IFAC

International Federation of Automatic Control

INTELLIGENT TUNING AND ADAPTIVE CONTROL

IFAC Symposia Series, 1991. Number 7

IFAC SYMPOSIA SERIES

Editor-in-Chief
Janos Gertler, Department of Electrical Engineering,
George Mason University, Fairfax, Virginia 22030, USA

JAAKSOO and UTKIN: Automatic Control (*11th Triennial World Congress*) (*1991, Nos. 1-6*)
DEVANATHAN: Intelligent Tuning and Adaptive Control (*1991, No.7*)
LINDEBERG: Safety of Computer Control Systems (SAFECOMP'91) (*1991, No.8*)

JOHNSON *et al*: Adaptive Systems in Control and Signal Processing (*1990, No.1*)
ISIDORI: Nonlinear Control Systems Design (*1990, No.2*)
AMOUROUX & EL JAI: Control of Distributed Parameter Systems (*1990, No.3*)
CHRISTODOULAKIS: Dynamic Modelling and Control of National Economies (*1990, No.4*)
HUSSON: Advanced Information Processing in Automatic Control (*1990, No.5*)
NISHIMURA: Automatic Control in Aerospace (*1990, No.6*)
RIJNSDORP *et al*: Dynamics and Control of Chemical Reactors, Distillation Columns and Batch Processes (DYCORD'89) (*1990, No.7*)
UHI AHN: Power Systems and Power Plant Control (*1990, No.8*)
REINISCH & THOMA: Large Scale Systems: Theory and Applications (*1990, No.9*)
KOPPEL: Automation in Mining, Mineral and Metal Processing (*1990, No.10*)
BAOSHENG HU: Analysis, Design and Evaluation of Man-Machine Systems (*1990, No.11*)
PERRIN: Control, Computers, Communications in Transportation (*1990, No.12*)
PUENTE & NEMES: Information Control Problems in Manufacturing Technology (*1990, No.13*)
NISHIKAWA *et al.*: Energy Systems, Management and Economics (*1990, No.14*)
DE CARLI: Low Cost Automation: Techniques, Components and Instruments, Applications (*1990, No.15*)
KOPACEK & GENSER: Skill Based Automated Production (*1990, No.16*)
DANIELS: Safety of Computer Control Systems 1990 (SAFECOMP'90) (*1990, No.17*)

Other IFAC Publications

AUTOMATICA
the journal of IFAC, the International Federation of Automatic Control
Editor-in-Chief: G.S. Axelby, 211 Coronet Drive, North Linthicum,
Maryland 21090, USA

IFAC WORKSHOP SERIES
Editor-in-Chief: Pieter Eykhoff, University of Technology, NL-5600 MB Eindhoven,
The Netherlands

Full list of IFAC Publications appears at the end of this volume

INTELLIGENT TUNING AND ADAPTIVE CONTROL

Selected Papers from the IFAC Symposium, Singapore,
15 - 17 January 1991

Edited by

R. DEVANATHAN

School of Electrical and Electronic Engineering,
Nanyang Technological University, Singapore

Published for the

INTERNATIONAL FEDERATION OF AUTOMATIC CONTROL

by

PERGAMON PRESS

OXFORD · NEW YORK · SEOUL · TOKYO

UK	Pergamon Press plc, Headington Hill Hall, Oxford OX3 0BW, England
USA	Pergamon Press, Inc., 395 Saw Mill River Road, Elmsford, New York 10523, USA
KOREA	Pergamon Press Korea, KPO Box 315, Seoul 110-603, Korea
JAPAN	Pergamon Press, 8th Floor, Matsuoka Central Building, 1-7-1 Nishi-Shinjuku, Shinjuku-ku, Tokyo 160, Japan

First edition 1991

Library of Congress Cataloguing in Publication Data

Intelligent tuning and adaptive control: proceedings of the IFAC
Symposium, Singapore, 15-17 January 1991
edited by R. Devanathan. - 1st ed. p. cm. - (IFAC symposia series: 1991. no.7)
Sponsored by: International Federation of Automatic Control
1. Intelligent control systems - Congresses. 2. Adaptive control systems - Congresses.
I. Devanathan, R. II. International Federation of Automatic Control
III. IFAC Symposium on Intelligent Tuning and Adaptive Control (1991: Singapore)
IV. Series
TJ217.5.1545 1991 629.8'36 - dc20 91-31032

British Library Cataloguing in Publication Data

Devanathan, R.
Intelligent tuning and adaptive control. -
(IFAC symposia series; 1991 v.7)
I. Title II.Series
629.836

ISBN: 9780080409351

These proceedings were reproduced by means of the photo-offset process using the manuscripts supplied by the authors of the different papers. The manuscripts have been typed using different typewriters and typefaces. The lay-out, figures and tables of some papers did not agree completely with the standard requirements: consequently the reproduction does not display complete uniformity. To ensure rapid publication this discrepancy could not be changed: nor could the English be checked completely. Therefore, the readers are asked to excuse any deficiencies of this publication which may be due to the above mentioned reasons.

The Editor

Transferred to digital print 2009
Printed and bound in Great Britain by CPI Antony Rowe, Chippenham and Eastbourne

IFAC SYMPOSIUM ON INTELLIGENT TUNING AND ADAPTIVE CONTROL

Sponsored by
International Federation of Automatic Control (IFAC)

Co-sponsored by
The Instrumentation and Control Society (ICS), Singapore
The Institute of Electrical and Electronic Engineers, Inc., Singapore Section

Organized by
The Instrumentation and Control Society (ICS), Singapore

CONTENTS

NEURAL NETWORK/SELF-TUNING APPLICATIONS

IMPLEMENTATION ISSUES

ADAPTIVE CONTROL

PLENARY SESSION

ROBOTIC CONTROL

KNOWLEDGE BASED CONTROL

INTELLIGENT TUNING/CONTROL

PLENARY SESSION

CONTROL OF DRIVES/SERVOS/APPLICATIONS

KNOWLEDGE BASED CONTROL/ADAPTIVE CONTROL

PREDICTIVE AND ROBUST CONTROL

ADAPTIVE CONTROL/KNOWLEDGE BASED CONTROL

DIRECTIONS IN INTELLIGENT CONTROL

K. J. Åström

*Department of Automatic Control, Lund Institute of Technology,
Box 118, S-221 00 Lund, Sweden*

Abstract. This paper outlines some different approaches to intelligent control based on fuzzy control, neural networks and expert systems. The key ideas behind these approaches are outlined and it is indicated how they may assist in making better control systems or better tools for design of control systems. The possibilities of obtaining controllers with significantly improved capabilities are also discussed.

Keywords: Auto-tuning, Control Design, Control Loop Auditing, Expert Control, Feedback, Fuzzy Control, Intelligent Control, Intelligent User Interfaces, Neural Networks, PID Control.

1. Introduction

In the early development of artificial intelligence there were strong ties to automatic control. Typical examples are the development of cybernetics, robotics and early learning systems. The fields did however diverge after a short period, one reason being that there were so many difficult problems in each subfield. It now appears that the fields are interacting again. An indication of this is that the word intelligence is increasingly appearing as an adjective to sensors, components, and control systems. The word also appears in titles of conferences like the one we are now attending. Reasons for this are increasing requirements on control systems and new demanding applications.

Control theory has for a long time focused on algorithms. Control problems can not be solved by algorithms alone and it is also necessary to add other elements like heuristics. There are many examples of this, control system design is one example, both control laws and design procedures are algorithmic. There are however many heuristic elements for example selection of appropriate methods, trade offs between different specifications etc. Typical examples are systems for autonomous vehicles, systems for industrial automation and process control systems with automatic tuning and adaptation.

Intelligence is a loaded word because of its strong association to human ability. Personally I share many views on AI, as expressed in [Penrose, 1989]. Even if AI fails to live up to human capabilities, I strongly believe that AI research has generated many useful ideas and techniques that can be used to make better control systems. To be specific, it is my opinion that automatic control can make very effective use of

Programming techniques
Programming environments
Methods for dealing with heuristics
Neural algorithms and hardware

I also believe the AI field can be stimulated by interaction with feedback control, e.g. by introducing dynamics and analysis [Birdwell et al., 1986], [Verbruggen and Åström, 1989].

This paper is organized as follows. Some specific approaches are discussed in Section 2, this includes fuzzy control, neural networks and expert control. Section 3 deals with implications for control system design, which is an area of significant current interest. In Section 4 we discuss impact on on line systems. Section 5 treats the particular case of single loop controllers, where several of the ideas discussed in the paper have been put to industrial use. By extrapolating from the properties of systems that are currently used we obtain a class of systems called autonomous controllers, which have interesting and useful properties.

2. Methodologies

Apart from the general methods discussed in the introduction there are three methodologies that are commonly used in intelligent control systems, they are fuzzy control, neural networks and knowledge-based systems [Zadeh, 1968]. These methodologies are briefly reviewed in this section.

Fuzzy Control

The idea of fuzzy sets is due to [Zadeh, 1968] and [Zadeh, 1988]. The key idea is to develop a framework to deal with imprecision. Instead of using the ordinary concept of set inclusion Zadeh introduced a function that expressed the degree of belonging to a given set as a function taking values in the range 0 to 1. The idea was first applied to control systems in [Mamdami and Assilian, 1974] and has since then had substantial following [Kaufmann and Gupta, 1985]. It leads to control systems, where the signals are quantized crudely into a few levels like very low, low, about right, high and very high. A control variable is then computed as a fuzzy variable using fuzzy logic. The fuzzy control variable is then mapped into a real number which is applied to the process ("defuzzification"). Very often the controller computes the change in the control variable. This means that the controller obtained becomes a nonlinear PI controller. Interest in fuzzy control has recently increased significantly because of good practical experiences in control of cement kilns, [Holmblad and Østergaard, 1981], [Haspel et al., 1987] and other areas, and the creation of the institute of fuzzy control in Japan [Sugeno, 1985], [Kaufman and Gupta, 1985], [Watanabe and Dettloff, 1987] [van der Rhee et al., 1990], [Harris and Moore, 1990]. A number of fuzzy controllers for simple processes like cameras, vacuum cleaners and washing machines have also appeared. One reason for this is that it is easy to implement the controllers in VLSI, because they only require crude quantization and simple logic. [Togai and Watanabe, 1986] The cement kiln systems are interesting, because in this case traditional approaches using standard control methodology were found to consume large engineering efforts and poor control performance. Fuzzy controllers are well suited for simple control tasks. Because of their simplicity they can be implemented with very little silicon surface. The problems of A/D and D/A conversion are simplified significantly because of the crude quantization used. Special chips to do the fuzzy logic are also appearing. They can easily be interfaced to sensors and actuators.

Fuzzy control algorithms can be expressed in statements like:

If
    the pendulum is upright
    and pendulum velocity is small
    and cart to the right
then
    move cart to the right

Fuzzy control is sometimes also called linguistic control. Notice that the above expression is very closely related to rule based expert systems. Software for fuzzy control have also been significantly influenced by the development of knowledge-based systems. Recent implementations of fuzzy controllers make it very easy to change the logic and experiment on-line. Fuzzy control laws are often derived by attempting to mimic the actions taken by an operator or an experienced process engineer.

There are several ways to implement fuzzy controllers.

Neural Networks

After a long period in the doldrums the interest in neural networks are now florishing [Rumelhart and McClelland, 1986], [DARPA, 1988], [Hecht-Nielsen, 1990]. There are several types of neural networks that can be used in control systems, the multilayer perceptron, Kohonen's self-organizing map and the Boltzmann machine.

A multi layer perceptron consists of a collection of neurons that are organized in layers with feedforward connections from one layer to another. The output of a neuron is given by

$$y = f\left(\sum_{i=1}^{n} w_i u_i\right)$$

where u_i is the i:th input, ω_i the weight associated with u_i, and f is a sigmoid function, e.g.

$$f(x) = \frac{1}{1+e^{-ax}}$$

It has been shown that large classes of continuous functions can be approximated by a multilayer perceptron. There are also learning algorithms, e.g., back propagation so that the function can be constructed by giving the network a number of values and arguments. Learning is performed by giving the network an input and a corresponding desired output. There are algorithms to adjust the weights so that the machine will give the desired response. The training is repeated for more input-output pairs. There are strong connections between training of neural networks and adaptive control [Åström and Wittenmark, 1989]. The algorithms commonly used to train a neural network can, e.g., be interpreted as special versions of model reference adaptive systems.

The multilayer perceptron can be used to implement a trainable nonlinear function. The perceptron was originally developed as a pattern recognition device. The network has also other uses in control engineering. The function can be used in static and dynamic modeling and as an element of a controller. A difficulty is that there are no systematic procedures to determine how many layers and nodes are required to approximate a function well. These problems are therefore determined by trial and error.

Kohonen's self-organizing map is another type of neural networks. See [Kohonen, 1989]. It consists of one layer of neurons that are connected to the input. The input is typically vector valued. There are also internal connections between the neurons in the layer. Connections between neighboring neurons are obtained by the learning algorithms which update neighboring neurons in the same way. The network works as follows. The neuron whose output is largest is selected and its weights are updated so that it becomes closer to the input vector. The neighboring neurons are updated similarly. Kohonen's network can be used to classify data. If a stream of vector signals are connected to the network the weights of the network will adjust themselves to give a representation of the distribution of the inputs. The connection of the neurons in the layer will also be reflected in an ordering of the neurons. Kohonen's network can thus be used to classify data and signals. One application is to classify dynamic responses of a system or the time distribution of control system performance. Neural networks represent a return to the analog computing paradigm. There are many interesting analog computing circuits. Brockett has recently written a number of papers where he gives differential equations with interesting computing capability. The differential equation given in [Brockett, 1988] that sorts lists is closely related to Kohonen's network.

The input-output relation of the multilayer perceptron is a static nonlinearity when the training is finished. There are more complicated

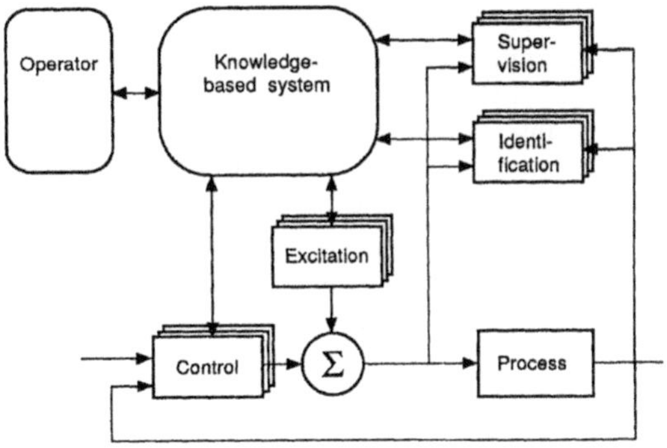

Figure 1. An expert control system.

networks which also are dynamic systems. Examples are the Hopfield network and the Boltzmann machine. In these networks there are arbitrary connections between the neurons. There may also be dynamics in the nodes. The Boltzmann machine can be used as a pattern recognition machine. In this way inputs and desired responses are given and the weights are adjusted until the machine responds in the desired manner. The Boltzmann machine can also be used as an associative memory and as an optimizer.

Applications of neural networks to control are given in [Narendra and Parthasarathy, 1990] and [Pao, 1990].

Expert Control

There have been several attempts to incorporate knowledge based systems into feedback loops. One reason for this is that conventional feedback systems contain a significant amount of logic that is conveniently described and implemented as rules. A block diagram of an expert control system is shown in Figure 1. This scheme was suggested in [Åström et al., 1986]. The system consists of an ordinary feedback loop with a process and a controller. There is a collection of different algorithms. There may be several different control algorithms as is indicated by the different layers in the figure. There are also algorithms for identification, diagnosis, and supervision. The identification algorithm may also generate perturbation signals to excite the process. The algorithms are coordinated by a knowledge based system. This decides what algorithm to use and it also interacts with the operator. The system in Figure 1 is very general. For example, it contains conventional adaptive control as a special case. The system also admits a nice separation of algorithms and logic.

It was shown in [Årzén, 1989a] that a blackboard system is a convenient architecture for implementing expert systems. A detailed description of an implementation is given in [Årzén,

1989a]. Alternative implementations are also found in papers at this conference. See [Sallé and Åström, 1991].

An interesting aspect of fuzzy control and neural networks is that they may be considered as sources for system components that provide useful functions like pattern recognition, classification, associative memory, learning and optimization. These functions can be combined with conventional control techniques using an expert system as an integration tool.

3. Control System Design

Control system design includes different tasks like modeling, analysis, simulation, design, implementation, and verification. These tasks are of central importance to control engineering. A good control system design is a combination of many different elements—control theory, algorithms, engineering judgement, rules of thumb, etc. It requires a good understanding of the system, its environment, specifications and design methods. Research in control theory has for a long time focused on detailed studies of idealized design problems that focus on one or two aspects of the problem. This has resulted in a proliferation of so called design methods. In some cases the design methods have also been implemented in some interactive computer software [Åström, 1983], e.g. as toolboxes in Matlab. See [Chiang and Safonov, 1988] and [Maciejowski, 1989]. This is a useful development but it still leaves the control system designer with many tasks for which there is little support. The proliferation of design methods is also somewhat bewildering to an engineer in industry. An indication of this is the large number of benchmark sessions that are appearing in our conferences, which attempts to provide insight into the properties of the different design methods.

Control system design is a natural application of AI methods. A number of tools for computer aided control engineering have been developed. An overview is presented in [Jamshidi and Herget, 1985]. The commercial packages CTRL-C, CC, $Matrix_X$ and PROMATLAB are all derived from the program Matlab developed by Cleve Moler. These programs are useful tools to simplify the algorithmic aspects of control design. There are however many other aspects of control system design where computer aids are useful. Tools for selecting methods and design procedures is one area. There is also a growing awareness that the user interface is a key issue. The interface should be such that the packages are easily handled by the inexperienced user and convenient and flexible for the experienced user. AI methods have proven very useful in the development of user interfaces they are probably also useful to describe design heuristics.

There are some experimental systems which attempt to incorporate AI methods in different ways. See [Trankel et al., 1986], [Taylor et al., 1990], [Marttinen and Telkka, 1990], [Munro, 1990] and [Rimvall et al., 1989].

There are many ways to use knowledge-based systems for control system design. An overview of several approaches are given in [James, 1987], [James, 1988]. The paper [James et al., 1987] describes design of lead/lag networks. In this paper the empirical design knowledge is described in a rule based format. The paper [Taylor, 1988] describes how knowledge based systems are used to provide user assistance in a large integrated design package. A computer algebra system for analysis and design of nonlinear systems is described in [Akhrif et al., 1987].

Control system design may be formalized as a high level problem-solving language [Åström, 1983]. This is a useful viewpoint because it is then possible to apply some formal methods. Ideas from natural language processing can also be applied. For example, since natural language is so ambiguous it is necessary to introduce context in order to understand it. The notion of script introduced by Roger Schank [Schank and Abelson, 1977] is one way to describe context. This is one way to encode common situations to allow interpretation of natural language sentences. Although it is quite difficult to describe context in an everyday conversation, it is easy to describe the context in control system design. In [Larsson and Persson, 1988a], [Larsson and Persson, 1988b] and [Larsson and Persson, 1991] it is described how scripts are used to develop an intelligent user interface for solving system identification problems. The interface contains knowledge about the problem domain (system identification) as well as about the particular software used (Idpac). An interesting aspect is that the user interface could be developed without changing one single line in the identification package, which was an old Fortran program.

Good use can also be made of computer science methods in the implementation area. If a controller is completely descibed in some formal manner the implementation of real time control code may be viewed as automatic code generation from a high level problem description. This viewpoint is taken in [Dahl, 1989], which describes a system which automatically generates Modula II code and appropriate routines interaction with a real time kernel and a man machine interface from simulation code in Simnon. An interesting aspect is that by using programming methodology from AI the translator could be written with a very modest effort of 2 man-months.

Some of the experimental systems have been tested in a teaching environment. I do not know of any system that is in routine industrial use. For this to happen we will probably have to wait for the next generation of CACE tools, which are more flexible and easier to interface with other programs.

There are many interesting future developments in this area. There are also many fundamental problems that must be resolved. One key issue is how to describe and represent design knowledge. This is a problem that arises also in many other areas of engineering. Another is to describe the properties of the different design methods. A third is to capture and describe good design heuristics. A good starting point would be to try to develop good empirics for design in parallel with the development of design methods. I believe that it is a good starting point to view the design problem as a high level problem-solving language. Another interesting possibility is to use the idea of auditing systems to check that a control design satisfies all specifications.

4. On-Line Systems

Industrial automation systems were traditionally based on relay boxes for interlocks, sequencing and controllers. With the introduction of microprocessors, these were implemented as programmable logic controllers (PLC) and direct digital controllers (DDC). Since the systems are implemented in the same technology, a natural merging of the techniques of logic, sequencing and algorithms is occurring. This has opened up many interesting possibilities of making systems with increased capabilities. For example, it is possible to process alarm signals to give more meaningful information. It is also possible to provide the alarm system with capabilities for inquiries. This is clearly an area that can benefit from use of methodologies from AI and feedback control. One observation is that logic and sequencing can be expressed very conveniently in rules. The use of an AI programming style admits system descriptions that are much more compact than those normally used for PLCs. For example, it is possible to have generic rules that apply to all processes of a certain type, allowing significant simplification in programming, modifications, and troubleshooting.

The merger of algorithms and logic is also noticeable for simple controllers. A recent standard proposal for a PID controller has 255 different modes. The reason for the large number of modes is that it is attempted to cover all possible situations. A much smaller number of the modes will be used in each specific application. An alternative implementation would be to incorporate a small knowledge based system in the controller that admits easy customization.

The expert control system discussed in Section 2 is one possibility to combine algorithms and heuristics. Interesting views, which combine conventional control with knowledge-based systems, are given in [Årzén, 1989b] and [Beck and Lauber, 1990]. Much fundamental work remains to be done to achieve an effective combination of algorithms and logic. Finding methods to analyze systems and reason in real time are key issues.

There is exploratory work in the process industry aimed at investigating the potential of combining feedback control and knowledge-based systems. Several system vendors have developed commercial systems intended to be used in combination with distributed systems for process control. Bailey has an embedded expert system shell in their distributed control system [Oyen et al., 1988]. Foxboro has provided an interface to a small commercial shell into their intelligent automation system. Probably the most sophisticated system for real time process control is G2 developed by Gensym corporation [Moore et al., 1987], [Moore et al., 1990]. Many vendors of distributed control systems have developed interfaces to this system. There are also a few other systems that are designed for real time control like Muse and Chronos. A comparison of some real time expert systems is given in [Sallé and Årzén, 1989].

The key element of G2 is an item. This can be rules, objects, procedures, graphs, buttons, textboxes etc. The items are organized hierarchically, they have all graphical representations that can be manipulated directly using mouse or keyboard operations. Objects are organized into a class hierarchy with single inheritance. Properties of an object are described by attributes that may be constants, variables, lists or other objects. There are flexible mechanisms for structuring the rule base. There are also convenient ways to give generic rules that apply to classes of objects. There are some facilities for dealing with the real time issues. It is possible to restrict the reasoning to part of the rule base through a focusing mechanism. Time tags can be associated with values. The inference engine will automatically request for variables that have become invalid and waits for new values. There are also mechanisms to save histories of old values for all variables and to reason about the past history.

Alarm analysis and diagnosis is a common application of the commercial systems.

5. Autonomous Controllers

Simple process controllers of the PID type are currently going through an interesting development [Bristol, 1977], [Kraus and Myron, 1984]. [Åström and Hägglund, 1988]. Features like automatic tuning, adaptation and gain scheduling are currently being incorporated even in single loop controllers. To achieve this, it is necessary to automate modeling as well as control design. Modeling has been automated both by conventional system identification methods [Åström and Hägglund, 1988] and with heuristical approaches based on pattern recognition [Bristol, 1977]. Control design has also been automated using both algorithmic and heuristic methods. The traditional way of tuning controllers is often based on heuristic rules of the Ziegler-Nichols type [Deshpande and Ash, 1981]. Recently there have been significant efforts to improve and extend such [Hang and Åström, 1988], [Anderson et al., 1988], [Åström et al., 1989].

A result of this development is that the instrument engineers now have algorithms that will help them tune the controller or will even tune the controller automatically. An interesting side effect is that it has also made gain scheduling easy to use. Combined with an auto-tuner, it is straightforward to generate a gain schedule semi-automatically in the following manner. A suitable scheduling variable is first determined. This could be controller output, measured signal or an external signal. The process is then operated in a number of points over the range of variation of the scheduling variable. Tuning is executed at the chosen operating points. The controller parameters obtained at the different operating conditions are stored. Interpolation may be used to obtain controller parameters at intermediate operating conditions. The schedule and the chosen operating points can be refined by plotting controller parameters to see if they are suitably spread over the operating range.

Many of the design choices in existing systems were restricted by the computational power that was economically feasible. With the current rate of increase of computational power and techniques that have already been proven in laboratory tests one can extrapolate the characteristics of future controllers. Natural next steps are to include diagnostics and loop auditing [Brandl and Jeffreys, 1988]. With such features we can talk about autonomous controllers, i.e. controllers that when connected to an unknown process will explore the features of the process and the disturbances to decide upon a suitable controller structure and perform the control functions. A system may have the following key components for analysis of the process.

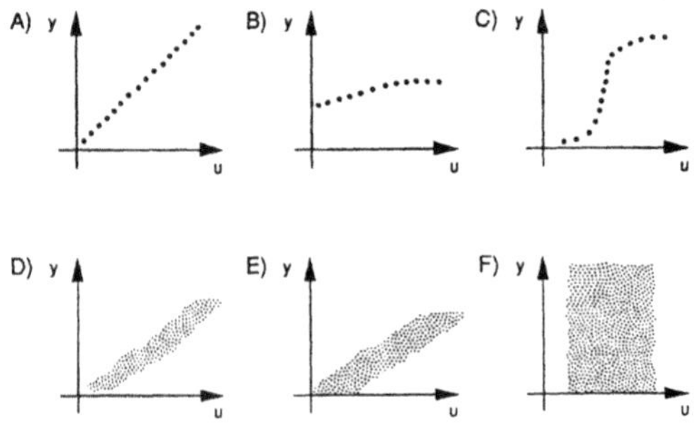

Figure 2. Example of static analysis.

Static Analyser
Transient Response Analyser
Relay Feedback Analyser
Frequency Response Analyser
Parameter Estimator
Noise Analyser

The static analyser gives the relation between the process inputs and outputs in steady state. It tells if the relation is linear or essentially nonlinear. It also gives an indication if the primary function of the feedback loop is regulation or servo following. See Figure 2. The behaviour, illustrated by curves A, B, and C, indicates a pure servo function, because there is a well defined relation between input and output. Curve F is a typical behaviour for a pure regulation problem. In this case there is no relation between input and output because of the disturbances acting on the plant. The behaviours shown in D and E are mixed cases. The behaviour in B indicates that the sensor has poor resolution and C indicates poor actuator sizing.

The transient response analyser investigates the transient responses both passively when there are natural disturbances and actively by introducing perturbations. The analyser determines the period and the damping of the dominant mode, static gain, settling time, rise time, etc.

The relay feedback analyser determines the frequency ω_{180} where the system has 180° phase lag and the gain k_{180} at that frequency. When applied to an open loop system this data can be used to determine parameters of a PID controller. See [Åström and Hägglund, 1988]. The frequency ω_{180} is the crossover frequency of the open loop system. Knowledge about this is useful for many purposes. When applied to a closed loop system, relay feedback can be used to determine the amplitude margin. Simple calculations show that this is given by

$$a_m = \frac{1 + k_{180}}{k_{180}}$$

where k_{180} is the gain obtained from the relay analyser.

The frequency response analyser determines the frequency response of a system by active perturbation. This is useful in order to determine dynamics accurately. The parameter estimator determines a parametric model of the system and the noise analyser characterizes the disturbances acting on the system.

The control design can be based on the following subsystems.

Controller Assessment
Control Design

The controller assessment determines the structure of a controller that will satisfy the specifications based on data obtained from the analysis and interaction with the operator. Different ways of doing this are discussed in [Åström, 1991].

When the system is running it is useful to have various monitoring routines like

Stability Assessment
Performance Assessment
Actuator Monitor

The stability assessment monitors the stability margins continuously. This can be done passively by tracking the loop transfer function at a few frequencies as is discussed in [Hägglund and Åström, 1989] and [Hägglund and Åström, 1990]. The performance monitor measures means, variances, max and mean of the control signal and the error. Actuators like valves are critical components in many cases. The actuator monitor determines if the actuator deteriorates by determining backlash and hysteresis.

Several of the components discussed above are running on existing systems.

6. Conclusions

There are opportunities to make control systems with significantly increased capability by combining methodologies from feedback control and artificial intelligence. In a simple setting this has been demonstrated by recently announced single loop controllers with capabilities for automatic tuning, gainscheduling and adaptation [Hägglund and Åström, 1991]. There are also interesting opportunities for better CACE systems with intelligent user interfaces, where heuristics and algorithms are combined. Such systems can be used off-line and possibly on-line. Ideas, methods and hardware are available, and it appears that control engineers have many possibilities to make useful and interesting systems. It is not clear if such systems are best labeled as intelligent.

7. References

Akhrif, O., G. L. Blankenship and L. S. Su (1987): "Computer algebra systems for analysis and design of nonlinear systems," *Proceedings of the American Control Conference*, Minneapolis, MN, pp. 547–554.

Anderson, K. L., G. L. Blankenship and L. G. Lebow, (1988): "A rule-based PID controller," *Proc. IEEE Conference on Decision and Control*, Austin, Texas.

Årzén, K. E. (1989a): "An architecture for expert system based feedback control," *Automatica*, **25**, 813–827.

Årzén, K. E. (1989b): "Knowledge-based control systems – Aspects on the unification of conventional control systems and knowledge-based systems," *Proc. 1989 Automatic Control Conference*, pp. 2233–2238.

Åström, K. J. (1983): "Computer-aided modeling, analysis and design of control systems – A perspective," *IEEE Control Systems Magazine*, **3**, May, 4–16.

Åström, K. J. (1991): "Assessment of achievable performance of simple feedback loops," *Int J of Adaptive Control and Signal Processing*, to appear.

Åström, K. J., and T. Hägglund (1988): *Automatic Tuning of PID Controllers*, Instrument Society of America, Research Triangle Park, NC.

Åström, K. J., and B. Wittenmark (1989): *Adaptive Control*, Addison Wesley, Reading, MA.

Åström, K. J., J. J. Anton, and K.-E. Årzén, (1986): "Expert control," *Automatica*, **22**, 277–286.

Åström, K. J., C. C. Hang and P. Persson (1989): "Towards intelligent PID control," *Preprints, IFAC Workshop on Artificial Intelligence in Real-Time Control, Shenyang, China*, Pergamon Press, Oxford, UK, pp. 38–43.

Beck, Th., and R. J. Lauber (1990): "Integration of an expert system into a real-time software system," *Preprints 11th IFAC World Congress*, Tallinn, Estonia, USSR.

Birdwell, J. D., J. R. Cockett and J. R. Gabriel (1986): "Domains of artificial intelligence relevant to systems," *Proceedings of the American Control Conference*, Seattle, WA.

Brandl, D., and S. A. Jeffreys (1988): "The computer assistant for control systems engineering in a historical perspective," *Proc. ISA Annual Conference*, Houston, TX, pp. 401–410.

Bristol, E. H. (1977): "Pattern recognition: an alternative to parameter identification in adaptive control," *Automatica*, **13**, 197–202.

Brockett, R. W. (1988): "Dynamical systems that sort lists, diagonalize matrices and solve linear programming problems," *Proceedings 27th IEEE Conference on Decision and Control*, Austin, Texas.

Chiang, R. Y., and M. G. Safonov (1988): *Robust-Control Toolbox*, The Mathworks, Inc.

Dahl, O. (1989): "Generation of structured Modula-2 code from a Simnon system description," TFRT-7415, Department of Automatic Control, Lund Institute of Technology, Lund, Sweden.

DARPA (1988): *Darpa Neural Network Study. October 1987 - February 1988*, AFCEA International Press, Fairfax, Virginia.

Deshpande, P. B., and R. H. Ash, (1981): *Computer Process Control*, ISA, Research Triangle Park, NC.

Hägglund, T., and K. J. Åström (1989): "An industrial adaptive PID controller," *IFAC Symposium on Adaptive Systems in Control and Signal Processing*, Glasgow, UK.

Hägglund, T., and K. J. Åström (1990): "A frequency domain approach to adaptive control," *Preprints 11th IFAC World Congress*, Tallinn, Estonia, USSR, pp. 265–270.

Hägglund, T., and K. J. Åström (1991): "Industrial adaptive controllers based on frequency response techniques," *Automatica*, to appear.

Hang, C. C., and K. J. Åström (1988): "Refinements of the Ziegler-Nichols tuning formula for PID auto-tuners," *Proc. ISA Annual Conference*, Houston, TX.

Harris, C. J., and C. Moore (1990): "Real time fuzzy based self-learning predictors and controllers," *Preprints 11th IFAC World Congress*, Tallinn, Estonia, USSR.

Haspel, D. W., C. J. Southan and R. A. Taylor (1987): "The benefits of kiln optimization using LINKman and high level control strategies," *World Cement*, **10**, No. 6.

Hecht-Nielsen, R. (1990): *Neurocomputing*, Addison-Wesley.

Holmblad, L. P., and J. Østergaard (1981): "Control of a cement kiln by fuzzy logic," *F.L. Smidth Review*, **67**, 3–11, Copenhagen, Denmark.

James, J. R. (1987): "A survey of knowledge-based systems for computer-aided control system design," *Proceedings of the American Control Conference*, Minneapolis, MN, pp. 2156–2161.

James, J. R. (1988): "Expert system shells for combining symbolic and numeric processing in CADS," *Preprints 4th IFAC Symposium on Computer Aided Design in Control Systems, CADCS'88*, Chinese Association of Automation, Beijing, P. R. China.

James, J. R., D. K. Frederick and J. H. Taylor (1987): "On the application of expert systems programming techniques to the design of lead-lag precompensators," *IEE Proceedings D: Ctrl Theory and Applications*, **134**, 137–144.

Jamshidi, M., and C. J. Herget (1985): *Computer-Aided Control Systems Engineering*, North-Holland.

Kaufmann, A., and M. M. Gupta (Eds.) (1985): *Introduction to Fuzzy Arithmetics: Theory and Applications*, Van Nostrand, Reinhold.

Kohonen, T. (1989): *Self-Organization and Associative Memory*, Springer-Verlag, Third edition.

Kraus, T. W., and T. J. Myron (1984): "Self-tuning PID controller uses pattern recognition approach," *Control Engineering*, June, 106–111.

Larsson, J. E., and P. Persson (1988a): "An intelligent help system for Idpac," *Proceedings of the 8th European Conference on Artificial Intelligence*, Technischen Universität München, Munich, FRG, pp. 119–123.

Larsson, J. E., and P. Persson (1988b): "The knowledge database used in an expert interface for Idpac," *Proceedings of the IFAC Workshop on Artificial Intelligence in Real-Time Control*, Swansea, Wales, UK.

Larsson, J. E., and P. Persson (1991): "An expert system interface for an identification program," *Automatica*, to appear.

Maciejowski, J. M. (1989): *Multivariable Feedback Design*, Addison-Wesley, UK.

Mamdani, E. H., and Assilian, S. (1974): "A case study on the application of fuzzy set theory to automatic control," *Proceedings IFAC Stochastic Control Symposium*, Budapest, Hungary.

Marttinen, A., and T. Telkka (1990): "A hierarchical process modelling environment," *Preprints 11th IFAC World Congress*, Tallinn, Estonia, USSR.

Moore, L. R., L. B. Hawkinson, M. Levin, A. G. Hoffman, B. L. Matthews and M. H. David (1987): "Expert system methodology for real-time process control," *Preprints, IFAC Workshop on AI in Real-Time Control, Swansea, UK*, Pergamon Press, Oxford, UK.

Moore, R., H. Rosenof and G. Stanley (1990): "Process control using a real time expert system," *Preprints 11th IFAC World Congress*, Tallinn, Estonia, USSR.

Munro, N. (1990): "Ecstasy — A control system CAD environment," *Preprints 11th IFAC World Congress*, Tallinn, Estonia, USSR.

Narendra, K. S., and K. Parthasarathy (1990): "Identification and control of dynamical systems using neural networks," *IEEE Trans. on Neural Networks*, **1**, 4–27.

Oyen, R. A., M. A. Keyes and M. P. Lukas (1988): "An expert system shell embedded in the control system," *Preprints, IFAC Workshop on AI in Real-Time Control, Swansea, UK*, Pergamon Press, Oxford, UK.

Pao, H. H. (1990): "Use of neural-net technology in control: A survey and a perspective," *Preprints 11th IFAC World Congress*, Tallinn, Estonia, USSR.

Penrose, R. (1989): *The Emperor's New Mind*, University Press, Oxford, UK.

Rimvall, M., H. Sutherland, J. H. Taylor and P. J. Lohr (1989): "GE's MEAD user interface – A flexible menu- and forms-driven interface for engineering applications," *IEEE Control Systems Society Workshop on Computer-Aided Control System Design (CACSD)*, Tampa, Florida.

Rumelhart, D. E., and J. L. McClelland (1986): *Parallel Distributed Processing – Explorations in the Microstructure of Cognition. Volume 1: Foundations*, MIT Press, Cambridge, Massachusetts.

Sallé, S. E., and K.-E. Årzén (1989): "A comparison between three development tools for real-time expert systems: Chronos, G2, and Muse," *IEEE Control Systems Society Workshop on Computer-Aided Control System Design (CACSD)*, Tampa, Florida.

Sallé, S. E., and K. J. Åström (1991): "Synthesis of a smart PID controller using expert system techniques," *IFAC International Symposium on Intelligent Tuning and Adaptive Control (ITAC 91)*, Singapore.

Schank, R. C., and R. P. Abelson (1977): *Scripts, Plans, Goals and Understanding*, Lawrence Erlbaum Associates, Hillsdale, NJ.

Sugeno, M. (Ed.) (1985): *Industrial Applications of Fuzzy Control*, Elsevier Science Publishers BV, The Netherlands.

Taylor, J. H. (1988): "Expert-aided environments for CAD of control systems," *Preprints 4th IFAC Symposium on Computer Aided Design in Control Systems, CADCS'88*, Chinese Association of Automation, Beijing, P. R. China.

Taylor, J. H., D. K. Frederick, C. M. Rimvall and H. A. Sutherland (1990): "Computer-aided control engineering environments: Architecture, user interface, data-base management, and expert aiding," *Preprints 11th IFAC World Congress*, Tallinn, Estonia, USSR.

Togai, M., and H. Watanabe (1986): "Expert systems on a chip: An engine for real-time approximate reasoning," *IEEE Expert*, **1**, 55–62.

Trankel, T. L., L. Z. Markosian and P. Sheu (1986): "Expert system architecture for control system design," *Proceedings of the American Control Conference*, Seattle, WA.

van der Rhee, F., H. R. van Nauta Lemke, and J. G. Dijkman (1990): "Knowledge based fuzzy modeling of systems," *Preprints 11th IFAC World Congress*, Tallinn, Estonia, USSR.

Watanabe, H., and W. Dettloff (1987): "Fuzzy logic inference processor for real time control: A second generation full custom design," *Proc 21st Asilomar Conference on Signals, Systems, and Computers*, Asilomar, California.

Verbruggen, H. B., and K. J. Åström (1989): "Artificial intelligence and feedback control," *Preprint, IFAC Workshop on Artificial Intelligence in Real-Time Control*, Shenyang, China, pp. 38–43.

Zadeh, L. A. (1968): "Fuzzy algorithms," *Information and Control*, **12**, 94–102.

Zadeh, L. A. (1988): "Fuzzy logic," *IEEE Computer*, April, 83–93.

Zhang, Z., H. Yang, M. Jia, L. Chen and X. Zhang (1990): "Application of expert control in plasma surface alloying furnace," *Preprints 11th IFAC World Congress*, Tallinn, Estonia, USSR.

INTELLIGENT SELF-TUNING PID CONTROLLER

H. Takatsu, T. Kawano and K. Kitano

Development & Engineering Dept., Process Automation Systems Div., Yokogawa Electric Corporation, Musashino-shi, Tokyo, Japan

ABSTRACT. A new self-tuning PID controller has been developed for distributed process control systems. The self-tuning PID controller is able to identify the process dynamics using a short period of process behaivior such as a setpoint change under the closed loop control condition, and to tune PID parameters based on the identified model parameters for both setpoint tracking and disturbance regulation characteristics, which makes it easy to set up the inital PID parameters and adapt the PID values to the process dynamics changes.
This paper describes the basic principles of the self-tuning controllers, and explains the functions to be implemented on. Finally, some application results illstrate the effectiveness of the self-tuning controllers.

KEYWORDS. Adaptive control; identification; intelligent control; PID control; self-tuning control

INTRODUCTION

Many different types of self-tunig PID controllers have been introduced into the industrial market since 1980. However, some self-tuning controllers need oscilation of process output; some need long duration of process output perturbations; some need long range observations of process output. These are disadvantages to be applied to the commercial practical plants regarded as the disturbances on plant operations. A new self-tunig control ler must satisfy the following users' needs:
(1) It helps operators and engineers to tune PID parameters without any special knowledge of the control theory and engineering.
(2) It automatically updates PID parameters in order to follow up changes in the plant dynamics quickly.
(3) It is available online in the commercial industrial plants without disturbing plant operations.

Based on the analyses of these users' needs, a new self-tuning PID controller has been developed to be implemented into the industrial process control systems. The new self-tuning controller mainly consists of a process identification part and a PID tuning part, and have realized the following features:
(1) Only a single small step change in the setpoint or control output is applied under the closed loop control condition to estimate a process model and compute the optimal PID parameters.
(2) The computed PID parameters are transfered to the control part when the certainty of the identified model is large enough.
(3) The estimated process model is displayed and utilized to detect if the process dynamics has changed and inform operators of the necessity of retunig in case of the monitoring mode.
(4) The controller may be easily combined with conventional control functions to provide more sophisticated control.

BASIC PRINCIPLES

Process Identification

Figure 1 shows the basic concept of the self-tunig controllers. The self-tuning controller always observes and collects the process input and output data. When the controller detects the process output move becomes larger than a specified level after removing the noises, signal trend and sensor failure, it starts the identification computation using the collected process input and output data.
The key technology is how to estimate the process model rapidly using a short duration of operating data in the closed loop control condition, because a self-tuning controller has to track the process dynamics change. Local identification such as a well known least square method is inadequate to the identification using a short duration of experiment. We have adopted an orignal global identification method. The difficult problems in on-line practical identification is the effects of unknown disturbances during the experiment. The proposed method can determine

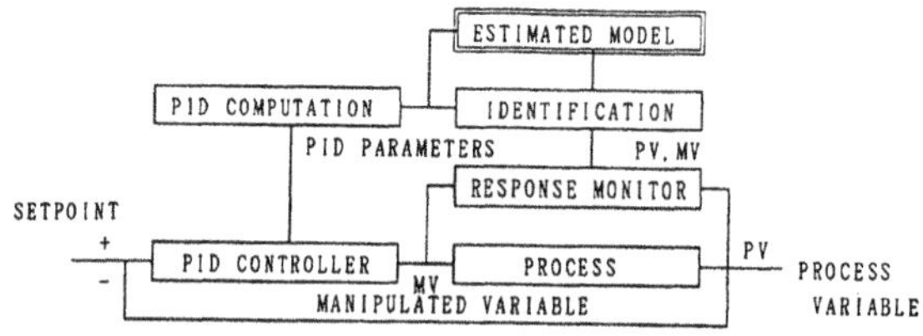

Fig. 1 Functional Block Diagram of Self-tuning Controller

a reasonable model with more than a given level of "model certainty", taking into account modelling errors, disturbances and so on. This method has the following features:
(1) The model is structured in the neighbourhood of the specified frequency domain to remove the effects of noises or disturbances.
(2) One kind of nonlinear programming techniques is adopted to obtain the model using a short period input and output data iteratively.
(3) PID parameters of the controller is adapted to the identified model only if the certainty of the identified model is large enough.
(4) Process behavior is observed every monitoring period and the identification is triggered again when the process move becomes large.

Figure 2 shows one of the model fitting trajectories of the adopted method in the parameter domain. The model is assumed to be a 'first-order plus time delay' model, which is a suitable model because our objective is PID tuning rather than model development. Contour lines in Fig. 2 indicate the sets of model parameters (gain, lag, delay) which give the same modeling errors. More optimal set of model parameters is sequentially obtained in order to minimize the modeling errors, using a set of process inputs and outputs iteratively.

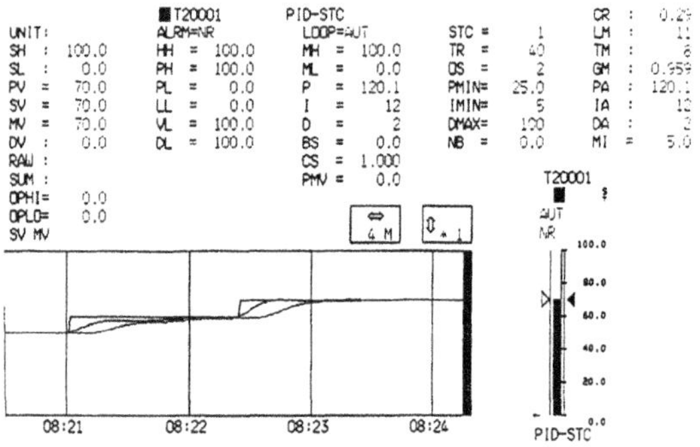

Fig. 3 Tuning Panel of Self-tunig Control

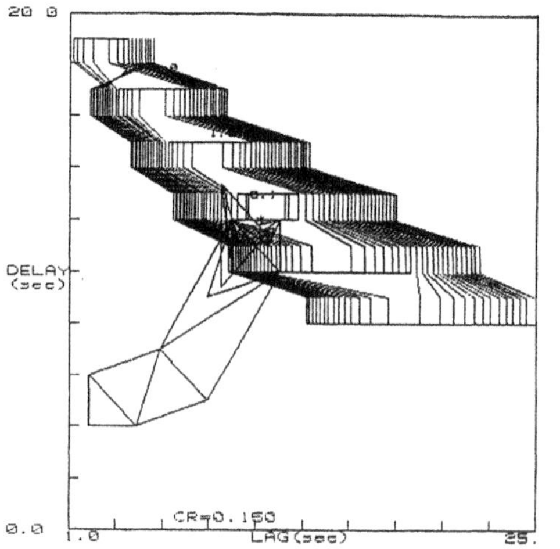

Fig. 2 Example of Non-linear Programming Application

PID Tuning

After the process identification, the optimal PID values are calculated using the estimated process model and specified controller type, based on the equations described below:

$K \cdot Kc = A(L/T)^2 + B(L/T) + C$
$Ti/T = D(L/T)^2 + E(L/T) + F$
$A, B, C, D, E, F = \text{function}(\, OS,\; CNT, \ldots\,)$
K: model gain, L: model delay,
T: model time lag, Kc: control gain,
Ti: integral time, OS: target response type, CNT: PID algorithm type

These equations are just like the Ziegler-Nichols method, but each coefficients have been decided after large amount of simulation runs. Generally speaking, as the optimal values for setpoint tracking and for disturbance regulation are of different form. The self-tunig controller has adopted two degrees of freedom structure in order to satisfy both control specifications.

Although PID parameters are usually computed based on the identified model and the user-specified response type, special tunig mechanizm is prepared for to stabilize the behavior rapidly. In the case of oscillating condition PID control gain is decreased temporarily; and vice versa in the case of overdamping the gain is increased.

Auto Startup

Figure 3 shows the controller tuning panel on the display terminals. The self-tuning controller has more parameters than conventional PID controllers, most of which are prepared for safety and easiness of plant operation. Auto startup mode sets up the initial values of these parameters automatically. In this mode step inputs are applied to the process under the manual mode and the subsequent time response data are used to obtain the process model and compute the initial PID values through the above described techniques.

Sampling perid and other parameters are also decided based on the response characteristics.

LOOP SUPERVISOR

Self-tuning control without loop supervising function is dangerous because it is difficult to distinguish process dynamics changes from the device failure such as sensors and actuators. In industrial applications most of contol loops are set out of self-tuning mode once the initial PID values have been set. In these situations, some warning systems are indispensable to inform operators of the necessity of retuning. The self-tuning controller always computes the ratio of process output and model output variances. As the variance ratio becomes large after the process dynamcs changes, it occurs a warning for a request of retuning.

SPECIFICATIONS

Self-tuning Operation Mode

Figure 4 is the flow chart of the self-tunig control. Four self-tunig modes are available for use.

(1) Auto Mode: The self-tunig controller always observes the process behavior, and tracks the process dynamics changes by the above described identification and PID tuning functions.

(2) Monitoring Mode: In this mode, the self-tunig control only displays the computed process model and PID values; it does not set the computed PID values to the control part. This mode is effective for validating the self-tunig function or checking the dynamics changes under practical operations.

(3) Auto startup Mode: Auto startup mode is effective to compute the initial PID parameters without any special pre-information. In this mode, an open loop step response is utilized for the identification and PID computation using the same algorithms.

(4) On-demand Mode: On-demand mode is available when setpoint changes are not preferable because of some operating constraints. When the on-demand tuning is requested, the step signal is applied to the process input only once. The controller estimates the process model using the subsequent closed-loop time response data.

Parameters for Self-tunig Control

Table 1 shows the parameters of self-tunig control and Fig.3 is the tuning panel of the controllers. 'TR' is specified to be an approximately 95% response time of the process, which is used to decide the sampling period and data collecting duration. 'OS' is the control target, which corresponds to the overshoot or rising time of the responses or control criterions shown in Table 2.

The estimated model is displayed in the form of an equivalent 'first lag plus time delay' model with an error of model certainty 'CR'. CR is normalized to the input signal amplitude and duration. If CR is less than 5%, the estimated model is adopted to be a reasonable model for PID computation.

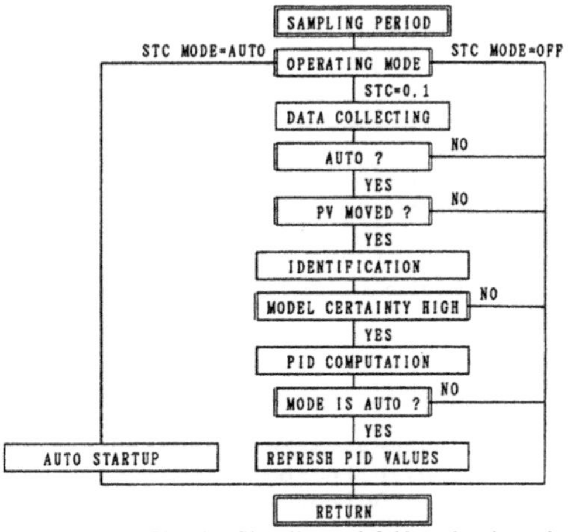

Fig. 4 Flow Chart of Self-tuning Control

TABLE 1 Self-tuning Control Parameters

SIMBOL	MANE	DESCRIPTION
STC	STC MODE .	To specify Self-tuning mode
TR	95% RESPONSE TIME	Approximate value of rising time. Used for sampling time decision.
NB	NOISE BAND	Noise amplitude to be added on PV.
PMIN	MIN OF PROPO RTIONAL BAND	Lower limit of proportional band for self-tuning
IMIN	MIN. OF INTE GRAL TIME	Lower limit of integral time for self-tuning
DMAX	MIN. OF DERI VATIVE TIME	Upper limit of derivative time for self-tuning
MI	TEST SIGNAL AMPLITUDE	Test signal amplitude applied to MV on auto start and ondemand mode
PA	NEW PROPORTI ONAL TIME	Newly computed proportional band
IA	NEW INTEGRAL TIME	Newly computed integral time
DA	NEW DERIVATI VE TIME	Newly computed derivative time
CR	MODELLING ERROR	Error of the identified model certainty
LM	MODEL DEAD TIME	Dead time (delay) of the identified model
TM	MODEL TIME CONSTANT	Time constant (lag)of the identified model
GM	MODEL GAIN	Gain of the identified model

TABLE 2 Control Target

OS	TARGET RESPONSE	CRITRION
0	NO OVERSHOOT	NO OVERSHOOT
1	OVERSHOOT SMALL (5%) SETTLING TIME LARGE	ITAE MINIMUM: $\text{Min.}\ \Sigma \lvert e \rvert t \cdot dt$
2	OVERSHOOT MIDIUM (10 %) SETTLING TIME MEDIUM	IAE MINIMUM: $\text{Min.}\ \Sigma \lvert e \rvert \cdot dt$
3	OVERSHOOT LARGE (15 %) SETTLING TIME SMALL	ISE MINIMUM: $\text{Min.}\ \Sigma e^2 \cdot dt$

IMPLEMENTATION

The self-tuning control function has been embedded into our single loop controllers and distributed process control systems. It is possible for all the control loops to have the self-tuning function, which helps the operators to tune each loops at the plant startup phase.
Combination of the self-tuning function with other control functions and sequence logics are effective for practical applications. For example, it is possible to be adapted to particular process control just like PH process for instance, by combining with nonlinear compensation, deadtime compensation and so on . Start and stop operations of self-tunig control according to the plant operating conditions is also available from other control functions.

SIMULATIONS

Figure 5 indicates a simulation result of self-tunig control. The simulation process is assumed to be as follows:

$$Gp(S) = \exp(-10S) / (1 + 10S)$$

In simulation, step changes are applied to setpoint iteratively. At point①, process output rises slowly because the initial values of proportional band and integral time are large. At point②, the new PID parameters are set up and the process output rises again after the identification using the process behavior between point① and ②. The optimal values are always obtained after point④ to produce exellent responses as a result.
Figure 6 is another example of simulation. The simulated process is an integral process descrived below:

$$Gp(S) = \exp(-10S) / 20S(1 + S)$$

After setpoint changes with large initial PID parameters, the satisfactory PID settings can be obtained.

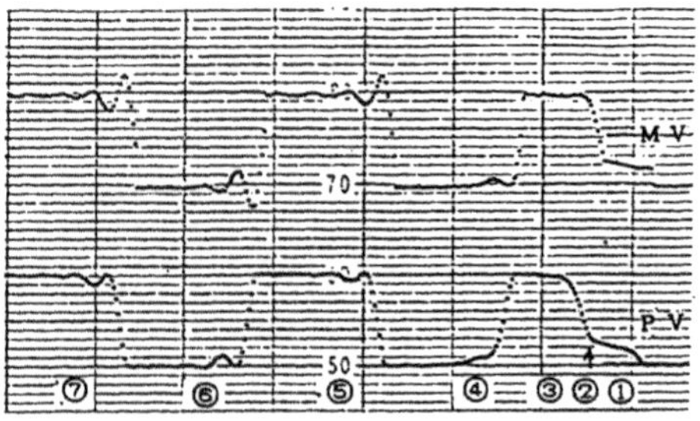

Fig. 5 Simulation Result for a first-lag with delay process

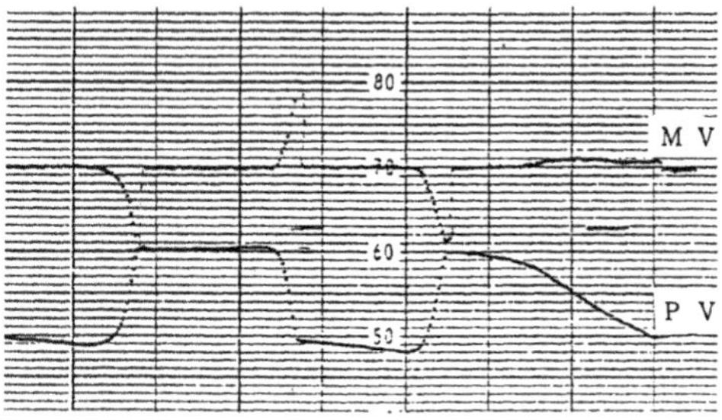

Fig. 6 Simulation Result for an integral process

FIELD TESTS

Some field tests were done in the pilot plant to evaluate the effectiveness of the self-tuning controller. The control target is the flow control and pressure control in a pipe line, each of them interacts each other. Figure 7a) is the step response of the pressure control loop in the setpoint change after auto startup. Figure 7b) shows the effects to the pressure control loop against the setpoint change of the flow control loop. Even though the process has noises and the interactions between control loops, the optimal PID parameters were computed online using the short period of operating data.

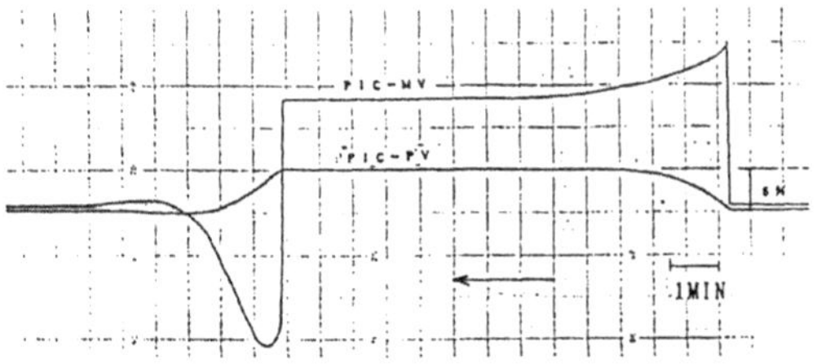

Fig. 7a) Step Response of Presure Control Loop

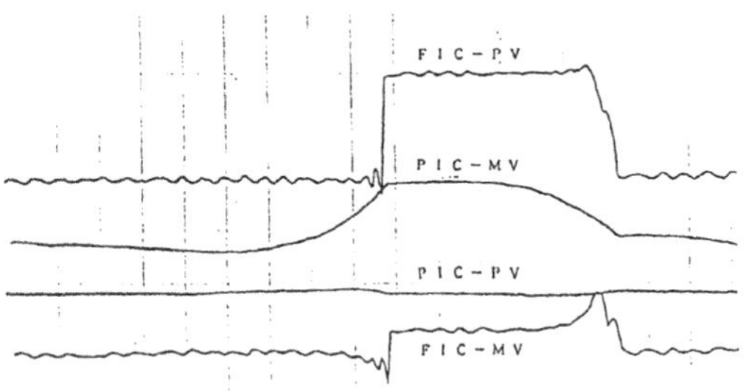

Fig. 7b) Effects of Setpoint Change of Flow Control Loop

CONCLUSIONS

A lot of control techniques have been embedded to the PID controller as "intelligence" in developing the new self-tuning control. One is a process model which is used to compute the PID parameters and detect the process dynamics changes. Another is the identification technique which enables the controller to identify the process dynamics from a short duration of operating data. The other is the PID computation rules to compute the optimal PID values according to the identified process model and the current process behavior. Prediction of future behaivior is also possible using the embedded process model.

REFERENCES

Himmelblau, D. M. (1972). Applied nonlinear progamming. McGraw-Hill.

Kitamori, T. and S. Sin(1990). Special Issue of Adaptive Control. Computrol, 32, Corona, Japan

Yuwana, M., and Seborg, D. E. (1982). A New Method for On-line Controller Tuning. AIchE Journal, 28, 434-440.

Ziegler, J. G. and N. B. Nichols (1942). Optimum Settings for Automatic Controllers. Trans. ASME, 64, 759.

AN ELEMENTARY PATTERN RECOGNITION SELF-TUNING PI-CONTROLLER

M. Klein, T. Marczinkowsky and M. Pandit

Dept. of Electrical Engineering, University of Kaiserslautern, P O Box 3049, D-6750 Kaiserslautern, Germany

Abstract. This paper deals with the modelling of an adaptive control system from an informational point of view in general and the design of an adaptive pattern recognition PI-controller in particular. The design starts off with a state-and-event oriented *overall* adaptation strategy so that the resulting adaptive contoller can be modelled as a finite state machine. The presented *closed-loop* adaptation strategy is based on pattern recognition control which yields essentially process model-free tuning methods. The three basic tasks in the design of adaptive pattern recognition controllers (APRC) are formulated and treated sepatately: (1) pattern description, (2) definition of control objectives in terms of patterns and (3) mapping the patterns to controller settings. In this context, the powerful Elementary Pattern Description (EPD) technique is presented solving the problem of information loss adherent to the other known pattern description techniques. A parameter-adaptive PI controller based on this EPD approach yields good results in simulation studies for linear time-varying stable minimum/nonminimumphase processes with dead time. The application to laboratory plants for the control of flow, pressure and temperature was also sucessful.

Keywords. Adaptive Control; Adaptation Strategy; Elementary Pattern; Expert Control; Intelligent Tuning; PI Control; Real Time Pattern Recognition.

INRODUCTION

Adaptive controllers based on pattern recognition and expert control have proved to be at least as sucessful as model-based adaptive controllers for process industy applications. Pattern recognition adaptive controllers perform the adaptation similar to a human operator: they evaluate the goodness of the control by observing the run of the process output variable, associate the run with performance characteristics and adjust the controller on the basis of empirical knowledge.

The design procedure and informational modelling of a self-tuning pattern recognition PI-controller for disturbed, nonlinear, stable SISO processes with dead time is illustrated. It is assumed that the set-point signal w(t) is piecewise constant. The *assumptions* concerning the processes to be controlled are: (A1) step response stability, (A2) a *static* nonlinearity is monotone and (A3) a *dynamic* nonlinearity can be approximated in the working-point by a linear first or second order model with dead time.

The necessary *a-priori-knowledge* of the plant operator should be: (K1) the range of the controller output signal $[u_{min}, u_{max}]$ and (K2) a rough measure for the transient time T_T of the process step response into a +/- 5% tolerance band around the steady state value.

ADAPTATION STRATEGY AND INFORMATIONAL MODELLING

An adaptive control system consists out of the basic control loop (Fig. 1) and the adaptation system. The chosen adaptation strategy determines which module or functional block of the adaptation system will be implemented. The most important modules (Fig. 2) of the adaptation system can be derived from the basic control problems: set-up into the working-point of the plant, modelling or identification, controller design, emergency treatment, operator I/O and the adaptation as *the* specific problem of adaptive systems.

The guideline of the subsequent design is basically the automation of already sucessfully used methods, i. e. the implementation of expert knowledge.

As shown in Fig. 3, there is a division between an open and closed-loop operation mode. After initialization and the operator input of the set-point w the plant is brought into its working-point - either automatically or manually. Due to this set-up procedure, the plant output y is now in an operator-defined region (tolerance band) around the set-point.

By noise measurement and process step response identification a rough modelling in the working-point is performed. An automatic graphical evaluation of the process step response yields a rough first order model with dead time in the three parameters gain K, lag time T_L and 63% rise time T_R. With these identified values (K,T_L,T_R) the controller sampling time $T_C=0.1*(T_L+T_R)$ and the starting values of the PI controller parameters (K_P,K_I) are computed according to known tuning rules. Then, the operation is changed from open to closed loop mode and the closed-loop adaptation module is activated.

Informational Modelling

The evolution of the design procedure is based on the questions: *When* (state/event), *why* (cause), *what* (model, controller, control signal) and *how* (method) the adaptation is performed? This results in an adaptation strategy which consists of appropriate decisions based on the above questions in any control situation.

This *state-and-event oriented approach* is related to automata theory. Thus, it makes sense to use the models and description methods known in this field. From this point of view, the adaptive controller can be modelled as a *finite state machine*. It is further helpful to divide the adaptive controller into two communicating parts where one part is composed of the *controlled* units or *slave modules* and the other part is the *controlling* unit or *master module*. This interpreted structure (Fig. 2) is the *master-slave state machine* and as a formal description method we choose here an *access graph* which shows the data communication between the action modules (Wendt 1979, 1983).

The actions of the slave modules can be *concurrent* so that a *petri net* is adequate for modelling their synchronization. In contrast to this, the action sequence of the master module is purely *sequential*, since it is a finite state machine. Thus, flow diagrams, flow tables or state graphs (Fig. 3) can be used for its description (Wendt, 1983).

Adapation Strategy

In the context with the master-slave state machine model of the adaptive controller, the notion of *adaptation strategy* shall now be introduced in a *wide* and *strict* sense.

The *overall adaptation strategy* (wide sense) deals with the interaction of all modules of the adaptation system. If the controller is modelled as a master-slave state machine (as we do here), the overall adaptation strategy is the program that is performed by the state graph processor of the master module (Fig. 2); thus it includes *all* the states in Fig. 3.

The *closed-loop adaptation strategy* (strict sense) concerns only the event-oriented adjustment of the controller in the closed-loop operation mode, i.e. only the states in the closed loop (Fig. 3). Within the implemented closed-loop adaptation strategy, four kinds (or states) of adaptation with specific intelligent tuning methods are separated: (1) Closing Tuning, (2) Set-Point-Change Tuning, (3) Critical-Event Tuning and (4) Iterative (Set-Point Change) Optimization Tuning.

The basic idea for these process model free intelligent tuning methods is the direct evaluation of the process output signal y(t) or the error e(t)=w(t)-y(t) and the adjustment of the controller parameters according to prescribed performance criteria on the basis of empirical knowledge. Since all this is done in terms of *patterns*, the topic now is pattern recognition control.

PATTERN RECOGNITION CONTROL

The development of adaptive pattern recognition control is derived from the analysis and description of the cognition and thinking of a human control engineer, particularly its associative and qualitative nature. This analysis yields that the design procedure can be divided into three basic tasks which can be treated separately: pattern description (Task 1), definition of control objectives in terms of patterns (Task 2) and mapping the patterns to controller settings (Task 3).

Pattern Description (Task 1)

The problem is to find an adequate description for a given measured signal which should fulfill the following objectives: (O1) easy to handle and understand, (O2) compact: data reduction, (O3) unique and (O4) precise and complete: full pattern information.

To illustrate the possible pattern description techniques we will now focus on some typical shapes for the error e(t)=w(t)-y(t) as depicted in Fig. 4.

As discussed in (Klein, Marczinkowsky, Pandit, 1990), the two approaches used hitherto are the *qualitative evaluation* and the *key value* techniques. The first method consists of a purely qualitative evaluation of the transient response characteristics which has been classified into several categories such as "too low monotone", "overshoot oscillatory" etc. (Porter, Jones, McKeown, 1987) or "error is small positive" and so on as done by Sripada, Fisher and Morris (1987). The second method takes into account the quantitative information of the signal in terms of easy to handle key values like damping, overshoot and time period (Bristol, 1986).

The Elementary Pattern Description Technique. In contradistinction to these methods of pattern description, starting from a small repertoire of elementary patterns and a set of rules, any given run of variable can be represented as a combination of suitably scaled elementary patterns.

As shown in Fig. 4, the complete measured signal ("holo-pattern") is decomposed into so-called *partial patterns*. Each partial pattern is completely and uniquely described by its *qualitative* shape form, the *elementary pattern* and three *quantitative* values, namely amplitude(=maximum), starttime and duration. These quantitative values can be regarded as *scaling factors* for the *elementary pattern* which is defined with an amplitude (ordinate) and duration (abscisse) of 1.

Assuming the *set of elementary patterns* consists of eight elements as shown in Fig. 5, one can write a *pattern formula* for the measured signal or holo-pattern in Fig. 4. Here, the *holo-pattern H* is composed of six partial patterns Pi, i=1,..,6 where each partial pattern P can be described by the following notation:

Partial Pattern = Pattern Type (amplitude, starttime, duration) or

$$Pi = Ti(Ai,Si,Di) \qquad (1)$$

where Ti is the corresponding Elementary Pattern Ej, j=0,..,7 and Ai,Si,Di are the amplitude, starttime and duration of the Partial Pattern Pi. So, we get the following formula for the Holo-Pattern H:

$$H = P1+P2+P3+P4+P5+P6 = \sum_{i=1}^{6} Ti(Ai,Si,Di)$$

$$= E1(1,0,18) + E2(0.94,18,36) + E5(-0.39,54,75) + E5(0.15,129,75) + E5(-0.05,204,75) + E5(0.02,279,76) \qquad (2)$$

Thus, the number of data necessary to describe the complete signal is only the number of partial patterns multiplied with 4 (or 3, since the starttime is redundant); in this case, 24 or 18 data.

The main problem now is to find the set of elementary patterns which is large enough to describe all practical relevant patterns but small enough for an efficient implementation in real-time.

For arriving at such a set of elementary patterns (as in Fig. 5), a large ensemble of runs of the error signals measured in simulation and in laboratory plants for process control were studied. Using these data the set in Fig. 5 with eight elements was generated.

Out of the holo-pattern formula H the measured signal can then be regenerated - apart from measurement noise which is automatically filtered by the Partial Pattern Recognition Procedure. Therefore, out of the holo-pattern formula H, (a) arbitrary defined key values can be computed and (b) a mapping to an arbitrary defined set of qualitative characteristics can be done.

Herewith, the pattern description problem is solved sufficiently well fulfilling the objectives (O1) through (O4).

Control Objectives (Task 2)

To judge the goodness of the control, performance criteria are needed. Depending on the purpose of the underlying control system and the application-relevant control situations, the number and kind of performance criteria are different. The causes for (desired or undesired) changes in the process output signal are manifold: e.g. deterministic or stochastic disturbances; changes of process/controller parameters, load or set-point. Nevertheless, for practical applications in the processing industry, it is often sufficient to tune the controller such that it exhibits a "good" transient and steady-state behaviour due to a set-point change.

The task now is to define adequate performance criteria in terms of patterns. Typical desired set-point change response shapes can be stated in a qualitative and/or quantitative way using the EPD notation (see also Fig. 5), e.g.

$$H_{D1} = E1 + E2 \quad (3)$$

$$H_{D2} = E1 + E2 + E5 \quad (4)$$

where H_{D1} is a desired response pattern without overshoot and H_{D2} has exactly one overshoot. E1 only occurs when the process has a dead time. For nonminimalphase processes, E7 is needed instead of E2. If additionally *optional* ({})and *choice* operators (|) are introduced,

$$H_{D3} = \{E1\} + (E2|E7) + \{E5\} \quad (5)$$

can be chosen to tolerate one overshoot and to admit processes with or without dead time and minimum or nonminimumphase behaviour. To specify the admissable values of e.g. transient times and overshoot, the three scaling factors (Ai,Si,Di) can be used. These criteria are relevant for the implemented Set-Point Change and Iterative Optimization Tuning.

In the context with intelligent tuning methods, the criteria for starting and breaking up the controller adaptation are very important. In the steady-state case, the desired partial pattern is of the type E0 or a special defined *noise pattern En,* i.e. the process output is within a noise-dependent tolerance-band around the set-point. Especially with the Critical-Event Tuning, E0 and En are used (together with a time condition) in the starting and breaking-up criterion, e.g. if the current partial pattern is E0 or En, no adjustment occurs.

For a sensible choice of these important threshold and timing conditions ("adaptation timing"), a rough process model is needed. This is the main reason for the noise and step response identification in the beginning.

Mapping Patterns to Controller Settings (Task 3)

For adapting the PI parameters on the basis of the holo-pattern identified, intelligent tuning methods with specific search algorithms and expert rules were implemented for each kind of adaptation (Fig. 3). Important items in the development of such tuning algorithms are the direction and step width of search for (K_P,K_I) and the adaptation timing. This knowledge may be attained through extended performance studies on simulated and original processes combined with a systematic data handling.

Thus, one helpful idea was the generation of so-called "(set-point change) process behaviour maps" (Fig. 6) which show the error signal e(t) (due to a set-point change) for a specific process as a function of the controller parameters (K_P,K_I). The underlying control loop is shown in Fig. 1. The empirical knowledge for the tuning methods corresponding to the control situation "set-point change", the Set-Point Change and Iterative Optimization Tuning, is essentially derived from the process behaviour maps of a large ensemble of representative test processes. In a similar way, for other control situations expert rules can be deduced. So, for the Critical-Event Tuning the influence of process and/or controller parameter changes as well as signal disturbances on e(t) can be studied. The change of operation mode from open to closed loop is treated with the Closing Tuning.

These expert rules, incorporating the empirical knowledge, have the following structure (with weights according to the probability of the corresponding action):

IF condition THEN action1 (weight1); action2 (weight2). e.g.

IF the signal is decreasing oscillatory AND the number of overshoots is greater than 1 THEN reduce the integral factor ki (weight=3).

The resulting *general informational structure* of an APRC based on the EPD technique is depicted in Fig. 7. The implemented APRC was tested with simulated and original processes. The results obtained are summarized in the following section.

PERFORMANCE RESULTS

Figure 8 shows the robustness of the implemented APRC (Fig. 8b) with respect to process parameter changes as compared to a fixed PI controller (Fig. 8a) with the same properly tuned starting values. To show the typical control behaviour, a second order process with damping d=0.7 is simulated. At t=0 sec, the process dead time, generally the most critical parameter, is changed from 0.2 sec to 0.6 sec together with the set-point w(t). This results in instability in the case of the fixed controller (Fig. 8a). The APRC rapidly adapts (crosses x in the run of the output variable y(t)) and optimizes the controller parameters with the subsequent set-point changes (Fig. 8b).

Figure 9 depicts the application of the APRC in a laboratory plant for the control of pressure with a process time constant of about 3 sec. At t=0 sec, the initially properly tuned controller parameters (K_P, K_I) are both decreased by the factor 10 together with a set-point change. The APRC exhibits its robustness against nonlinearity, noise and parameter changes (process and/or controller) and optimizes the control with the subsequent set-point changes.

For all test processes investigated, including nonminimalphase and oscillatory systems as well as laboratory plants (flow, pressure, temperature), the APRC has shown to perform better or at least as well as a fixed tuned controller without exception.

CONCLUSIONS

The state-and-event oriented adaptation strategy results in the master-slave state machine model for the adaptive controller. Thus, methods known from modern information technique and automata theory can easily be transferred to our field.

The EPD approach forms a basis for a better or at least a more transparent controller adaptation, since the complete information of the measured signal can be used in an easy to handle notation. This yields potential advantages concerning the design procedure and its result, the final implementation, which is only partially possible when restricted to the other pattern description techniques. Future work will concentrate on the analysis of the mapping between the holo-pattern formula (and/or derived information) and the controller parameters to exploit the information contents obtained by the EPD approach. Further, applications to industrial plants are planned.

REFERENCES

Bristol, E.H. (1986). The EXACT pattern recognition adaptive controller, a user-oriented commercial sucess. in: Narendra, K.S. (1986). *Adaptive and learning systems.* Plenum Press , New York, USA, 149-163.

Klein M., T. Marczinkowsky, M. Pandit (1990). An adaptive controller based on the elementary pattern description technique. *Proc. of BILCON '90, Bilkent Int. Conf.,* Ankara, Turkey, 2-5 July 1990, 711-717.

Porter B., A.H. Jones, C.B. McKeown (1987). Real-time expert tuners for PI-controllers. *IEE Proc.,* Vol. 134, Pt. D, No. 4, 260-263.

Sripada N.R., D.G. Fisher, A.J. Morris (1987). AI application for process regulation and servo control. *IEE Proc.,* Vol. 134, Pt. D, No. 4, 251-259.

Wendt S. (1979). The programmed action module: an element for system modelling. *Digital Processes,* 5, 213-222.

Wendt S. (1983). Nachrichtenverarbeitung. Vol. 3 of K. Steinbuch, W. Rupprecht: *Nachrichtentechnik.* Springer Verlag, Heidelberg, F.R.G.

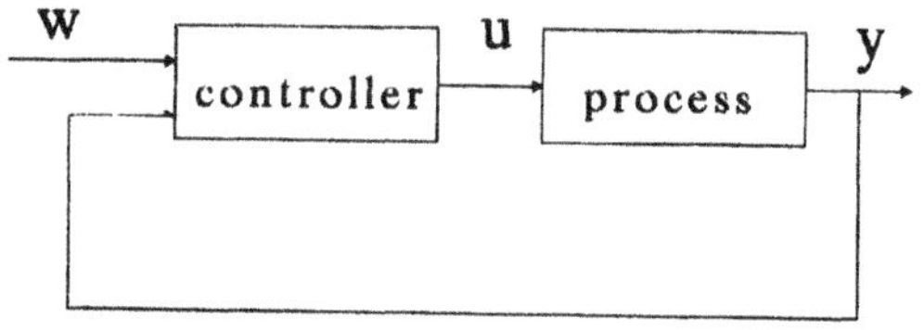

Fig. 1. Basic Control Loop.

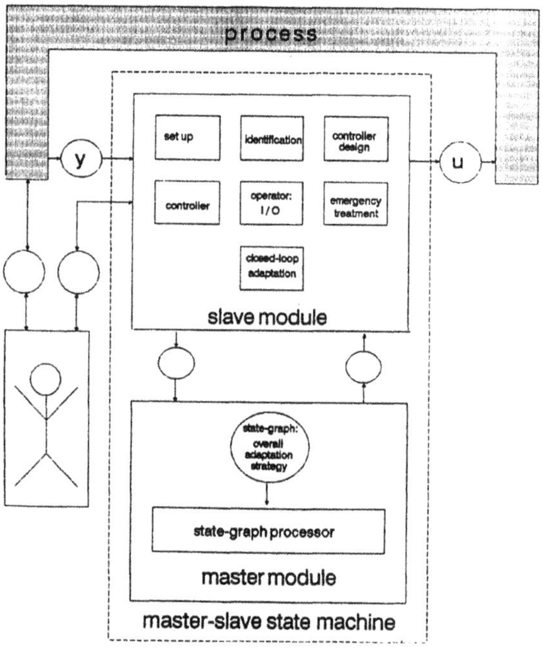

Fig. 2. Master-Slave State Machine Model of the Adaptive System.

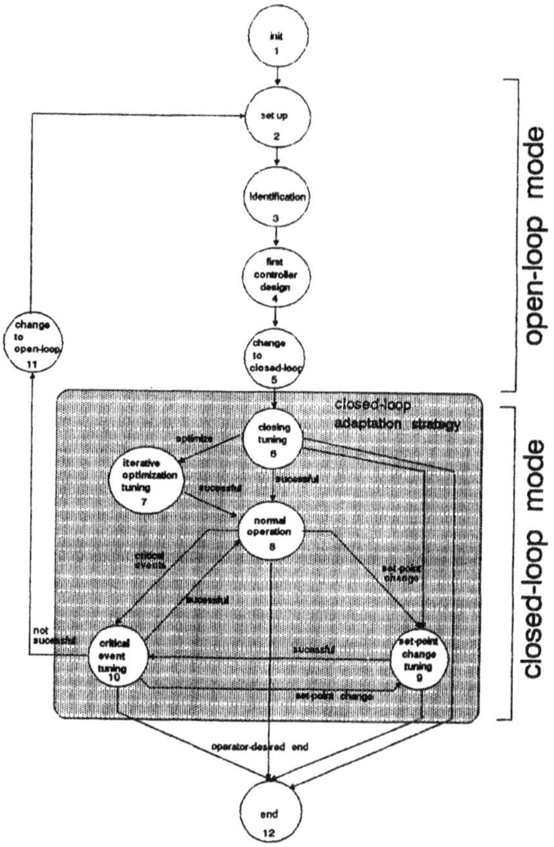

Fig. 3. State Graph of the Overall Adaptation Strategy.

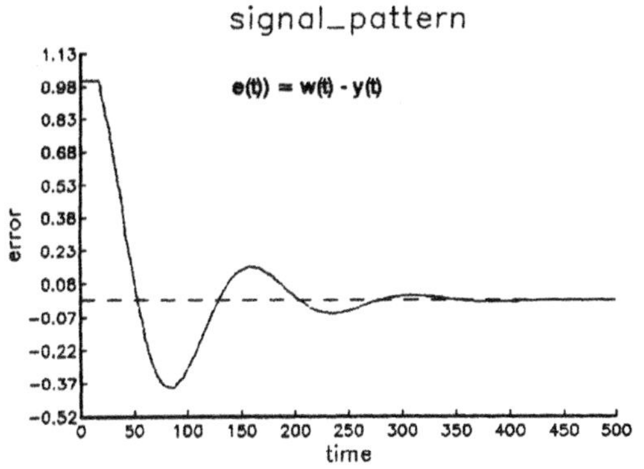

Fig. 4. Typical measured Signal.

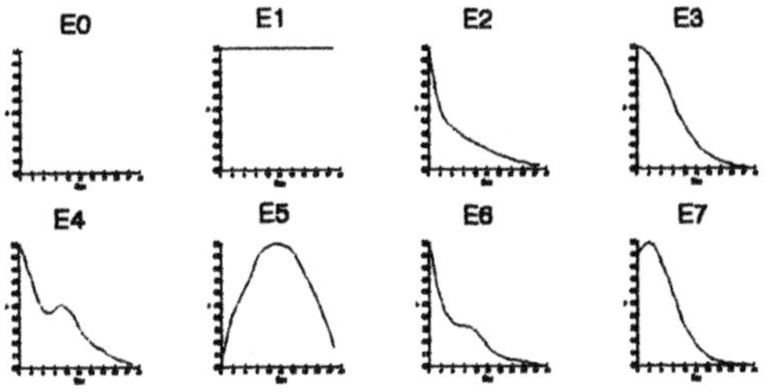

Fig. 5. Possible Set of Elementary Patterns.

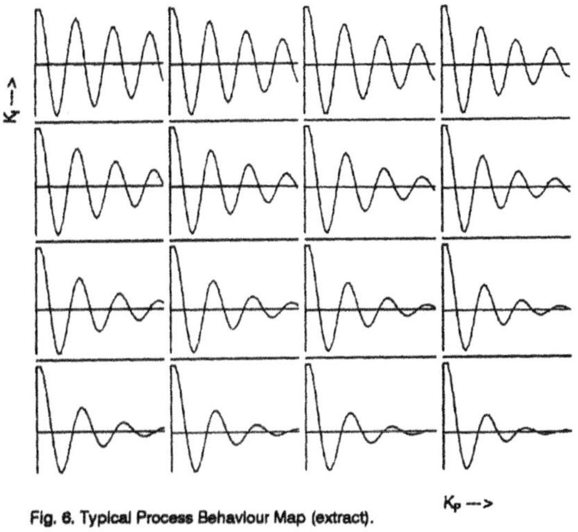

Fig. 6. Typical Process Behaviour Map (extract).

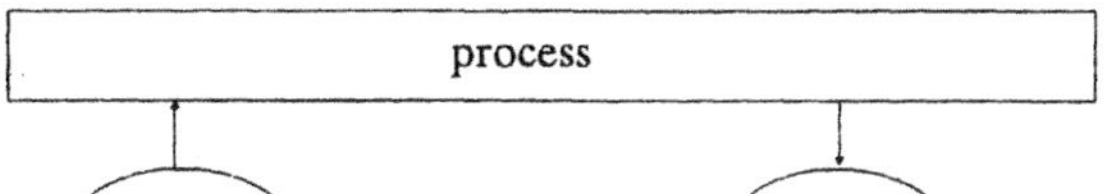

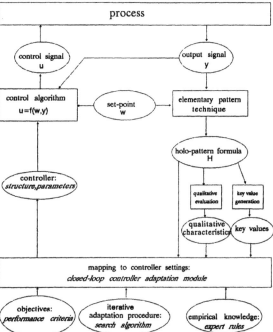

Fig. 7. Access Graph of an APRC based on the EPD Technique.

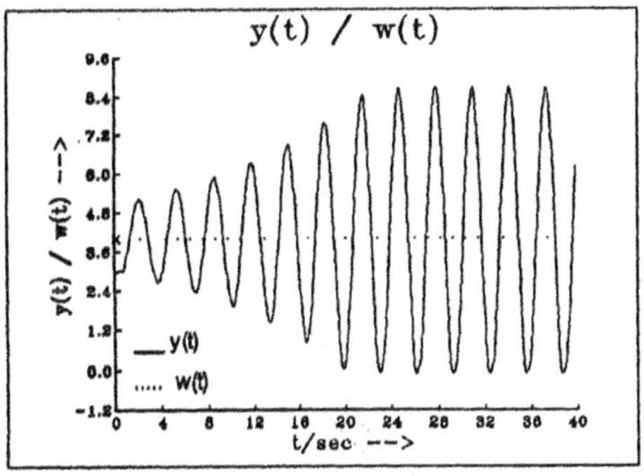

Fig. 8a. Fixed Controller.

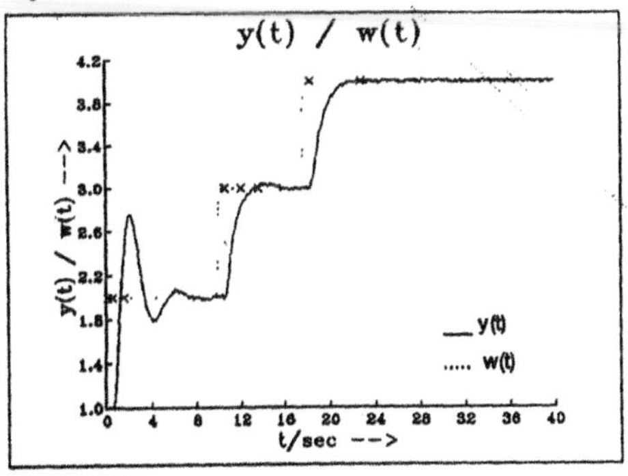

Fig. 8b. APRC.

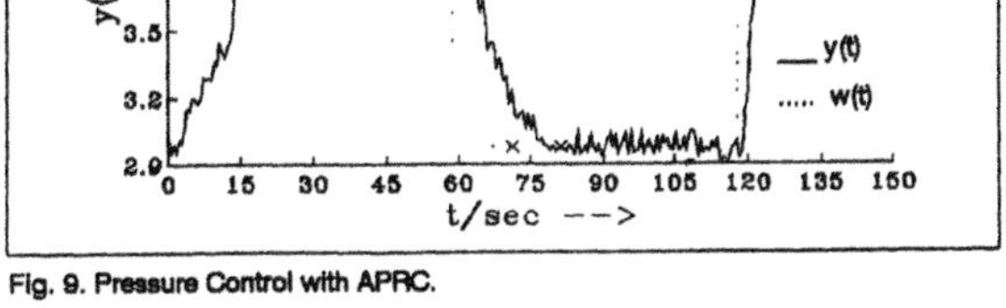

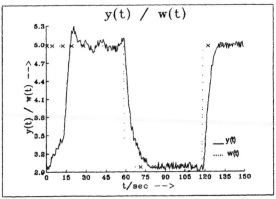

Fig. 9. Pressure Control with APRC.

A DESIGN OF HYBRID ADAPTIVE CONTROL HAVING COMPUTATIONAL DELAYS

S. Shin

*Institute of Information Sciences and Electronics,
University of Tsukuba, Ten-nodai, Tsukuba City 305, Japan*

Abstract. This paper presents a design of hybrid adaptive control which affords computational delays caused by updating estimators for unknown parameters. This is realized by making an artificial phase difference between frame periods of the control and the estimation. This difference is used for the computation of updating estimators with a similar adaptive law as the discrete-time one, although the control is performed in the continuous-time. The global stability of the proposed hybrid adaptive control is also presented briefly here.

Keywords. Model reference adaptive control; hierarchially intelligent control; digital control; microcomputer-based control; recursive estimation; convergence analysis; stability.

INTRODUCTION

There are some literatures (Elliotto, 1982; Miyasato and Oshima, 1987; Narendra, Khalifa and Annaswamy, 1985; Shin, 1989) concerning the hybrid adaptive control which has a continuous-time control law and a discrete-time like adaptive law.

One of main features of this control system is suitable for the digital control. The estimators are updated not so frequently compared with the continuous-time adaptive control and the computation burden of updating estimators is less in the hybrid adaptive control.

However, there remains another aspect if we want to reduce the computational burden. That is, usual hybrid adaptive controls does not afford computational delay for updating estimator as shown in **Fig. 1**. At time t_k, an adaptive law updates estimators based on the input and the output of the controlled object up to this time t_k. At the same time, the controller needs this newly updated estimators for the control. Therefore, a digital computer used for this control has no time for computing.

In the practical situation, the computer needs a little bit time for this calculation. Therefore, we propose here the new hybrid adaptive control which has a phase difference between frame periods of the control and the estimation as shown in **Fig. 2**. This control may require two CPUs (Central Processing Units) as shown in **Fig. 3**. One is used for control. It gathers the input and the output data and generates the control input continuously. Another is used for estimation. It updates estimators based on the gathered input and output of the controlled object up to time t_{ek} and transfers this updated estimators to the CPU for control till time t_{ck}, where $t_{ck} > t_{ek}$. Therefore, the proposed control affords computational time for updating estimators.

In the following sections, we present a detailed description and a brief proof of the global stability of the proposing hybrid adaptive control.

SYSTEM DESCRIPTION

A controlled object considered here is a continuous-time single-input and single-output n-th order linear system described by,

$$y(t) = \mathbf{c}^T\mathbf{x}(t) \qquad (1\ a)$$

$$s\mathbf{x}(t) = A\mathbf{x}(t) + \mathbf{b}u(t) \qquad (1\ b)$$

where ^{T} means the transpose of vectors or matrices and s is the time differential operator. Scalars, $y(t) \in R$ and $u(t) \in R$ respectively, are the output and the input of the controlled object and n-th dimensional vector $\mathbf{x}(t) \in R^n$ is a state of the controlled object, where R means the field of real numbers. Elements of matrix $A \in R^{n \times n}$ and vectors $\mathbf{c}, \mathbf{b} \in R^n$ are unknown here. However, the system degree n and the relative degree m;

$$m = \min\{i \mid \mathbf{c}^T A^{i-1}\mathbf{b} \neq 0\} \qquad (2)$$

is assumed to be known. This controlled object is also assumed to be controllable, observable, and stably invertible for ensuring the global stability of the designed adaptive control system.

The reference model to be followed is,

$$y_m(t)= c_m^T x_m(t) \qquad (3\ a)$$

$$s x_m(t)= A_m x_m(t)+b_m r(t) \qquad (3\ b)$$

where $y_m(t)\in R$, $r(t)\in R$ and $x_m(t)\in R^n$ respectively are the output, the input, and the state of the reference model. This model must be stable and r(t) is a piecewise continuous bounded function. A matrix A_m and vectors c_m, b_m have their own suitable sizes. However, the relative degree m_m of the reference model;

$$m_m= \min \{i \mid c_m^T A_m^{i-1} b_m \neq 0\} \qquad (4)$$

must satisfy,

$$m_m \geq m \qquad (5)$$

for avoiding differentiations of r(t) as described in the later.

Since the relative degree m is defined by (2), we get,

$$y(t)= c^T A^{i-1} x(t), \text{ for } \forall i= 1,2, \cdots, m \qquad (6\ a)$$

$$y(t)= c^T A^m x(t)+c^T A^{m-1} b u(t) \qquad (6\ b)$$

from successive differentiations of (1 a) and substitutions of (1 b). With a stable polynomial $q(s)$ with degree m and (6), the controlled object can be represented as,

$$q(s)y(t)= f^T x(t)+gu(t), \qquad (7)$$

where,

$$f= \{c^T q(A) b\}^T \qquad (8\ a)$$

$$g= c^T A^{m-1} b \qquad (8\ b)$$

It is needless to say that the definition of the relative degree, (2), gurantees $c^T A^{m-1} b$ is not zero.

In order to eliminate direct use of the state, we introduce here an observer written as,

$$x(t)= K_1 z_1(t)+K_2 z_2(t)+\varepsilon_0(t) \qquad (9\ a)$$

$$s z_1(t)= A_0 z_1(t)+b_0 u(t) \qquad (9\ b)$$

$$s z_2(t)= A_0 z_2(t)+b_0 y(t) \qquad (9\ c)$$

where $z_1, z_2 \in R^n$ and $A_0 \in R^{n\times n}$, $b_0 \in R^n$ should be so specified that A_0 is stable and (A_0, b_0) becomes a controllable pair. A vector $\varepsilon_0(t)\in R^n$ is a decaying term proportional to $\exp(A_0 t)$. Matrices $K_1 \in R^{n\times n}$ and $K_2 \in R^{n\times n}$ are unknown here since they depend on the system matrices A, **b**, **c**, whose elements are assumed to be unknown.

Combining (7) and (9 a), we get a nonminimum representation of the controlled object as follows,

$$q(s)y(t)= k_1^T z_1(t)+k_2^T z_2(t)+gu(t)+\varepsilon_1(t) \qquad (10)$$

where $\varepsilon_1(t)$ is an exponentially decaying term and,

$$k_i= \{f^T K_i\}^T, \quad i= 1,2 \qquad (11)$$

From this representation of the controlled object, we can easily derive the control law which makes y(t) follow $y_m(t)$. That is,

$$\begin{aligned} u(t) &= g^{-1}\{-k_1^T z_1(t)-k_2^T z_2(t)+q(s)y_m(t)\} \\ &= g^{-1}\{-k_1^T z_1(t)-k_2^T z_2(t) \\ &\quad +q(s)c_m^T(sI-A_m)^{-1} b_m r(t)\}. \end{aligned} \qquad (12)$$

The condition (5) guarantees r(t) is differential free. In order to verify viability of this control law, u(t) in the right hand side of (10) is substitute by (12). Then, we get,

$$q(s)e(t)= \varepsilon_1(t) \qquad (13)$$

where the output error e(t) is defined by,

$$e(t)= y(t)-y_m(t) \qquad (14)$$

The stability of q(s) and (13) ensure,

$$\lim_{t\to\infty} e(t)= 0 \qquad (15)$$

Then, the control law turns out to be suitable for the model reference control.

ADAPTIVE CONTROL

The control law, (12), is not realizable since the controlled object considered here is not perfectly known. The unknown parameters, g and k_i in (12), should be estimated.

In order to eliminate differentiations of y(t) in the left hand side of (10), we multiply $q(s)^{-1}$ both sides of (10). Then, we get,

$$y(t)= k_1^T \zeta_1(t)+k_2^T \zeta_2(t)+g\upsilon(t)+\varepsilon_2(t) \qquad (16)$$

where $\varepsilon_2(t)$ is an exponentially decaying term and,

$$\zeta_i(t)= q(s)^{-1} z_i(t), \text{ for } i= 1,2 \qquad (17\ a)$$

$$\upsilon(t)= q(s)^{-1}u(t) \tag{17 b}$$

Since $\mathbf{k}_1, \mathbf{k}_2$, and g are unknown in (10), they will be estimated with the following estimation model,

$$y_e(t)= \mathbf{k}_{e1}(t)^T\zeta_1(t)+\mathbf{k}_{e2}(t)^T\zeta_2(t)+g_e(t)\upsilon(t) \tag{18}$$

where $\mathbf{k}_{e1}(t)\in R^n$, $\mathbf{k}_{e2}(t)\in R^n$, and $g_e(t)\in R$ are estimators for corresponding unknown values and $y_e(t)\in R$ is an estimator for y(t). Furthermore, since we consider here the hybrid adaptive control based on the timing shown in Fig. 3,

$$\mathbf{k}_{ei}(t)= \mathbf{k}_{ei}(t_{ek}), \quad \text{for } \forall t\in [t_{ek}, t_{ek+1}], \text{ for } i= 1,2 \tag{19 a}$$

$$g_e(t)= g_e(t_{ek}), \quad \text{for } \forall t\in [t_{ek}, t_{ek+1}] \tag{19 b}$$

where t_{0k} should be less or equal to t_{ek}.

Based on the estimation error $e_e(t)$ defined by,

$$e_e(t)= y(t)-y_e(t) \tag{20}$$

these estimators are updated after time t_{ek+1} till t_{ck+1} by,

$$\mathbf{k}_{ei}(t_{ck+1})= \mathbf{k}_{ci}(t_{ek})+\alpha_k b_k \int_{t_{ek}}^{t_{ek+1}} \zeta_i(\tau)e_e(\tau)d\tau \quad \text{for } i= 1,2 \tag{21 a}$$

$$g_e(t_{ck+1})= g_e(t_{ek})+\alpha_k b_k \int_{t_{ek}}^{t_{ek+1}} \upsilon(\tau)e_e(\tau)d\tau \tag{22 b}$$

$$0< \alpha_k < 2 \tag{22 c}$$

$$b_k= [\lambda+ \int_{t_{ek}}^{t_{ek+1}} \{ \sum_{i=1}^{2} \zeta_i(\tau)^T\zeta_i(\tau)+\upsilon(\tau)^2\}d\tau]^{-1} \tag{22 d}$$

where λ is a positive number. This is the same adaptive law used by Narendra, Khalifa and Annaswamy (1985), except the integral interval.

The control law is defined by,

$$u(t)= g_c(t)^{-1}\{-\mathbf{k}_{c1}(t)^T\mathbf{z}_1(t)-\mathbf{k}_{c2}(t)^T\mathbf{z}_2(t) +q(s)y_m(t)\}$$

$$= g_c(t)^{-1}\{-\mathbf{k}_{c1}(t)^T\mathbf{z}_1(t)-\mathbf{k}_{c2}(t)^T\mathbf{z}_2(t) +q(s)\mathbf{c}_m{}^T(sI-A_m)^{-1}\mathbf{b}_m r(t)\} \tag{23}$$

where,

$$\mathbf{k}_{ci}(t)= \mathbf{k}_{ei}(t_{ek}), \quad \text{for } \forall t\in [t_{ck}, t_{ck+1}], \text{ for } i= 1,2 \tag{24 a}$$

$$g_c(t)= g_e(t_{ek}), \quad \text{for } \forall t\in [t_{ck}, t_{ck+1}] \tag{24 b}$$

It can be seen (19) and (24) satisfy the timing shown in Fig. 3.

ANALYSIS

In this section, we show the stability and the convergence of the output error of the hybrid adaptive controller designed in the section 3.

Let a performance measure v(k) for adaptation be,

$$v(k)=\sum_{i=1}^{2}\mathbf{k}_{di}(t_{ek})^T\mathbf{k}_{di}(t_{ek})+g_d(t_{ek})^2 \tag{25}$$

where,

$$\mathbf{k}_{di}(t_{ek})= \mathbf{k}_i-\mathbf{k}_{ei}(t_{ek}), \quad \text{for } i= 1,2 \tag{26 a}$$

$$g_d(t_{ek})= g-g_e(t_{ek}) \tag{26 b}$$

In the following analysis, exponential decaying terms are omitted. Difference of v(k) along (21) becomes,

$$v(k+1)-v(k)$$

$$= -2\alpha_k b_k \int_{t_{ek}}^{t_{ek+1}} \{\sum_{i=1}^{2}\mathbf{k}_{di}(t_{ek})^T\zeta_i(\tau) +g_d(t_{ek})\upsilon(\tau)\}e_e(\tau)d\tau$$

$$+(\alpha_k b_k)^2 \sum_{i=1}^{2}[\{ \int_{t_{ek}}^{t_{ek+1}} \zeta_i(\tau)e_e(\tau)d\tau\}^T \cdot \int_{t_{ek}}^{t_{ek+1}} \zeta_i(\tau)e_e(\tau)d\tau]+\{ \int_{t_{0k}}^{t_{ek+1}} \upsilon(\tau)e_e(\tau)d\tau\}^2$$

$$\le -2\alpha_k b_k \int_{t_{ek}}^{t_{ek+1}} e_e(\tau)^2d\tau +(\alpha_k b_k)^2 \int_{t_{ek}}^{t_{ek+1}} e_e(\tau)^2d\tau(b_k{}^{-1}-\lambda)$$

$$\le -\alpha_k(2-\alpha_k)b_k \int_{t_{ek}}^{t_{ek+1}} e_e(\tau)^2d\tau \tag{27}$$

Since α_k is restricted as (22 c), the difference is semi negative definite. Then, the semi positive definite function v(k) converges to a non-negative constant as t increased. This means the uniform boundness of estimators $k_{ei}(t_{ek})$, $g_e(t_{ek})$ and,

$$\lim_{t\to\infty} b_k \int_{t_{ek}}^{t_{ek+1}} e_e(\tau)^2 d\tau = 0 \quad (28\text{ a})$$

$$\lim_{t\to\infty} \{\mathbf{k}_{ei}(t_{ek+1})-\mathbf{k}_{ei}(t_{ek})\} = \mathbf{0} \quad (28\text{ b})$$

$$\lim_{t\to\infty} \{g_e(t_{ek+1})-g_e(t_{ek})\} = 0 \quad (28\text{ c})$$

It is needless to say that differences, $\mathbf{k}_{ei}(t_{ek+1})-\mathbf{k}_{ei}(t_{ek})$ and $g_e(t_{ek+1})-g_e(t_{ek})$ are uniformly bounded. From the analysis given by Narendra, Khalifa and Annaswamy (1985), (28 a) means,

$$\lim_{t\to\infty} b(t)|e_e(t)| = 0 \quad (29)$$

where,

$$b(t) = [\sup_{\tau\le t}\{\zeta_i(\tau)^T\zeta_i(\tau)+\upsilon(\tau)^2\}]^{-1} \quad (30)$$

From (14), (18), (20), and (23), the estimation error $e_e(t)$ and the output error $e(t)$ are related as,

$$e(t) = e_e(t) - q(s)^{-1}\{g_c(t)u(t)\} + g_e(t)\upsilon(t) + \sum_{i=1}^{2}[q(s)^{-1}\{\mathbf{k}_{ci}(t)^T\mathbf{z}_i(t)\} - \mathbf{k}_{ei}(t)^T\zeta_i(\tau)] \quad (31)$$

Then, we get,

$$\begin{aligned} e(t) = {} & e_e(t) - [q(s)^{-1}\{g_c(t)u(t)\} - g_c(t)q(s)^{-1}\{u(t)\}] \\ & - \sum_{i=1}^{2}[q(s)^{-1}\{\mathbf{k}_{ci}(t)^T\mathbf{z}_i(t)\} - \mathbf{k}_{ci}(t)^T q(s)^{-1}\{\mathbf{z}_i(t)\}] \\ & + [g_c(t)-g_e(t)]q(s)^{-1}\{u(t)\} \\ & + \sum_{i=1}^{2}[\mathbf{k}_{ci}(t)-\mathbf{k}_{ei}(t)]^T q(s)^{-1}\{\mathbf{z}_i(t)\} \end{aligned} \quad (32)$$

Therefore,

$$\begin{aligned} b(t)|e(t)| = {} & b(t)|e_e(t)| \\ & + b(t)|q(s)^{-1}\{g_c(t)u(t)\} - g_c(t)q(s)^{-1}\{u(t)\}| \\ & + b(t)|\sum_{i=1}^{2}[q(s)^{-1}\{\mathbf{k}_{ci}(t)^T\mathbf{z}_i(t)\} \\ & - \mathbf{k}_{ci}(t)^T q(s)^{-1}\{\mathbf{z}_i(t)\}]| \\ & + b(t)|[g_c(t)-g_e(t)]q(s)^{-1}\{u(t)\}| \\ & + b(t)|\sum_{i=1}^{2}[\mathbf{k}_{ci}(t)-\mathbf{k}_{ei}(t)]^T q(s)^{-1}\{\mathbf{z}_i(t)\}| \end{aligned} \quad (33)$$

From (19), (24), (28 b), and (28 c),

$$\lim_{t\to\infty} \{\mathbf{k}_{ci}(t)-\mathbf{k}_{ei}(t)\} = \mathbf{0} \quad (34\text{ a})$$

$$\lim_{t\to\infty} \{g_c(t)-g_e(t)\} = 0 \quad (34\text{ b})$$

This means the 4th and the 5th terms of the right hand side of (33) are bounded and go to zero as t increased since $b(t)q(s)^{-1}\{u(t)\}$ and $b(t)\|q(s)^{-1}\{\mathbf{z}_i(t)\}\|$ are bounded uniformly with respect to time t, where $\|\bullet\|$ is Euclid norm. From the lemma given by Narendra, Khalifa and Annaswamy (1985), the 2nd and the 3rd terms of the right hand side of (33) are also bounded and go to zero as t increased. After all, (29) can be rewritten as,

$$\lim_{t\to\infty} b(t)|e(t)| = 0 \quad (35)$$

This and the stably invertibility of the controlled object lead to the boundness of all variables in the adaptive control system considered here and,

$$\lim_{t\to\infty} e(t) = 0 \quad (36)$$

from the lemma given by Narendra, Khalifa and Annaswamy (1985).

SIMULATION

A numerical simulation is performed on the 2nd order linear sytem where,

$$A = \begin{pmatrix} -0.3 & 0 \\ 1 & -0.3 \end{pmatrix}$$

$$\mathbf{b} = (1\ 0)^T, \quad \mathbf{c} = (0,1)^T \quad (37)$$

Then, the relative degree becomes 2. The reference model and the reference input respectively are,

$$y_m(t) = (s+1)^{-2} r(t) \quad (38\text{ a})$$

$$r(t) = 1.5\sin(2\pi t/5) \quad (38\text{ b})$$

The other settings are,

$$A0 = \begin{pmatrix} -1 & 0 \\ 1 & -1 \end{pmatrix} \quad (39\text{ a})$$

$$b_0 = (1\ 0)^T \quad (39\text{ b})$$

$$q(s) = (s+1)^2 \tag{39 c}$$

$$t_{ek} - t_{ek-1} = t_{ck} - t_{ck-1} = 1, \text{ for all } k \tag{39 d}$$

$$t_{ck} - t_{ek} = 0.2, \text{ for all } k \tag{39 e}$$

Figure 4 is a result of the simulation. It can be shown that the proposing adaptive control system works well.

CONCLUSION

We present here a design and an analysis of the hybrid adaptive control affording computatiohal delay. This work is partially supported by University of Tsukuba Project Research.

REFERENCES

Elliott, H.(1982). Hybrid adaptive control of continuous time system. *IEEE Tr. Automat. Contr.*, AC-27, 419-426.

Miyasato, Y., and Y. Oshima (1987). A design of learning control systems, *Tr. SICE*, 23, 576-583

Narendra, K. S., I. H. Khalifa, and A. M. Annaswamy (1985). Error models for stable hybrid adaptive systems. *IEEE Tr. Automat. Contr.*, AC-30, 339-347.

Shin, S. (1989). Hybrid model reference adaptive control based on weighting functions, *Preprint of the 18th SICE Symposium on Control Theory*, Kobe, JAPAN, 99-102

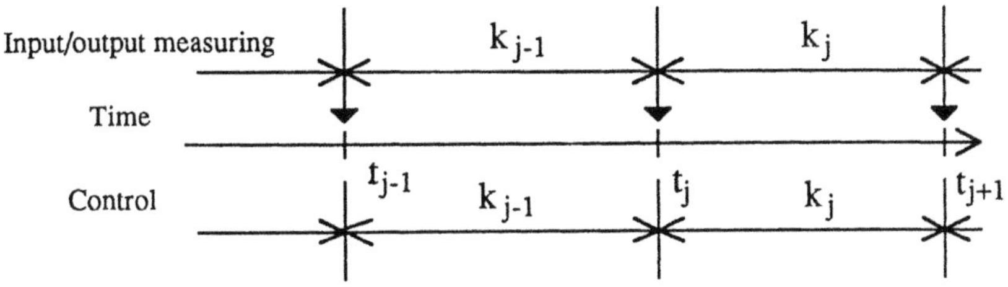

Fig. 1. Update timing used for usual hybrid adaptive control.

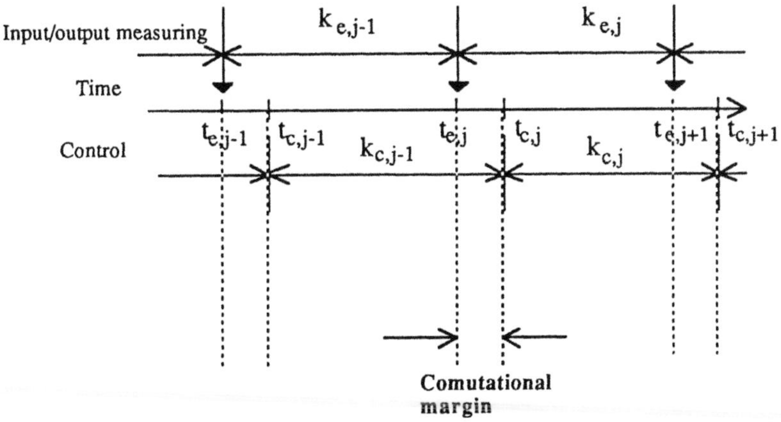

Fig. 2. Update timing used for proposed hybrid adaptive control.

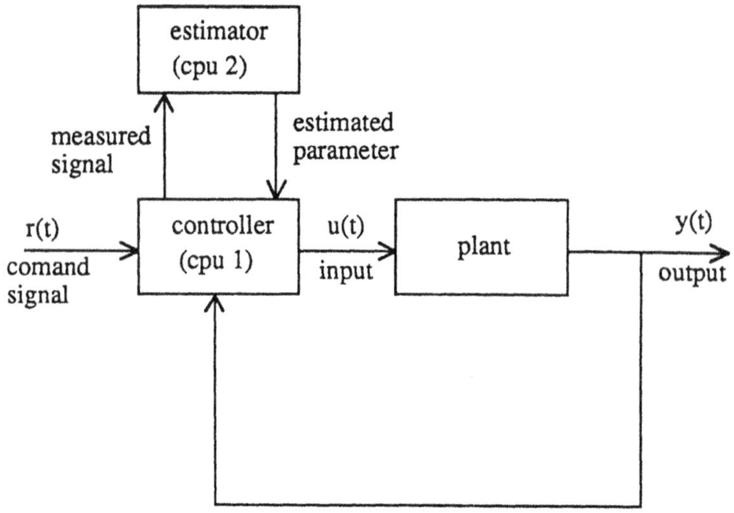

Fig. 3. Structure of the proposing hybrid adaptive control system. The controller measures the output and synthesizes the input. Based on these values, the estimator updates estimators and put them in the controller.

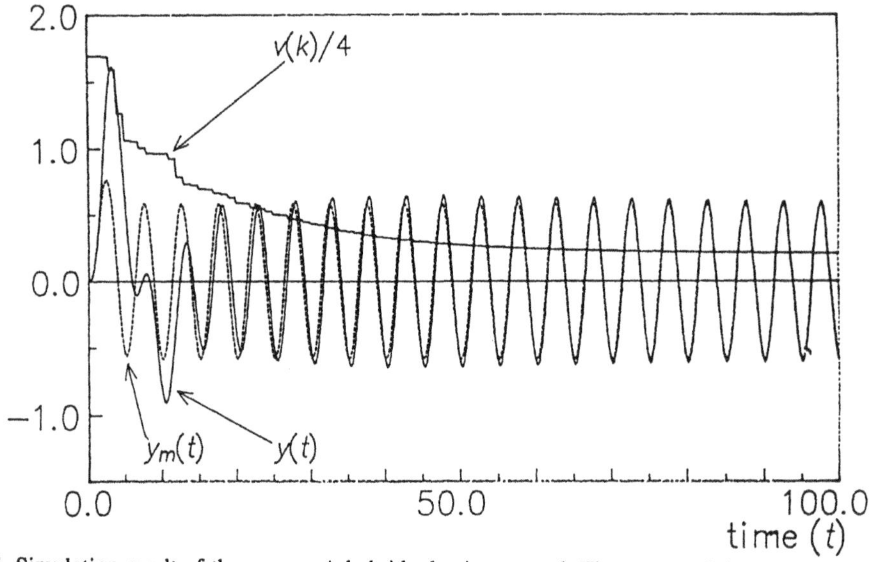

Fig. 4. Simulation result of the proposed hybrid adaptive control. The output of the controlled object, y(t) becomes asymptotically to follow the output of the reference model and the parameter estimation error v(t) decreases stepwise since discrete-time like adaptive law is used.

ROBUSTNESS OF THE COMBINED MRAC USING KNOWLEDGE OF A BOUND ON $\| \theta^* \|$

M. A. Duarte

Dept. of Electrical Engineering, University of Chile, Casilla 412-3, Santiago, Chile

Abstract. The combined model reference adaptive control (CMRAC) scheme, that brings together direct and indirect adaptive control methods has been extensively studied in the ideal case. Under non ideal conditions this algorithm has also been tested and quite good performance was also detected. In this paper the CMRAC is modified in order to include the knowledge of a bound on the norm of the desired controller parameter vector, which is somehow related to the actual plant parameters. This information is used to modify both control adaptive laws as well as identification adaptive laws. It is shown that the modified CMRAC gives a stable adaptive scheme in the presence of bounded external disturbances. Thus, CMRAC suitably modified is also useful to face this kind of robust adaptive control problems. The study is done for the continuous-time case and a first order plant is treated in detail. Extension of the method to high order plants is also included. A comparison with the direct MRAC using this kind of extra information indicates that transient response is improved if the modified CMRAC is used. The main contribution of the paper is the presentation of a new robust MRAC method based on the combination of direct and indirect adaptive control philosophies, which is globally stable in the presence of bounded external perturbations acting on the input and/or output of the plant. The advantage over the direct MRAC lies on the fact that a better transient response can be obtained with the proposed method due to flexibility in the choice of design parameters and adaptive gains.

Keywords. Adaptive control, adaptive systems, model reference adaptive control, robust adaptive control, direct and indirect adaptive control, combined adaptive control.

INTRODUCCION

Robust model reference adaptive control is an area of active research in automatic control. Several new and important methods, robust in the presence of external perturbations have been proposed by different authors (Kreisselemeier and Anderson 1986; Ioannou and Tsakalis 1986; Narendra and Annaswamy 1985, 1986, etc). One group of methods modify the standard adaptive laws whereas other simply keep the same adaptive laws as in the ideal case. Peterson and Narendra (1982), Samson (1983) and Egart (1979) introduce the concept of dead zone to modify adaptive laws to provide stability of the overall adaptive system in presence of external disturbances. Ioannou and Kokotovic (1983) use the so called $\sigma\theta$ modification in the adaptive law so that stability of the adaptive system is achieved when external perturbations act on the plant. Narendra and Annaswamy (1987) proposed the $|e|\theta$ modification with better results than the $\sigma\theta$ modification and again global stability is guaranteed in the presence of bounded external perturbations. Finally, Kreisselmeier and Narendra (1982) use the knowledge of a bound on the desired controller parameter θ^* to meet the same objective. All the previous approaches use a direct controller and control parameters are adjusted based on tracking error and modified adaptive laws.

One the other hand a combined approach to model reference adaptive control has been recently introduced in control literature by Duarte and Narendra (1989, 1989a, 1988). This method combines the direct and indirect philosophies to model reference adaptive control and exhibits some interesting characterists that make the method an attractive tool to study robust adaptive control problems (Duarte 1989b). The method has also been tested under non ideal conditions (Narendra and Duarte 1988, 1989) and a better transient bebaviour of the combined method with respect to the component methods has also been observed. These facts have motivated the analysis of the combined model reference adaptive control (CMRAC) when modifications on the adaptive laws are included. The $|e|\theta$ modification (Narendra and Annaswany 1987) has been used in Duarte (1990a, 1991) and the dead zone type of ideas (Kreisselmeier and Narendra 1982) has been used in Duarte (1990).

In this paper, the combined MRAC is modified to include knowledge of a bound on $\|\theta^*\|$. With this modification the CMRAC results in a robust scheme with respect to bounded external perturbations where the knowledge of a bound on the external perturbations is not needed.

First order plants are treated in detail for the sake of clarity. Later, analysis of nth order plants is done. Simulation results for first order plants are included to compare the behaviour of the CMRAC and the direct MRAC.

CMRAC OF FIRST ORDER PLANTS

All the concepts involved in the proposed method are better understood studying a first order plant. In the next section these concepts will be extendend to n^{th} order plants. Here, first order plants are treated in detail to realize all the ingredients of the method.

Let consider a first order plant described by equation

$$\dot{y}_p(t) = -a_p y_p(t) + u(t) + p(t), \quad (1)$$

where $u(\cdot)$, $y_p(\cdot)$: $R^+ \rightarrow R$ are the input and output of the plant respectively. $p(\cdot)$: $R^+ \rightarrow R$ is an external perturbation affecting plant input and it is assumed that $p(t)$ is a bounded time function. $a_p \in R$ is a constant unknown plant parameter.

Let the model reference be defined by the differential equation

$$\dot{y}_m(t) = -a_m y_m(t) + r(t), \quad (2)$$

with am > 0, i.e. the reference model is asymptotically stable and it is assumed that $r(\cdot)$: $R^+ \rightarrow R$ is a bounded piecewise continuous time function. The control law to be used for a first order plant is as follows:

$$u(t) = \theta(t)y_p(t) + r(t), \quad (3)$$

where $\theta(\cdot)$: $R^+ \rightarrow R$ is a controller adjustable parameter. The way in which this parameter is to be adjusted will be defined later.

The identification model to be used in the combined MRAC method has the form:

$$\dot{\hat{y}}_p(t) = a_i e_i(t) - \hat{a}_p(t)y_p(t) + u(t), \quad (4)$$

where $a_i > 0$, $\hat{y}_p(t)$, $\hat{a}_p(t)$ are the estimates of $y_p(t)$ and a_p at time 't' and $e_i(t) = y_p(t) - \hat{y}_p(t)$ is the identification (output) error.

It can be easily verified that the desired controller parameter θ^*, the true plant parameter a_p and the model reference parameter a_m satisfy the following relationship

$$\theta^* = a_p - a_m.$$

Based on this equation closed-loop estimation error $\epsilon(t)$ is computed by replacing true parameters by their estimates (Duarte and Narendra 1989,1989a),

$$\epsilon(t) = \theta(t) - \hat{a}_p(t) + a_m. \quad (5)$$

If we define the parameter errors as $\emptyset(t) = \theta(t) - \theta^*$ and $\delta(t) = \hat{a}_p(t) - a_p$, together with the control (tracking) error $e_c(t) = y_p(t) - y_m(t)$, it is possible to write the following set of equations describing the dynamical behaviour of control, identification and closed-loop estimation errors

$$\dot{e}_c(t) = -a_m e_c(t) + \emptyset(t)y_p(t) + p(t), \quad (6)$$

$$\dot{e}_i(t) = -a_i e_i(t) + \delta(t)y_p(t) + p(t), \quad (7)$$

$$\epsilon(t) = \emptyset(t) - \delta(t). \quad (8)$$

This set of equations allows to analyse the adaptive system and it has the same form as in the ideal case, except for the perturbation term $p(t)$.

In order to guarantee the stability of the averall adaptive system, adaptive laws used to adjust $\theta(t)$ and $\hat{a}_p(t)$ have the form:

$$\dot{\theta}(t) = \begin{cases} -e_c(t)y_p(t)-\theta(t)\left(1 - \dfrac{|\theta(t)|}{\theta^*_M}\right)^2-\epsilon(t) & \text{if } |\theta(t)|>\theta^*_M \\ -e_c(t)y_p(t) - \epsilon(t) & \text{otherwise} \end{cases} \quad (9)$$

$$\dot{\hat{a}}_p(t)= \begin{cases} -e_i(t)y_p(t)-\hat{a}_p(t)\left(1 - \dfrac{|\hat{a}_p(t)|}{a_{pM}}\right)^2+\epsilon(t) & \text{if } |\hat{a}_p(t)|>a_{pM} \\ -e_i(t)y_p(t)+\epsilon(t) & \text{otherwise} \end{cases} \quad (10)$$

Now we can state the following theorem concerning the adaptive control of first order plants in the presence of bounded external disturbances.

<u>Theorem</u>

Let consider the plant (1), the reference model (2) the identification model (4) and the controller defined by equations (3), (9) and (10). If the external perturbation $p(t)$ is uniformly bounded i.e. $|p(t)| \leq p_0$ for all $t \geq t_0$, then the adaptive system is unifomly stable and all the signals remain bounded for all $t \geq t_0$.

<u>Proof</u>

Let consider the equivalent description of the adaptive system given by the set of equations (6)-(10) and the Lyapunov function candidate defined by

$$V(e_c,e_i,\emptyset,\delta) = \tfrac{1}{2}(e_c^2 + e_i^2 + \emptyset^2 + \delta^2). \quad (11)$$

The time derivative along the trajectory defined by equations (6)-(10) yields

$$\dot{V} = -a_m e_c^2+pe_c-\theta\emptyset f(\theta)-a_i e_i^2+pe_i-\hat{a}_p\delta f(\hat{a}_p)-\epsilon^2. \quad (12)$$

Analyzing terms $\theta\emptyset f(\theta)$ and $\hat{a}_p\delta f(\hat{a}_p)$, considering that $|\theta(t)|>\theta^*_M$ and $|\hat{a}_p(t)|>a_{pM}$, and according to the definition of $f(\cdot)$ function, it can be concluded that $\dot{V} \leq 0$ in the region D^c where D is defined as follows:

$$D=\{(e_c,e_i,\emptyset,\delta) \in R^4 \ / \ |e_c|\leq \frac{p_0}{a_m} \text{ and } |e_i|\leq \frac{p_0}{a_i}\} \quad (13)$$

Since V is positive definite and $\dot{V}$ is negative semi-definite outside the bounded region D (Narendra and Annaswamy 1989a) it follows that all solutions of the adaptive system are bounded.

SIMULATION RESULTS

To compare the behaviour of the combined MRAC with that obtained from the direct MRAC under external perturbations, several computer simulations of first order plants were performed. The unstable plant defined by $\dot{y}_p = 4y_p + u + p$, with the initial condition $y_p(0) = 1$ was chosen.
The stable reference model with zero initial condition defined by $\dot{y}_m = -y_m + r$, $y_m(0) = 0$ was used in simulations. Finally, the identification model described by $\dot{\hat{y}}_p = e_i + \hat{a}_p y_p + u$, with $\hat{y}_p(0) = 0$, was utilized. All the adaptive gains were unity and all other initial conditios are zero.

Figures 1 and 2 show the case of a constant perturbation $p(t) \equiv 2$ and zero reference input $r(t) \equiv 0$. It is interesting to point out that the bias on tracking error present in the direct MRAC is eliminated by using the CMRAC. Also, control range is smaller in the CMRAC than in the direct MRAC.

Same results are shown in Figures 3 and 4 when the same constant external perturbation is affecting the plant $p(t) \equiv 2$ and a sinusoidal reference input is used, $r(t) \equiv 15\cos(t)$. Again, better transient behaviour of the CMRAC is observed with respect to the direct MRAC under same conditions.

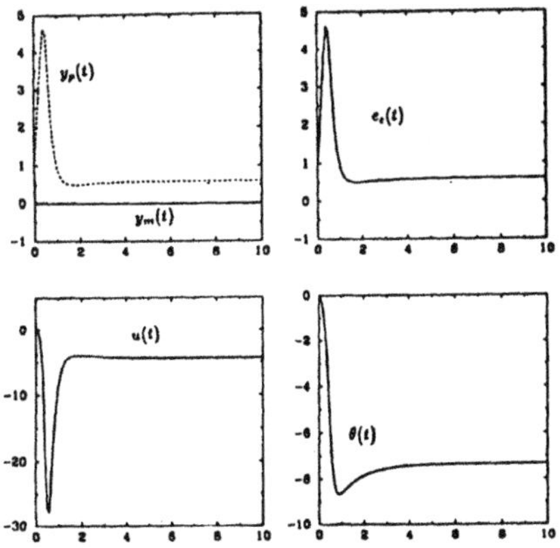

Figure 1. Direct MRAC with constant perturbation and zero reference input.

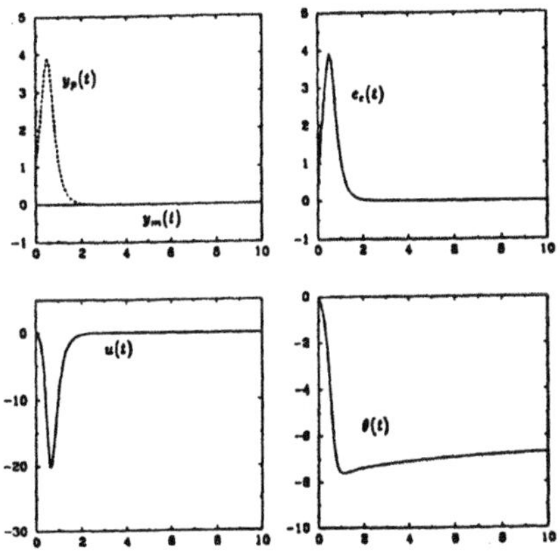

Figure 2. Combined MRAC with constant perturbation and zero reference input.

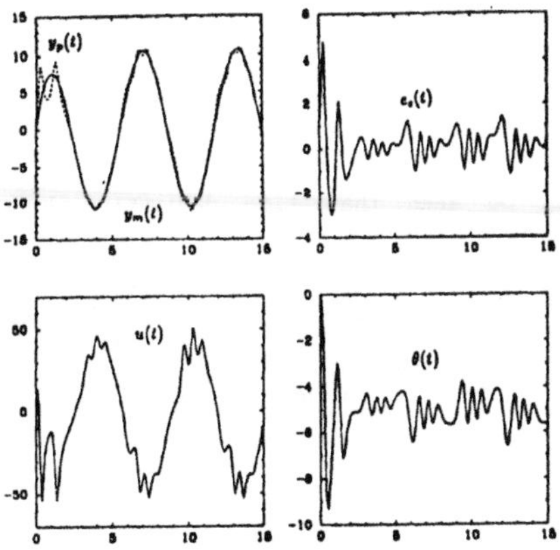

Figure 3. Direct MRAC with constant perturbation and sinusoidal reference input.

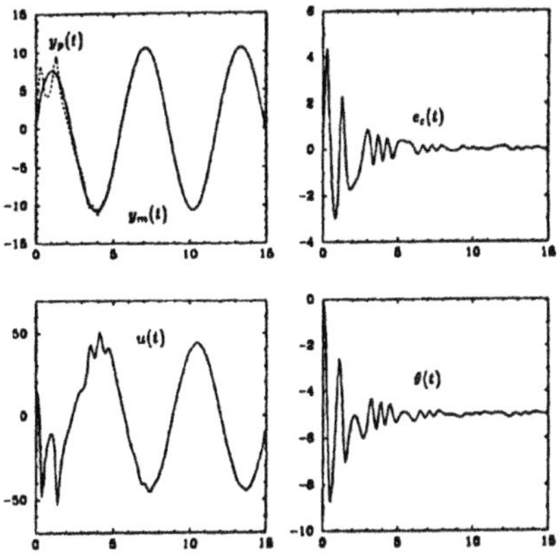

Figure 4. Combined MRAC with constant perturbation and sinusoidal reference input.

CMRAC OF GENERAL N^{th} ORDER PLANTS

Let consider an nth order plant defined by

$$\dot{x}_p(t) = A_p x_p(t) + b_p u(t) + d_p p_1(t), \qquad (14)$$

$$y_p(t) = c_p^T x_p(t) + p_2(t), \qquad (15)$$

where $A_p \in R^{n\times n}$; x_p, b_p, d_p, $c_p \in R^n$ and u, y_p, p_1, $p_2 \in R$. $p_1(\cdot)$, $p_2(\cdot)$: $[t_0,\infty) \rightarrow R$ are two bounded piecewise continuos functions of time. In addition it is assumed that $p_1(t)$ is differentiable. The corresponding transfer function is given by

$$H_p(s) = c_p^T(sI-A_p)^{-1}b_p = k_p Z_p(s)/R_p(s), \qquad (16)$$

where $Z_p(s)$ and $R_p(s)$ are two monic coprime polynomials of degree m and n respectively.

Roots of polynomial $Z_p(s)$ are all in the open left half of the complex plane. It is assumed that m, n and the sign of k_p are known but the parameters of the plant are unknown. Assumptions on the knowledge of the sign of k_p and the order n are not really needed. Nussbaum gain (Nussbaum 1983, Mudget and Morse 1985) can be used when the sign of k_p is not known and only an upper bound on the order n rather than the exact value is needed (Narendra and Annaswamy 1989a). The model reference is defined by the transfer function

$$H_m(s) = k_m Z_m(s)/R_m(s), \qquad (17)$$

where $Z_m(s)$ and $R_m(s)$ are known monic Hurwitz polynomials of degree m and n respectively. k_m is a known positive constant. The transfer function $H_m(s)$ relates the reference input $r(t)$ with the reference output $y_m(t)$ and $r(\cdot)$: $[t_0,\infty]$ R is a bounded piecewise continuos function of time. That is to say

$$y_m(t) = H_m(s)r(t). \qquad (18)$$

The input to the plant is defined by

$$u(t)=\theta^T(t)w(t), \qquad (19)$$

with $\theta(t), w(t) \in R^{2n}$. The control parameter vector is defiend as $\theta(t)=[k(t),\theta_1^T(t),\theta_0(t),\theta_2^T(t)]^T$ with k, $\theta_0 \in R$; $\theta_1,\theta_2 \in R^{n-1}$ and the control sensitivity vector is $w(t)=[r(t),w_1^T(t),y_p(t), w_2^T(t)]^T$ where $\dot{w}_1(t) = \Omega w_1(t)+lu(t)$ and $\dot{w}_2(t) = \Omega w_2(t)+ly_p(t)$.

(Ω, l) is any controllable pair with $\Omega \in R^{(n-1)\times(n-1)}$ an asymptotically stable matrix.

It can be seen that the tracking error $e_c(t)=y_p(t)-y_m(t)$ is given by the following relationship (Narendra and Annaswamy 1989a)

$$e_c(t)=k_p/k_m\, W_m(s)(\Phi^T(t)w(t)) + p_3(t), \tag{20}$$

where $\Phi(t)=\theta(t)-\theta^* \in R^{2n}$ and $p_3(t) \in R$ is the effect produced by the external pertubations $p_1(t)$ and $p_2(t)$ acting on the plant.

Another way of writting equations decribing the dynamical behaviour of $e_c(t)$ is given by

$$\dot{e}(t) = Ae(t) + b[\Phi^T(t)w(t)] + p_4(t), \tag{21}$$

$$e_c(t) = h^Te(t), \tag{22}$$

where $e(t)=x(t)-x_m(t)$, $x(t)=[x_p^T(t),w^T(t)]^T$, $x_m(t)=[x_p^{*T}(t),w_1^{*T}(t),w_2^{*T}(t)]^T$. $p_4(t) \in R^{3n-2}$ is an equivalent perturbation replacing the effects of $p_1(t)$ and $p_2(t)$. $x_m(t)$ is the augmented state of the reference model given by

$$\dot{x}_m(t) = Ax_m(t) + br(t), \tag{23}$$

$$y_m(t) = h^Tx_m(t). \tag{24}$$

Since this is a general case and $n^* = n - m$ is arbitrary, it is necessary to introduce the concept of augmented error $\epsilon_c(t)$ as follows:

$$\epsilon_c(t)=(k_m/k_p)\Phi^T(t)\bar{w}(t) + \emptyset_{k1}(t)e_2(t), \tag{25}$$

where $\emptyset_{k1}(t)=k_1(t)-k_1^*=k_1(t)-(k_p/k_m) \in R$, $e_2(\cdot)$: R^+ R is defined as

$$e_2(t)=\theta^T(t)W_m(s)I_{2n}w(t) + W_m(s)\theta^T(t)w(t), \tag{26}$$

where I_{2n} is the identity matrix of order 2n and $\bar{w}(t) \in R^{2n}$ is such that

$$\bar{w}(t) = W_m(s)I_{2n}w(t).$$

The identification model to be used is given by

$$\hat{y}_p(t)=(\hat{k}_p(t)/k_m)[\hat{\beta}^T(t)\bar{w}_1(t)+\bar{u}(t)]+\hat{\alpha}^T(t)\bar{w}_2(t)+ +\hat{\alpha}_{n-1}(t)\bar{y}_p(t), \tag{27}$$

which gives the following output error $e_i(t)=y_p(t)-\hat{y}_p(t)$

$$e_i(t)=(1/k_m)\mu_{kp}(t)[\hat{\beta}^T(t)\bar{w}_1(t)+\bar{u}(t)]+ +(k_p/k_m)\mu_\beta{}^T(t)\bar{w}_1(t)+\mu_\alpha{}^T(t)\bar{w}_2(t)+ +\mu_{\alpha n-1}(t)\bar{y}_p(t), \tag{28}$$

where $\bar{y}_p(t)=W_m(s)y_p(t) \in R$, $\bar{w}_1(t)=W_m(s)I_{n-1}w_1(t) \in R^{n-1}$, $\bar{w}_2(t)=W_m(s)I_{n-1}w_2(t) \in R^{n-1}$, $\bar{u}(t)=W_m(s)u(t) \in R$. I_{n-1} is the identity matrix of order n-1.

$\hat{k}_p(t),\hat{\alpha}_{n-1}(t) \in R$ and $\hat{\beta}(t),\hat{\alpha}(t) \in R^{n-1}$ are the estimates at time 't' of plant parameters k_p, $\alpha_{n-1} \in R$, and β, $\alpha \in R^{n-1}$ respectively. Parameters k_p, α_{n-1}, β and α are linearly related to parameters k_p, a_{n-1}, a and b which contain the coefficients of the polynomials $R_p(s)$ and $Z_p(s)$.

Parameter errors are defined as $\mu_{kp}(t)= \hat{k}_p(t)-k_p \in R$, $\mu_{\alpha n-1}(t)=\hat{\alpha}_{n-1}(t)-\alpha_{n-1} \in R$, $\mu_\beta(t)= \hat{\beta}(t)-\beta \in R^{n-1}$, $\mu_\alpha(t)=\hat{\alpha}(t)-\alpha \in R^{n-1}$, $\Phi_k(t)=k(t)-k^* \in R$, $\Phi_{\theta 1}(t)=\theta_1(t)-\theta^*_1 \in R^{n-1}$, $\Phi_{\theta 0}(t)=\theta_0(t)-\theta^*_0 \in R$ and $\emptyset_{\theta 2}(t)=\theta_2(t)-\theta^*_2 \in R^{n-1}$.

Closed-loop estimation errors can be computed as follows (Duarte and Narendra 1989,1989a):

$$\epsilon_{\theta 1}(t)=\hat{\beta}(t)+\theta_1(t), \tag{29}$$

$$\epsilon_k(t)=k(t)\hat{k}_p(t)-k_m, \tag{30}$$

$$\epsilon_{\theta 0}(t)=\theta_0(t)\hat{k}_p(t)+\hat{\alpha}_{n-1}(t)k_m, \tag{31}$$

$$\epsilon_{\theta 2}(t)=\theta_2(t)\hat{k}_p(t)+\hat{\alpha}(t)k_m, \tag{32}$$

or equivalently

$$\epsilon_{\theta 1}(t)=\mu_\beta(t)+\Phi_{\theta 1}(t), \tag{33}$$

$$\epsilon_k(t)=k(t)\mu_{kp}(t)+k_p\Phi_k(t), \tag{34}$$

$$\epsilon_{\theta 0}(t)=\theta_0(t)\mu_{kp}(t)+k_p\Phi_{\theta 0}(t)+k_m\mu_{\alpha n-1}(t), \tag{35}$$

$$\epsilon_{\theta 2}(t)=\theta_2(t)\mu_{kp}(t)+k_p\Phi_{\theta 2}(t)+k_m\mu_\alpha(t). \tag{36}$$

Adaptive laws to adjust controller and identifier parameters are given by

$$\dot{\hat{\alpha}}(t)=-e_i(t)\bar{w}_2(t)/N_i(t)-k_m\epsilon_{\theta 2}(t)-\hat{\alpha}(t)f(\hat{\alpha}), \tag{37}$$

$$\dot{\hat{\alpha}}_{n-1}(t)=-e_i(t)\bar{y}_p(t)/N_i(t)-k_m\epsilon_{\theta 0}(t)- -\hat{\alpha}_{n-1}(t)f(\hat{\alpha}_{n-1}), \tag{38}$$

$$\dot{\hat{\beta}}(t)=-sgn(k_p)e_i(t)\bar{w}_1(t)/(N_i(t)k_m)-\epsilon_{\theta 1}(t)- -\hat{\beta}(t)f(\hat{\beta}), \tag{39}$$

$$\dot{\hat{k}}_p(t)=-e_i(t)[\hat{\beta}^T(t)\bar{w}_1(t)+\bar{u}(t)]/(N_i(t)k_m)- -k(t)\epsilon_k(t)-\theta_0(t)\epsilon_{\theta 0}(t)-\theta_2{}^T(t)\epsilon_{\theta 2}(t)- -\hat{k}_p(t)f(\hat{k}_p), \tag{40}$$

$$\dot{k}(t)=-sgn(k_p)[\epsilon_c(t)y_m(t)/(N_c(t)k_m)+\epsilon_k(t)]- -k(t)f(k), \tag{41}$$

$$\dot{\theta}_1(t)=-sgn(k_p)\epsilon_c(t)\bar{w}_1(t)/(N_c(t)k_m)-\epsilon_{\theta 1}(t)- -\theta_1(t)f(\theta_1), \tag{42}$$

$$\dot{\theta}_0(t)=-sgn(k_p)[\epsilon_c(t)\bar{y}_p(t)/(N_c(t)k_m)-\epsilon_{\theta 0}(t)]- -\theta_0(t)f(\theta_0), \tag{43}$$

$$\dot{\theta}_2(t)=-sgn(k_p)[\epsilon_c(t)\bar{w}_2(t)/(N_c(t)k_m)-\epsilon_{\theta 2}(t)]- -\theta_2(t)f(\theta_2). \tag{44}$$

where $N_c(t)=1+y^2{}_m(t)+\bar{w}^T{}_1(t)\bar{w}_1(t)+\bar{y}^2{}_p(t)+\bar{w}^T{}_2(t)\bar{w}_2(t)+e^2{}_2(t) \in R$ and $N_i(t)=1+(\hat{\beta}^T(t)\bar{w}_1(t)+ \bar{u}(t))^2+\bar{w}^T{}_1(t)\bar{w}_1(t)+\bar{y}^2{}_p(t)+\bar{w}^T{}_2(t)\bar{w}_2(t) \in R$ are normalization factors. The function $f(\cdot)$ is defined as

$$f(x)=\begin{cases}(1+\|x\|/x^*_M)_2 & \text{if } \|x\| > x^*_M\\ 0 & \text{elsewhere}\end{cases}$$

where $\|x(t)\| \leq x^*_M$ for all $t \geq t_0$ (x^*_M is a known bound on the norm of $x(t)$).

Stability proof of the adaptive system is done by using Lyapunov theory and extensions. Let consider the Lyapunov function candidate

$$V(e,e_i,\Phi,\mu) = \tfrac{1}{2}\,(|k_p|\Phi_k{}^2 + |k_p|\Phi^T{}_{\theta 1}\Phi_{\theta 1} + |k_p|\Phi^2{}_{\theta 0} + + |k_p|\Phi^T{}_{\theta 2}\Phi_{\theta 2} + |kp|\mu^T{}_\beta\mu_\beta + \mu^T{}_\alpha\mu_\alpha + + \mu^2{}_{kp} + \mu^2{}_{\alpha n-1}). \tag{45}$$

It can be seen that time derivative along the trayectory of the adaptive system is negative semi-definite . Since $V > 0$ and $\dot{V} \leq 0$, boundedness of $\mu_{kp}(t)$, $\mu_{\alpha n-1}(t)$, $\mu_\beta(t)$, $\mu_\alpha(t)$, $\Phi_k(t)$, $\Phi_{\theta 1}(t)$, $\Phi_{\theta 0}(t)$ and $\Phi_{\theta 2}(t)$ immediately follows. Boundedness of the rest of the signals in the adaptive system follows using arguments given by Kreisselmeier and Narendra (1982) and Narendra and Annaswamy (1989a).

CONCLUSIONS

A new MRAC, robust in the presence of bounded external perturbation has been presented in this paper. The method uses the combined MRAC recently developed by Duarte and Narendra (1989,1989a) for ideal conditions and it is modifed by using the results by Kreisselmeier and Narendra (1982) which uses a knowledge of a bound on $\|\theta^*\|$.

Global stability of the method has been proved for first order and general n^{th} order plants. Explicit knowledge of bounds on external perturbations affecting the system is not needed.

Simulation results indicate that CMRAC present better transient behaviour than direct MRAC if used alone. In some cases bias can be removed from tracking error and less restrictive conditions on the control input can be obtained by using the CMRAC.

Currently, research on using the combined MRAC incorporating dead zone, $|e|\theta$ and $\sigma\theta$ modifications on adaptive laws is under way .

ACNOWLEDGEMENTS

This work has been partially supported by FONDECYT under Project FONDECYT No. 01130/89.

REFERENCES

Duarte M.A. (1991). Combined model reference adaptive control using the $|e|\theta$ modification. Submitted to 9 IFAC/IFORS Symposium on Identification and System Parameter Estimation, Budapest, Hungary, July 1991.

Duarte M.A.(1990). Robust adaptive control in the presence of external perturbations (In Spanish). Accepted to be presented at IV Latinamerican Congress on Automatica, Puebla, México, November 1990.

Duarte M.A.(1990a). A robust model reference adaptive control scheme.(In Spanish) Proceedings of II Chilean Symposium on System Identification, Parameter Estimation and Adative Control. July 1990, Santiago, Chile: Department of Electrical Engineering, University of Chile, ACCA, IEEE-Chile, pp. 25-30.

Duarte M.A. and K.S. Narendra (1989). Combined direct and indirect approach to adaptive control. IEEE Transactions on Automatic Control, v. AC-34, No. 10, pp. 1071-1075.

Duarte M.A. and K.S. Narendra (1989a). A new approach to model reference adaptive control. International Journal of Adaptive Control and Signal Processing, v. 3, No. 1, pp. 53-73.

Duarte M.A. (1989b). A combined approach of direct and indirect model reference adaptive control (In Spanish). Proceedings of the VIII Chilean Conference on Electrical Engineering. Concepción, Chile: Departament of Electrical Enegineering, University of Bío-Bío, pp. 442-446.

Duarte M.A. and K.S. Narendra (1988). Combined direct and indirect adaptive control of plants of relative degree one. Proceedings of the 22nd Princeton Conference on Information Sciences and Systems. Princeton, New Jersey, USA: Department of Electrical Engineering, Princeton University, v. 1, pp. 400-404.

Egart B. (1979). Stability of adaptive controllers. Springer-Verlag, Berlin, 1979.

Ioannou P.A. and K.S. Tsakalis (1986). A robust direct adaptive controller. IEEE Transactions on Automatic Control, v. AC-31, No. 11, pp. 1035-1043.

Ioannou P.A. and P.V. Kokotovic (1983). Adaptive Systems with Reduced Models. Springer-Verlag, New york.

Kreisselmeier G. and B.D.O. Anderson (1986). Robust model reference adaptive control. IEEE Transactions on Automatic Control, v. AC-31, No. 2, pp. 127-133.

Kreisselmeier G. and K.S. Narendra (1982). Stable model reference adaptive control in the presence of bounded disturbances. IEEE Transactions on Automatic Control, v. AC-27, pp. 1169-1175.

Narendra K.S. and M.A. Duarte (1989). Application of robust adaptive control using combined direct and indirect methods. International Journal of Adaptive Control and Signal Processing, v. 3, No. 2, pp. 131-142.

Narendra K.S. and A.M. Annaswamy (1989a). Stable adaptive systems. Prentice Hall, New Jersey, 1989.

Narendra K.S. and M.A. Duarte (1988). Robust adaptive control using direct and indirect methods. Proceedings of the 7th American Control Conference. Atlanta, Georgia, USA: American Automatic Control Council, v. 3, pp. 2429-2433.

Narendra K.S. and A.M. Annaswamy (1987). A new adaptive law for robust adaptive control without persistent excitation. IEEE Transactions on Automatic Control, v. AC-32, No. 2, pp. 134-145.

Narendra K.S. and A.M. Annaswamy (1986). Robust adaptive Control in the presence of bounded disturbances. IEEE Transaction on Automatic Control, v. AC-31, pp. 306-315.

Narendra K.S. and A.M. Annaswamy (1985). Robust adaptive Control". in K.S. Narendra, Ed. Adaptive and Learning Systems: Theory and Applcations, pp. 3-31, Plenum Press, New York.

Nussbaum R.D. (1983). Some remarks on a conjecture in parameter adaptive control. Systems and Control Letters, v. 3, pp. 243-246.

Mudget D.R. and A.S. Morse (1985). Adaptive stabilization of linear systems with unknown high-frequency gain. IEEE Transactions on Automatic Control, v. AC-30, pp. 549-554.

Peterson B.B. and K.S. Narendra (1982). Bounded error adaptive control. IEEE Transactions on Automatic Control, v. AC-27, pp. 1161-1168.

Samson, C. (1983). Sability analysis of adaptively controlled systems subject to bounded disturbances. Automatica, v. 19, No. 1, pp. 81-86.

A NEW SOLUTION TO ADAPTIVE INVERSE CONTROL

B. Farhang-Boroujeny* and K. Ayatollahi**

**Electrical Engineering Dept., National University of Singapore, Singapore 0511*
***Esfahan Nuclear Technology Center, Isfahan, Iran*

Abstract. The inverse control is a solution to the adaptive control problem, in which the controller approximates the inverse of the plant model. In other words, the controller transforms the input sequence to a sequence that when applied to the plant input, the plant output will be the same as the input sequence, or at least the same as its delayed replica. In this paper the previous works on the subject are extended by introducing a nonadaptive feedback from the plant output to the controller input. It is shown that this extension results in more robust implementation. The stability of the proposed scheme is also addressed and solutions are given to overcome the instability problems that may arise under some rare conditions.

Keywords: Closed-loop systems; controllers; adaptive control; inverse modeling; non-minimum systems; stability.

INTRODUCTION

Self-tuning adaptive controllers have been widely considered in the literature (Astrom, 1987; Clarke and co-workers, 1985; Goodwin and Sang Sin, 1984). Figure 1 shows a general structure of an adaptive self-tuner. The plant input and output sequences are monitored to estimate its parameters. The estimated parameters are then used to choose (directly or indirectly) the parameters of the controller. Examples are the minimum-variance (MV) self-tuner of Clarke, Kanjilal and Mohtadi (1985), and the pole-placement (PP) self-tuners of Lelic and Zarrop (1985). Lelic and Zarrop (1985) calculate the controller parameters so that the output $\{y_k\}$ obeys the input $\{x_k\}$ in some predefined manner. In the MV self-tuners the design and the controller blocks of Fig. 1 are mixed. Then, by choosing a proper cost function, a control law is developed to predict $\{u_k\}$ properly (Astrom, 1987; Clarke and co-workers, 1985).

An alternative approach to the solution of adaptive control problem is the scheme proposed by Martin-Sanchez (1985). He argues that if the plant model is known, and is reversible, then by application of the inverse of the plant model, the input sequence $\{x_k\}$ may be converted to the sequence $\{u_k\}$ which when applied to the plant input, the plant output will be equal to $\{x_k\}$. The reversibility problem of the plant model, which arises when it is non-minimum phase, is solved by Widrow and Stearns (1985), by introducing a delay between the input sequence and its replica at the output of the plant. They have used a finite impulse response (FIR) digital filter structure for the controller, and have proposed a least mean square (LMS) algorithm for the controller adjustment.

In this paper the Widrow and Stearns scheme is extended by introducing a feedback from the output of the plant to input of the controller. We show that by this approach the inaccuracy of the inverse model is made tolerable to some extent, and a more robust system is realized. We consider only the single input single output systems, and we assume that the original plant to be controlled is stable, but it may be non-minimum phase.

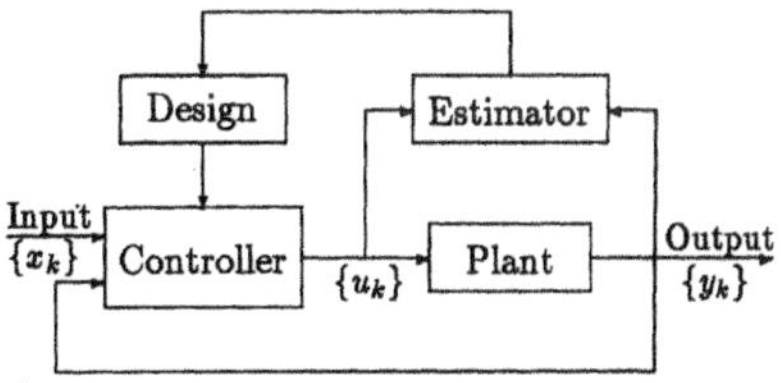

Fig. 1. The general structure of an adaptive self-tuner.

NEW SCHEME

Figure 2 shows the inverse control scheme proposed in by Widrow and Stearns (1985). The drift compensator is included for compensation of the plant output deviation from its nominal (desired) value, which may be due to the existence of a nonzero bias at the plant output, or inaccuracy of the controller in realizing the inverse of the plant. In the latter case any variation in the input signal changes the output drift. If the structure of Fig. 2 with the LMS algorithm suggested by Widrow and Stearns (1985) is used to compensate the considered drift, the resulted system may be unacceptably slow. For the LMS algorithm it takes at least tens of data samples to converge. Faster algorithms, such as Kalman filtering (Goodwin and Sang Sin, 1984), may also be too slow for this purpose, if the system is noisy. It is

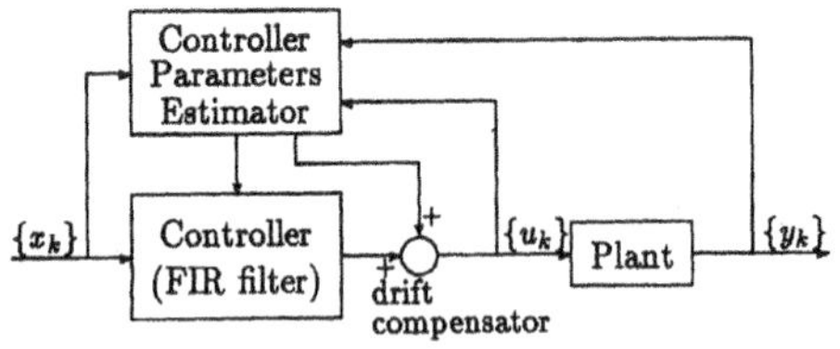

Fig. 2. The inverse modeling approach of Widrow and Stearns.

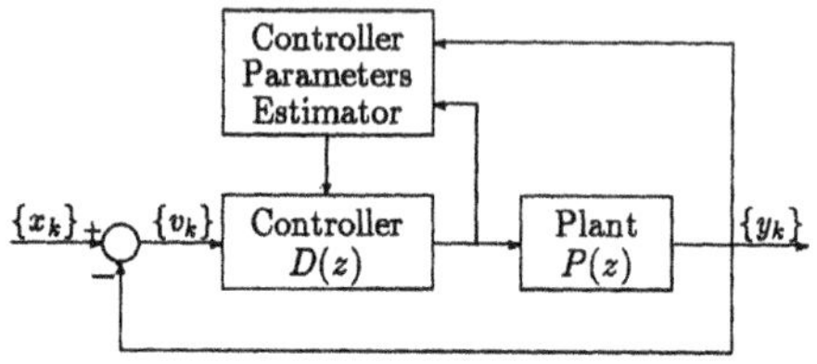

Fig. 3. Adaptive inverse control with feedback to compensate the output drifts.

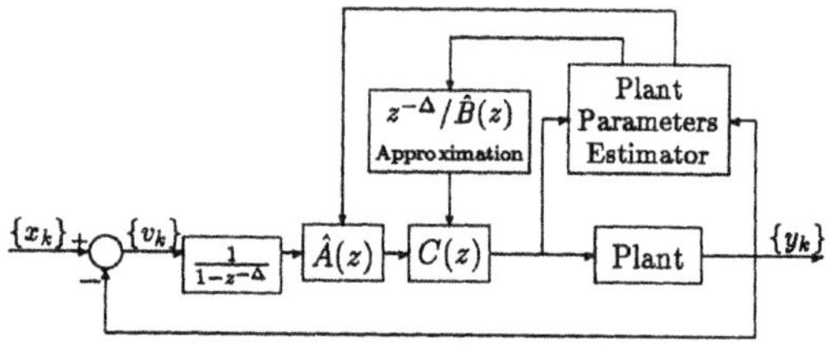

Fig. 4. The complete block diagram of the proposed scheme.

because averaging is required for estimates made in noisy environments.

Figure 3 shows our proposed structure of the inverse controllers. The controller, D(z), should be chosen so that the output sequence $\{y_k\}$ be a delayed replica of the input sequence $\{x_k\}$. In term of the z-transforms D(z) should be chosen so that

$$\frac{Y(z)}{X(z)} = z^{-\Delta} \tag{1}$$

where $X(z)$ and $Y(z)$ are, respectively, the z-transforms of the sequences $\{x_k\}$ and $\{y_k\}$, and Δ is a properly chosen delay.

Referring to Fig. 3 and using (1) we obtain

$$D(z) = \frac{1}{1-z^{-\Delta}} \cdot \frac{z^{-\Delta}}{P(z)} \tag{2}$$

Of course, the above controller can not be realized in its present form, if the plant, $P(z)$, is non-minimum phase. To overcome this problem we may proceed as follows.

The first term at the right-hand-side of (2) can be easily realized by a simple recursion which adds delayed samples of the sequence $\{v_k\}$ of Fig. 3. The Widrow and Stearns (1985) technique may be used to realize the last term of (2) by employing an FIR approximation of the two-sided expansion of $z^{-\Delta}/P(z)$, in which the right-sided expansion corresponds to the stable zeros of $P(z)$ and the left-sided expansion corresponds to its unstable (non-minimum phase) zeros.

As a result cancellation of the non-minimum phase poles and zeros, which could result in instability of the whole system, if $z^{-\Delta}/P(z)$ was implemented directly, is prevented. Of course in practice $P(z)$ is not available and, therefore, must be replaced by its estimate, $\hat{P}(z)$, which can be obtained by application of a proper identification algorithm (Goodwin and Sang Sin, 1984).

Assuming an ARMA model for the plant, and having its estimate, $\hat{P}(z) = \hat{B}(z)/\hat{A}(z)$, we are interested in implementing:

$$\frac{z^{-\Delta}}{\hat{P}(z)} = \hat{A}(z) \cdot \frac{z^{-\Delta}}{\hat{B}(z)} \tag{3}$$

$\hat{A}(z)$ çan be readily realized as it is a simple linear combiner. Therefore, the next problem to be solved is approximation of $z^{-\Delta}/\hat{B}(z)$ by a FIR filter. That is two-sided expansion of $z^{-\Delta}/\hat{B}(z)$ within a limited length.

Figure 4 gives the complete block diagram of our new scheme.

POLYNOMIAL INVERSION

We are given a polynomial

$$\hat{B}(z) = \hat{b}_0 + \hat{b}_1 z^{-1} + \hat{b}_2 z^{-2} + \cdots + \hat{b}_M z^{-M} \tag{4}$$

and we are interested in finding the coefficients of the polynomial

$$C(z) = c_0 + c_1 z^{-1} + c_2 z^{-2} + \cdots + c_N z^{-N} \tag{5}$$

so that

$$\hat{B}(z)C(z) \approx z^{-\Delta} \tag{6}$$

with best approximation. Different criteria may be chosen to evaluate this approximation. We choose $C(z)$ so that the output of the system $\hat{B}(z)C(z)$ has minimum variance with respect to its desired value, which should be equal to a delayed replica of its input. If the input signal to this system is assumed to be white, it will be equivalent of minimizing the following cost function.

$$\xi = \|\hat{E}(z)\| \tag{7}$$

where $\hat{E}(z) = z^{-\Delta} - \hat{B}(z)C(z)$, and $\|\hat{E}(z)\|$ means Euclidian or L_2 norm of $\hat{E}(z)$ which is defined as follows:

$$\|\hat{E}(z)\| = \sqrt{\frac{1}{2\pi}\int_0^\infty |\hat{E}(e^{j\omega})|^2 d\omega} = \sqrt{\sum \hat{e}_i^2} \tag{8}$$

where $\{\hat{e}_i\}$ are the coefficients of the polynomial $\hat{E}(z)$.

Other criterions that may be used are the L_1 and L_∞ norms of $\hat{E}(z)$. Although these criterions might be preferred to the L_2 norm, with some respects, they cannot be easily implemented in practice, as there are no simple solutions to them.

The minimization of the L_2 norm is straightforward. One may argue that choosing the coefficients $\{c_i\}$ is a Wiener filtering problem. A white noise is passed through a moving average (MA) system with impulse response $\{\hat{b}_i\}$. The output sequence from the MA system should be processed by a FIR filter whose tap gains, $\{c_i\}$, should be adjusted to minimize the mean square error (MSE) between the output of the latter filter and a delayed replica of the input to the

MA system; see Fig. 5. The Wiener equation governing the optimum tap gains of the FIR filter reads:

$$RC_{opt} = P \tag{9}$$

where $R = E\{Z_k Z_k^t\}$, $P = E\{d_k Z_k\}$, $Z_k = [z_k z_{k-1} \cdots z_{k-N}]^t$, $d_k = \nu_{k-\Delta}$ and $E\{\cdot\}$ means statistical expectation.

It is straightforward to show that for white ν_k with average power of one,

$$r_{ij} = \sum_{l=0}^{M-|i-j|} \hat{b}_l \hat{b}_{l+|i-j|}; \qquad 0 \le i,j \le N \tag{10}$$

and

$$p_i = \hat{b}_{\Delta-i}; \qquad \max(0, \Delta - M) \le i \le \min(\Delta, N) \tag{11}$$

where r_{ij} and p_i are the ijth element of R and ith element of P, respectively.

There are many noniterative efficient algorithms for solving the Wiener equations. Most of these algorithms have complexity of $2n^2$ mathematical operations, where n is the number of the Wiener equation coefficients (unknowns), and one operation is equivalent of one multiplication/division plus one addition/subtraction. For the considered system $n = N + 1$ and direct solution of the Wiener equation has the complexity of the order of $2(N+1)^2$.

Another algorithm which also finds the optimum coefficients of the inverse polynomials is given by Farhang-Boroujeny (1986), where the same problem is encountered in channel equalization. This is an indirect iterative solution of the present Weiner equations. It calculates the coefficients $\{c_i\}$ by direct manipulation of the coefficients $\{\hat{b}_i\}$. Figure 6 gives a summary of that. Inspection of this algorithm shows that the complexity of each of its iterations is $2(M+1)(N+1)$. This does not include the initial calculation of $\hat{E}(z)$, as in the case of noniterative solution of the Wiener equation, also, some complexity requires to calculate the correlation coefficients from $\{\hat{b}_i\}$. Thus, as in practice N is usually larger than M (most of times by a factor of 2 or more), we may conclude that the iterative algorithm of Fig. 6 is simpler than the conventional noniterative algorithms by a factor of $M + 1/N + 1$. This is at the expense of requiring some iterations. However, as in tracking mode we are usually interested in updating the controller coefficients frequently, this does not turn out to be a serious problem. This is similar to the situation that occurs with the Riccati equation solution in the LQG self-tuners, where to keep the complexity of the controller lower, only one iteration of that is performed during each sampling interval.

STABILITY

A fundamental question that arises when a new scheme is suggested is whether it is stable or not. If it is stable, is it unconditionally stable or its stability can be assured under some particular circumstances. In this section we discuss the stability of the proposed scheme and give solutions to the instability problem that might arise under some rare conditions. Our analysis is limited to a deterministic case where we assume that the plant is fixed and an accurate estimate of its parameters are available. To simplify the discussion, further, we assume that $\hat{A}(z) = A(z)$ and $\hat{B}(z) = B(z)$; that is, perfect estimation of the plant model is available.

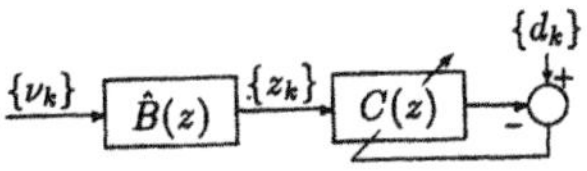

Fig. 5. Wiener filtering interpretation of the $C(z)$ approximation.

Under the above conditions, we may argue that as our criterion in the design of the controller is to force all of the closed loop poles toward the origin, we expect the resultant closed loop system to be stable, and robust with this respect. Of course, this is a subjective argument. In practice, precautions should be taken to prevent any possible instability of the system. We may also notice that the cancellation of poles of the plant by $\hat{A}(z)$ in the closed loop will result in a set of poles that are very close to the poles of the plant. However, if the plant is assumed to be stable, this will not be a problem (as far as the input-output system function of the closed loop is concerned), as the latter poles will be canceled by the zeros of $\hat{A}(z)$ that will appear in the numerator of the system function of the closed loop system as well.

To explore the stability of the proposed scheme more carefully we replace $\hat{A}(z)$ by $A(z)$ in Fig. 4. It gives

$$G(z) = \frac{Y(z)}{X(z)} = \frac{C(z)B(z)}{1 - z^{-\Delta} + C(z)B(z)} \tag{12}$$

If $C(z)$ could be chosen to be an accurate approximation to $z^{-\Delta}/B(z)$, then, $C(z)B(z)$ would be equal to $z^{-\Delta}$ and, as expected, $G(z) = z^{-\Delta}$. Moreover, for further reference, we may notice that the dc gain of G(z), i.e., G(1), is always equal to one, irrespective of the accuracy of the approximation $C(z)B(z) \approx z^{-\Delta}$. This is because of the infinite dc open loop gain of the proposed system, due to the term $1/(1 - z^{-\Delta})$. An exception to the latter result is the case when $B(z)$ has a zero at $z = 1$. In this case, the pole of $1/(1 - z^{-\Delta})$ at $z = 1$ is canceled by the latter zero, and the infinite dc open loop gain is not fulfilled. Therefore, we should admit that the proposed scheme cannot be used for controlling the plants that have zero dc gain.

Unfortunately, $C(z)$, with its limited length, may not be a good approximation to $z^{-\Delta}/B(z)$ and, therefore, the terms $z^{-\Delta}$ and $C(z)B(z)$ may not be canceled out in the denominator of $G(z)$. The stability problem can then be formulated. Choosing the polynomial $C(z)$ according to the criterion given in the previous section, the resulted closed loop system will be stable if the zeros of

$$1 + E(z) = 0 \tag{13}$$

where $E(z) = C(z)B(z) - z^{-\Delta}$, stay within the unity circle.

The Nyquist stability criterion is conventionally used to investigate both the absolute and relative stabilities of linear closed loop systems from a knowledge of their open loop frequency response characteristics. The phase and gain margins are the commonly used criteria for the relative stabilities of the latter systems. The gain margin can be conveniently used to evaluate the relative stability of the closed loop systems that we encounter in this paper. Considering the definition of gain margin, we may conclude that a lower limit to the gain margin is $-20\log\{\max|E(e^{j\omega})|\}$. We may further notice that

$$\max|E(e^{j\omega})| \le \sum_i |e_i|$$

Initialize $\{c_i\}$

Obtain the error terms $\{\hat{e}_i\}$ by convolving the sequences $\{\hat{b}_i\}$ and $\{c_i\}$

for $i = 0$ to N
begin
$$\delta_i = -\sum_{j=0}^{M} \hat{b}_j \hat{e}_{i+j} / \|\hat{B}\|^2$$
$c_i = c_i + \delta_i$
Renew the error terms $\{\hat{e}_i\}$
end.

Fig. 6. Polynomial inversion algorithm.

Thus, $-20\log(\sum|\epsilon_i|)$ gives a lower limit to the gain margin of the proposed system. Moreover, if the polynomial inversion algorithm of Farhang-Boroujeny (1986) is used, $\{\epsilon_i\}$ will be available and $\sum|\epsilon_i|$ can be readily evaluated.

After stability analysis of the system we may notice that it is either absolutely or relatively unstable. A simple approach to overcome the instability problem is to add a proper factor, $0 < \alpha < 1$, to $E(z)$ in (13). This can be achieved by adding similar factor to $z^{-\Delta}$ and $C(z)$ in (12) or, equivalently, in Fig. 4. If $-20\log(\sum|\epsilon_i|)$ is considered as the present gain margin, and the gain margin is required to be more than γ dB, the following equation should be used to calculate α.

$$\alpha = \frac{10^{\gamma/20}}{\sum|\epsilon_i|} \tag{14}$$

Adding the coefficient α, we obtain

$$G(z) = \frac{\alpha C(z)B(z)}{1 + \alpha[C(z)B(z) - z^{-\Delta}]} \tag{15}$$

Then, the dc gain of $G(z)$ is

$$G(1) = \frac{\alpha C(1)B(1)}{1 - \alpha + \alpha C(1)B(1)} \tag{16}$$

which for the values of α in the range 0 to 1 is not anymore equal to one. It means that the plant output does not follow the exact replica of the reference input. However, as $C(z)$ and the estimate of $B(z)$ are available, it is possible to evaluate $G(1)$ and multiply the set point by $1/G(1)$, before its application as input to the proposed system. Nevertheless, as $G(1)$ is not precise ($\hat{B}(z)$ is used instead of $B(z)$), the latter scaling is not accurate and, thus, the plant output does not follow the desired set point exactly. Therefore, addition of the coefficient α to prevent the instability of the proposed scheme is not without its woes, and it should be prevented if possible.

If sufficient computational resources are available, one may try to obtain a better estimate of gain margin and/or phase margin by proper processing of the available information. For example, the DFT of the sequence $\{\epsilon_k\}$ may be calculated to obtain sufficient points of the Nyquist plot.

Accuracy of the inversion of $B(z)$ highly depends on the locations of its zeros. If the plant is badly conditioned, so that, it has zero(s) on or close to the unity circle, $C(z)$ can hardly realize the high peak(s) required in $1/B(e^{j\omega})$. On the other hand, if the plant is well conditioned, so that, its zeros are all far from the unity circle, $C(z)$ can easily realize $z^{-\Delta}/B(z)$. Moreover, if a priori knowledge of the plant is available, it may be used to design a more efficient controller. For example, if the plant is known to be well conditioned, we may be sure of the stability of the proposed scheme and, therefore, there will be no need to test the gain margin. On the other hand, if we know that the plant is not well conditioned, but we have a priori knowledge of location(s) of its badly conditioned zero(s), it is possible to use the following procedure to redesign the proposed controller:

To deal with the badly conditioned plants, the term $z^{-\Delta}$ of equation (1) and Fig. 4 may be replaced by a polynomial $F(z)$, whose zeros approximately match the badly conditioned zeros of $B(z)$. Then $C(z)$ should be chosen so that

$$C(z)B(z) \approx F(z) \tag{17}$$

This approximation can be easily fulfilled if $F(z)$ includes all of the badly conditioned zeros of $B(z)$(at least approximately), since, then, there will be no sharp peak in $\frac{F(e^{j\omega})}{B(e^{j\omega})}$. Furthermore, to force the dc open loop gain of the system to be infinite, $F(z)$ should be properly scaled so that $F(1) = 1$. Having done these, and assuming that the approximation (17) is well fulfilled, the system function of the closed loop system will be approximately equal to $F(z)$. It means that $F(z)$ gives the impulse response of the system. Therefore, in choosing $F(z)$ the latter fact should be also carefully considered. In the next section we present some numerical results to show the efficiency of using $F(z)$, instead of $z^{-\Delta}$, and also its selection when some prior information of the plant is available.

COMPUTER RESULTS

Simulation results of the proposed scheme are discussed in this section. Figure 7 shows a schematic diagram of the simulated plant. The sequence $\{\zeta_k\}$ is the plant disturbance which is either assumed to be zero or a step.

Figure 8 compares the new scheme with the scheme proposed by Widrow and Stearns (1985) for a plant with

$$P_1(z) = \frac{z^{-3}(-0.02 + 0.03z^{-1})}{1 - 1.6z^{-1} + 0.62z^{-2}} \tag{18}$$

Superiority of the suggested scheme, under the conditions that Δ is not properly chosen and, therefore, $C(z)$ is not able to approximate $z^{-\Delta}/\hat{B}(z)$, is clear from these results. The disturbance $\{\zeta_k\}$ is assumed to be zero.

Figures 9, 10 and 11 give further results of the proposed scheme with $\{\zeta_k\}$ chosen to be a step sequence beginning at 110th iteration of each test. The results given in Fig. 9 correspond to $P_1(z)$ and the following two plants are used to obtain the results of Figs. 10 and 11, respectively,

$$P_2(z) = \frac{z^{-1}(1 + 2z^{-1})}{1 - 0.5z^{-1}} \tag{19}$$

and

$$P_3(z) = \frac{z^{-2}(1 + z^{-1})}{1 - 0.5z^{-1}} \tag{20}$$

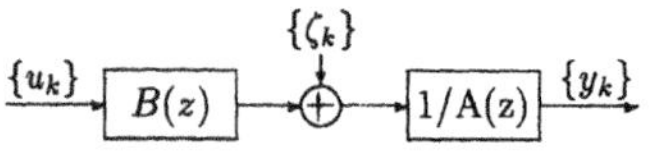

Fig. 7. Plant simulator.

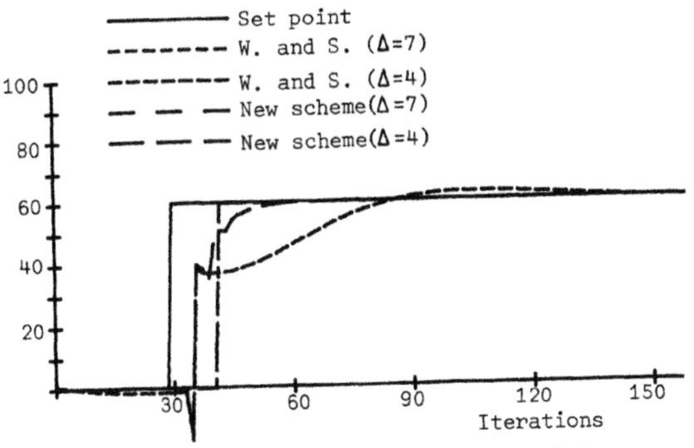

Fig. 8. Comparison of the Widrow and Stearns scheme with the new scheme for two values of Δ. The results of two schemes for Δ = 7 are overlapped.

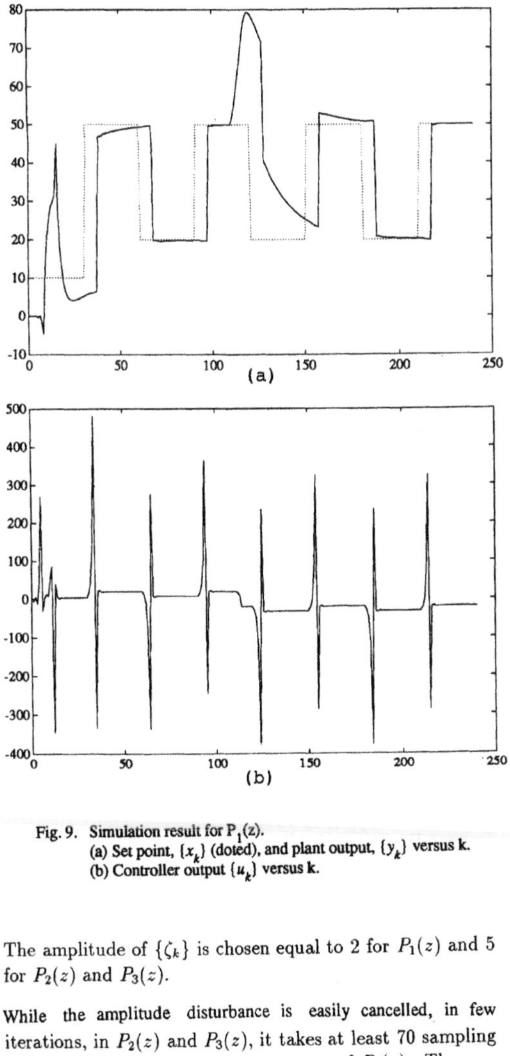

Fig. 9. Simulation result for $P_1(z)$.
(a) Set point, $\{x_k\}$ (doted), and plant output, $\{y_k\}$ versus k.
(b) Controller output $\{u_k\}$ versus k.

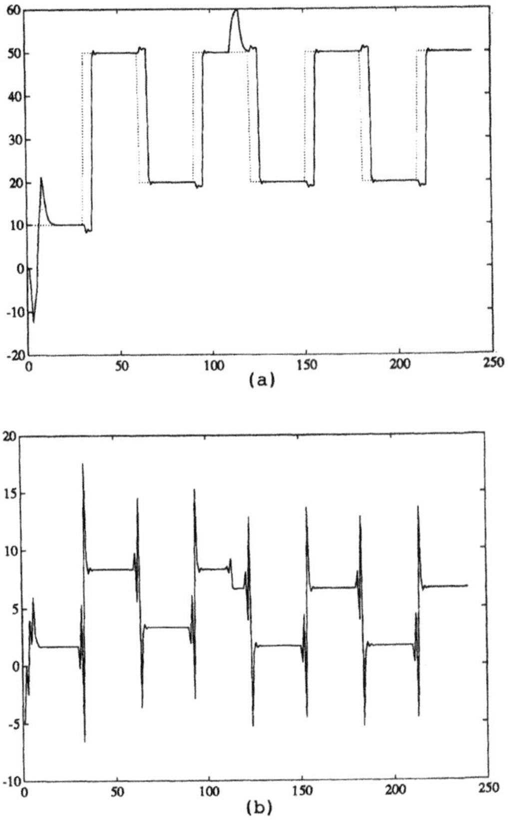

Fig. 10. Simulation result for $P_2(z)$.
(a) Set point, $\{x_k\}$ (doted), and plant output, $\{y_k\}$ versus k.
(b) Controller output $\{u_k\}$ versus k.

The amplitude of $\{\zeta_k\}$ is chosen equal to 2 for $P_1(z)$ and 5 for $P_2(z)$ and $P_3(z)$.

While the amplitude disturbance is easily cancelled, in few iterations, in $P_2(z)$ and $P_3(z)$, it takes at least 70 sampling intervals before it dies away at output of $P_1(z)$. The reason for this annoying result may be found to be due to a badly conditioned pole (very close to unity circle; z = 0.94) in $P_1(z)$ and its cancellation by the controller. This is in fact a drawback of the proposed and the previously developed inverse control schemes of Martin-Sanchez (1976) and Widrow and Stearns (1985). However, if the plant disturbance is low, this is not going to be a serious problem as none of the poles of the plant affect the system function of the closed loop system. Of course, we are assuming that the plant estimation is accurate enough.

Note that $P_3(z)$ has a zero over the unity circle and, therefore, the approximation $C(z) \approx z^{-\Delta}/B(z)$ cannot be accurate. However, despite the latter fact, the closed loop system (with $\alpha = 1$) has remained stable.

To further investigate the performance of the proposed scheme, when the plant has badly conditioned zeros, a few different examples of $B(z)$ and their corresponding $\sum |e_i|$ (after optimum setting of $C(z)$) are given in Table 1. All results are obtained for $N = 11$. Also, included in this table are the results of the cases when a proper $F(z)$ is used instead of $z^{-\Delta}$. The effectiveness of application of $F(z)$ can be seen from these results. Fig. 12 gives, as an example, the simulation result of $P_3(z)$ with $F(z) = 0.5z^{-3} + 0.5z^{-4}$.

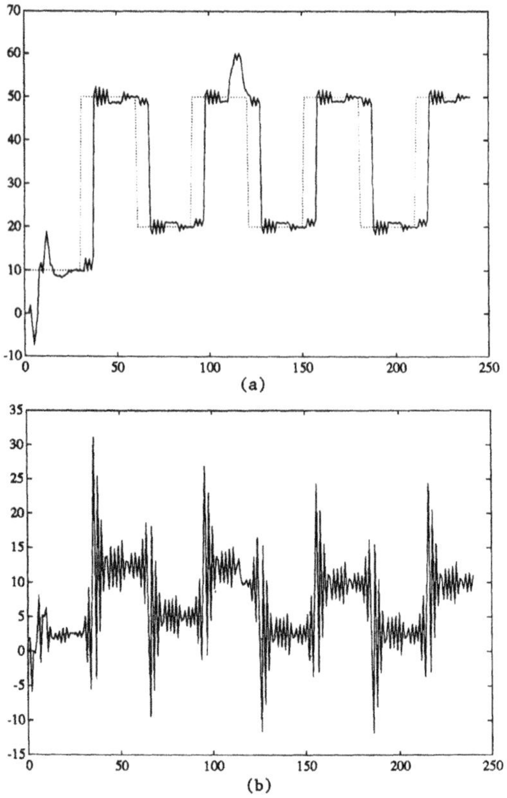

Fig. 11. Simulation result for $P_3(z)$.
(a) Set point, $\{x_k\}$ (doted), and plant output, $\{y_k\}$ versus k.
(b) Controller output $\{u_k\}$ versus k.

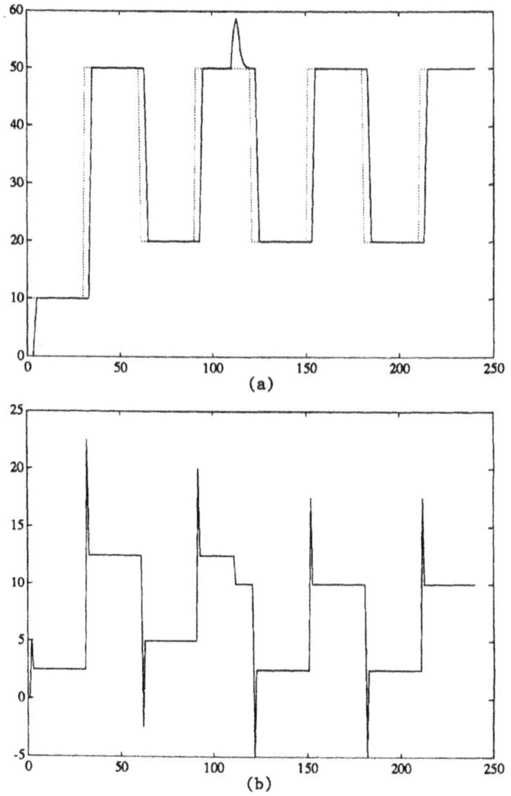

Fig. 12. Simulation result for $P_3(z)$ with $F(z) = 0.5z^{-3} + 0.5z^{-4}$.
(a) Set point, $\{x_k\}$ (doted), and plant output, $\{y_k\}$ versus k.
(b) Controller output $\{u_k\}$ versus k.

TABLE 1 Results of $B(z)$ Inversion

$B(z)$	$\sum \lvert e_i \rvert$	
	$F(z) = z^{-6}$	$F(z) = 0.5z^{-5} + 0.5z^{-6}$
$(1+0.6z^{-1})(1+2z^{-1})$	0.097	—
$(1+0.7z^{-1})(1+1.5z^{-1})$	0.32	—
$(1+0.8z^{-1})(1+1.3z^{-1})$	0.63	0.088
$(1+0.9z^{-1})(1+1.1z^{-1})$	0.93	0.109
$1+Z^{-1}$	1.00	0
$(1+Z^{-1})^2$	1.00	0.115
$(1+Z^{-1})^3$	1.32	0.126

CONCLUSIONS

It was shown that the inverse modeling approach of Widrow and Stearns (1985) could be made more robust with respect to the inaccurate estimation of the inverse model, by introducing a fixed feedback from the plant output to the controller input. Efficient techniques for approximating the inverse of a polynomial in form of a finite length FIR filter, which was required in the process of the controller design, were addressed. The stability of the proposed scheme was also discussed and solutions were given to this problem. The suggested scheme was tested by computer simulations and it was observed that as long as the plant was well conditioned, in the sense that its poles and zeros were not very close to the unity circle, it could be easily controlled, with sufficient robustness to the load disturbance.

ACKNOWLEDGMENT

The authors wish to thank Dr. T.T. Tay of Electrical Engineering Department of National University of Singapore for many helpful discussions during preparation of this paper.

REFERENCES

Astrom, K.J.(1987). Adaptive Feedback Control. *IEEE Proc.*, *75*, pp.185-217.

Clarke, D. W., P. P. Kanjilal, and C. Mohtadi (1985). A Generalized LQG Approach to Self-tuning Control, Part I. Aspects of Design, and Part II. Implementation and Simulation. *Int. J. Control*, *41*, pp.1509-1544.

Farhang-Boroujeny, B.(1986). Application of Gauss-Seidel Method to Channel Equalization. *IEEE Montech'86 Proc., Conference on Antennas and Commun.*, pp.345-348.

Goodwin, G. C. and K. Sang Sin (1984). *Adaptive Filtering Prediction and Control*. Prentice-Hall, Englewood Cliffs, N.J.

Lelic, M. A. and M. B. Zarrop(1987). Generalized Pole-placement Self-tuning Controller, Part 1. Basic Algorithm. *Int. J. Control*, *46*, pp.547-568.

Martin-Sanchez, J. M. (1975). A New Solution to Adaptive Control. *IEEE Proc.*, *64*, pp.1209-1218.

Sugiyama, Y.(1986). An Algorithm for Solving Discrete-time Wiener-Hopf Equations Based on Euclid's Algorithm. *IEEE Trans. Inform. Theory*, *IT-32*, pp.394-409.

Widrow, B. and S.D. Stearns(1985). *Adaptive Signal Processing*. Prentice-Hall, Englewood Cliffs, N.J.

CONTINUOUS-TIME GENERALIZED PREDICTIVE ADAPTIVE CONTROL

D. Matko

Faculty of Electrical and Computer Engineering Ljubljana, Trzaska 25, Yugoslavia

Abstract. The paper describes an approach to the continuous time adaptive control systems, which are based on the continuous time generalized predictive (GP) control. An alternative procedure for GP controller design is represented, which is based on a state observer - Kalman filter. First the design procedure is shown for discrete time systems. In the first step the discrete time Kalman filter for processes disturbed by noise which is usually used at the GPC design is developed. The second step is to solve a time variant Riccate type equation which is obtained if the standard optimal state controller cost function is made equal to the GPC cost function by introducing time variant costing coefficients. It is shown that using the proposed procedure GPC for any costing and control horizon can be designed and that GPC is an optimal state controller with a special representation of the Kalman filter what extends previous results showing that the linear quadratic Gaussian (LQG) control is a special case of the GPC. The idea of GPC design as a state controller with an Kalman filter can be carried over to the continuous time domain. The continuous time GPC criterion function for regulation can be made equal to the standard criterion function of the continuous time state controller by an appropriate choice of the standard quadratic performance criterion matrices in the same way as with discrete time systems. The control law which is believed to be an equivalent of the continuous time GPC is then given by standard state controller feedback law and can be combined with an continuous time identification procedure to yield adaptive control system. Simulation results with the nonminimum phase proces illustrate the proposed adaptive controll system design procedure.

Keywords. Adaptive control, Computer-aided design, Kalman filters, Linear optimal control, Predictive control.

INTRODUCTION

The paper represents an approach to continuous time adaptive control system, which is based on the continuous time generalized predictive control. Generalized predictive control (GPC) was developed in discrete-time framework, where a special form of the noise filter (noise spectral density) was supposed and a quadratic cost function was optimized, Clarke, Mohtadi, Tuffs (1987a,b). Continuous time counterpart based on emulators was developed later and is based on the mathematical analogy to the discrete predecessor, Gawthrop, Demircioglu (1989). In this paper an alternative procedure for GP controller design is represented, which is based on a state observer - Kalman filter and is applicable also in continuous-time systems. First the Kalman filter for processes disturbed by noise which is usually used at the GPC design (white noise filtered by a filter of the form $C(z^{-1})/A(z^{-1})$ is developed. The second step is to solve a time variant Riccati type equation which is obtained if the standard optimal state controller cost function is made equal to the GPC cost function by introducing time variant costing coefficients. It is shown that using the proposed procedure GPC for any costing (N_1,N_2) and control (N_u) horizon can be designed and that with respect to feedback properties GPC is equivalent to the optimal state controller with a special representation of the Kalman filter. This extends previous results showing that the linear quadratic Gaussian (LQG) control is a special case of the GPC for N_2, N_u tending to infinity. The same procedure is then applied to continuous time systems yielding the continuous time GPC controller. A modification of the cost function penalizing the control effort enables elimination of constant disturbances. Various possibilities leading to integral and proportional - integral type of controllers are reviewed. Simulation results illustrate the proposed controller design procedure.

DISCRETE TIME GENERALIZED PREDICTIVE CONTROL

The generalized predictive controller (GPC) will be reviewed here shortly for zero mean disturbances and zero reference signal (w=0). This simple GPC is usually not applicable in the practice because of the violation of the zero mean disturbances superposition. The original GPC, Clarke et all, (1987a,b) will be derived later from the simple GPC. Also only feedback properties of the controller will be discussed, the noncausal prefilter will be omitted.

The disturbed process shown in Fig. 1 is supposed to be described by the following difference equation

$$A(z^{-1})y(z)=B(z^{-1})\ z^{-d}u(z)+C(z^{-1})v(z) \tag{1}$$

where

$$A(z^{-1}) = 1+a_1z^{-1}+a_2z^{-2}+\dots+a_mz^{-m} ,$$

$$B(z^{-1}) = b_1z^{-1}+b_2z^{-2}+\dots+b_mz^{-m} ,$$

$$C(z^{-1}) = 1+c_1z^{-1}+c_2z^{-2}+\dots+c_mz^{-m}$$

and $y(z)$, $u(z)$ and $v(z)$ are process output, process input and zero mean white noise respectively.

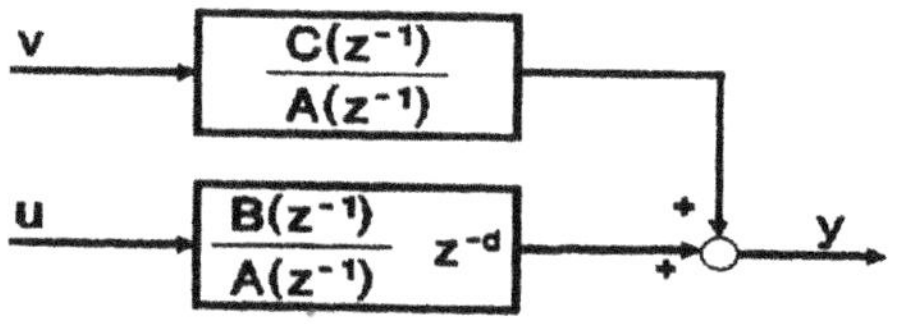

Fig.1: The disturbed process

The idea of the generalized predictive controller is to predict the process output for a set of steps, and to minimize the cost function

$$I=E\left\{\sum_{i=N_1}^{N_2}[y(k+i)]^2+\sum_{i=1}^{N_u}r_{GPC}(i)[u(k+i-1)]^2\right\} \tag{2}$$

The first term of this cost function penalizes the process output which is in this case equivalent to the system error $e(k)=w(k)-y(k)$ on a prediction horizon stretching from N_1 to N_2, where N_1 and N_2 are so called minimum and maximum costing horizons respectively. The second term of Eq. (2) takes into account the manipulated variable $u(k)$. The upper limit $N_u<N_2$ represents the so called control horizon, i. e. the maximum number of nonzero manipulated variable values; the manipulated variables beyond N_u are supposed to be zero. The weighting factor $r_{GPC}(i)$ can be variable to penalize the manipulated variables differently, but for simplicity a constant $r_{GPC}(i) = r_{GPC}$ will be used.

The minimization of I results in the GPC control algorithm

$$u = -[1\ 0\ \dots\ 0]\ [\underline{G}^T\underline{G}+r\underline{I}]^{-1}\underline{G}^T[\underline{L}_1\underline{u}^-+\underline{L}_2\underline{y}^-] \tag{3}$$

where $\underline{G}$ is the $(N_2-N_1) \times N_u$ matrix of the coefficients representing the process step response, $\underline{L}_1$ the $(N_2-N_1) \times (m+d-1)$ and $\underline{L}_2$ the $(N_2-N_1) \times m$ matrix consisting of coefficients obtained at the prediction of process output by the solution of Diophantine equation, Clarke, Mohtadi, Tuffs (1987). $\underline{y}^-$ is the vector of the present and the past process outputs

$$\underline{y}^- = [y(k)\ \ y(k-1)\ \dots\ y(k-m)]^T , \tag{4}$$

and $\underline{u}^-$ the vector of the past manipulated variables

$$\underline{u}^-= [u(k-1)\ u(k-2)\dots u(k-d-m+1)]^T \tag{5}$$

THE STANDARD APPROACH TO OPTIMAL CONTROL

The standard optimal state controller for univariable processes which are written in the state variable form

$$\underline{x}(k+1) = \underline{A}\ \underline{x}(k) + \underline{b}\ u(k) \tag{6}$$

where $\underline{A}$ is the process matrix and $\underline{b}$ the control vector, generates a manipulated variable $u(k)$ from the state variable vector $\underline{x}(k)$ so that the quadratic performance criterion

$$I = \sum_{i=0}^{N_3}[\ \underline{x}^T(i)\ \underline{Q}\ \underline{x}(i) + r_{SC}^2u(i)] \tag{7}$$

is minimized. Here $\underline{Q}$ is a positive semidefinite and symmetric matrix and r_{SC} is a positive scalar. The resulting controller is

$$u(N_3-j)=-(\underline{b}^T\underline{P}_{N_3-j+1}\underline{b}+r_{SC})^{-1}\underline{b}^T\underline{P}_{N_3-j+1}\underline{A}\ \underline{x}(N_3-j) \tag{8}$$

where $\underline{P}_{N_3-j+1}$ is the solution of the matrix Riccati equation

$$\underline{P}_{N_3-j} = \underline{Q}+\underline{A}^T\underline{P}_{N_3-j+1}[\underline{I}-\underline{b}(r+\underline{b}^T\underline{P}_{N_3-j+1}\underline{b})^{-1}\underline{b}^T\underline{P}_{N_3-j+1}]\underline{A} \tag{9}$$

with $\underline{P}_{N_3} = \underline{Q}$ as the final matrix. State vector $\underline{x}$ is usually not measurable, so an observer or in case of disturbances Kalman filter must be used. With standard Kalman filter the dynamic process is supposed to be represented by the following vector difference equations

$$\underline{x}(k+1) = \underline{A}\ \underline{x}(k)+\underline{b}\ \underline{u}(k)+\underline{f}\ v(k) \tag{10}$$

$$y(k) = \underline{c}^T\underline{x}(k) + n(k) \tag{11}$$

where $\underline{c}$ is output vector, $\underline{f}$ is the input noise vector, v(k) and n(k) (zero mean and covariances V and N respectively) are uncorrelated white noise sequences.

The well known Kalman filter estimates the state vector $\underline{x}(k)$ according to the measurements of the outputs $\underline{y}(k)$. The aposteriori estimated state vector is denoted by $\hat{\underline{x}}$. The recursive estimation algorithm is given by the following procedure Isermann (1981):

$$\underline{x}^*(k+1) = \underline{A}\,\hat{\underline{x}}(k) + \underline{b}\,u(k) \qquad (12)$$

$$\hat{\underline{x}}(k) = \underline{x}^*(k)+\underline{K}(k)[y(k)-\underline{c}^T\underline{x}^*(k)] \qquad (13)$$

If stationary noise is supposed, the steady state solution of the Kalman filter is used. The vector $\underline{K}$ is in this case constant

$$\underline{K} = \bar{\underline{P}}_K\underline{c}\,[\underline{c}^T\bar{\underline{P}}_K\underline{c} + \underline{N}]^{-1} \qquad (14)$$

where $\bar{\underline{P}}_K$ is the solution of the stationary matrix Riccati equation

$$\bar{\underline{P}}_K = \underline{f}\underline{V}\underline{f}^T+\underline{A}\bar{\underline{P}}_K[\underline{I}-\underline{c}(\underline{c}^T\bar{\underline{P}}_K\underline{c}+\underline{N})^{-1}\underline{c}^T\bar{\underline{P}}_K]\underline{A}^T \qquad (15)$$

AN ALTERNATIVE APPROACH

GPC and optimal state controller are with respect to feedback properties identical if the criterion functions (2) and (7) are chosen to be equal and if the Kalman filter is designed for noise filter C/A as supposed with GPC in Fig.1. First the corresponding Kalman filter will be designed and its relation with optimal prediction of polynomial systems by means of Diophantine equation will be discussed.

It is well known that the closed loop poles of the Kalman filter are the m roots within the unit circle of the 2m-th order equation

$$N + H(z^{-1})\ V\ H(z) = 0 \qquad (16)$$

where

$$H(z) = \underline{f}^T(z\underline{I}-\underline{A}^T)^{-1}\underline{c} = \underline{c}^T(z\underline{I}-\underline{A})^{-1}\underline{f} \qquad (17)$$

is the transfer function from the noise v to the process output. These m roots have also an other representation described below.

The spectral density of the noise on the output of the disturbed process described by Eqns. (10) and (11) is given in this case by the following equation

$$\Phi_y(\omega) = \frac{1}{2\pi}\,[N+H(z^{-1})VH(z)]_{z\,=\,e^{j\omega T}} \qquad (18)$$

From the spectral factorisation theorem (Åström, Wittenmark, 1984) follows that a noise with the same spectral density may be produced by filtering the white noise on a linear system with transfer function

$$G_v(z) = \frac{C(z^{-1})}{A(z^{-1})} = \frac{1 + c_1z^{-1} + \ldots + c_mz^{-m}}{1 + a_1z^{-1} + \ldots + a_mz^{-m}} \qquad (19)$$

The spectral density of the equivalent linear system output noise is given by

$$\Phi_y(\omega) = \frac{1}{2\pi}\left[\frac{C(z^{-1})\ C(z)}{A(z^{-1})\ A(z)}\,\sigma_v^2\right]\Bigg|_{z\,=\,e^{j\omega T}} \qquad (20)$$

where σ_v^2 is the covariance of the linear system input noise. It is obvious that the "stable" n roots of the Eq. (18) correspond to the numerator of the equivalent linear system transfer function $C(z^{-1})$. This fact is important for the practice and enables the design of Kalman filters for processes disturbed by noise as supposed at the GPC design and shown in Fig. 1. The Kalman filter characteristic equation has to be chosen in this case equal to the noise filter numerator polynomial $z^mC(z^{-1})$.

If the process is represented in observable canonical form, the matrix $\underline{AK}$ of the corresponding Kalman filter is given by

$$\underline{AK} = \underline{d}-\underline{a} = \begin{bmatrix} d_1 \\ d_2 \\ \vdots \\ d_m \end{bmatrix} - \begin{bmatrix} a_1 \\ a_2 \\ \vdots \\ a_m \end{bmatrix} = \begin{bmatrix} d_1 - a_1 \\ d_2 - a_2 \\ \vdots \\ d_m - a_m \end{bmatrix} \qquad (21)$$

Fig. 2 illustrates this case and it can be seen that the Kalman filter error $\tilde{y}$ tends to the v(k) as $t \to \infty$. Kalman filter reconstructs the noise v(k) and uses the reconstructed signal to drive the parallel model of the disturbed process. Using input output representation of the process state, the equivalence of the above Kalman filter representation and the output prediction can be established. Optimal output prediction using the Diophantine equation represents the discrete Kalman filter equivalent for polynomial systems, Lewis (1986).

Now the criterion function equivalence will be established. The criterion function (7) of the state controller can be made equal to the GPC criterion function (2) if for the process represented in observable canonical form the matrix $\underline{Q}$ is chosen to be

$$\underline{Q} = \begin{bmatrix} 0 & 0 & \ldots & 0 \\ 0 & 0 & \ldots & 0 \\ & & \vdots & \\ 0 & 0 & \ldots & 1 \end{bmatrix} \quad \text{for } N_1 \le i \le N_2 \qquad (22)$$

Q = 0 othervise

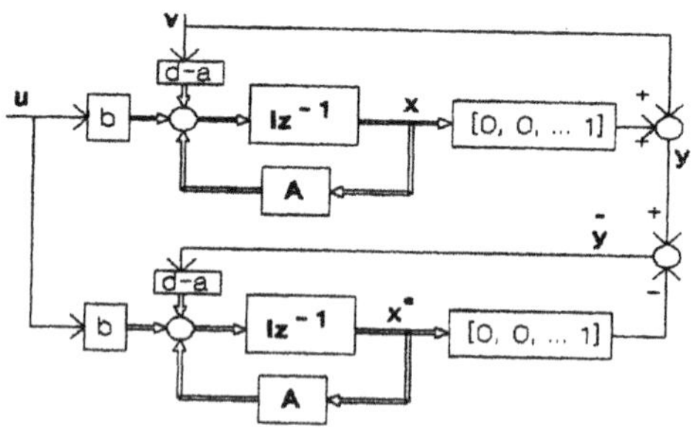

Fig. 2: Illustration of the Kalman filter.

and

$$r_{SC}=r_{GPC} = r \text{ for } 1 \le i \le N_u$$
$$r_{SC} = m \text{ for } i > N_u \qquad (23)$$

Without loss of generality $N_1 = N_2$ can be chosen. The procedure for the optimal state controller design is now as follows:

a. If $N_2 > N_u$ solve the equation

$$\underline{P}_{N_2-j} = \underline{Q} + \underline{A}^T \underline{P}_{N_2-j+1} \underline{A} \qquad (24)$$

for $j = 1$ to N_2-N_u with final condition $\underline{P}_{N_2}=\underline{Q}$

The result of this step is the matrix $\underline{P}_{N_u}$. If $N_2=N_u$ then $\underline{P}_{N_2}=\underline{P}_{N_u} = \underline{Q}$ is set

b. Use $\underline{P}_{N_u}$ as the final condition of the equation

$$\underline{P}_{N_2-j} = \underline{Q}_{N_2-j}+\underline{A}^T\underline{P}_{N_2-j+1}[\underline{I}-\underline{b}(r+\underline{b}^T\underline{P}_{N_2-j+1}\underline{b})^{-1}\underline{b}^T\underline{P}_{N_2-j+1}]\underline{A} \qquad (25)$$

where

$$Q_{n_2-j} = Q \quad \text{if } j \le N_2 - N_1 \qquad (26)$$
$$Q_{n_2-j} = 0 \quad \text{if } j > N_2 - N_1$$

and solve it for $j = N_2 - N_u + 1$ to N_2. The resulting matrix $\underline{P}_1$ is denoted as $\underline{P}$ and is used in control law synthesis.

If the receding horizon is introduced as with GPC, the control law (8) becomes

$$u = [\underline{b}^T\underline{P}\ \underline{b} + r]^{-1}\underline{b}^T\underline{P}\ \underline{A}\ \underline{x} \qquad (27)$$

Since the criterion functions of the GPC and optimal state controller are identical and optimal output prediction of the Diophantine equation represents the discrete Kalman filter equivalent for polynomial systems, GPC is identical to the optimal state controller with Kalman filter. This equivalence was established by numerous examples.

If nonzero mean disturbances or reference signals appear the criterion function (2) and (7) can not be used because of the required nonzero steady state value of the control variable. Filtered control variable must be used in criterion function in this case. The simplest filter leading to original GPC of Clarke, Mohtadi, 1987 is

$$u^*(z) = (1 - z^{-1})u(z) = \Delta u \,. \qquad (28)$$

Drifting disturbances are introduced by modeling the noise term in Eq. (1) by $(C(z^{-1}/\Delta)v(z)$ leading to the process equation

$$\Delta A(z^{-1})y(z) = B(z^{-1})z^{-d}u^*(z)+C(z^{-1})n(z) \qquad (29)$$

If now the described controller is designed for process (29) the original GPC is obtained. Eq. (28) represents the introduction of a serial PI term to the process transfer function, Isermann (1987). Instead of simple filter (28) any filter with DC component suppression can be used and the resulting controller is of integral type.

The idea of GPC design as a state controller with an Kalman filter can be carried over to the continuous time domain. For the process written in the state space representation

$$\dot{\underline{x}} = \underline{A}\ \underline{x} + \underline{b}\ u + \underline{f}\ v$$
$$y = \underline{c}^T\underline{x} + n \qquad (30)$$

where v and n (zero mean and covariances V and N respectively) are white noise processes uncorrelated with $\underline{x}(0)$ and with each other, the steady state Kalman filter estimate is given by

$$\dot{\hat{\underline{x}}} = \underline{A}\ \hat{\underline{x}} + \underline{b}\ u + \underline{K}(y - \underline{c}^T\hat{\underline{x}}) \qquad (31)$$

where for stationary noises the Kalman gain

$$\underline{K} = \underline{P}_K\underline{c}^T N^{-1} \qquad (32)$$

is constant with $\underline{P}_K$ as solution of the matrix Riccati equation

$$\underline{A}\underline{P}_K + \underline{P}_K\underline{A}^T + \underline{f}V\underline{f}^t - \underline{P}_K\underline{c}N^{-1}\underline{c}^T\underline{P}_K = 0 \qquad (33)$$

Noise with the same spectral density as the noise produced by (30) can be obtained also by filtering white noise with a phase minimal stable filter C(s)/A(s) with the same numerator and nominator degrees. The procedure of the Kalman filter design is then analogous to the discrete time Kalman filter design explained in Fig. 2. If however the degree of C(s) is supposed to be less than the degree of A(s) Kalman filter with no measurement noise, Lewis 1986, must be applied. Given example will illustrate this approach.

The continuous time GPC criterion function for regulation (no reference signal)

$$I = \int_{T_1}^{T_2} y^2(t)\, dt + \int_0^{T_u} r\, u^2(t)\, dt \tag{34}$$

can be made equal to the standard criterion function of the continuous time SC by an appropriate choice of the standard quadratic performance criterion matrices in the same way as with discrete time systems (Eqns. 22,23). The control law which is believed to be an equivalent of the continuous time GPC is then given by standard SC feedback law

$$u = r^{-1} \underline{b}^T \underline{P}\, \hat{\underline{x}} \tag{35}$$

with $\underline{P}$ as the result of the following procedure:

a. solve the equation

$$\dot{\underline{P}}(t_2-t) = -\underline{P}(t_2-t)\,\underline{A} - \underline{A}^T\underline{P}(t_2-t) - \underline{Q} \tag{36}$$

with $\underline{P}(t_2) = 0$ for t going from 0 to t_2-t_u.

b. Use $P(t_u)$ as the final condition for the matrix Riccati equation

$$\dot{\underline{P}} = -\underline{P}\,\underline{A} - \underline{A}^T\underline{P} + \underline{P}\underline{b}r^{-1}\underline{b}^t\underline{P} - \underline{Q} \tag{37}$$

and solve it for t going from t_2-t_u to t_2. The arguments t_2-t in Eq.(37) were omitted due to simpler notation. $\underline{Q}$ is given analogously to the discrete case with Eq.(22) for $t_1 \le t \le t_2$ and zero otherwise.

If nonzero mean disturbances or reference signals appear, the control variable u in criterion function (34) must be replaced by a filtered variable u^*. Used filter must suppress DC components. The process transfer function (and consequently the matrix $\underline{A}$ and vector $\underline{b}$ must be extended by the inverse of the filter transfer function. Simple possibilities are

$$u^* = s\,u \qquad \text{and} \qquad u^{**} = \frac{s}{s+1}\,u \tag{38}$$

leading to I and PI controllers respectively. The described procedure can be combined with continuous time identification procedure in adaptive control systems, but this is not the scope of this paper.

The procedure was tested on the nonminimum phase process

$$G_p(s) = \frac{-0.2s+1}{(s^2+1)}\,. \tag{39}$$

For the Kalman filter design the noise filters

$$G_{n1}(s) = \frac{1}{s} \qquad G_{n2}(s) = \frac{s^2+s+0.2}{s(s^2+1)} \tag{40}$$

were supposed. The extended process transfer functions yielding I and PI controllers are

$$G_{p1}(s) = \frac{u}{y^*} = \frac{-0.2s+1}{s(s^2+1)} \tag{41}$$

$$G_{p2}(s) = \frac{u}{y^{**}} = \frac{(-0.2s+1)(s+1)}{(s^2+1)}\,. \tag{42}$$

For Kalman filter the disturbed process was written in the state space representation (30) with

$$\underline{A} = \begin{bmatrix} 0 & 0 & 0 \\ 1 & 0 & -1 \\ 0 & 1 & 0 \end{bmatrix} \quad \underline{b}_1 = \begin{bmatrix} 1 \\ -0.2 \\ 0 \end{bmatrix} \quad b_2 = \begin{bmatrix} 1 \\ 0.8 \\ -0.2 \end{bmatrix}$$

$$\underline{f}_1 = \begin{bmatrix} 1 \\ 0 \\ 1 \end{bmatrix} \quad \underline{f}_2 = \begin{bmatrix} 0.2 \\ 1 \\ 1 \end{bmatrix} \quad c^T = [0 \quad 0 \quad 1] \quad n = 0 \tag{43}$$

where b_1, b_2 correspond to G_{p1}. G_{p2} and f_1, f_2 to G_{n1}, G_{n2} respectively. Kalman filter with no measurement noise is shown in Fig. 3. where K is given by

$$\underline{K} = (\underline{P}\underline{A}^T + \underline{f}V\underline{f}^T)\,\underline{c}\,(\underline{c}^T\underline{f}V\underline{f}^T\underline{c})^{-1} \tag{44}$$

with $\underline{P}$ as the solution of the Riccati equation

$$\underline{A}\underline{P} + \underline{P}\underline{A}^T + \underline{f}V\underline{f}^T - \underline{K}(\underline{c}^T\underline{f}V\underline{f}^T\underline{c})\underline{K}^T = 0, \tag{45}$$

Lewis 1986. The solution of this equation is trivial $P = 0$ yielding $\underline{K} = \underline{f}/(\underline{c}^T\underline{f}) = \underline{f}$

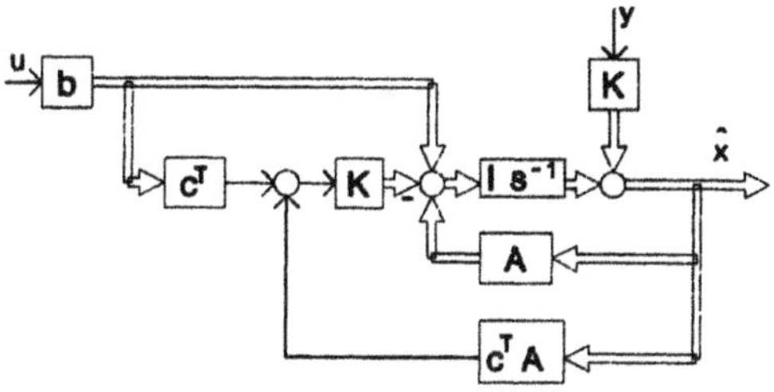

Fig.3: Kalman filter for noise filter (40)

The resulting feedback gain obtained by minimization of criterion (34) for $t_1=0$, $t_2=t_u=3$ and $r=0.01$ is

$$r^{-1}\underline{b}^T\underline{P} = [5.85 \quad 9.02 \quad 4.04] \tag{46}$$

The simulation results for I and PI controllers with noise filter G_{n1} and with square wave reference input without any noise are shown in Figs. 4 and 5. The corresponding results for the I controller design for the noise filter G_{n2} are shown in Fig. 6. It is obvious that the close loop step response with the controller designed for the coloured noise $n_2 = G_{n2}\, v$ is not satisfactory (steady state is reached after 20 s). This feature is

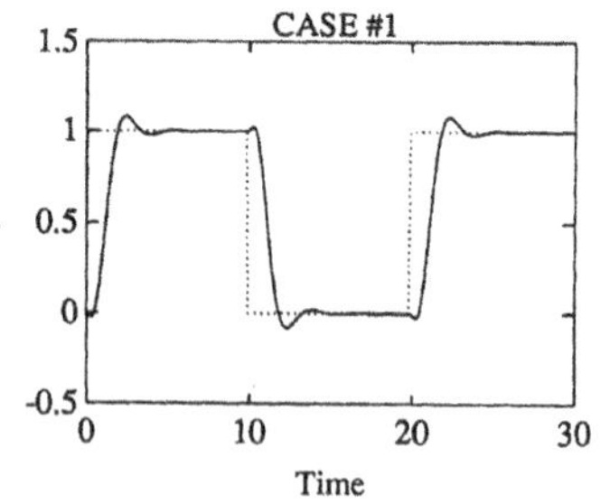

Fig. 4. Simul. results I contr.with G_{n1}

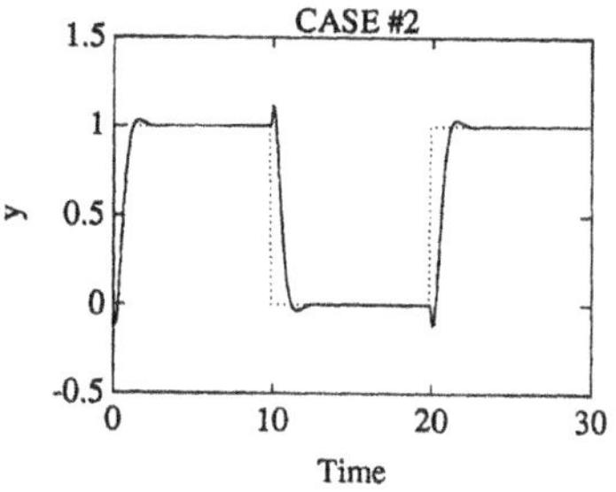

Fig. 5. Simul. results PI contr.with G_{n1}

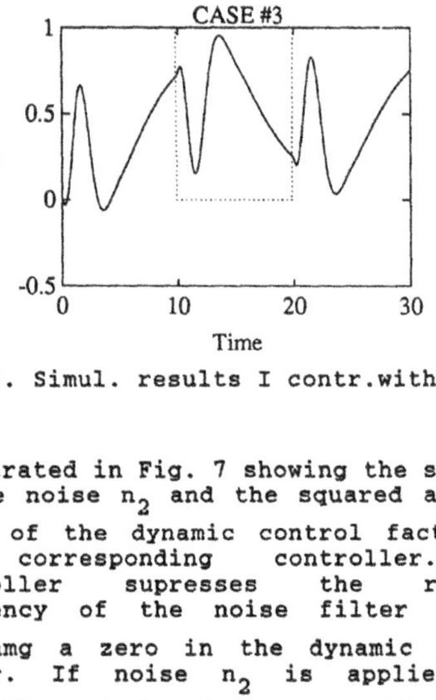

Fig. 6. Simul. results I contr.with G_{n2}

illustrated in Fig. 7 showing the spectrum of the noise n_2 and the squared absolute value of the dynamic control factor for the corresponding controller. The controller supresses the resonant frequency of the noise filter G_{n2} by settinmg a zero in the dynamic control factor. If noise n_2 is applied, the controller designed for n_2 reduces the output standard deviation from 0.65 to 0.35 while the controler designed for the noise filter 1/s increases it to 0.70

CONCLUSION

Equivalence with respect to feedback characteristics of generalized predictive controller and state controller with modified representation of Kalman filter is shown. The idea is then carried over to continuous time systems and a controller is derived which is belived to be the continuous time equivalent of the GPC.

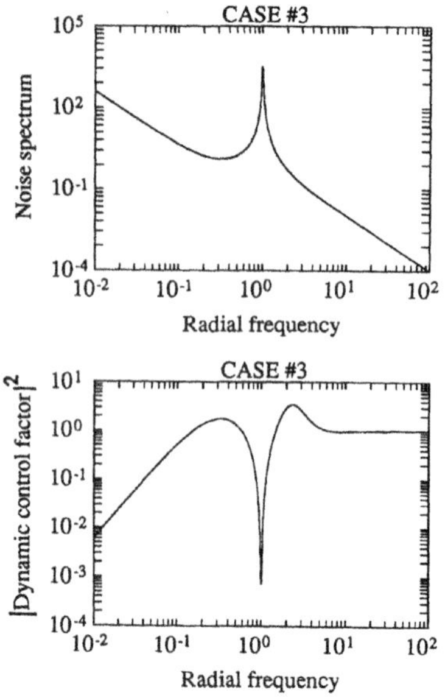

Fig. 7. Noise spectrum and the dynamic control factor (I contr.with G_{n1})

REFERENCES

Åström, K.J. and B. Wittenmark (1984). *Computer controlled systems.* Englewood - Cliffs, N.J., Prentice Hall.

Athans, M. and P.L. Falb (1966). *Optimal control*. New York, Mc Graw - Hill.

Clarke, D.W., C.Mohtadi and P.C. Tuffs (1987a). *Generalized predictive control - part I. The basic algorithm.* Automatica 23, 137 - 148.

Clarke, D.W., C. Mohtadi and P.C. Tuffs (1987b). *Generalized predictive control - part II. Extensions and interpretations.* Automatica 23, 149 - 160.

Clarke, D.W. and C. Mohtadi (1987). *Properties of generalized control.* Prep. 10.th IFAC Congress Munich, Vol 10. 63 - 74.

Gawthrop,p.J. and Demircioglu H (1989). *Continuous - time generalized predictive control.* IFAC Symposium Adaptive Systems in Control and Signal processing, Glasgow

Kwakernaak, H. and R. Sivan (1972). *Linear optimal control systems.* New York, Wiley-Interscience.

Lewis, F.L. (1987). *Optimal estimation.* New York, Wiley-Interscience.

Schumann, A. (1989). *Advantages of the Input - Output - Representation of Predictive Control Algorithms.* IFAC Symposium Adaptive Systems in Control and Signal processing, Glasgow

ADAPTIVE SYSTEMS IN PROCESS ENGINEERING

M. T. Tham, M. J. Willis, A. J. Morris and G. A. Montague

Dept. of Chemical and Process Engineering, University of Newcastle-upon-Tyne, Newcastle-upon-Tyne, NE1 7RU, UK

Abstract. Potential applications of adaptive systems to solve process engineering problems are presented. The application domains include control, optimisation and estimation. Results of evaluation studies on industrial plants, as well as nonlinear simulations are used as illustrative examples. Implementation issues are also discussed in detail.

Keywords. Adaptive systems, multivariable control, optimisation, estimation, inferential control.

INTRODUCTION

In the field of control engineering, the past two decades would be noted for the ardent study into adaptive control and estimation methodologies. This activity was not confined to 'ivory-towered' institutions of higher learning and research establishments. In fact, almost all major control systems manufacturers currently offer some form of adaptive/self-tuning devices (eg. Hardie, 1990). However, despite the embellishments about the advantages of adaptive schemes, it is dubious whether practising engineers have embraced the technology with the same enthusiasm and confidence as their proponents. This could perhaps be due to a poor appreciation for the capabilities of the various algorithms. In the tough world of business, effectiveness is more precious than promises of benefits through application of advanced technology. The failure of an evaluation study will not only quickly discourage the adoption of the algorithm under consideration, but will also cast doubt about the practicality of similar schemes. Although there are countless articles on adaptive techniques, they tend to be theoretically biased with little attention paid to implementation aspects, and this could be a root cause of the problem. Nonetheless, procedures which are capable of coping with changes in process dynamics still represent alluring strategies for process operation and supervision.

This contribution reiterates the assertion that adaptive systems can indeed improve process operations. Additionally, it is demonstrated that the role of adaptive systems is not confined to closed loop control. However, only the salient features of the algorithms considered will be described since the details can be found in the references. Instead, application examples are used to focus the discussion on problem formulation and the flexibility offered by adaptive systems. Implementation aspects are discussed, where it is emphasised that benefits from the deployment of adaptive systems can only be realised if they are applied in a discerning manner, and with a full appreciation of the problem at hand.

ADAPTIVE CONTROL

Adaptive systems came to the fore with studies into self-tuning controllers (STCs). Their appeal include easing the tedium of controller tuning; the ability to optimise controller parameters; auto-tuning of dynamic feedforward compensation terms; accommodation of nonlinear and time-varying dynamics; and in some cases, providing dead-time compensation. Many control equipment producers now offer self-tuning/adaptive devices as part of their product portfolio. These are generally single-input single-output (SISO) STCs with PID structures, the differentiating factor being the manner by which the PID parameters are updated. However, it is questionable whether SISO STCs could perform to 'expectations' in an environment where there are typically numerous control loops. Indeed, it has been shown that the application of SISO STCs in a multivariable environment can yield undesirable results (Peel and co-workers, 1985). It is suggested that in process control, a major advantage of adaptive systems is in facilitating the design of model-based multivariable control strategies. Consider, for example, the familiar Generalised Minimum Variance (GMV) strategy of Clarke and Gawthrop (1979) and the multivariable model:

$$\mathbf{A}\mathbf{y}(t) = \mathbf{B}\mathbf{u}(t) + \mathbf{L}\mathbf{v}(t) + \mathbf{C}\mathbf{e}(t) \qquad (1)$$

where **A**, **B**, **L** and **C** are (2x2) polynomial matrices in the back shift operator q^{-1}; **y**, **u**, **v** and **e** are (2x1) column vectors of outputs, inputs, measurable and random disturbances respectively. The form of the representation is particularly important in model based multivariable STCs (MVSTCs): the number of parameters to be estimated, and hence the time required for adaptation, increases with the order and 'size' of the system. Thus, parameter redundancy should be avoided, whilst the structure should be amenable to the incorporation of *a priori* information. Experience has shown that a controller parameterised with:

$$\mathbf{A} = \mathrm{diag}[A_1, A_2],\ \mathbf{C} = \mathrm{diag}[C_1, C_2]$$

$$\mathbf{L} = \mathrm{diag}[q^{-l_1}L_1, q^{-l_2}L_2]$$

$$\mathbf{B} = \begin{bmatrix} q^{-k_{11}}B_{11} & q^{-k_{12}}B_{12} \\ q^{-k_{21}}B_{21} & q^{-k_{22}}B_{22} \end{bmatrix}$$

provides the most flexible framework for applications to chemical processes. A_i and C_i are monic polynomials; k_{ij} denotes the delays between manipulated inputs and outputs; while l_i denote the delays between disturbances and outputs. The model thus assumes that loop interactions occur in a feedforward manner, ie. from inputs to outputs, and that the random disturbances affecting each loop are independent. The control objective is to minimise:

$$J_1 = ||\mathbf{y}^*(t) - \mathbf{w}(t)||^2 + ||\mathbf{Q'u}(t)||^2 \qquad (2)$$

where $||\mathbf{x}||^2 = \mathbf{x}^T\mathbf{x}$; **Q'** is a diagonal matrix of control weightings, and the vectors of delay-step-ahead predictions and set-points are respectively:

$$y^*(t) = [y^*_1(t+k_{11}/t), y^*_2(t+k_{22}/t)]^T$$

$$w(t) = [w_1(t), w_2(t)]^T$$

$y^*(t)$ can be derived as:

$$y^*(t) = Fy(t) + Gu(t) + Dv(t) + Hy^*(t-1) \quad (3)$$

$$G_{ij} = E_iB_{ij}, \quad D = EL, \quad H = (I - C)q$$

using the identity:

$$C = EA + F \quad (4)$$

$$F = diag[q^{-k}{}_{11}F_1, q^{-k}{}_{22}F_2]$$

$$E = diag[E_1, E_2], \quad deg(E_i) = k_{ii} - 1$$

where E_i are also monic polynomials. Substituting Eq.(3) into Eq.(2) and minimisation w.r.t. $u(t)$ will yield the following control law for the MVSTC:

$$w(t) - y^*(t) - Qu(t) = 0 \quad (5)$$

$$Q = diag[Q_1, Q_2], \quad Q_i = q'_{i,o}Q'_i/b_{ii,o}$$

Here the weighting, Q_i, are specified to have inverse PI forms (Morris and co-workers, 1982). By adopting the above model structure, the MVSTC has thus been reduced to a set of interconnected multi-input single-output STCs, where loop interactions are decoupled in a feedforward manner. This process description is also applicable to other MVSTC strategies (eg. Montague and co-workers, 1986; Wilkinson and co-workers, 1990). Although not the most general of process description, the attendant advantages are many: controller design becomes simpler and clearer; the decoupling mechanism is easily visualised; there are fewer parameters; the MVSTC can be parameterised initially using the results of open-loop step tests; commissioning can be performed on a loop-by-loop basis and different sampling rates can be specified for different loops (Morris and co-workers, 1982,1987; Tham, 1985).

The MVSTC has been successfully applied to a 10 stage pilot plant column which separates a 50-50 wt.% methanol-water feed mixture. Top and bottom product streams are simultaneously regulated to 95 wt.% and 5 wt.% methanol, using reflux and steam flowrates as the respective manipulated inputs. Bottoms composition is measured by a gas chromatograph (GC), which has a cycle time of 3 min. This measurement delay, coupled with highly nonlinear dynamics, can cause problems using conventional control, especially when the column is disturbed by large changes in feed flowrate. The UD-factored recursive least-squares (RLS) algorithm (RUD) (eg. Ljung and Soderstrom, 1984) was used to estimate the parameters of the output predictors for the composition loops. Using 3 min. sampling intervals for both loops, the regulatory performances of the MVSTC against a +25% and a -25% change in feed flowrate are shown in Fig.1. Under these conditions, the column is at its limits of operability. However, it can be observed that good control is achieved. Changing the top composition sampling time to 1 min. (the bottom composition sampling interval is still restricted to the GC cycle time of 3 min.), and subjecting the column to the same disturbances, led to the improved responses shown in Fig.2. As different loop sampling intervals are permissible, each may be specified to match individual loop dynamics instead of having to make a compromise between overall system requirements.

ADAPTIVE OPTIMISATION

Clearly, the self-tuning approach can also be used to directly provide for adaptive process supervision and optimisation. Consider the relationship:

$$y = f(x_i, \Theta_i, i=1,..m) \quad (6)$$

The variable of interest, y, is dependent on the data $\{x_i, i=1,..m\}$, and parameters $\{\Theta_i, i=1,..m\}$. Although f may be highly nonlinear, if the parameters of Eq.(6) can be identified, then a suitable technique may be used to optimise y. For example, RLS type algorithms may be employed to identify a nonlinear model which is linear-in-the-parameters, from which the steady states are extracted for use with a numerical optimisation technique to provide on-line adaptive optimising control (AOC) policies (eg. Bamberger and Isermann, 1978; Garcia and Morari, 1984; Golden and Ydstie, 1987; Lee and Lee, 1985). However, for a certain class of process description, the use of an 'additional' optimisation algorithm may not be necessary.

Consider the control of fixed bed chemical reactors (FBRs) where the usual objective is to keep product quality within specifications as well as maintain catalyst bed temperatures (eg. Jorgenson, 1989; McGreavy, 1983). The FBRs considered here employ multiple temperature controllers to influence reactor conversion. Knowing the process dynamics, the effects of temperatures and other process variables on reaction states could be precisely determined, and the scheme should yield good control performances. However, FBRs are typically nonlinear, non-stationary and distributed parameter systems with time-varying characteristics. Such attributes lead to difficulties in selecting the appropriate temperature set-points. A simple solution to this problem is to partitioning the FBR into temperature zones. Assuming that the flowrate, temperature and composition of the feed are relatively constant, Eq.(6) may be specifically expressed as:

$$C^s{}_{out} = x^T\Theta \quad (7)$$

$$x^T = [T^s{}_1, T^s{}_2, .., T^s{}_m], \quad \Theta^T = [\Theta_1, \Theta_2, .., \Theta_m]$$

$$T^s{}_i = T_i - T_{i,ss}, \quad C^s{}_{out} = C_{out} - C_{out,ss}$$

C_{out} = conversion; $\{T_i, i=1,..m\}$ = zone temperatures; $\{\Theta_i, i=1,..m\}$ = model parameters. Note that the variables in Eq.(7) are expressed as deviations about a steady-state condition (denoted by the subscript 'ss'). This is useful since the steady state model response will be independent of other unknown or unmeasured variables. The coefficients, $\{\Theta_i, i=1,..m\}$, are assumed unknown and, in fact, are expected take on different values at different operating levels. However, since Eq.(7) is linear-in-the-parameters, $\{\Theta_i, i=1,..m\}$ can be estimated on-line using RUD. The identified model is then used to calculate the 'best' set of temperature set-points which will achieve the desired feed conversion, C_d. This problem is posed as:

$$\min_{T^s_i} (C_d - C_{out})^2 \quad (8)$$

ie. the strategy aims to minimise the square of the deviations between C_{out} and C_d. Excessive changes in temperature set-points may be penalised by adopting a GMV type cost function. As the model is also linear in T_i, and noting that the RUD is basically a recursive minimisation algorithm, Eq.(8) may thus be solved using the RUD. Since a separate optimisation technique is not needed, implementation of the strategy is significantly simplified. The procedure can be summarised as follows:

Step 1. Measure product composition and reactor bed temperatures. Calculate conversion, C_{out}, and form data vector x, ie.

$$x^T = [T^s{}_i, i=1,..m], \quad T^s{}_i = T_i - T_{ssi}, \ i=1,..m$$

Step 2. Estimate model parameters, Θ, using RUD and assign $\hat{\Theta}$ to z, (where '^' denotes estimates).

Step 3. Given the 'data', z, use RUD to minimise Eq.(8) w.r.t. temperature set-points, s:

$$\frac{\partial(C_d - x^T\hat{\Theta})^2}{\partial s} = 0, \quad s^T = [\tilde{T}^s_i, i=1,..m]$$

Step 4. Send the set-points, $\{T_i = \tilde{T}^s_i + T_{ssi}, i=1,..m\}$, to the temperature controllers, wait for sampling time and repeat procedure from step (1).

The optimisation problem is thus solved by 'inverting' the parameter estimation problem. The applicability of this AOC scheme to FBRs and bioreactors have already been discussed in detail (Kambhampati and co-workers, 1990; Willis and Tham, 1989). Presented here are the results of an application to a nonlinear simulation of an endothermic FBR, which employs two heaters to maintain catalyst bed temperature, ie. there are two temperature zones.

Initially, and without invoking the optimisation stage, the model parameters were estimated by superimposing random sequences on temperature set-points to perturb the system. Only when it was deemed that the parameters have converged was the optimisation stage activated. Improved tracking of changing process dynamics was achieved using a variable forgetting factor. In the optimisation stage however, quick convergence to final values could only be achieved by re-initialising the covariance matrix of the RUD at each new conversion demand. The dotted line in Fig.3 shows the offset that would occur if the reactor was controlled to a fixed temperature profile, when the system was disturbed by a 10 deg.K decrease in feed preheat temperature. Using the AOC, a new temperature profile was automatically calculated to remove this offset. The optimality of the scheme is demonstrated in Fig.4, which compares the temperature trajectories calculated by the above procedure and a corresponding strategy which used the method of steepest descent, in response to a demand for increased conversion (Fig.5). It can be observed that although the routes to the optimum were different, the final temperature values were almost identical. Moreover, faster convergence to the final values were achieved using the proposed method.

ADAPTIVE INFERENTIAL MEASUREMENT AND CONTROL

If relevant process data is not available at a suitable frequency, and at regular intervals, then regardless of the control or optimisation policy, satisfactory performance will not be achieved. This problem is commonly encountered in direct product quality control situations where there is a strong reliance on the use of either composition analysers or laboratory analyses, eg. separation columns, chemical and polymerisation reactors and bioreactors. Due to the difficulties caused by the measurement delays and possibly irregular samples, 'inferential control' is usually applied. Here, control of the (primary) variable is achieved by controlling another (secondary) output variable which can easily measured, ie. the secondary output is used to 'infer' the state of the primary output. The product composition control of industrial columns, for example, is commonly achieved by controlling the liquid temperature of a specific tray. In practice, however, temperature-composition relationships are usually nonlinear and time-varying. Thus, good temperature control does not imply good composition control and inferential control schemes often require 'correction' using values of the primary variable whenever it is available. Unfortunately, the resulting 'parallel cascade' strategy (Luyben, 1974), can be difficult to tune.

Another approach is the application of soft-sensors, viz. algorithms that can estimate 'difficult-to-measure' variables, by inference from relatively 'easy' measurements (eg. Joseph and Brosilow, 1978; Morari and Stephanopoulos, 1980; Wright and co-workers, 1977). The estimates can subsequently be used for feedback control. In general, the use of observers and extended Kalman filters predominate, requiring the existence of accurate mechanistic models. This requirement can, however, be relaxed by designing the estimators within an adaptive framework. This may be approached by either adopting an input-output or a state space process representation. Both routes, however, lead to similar final estimator forms (Guilandoust and co-workers, 1987,1988):

$$y(t) = \Theta^T\phi(t\text{-}d) + e(t) \qquad (9)$$

where: $\Theta^T = [\alpha_d,..., \alpha_{nd}, \beta_1,..., \beta_{nd}, \tau_o,..., \tau_{nd}]$

$$\phi(t\text{-}d) = [-y(t\text{-}d),..-y(t\text{-}2d),..-y(t\text{-}nd), u(t\text{-}m\text{-}d\text{-}1),...u(t\text{-}m\text{-}nd\text{-}1), v(t\text{-}d),...v(t\text{-}nd)]$$

where y(t) is the primary output; u(t) and v(t) are the manipulated input and secondary output respectively; 'd' is the measurement delay and 'm' is the process delay. Both delays are expressed as integer multiples of the sampling interval of u(t) and v(t), which is smaller than the sampling time for y(t). 'n' is the order of the estimator which is specified by the user, and e(t) is the estimation error. When a measurement of y(t) becomes available, the parameter vector Θ is estimated using RUD based on Eq.(9). The *a posteriori* value of $\hat{y}(t)=\hat{\Theta}^T\phi(t\text{-}d)$ is then calculated and inserted in $\phi(t)$ for use in:

$$\begin{aligned}\hat{y}(t+d) &= \hat{\Theta}^T\phi(t)\\ &= -\alpha_d\hat{y}(t) - \alpha_{2d}\hat{y}(t\text{-}d) .. - \alpha_{nd}\hat{y}(t\text{-}(n\text{-}1)d)\\ &\quad + \beta_1 u(t\text{-}m\text{-}1) + .. + \beta_{nd}u(t\text{-}m\text{-}(n\text{-}1)d\text{-}1)\\ &\quad + \tau_o v(t) + .. + \tau_{nd}v(t\text{-}(n\text{-}1)d)\end{aligned}$$

to provide estimates of y(t) at those time instants when its measurements are not available, ie. at the sampling rate of u(t) and v(t). The parameters of Eq.(9) are again updated when a new value of y(t) becomes available.

AIEs have been successfully applied to several industrial processes. Due to commercial confidentiality, the vertical axes of the plots to be presented will not be annotated.

Continuous fermentation. In a continuous industrial fermenter, product specifications require that biomass concentration, X, be tightly controlled. The current control strategy is based upon a 4 hourly sample of X, and the dilution rate, D, adjusted upon return of the laboratory assays of X. Although a sample and control interval of 1 hr. would lead to more efficient control, it is not feasible to sample X at this frequency. However, D and carbon dioxide, CO_2, evolution rate (CER), are related to biomass production, and can be sampled frequently with little delays. Thus, hourly measurements of CO_2 and D, supported by infrequent biomass assays, could be used by an AIE to provide hourly estimates of X. Figure 6 compares the laboratory assays (step-like response) and estimates of X, over a 470 hour period of continuous operation. It can be seen that the AIE provided very good estimates of biomass transient behaviour. Although the assay indicated a large increase in X at time 248 hr., the AIE continued to predict a fall. Other 'suspect' biomass assays can also be observed at times 58, 100 and 183 hours respectively. However, the performance capability of the AIE was not compromised. A major plant disturbance occurred at around time 230 hr., causing a rapid fall in CER. Nevertheless, the AIE predicted the down-turn well, with only a slight undershoot. This is due to continuing estimator parameter adaptation to the new operating conditions. Note that if a non adaptive inferential estimator was to be employed, the estimates would show an offset. This is demonstrated in Fig.7, where the RUD was switched off at time 250 hr.

High purity distillation. In addition to nonlinear dynamics, acceptable product composition control of high purity distillation columns is often hampered by measurement delays. In one such column, an industrial demethaniser, top product composition is controlled by manipulating reflux flowrate. Except for top product composition, which is measured by an analyser at 20 min. intervals, all other process variables are available every 5 min. A potential for improved control exists if composition could be regulated at 5 min. intervals. This may be realised via an AIE, using column overheads vapour temperature and reflux flowrate, to provide composition estimates at 5 min. intervals. Figure 8 shows the results of a 167 hr. segment from a 21 day continuous evaluation study, comparing product composition with its estimates. Good estimates were obtained and degradation in performances were observed only during large process transients, as shown in the corresponding plot of estimation errors (Fig.9). This is because the AIE had to 'tune' to frequent changes in operating conditions. Under steadier operating conditions, it can be observed that there were large reductions in estimation errors, which were generally within the tolerances demanded by plant personnel.

Adaptive Inferential Control. If 'fast' and accurate primary output estimates are available, they can be used as feedback signals for control. While, the use of a secondary variable is, strictly speaking, not necessary, given a suitable choice of secondary variable, the resulting adaptive inferential control (AIC) strategy will inherently provide for 'anticipatory' regulation in a feedforward sense. This is because disturbance effects will manifest first in secondary variable responses, and hence will be reflected in primary output estimates. This is illustrated by application to a nonlinear simulation of the pilot scale column mentioned previously. Steam flowrate is used to regulate bottom product composition against a sequence of changes in feed flowrate: +25% from steady state; -25% back to steady state; -25% from steady state; and finally a +25% back to steady state. The liquid temperature of the stage immediately below the feed tray was chosen as the secondary variable. Bottom composition is measured by a GC at 3 min. intervals while temperature and steam flowrate are measured every 0.5 min.

The performances of the conventional scheme (using composition measurement) and the AIC scheme (using composition estimates) are compared in Fig.10. Both approaches employed fixed parameter PI controllers and their settings were tuned to obtain responses with minimum integral of absolute error. As expected, the regulatory capabilities of the AIC scheme is clearly superior, due to its ability to anticipate disturbance effects. Moreover, making use of the estimates for control also compensates for the effects of measurement delays. Thus higher controller gains may be used, leading to tighter control. This is highlighted by the PI settings used in the AIC scheme, viz. K_c=-175 and T_I=2.5 min., compared to K_c=-60 and T_I=20 min., used in the conventional strategy.

Summary. The adoption of an adaptive approach to inferential estimator design has resulted in an algorithm which is highly 'portable', ie. the AIE was not process specific. During large process transients, slight deteriorations in performances were observed, and is to be expected given the highly nonlinear nature of the processes considered. However, offsets between estimates and primary variable measurements were not observed, and this is in marked contrast to the performances of other inferential schemes (Patke and co-workers, 1982).

The above results, together with those of Guilandoust and co-workers (1987,1988), and Tham and co-workers (1989), demonstrate the potential of AIEs for improving the product quality control of processes faced with measurement delay problems. Additionally, it is suggested that AIC schemes incorporating fixed parameter PI(D) controllers will out-perform STC strategies which rely solely on feedback of delayed output measurements. An added advantage with an AIC scheme is that the estimates can always be validated based upon an engineering appreciation of process capabilities, before they are used for control. In the case of a STC, however, it will be difficult to ascertain whether the calculated control is tenable.

IMPLEMENTATION ASPECTS

In applying STC and AOC strategies, the guidelines given by Astrom and Wittenmark (1989) should ensure success, eg. the use of pre- and post-sampling filters; choice of sampling rates; monitoring of estimator performance; application of 'dead-zones' and 'leakages'; rate limiting and bumpless transfer of control signals. Additionally, the existence of process constraints must be made known to the algorithms. For instance, calculated control must be 'clipped' so that physical limits are never exceeded, otherwise the estimator will be corrupted by inconsistent data. In multivariable control, however, clipping may lead to undpredictable results. In particular, interaction decoupling may not be achieved. Thus, process constraints must be explicitly taken into account when solving for the multivariable control law (Wilkinson and co-workers, 1990).

Process signals are also usually corrupted by noise which should, ideally, be attenuated by analog filters prior to sampling. However, there may be instances such as an evaluation study, where the case for new instrumentation may not be fully justifiable. Noise filtering may therefore have to be performed digitally, and probably at low sampling rates. While it may be of little consequence in a control situation, the phase shift properties of digital filters can reduce the predictive capability of AIEs and lead to poor performances in terms of inferential measurement. Although a delay may be used to model and hence compensate for the phase shift, estimator transient behaviour may suffer. Thus, minimum phase shift filters should be applied. In the above AIE applications, process noise were attenuated by a linear trend 'smoother' (eg. Gardner, 1985). This algorithm uses gradient information and hence its transfer function is similar to that of a lead-lag compensator, with one tunable parameter, viz. the smoothing constant.

Central to the functionality of all adaptive systems is the parameter estimator which provides the respective strategies with their 'learning' capabilities. It is therefore imperative to ensure that only reliable data is used, ie. they are not discordant points or 'outliers'. Outliers are not common in laboratory environments but occur quite frequently in industrial settings. They generate deceptively large estimation errors. Since least-squares techniques seek to minimise the square of estimation errors, the consequence is grossly biased parameter estimates and hence poor system performances. Although low pass filters with narrow bandwidths may be used to mitigate the effects, they also tend to obliterate true process transients. Thus, the ideal solution is not to apply indiscriminate global filtering, but to first test whether a data point is valid, and if it is not, 'reconstruct' the value, eg. using a smoothing algorithm. In an off-line situation, and using various statistical criteria, outliers can easily be editted from the data set (eg. Barnett and Lewis, 1979). Unfortunately, these techniques are not suited for on-line implementation since they require future data points. Nevertheless, a heuristic method may be devised, based upon first and second statistical moments calculated from data histories, as well as knowledge of process capabilities, to perform on-line outlier detection and data reconstruction. This was indeed performed in the AIE applications to the industrial processes presented above. A more integrated solution to the problem of outliers would be to use such a logic filter in conjunction with a parameter estimator derived based upon theories of 'robust statistics' (eg. Huber, 1977).

To promote consistency of estimation in the presence of unmodelled static effects, zero mean data should be employed. Although the use of 'incremental' data has been advocated, this operation amplifies high frequency components of the data, eg. process noise. Moreover, since the models used in the design of adaptive systems are usually lower order approximations, an incremental approach could also accentuate process-model mismatch, resulting in drifting and even unstable parameter estimates. The problem can be alleviated if the signals are low-pass filtered before differencing, ie. 'mean-deviational' data (Tham, 1985). This is equivalent to band-pass filtering the data to remove undesirable high and low frequency components. It is also good practice to 'scale' the data such that they are of the same order of magnitude. A non process specific scaling procedure is to divide the band-pass filtered data by the standard deviation of the data sequence. The standard deviation is calculated using the intermediate low-pass filtered signal as the mean. This data 'standardisation' procedure ensures that each variable is given equal emphasis in the regression relationships, and that the data is always bounded (between ± 1), a necessary condition for stable parameter estimation. The use of standardised data also leads to a better conditioned covariance matrix in RLS based parameter estimators.

CONCLUDING REMARKS

Although the above tasks may appear numerous, increasing computational requirements significanlty, they are relatively easy to perform and are integral to the successful implementation of adaptive techniques. Some of these practical issues have been recognised and adaptive devices with enhanced monitoring functions are beginning to appear on the market. However, it is in ensuring the integrity of the algorithms to specific process upsets and engineering problems that will be most demanding. For instance, instrumentation re-calibration, hard-wiring faults, plant trips and operator intervention can cause the estimator to converge to erroneous values. These events may not be accompanied by specific process behaviour and thus may not be easy to identify, and the appropriate action taken, eg. "continue when normal process operation is restored", or "re-initialise and restart". Nevertheless, a minimum requirement for practicable adaptive systems is that they should be equipped to recognise such occurrences so that the parameter estimator could at least be switched off.

This leads to the contentious question of whether an adaptive system should be left in the adaptive mode at all times. To accrue the full potential of their application, this should indeed be approach to adopt. Unless proper monitoring of the system can be achieved, ie. of both the algorithm and the process, and all major cause and effects identified, it would be deleterious to follow such a course of action. For example, in the above on-line applications, especially the demethaniser column, the respective algorithms have been operated in the adaptive mode for many months without external supervision. However, the algorithms have been subjected to rigorous off-line tests, including the use of real-time simulations, and a process analysis performed on the candidate processes. The former is to obtain an unequivocal understanding of algorithmic capabilities, while the latter is to gather process knowledge for incorporation into the algorithm; identify potentially practical difficulties and notably, to inculcate an engineering appreciation of the problem. Based on the current state-of-the-art, it is therefore suggested that the full benefits of adaptive systems can only realised with in-house developments.

ACKNOWLEDGEMENTS

The authors gratefully acknowledge the support of the Dept. of Chemical and Process Engnrg., Uni. Newcastle upon Tyne; ICI Engnrg., and the UK SERC (MJW).

REFERENCES

Astrom, K.J. and B. Wittenmark (1989). Adaptive Control, Addison-Wesley.

Bamberger, W. and R. Isermann (1978). Adaptive on-line steady-state optimisation of slow dynamic processes. Automatica, 14, 223-230.

Barnett, V. and T. Lewis (1979). Outliers in Statistical Data. John Wiley.

Clarke, D.W. and P.J. Gawthrop (1979). Self-tuning control. Proc. IEE, 126, 6, 633-640.

Garcia, E. C. and M. Morari (1984). Optimum operation of integrated processing systems. Part II Closed loop on-line optimizing control. AIChE Journal, 30, 226-234.

Gardner, E.S. (1985). Exponential smoothing: the state of the art. J. of Forecasting, 4, 1, 1-28.

Golden, M. and E. Ydstie (1987). Nonlinear adaptive optimisation of continuous bioreactors. Proc. 10th IFAC World Congress, Munich, Germany.

Guilandoust, M.T., A.J. Morris and M.T. Tham (1987). Adaptive inferential control. Proc. IEE Pt.D, 134, 3, 171-179.

Guilandoust, M.T., A.J. Morris and M.T. Tham (1988). An adaptive estimation algorithm for inferential control. I & EC Research, 27, 1658-1664.

Hardie, I. (1990). Auto-tuned control. Control and Instrumentation, March.

Huber, P.J. (1977). Robust Statistical Procedures. SIAM.

Jorgensen, S.B. and N. Jensen (1989). Dynamics and control of chemical reactors - selectively surveyed. Proc. IFAC Symp. DYCORD'+89, Maastricht, The Netherlands, 21-23 Aug, 359-370.

Joseph, B. and C. Brosilow. (1978). Inferential control of processes. Parts I and III, AIChE Journal, 24, 3, 485-491 and 500-509.

Kambhampati, C., M.T. Tham, G.A. Montague and A.J. Morris (1990). Multivariable adaptive optimisation of the performance of fermentation process. Proc. ACC, San Diego, May, 2673-2678.

Lee, K.S. and W.K. Lee (1985). On-line optimising control of a non-adiabatic fixed bed reactor. AIChE Journal, 31, 4, 667-675.

Ljung, L. and T. Soderstrom. (1983). Theory and Practice of Recursive Identification. MIT Press.

Luyben, W.L. (1973). Parallel Cascade Control. Ind. Eng. Chem. Fundam., 12, 4, 463-467.

McGreavy, C. (1983). On-line computer control systems for chemical reaction processes. Computers and Chemical Engineering, 7, 4, 529-566.

Montague, G.A., M.T. Tham and A.J. Morris (1986). A comparison of multivariable long range predictive control in a highly nonlinear environment. Proc. ACC, Seattle, 721-727.

Morari, M., and G. Stephanopoulos. (1980). Minimizing unobservability in inferential control schemes. Int. J. Control, 31, 2, 367-377.

Morris, A.J., Y. Nazer and R.K. Wood (1982). Multivariate self-tuning control. Optimal Control Applications and Methods, 13, pp363-387.

Morris, A.J., M.T. Tham and R.K. Wood (1987). Multivariable and multirate self-tuning controllers. In Encyclopedia of Systems and Control. (ed. M. Singh), Pergamon Press, Oxford.

Patke, N. G., P.B. Deshpande and A.C. Chou. (1982). Evaluation of inferential and parallel cascade schemes for distillation control. Ind. Eng. Chem. Process Des. Dev., 21, 266-272.

Peel, D., A.J. Morris and M.T. Tham (1985). Univariate self-tuning control in a distributed control environment. Proc. IEE Conf. 'Control 85', Cambridge.

Tham, M.T. (1985). Some Aspects of Multivariable Selftuning Control. PhD Thesis, Uni. Newcastle upon Tyne, UK.

Tham, M.T., A.J. Morris and G.A. Montague (1989). Soft-sensing: a solution to the problem of measurement delays. Chemical Engineering Research and Design, 67, 6, 547-554.

Wilkinson, D.J., A.J. Morris and M.T. Tham (1990). Multivariable constrained generalised predictive control. Proc. ACC, San Diego, May, 1260-1265.

Willis, M.J. and M.T. Tham (1989). A supervision scheme for fixed bed tubular reactors. Proc. IFAC Symp. DYCORD+'89, Maastricht, The Netherlands, Aug. 21-23.

Wright, J.D., J.F. MacGregor, A. Jutan, J.P. Tremblay and A. Wong. (1977). Inferential control of an exothermic packed bed tubular reactor. Proc. JACC, 1516-1522.

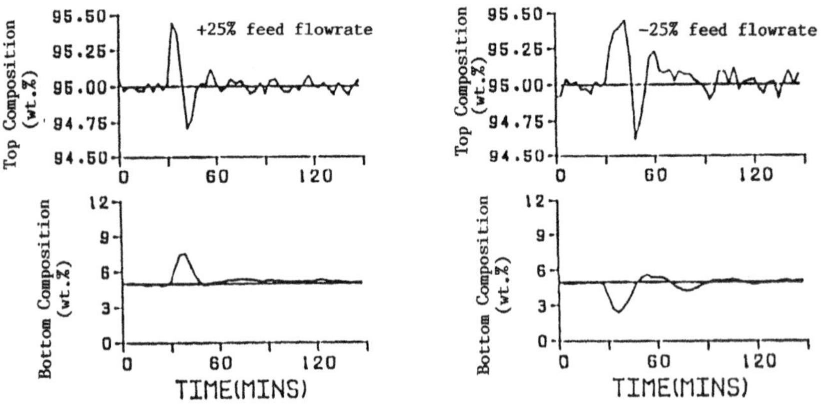

Fig. 1. GMV multivariable self-tuning control of pilot plant distillation column.

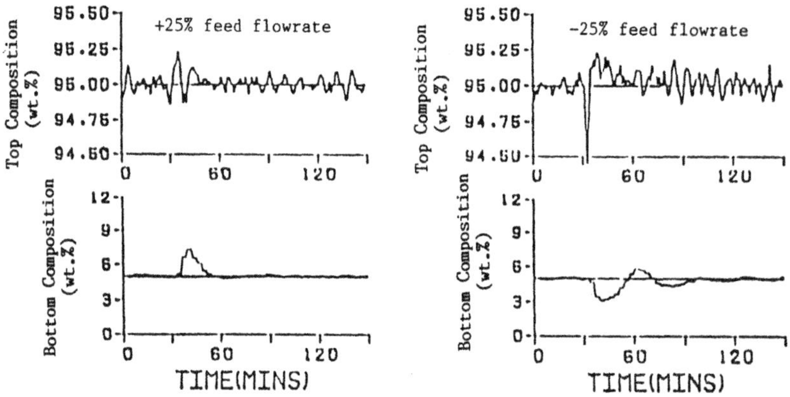

Fig. 2. GMV multivariable-multirate self-tuning control of pilot plant distillation column.

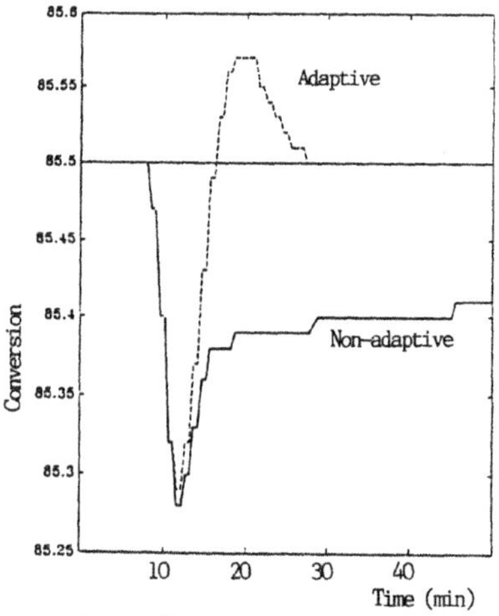

Fig. 3. Conversion response to change in feed temperature.

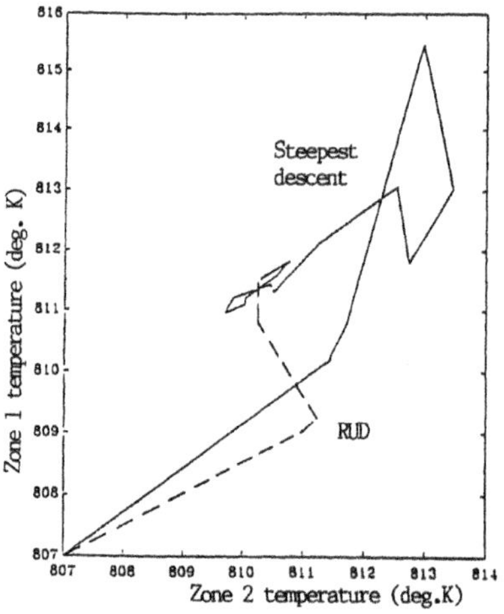

Fig. 4. Routes to optimum temperatures.

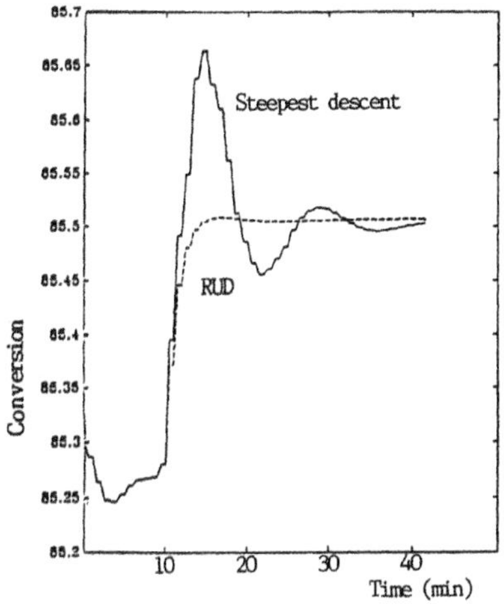

Fig. 5. Response to change in conversion demand.

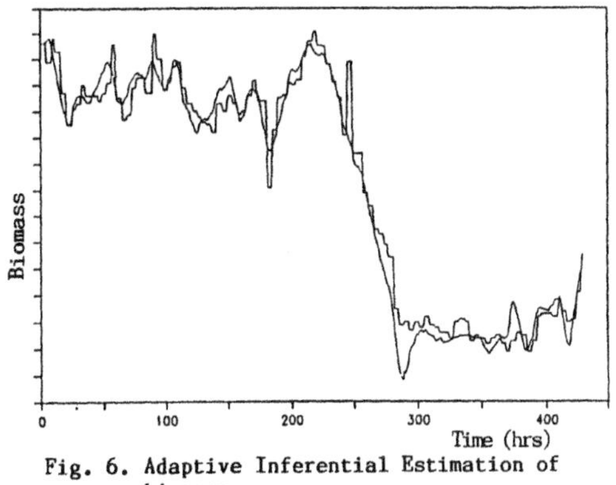

Fig. 6. Adaptive Inferential Estimation of biomass.

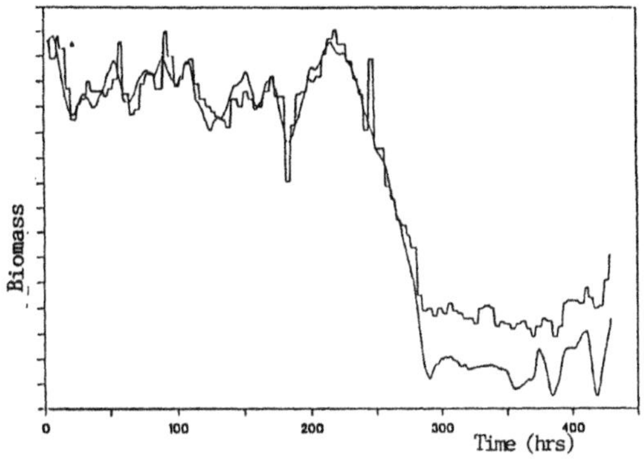

Fig. 7. Estimation of biomass (adaptation switched off at time 250 hr.)

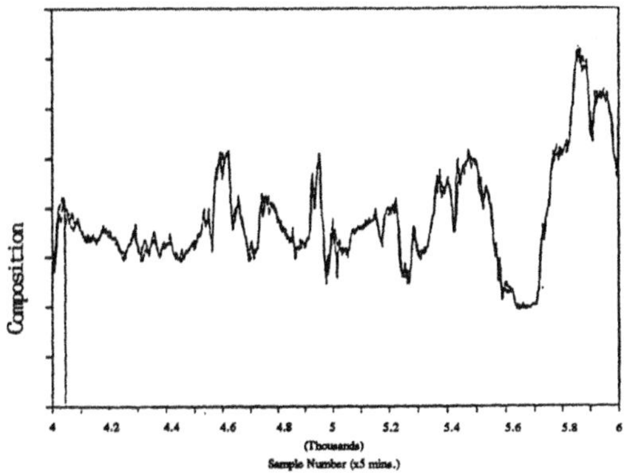

Fig. 8. Adaptive Inferential Estimation of composition of an industrial demethaniser.

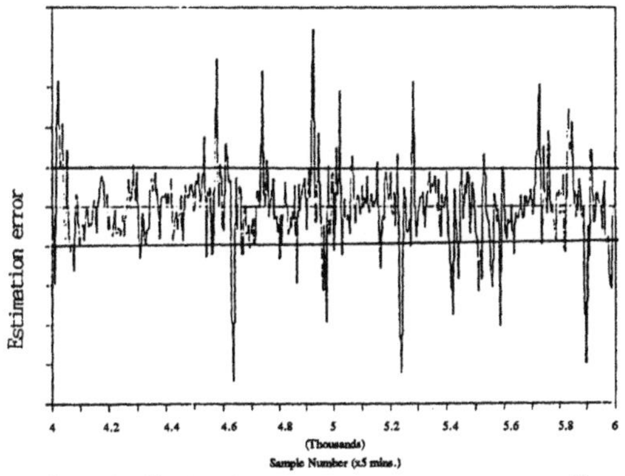

Fig. 9. Estimation error sequence corresponding to Fig. 8.

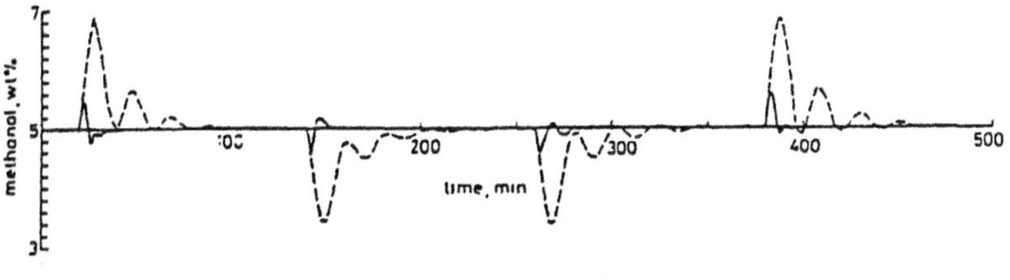

Fig. 10. Comparison of Adaptive Inferential Control and conventional PI control of bottom product composition.

——————— Adaptive Inferential Control
- - - - - - - Conventional PI Control

PROCESS CONTROL UNDER VARIABLE FLOW AND VOLUME

A. J. Niemi

Helsinki University of Technology, SF-02150 Espoo, Finland

Abstract. Models of continuous flow vessels with variable volume and flow are studied and their control algorithms developed. It is shown that the weighting functions of many vessels which are time-variable under such conditions, are invariant with regard to new, composite variables which are introduced. The validity of the basic assumption on invariability of the flow pattern under flow and volume variations is shown valid for several, analytically known cases. The approach is extended to arbitrary, non-analytic response functions obtained by measurements. Automatic controllers are then introduced for processes with variable liquid volume and flow. Both the feedforward and feedback controllers developed in terms of the new variables have constant parameters and eliminate correctly the effects of the disturbance variables in control of concentration or other material property. The corresponding gain scheduling controllers are also presented which operate in terms of the time. Both types of controllers appear easy to implement in hardware and software. Industrial buffer vessels, natural water reservoirs and physiological systems are stated as potential applications of modelling and control, and extensions to chemical reactors are discussed.

Keywords: Time-varying systems, Invariance, Process control, Modeling, Continuous flow systems, Fluid composition control, Feedback, Three-term control, Feedforward.

1 INTRODUCTION

A continuous flow vessel is characterized by the probability disribution of residence of material passing it. Probability density function models are available for many types of vessels in the literature for constant values of the parameters, i.e. of the liquid volume and volumetric flow. If such a constant parameter vessel is provided with feedforward or feedback control and the residence time distribution is known, the controller can be tuned with well known methods.

In a more general case of the continuous flow vessel, its parameters are variable. E.g. the flow through an industrial process vessel is submitted to changes both for random reasons, like for disturbances from other vessels connected serially with it, and intentionally, when the production is increased or decreased. A method has been developed earlier for incorporation of the unsteady flow in the control, assuming the liquid volume constant (Niemi 1981).

Also the amount of material, like the volume of liquid, is often variable which automatically implies a variation of at least one of the flows into or out from the vessel. Thus e.g. a buffer vessel usually serves the smoothing of both the flow and quality variations of the entering material. In such a case, variation of both flow and volume parameters in addition to that of inlet concentration, should be counteracted in control of the output concentration of a material component.

Variable parameter residence time distributions and weighting functions cannot be expressed as functions of a single time variable, and therefore the constant parameter models do not apply to description of processes which are subject to flow and volume variations. On the other hand, since the measurements of these parameters are standard operations in the industry, their values are continuously available. It will be explained in the following, how this information can be incorporated in the process models. Likewise it will be studied, how such models are used for computation of the output of the time-variable process. Finally, it will be shown, how standard methods developed for constant parameter control of time-invariant processes are extended to control of vessels with variable flow and volume.

2 CONTROL OF CONSTANT PARAMETER VESSEL

If the flow and volume of a continuous flow process are constant, its residence time distribution or weighting function of the input/output concentration process is time-invariant. Thus the methods of classical and modern control theory apply to design of its feedforward and feedback controls which may be based on a time series model obtained by measurements or its analytical approximation or a transformed model.

The model is preferably expressed as a function of a normalized, dimensionless time variable $\theta = t/\tau$ where the mean residence time is $\tau = V/Q$ (V volume, Q volumetric flow). If the distribution is known at one steady state and the process is invariant with regard to θ within a range of values of V and Q, the model can be converted to conform with other steady states and then used for tuning of a controller operable around such a new state.

Alhough this method is simple in principle, it appears that it is not much used to tuning of controllers for new points of operation. The few examples of its application include the control of grinding of minerals under different, constant feeds in which case the residence time distribution has been additionally superimposed with linear kinetics of grinding, in order to obtain the weighting function of the particular process.

3 PROCESS WITH VARIABLE FLOW AND CONSTANT VOLUME

The above assumption on invariability of the flow pattern represented by the dimensionless residence time distribution can be analogously extended to such processes which are subject to unsteady flow. Under such conditions, the process and model are time-variable, but it has turned out that they are brought to a constant parameter form through introduction of a new variable (Niemi 1977, 1981)

$$z = \frac{1}{V}\int_{\nu}^{t} Q(\mu)\,d\mu \qquad t \geq \nu \qquad (1)$$

The defined new variable z is thus proportional to the amount of liquid which passes the plant during an interval $\nu \to t$ of time. As it is introduced to the residence time or weighting function models, the time t and its explicit function Q(t) are eliminated from them.

Analysis of many mathematically known flow models has shown that they can be consistently described by functions of the new variable (1). The weighting functions p of the perfect mixer (2) and the plug flow or pure time delay (3) are basic examples. An arbitrary number of other examples can be developed by combining them e.g. serially or by means of recycling loops (Niemi 1977).

$$p(z)dz = e^{-z}\,dz \qquad (2)$$

$$p(z)dz = \delta(z-1)dz \qquad (3)$$

Models of many stationary flow systems of other types are found in the literature, mostly as functions of the normalized time (e.g Levenspiel 1962). Since θ (see Chapter 2) is under constant flow equal to the value of the expression (1), θ in such models may be substituted by z (1), in order to extend them to variable flow conditions. Application of the method is not limited to analytically known distributions, and arbitrary continuous flow systems can be modelled as functions of this variable, subject to the assumption of invariability of the flow pattern.

Various test procedures are available for determination of unknown residence time distributions or weighting functions, like deterministic test signals produced with tracers, or stochastic identification. During such a test both the flow and concentrations are measured and recorded, and the value of the variable z calculated corresponding to each concentration reading. The weighting function is then obtained as a function of z (1), irrespective if the flow during the test is constant or variable.

A study of the process models shows that, if the residence time distribution and weighting function of a vessel are expressed as functions of the integrated flow variable, they have constant parameters and are equal, although the corresponding time-functions have variable parameters and differ from each other (Niemi 1977). The constant parameter presentation results in an easy computation of the output concentration by convolution, assuming the inlet concentration and the flow are measured continuously or at intervals.

3.1. Control of Variable Flow Process

Since accordingly the output concentration can be computed on the basis of input concentration at a variation of flow without decrease of accuracy, the computed signal can be used for production of the compensating feedforward control action at the output of the process in open loop. The control result is then continuously equal to that result which is reached by the corresponding conventional, constant parameter feedforward controller operating in terms of time, only under that constant flow for which it was tuned.

Also a feedback controller can be constructed or programmed to operate with z (1) as the argument instead of time. Since the process parameters are constant in the z-domain representation also under unsteady flow, it is appropriate to adopt a controller which has constant parameters. All conventional methods of linear feedback control can then be used for design and tuning of the controller (Niemi 1981).

Operation of the feedback controller is based, in addition to the measured output signal, on measurement of flow and use of the latter signal in open loop. The control signal is brought to the actuator in the same manner as that from a conventional controller. At the computation of the control signal, one may alternatively keep to functions of time, without a use of the variable z (1), whereby the controller will have variable parameters in order to reach the same independence of flow variations (Niemi 1981, 1982).

Åström and Wittenmark (1989) discuss such a gain scheduling control in which the sampling period is inversely proportional to the flow rate. Their simulations show that the feedback control loop operates in the same manner at different values of the flow. As another alternative, the sampling at constant increments of z (1) can be synchronized with a positive displacement meter instead of clock, resulting in complete neglection of the time scale in control (Niemi 1977).

4. PROCESSES WITH VARIABLE VOLUME AND FLOW.

Variation of the volume (V) of liquid in a vessel implies necessarily the existence of a difference between its feed (Q_{in}) and outlet (Q_o) flow rates. These three variables are contained in the dynamic overall mass balance of the vessel (4) which includes them as functions of time. Thus any two of them may change independently and the third one is determined by these; i.e. if two variables are measured physically, the third one can be computed. Especially the liquid volume is easily obtained indirectly, through measurement of the level height.

$$\frac{dV(t)}{dt} = Q_{in}(t) - Q_o(t) \qquad (4)$$

As the flow patterns of continuous flow vessels and their invariability under volume changes are studied in more detail, it turns out that the vessels have to be classified in two groups. The first group comprises such vessels in which the movement of liquid can be represented by an axial velocity profile of the elements of its cross-section, and the other one the perfectly mixing vessels.

4.1. Flow Processes with Distinct Velocity Profiles

The axial velocity of liquid is distributed around an average velocity which, during a volume change, is different at different cross sections. In the special case of a plug flow vessel, the distribution is reduced to an impulse, i.e. all elements of the same cross-section have an equal speed. A liquid element which has entered such a vessel at time v will be found at time t (t > v) at the relative distance expressed by Eq. (5) from the entrance. This is equivalent to the ratio of the volume of material entered between v and t to the total volume at time t (Fig. 1). A more detailed study yields this result as the solution of an integral equation (Niemi 1988, 1990). - The length L of the vessel has been set constant for convenience.

$$\frac{x_{av}(t)}{L} = \frac{1}{V(t)} \int_{v}^{t} Q_{in}(\mu)\, d\mu = z_1 \qquad (5)$$

If this relative displacement is adopted as a variable (z_1), the model of the plug flow vessel of variable volume is consequently (δ unit impulse function):

$$p(z_1)\,dz_1 = \delta(z_1 - 1)\,dz_1 \qquad (6)$$

In practical vessels, the velocity is continuously distributed and different

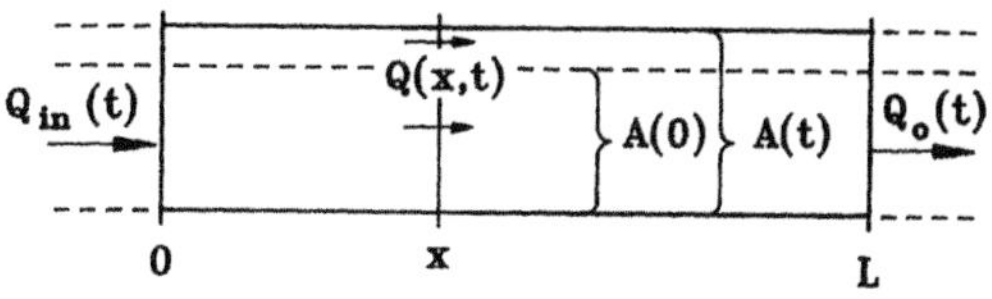

Fig. 1. Continuous flow vessel of fixed length L and variable volume V(t); cross section A(t) independent of distance x.

liquid elements move over different distances x between the times v and t. E.g. the elements found at the far end of the vessel at time t have entered it at different times v and are therefore members of different distributions of travel $x(v,t)$ which are characterized by a different value of $x_{av}(v,t)$ each. The distribution of material, in a sample taken from the outlet at time t after it has entered at different times v and passed the fixed length L of the vessel, with regard to x_{av} can now be derived. The end conditions are disregarded here by assuming the studied vessel to be a limited part of a vessel of infinite length (Fig. 1).

Thus the residence of liquid in the vessel can also be described by a distribution with regard to z_1 (5), and the content of a component of material which is carried in the elements of incoming liquid and is proportional to its concentration, is correspondingly weighted (Niemi 1988, 1990). The assumption of an invariable flow pattern means here an invariable velocity profile or distribution of the relative velocity $p(v/v_{av})$ in all cross sections, at all times, producing an equal distribution $p(x/x_{av})$ of relative travel of the liquid elements. From the latter, the corresponding residence distribution $p(z_1)$ of a vessel of fixed length can be derived corresponding, further on, to the relative velocity profile. The profile may well be invariable despite of changes of flow and volume.

It has been shown that many continuous flow processes of variable volume which are important in the industry can be described by means of residence distributions which are based on the presented variable z_1 and on the elimination of the explicit time variable by its introduction. The plug flow and laminar flow vessels are examples of such processes, and likewise that of Gaussian velocity profile and, on additional conditions, the dispersion flow vessel. The description can be generalized to such a vessel of arbitrary flow pattern in which the relative velocity profile may be different in different cross sections, like in a neighbourhood of its physical ends, but in which it is continuously invariable in each fixed cross section. Such process needs not to be described by a closed mathematical model of relative travel or residence, and its flow pattern, i.e. typically its residence distribution, may be known e.g. as numerical data obtained experimentally. The variables constituting z_1 and the residence distribution in terms of the latter are relatively easy to measure as compared e.g. with determination of relative travel and its distribution.

As an example of analytically known cases, the laminar type of flow will be discussed. The relative velocity profile (7) of the laminar flow is known to be independent of the absolute velocity (r distance from axis).

$$\frac{v}{v_{av}} = 2\left(1 - \frac{r^2}{r_m^2}\right) \qquad 0 \le r \le r_m \tag{7}$$

The corresponding distribution of relative velocity is constant (= 1/2 for $0 \le v/v_{av} \le 2$, otherwise 0). This model is now considered valid, even if v_{av} and r_m vary, i.e. the profile is assumed to follow changes of inlet and outlet flows and the volume without inertia. - Changes of volume may be due to expansion or contraction of the pipe under variable pressure.

The distribution of relative travel, and that of average travel for a section of length L of the laminar flow pipe are then obtained, and from the latter the distribution of z_1.

$$p(z_1)dz_1 = \frac{1}{2z_1^2} dz_1 \qquad z_1 \ge \frac{1}{2} \tag{8}$$

As this model is applied to computation of concentrations, these are to mean the average concentration in an infinitesimally short axial section of the pipe. Differing from this general understanding of the present paper, and in order to illustrate the relationship of the model with process physics, a semi-infinite pipe is considered in the following which has its inlet end at x = 0. The radial distribution of the entering liquid element is then not considered constant at the inlet, but proportional to the velocity profile of the flowing medium, corresponding to a feed to the pipe from a tiny perfect mixer in which the concentration is variable. The weighting function of a finite length L of the pipe is in such a case:

$$p(z_1)dz_1 = \frac{1}{2z_1^3} dz_1 \qquad z_1 \ge \frac{1}{2} \tag{9}$$

This result, and corresponding results for vessels with Gaussian velocity profile and with dispersion flow, have been derived separately; in the last case additional assumptions are needed (Niemi 1988, 1990). The effect of end conditions, especially on dispersion models, has been studied in detail elsewhere (Kreft, Zuber 1978).

4.2. Perfect Mixer

The perfect mixer is represented by an ordinary differential equation which can also be written in terms of variable coefficients (C_{in} input, C output concentration).

$$\frac{d[V(t)C(t)]}{dt} = Q_{in}(t)C_{in}(t) - Q_o(t)C(t) \tag{10}$$

If also Eq. (4) is observed, this model can be brought to a solved form (Nir 1973). The corresponding weighting function p is expressed consistently as a function of another, single new variable z_2 (Niemi 1988).

$$p(z_2)dz_2 = e^{-z_2} dz_2 \tag{11}$$

$$z_2 = \int_v^t \frac{Q_{in}(\mu)}{V(\mu)} d\mu \tag{12}$$

In terms of this variable, Eq. (10) obains a very simple form.

$$\frac{dC}{dz_2} = C_{in} - C \tag{13}$$

Model of a series of variable volume mixers consists of successive convolutions of single mixer models. It is therefore less compact than the corresponding models of sets of constant volume mixers (Niemi 1990).

4.3. Discussion of Variable Volume Models

The new variables introduced, (5) and (12), have different structures. This is related to the different structures of flow patterns of their physical counterparts which makes that essentially different methods had to be used for their derivation. Cases of both types are found in the industry and nature, and the model type has to be chosen on the basis of test or experience.

Like in the case of variable flow and constant volume (z in Chapter 3), the variables z_1 and z_2 may be substituted in the constant parameter models found in the literature, in order to extend these to variable volume conditions. Functional form of models and other parameters than volume and flow are thereby assumed constant. This assumption is equivalent to that on the invariability of flow pattern, and subject to this assumption, the variables can be applied to arbitrary processes and models obtained by measurements. The variable z_1 is used more generally for arbitrary flow patterns while z_2 requires that the process or its element being modelled is similar to a perfect mixer.

From the constant parameter representation, it follows that the output concentration C of a variable volume system can be computed by convolution. This requires that both the volume and the inlet and outlet flow rates, or any two of them, are known by measurements made continuously or at intervals during a sufficient time, in addition to the concentration input data. Correspondingly, both of the new variables can be written in the form of a difference (14,15) of form $z-\zeta$ to serve as the argument of the weighting function in the convolution integral (16), while the meanings of z and ζ in this appear from Eqs (14) and (15) (η is origin of time). - Similar presentations apply also to the models and variables of the constant volume system.

$$z_1 = \frac{1}{V(t)}\int_\eta^t Q_{in}(\mu)\,d\mu - \frac{1}{V(t)}\int_\eta^v Q_{in}(\mu)\,d\mu \tag{14}$$

$$z_2 = \int_{\eta}^{t} \frac{Q_{in}(\mu)}{V(\mu)} d\mu - \int_{\eta}^{v} \frac{Q_{in}(\mu)}{V(\mu)} d\mu \qquad (15)$$

$$C(z) = \int_{-\infty}^{z} C_{in}(\zeta) p(z-\zeta) d\zeta \qquad \zeta \le z \qquad (16)$$

Integration with regard to ζ is not affected by the presence of z in the integrand. The result is therefore always unambiguous although the dependence of z_1 on t (5) is not necessarily uniquely invertible, differing from those of z (1) and z_2 (12) on t. A further use of the output concentration $C(z_1)$ therefore generally requires a storing of z_1 as a function of time t.

Practical computation can alternatively be based on the corresponding time-variable form, which yields the output concentration directly for different values of t and which is written below for the case of Section 4.1 only (17). Also then it is essential, that only one function p which is a function of one variable needs to be stored in the memory, in addition to the measured (or computed) values of C_{in}, V and Q_{in}.

$$C(t) = \int_{-\infty}^{t} C_{in}(v) p\left[\frac{1}{V(t)} \int_{v}^{t} Q_{in}(\mu) d\mu\right] \frac{Q_{in}(v)}{V(t)} dv \qquad (17)$$

5. FEEDFORWARD CONTROL UNDER VARIABLE VOLUME

It has been shown above that the concentration process in a variable volume system can be described by a model which takes into account the variable flow rates and volume, two of which may vary independently. Because the model is linear in concentration and has constant parameters, a feedforward control signal may now be obtained through convolution similarly to Eq. (16). If the assumption on an invariable flow pattern is satisfied also under variation of volume, and if the input concentration and two of the stated other independent process variables are measured, the results will be equal to those produced by the corresponding conventional feedforward controller under constant flow and volume, or by the feedforward controller of Sec. 3.1 under constant volume. The essential difference lies in the fact that in the latter cases the control results deteriorate, if the stated quantities get variable, while in the control according to the presented method they are not affected by such variations.

In an ideal case of traditional feedforward control, the effects of one disturbance variable are compensated completely. The ideal controller of the type presented now eliminates, for its part, fully the effects of three disturbance variables or, in addition to the same compensation, adapts accurately to two other disturbance variables (Fig. 2). If the control algorithm is programmed as a function of time, a gain scheduling controller with variable coefficients is obtained as shown by the model (17) as an example. In all types of feedforward control, the accuracy depends on correctness of the weighting function model used.

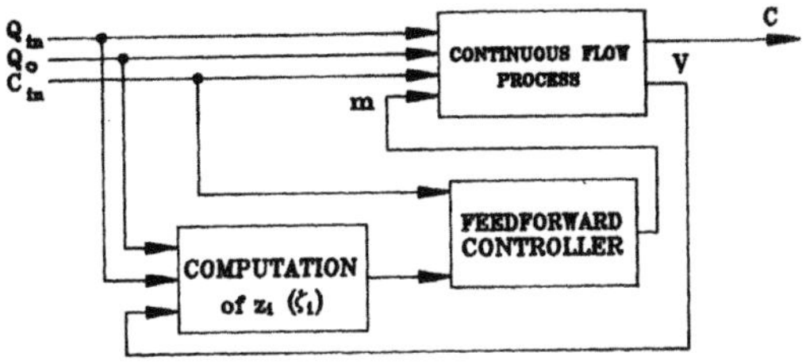

Fig. 2. Feedforward control of plant with variable volume and flow rates.

The presented control procedure applies both to vessels with distinct velocity profile and to perfect mixers. Because the input/output models produced by different flow patterns may appear similar, it must be checked in the case of an experimental model to which one of the two types the process belongs. In both cases, the same process variables are measured and used in the argument of the appropriate process model.

6. FEEDBACK CONTROL UNDER VARIABLE VOLUME

Because the time-variable process can be converted, subject to the assumptions made, to a linear constant parameter plant, its feedback control can be based on the variable z_1 or z_2 and the controller designed as one with constant parameters. After a transition from time to the appropriate z-variable, all coventional control design methods can be used including those of the frequency domain which require a further analytical or numerical transformation. The results which are reached with such a control under variable volume and flow, will be equal to those produced by the corresponding conventional controller under constant flow rate, or by the feedback controller of Sec. 3.1 under constant volume.

E.g. the model of the constant parameter PID controller which eliminates the effects of two disturbances (volume and flow) while tuned to counteract a third disturbance (concentration) is accordingly (e control deviation, m actuating variable):

$$m(z_i) = K_p e(z_i) + K_I \int_0^{z_i} e(\zeta_i) d\zeta_i + K_D \frac{de(z_i)}{dz_i} \qquad i = 1,2 \qquad (18)$$

The corresponding time domain controllers with gain scheduling are obtained by means of the expressions of the variables; Eq. (19) using Eq. (5) and Eq. (20) using Eq.

(12). The last term of Eq. (19) follows from the basic form of derivative response.

$$m(t) = K_p e(t) + \frac{K_I}{V(t)} \int_0^t e(\nu) Q_{in}(\nu) d\nu + $$

$$+ K_D \frac{de(t)}{dt} \cdot \frac{V(t)}{Q_{in}(t)} \tag{19}$$

$$m(t) = K_p e(t) + K_I \int_0^t e(\nu) \frac{Q_{in}(\nu)}{V(\nu)} d\nu + $$

$$+ K_D \frac{de(t)}{dt} \cdot \frac{V(t)}{Q_{in}(t)} \tag{20}$$

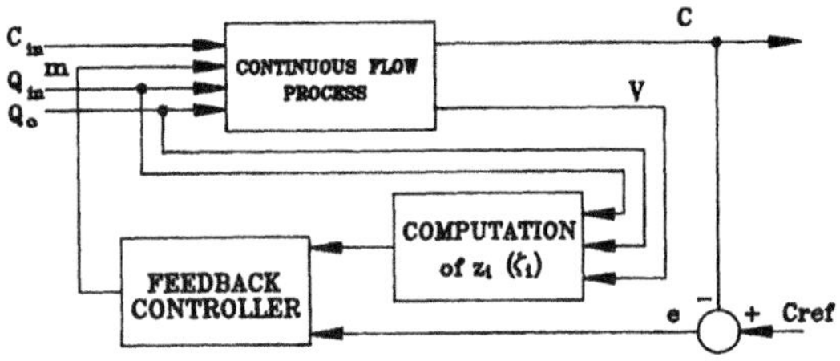

Fig. 3. Feedback control of plant with variable volume and flow rates.

Fig. 3 shows the structure of the control loop; the inlet (Q_{in}) and outlet (Q_o) flows appear as independent, measured variables, similarly to Fig. 2. For computation of z_i, two of the inputs of related block are sufficient, but the volume V or the liquid level has to be measured at least timewise, in order to eliminate the accumulation of errors in computation according to Eq. (4). The controller of Fig. 3 is e.g. one of the type (18). An implementation of the system in hardware and software is logical on the basis of the diagram.

7. DISCUSSION AND CONCLUSIONS

The described methods relate to control of concentration or other material property under variable volume and flow. These variables are incorporated in the new, combined variables introduced with which the operation of the controller is synchronized. Two such new variables are presented; one for vessels with a distinct flow pattern and the other for perfect mixers. Their use results in constant parameter process models and, further on, in constant parameter controllers which can be designed by standard methods. The resulting controllers adapt to changes of volume and flow through the stated combined variables. If the controllers are developed to the alternative form of presentation in time domain, gain scheduling controllers are obtained with parameters which are continuously updated by signals from the flow and level transmitters.

Assuming the flow pattern of the process is insensitive to changes of flow and volume, such controllers yield under variable conditions equal results as those delivered by conventional controllers under constant flow conditions. The needed computations are straightforward and the measurements, communications and manipulations easy to implement. If a conroller requires a storage of the process model, this will be a function of a single variable only, despite of variations of flow and volume.

Potential applications of the approach are control of industrial buffer vessels and of natural and artifical water reservoirs, and modeling of physiological systems. The range of validity of the basic assumption of invariability of the flow pattern may then have to be checked in specific applications. However, one may conclude that wherever the same time-invariable model has been proven valid under different stationary conditions (see Sec. 2), the corresponding model which uses a variable introduced will be valid in the same range also under dynamic changes of process variables, possibly excluding the fastest dynamics. The presented feedback controller may be tuned similarly to the conventional PID controller, even if the process model is only rudimentarily known.

The method will work accurately also in control of such chemical reactors in which fast, essentially time-independent reactions take place. Same effects are reached in feedforward control of such processes which involve time-dependent linear, e.g. kinetic or thermal phenomena. The z-variables are then used only for storing of the p-function and the computations are mainly run in time domain using a model in which the function p (e.g. in Eq. (17)) is superimposed with the models of the latter process components. Also in feedback control of such processes, the effects of variable volume and flow are compensated effectively, although theoretically not completely.

The parameters of the flow pattern have been assumed constant in the presented study. If their values are not known they can be determined by tracer testing or estimated by e.g. stochastic identification, if the volume and flow are also constant and if the primary input and output variables (concentrations etc.) change in a random manner and are measured continuously. Variations of volume and flow have so far made a time-based identification effectively difficult, but if they are measured and the appropriate new variable of the presented type is used as the argument, no difficulty is foreseen, and the z-based identification is a subject of further study. It appears that even the assumption of an invariable flow pattern can then be relaxed and its parameters tracked, if they change slowly enough, despite of volume and flow variations.

REFERENCES

Åström, K.J. and B. Wittenmark (1989). Adaptive Control. Addison-Wesley, Reading, Mass.

Kreft, A. and A. Zuber (1978). On the physical meaning of the dispersion equation and its solutions for different initial and boundary conditions. Chem. Eng.Sc., 33, 1471-1480.

Levenspiel, O. (1962). Chemical Reaction Engineering. Wiley, New York.

Niemi, A. J. (1977). Residence time distributions of variable flow processes. Int. J. Appl. Radiation and Isotopes, 27, 855-860

Niemi, A. J.. (1981). Invariant control of variable flow processes. Proc. 8th IFAC World Congress, Pergamon Press, Oxford, 2687-2692.

Niemi, A.J. (1982). Method an apparatus for the incorporation of varying flow in the control of process quantities. U.S. Patent No 4358821 (UK No 2051424).

Niemi, A. J. (1988). Variable parameter model of the continuous flow vessel. Mathl Comput. Modelling, 11, 32-37.

Niemi, A.J. (1990). Processes with variable flow and volume. In Radioisotope Tracers in Industry, IAEA, Vienna.

Nir, A. (1973). Tracer relations in mixed lakes in non-steady state. J. Hydrol., 19, 33-41.

TURBINE GENERATOR EXCITATION CONTROL USING ADVANCED DIGITAL TECHNIQUES

K. J. Zachariah*, P. A. L. Ham*, J. W. Finch and M. Farsi****

**NEI Parsons Ltd, Newcastle-upon-Tyne, UK*
***University of Newcastle-upon-Tyne, UK*

Abstract. The use of a self-tuning strategy for excitation control of turbine generators is described. Schemes to improve the robustness of parameter estimation in the self-tuning algorithm are presented. A software simulator and a laboratory model turbogenerator which have been used to evaluate the performance of the self-tuning excitation controller are briefly described. Some of the results obtained are documented.

Keywords. Self-tuning AVR; Advanced turbine generator excitation control; GPC; Robust parameter estimation; Micro-alternator system.

INTRODUCTION

Excitation control systems for large turbine generators can greatly benefit by the use of digital technology. The major advantages of using digital technology over conventional analogue systems are greater flexibility of control, improved operator interface, enhanced fault diagnostic capability and ease of upgrading. It is for these reasons that Power Utilities are now keen on specifying digital excitation control systems for modern power plant.

The Automatic Voltage Regulator (AVR) is the heart of the excitation control system for the turbogenerator. For an isolated machine, the primary function of an AVR is to regulate the terminal voltage; when connected to a large power system, it contributes to the control of overall system voltage, but acquires further roles in respect of the control of reactive power and contribution to stability for that machine. The present generation of AVR's, developed by NEI Parsons (Hingston, Ham and Green, 1989), is modelled on existing analogue electronic excitation control systems. These are fixed-parameter controllers which are generally based on lag-lead compensation techniques.

In recent years there has been considerable interest in the application of adaptive/self-tuning control theory to the design of AVR's (Kanniah, Malik and Hope, 1984; Wu and Hogg, 1988; Finch, Zachariah and Farsi, 1989). The need for more effective excitation control with the objective of extending the operational margins of stability is the key motivating factor for this interest.

The design of an AVR assumes that the turbogenerator can be considered to be linear or can be linearized around the nominal operating point. Turbine generators, however, are non-linear systems and are continuously subjected to load variations of random magnitude and duration. In order to improve the overall performance of the turbogenerator under varying operating conditions, the use of an adaptive/self-tuning control strategy is a possible solution. Since the parameter estimator of a self-tuning controller can track the varying dynamics of the system, the desired performance over the complete range of the turbogenerator can be achieved.

The paper describes the implementation and evaluation of a self-tuning AVR. This work is being carried out as part of an on-going research and development programme between NEI Parsons Ltd. and the University of Newcastle upon Tyne, U.K. The Generalised Predictive Control (GPC) strategy (Clarke, Mohtadi and Tuffs, 1987) has been used for the design of the controller. Schemes to improve the robustness of the parameter estimator of the self-tuning algorithm have been employed. The self-tuning AVR has been evaluated using a software simulator as well as a laboratory model turbogenerator.

THEORY

The GPC strategy which is employed in the design of the self-tuning AVR is based on a 'Controlled Auto-Regressive Integrated Moving Average' (CARIMA) model of the system. This particular discrete time model, in this case representing the excitation system of the turbine generator linearized around the operating point, is defined as:

$$A(z^{-1})y(t) = B(z^{-1})u(t-1) + C(z^{-1})e(t)/\Delta \qquad (1)$$

where $A(z^{-1})$, $B(z^{-1})$ and $C(z^{-1})$ are polynomials in the backward shift operator z^{-1} with appropriate dimensions. The terminal voltage of the generator and the field current demand are the plant output $y(t)$ and input $u(t)$ respectively. $C(z^{-1}).e(t)/\Delta$ models the disturbance acting on the system where $e(t)$ is an uncorrelated random sequence having finite variance. Δ is the differencing operator $(1-z^{-1})$. In this case, $C(z^{-1})$ is assumed to be 1 for simplicity.

GPC is a predictive controller which can take different forms by extending the basic theory. The type that is chosen for this application is the one that has an approximate closed-loop model following feature. The control law minimizes a cost function, J, which is defined as:

$$J = \sum_{j=1}^{N_y} [\phi(t+j) - w(t+j)]^2 + \sum_{j=1}^{N_u} \lambda[\Delta u(t+j-1)]^2 \qquad (2)$$

where Ny and Nu are the maximum output prediction horizon and the control horizon respectively. Nu specifies a horizon beyond which control input increments are assumed to be zero thereby setting the liveliness of the controller. λ is a control weighting factor which can be used as an additional detuning knob. The future output reference is represented by $w(t+j)$ and $\phi(t+j)$ is an auxiliary output derived from the normal plant output as:

$$\phi(t+j) = P(z^{-1})y(t+j) \qquad (3)$$

$P(z^{-1})$ is a transfer function of the form

$$P(z^{-1}) = p\ Pn(z^{-1})/Pd(z^{-1}) \qquad (4)$$

where $Pn(z^{-1})$ and $Pd(z^{-1})$ are monic polynomials in z^{-1} and p is a scalar quantity chosen such that $P(z^{-1})$ equals unity in steady state. The transfer function $P(z^{-1})$ is related to the model of the closed-loop, $M(z^{-1})$ such that

$$M(z^{-1}) \simeq 1/P(z^{-1}) \qquad (5)$$

Minimizing the cost function J with respect to Δu leads to the following control law (Clarke, Mohtadi and Tuffs, 1987):

$$\Delta u = (G^TG+\lambda I)^{-1}G^T(w-s) \qquad (6)$$

where w and s are Ny vectors containing the future voltage reference sequence and the components of future auxiliary plant output ϕ which are known at time t respectively. G is a Ny x Nu matrix and Δu is a Nu vector containing a suggested sequence of future control actions.

Since GPC is a receding horizon control law only u(t), which is the first element of the Δu vector, need be calculated at time t. Equation (6) can thus be simplified to:

$$u(t) = u(t-1) + g^T\ (w-s) \qquad (7)$$

in which g is a Ny vector containing the first row of the Nu x Ny matrix $(G^TG+\lambda I)^{-1}G^T$.

GPC is an explicit self-tuning algorithm and the plant parameters $A(z^{-1})$ and $B(z^{-1})$, which are estimated on-line, are used to calculate the vectors g and s. This is done by using a recursion technique (Clarke, Mohtadi and Tuffs, 1987) and although the calculation is relatively time consuming the right choice of Ny and Nu can reduce the computational burden considerably.

IMPLEMENTATION OF THE SELF-TUNING AVR

The self-tuning AVR is implemented on a VMEbus based microcomputer system. It consists of a 32-bit CPU board with a M68020 microprocessor supported by a hardware floating point unit and peripheral boards for analog and digital interface. The software is written in the high-level language 'C' which is considered most suitable for computationally intense real-time control applications such as this.

Care has been taken to ensure that similar hardware to that of the digital AVR (Hingston, Ham and Green, 1989) was chosen for implementation of the self-tuning controller. The control algorithm has been written in such a way that it will interface with the existing software structure of the digital AVR with little difficulty. This will enable the digital AVR system to be upgraded at a later date to possess adaptive control capability.

The self-tuning controller employs a recursive least-squares (RLS) parameter estimator with an upper-diagonal factorised covariance update to obtain the coefficients of the digital model of the plant. The CARIMA plant model requires the use of differential data in the parameter estimator. This can lead to poor estimates of parameters in situations where the plant is in a steady-state condition. Hence a CARMA Plant model, with the input output data prefiltered using a high pass filter, is used instead. This approach eliminates the undesirable effects of offsets and load disturbances on the estimator, but allows a more effective use of the low frequency dynamic information from the plant. The prefiltering is achieved by routing the input output data of the plant through the sampled data version of a 'washout' function of the form $sT_w/(1+sT_w)$ where T_w is the washout time constant. A value of 10 seconds is used for T_w during the tests.

A variable forgetting factor scheme (Fortescue, Kershenbaum and Ydstie, 1981) is used to improve parameter tracking and prevent wind-up of the estimator. The information content in the estimator is thus kept constant which increases its robustness. The memory length of the estimator and the expected noise variance of the data, which relate to the variable forgetting factor scheme, are set to 3000 samples and 10^{-5} respectively.

The parameter estimator is started with zero initial estimates. In order to reflect the uncertainty in the initial parameter values, the covariance matrix is initialised with a large value of 1000.I, where I is an identity matrix of the required size. Random values of control input are applied for the first few sample periods to enable the estimator to produce reasonable parameter values before invoking the control law calculation.

A third order plant model is assumed within the algorithm and the sampling period is chosen to be 20 milliseconds which complies with the bandwidth generally specified for the AVR. The output prediction horizon (Ny) of the controller is set to 10 and the control horizon (Nu), which determines the liveliness of the controller, is set to 2. The control weighting factor λ, which acts as a fine-tuning knob, is set to 2. The transfer function $P(z^{-1})$, which is used to approximately specify the closed-loop model, is set such that $Pn = 1-0.8z^{-1}$, $Pd = 1$ and $p = 5$.

The facility for invoking a 'random walk' to increase the dynamic information content in the input output data of the plant is incorporated in the parameter estimator. This is a useful feature and can be used to 'excite' the plant when the covariance matrix indicates low information content which can be monitored through its 'trace'. Another situation where a 'random walk' can be performed is when it is known that the dynamics of the system have changed considerably, for example, when the main circuit breaker of the generator has changed state.

When the turbine generator is operating without any perturbation or is subjected to severe disturbances, it is advisable that the parameter estimator is 'frozen' to improve its robustness. Severe disturbances, which can occur on the machine due to faults in the power system, are transient in nature and typically persist for only a few machine cycles. The system dynamics will be considerably different during these abnormal conditions, but since the faults generally are cleared very quickly, it is better for robustness in the estimator that it is stopped during these 'glitches'.

The dynamic information content in the input output data of the plant can be checked against a 'window' to 'freeze' the estimator. A measure of the information content can be obtained from x^TPx where x is the data vector and P is the covariance matrix of the parameter estimator. When the dynamic information is outside the 'window', the estimator is 'frozen'. It is worth mentioning at this point that this feature is not invoked at start-up of the estimator nor during the 'random walk'.

EVALUATION OF THE CONTROLLER

The performance of the self-tuning AVR was evaluated using a non-linear computer simulation model as well as a laboratory model turbogenerator system both of which are widely used for this type of work. A brief description of these follows.

Simulation Model

The non-linear computer model of the turbine generator is formulated in state space form and is designed mainly for fault simulation studies. It has 10 states which adequately model the dynamics of the plant, although various levels of complexity of the model are possible.

The basic dynamics of the synchronous machine are represented in the model using flux linkages of the direct and quadrature axis armature and rotor circuits. The equations relating generator fluxes, currents and voltages are derived using the two-axis theory based on Park's transformation (Adkins and Harley, 1975). Five state variables represent the electromagnetic dynamics of a modified generator with the transmission system and unit transformer impedances absorbed into it.

The equations representing the conversion of mechanical to electrical torque are derived in terms of the rotor position, speed and acceleration of the generator and these occupy two states in the model. The various turbine stages are not separately considered for modelling and hence a simple two-time constant approximation of the governor and turbine is used. A rotating exciter supplying diodes is represented by one state variable in the model.

Within the simulator, constant values of reactances are assumed for the machine, the varying saturation during and following a fault being neglected. Reactances are taken to be on-load pre-fault values. Space harmonics in the flux wave are neglected and the capacitance in the generator windings and transmission system is not modelled. The simple fourth-order Runge-Kutta numerical integration method is used for digital simulation.

Laboratory Model Turbogenerator

A block schematic of the system is given in Fig. 1. It consists of a 3-phase, 4 pole micro-alternator whose field is driven through a Time Constant Regulator (TCR). The TCR enables the field time constant of the micro-alternator to be modified such that the system represents a full size machine. The TCR is set to give an open-circuit field time constant of 5 seconds during the tests. The micro-alternator can be synchronised to the busbar via the step-up transformer and transmission line simulators. Measurement of machine terminal volts, active and reactive power, frequency, rotor angle etc. are available as part of the arrangement.

The prime mover torque is supplied by a DC motor with its speed and torque controlled by an electronic turbine and governor simulator. The simulator represents the droop characteristics of the governor as well as the dynamics of the steam valves and the reheat system of the turbine. The output of the simulator is a signal representing the total mechanical torque of the turbogenerator and this signal is used to drive a thyristor convertor which regulates the armature current of the DC motor. Since the field current of the motor is supplied from a constant current source, the mechanical torque developed by it is directly proportional to the armature current. The effect of the relatively large losses associated with the micromachine has been dynamically compensated for in the mechanical torque signal (Auckland and Shuttleworth, 1981).

MICRO-ALTERNATOR AND SIMULATION RESULTS

It is important that the self-tuning AVR performs satisfactorily under normal operation of the generator as well as abnormal/fault conditions. Performance under normal conditions of the machine was first analysed using the laboratory model turbogenerator. This was followed by simulation studies to evaluate the self-tuning controller's response during abnormal conditions of the machine. Finally, the simulation results were verified by applying faults on the micro-alternator system with the self-tuning AVR in operation. Some of the earlier results obtained have been reported elsewhere (Zachariah, Ham, Finch and Farsi, 1988; Finch, Zachariah and Farsi, 1989), while this section briefly documents the more recent ones.

Fig. 2 gives the response of the closed-loop system with the self-tuning AVR during manual synchronisation of the generator to the busbar with subsequent ramping of the voltage set point towards a target operating point. Although the change in the turbogenerator dynamics from open-circuit to on-load condition is appreciable, the controller performs satisfactorily, as can be seen from the Fig. 2.

The performance of the self-tuning AVR was evaluated under severe fault conditions by applying a 3-phase to neutral short circuit for 125 milliseconds at the generator end of the transmission line. This test was initially conducted using the software simulator. A study of the pattern of oscillations of the rotor angle following a fault gives a good indication of whether the system is transiently stable under these conditions. Fig. 3 gives the response of the system during this test and indicates that the self-tuning AVR copes well with this severe disturbance.

A similar 3-phase short circuit was applied to the micro-alternator with the self-tuning controller in service. Fig. 4 documents the effect of the test on the rotor angle, terminal volts as well as active and reactive power. The damping of the resulting rotor angle oscillations and the recovery of the terminal voltage are found to be acceptable.

CONCLUSION

Results of some of the recent trials with the self-tuning AVR have been documented. The controller performs satisfactorily under normal conditions of the generator as well as during severe disturbances. The modifications to the parameter estimation algorithm to improve its robustness have been found to work well. The trials conducted using the self-tuning AVR based on the GPC strategy indicate its potential for turbine generator excitation control.

ACKNOWLEDGEMENT

The authors wish to thank NEI Parsons Ltd., for support of this research project and for giving permission to publish this work.

REFERENCES

Hingston, R.S., Ham, P.A.L., and Green, N.J. (1989). Development of a digital excitation control system. IEE EMD'89 Conference, pp.125-129, London, UK.

Kanniah, J., Malik, O.P., and Hope, G.S. (1984). Excitation control of synchronous generators using adaptive regulators, Part 1 - Theory and simulation results, Part 2 - Implementation and test results. IEEE Trans. on PAS, Vol. PAS 103, No. 5, pp.897-910.

Wu, Q.H., and Hogg, B.W. (1988). Robust self-tuning regulator for a synchronous generator. Proc. IEE, Vol. 135, Pt.D, No. 6, pp.463-473.

Finch, J.W., Zachariah, K.J., and Farsi, M. (1989). Generalised predictive control applied to a turbogenerator voltage regulator. IEE EMD'89 Conference, pp.130-134, London, UK.

Clarke, D.W., Mohtadi, C., and Tuffs, P.S. (1987). Generalised Predictive Control, Part 1 - The basic algorithm, Part 2 - Extensions and interpretations. Automatica, 23, (2), pp.137-160.

Fortescue, T.R., Kershenbaum, L.S., and Ydstie, B.E. (1981). Implementation of self-tuning regulators with variable forgetting factors. Automatica, 17, (6), pp.831-835.

Adkins, B., and Harley, R.G. (1975). The general theory of a.c. machines - Applications to practical problems. Chapman and Hall, London.

Auckland, D.W., and Shuttleworth, R. (1981). Compensation systems for a micromachine model. Proc. IEE, Vol. 128, Pt. C, No. 1, pp.12-17.

Zachariah, K.J., Ham, P.A.L., Finch, J.W., and Farsi, M. (1988). A self-tuning voltage regulator for turbogenerators. IEE CONTROL 88 Conference, pp.366-370, Oxford, UK.

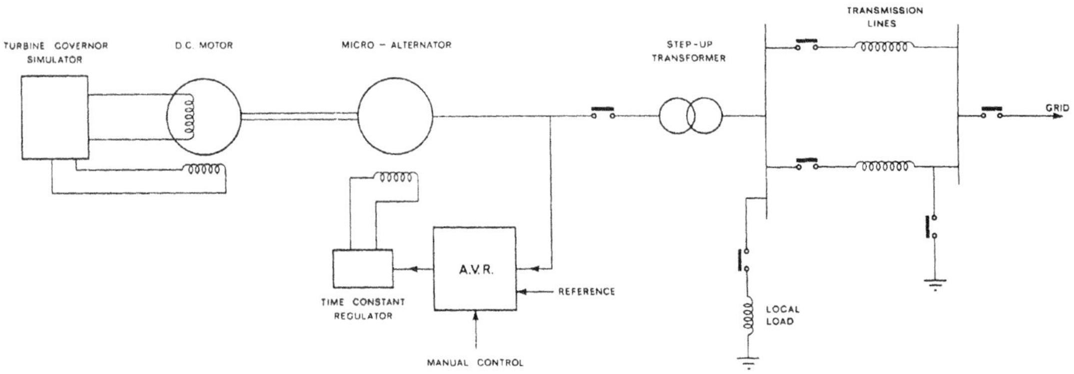

FIG. 1. Laboratory model turbogenerator

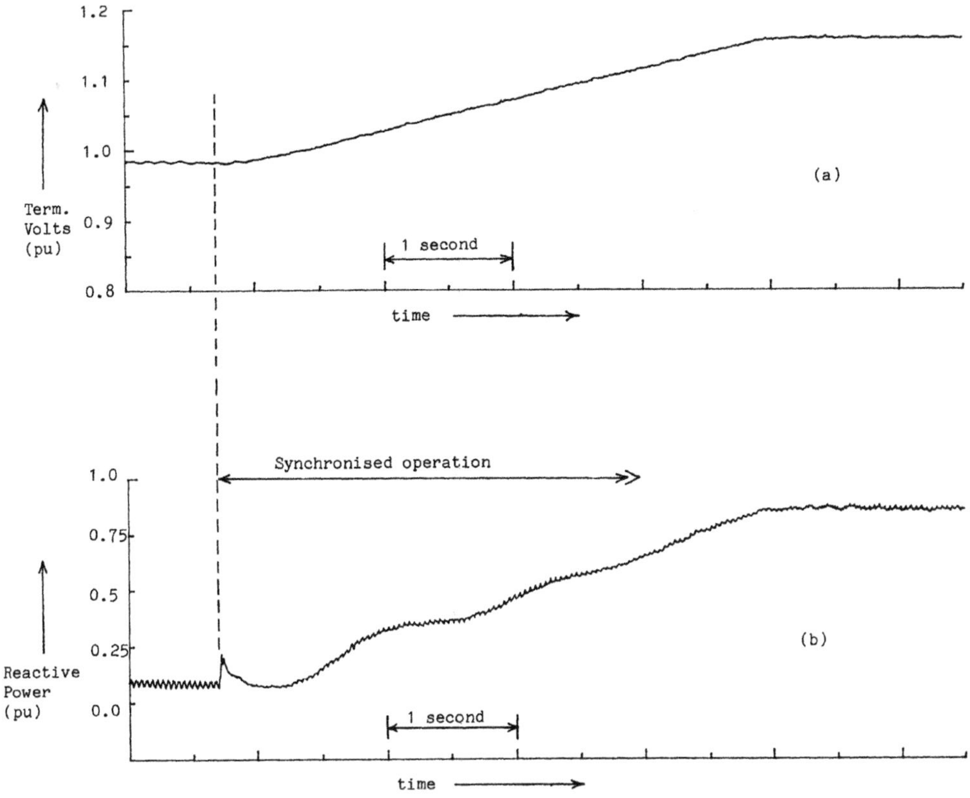

FIG. 2. System response during synchronisation and subsequent operating point change (Micro-alternator)

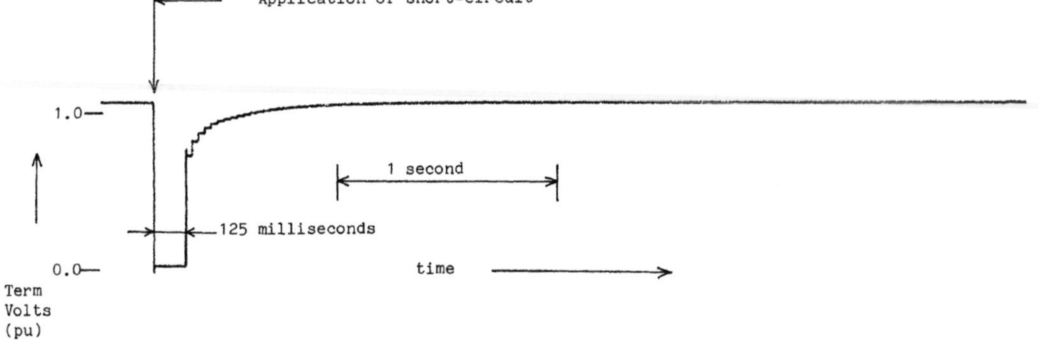

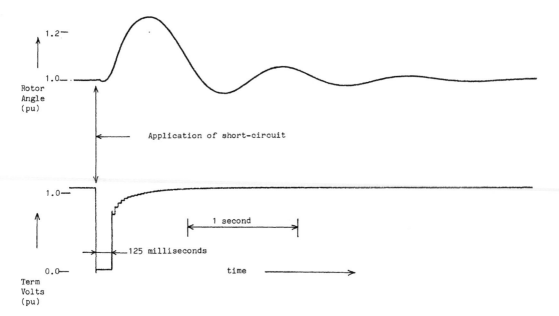

FIG. 3. System response to a 3 phase to neutral short-circuit for 125 milliseconds. (Simulation)

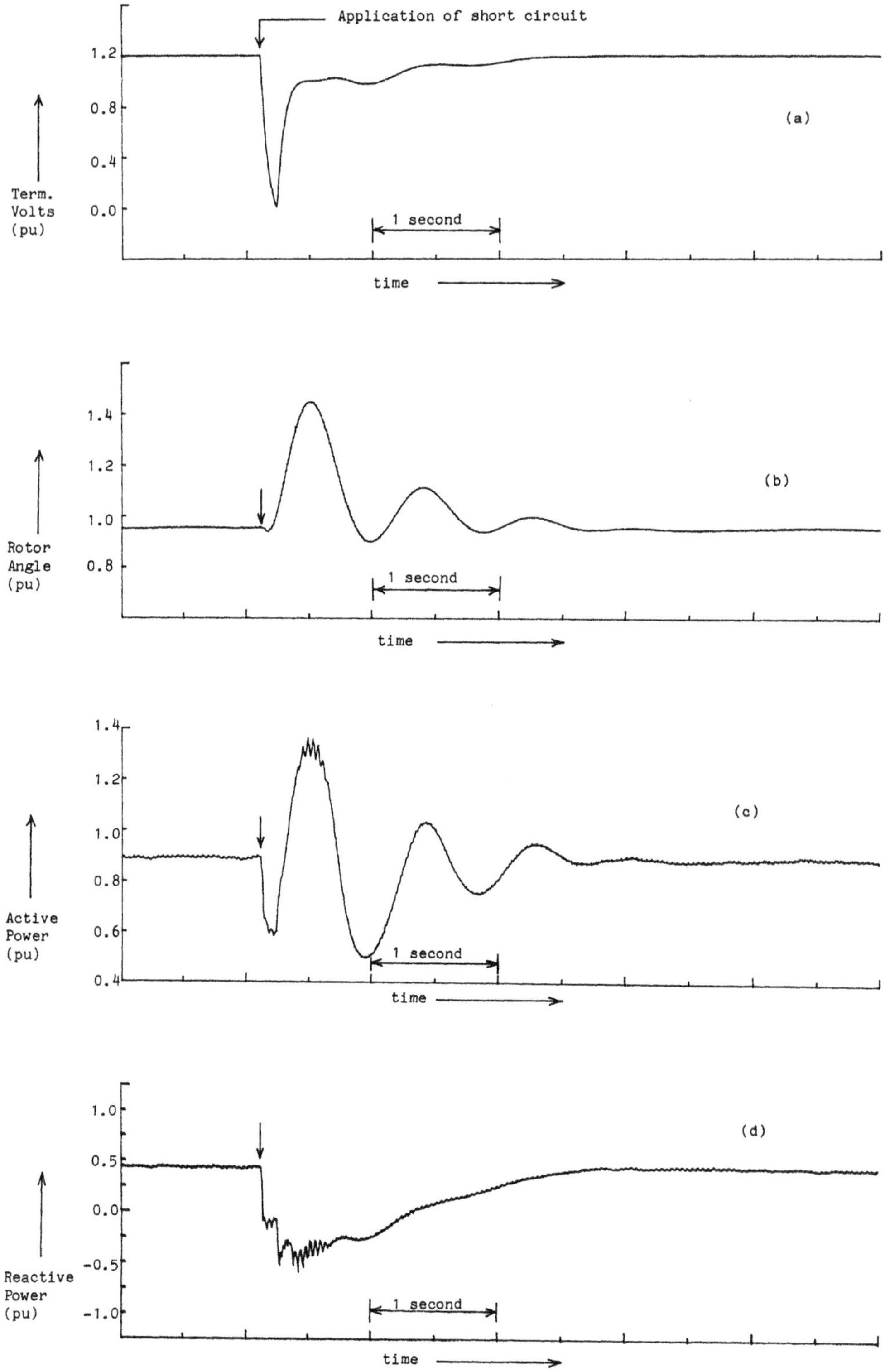

FIG. 4. System response to a 3 phase to neutral short-circuit for 125 milliseconds. (Micro-alternator)

NON-LINEAR PREDICTIVE CONTROL

M J. Willis, M. T. Tham, G. A. Montague and A. J. Morris

Dept. of Chemical and Process Engineering, University of Newcastle-upon-Tyne, Newcastle-upon-Tyne, NE1 7RU, UK

Abstract. In this contribution, an approach to nonlinear predictive control is discussed. The philosophy involves the use of an on-line optimisation algorithm to determine the future inputs that will minimise the deviations between desired and predicted outputs. Control, however, is implemented in a receding horizon fashion. The proposed procedure permits the use of a variety of model types. In particular, nonlinear models may be incorporated into the scheme without the need for linearisation. The advantage of this feature is demonstrated by simulated examples. It is shown that the use of a nonlinear model within the procedure leads to superior performance capabilities compared to that obtained when an equivalent linearised model is used. This highlights the well known model based control edict: the more accurate the process model, the better the achievable control performance.

Keywords. Predictive control; nonlinear systems; neural nets; optimisation; models.

INTRODUCTION

In the chemical process industry, although most systems possess nonlinear characteristics, control algorithms are generally synthesised based upon linear systems considerations. In many cases, the use of a linear control algorithm may be adequate. However, in some process situations the use of a nonlinear control strategy may impart operational advantages. This is because a linear model may be restricted in the amount of information it can contain about a process, thus compromising achievable control quality. Particular instances where this may be true are for highly nonlinear continuous processes or for batch processes that operate over a wide dynamic region. Under these circumstances, the application of a nonlinear control algorithm may be appropriate.

The use of a nonlinear process representation within a model based control strategy was investigated by Lee and Sullivan (1988). The utility of the technique was demonstrated in an application to a forced circulation single-stage evaporator which is a nonlinear interacting process (Lee and co-workers, 1989). Their strategy, Generic Model Control, bears strong similarities to the Internal Model Control (IMC) approach exhorted by Garcia and Morari (1982). IMC was also used as a basis in the development of a nonlinear control algorithm by Economou and Morari (1986). Their method made use of a Newton-type algorithm to provide a tractable solution for the control law, and was shown to perform well in a simple but realistic example. The development did not consider process constraints, although this was addressed in a later contribution (Li and Biegler, 1988).

Another approach to the development of a nonlinear controller, is the nonlinear inferential control policy of Parrish and Brosilow (1988). The control algorithm was developed from a discrete representation of a nonlinear process model. It was shown that a nonlinear controller results if the predicted, one-step ahead, process output was replaced by the desired output and the resulting equations rearranged to solve for the corresponding manipulated input. The performance of this nonlinear control algorithm was compared with that obtained using Proportional-Integral (PI) control by applications to a simulated pH process and a heat exchange system. The results showed that significant improvements in performance can be achieved using the nonlinear controller. The sensitivity of the algorithm to modelling errors was alleviated by the use of a low pass filter. This filter essentially attenuates the effect of process-model mismatch which normally occurs at the high-frequency end of the dynamic spectrum. The result is that control action is desensitised, with an accompanying degradation in system performance.

Recently, the use of an optimisation routine to solve for the required manipulated input within a model based control strategy has received attention, (eg. Eaton and Rawlings, 1990). Although the approach adopted in this paper is similar, the technique was developed independently.

AN APPROACH TO NONLINEAR PREDICTIVE CONTROL

In the majority of modern control philosophies, the controller is model based. Consequently, control performance is directly dependent on the nominal model used for controller synthesis. Within the context of nonlinear control, however, nonlinear models may assume a plethora of different forms. Thus, if the structure of the nonlinear controller is highly inextricably linked to the structure of the model, then the practicality of the algorithm will be severely diminished. The ability to incorporate various model forms without having to modify the structure of the controller, therefore, addresses an important implementation issue. This paper demonstrates the applicability of an approach to nonlinear predictive control, which permits the use of any discrete model forms.

The predictive control algorithm to be described is essentially a nonlinear variant of the Generalised Predictive Control (GPC) algorithm discussed in Clarke and Mohtadi (1987). GPC is a linear, explicit technique which provides an analytical solution to the following cost function:

$$J = \sum_{n=N_1}^{N_2} [w(t+n) - y(t+n)]^2 + \sum_{i=0}^{N_u} [qu(t+i)]^2 \quad (1)$$

where y(t), u(t) and w(t) are the controlled output, manipulated input and set-point sequences respectively. N_1 is the minimum output prediction horizon while N_2 is the maximum output prediction horizon. N_u is the control horizon and q is a weighting which penalises excessive changes in manipulated input. The method uses the linear Controlled Auto-Regressive Integrated Moving Average model, leading to a controller with inherent integrating properties. Here, an implicit development of the prediction equations is combined with an iterative solution procedure.

In the cost function, the term y(t+n), $n = N_1,...,N_2$, represents a 'sequence' of future process output values, and hence unknown values. Thus, in order to minimise the cost function, the sequence of process outputs has to be replaced by their respective n-step-ahead predictions.

Nonlinear n-step-ahead predictors

For a linear discrete representation, it is easy to show that prediction equations may be developed implicitly (eg. see Peterka, 1984; Albertos and Ortega, 1989). With nonlinear systems, this is not possible and an explicit approach is required. Consider the following continuous time process representation:

$$\mathbf{x}(\tau) = F(\mathbf{x}(\tau), \mathbf{u}(\tau))$$
$$\mathbf{y}(\tau) = G(\mathbf{x}(\tau)), \qquad \tau \geq 0 \qquad (2)$$

where $\mathbf{x}(\tau)$ represents an l-dimensional state; $\mathbf{u}(\tau)$ and $\mathbf{y}(\tau)$ are m-dimenional process input and output respectively, and 'τ' is time. The functions $F{:}R^{l \times m} \quad R^l$, and $G{:}R^l \quad R^m$ are smooth and continuously differentiable mappings. The corresponding discrete representation of Eq.(2) is given by:

$$\mathbf{x}(t) = F(\mathbf{x}(t-1), \mathbf{Du}(t))$$
$$\mathbf{y}(t) = G(\mathbf{x}(t)) \qquad (3)$$

where **D** is a diagonal matrix which defines the discrete time-delays between each pairing of the elements in **x** and **u**. At each sampling instant, Eq.(3) can be used to generate a one step ahead prediction of the states, $\mathbf{x}(t+1|t)$, and hence $\mathbf{y}(t+1|t)$, simply by using the current state x(t) in the function F. Similarly, inserting $\mathbf{x}(t+1|t)$ into F yields $\mathbf{y}(t+2|t)$. Thus, by successive substitutions of the previous prediction $\mathbf{x}(t+n-1|t)$, $n = N_1,..,N_2$, a sequence of n-step-ahead output predictions, $\mathbf{y}(t+n|t)$, may be obtained in a recursive manner, ie.

$$\mathbf{x}(t+n|t) = F(\mathbf{x}(t+n-1|t), \mathbf{Du}(t+n))$$
$$\mathbf{y}(t+n|t) = G(\mathbf{x}(t+n|t)) \qquad (4)$$

Note that in generating the predictions, known values of **x** and **u** will be used where appropriate.

Solution of the objective function

After the future outputs, y(t+n), have been predicted, and if the future set-points are known, then the future controls which will minimise the cost function given by Eq.(1) may be determined. Since the functions F and G may be nonlinear, an analytical solution of the cost function may not be possible. However, by adopting a numerical optimisation approach, a generalised solution technique results which can be used to provide the 'optimal' solution regardless of the form of F and G.

Most optimisation algorithms employ some form of search technique to scan the feasible space of the objective function until an extremum point is located (eg. Luenberger, 1973; Edgar and Himmelblau, 1989). The search is generally guided by calculations on the objective function and/or the derivatives of this function. The various procedures available may be broadly classified as either 'gradient based' or 'gradient free'. In an on-line situation, where process measurements are often corrupted by noise, the use of gradient based methods may not be feasible due to their susceptibility to discontinuities.

In the case of gradient free methods, Fletcher (1980) showed that the most efficient technique was due to Powell (1964). This method locates the extremum of a function using a sequential unidirectional search procedure. Starting from an initial point, the search proceeds according to a set of conjugate directions generated by the algorithm until the extremum is found. This technique was therefore chosen as the basis of the nonlinear predictive control philosophy advocated in this paper.

Ensuring offset free response

It is also required that set-points are tracked with zero errors. Since control is based upon the predictions of future outputs obtained from a nominal model of the process, offsets may occur due to model deficiencies, eg. not taking into account the effects of disturbances, and mismatch between the gains of the process and the model. Following IMC and the Dynamic Matrix Control, (Cutler and Ramaker, 1979), the discrepancy between model and process responses can be estimated at each sample instant as:

$$d(t) = y(t) - y(t|t-1) \qquad (5)$$

where y(t) is the actual process output at time t and $y(t|t-1)$ is the corresponding process model output. This estimate is then used to 'correct' the predictions obtained from the model, ie.

$$y_c(t+n|t) = y(t+n|t) + d(t) \qquad \forall n = N_1,..N_2 \qquad (6)$$

where the subscript 'c' denotes a corrected value. It is the sequence of corrected predictions that are then used in the cost function, Eq.(1), and has the effect of including integrating action within the control scheme. Additionally, in common with most predictive control strategies, an assumption is made concerning future control moves, viz. beyond the control horizon, N_u, it is assumed that $u_j = u_{j-1}$, $\forall j = N_u+1,...N_2$.

Summary of algorithm

The implementation procedure of an unconstrained predictive control algorithm within the proposed framework is summarised below:

Step I. Sample process output, y(t), and calculate the error, d(t), between y(t) and $y(t|t-1)$.

Step II. The process model is used to generate predictions of process outputs $y(t+n|t)$, $n = N_1,....N_2$

Step III. Correct the predictions from Step II using the result of Step I, to obtain $y_c(t+n|t)$, $n = N_1,....N_2$.

Step IV. Formulate cost function:

$$J = \sum_{n=N_1}^{N_2} [w(t+n) - y_c(t+n|t)]^2 + \sum_{i=0}^{N_u} [qu(t+i)]^2$$

Step V. An optimisation algorithm is used to determine the sequence of controls:

$$u(t+i), i = 0,...,N_u$$

Step VI. Implement u(t), wait till next sample and repeat procedure from Step I.

ILLUSTRATIVE EXAMPLES

Here, a Hammerstein model; a discretised mechanistic model and an artificial neural network based model will be used within the above procedure to illustrate the utility of the proposed nonlinear predictive control strategy.

Use of a Hammerstein model

This representation assumes that process characteristics can be described by a nonlinearity in series with linear dynamics. Consider the following deterministic representation of a Hammerstein model:

$$A(q^{-1})y(t) = q^{-k}B(q^{-1})u_{NL}(t) \quad (6)$$

where $u_{NL}(t)$ may be typically defined by:

$$u_{NL}(t) = \sum_{i=0}^{n} \alpha_i u^i(t) \quad (7)$$

and n is the order of the 'nonlinearity'. A and B are polynomials in the backward shift operator, q^{-1}, and k is the system time delay. Equation (6) may be expanded as:

$$y(t) = \sum_{i=1}^{N_a} a_i y(t-i) + \sum_{i=0}^{N_b} b_i u_{NL}(t-k-i) \quad (8)$$

where N_a and N_b are the degrees of the polynomials A and B. The n-step-ahead predictors are obtained by recursively propagating Eq.(8) forward in time, as described previously. These may be written as:

$$y(t+n|t) = \sum_{i=1}^{N_a} a_n y(t+n-i|t) + \sum_{i=0}^{N_b} b_i u_{NL}(t+n-k-i)$$

Where actual (past) values of outputs are known, these are used in place of the corresponding estimates. For predictions up to and including the process time delay, the control input values are known. To predict beyond the process time delay, unknown future control inputs are required, and these are determined by the optimisation procedure.

Consider the process studied by Bamberger and Isermann (1978):

$$A(q^{-1})y(t) = q^{-k}B(q^{-1})u_{NL}(t)$$

where $u_{NL}(t) = 1 + 2.0u(t) - 0.25u(t)^2$

The time-delay and the coefficients of the A and B polynomials are defined as follows:

$a_1 = -1.425, a_2 = 0.496$
$b_0 = -0.102, b_1 = 0.173, k=1$

The parameters of the nonlinear predictive controller were chosen to be $N_1 = 1$, $N_2 = 10$, $N_u = 1$ and $q = 0$. The control objective was to follow a series of set-point changes in the form of an asymmetrical rectangular wave. In developing the control law, it was assumed that the process was known exactly. Figure 1 shows the closed loop response of the system. As may be observed, excellent servo control characteristics were obtained. Note also that the process in this example is nonminimum phase (NMP). Hence, as with the standard GPC philosophy a stable response was obtained by setting the prediction horizon beyond this NMP behaviour.

Use of a mechanistic model

For the sake of illustration, consider the ideal continuous stirred tank reactor (CSTR) where a reversible exothermic reaction takes place. The temperature of the feed stream is the manipulated variable while the controlled variable is the concentration of the product stream. This reactor system, described in Economou and Morari (1986), has a well defined maximum conversion. As a result, when the operating point changes from one side of the maximum to the other, the gain of the system changes sign. Moreover, the temperature of the feed stream is constrained to lie between 250 deg. K and 650 deg. K.

In implementing the nonlinear predictive controller, only the process description needs to be modified for each new application. Thus, the above Hammerstein model is now replaced is by a set of discrete reactor model equations. The latter was obtained by applying the backward difference mapping (Euler approximation) to transform the continuous time differential equations into a set of discrete difference equations. For simplicity, it was also assumed that 'instantaneous' measurements of composition were available.

The predictive controller was designed using: $N_1 = 1$; $N_2 = 10$; $q = 0$ and $N_u = 1$. The differential equations describing the reactor were discretised with a step size of one second. A series of set-point changes were used to assess the performance of the nonlinear predictive controller over a 'wide' operating range. As may be observed in Fig.2, very acceptable set-point tracking was achieved. When the required set-point was above the maximum achievable product composition of 0.5085 mol/l, it may be seen that the process was regulated to the process maximum. The corresponding feed and reaction mass temperatures are shown in Fig. 3. Note that in this application, the input constraints were implemented by simply 'clipping' the calculated control moves. For a single-input single-output system, and for $N_u = 1$, the solution is identical to that obtained when the constraints are explicitly included as part of the optimisation problem. From the feed temperature plots, it may also be observed that the gain of the system changed signs several times. However, this did not compromise controller performance and the system remained stable.

It is evident that the utility of a mechanistic model within a model based control strategy depends on the accuracy of the particular model. The advantage of the proposed technique, compared to other philosophies reliant upon a mechanistic model for control algorithm synthesis, is that the mechanistic model does not require linearisation in realising the control scheme.

Use of an artificial neural network model

The use of artificial neural networks (ANNs) in process model development has been shown to be a cost effective approach. They provide a mapping between sets of inputs and outputs, without requiring the form of the mapping functions to be specified. Thus, ANNs are capable of modelling a large class of nonlinear dynamic systems. If the neural network based models could be incorporated as an integral component of a control strategy, then the resulting controller would be applicable to an equally large class of nonlinear systems. This is particularly attractive in situations where a physico-chemical model of the process to be controlled is not available. Another important implication is that a generic nonlinear modelling and control philosophy can be realised. The application and utility of ANN's within the domain of process estimation and control have been discussed extensively elsewhere, (eg. see Willis and co-workers, 1990, 1991; Di Massimo and co-workers, 1990), and thus will not be repeated.

Here, the performance of the proposed predictive control scheme, in conjunction with a neural network based model, is compared with the performances of conventional PI control, and with a linear predictive control strategy. The process being considered is a nonlinear simulation of the 10 stage pilot plant distillation column, installed at the University of Alberta, Canada. The top composition loop is controlled by manipulating reflux flowrate, using a PI controller. The objective is to control the bottom product composition using the steam flowrate to the reboiler as manipulated input.

First, the relationships between bottom product composition and steam flowrate, were established by the use of both a linear model and a nonlinear neural network model. Both descriptions also took into account the effects of reflux changes (interaction) and feed flowrate (disturbance) on bottom product composition. In developing the linear model, simple first order plus deadtime transfer functions were used to describe the dynamics of each input-output pair, whilst the neural network employed a structure with three input neurons, two hidden layers each containing four neurons, and one output neuron. (For a detailed definition of the terminology and a discussion of the use of ANNs as a modelling tool, please refer to the companion paper by Willis and co-workers, (1991)). After identification of the respective models, they were incorporated within the proposed control scheme to provide, respectively, a linear and a nonlinear predictive controller (LPC and NLPC). Note that the LPC is basically a variant of the linear GPC algorithm of Clarke and Mohtadi (1987), where the cost function, Eq.(1), is solved using an iterative optimisation procedure instead of an analytical approach.

The output prediction horizons for both predictive controllers were set as $N_1 = 1$ and $N_2 = 7$, while the control horizons were chosen to be $N_u = 1$. A set of n-step-ahead output predictions were obtained following the procedure outlined above for the Hammerstein model application. In addition, it is noted that the maximum prediction horizon is actually larger than the delays in the response of bottom product composition to feed flowrate disturbances. Hence, in realising the predictor equations, future values of the feed flowrate would be required. However, a causal implementation is made possible by assuming that feed flowrate remains constant at the last sampled value. The n-step-ahead predictions using the neural network model were obtained in a similar manner.

The disturbance rejection properties of the LPC, the NLPC and a 'well-tuned' conventional Proportional + Integral (PI) controller were compared and evaluated. It is required that bottom product composition is maintained at 5 wt.% methanol, when the column is disturbed by step changes in feed flowrate. The integral of the absolute set-point tracking error (IAE) was used to quantify the performance characteristics of the three controllers.

Figure 4 shows the performance of the above controllers. It can be observed that the PI algorithm resulted in the worst control performance, yielding an IAE of 78.1. This is primarily because the algorithm does not provide feed forward compensation against the feed flowrate changes. Better disturbance rejection characteristics were observed when the LPC strategy was employed. However, bearing in mind that the nominal model used in formulating the LPC also accounts for the effects of disturbance and loop interactions, the recorded IAE value of 62.1 does not represent a significant improvement over the performance of the well tuned PI strategy. The best regulatory performance was obtained When the NLPC was applied as is evident from Fig.7. The corresponding IAE value in this application was 5.7, showing an order of magnitude improvement in closed loop performance. This superior disturbance rejection behaviour is mainly attributed to the ability of the neural network model to represent the dynamic characteristics of the process with better accuracy. Since the performance of model based controllers is directly dependent on the accuracy of the nominal model used in controller synthesis, superior control is achieved.

CONCLUDING REMARKS

An approach to model based nonlinear predictive control has been described and its utility demonstrated. Using the proposed procedure, the structure of the controller is made independent of the structure of the model. This imparts implementation flexibility, since any model form, linear or nonlinear, may be used. Thus unlike model specific controller algorithms, the formulation can be applied to a wide class of processes, limited only by the availability of the process model. The ability to accommodate such a wide variety of model types is only made possible by the use of a numerical optimisation technique to minimise the predictive control objective function. However, as the solution procedure is iterative, the algorithm may only be applicable in cases where time is not a critical parameter. Nevertheless, as chemical systems are often characterised by relatively large time constants and correspondingly large sample times this may not prove too restrictive.

Current work is aimed at the development of an equivalent policy for multivariable control applications. It is envisaged that the inclusion of process constraints will feature significantly in the strategy. It is also suggested that the use of multivariable neural network based models appear to hold significant promise in attempting to realise a generic nonlinear predictive control philosophy.

ACKNOWLEDGEMENTS

The authors gratefully acknowledge the support of the Dept. of Chemical and Process Engineering, University of Newcastle upon Tyne, UK.

REFERENCES

Albertos, P. and R. Ortega (1989). On generalised predictive control: two alternative formulations. Automatica, 25, 5, 753-755.

Bamberger, W. and R. Isermann (1978). Adaptive on-line steady-state optimisation of slow dynamic processes. Automatica, 14, 223-230.

Clarke, D.W., C. Mohtadi, and P.S. Tuffs (1987). Generalised predictive control: Part I: the basic algorithm. Automatica, 23, 137-148.

Cutler, C.R. and B.L. Ramaker (1980). Dynamic matrix control: a computer control algorithm. Preprints JACC, San Francisco.

Di Massimo, C., M.J., Willis, G.A. Montague, C. Kambhampati, A.G. Hofland, M.T. Tham and A.J. Morris (1990). On the applicability of neural networks in chemical process control. (To be presented at the AIChE Annual Meeting, Chicago).

Eaton, J.W. and J.B. Rawlings (1990). Feedback control of chemical processes using on-line optimization techniques. Comp. Chem. Engng., 14, 4/5, 469-479.

Economou, C.G. and M. Morari (1986). Internal model control 5: extension to nonlinear systems. Ind. Eng. Chem. Process Des. Dev., 25, 403-411.

Edgar, T.F. and D.M. Himmelblau (1989). Optimization of chemical processes. McGraw-Hill.

Fletcher, R. (1980). Practical methods of optimization. Volume 1. John Wiley.

Garcia, G.E. and M. Morari (1982). Internal model control 1: a unifying review and some new results. Ind. Eng. Chem. Process Des. Dev., 21, 308-323.

Lee, P.L., R.B. Newell, and G.R. Sullivan (1989). Generic model control- a case study. Can. J. Chem. Eng., 67, 478-484.

Lee, P.L. and G.R. Sullivan (1988). Generic model control (GMC). Comput. Chem. Eng., 12, 6, 573-580.

Li, W.C. and L.T. Biegler (1988). Process control strategies for constrained nonlinear systems. Ind. Eng. Chem. Process Des. Dev., 1421-1433.

Luenberger, D.G. (1973). Linear and Nonlinear Programming. Addison-Wesley.

Parrish, J.R. and C.B. Brosilow (1989). Nonlinear inferential control. AIChE Journal, 34, 633-644.

Peterka, V. (1984). Predictor based self-tuning control. Automatica, 20, 39-50.

Powell, M.J.D. (1964). An efficient method for finding the minimum of a function of several variables without calculating the derivatives. Comput. J., 7, 155-162.

Willis, M.J., C. Di Massimo, G.A. Montague, M.T. Tham and A.J. Morris (1990). On neural networks in chemical process control. (Submitted to the Proc. IEE, Pt.D.)

Willis, M.J., G.A. Montague, M.T. Tham, and A.J. Morris (1991). Inferential measurement using artificial neural networks. (to be presented at IFAC Conf., ITAC'91, Singapore, Jan.)

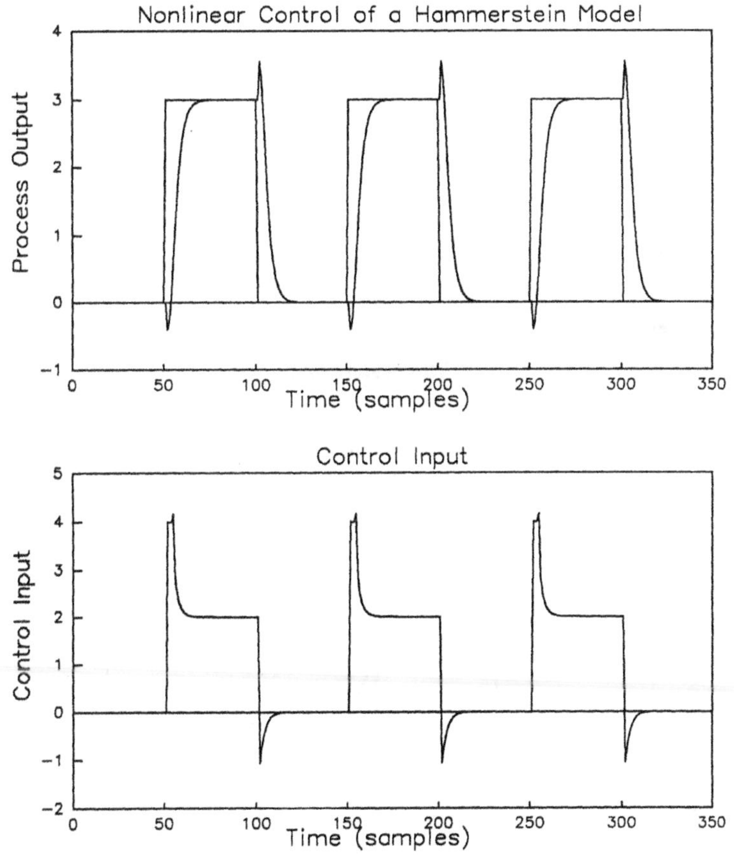

Fig.1. Nonlinear Predictive Control of a Hammerstein model

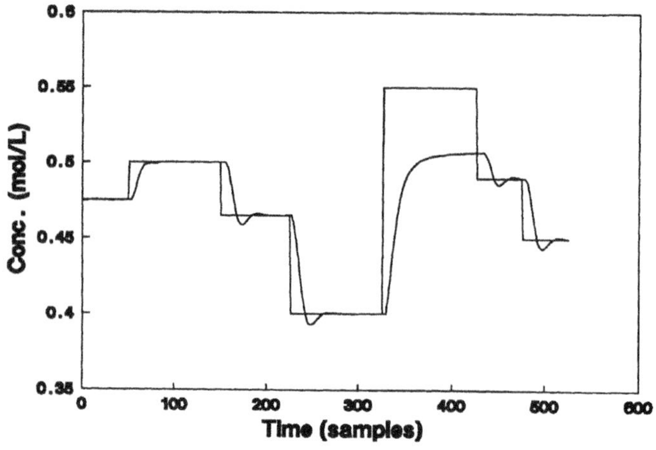

Fig.2 Control of Exothermic CSTR Product Composition

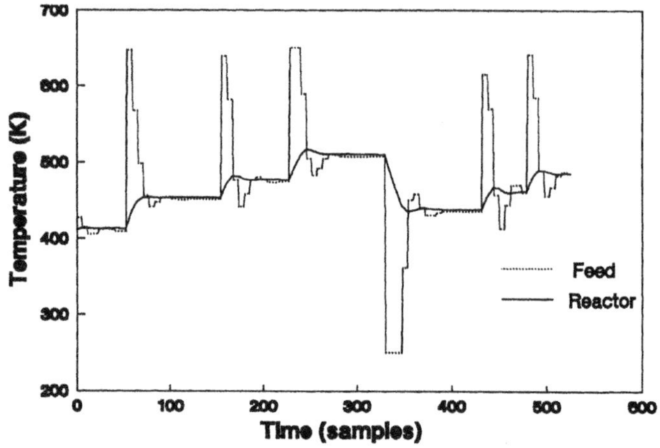

Fig.3. CSTR Feed and Reactor Temperatures

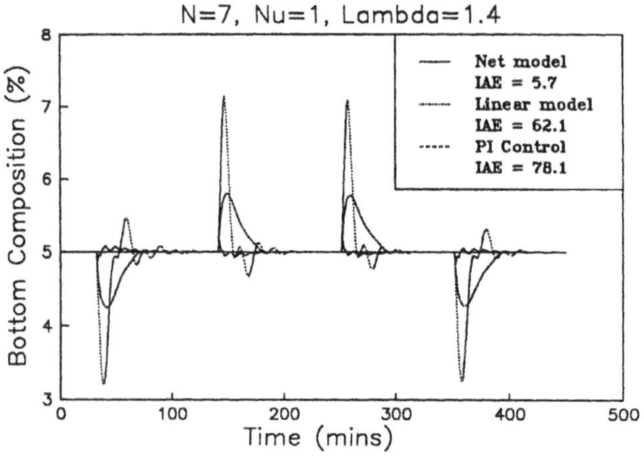

Fig.4 Comparison of Regulatory Performances of a PI; LPC and a Neural Network based NLPC

TUNING OF PID CONROLLERS: SURVEY OF SISO AND MIMO TECHNIQUES

H. N. Koivo and J. T. Tanttu

*Tampere University of Technology, Control Engineering Laboratory,
P O Box 527, SF-33101, Tampere, Finland*

Abstract.

A survey of different tuning methods of a PID controller is given. The scalar case is limited to most recent developments only. In multivariable case the different approaches for unknown plants are reviewed. The basic results have been extended to time-delay systems and distributed parameter systems. Decentralized control tuning, adaptive and expert system directions are also briefly discussed.

Key words. PID control; Multivariable control systems; Expert systems; Decentralized control; Adaptive control.

INTRODUCTION

The perennial workhorse of industrial controls is a proportional-integral-derivative (PID) controller. Well over ninety percent of all existing control loops are PID controllers. Since their number is so overwhelming, tuning of PID controllers is an important issue both in starting up new plants and in everyday operation. It is therefore surprising that only in the 1980's has automatic tuning of PID controllers received more attention, witnessed by a number of excellent surveys for tuning of SISO (single input - single output) PID-controllers (Åström and Hägglund, 1984, 1988; Clarke, 1986; McMillan, 1983).

Several different design methods for multivariable PI(D) controllers have also been suggested - e.g. Smith and Davison (1972), Solheim (1974), Seraji and Tarokh (1977), Owens (1978) - the tuning in the MIMO (multiple input -multiple output) case has received much less interest, although from the industrial point of view it is equally important. Multivariable PI(D) controllers have been used in a number of applications, among them heat exchanger (Davison, Taylor, Wright, 1980), paper quality variable control (Kunde, Chen, Mathre, 1983), robot control (Koivo and Sorvari, 1985) artificial heart (Wang, Lu, and McInnis, 1987), and paper machine head box (Jussila. Tanttu, Piirto, and Koivo, 1989).

The objective of this paper is to survey the existing results for tuning both scalar and multivariable PID controllers.

In another paper of this conference (Lieslehto, Tanttu) a comparison is made between four different multivariable PI controller tuning methods. Lieslehto, Tanttu, and Koivo (this conference) consider an expert system, which performs the steps from input -output pair selection to controller tuning . So in this paper we concentrate more on the tuning methods proper, although the previous steps of the design process i.e. modelling, dividing the process to be controlled into manageable blocks, selecting the input-output pairs for SISO control, and the interaction analysis to decide whether multivariable control should be used, are very important.

SCALAR PID CONTROLLER TUNING

The ideal SISO PID controller is of the form

$$u(t) = K_P e(t) + K_I \int_0^t e\,dt + K_D \frac{de}{dt}$$

or equivalently

$$u(t) = K\left[e(t) + \frac{1}{T_I}\int_0^t e\,dt + T_D \frac{de}{dt}\right]$$

where u is the control variable, $e = y_{ref} - y$ the error, y the output, , and y_{ref} the reference signal. The gains K_P, K_I, and K_D are real numbers. The latter form with proportional gain K, integral time T_I, and derivative time T_D is often preferred.

Because of the space limitations the review on the SISO tuning methods will be limited to some recent developments only. The reader is assumed to be familiar with the classical tuning methods.

The emphasis in 1980's has been on automatic tuning of scalar PID controllers. Automatic tuning is not a straightforward exercise in applying e.g. standard optimization techniques, but a rather complicated problem in data analysis, modelling, and control. The steps are roughly as follows: when to start tuning procedure, how long a data window should be used, is process delay significant, what is the model to be used, is the modelling (identification) performed in open loop or closed loop, is tuning a one shot effort or continuous (self tuning).

Some tuning methods rely on examining the process data (Kraus and Myron, 1984) to initiate tuning. Others generate their own excitation e.g. with a relay technique (Åström and Hägglund, 1984, 1988). A separate time delay estimation seems to be required in many industrial problems where the delay is unknown or time varying. Efforts in this direction have been suggested in Hansen (1983) and de Keyser (1986).

Dumont, Martin-Sánchez, and Zervos (1988) make an interesting comparison of auto-tuned regulators and adaptive predictive control system on an industrial bleach plant. Other comparisons have been made by Kaya and Titus (1988) on four self tuning control products and by Minter and Fisher (1988) on academic vs. industrial adaptive controllers.

An interesting development in tuning scalar PID controllers has been the use of expert systems: Åström, Anton, and Årzén (1986), Porter, Jones, and McKeown (1987), Sripada, Fisher, and Morris (1987), Lieslehto, Tanttu, and Koivo (1988), Thompson and McCluskey (1988). Robust control has also influenced PID tuning witnessed by Pohjolainen and Mäkelä (1988) and Rivera and Morari (1990). A fresh look into tuning problem has been taken in Zervos, Bélanger, and Dumont (1988).

A more detailed discussion of scalar PID tuning is provided in Koivo and Tanttu (1990).

MULTIVARIABLE PID CONTROLLER

Analogously to scalar PID controller the multivariable controller takes the form

$$u(t) = K_P e(t) + K_I \int_0^t e\,dt + K_D \frac{de}{dt}$$

or the corresponding transfer function

$$G_{PID}(s) = K_P + K_I \frac{1}{s} + K_D s$$

where u is the control vector, $e = y_{ref} - y$ the error vector, y the output vector, and y_{ref} the reference signal. The gains K_P, K_I, and K_D are matrices of appropriate dimensions. The multivariable PID structure has been suggested by many authors among them Davison and Smith (1971 and 1972), Fond and Foulard (1971), Johnson (1968), Porter (1970), and Rosenbrock (1971). The design of multivariable PID controller for linear plants has further been considered from a number of different viewpoints.

The idea of controlling unknown multivariable plants with a PID controller and determining the gain matrices based on a number of measurements is not so well established. What make this significantly harder than the scalar case are interactions between different variables. Can simple tuning rules for multivariable PID controller tuning be established? And further, can these be automated? Although scalar controllers are used in the majority of industrial applications, there are 5-10% control loops which cannot be controlled by SISO PID controllers. These require multivariable controllers.

From the scalar case one can deduce that it is important how the unknown system is modelled. Stability of the plant is essential. One of the first contributions to the tuning is by Davison (1976). He considers an unknown linear time-invariant plant, which is open-loop stable. Limiting his more general results to the multivariable PID tuning the system considered is

$$\dot{x} = A x + B u$$

$$y = C x$$

$$e = y_{ref} - y$$

where x is the state vector, u is the control vector, y is the output vector, and y_{ref} the reference signal. The assumptions made are:

1. Matrix A is asymptotically stable.
2. The control input u can be excited and the output y, which is to be regulated, can be measured. This implies that the number of inputs and outputs are known.
3. The order of the plant and the matrices A, B, and C are unknown.

Problem. It is desired to find a multivariable PID controller for the system so that asymptotic regulation occurs for all constant reference signals and that the response is as "good" as possible.

The problem definition is vague mathematically regarding the goodness of the response. Of course, an optimal control problem could be formulated to determine the "best" gain parameters. In tuning a practical engineering approach is used, as is also the case e.g. in multivariable frequency domain techniques by Rosenbrock (1974) and MacFarlane and Belletrutti (1973). Tuning controllers is based on experiments, e.g. step responses, and the human judgement is used to evaluate the performance of the controller.

Multivariable feedback design in frequency domain (Rosenbrock, 1974; MacFarlane and Belletrutti, 1972) is based on the idea of constructing a controller which approximately decouples the closed-loop plant both at zero frequency and at high frequencies. Consider the open-loop transfer function

$$G(s) = C(sI - A)^{-1} B$$

It is easy to see that at zero frequency, $s = 0$, $G(0) = C(-A)^{-1}B$. Since the integral controller takes care of the control at steady-state, or zero frequency, the appropriate choice for the integral gain should be inverse of the plant at steady state:

$$K_I = \varepsilon G^{-1}(0) = \varepsilon[C(-A)^{-1}B]^{-1}$$

where ε is a fine tuning parameter. This has been suggested e.g. by Rosenbrock (1971) and Davison (1976).

Proportional controller should be effective at high frequencies. If the transfer function is developed into Taylor series, when s is large then

$$G(s) = \frac{CB}{s} + \frac{CAB}{s^2} + \frac{CA^2B}{s^3} + \ldots$$

Taking only the first term into account, the proportional gain K_P is proposed to be (Penttinen and Koivo, 1979, 1980)

$$K_P = \delta\,(CB)^{-1}$$

where δ is the fine tuning constant. A similar suggestion has also been made by Mayne (1979).

The derivative gain matrix is harder to pick. If it is kept small ($\|K_D\|$ small), a reasonable choice for K_D is to use the same matrix as in proportional gain (Koivo, 1980):

$$K_D = \gamma\,(CB)^{-1}$$

where γ is the fine tuning constant.

Davison, Taylor, and Wright (1980) use as the proportional gain matrix the same as in the integral gain matrix

$$K_P = \delta\, G^{-1}(0) = \delta\left[C(-A)^{-1}B\right]^{-1}$$

Thompson's (1982) proportional gain matrix is based on arguing that CB is not easy to determine accurately or it might be nonsingular. Therefore let

$$K_P = \delta\left[C\,A^{-1}(e^{A\tau}-1)\,B\right]^{-1}$$

Here τ is the so-called average e-folding time of the system. It is likely that this matrix will decouple the plant in over its midrange frequency. Koivo and Pohjolainen (1981) give a similar gain matrix to avoid rank deficiency

$$K_P = \delta\left[C\,e^{At_0}B\right]^{-1}$$

For derivative gain Thompson (1982) proposes the matrix

$$K_D = \gamma I$$

He argues that the effect of D-part is minor compared with the integral and proportional gains at high frequencies. So although it is unlikely that this matrix gain will decouple the plant at high frequencies, both proportional and integral gains will take care of decoupling.

Since the initial assumption is that the plant is stable, the controlled plant must remain stable. This has been proved by Davison (1976) for I controller and by Penttinen and Koivo (1980) for PI controller.

The previous controller gain matrices can be classified as "low" gains. Owens and Chotai (1982) approach the tuning problem from a different angle using "high" controller gain. The plant assumptions are that the plant may be unstable but minimum phase.

Tuning of multivariable discrete time PI controllers for unknown systems has been discussed by Peltomaa and Koivo(1983).

EXPERIMENTAL DETERMINATION OF THE GAIN MATRICES

Since the plant is unknown, the matrices CB and $-CA^{-1}B$ have to be determined experimentally. Algorithms 1 and 2 outline the procedure:

Algorithm 1. (determining CB)

Step 1. Let $u(t) = 0$. When the steady-state ($y(\infty) = 0$) is achieved, choose a constant input $u(t) = u_1$ ($\neq 0$) and apply it to the plant. Then .

$$\dot{y}(0) = \dot{y}_1 = CB\,u_1$$

From the corresponding step response $\dot{y}_1$ can be computed, e.g., graphically.

Step 2. Repeat step 1 with a constant control u_2, which is linearly independent of u_1. Then

$$\dot{y}(0) = \dot{y}_2 = CB\,u_2$$

Compute $\dot{y}_2$ from the corresponding step response.

Step m. Repeat step 1 with a constant control u_m, which is linearly independent of $u_1, u_2, \ldots, u_{m-1}$. Then .

$$\dot{y}(0) = \dot{y}_m = CB\,u_m$$

Compute $\dot{y}_m$ from the corresponding step response. As a result

$$CB = [\,\dot{y}_1, \dot{y}_2, \ldots, \dot{y}_m\,]\,[\,u_1, u_2, \ldots, u_{m-1}\,]^{-1}$$

Algorithm 2. (determining $G(0) = -CA^{-1}B$)

Step 1. Apply a constant input $u(t) = u_1$ ($\neq 0$) to the open-loop plant. The system is stable, so a steady-state solution $y(\infty) = y_1$ exists. Measure y_1.

Step 2. Repeat step 1 with a constant control u_2, which is linearly independent of u_1. Measure y_2.

Step m. Repeat step 1 with a constant control u_m, which is linearly independent of $u_1, u_2, \ldots, u_{m-1}$. Measure y_m.

As a result

$$G(0) = [\,y_1, y_2, \ldots, y_m\,]\,[\,u_1, u_2, \ldots, u_{m-1}\,]^{-1}$$

EXTENSIONS TO OTHER PLANTS

Many processes have time delays. In SISO cases PID controllers are frequently used to control plants with time delays. Some of the tuning procedures discussed above have been extended for such plants. Koivo and Pohjolainen (1981, 1985) consider systems, which have time delays in control variable. Jussila and Koivo (1985, 1987) treat the case where the delays may also occur in the state variables.

Distributed parameter systems have first been treated in Pohjolainen and Koivo (1979), where tuning of I controller is discussed. Pohjolainen (1982) discusses PI controller for infinite dimensional systems. Kobayashi (1986) considers tuning of a discrete I controller for parabolic systems and a digital PI controller for exponentially stable distributed parameter systems. Logemann and Owens (1987) using frequency domain theory treat the low-gain PI control for a wide class of infinite-dimensional systems.

DECENTRALIZED AND EXPERT TUNING

Davison (1978) first addressed a very important issue of tuning decentralized controllers. He later applied these ideas to load and frequency control of a power system with Tripathi (1978, 1980). Some issues related to this have been treated by Lunze (1989). Adaptive and expert systems in multivariable PID control tuning have been developed by number of authors: Bristol and Hansen (1988), Jones (1988), Jones and Porter (1985), Koivo and Sorvari (1985), Lieslehto, Tanttu, and Koivo (this symposium), Porter and Khaki-Sedigh (1987), Tanttu (1984, 1987), Wang and Owens (1988). Piirto and Koivo (1989) demonstrate in an advanced portable control station how a multivariable PI(D) controller can be implemented and used in real time control of processes.

Multiloop applications are discussed e.g. by Gawthrop and Nomikos (1989).

CONCLUSIONS

Although tuning of MIMO PID controllers has not progressed to the stage of SISO PID control, strong inroads have been made into this very important practical issue. The PID control structure seems to be a good candidate for robust control also in multivariable case. Its simplicity and demonstrated usefulness make the tuning problem an interesting challenge. Different approaches to this problem are surveyed. Decentralized, adaptive and expert system viewpoints to MIMO PID tuning have also been reviewed in the paper.

REFERENCES

Åström, K.J. and **T. Hägglund** (1988). *Automatic tuning of PID controllers*, ISA, Research Triangle Park, N.C.

Åström, K.J. and **B. Wittenmark** (1984). *Computer Controlled Systems*, Prentice-Hall, Englewood Cliffs, New Jersey.

Åström, K.J., J.J. Anton, and K.E. Årzén (1986). Expert control. *Automatica* **22**, 277-286.

Bristol, E.H. and P.D. Hansen (1987). Moment projection feedforward control adaptation. *American Control Conference*,. Minneapolis, Minnesota.

Clarke, D. (1986). Automatic tuning of PID regulators. *Expert Systems and Optimization in Process Control*, Technical Press, Aldershot, England.

Davison, E.J. (1976). Multivariable tuning regulators: The feedforward and robust control of general servomechanism problem. *IEEE Trans. Aut. Control* **AC-21**, 35-47.

Davison, E.J. (1978). Decentralized robust control of unknown systems using tuning regulators. *IEEE Trans. Aut. Control* **AC-23**, 276-289.

Davison, E.J. and **H.W. Smith**.(1971). Pole assignment in linear time-invariant multivariable systems with constant disturbances. *Automatica* **7**, 489-498.

Davison, E.J., P.Taylor and **J. Wright** (1980). On the application of tuning regulation to obtain a robust feedforward-feedback controller of an industrial heat exchanger. *IEEE Trans. Aut. Control* **AC-25**, 361-375.

Davison, E.J. and **N.K. Tripathi** (1978). The optimal decentralized control of a large power system: Load and frequency control. *IEEE Trans. Aut. Control* **AC-23**, 312-325.

Davison, E.J. and **N.K. Tripathi** (1980). Decentralized tuning regulators: an application to solve the load and frequency control problem for a large power system. *Large Scale Systems* **1**, 3-15.

De Keyser R.M.C. (1986). Adaptive dead-time estimation. *IFAC Workshop on Adaptive Systems in Control and Signal Processing*. Lund, Sweden, 209-213.

Dumont, G.A., J.M. Martin-Sánchez and C:C. Zervos (1989). Comparison of an auto-tuned PID regulator and adaptive predictive control system on an industrial bleach plant. *Automatica* **25**, 33-40.

Fond, M. and **C. Foulard** (1971). Linear multivariable control in the presence of unknown non-zero mean value perturbations. *IFAC Symposium on Multivariable Control*.paper 2.2.2, Düsseldorf, West Germany.

Gawthrop, P.J. and **P.E. Nomikos** (1990). Automatic tuning of commercial PID controllers for single-loop and multiloop applications. *IEEE Control Systems Magazine* **10-1**, 34-42.

Hansen, P.D. (1983). Robust identification for a self-tuning controller. *Proc. of the Third Yale Workshop on Applications of Adaptive Systems Theory*. New Haven, Conneticut, 47-56.

Johnson, C.D. (1968). Optimal control of the linear regulator with constant disturbances. *IEEE Trans. Aut. Control* **AC-13**, 416-421.

Jones, A.H.(1988). Design of adaptive digital set-point tracking PI controllers incorporating expert tuners for multivariable plants. *IFAC Symposium on Adaptive Systems in Control and Signal Processing*. Glasgow, United Kingdom, 579-586.

Jones, A.H. and B.Porter (1985). Design of adaptive digital set-point tracking PID controllers incorporating recursive step-response matrix identifiers for multivariable plants. *IEEE Conf. on Decision and Control*, Ft. Lauderdale, Florida.

Jussila, T.T. and **H.N. Koivo**.(1987). Tuning of multivariable PI-controllers for unknown delay-differential systems. *IEEE Trans. Aut. Control* **AC-32**, 364-368.

Jussila, T.T., J.T. Tanttu, M. Piirto and **H.N. Koivo**.(1987). An extended self-tuning multivariable PI-controller. *IFAC Symposium on Adaptive Systems in Control and Signal Processing*. Glasgow, United Kingdom, 375-380.

Kaya. A. and S. Titus (1988). A critical performance evaluation of four single loop self tuning control products. *American Control Conference*. Atlanta, Georgia, 1659-1664.

Kobayashi, T. (1988). A digital PI-controller for distributed parameter systems. *SIAM J. Control and Optimization* **26**, 1399-1414.

Kobayashi, T. (1986). Design of robust multivariable tuning regulators for infinite-dimensional systems by discrete-time input-output data. *Int. J. Control* **44**, 89-90.

Koivo, H.N. (1979). Robust controllers and their tuning. *Acta Polytechnica Scandinavica* **Ma31**. 111-119.

Koivo, H.N. (1980). Tuning of multivariable PID controller for unknown systems. *Proc. 19th EEE Conference on Decision and Control*. Albuqueque, New Mexico.

Koivo, H.N. and **S. Pohjolainen** (1981). Tuning of multivariable PI controllers for unknown systems with input delay. *Proc.of the 8th IFAC World Congress*. 951-956.

Koivo, H.N. and **S. Pohjolainen** (1985). Tuning of multivariable PI controllers for unknown systems with input delay. *Automatica* **21**, 81-91.

Koivo, H.N. and **J. Sorvari** (1985). Tuning of multivariable controller for a robot manipulator. *IEEE conf. on Robotics and Automation*. St. Louis, Missouri, 290-294.

Koivo, H.N. and **J. Sorvari** (1985). On-line tuning of a multivariable PID controller for robot manipulators. *IEEE Conf. on Decision and Control*, Ft. Lauderdale, Florida.

Koivo, H.N. and **J.T. Tanttu** (1990). Tuning of PID controllers: Survey of SISO and MIMO techniques. Report-3, Control Engineering Laboratory,Tampere University of Technology, Tampere Finland.

Kraus, T.W. and T.J. Myron (1984). Self-tuning PID controllers based on a pattern recognition approach. *Control Engineering*, 106-111.

Kunde, F.K., T-S. Chen and J.L. Mathre (1983). Tuning weight and moisture control automatically. *Tappi J.* **66**, 35-38.

Lieslehto, J., Tanttu J.T. and **H.N. Koivo** (1988) An expert system for tuning PID controllers. *The Fourth IFAC Symposium on Computer Aided Design in Control Systems*. Beijing, P:R: China, 387-391.

Lieslehto, J., Tanttu J.T. and **H.N. Koivo** (1990) An expert system for the multivariable controller design. This Symposium.

Logemann, H. and D. Owens (1987). Robust high-gain feedback control of infinite-dimensional minimum-phase systems. *IMA Journal of Mathematical Control and Information***4**, 195-220.

Logemann, H. and D. Owens (1987). Low-gain feedback control of unknown infinite-dimensional systems. *Research report no. 7*, University of Strathclyde, Glasgow, United Kingdom.

Lunze, J. (1989). *Robust Multivariable Feedback Control*. Prentice Hall, New York.

MacFarlane, A.G.J. and **J.J. Belletrutti**.(1973). The characteristic locus design method. *Automatica* **9**, 575-588.

Mayne, D. (1979). Sequential design of linear multivariable systems. *Proc. IEE* **126**, 568-572.

McMillan, G.K. (1983). *Tuning and Control Loop Performance*. ISA Monograph, Research Triangle Park, North Carolina.

Minter, B.J. and D.G. Fisher (1988). A comparison of adaptive controllers: Academic vs industrial. *American Control Conference*. Atlanta, Georgia, 1653-1658.

Owens, D. (1978). *Feedback and Multivariable Systems*. Peter Peregrinus, Stevenage, United Kingdom.

Owens, D. (1985).A loop tuning condition for multivariable process control using steady state step response data. *Control - Theory and Advanced Technology* **1**, 267-274.

Owens, D. and A. Chotai (1982).High performance controllers for unknown multivariable systems. *Automatica***18**, 583-587.

Owens, D. and A. Chotai (1981). Controllers for unknown multivariable systems using monotone modelling errors. *Proc. IEE* **129, Pt D**, 106-107.

Owens, D. and A. Chotai (1983). Robust controller design for linear dynamic systems using approximate models. *Proc. IEE* **130, Pt D**, 45-54.

Owens, D. and A. Chotai (1986). Approximate models in multivariable process control - An inverse Nyquist array and robust tuning regulator interpretation. *Proc. IEE* **133, Pt D**, 1-12.

Piirto, M. and **H.N. Koivo** (1989). Advanced portable control station. *IFAC Conference on Low Cost Automation*. Milano, Italy.

Peltomaa, A., and **H.N. Koivo** (1983). Tuning of multivariable discrete time PI controller for unknown systems. *Int. J. Control* **38**, 735-745.

Penttinen, J. and **H.N. Koivo** (1979). Multivariable tuning regulators for unknown systems. *Proc. of the 17th Annual Allerton Conference on Communication, Control and Computing*, Monticello, Illinois..

Penttinen, J. and **H.N. Koivo** (1980). Multivariable tuning regulators for unknown systems. *Automatica***16**, 393-398.

Pohjolainen, S. (1987). Robust multivariable PI-controller for infinite dimensional systems. *IEEE Trans. Aut. Control* **AC-27**, 17-30.

Pohjolainen, S. (1987). On optimal tuning of a robust controller for parabolic distributed parameter systems. *Automatica***23**, 719-728.

Pohjolainen, S. and H.N. Koivo (1979). Robust controller for distributed parameter systems. *Proc. the IV Symposium Über Oper. Res.* Saarbrücken, West Germany.

Pohjolainen, S. and T. Mäkelä (1988). How to improve a given linear controller using H_∞-methods. *IMACS World Congress*. Paris, France.

Porter, B. (1981). Design of error-actuated controllers for linear multivariable plants. *Electronic letters* **17**, 106-107.

Porter, B. (1982). Design of tunable set point tracking controllers for linear multivariable plants. *Int. J. Control* **35**, 1107-1115.

Porter, B., A.H. Jones and C.B. McKeown (1987). Real-time expert tuners for process control. *IEE Proc.* **134, Pt.D**, 2260-263.

Porter, B. and A. Khaki-Sedigh (1987). Singular perturbation analysis of the step-response matrices of a class of linear multivariable systems. *Int. J. of System Science* **18**, 205-211.

Porter, B. and A. Khaki-Sedigh (1989). Robustness properties of tunable digital set-point tracking PID controllers for linear multivariable plants. *Int. J. Control* **49**, 777-789.

Porter, B. and H.M. Power (1981). Controllability of multivariable linear systems incorporating integral feedback. *Electronic letters* **7**, 689-692.

Rivera, D.E. and M. Morari (1990). Low order SISO controller tuning methods for the H_2, H_∞ and μ objective functions. *Automatica* **26**, 361-369.

Rivera, D.E., M. Morari and **S. Skogestad** (1986). Internal model control. 4. PID controller design, *Ind.Eng.Chem.Process Des.Dev.*, **25**, 252-265.

Rosenbrock, H.H. (1971). Progress in the design of multivariable control systems. *Trans. Inst. Measure. Control*, **4**, 9-11.

Seraji, H. and **M. Tarokh** (1977). Design of PID controllers for multivariable systems. *Int J. Control* **26**, 75-83.

Smith H.W. and **Davison, E.J.** (1972). Design of industrial regulators. *Proc.IEE* **119**, 1210-1216.

Solomon, A. and E.J. Davison (1986). Control of unknown linear systems using gain-scheduling tuning regulators. *American Control Conference*, Boston, Mass.

Sripada, N.R., D.G. Fisher and A.J. Morris (1987). AI application for process regulation and servo control. *IEE Proc. Expert Systems in Engineering* **134, Pt.D**, 251-259.

Tanttu J.T. (1984). A multivariable PI controller for DARMA processes. Report 1-84, Control Engineering Laboratory, Department of Electrical Engineering, Tampere University of Trechnology, Tampere Finland.

Tanttu J.T. (1987). A comparative study of three multivariable self-tuning controllers. Ph.D. thesis, Tampere University of Technology Publications 44, Tampere Finland.

Tanttu J.T. (1988). Tuning of the scalar PID controller. Report 1-88, Control Engineering Laboratory, Department of Electrical Engineering, Tampere University of Technology, Tampere Finland.

Thompson, S. (1982). Multivariable PID controller for unidentified plant. *ASME Journal of Dynamic Systems, Measurement, and Control* **104**, 270-274.

Thompson, S. and E.G. McCluskey (1988). An expert adaptive PID-controller. *IFAC Symposium on Adaptive Systems in Control and Signal Processing*. Glasgow, United Kingdom, 553-558.

Wang, J.C., P.C. Lu and B.C. McInnis (1987). A microcomputer-based control system for the total artificial heart. *Automatica* **23**, 275-286.

Wang, L. and D.H. Owens (1988). Multivariable first order adptive controllers. *IFAC Int. Symp. ADCHEM'88 Adaptive Control of Chemical Processes*. Lyngby, Copenhagen, Denmark, 191-196.

Zervos, C., P.R. Bélanger and G. Dumont (1988). On PID controller tuning using orthonormal series identification. *Automatica* **24**, 165-175.

Yuwana, M., and D. Seborg (1982). A new method for on-line controller tuning. *AIChE J.* **28**, 434-440.

PARAMETER ESTIMATION USING ARTIFICIAL NEURAL NETS

A. P. Loh and T. H. Lee

Dept. of Electrical Engineering, National University of Singapore, 10 Kent Ridge Crescent, Singapore 0511

Abstract. Artificial neural nets have become increasingly popular in solving optimization problems such as the Travelling Salesman Problem, A/D converter realization, linear programming and pattern recognition. In many of these optimization problems, the neural net structure is based on the Hopfield model and each neuron is assumed to have two distinct outputs, 0 or 1. In this paper, a continuous Hopfield net is used to formulate and solve a least squares minimization problem associated with the parameter estimation of a physical system. The algorithm shows good convergence and a certain amount of "recursion" can be incorporated which makes it useful for real time applications.

Keywords. Neural net; Hopfield model; Least squares estimation; Recursion; Real time.

INTRODUCTION

In recent years, artificial neural nets have become increasingly popular in solving optimization problems such as the Travelling Salesman Problem (TSP), A/D Converter realization, linear programming and pattern recognition. Their computational ability is derived from the highly interconnected networks of simple analog processors which are modelled around aspects of neurobiology. One of the biggest advantages of the neural net is the speed at which it can solve some classic combinatorial optimization problems like the TSP (Giarratano, Villareal, Savely, 1990). The problem is to compute the shortest route through a given network when starting from a certain city. A neural net took less than 0.1 second to solve both the ten- and thirty-city problems on a microcomputer compared to 1 hour using a conventional searching algorithm on a large mainframe. In addition, an electronic circuit was constructed to solve these problems in 1 μsec (Hopfield, Tank, 1985).

Neural nets may take different forms e.g. single layer adalines or multi-layered perceptrons. In each of these, it is characterised by a system of nodes called neurons which are connected to one another via weights. The neurons themselves are simple non-linear processing elements implementing monotonically increasing but bounded functions often referred to as sigmoid functions. Changing the weights between neurons will alter the behaviour of the whole network. Hopfield and Tank (1986) has shown that by appropriately choosing these weights, the network can be made to solve the optimization problems mentioned above.

In many of these problems, each neuron is assumed to have two distinct outputs, 0 or 1. Using the original Hopfield model, the objective function is chosen to be

$$E = -0.5V^tTV - I^tV \qquad (1)$$

where T is the symmetric connection matrix for all the neurons,
V is the vector of neural outputs,
I is the vector of inputs to the neurons,
$(.)^t$ denotes the transpose.

The characteristic of each neuron is such that

$$V_j = \begin{cases} 0 & \text{when } TV + I < 0 \\ 1 & \text{elsewhere.} \end{cases} \qquad (2)$$

It can be verified that the algorithm (2) will lead to a continuously decreasing objective function, E, which forms the basis for the optimization.

There has been an extension of the original Hopfield model to continuous working nodes with neurons taking arbitrary (but bounded) values. The objective function is similar except that the energy surface is continuous with all neurons, V_j operating continuously as well. The dynamics of the ith neuron is given by the differential equation:-

$$dx_i/dt = \sum_{j=1}^{m} T_{ij}V_j + I_i \quad i = 1, 2, \ldots, m \qquad (3)$$

$$V_i = g_i(x_i) \qquad (4)$$

where x_i is a state of the ith neuron,
V_i is the output of the ith neuron,
$g_i(.)$ is the map of the state to the output and
m is the number of neurons.

Under certain conditions of g_i's and T, the above dynamical system will lead to a monotonically decreasing E as the system evolves in time.

In this paper, a neural network with the above structure is proposed for implementing a least squares estimator for a linear system. The algorithm is also valid for systems whose models are linear in the parameters. The problem formulation is given in the next section with simulation results to follow. It will be shown how "recursion" can be incorporated, thus extending its usefulness to real time or on-line applications.

THE ESTIMATOR

Consider a physical system whereby the input-output characteristics can be described in the following simple form :

$$y(t) = \beta(t-1)^t\theta \quad (5)$$

where $y(t)$ denotes the system output at time t, $\beta(t-1)$ denotes a vector that is a linear or nonlinear function of past outputs and inputs:

$$Y(t-1) = [y(t-1), y(t-2), \ldots]^t$$
$$U(t-1) = [u(t-1), u(t-2), \ldots]^t$$

θ denotes the unknown parameter vector.

(5) represents a wide class of linear and nonlinear deterministic dynamical systems (Goodwin, Sin, 1984). Suppose we have N sets of input-output pairs. Then, collecting them together, we have

$$\Gamma(N) = \Phi(N)\theta \quad (6)$$

where $\Gamma(N) = [y(t), y(t+1), \ldots, y(t+N-1)]^t$
$\Phi(N) = [\beta(t-1), \beta(t), \ldots, \beta(t+N-2)]^t$

The least squares parameter estimation problem becomes finding θ such that the objective function

$$E' = (\Gamma(N) - \Phi(N)\theta)^t(\Gamma(N) - \Phi(N)\theta) \quad (7)$$

is minimized. The minimization of E' is equivalent to minimizing

$$E = -\theta^t\Phi(N)^t\Gamma(N) - \Gamma(N)^t\Phi(N)\theta + \theta^t\Phi(N)^t\Phi(N)\theta \quad (8)$$

Comparing (8) and (1), therefore

$$T = -2\Phi(N)^t\Phi(N),\ I = -2\Phi(N)^t\Gamma(N) \text{ and } V = \theta \quad (9)$$

Hence, the weights matrix is a function of the inputs and outputs of the system to be estimated and so is I. As long as the inputs are persistently exciting (Astrom, Wittenmark, 1989), T will be negative definite and the minimization problem will be well posed since -T then becomes positive definite.

Having now defined the objective function in the context of a Hopfield net, we now turn to the dynamics of the neurons themselves. From (3) and (4)

$$[dx/dt] = TV + I \quad (10)$$
$$V = G(x) \quad (11)$$

where $[dx/dt] = [dx_1/dt, dx_2/dt, \ldots, dx_m/dt]^t$, m is also the number of parameters to be estimated, $x = [x_1, x_2, \ldots, x_m]^t$ - vector of states and $G(x)$ = diagonal matrix of $g_i(x_i)$, i = 1, ..., m.

Hence (10) becomes

$$[dx/dt] = TG(x) + I \quad (12)$$

and the stability of (13) depends critically on TG(x).

If however, $G(x) = Dx$ where D is a diagonal matrix of positive constants, then $TG(x) = TDx$ and TD will be negative definite, with eigenvalues being strictly negative. Thus the stability of the network is assured by choosing $g_i(x_i)$ to be monotonically increasing linear functions of the states. This is sufficient to ensure stability and convergence of the algorithm. In general, $g_i(x_i)$ need only be monotonically increasing. If bounds are placed on them, the bounds must include the bounds of the parameters to be estimated. Otherwise, the algorithm will lead to a local minimum with at least one of the parameters taking the value of the corresponding bound. In our case, $g_i(x_i)$ is not bounded.

Hence from (12) and (11),

$$[dx/dt] = TDx + I \quad (13)$$
$$V = Dx \quad (14)$$

Thus, the solution to the least squares minimization problem has been transformed to one of solving for the output response (implicitly, to a unit step input) of a dynamical system given by the equations above.

Observe that the right hand side of (13) is the gradient of the objective function with respect to the output of the neurons or the unknown parameters. At the minimum, $[dx/dt] = 0$, implying that the solution occurs at the steady state of (13) and (14).

Recursion

Consider at time = t_o sec and the first set of input-output data is collected. At this point, the weights matrix, denoted by T_o, is given by

$$T_o = -2\beta(t_o - T_s)\beta(t_o - T_s)^t \quad (15)$$

and

$$I_o = -2\beta(t_o - T_s)y(t_o) \quad (16)$$

At time = $(t_o + kT_s)$ sec, the weights matrix can be computed as

$$T_k = \lambda T_{k-1} - 2\beta(t_o + (k-1)T_s)\beta(t_o + (k-1)T_s)^t \quad (17)$$

$$I_k = \lambda I_{k-1} - 2\beta(t_o + (k-1)T_s)y(t_o + kT_s) \quad (18)$$

where λ is a forgetting factor with value between 0 and 1.

At each sampling instant, the matrices are updated as in (17) and (18) and the steady state solutions to (13) and (14) solved to obtain the estimate for the parameters. As more data are collected, the weights matrix, T becomes more negative definite, ensuring the rapid convergence of the solution.

Although the procedure is not recursive with respect to the parameters themselves, the update of the relevant matrices can be carried out recursively without too much computational effort. In the event that there is a drift of the parameter values during the estimation process, the algorithm also works by incorporating a forgetting factor in the data matrices as shown in (17) and (18). This is demonstrated in one of the simulation results.

SIMULATION RESULTS

Consider the following system :

$$y(t) = 0.1u(t-1)+0.2u(t-2)+0.5u(t-3)+1.5u(t-4)$$

The inputs u(t) was generated as a random sequence. A forgetting factor of 0.98 was used for the identification of the above system. Two different gain matrices were used to demonstrate the relationship between convergence and the condition of TD.

The solution was reached by continuously solving (14) and (15) as more data is accumulated. Fig. 1 shows the estimates as they evolve in time for D = diag(20) i.e. diagonal matrix with elements = 20. Fig. 2 is an illustration for the same system but with D = diag(2).

Fig. 3 demonstrates the ability of the algorithm

to track the parameters as they change. The system in this case is as follows:

$$y(t) = 0.1u(t-1) + 0.9u(t-2)$$

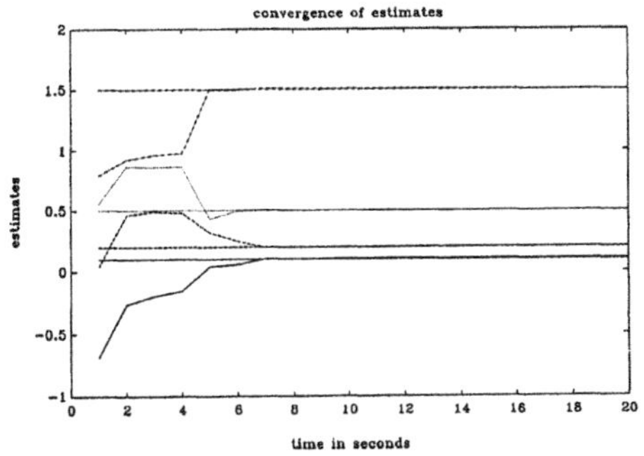

Fig. 1 : Actual and Estimated Parameters

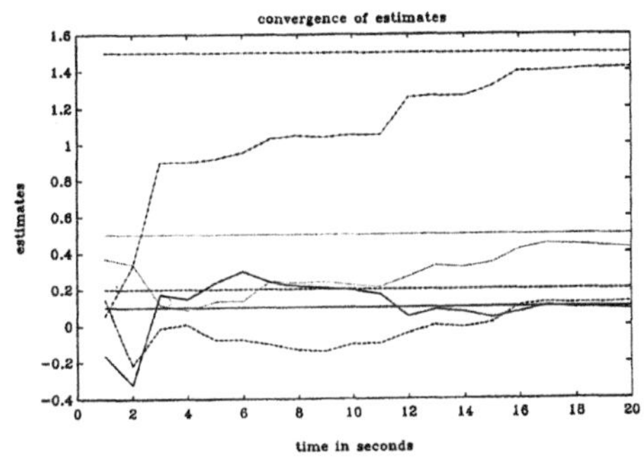

Fig. 2 : Actual and Estimated Parameters

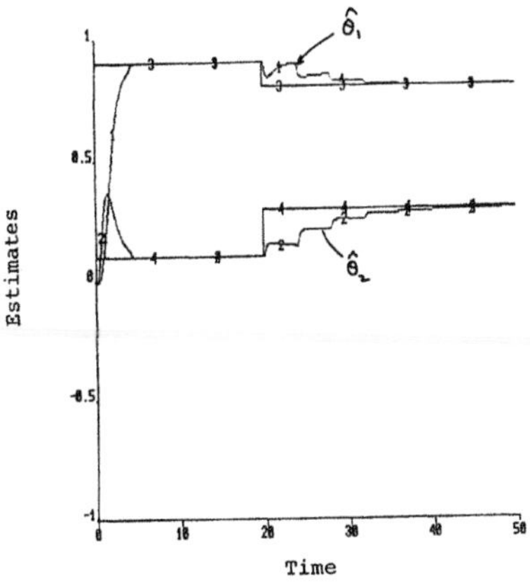

Fig. 3 : Actual and Estimated Parameters

DISCUSSION

The structure of the neural net proposed is based on the continuous Hopfield model with the number of neurons equal to the number of parameters to be estimated. The outputs of the neurons are the estimated parameters themselves while the weights between them are functions of the inputs and outputs of the process to be estimated. The neural net eventually learns about the system without any supervised training. This is in contrast to many neural nets where the weights are the variables to be adjusted with each training set of input and output. One such net is the adaline which is extremely well suited to our problem. In the adaline, the information is coded in the weights of the neurons as opposed to our algorithm where the information is coded in the output of the neurons themselves. The latter offers the advantage that the stability of the net is far more predictable. In the former, the stability with respect to the weights is difficult to analyse. Also, by choosing a large gain matrix, we can accelerate convergence. Again, in the former, there is no such guarantee and the whole training process may be quite tedious.

In comparison with the conventional recursive least squares method, it is obvious that the algorithm does not experience the difficulties encountered in matrix inversion. It however requires a set of differential equation solvers. Also, if there are any a priori information on the bounds of the parameter values, this can be helpful in ensuring better convergence by incorporating these bounds in the sigmoid functions.

Although the algorithm has been proposed in the framework of a neural net, it is really nothing more than the steepest descent method in conventional optimization where the gradient directions are used with infinitesimal step lengths leading to a continuous scheme. The descent path is a function of time with descent steps also taken in the time domain.

CONCLUSION

We have shown how a parameter estimation problem can be posed in the framework of a continuous neural net. It applies to any deterministic system with models that are linear in the parameters. One of its advantages is that stability of the solution is guaranteed provided the inputs are persistently exciting and the gain matrix is appropriately chosen. The algorithm is recursive to a certain extent and together with strong convergence, it is useful as an on-line system identification tool.

REFERENCES

Giarratano, J. C., Villarreal, J. A., Savely, R. T. (1990). Future Impacts of Artificial Neural Systems on Industry. ISA Transactions, 29 No 1, 9 - 14.

Hopfield, J. J., Tank, D. W. (1985). Neural Computations of Decisions in Optimization Problems. Biological Cybernetics, 52, 141 - 152.

Tank, D. W., Hopfield, J. J. (1986). Simple Neural Optimization Networks : An A/D Converter, Signal Decision Circuit and a Linear Programming Circuit. IEEE Trans on Circuits and Systems, 33 No 5, 533 - 541.

Goodwin, G. C., Sin, K. S. (1984). Adaptive Filtering, Prediction and Control. Prentice Hall.

Astrom, K. J., Wittenmark, B. (1989). Adaptive Control. Addison Wesley.

INFERENTIAL MEASUREMENT VIA ARTIFICIAL NEURAL NETWORKS

M. J. Willis, C. Di Massimo, G. A. Montague, M. T. Tham and A. J. Morris

Dept. of Chemical and Process Engineering, University of Newcastle-upon-Tyne Newcastle-upon-Tyne, NE1 7RU, UK

Abstract. In this article, the use of an artificial neural network to provide a cost efficient model of two distillation processes is considered. First the concepts involved in the formulation of artificial neural networks are discussed. The suitability of the technique to provide estimates of the product composition from an industrial distillation tower is then demonstrated. Measurements from established instruments (ie overheads temperature) are used as secondary variables for estimation of the 'primary' composition variables. The advantage of using these estimates for feedback control is then demonstrated.

Keywords. Neural nets; distillation columns; inferential control; nonlinear systems; models.

INTRODUCTION

The use of artificial neural networks (ANNs) as a possible solution strategy for problems which require complex data analysis is not new. Over the last 40 to 50 years intensive research has been devoted to this cause in an attempt to develop an algorithmic equivalent of the human learning process. The principal motivation behind this research is the desire to achieve the sophisticated level of information processing that the brain is capable of. Whilst the function of single neurons is relatively well understood, their collective role within the conglomeration of cerebrum elements is less clear and a subject of avid postulations. Consequently, the architecture of an ANN is based upon a primitive understanding of the functions of the biological neural system. Even if neuro-physiology could untangle the complexities of the brain, due to the limitations of current hardware technology, it would be extremely difficult, if not impossible, to emulate exactly its immensely distributed structure. Thus, rather than accurately model the intricacies of the human cerebral functions, ANNs attempt to capture and utilise the connectionist philosophy on a more modest and manageable scale.

Within the domain of process engineering, process design; simulation; process supervision, control and estimation; fault detection and diagnosis their is a reliance upon the effective processing of unpredictable and imprecise information. To tackle such tasks, current approaches tend to be based upon some 'model' of the process in question. The model can either be qualitative knowledge derived from experience; quantified in terms of an analytical (usually linear) process model, or a loosely integrated combination of both. Although the resultant procedures can provide acceptable solutions, there are many situations in which they are prone to failure because of the uncertainties and the nonlinearities intrinsic to many process systems. These are, however, exactly the problems that a well trained human decision process excels in solving. Thus, if ANNs fulfil their projected promise, they may form the basis of improved alternatives to current engineering practice. Indeed, applications of artificial neural networks to solve process engineering problems have already been reported.

For instance, the use of neural network based models (NNMs) for the on-line estimation of process variables was considered by Montague et al (1989). Lant et al (1990) further discussed the relative merits of process estimation using an adaptive linear estimator (Guilandoust et al, 1987), and an estimator based on an NNM. The applicability of neural networks for improving process operability was investigated by Di Massimo et al (1990). Hoskins and Himmelblau (1988) described the desirable characteristics of neural networks for knowledge representation in chemical engineering processes. Their paper illustrated how an artificial neural network could be used to learn and successfully discriminate amongst faults. Additionally, fault diagnosis and the development of intelligent control systems was discussed in Bavarian (1988). Birky and McAvoy (1989) presented an application where a neural network was used to learn the design of control configurations for distillation columns. They were able to demonstrate that their approach was an effective and efficient means to extract process knowledge.

In some situations, techniques based on the use of an NNM may offer significant advantages over conventional model based techniques. For instance, if the NNM is sufficiently accurate, it could theoretically be used in place of an on-line analyser. Indeed, such a philosophy may be used to provide more frequent measurements than could be achieved by hardware instrumentation. This is advantageous from the control viewpoint as the feedback signals will not be subject to measurement delays. Consequently, significant improvements in control performance can be expected. A tentative exposition of an inferential controller comprising a neural network model and a simple PI algorithm is attempted here. The NNM is used to provide estimates of 'difficult-to-measure' controlled variables by inference from other easily measured outputs. These estimates are then used for feedback control.

PROCESS MODELLING VIA ARTIFICIAL NEURAL NETWORKS.

In almost all of the work cited above, an NNM is used in place of a conventional model. The accuracy of the NNM model may be influenced by altering the topology (structure) of the network. It is the topology of the network, together with the neuron processing function, which impart to an ANN, its powerful signal processing capabilities. Although a number of ANN architectures have been proposed (see Lippmann, 1987), the 'feedforward' ANN (FANN) is by far the most widely applied. Indeed, Cybenko (1989) has recently claimed that any continuous function can be approximated arbitrarily well on a compact set by a FANN, comprising two hidden layers and a fixed continuous non-linearity. This result essentially states that a FANN could be confidently used to model a wide range of non-linear relationships. The implications of this statement are therefore considerable. In view of this, subsequent discussions will therefore be restricted to FANNs.

Feedforward Artificial Neural Networks

The architecture of a typical FANN is shown in Fig.1. The nodes in the different layers of the network represent 'neuron-like' processing elements. There is always an input and an output layer. The number of neurons in both these layers depend on the respective number of inputs and outputs being considered. In contrast, hidden layers may vary from one to any finite number, depending on specification. The number of neurons in each hidden layer is also a user specification. It is this hidden layer structure which essentially defines the topology of a FANN.

The neurons in the input layer do not perform data processing functions. They merely provide a means by which scaled data is introduced into the network. These signals are then 'fed forward' through the network via the connections, through hidden layers, and eventually to the final output layer. Each interconnection has associated with it, a weight which modifies the strength of the signal flowing along that path. Thus, with the exception of the neurons in the input layer, inputs to each neuron is a weighted sum of the outputs from neurons in the previous layer.

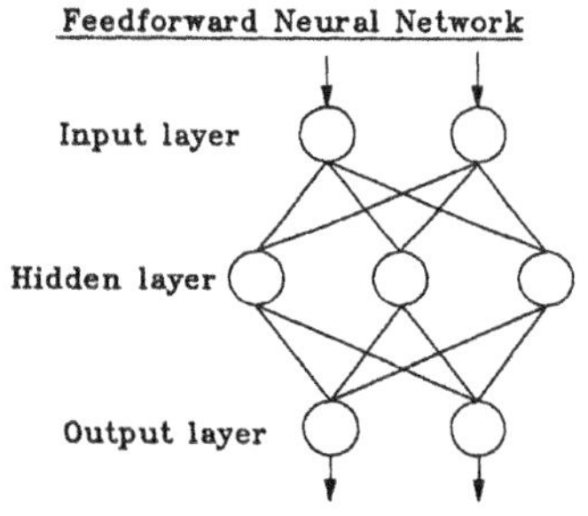

Fig. 1. Schematic of Feedforward Neural Network Architecture.

For example, if the information from the ith neuron in the j-1th layer, to the kth neuron in the jth layer is $I_{j-1,i}$, then the total input to the kth neuron in the jth layer is given by:

$$\alpha_{j,k} = d_{j,k} + \sum_{i=1}^{n} w_{j-1,i,k} I_{j-1,i} \qquad (1)$$

where $d_{j,k}$ is a bias term and $w_{j-1,i,k}$ is the weight which are associated with each interconnection. The output of each node is obtained by passing the weighted sum, $\alpha_{j,k}$, through a nonlinear operator. This is typically a sigmoidal function, the simplest of which has the mathematical description:

$$I_{j,k} = 1/(1+\exp(-\alpha_{j,k})) \qquad (2)$$

Within an ANN, this function provides the network with the ability to represent nonlinear relationships. Additionally, note that the magnitude of bias term in Eq.(1) effectively determines the co-ordinate space of the input-output mapping. This implies that the network has a greater potential to characterise the structure of the nonlinearities: a highly desirable feature.

To develop a process model using the neural network approach, the topology of the network must first be declared. The convention used in referring to a network with a specific topology follows that adopted by Bremmerman and Anderson (1989). For example, a FANN with 3 input neurons, 2 hidden layers with 5 and 9 neurons respectively, and 2 neurons in the output layer will be referred to as a (3-5-9-2) network. A (2-10-1) network, thus refers to a FANN with 2 neurons in the input layer, 10 neurons in 1 hidden layer, and 1 neuron in the output layer.

Algorithms for Network Training (Weight Selection)

Having specified the network topology, a set of input-output data is used to 'train' the network, ie. to determine the weights (and the bias terms) associated with each interconnection. The data is propagated forward through the network to produce an output which is compared with the corresponding output in the data set, hence generating an error. This error is minimised by adjusting the network weights and the bias terms, and may involve many passes through the training data set. When no further decrease in error is possible, the network is deemed to have 'converged', and the last set of weights are retained as the parameters of the NNM. For a given topology, therefore, the magnitudes of the weights define the characteristics of the network.

A numerical search technique is applied to determine these weights as this procedure is usually not amenable to analytical solution. Clearly, this may be regarded as a nonlinear optimisation problem, where the objective function for the optimisation is written as:

$$V(\Theta,t) = \tfrac{1}{2} \Sigma\, E(\Theta,t)^2 \qquad (3)$$

where 'Θ' is a vector of network weights, 'E' is the output prediction error and 't' is time.

A popular solution technique is the 'back-error propagation' algorithm (Rumelhart and McClelland, 1986), which is a distributed gradient descent technique. In this approach, weights in the jth layer are adjusted by making use of locally available information, and a quantity which is 'back-propagated' from neurons in the $j+1^{th}$ layer. This simple procedure only makes use of the Jacobian of the objective function to determine the search direction. A potentially more appealing technique, therefore, would be a Newton-like algorithm which includes second derivative information. Indeed, recently Sbarbaro (1989) employed an approach similar to that utilised in 'classical' recursive system identification algorithms (eg. Ljung and Soderstrom, 1983), and showed that faster network convergence could be achieved. The increase in computational overheads, due to the matrix inversion, is balanced by improved convergence speeds.

An alternative approach, termed the 'chemotaxis algorithm', was proposed by Bremermann and Anderson (1989). Postulating that weight adjustments occur in a random manner, and that weight changes follow a multivariate Gaussian distribution with zero-mean, their algorithm adjusts weights by adding Gaussian distributed random values to old weights. The new weights are accepted if the resulting prediction error is smaller than that obtained using the previous set of weights. This procedure is repeated until the reduction in error is negligible.

In this contribution, attention will be confined to this chemotaxis algorithm: a comparison between the back-error propagation and the chemotaxis algorithm may be found elsewhere, (Willis et al, 1990).

The Chemotaxis Algorithm (Bremmermann and Anderson, 1989)

- Step I — Initialise weights with small random values
- Step II — Present the inputs, and propagate data forward to obtain the predicted outputs
- Step III — Determine the cost of the objective function, E_1, over the whole data set.
- Step IV — Generate a Gaussian distributed random vector.
- Step V — Increment the weights with random vector.
- Step VI — Calculate the objective function, E_2, based on the new weights.
- Step VII — If E_2 is smaller than E_1, then retain the modified weights, set E_1 equal to E_2, and go to Step V. If E_2 is larger than E_1, then goto Step IV.

Note that during the minimisation, the allowable variance of the increments may be adjusted to assist network convergence.

Dynamic Neural Networks

The ANNs discussed above, merely perform a non-linear mapping between inputs and outputs. Dynamics are not inherently included within their structures, whilst in many practical situations, dynamic relationships exist between inputs and outputs. As such, the ANNs may fail to capture the essential characteristics of the system. Although, dynamics can be introduced in a rather inelegant manner by making use of time histories of the data, a rather more attractive approach is inspired by analogies with biological systems. Studies by Holden (1976) suggest that dynamic behaviour is an essential element of the neural processing function. It has also been suggested that a first-order low-pass filter may provide the appropriate representation of the dynamic characteristics (Terzuolo et al, 1969). The introduction of these filters is relatively straightforward, with the output of the neuron being transformed in the following manner:

$$y^f(t) = \Omega y^f(t-1) + (1-\Omega)y(t) \quad 0 \le \Omega \le 1 \qquad (4)$$

Suitable values of filter time constants cannot be specified a priori, and thus the problem becomes one of determining 'Ω' in conjunction with the network weights. The chemotaxis approach does not require modification to enable incorporation of filter dynamics: the filter time constants are determined in the same manner as network weights.

INFERENTIAL MEASUREMENT

There are many industrial situations where infrequent sampling of the 'controlled' process output can present potential operability problems. Linear adaptive estimators have been employed to provide 'fast' inferences of variables that are 'difficult to measure' (Tham et al, 1989). Although results from industrial evaluations have been promising, it is suggested that, due to their ability to capture nonlinear characteristics, the use of NNMs may provide improved estimation performances. The following sections present the results from a recent evaluation study.

Estimation of product composition in a high purity distillation.

Distillation columns, especially those with high purity products, can exhibit highly nonlinear dynamics, resulting in controllability problems. These difficulties may be exacerbated by the measurement delays due to the use of on-line composition analysers. The column being considered here is an industrial demethaniser, where the control objective is to regulate the composition of the high purity top product stream by varying reflux flow rate. Apart from top product composition, which is subject to an analyser cycle time of 20 mins., all other process variables are available at 5 mins. intervals. The aim is to use fast secondary measurements, here, column overheads vapour temperature and reflux flow rate, to provide an estimate of a slow primary measurement (top product composition) every 5 mins. Since column vapour temperature can be overly sensitive to disturbances which do not necessarily affect product composition, the use of a tray liquid temperature would be preferable. However, this choice of secondary variable was dictated by availability at the time of the tests.

The neural estimator was based on a (2-9-9-1) network. Additionally, 1st-order low-pass filters were associated with each neuron and the filter constants were determined simultaneously with network weights. The resulting NNM produced the results shown in Fig. 2. Clearly, good estimates have been achieved.

INFERENTIAL CONTROL VIA ANNs.

With the availability of 'fast' and accurate product quality estimates, the option of closed loop 'inferential' control instantly becomes feasible. The effectiveness of such a strategy has been demonstrated by Guilandoust et al (1987, 1988), where adaptive linear algorithms were used to provide inferred estimates of the controlled output for feedback control. Here, the practicality of an NNM based inferential control scheme is explored via nonlinear simulation.

The process under consideration is a nonlinear simulation of the 10 stage pilot plant column, installed at the University of Alberta, Canada. It separates a 50-50 wt.% methanol-water feed mixture which is introduced at a rate of 18.23g/s into the column on the 4th tray. Bottom product composition is measured by a gas-chromatograph which has a cycle time of 3 mins. The control objective is to use steam flow rate to regulate bottom product composition to 5 wt.% methanol, when the column is disturbed by step changes in feed flow rate.

The training data set consists of paired values of the input variable, ie. temperature of the liquid on tray 2 (the 2nd tray from the reboiler) and the output variable, ie. the corresponding bottom product composition. The data was collected at 3 mins. intervals, this being the cycle time of the gas chromatograph. The chemotaxis algorithm was used to determined both the weights and filter time constants of a (1-4-4-1) network.

Fixed parameter PI controllers were fine-tuned to obtain responses with minimum integral of absolute error. Comparative performances of PI feedback control and NNM based inferential control, under major disturbance conditions, were assessed by subjecting the column to a sequence of step disturbances in feed flow rate, viz. a 10% decrease from steady state followed by a 10% increase back to steady state. As reported in Guilandoust et al (1987,1988), the responses of bottom product composition and tray liquid temperatures, exhibit quite different gains and time-constants, depending on the direction of inputs. Thus, the NNM must be able to characterise these nonlinear effects in order to provide accurate composition estimates. Moreover, for the inferential control technique to be practicable, the NNM is required to provide composition estimates at a rate faster than that obtainable from the gas chromatograph. This is achieved by introducing into the NNM, tray 2 liquid temperatures sampled every 0.5 mins.

The inferential controlled response, using a PI controller (K_c=2.2 and T_I=20), together with composition estimates, are shown in Fig. 3. It can be observed that although the estimate and actual composition compare favourably, there is a difference between the two values. This is because the neural estimator is operating in the open loop, ie. the estimates are not corrected by feedback of measured composition. Thus, if the estimates are used for control, offsets between the actual process output and the desired value will occur. However, offset free inferential control can be achieved by treating the elimination of estimation errors as a control problem. Here, whenever a new composition value is available, ie. every 3 mins., the estimation error is calculated and used to correct subsequent estimates. This technique is based on the Internal Model Control (IMC) strategy of Garcia and Morari (1982). The success of this modification is demonstrated in Fig.3, where it may be observed that it is the actual composition that is at setpoint.

Fig. 4 demonstrates the effect of using the actual (analyser delayed) composition values for control. The oscillatory responses observed are a direct consequence of using the same control settings as in the previous case. Clearly these parameters are unsuitable for a feedback signal which suffers from analyser delays. Reducing the proportional gain to 1.57 resulted in the response shown in Fig. 5, where the response shown in Fig. 3 has also been superimposed. Although a much improved performance has been achieved, comparison with the inferential controlled response will show that there is an increase in peak overshoot of more than 1.0 wt.% methanol.

The performance of the NNM based inferential control scheme is clearly better. Using the output of the neural estimator for feedback control may be regarded as implicitly providing deadtime compensation in the closed loop. This then permits the use of higher gain control. Moreover, as changes in feed flow rate affect tray 2 temperature before bottom product composition, the estimates will contain information about the impending effects of the disturbance. As a result, disturbance rejection responses are superior to that obtained using conventional control.

CONCLUDING REMARKS

In this contribution, the applicability of neural networks as a potential aid in the development of an inferential estimator was explored. Of particular interest was the ability of the neural estimator to provide a 'fast' inference of a 'difficult to measure' process output, from other easily measured variables. It was demonstrated how significant improvements in process regulation could be achieved if the estimates produced by the neural estimator were used as feedback signals for control. This is possible because the use of secondary variables means that the effects of load disturbances are anticipated in a feedforward sense.

ACKNOWLEDGEMENTS

The support of the Dept. of Chemical and Process Engnrg., Uni. of Newcastle upon Tyne; and ICI Engnrg. are gratefully acknowledged.

REFERENCES

Bavarian, B. (1988). Introduction to neural networks for intelligent control, IEEE Control Systems Magazine, April, 3-7.

Birky, G.J. and McAvoy, T.J. (1989). A neural net to learn the design of distillation controls. Preprints IFAC Symp. Dycord+89, Maastricht, The Netherlands, Aug. 21-23, pp205-213

Bremermann, H.J. and Anderson, R.W. (1989). An alternative to Back-Propagation: a simple rule for synaptic modification for neural net training and memory. Internal Report, Dept. of Mathematics, Uni. of California, Berkeley.

Cybenko, G. (1989). Continuous value neural networks with two hidden layers are sufficient. Internal report, Dept. of Comp. Sci. Tufts Univ. Medford.

Di Massimo, C., Willis, M.J., Montague, G.A. Kambhampati, C., Hofland, A.G., Tham, M.T. and Morris, A.J. (1990). On the applicability of neural networks in chemical process control. To be presented at AIChE Annual Meeting, Chicago.

Garcia, C. and Morari, M. (1982). Internal Model Control, Pt. 1, A unifying review and some new results. Ind. Eng. Chem. Process Des. Dev.,21, 308-323.

Guilandoust, M.T., Morris, A.J. and Tham, M.T. (1987). Adaptive Inferential Control. Proc.IEE,Pt.D. 134.

Guilandoust, M.T., Morris, A.J. and Tham, M.T. (1988). An Adaptive Estimation Algorithm for Inferential Control. Ind. Eng.Chem. and Res., 27, 1658-1664.

Holden, A.V. (1976). Models of the stochastic activity of neurones, Springer Verlag.

Hoskins, J.C. and Himmelblau (1988). Artificial Neural Network models for knowledge representation in Chemical Engineering. Comput. Chem. Engng, 12, 9/10, 881-890.

Lant, P.A., Willis, M.J., Montague, G.A., Tham, M.T. and Morris, A.J. (1990). A Comparison of Adaptive Estimation with Neural based techniques for Bioprocess Application. Preprints ACC, San Diego.pp 2173-2178.

Lippmann, R.P. (1987). An Introduction to Computing with Neural Nets. IEEE ASSP Magazine, April.

Ljung, L. and Soderstrom, T. (1983). Theory and Practice of Recursive Identification', MIT Press.

Montague, G.A., Hofland, A.G., Lant, P.A., Di Massimo, C., Saunders, A., Tham, M.T. and Morris, A.J. (1989). Model Based Estimation and Control: Adaptive filtering, Nonlinear observers and Neural networks', Proc. 3rd-Int. Symp. 'Control for Profit', Newcastle-upon-Tyne.

Rumelhart, D.E. and McClelland, J.L. (1986). Parallel Distributed Processing: Explorations in the Microstructure of Cognition. Vol.1: Foundations, MIT Press, Cambridge.

Sbarbaro, D. (1989). Neural nets and nonlinear system identification. Control Engineering Report 89.9., Uni. Glasgow.

Terzuolo, C.A., McKeen, T.A., Poppele, R.E. and Rosenthal, N.P. (1969). 'Impulse trains, coding and decoding'. In Systems analysis to neurophysiological problems, Ed. Terzuolo, University of Minnesota, Minneapolis, pp86-91.

Tham, M.T., Morris, A.J. and Montague, G.A. (1989) Soft sensing: A solution to the problem of measurement delays. Chem. Eng. Res. and Des., 67, 6, 547-554.

Willis, M.J., Di Massimo C., Montague, G.A., Tham, M.T. and Morris, A.J. (1990). On Artificial Neural Networks in Process Engineering, Submitted to Proc. IEE. Pt. D

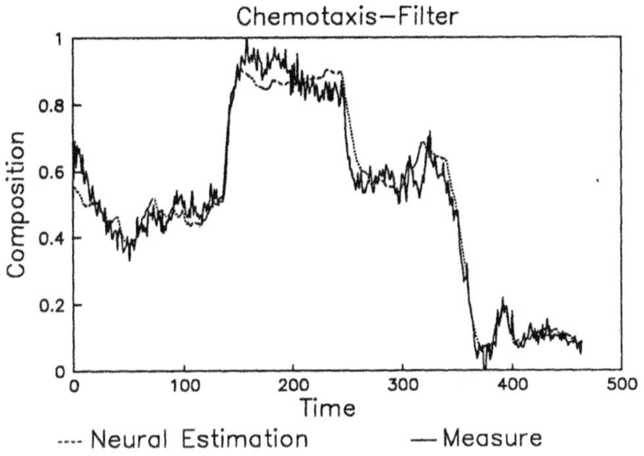

Fig.2. Neural Network Estimator Applied to High Purity Column.

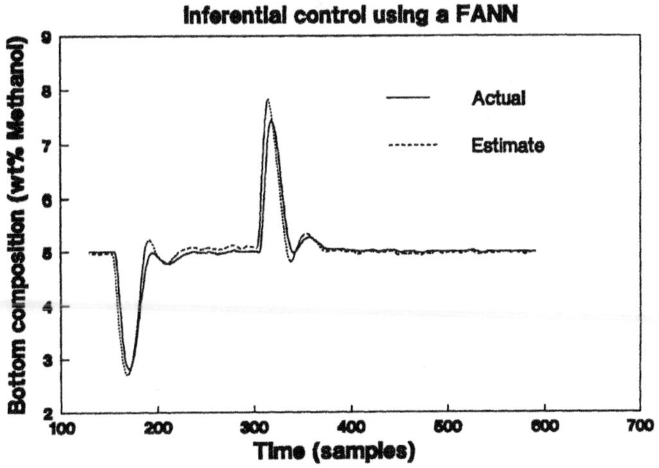

Fig.3 Inferential Control using a Neural Network.

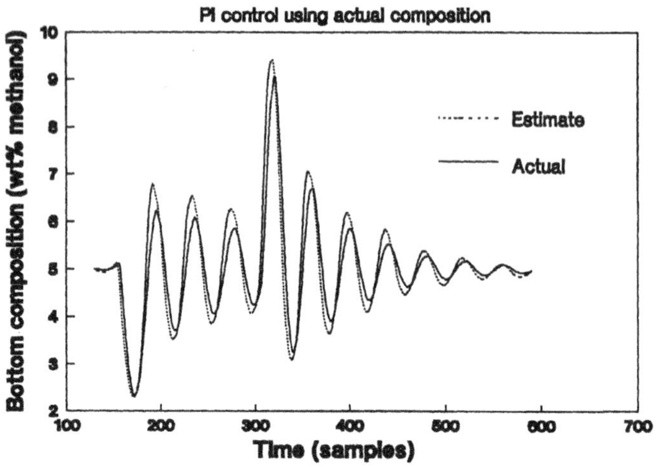

Fig.4 Control using analyser delayed value.

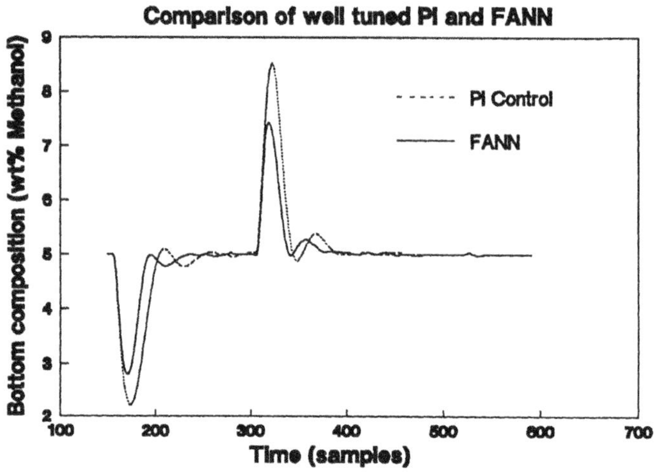

Fig.5 Detuned control using analyser delayed value.

ROBOT MOTION PLANNING AND CONTROL USING NEURAL NETWORKS

M. H. Ang, Jr.*, G. B. Andeen and V. Subramaniam*****

**Dept. of Mechanical and Production Engineering, National University of Singapore, Singapore 0511*
***SRI International, Menlo Park, CA 94025, USA*
****Dept. of Mechanical Engineering, Massachusetts Institute of Technology, Cambridge, MA 02139, USA*

Abstract. A method for motion path selection and control using a neural network that is especially applicable for manipulators with compliant limbs is proposed. The "trainability" of the neural network allows the design of path planners and controllers that have smooth force/torque-time profiles. For nonlinear control problems, performance is much better if the overall control function is decomposed into an acceleration controller neural network that outputs the required acceleration, and a lineariser neural network which compensates for nonlinearities. The feasibility of our method has been verified through simulations for both linear and nonlinear problems of one degree-of-freedom. Point-to-point control and smooth motion has been achieved with some overshoot.

Keywords. Robots; neural networks; motion control; motion planning; control applications; robot control.

NEURAL NETWORKS IN ROBOTICS

Neural networks offer several advantages over conventional computing architectures. Calculations can be carried out in parallel and any desired input-output behavior can be emulated through training. Neural networks can be thought of as "black boxes" which map input to output. The mapping is done without the need for explicit rules for computing the output from the input. In fact, there may be no analytic function that evaluates the output from the input. Instead, the neural network learns the desired mapping by a process called *training*. Training involves presenting the network with a set of input-output pattern pairs. The training set may consist of the desired behavior that the neural network should emulate and for which a desired mapping is achieved. Or the training set may be the actual inputs and outputs of a physical system, from which the neural network represents and learns a model of the system.

Many uses of neural networks have been uncovered in robotics. They have been employed to solve the inverse kinematics problem of serial robotic manipulators (Guez and Ahmad, 1988). The end-effector position is the input while the joint position is the output of the neural network. In this problem, the inverse mapping (inverse kinematics) from end-effector position to joint position is difficult while the forward mapping (forward kinematics) from joint position to end-effector position is straightforward to compute. Training patterns are easily generated through forward kinematics computations.

We find neural networks attractive for robot control. Their inherently parallel architecture makes real-time robot control feasible and their trainability makes them suitable for controlling plants with unknown models and for adaptive control. Furthermore, their learning ability allows the realisation of any desired controller behavior.

One control application involves using the neural network to emulate the system dynamics and to feedforward the systems dynamics as shown in Fig. 1a. The input in this case is a small change from the present position as in following a path as well as the present state of the robot. The neural network then provides a correction to the force from the controller. The neural network is continually trained over the space of small changes as the "ordinary" feedback controller is functioning. Gradually the neural network takes over the control of the robot as it makes more nearly exact choices and the feedback control functions less. This kind of adaptive control is suggested by (Miller and others, 1987).

Yet another technique is to use the neural network to make the performance of the system appear linear to the controller as shown in Fig. 1b (Ang and Andeen, 1990). The controller implements the *computed-torque* algorithm to provide the desired acceleration signal to the neural network. The neural network performs the inverse dynamics evaluation to provide the force required to give the desired acceleration taking into consideration nonlinearities based on the state feedback to the network. Such a scheme requires the network to be pretrained to learn the robot dynamic model offline. Training can also be done while the manipulator is operating if the network is set up to give a signal to correct the linear signal as shown in Fig. 1c.

OUR APPROACH TO THE ROBOT MOTION CONTROL PROBLEM

The robot motion control problem involves the computation of the trajectory of actuating joint forces/torques required to accomplish the desired robot motion. The desired motion may be to move to a final position at zero velocity from the initial position at rest, or it may be that of following a

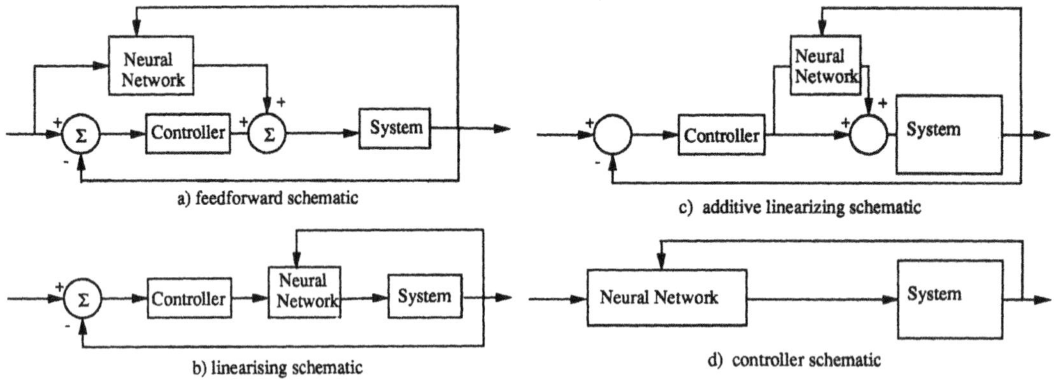

Fig. 1. Neural networks in robot motion control.

specific motion trajectory. We choose to be concerned with smoothness of the applied forces/torques rather than with path following. Our emphasis is on accomplishing a move from an initial to a final position with smooth force/torque–time profiles. Smooth force/torque–time profiles will excite a minimum of resonances in the manipulator, an increasing concern as manipulator designs become more lightweight and compliant (Bayo and Paden, 1987). Instead of trying to follow a path (which must be differentiated twice to arrive at a force/torque, and which may demand large force/torque spikes when a small change is made in the path), the path follows from the kind of force/torque profile.

NEURAL NETWORK CONTROLLER

We are exploring the feasibility of having neural networks take over the entire control function as shown in Fig. 1d. The concept may be similar to the feedforward compensator, but there is no feedback through an alternative compensator. We are not limited to small changes in reference values. When the moves are large, the neural network essentially plans the trajectory. A wide variety of desired trajectory planning behaviors can be achieved depending on how the neural network is trained. The neural network serves as the robot controller and can therefore be trained according to the desired controller performance we want.

The neural network is trained using desired force/torque-time trajectories. The force/torque trajectory provides a move whose characteristics are recorded and used for learning. Given two points along a path that has resulted from a training trajectory, the network selects the force profile that produced that motion path, and selects the actual force/torque to be used. Nearby paths, not specifically programmed, can also be followed. The controller responds to the updated position and provides another force/torque profile and value. The controller responds to the disturbances essentially by finding the new force trajectory from the disturbed state to the desired state.

Force/Torque–Time Profiles

The force/torque-time trajectories are designed to be smooth "sinusoidal-like" functions. The magnitude of the force/torque F gradually increases at the start of the move and then decreases towards the end of the move. We have found it useful to take functions that can be described easily in terms of a few parameters such as amplitude and move time. For linear motion control problems starting from zero initial conditions, the following third order polynomial would qualify for a smooth force/torque time-trajectory:

$$F = t^3 - \frac{3P}{2}t^2 + \frac{P^2}{2}t \tag{1}$$

for $0 \leq t \leq P$, and $F = 0$ elsewhere as shown in Fig. 2. Typical motions resulting from this family of trajectories are shown in Fig. 3. P is the sole parameter that describes the trajectory and can be thought of as the move time. Note that $F = 0$ at $t = 0$ and $t = P$. Increasing the value of P gives force trajectories of increasing magnitude and duration. It is important to avoid multiple trajectory solutions that would bring the current state (say state B) to the same final state, such as shown in Fig. 4. It can be shown that the trajectories (1) in Fig. 3 avoid the multiple solution problem (Fig. 4) since the trajectories do not cross each other.

Obviously for nonlinear systems where a holding force/torque may be required at the start and the end of the move, the trajectory (1) needs to be offset by the corresponding holding force/torque profile.

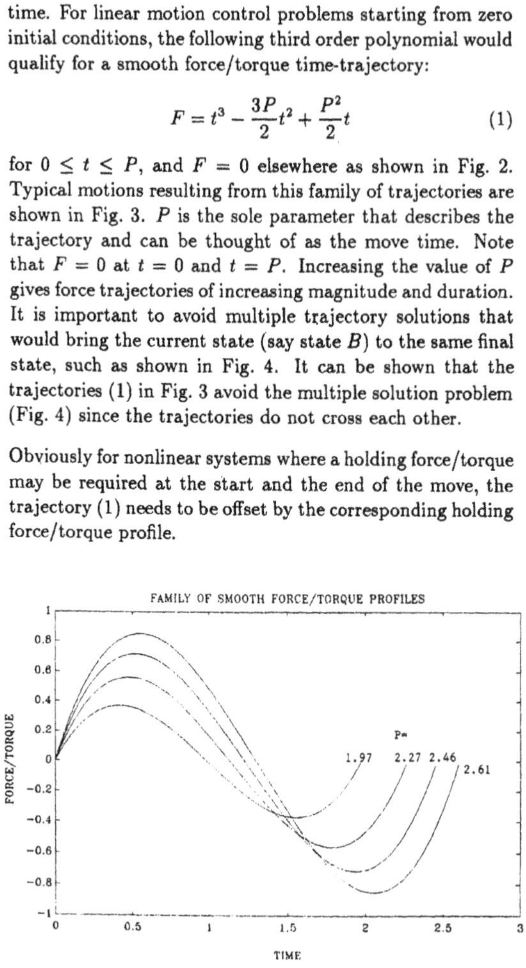

Fig. 2. Desired smooth force/torque trajectory.

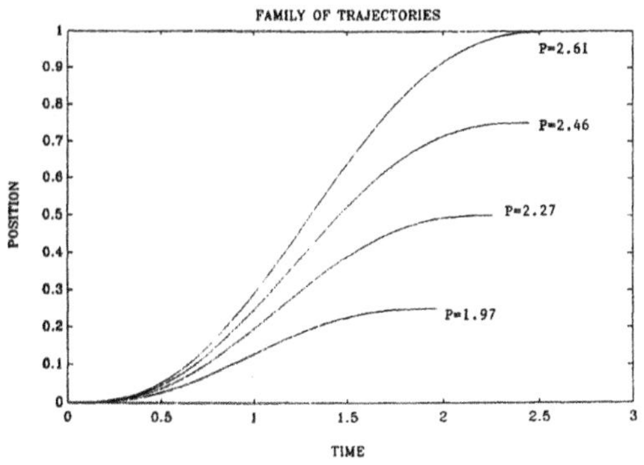

Fig. 3. Family of motion trajectories.

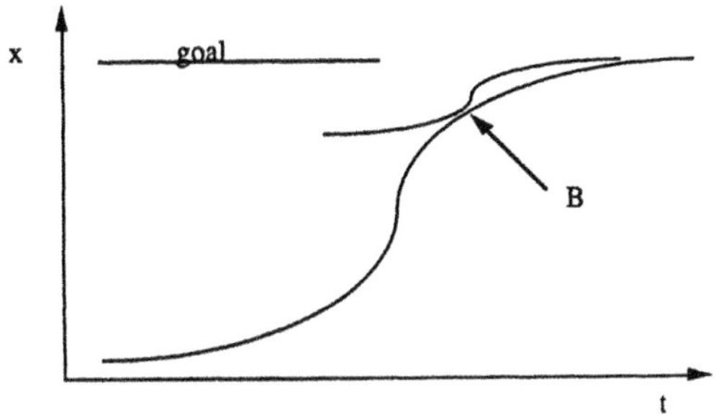

Fig. 4. Multiple Solutions that arrive at the same final state.

Training the Neural Network

The neural network controller is trained to provide smooth actuating force/torques like (1). The input to the network is the present state (position and velocity) and the final position at the end of the move; the output of the network is the force/torque signal (Fig. 5a). As the neural network is executing control functions, the force/torque signal should follow (1) with P as the only degree-of-freedom. The trajectory (1) is applied to the system (Fig. 5b) to generate the resulting motion trajectory. Two states at $t = t_A$ and $t = t_B$ ($t_B > t_A$) are randomly selected from the resulting motion trajectory (Fig. 3). These states together with the force/torque at $t = t_A$ form one training pattern. One trajectory (1) therefore generates a set of training data. Several trajectories (using different P's) can be used to generate several sets of training patterns. The training is of course done off line. Once the training has achieved a tolerable accuracy level, the neural network can then be used online for control.

Linear System Simulation: One Mass Model with No Gravity

We consider first the linear problem of moving a cart of mass 1 kg (force=acceleration) on a horizontal rail with no gravity, as shown in Fig. 6. The initial position is $x = 0$ with zero initial velocity and the final position is x_f, also at zero velocity. For the force/torque trajectory (1), the corresponding motion can be obtained by integrating the trajectory twice to obtain:

$$x = \frac{t^5}{20} - P\frac{t^4}{8} + P^2\frac{t^3}{12} \qquad (2)$$

The value of P can be computed from the final position using (2) since $x = x_f$ at $t = P$:

$$P = (120x_f)^{0.2} \qquad (3)$$

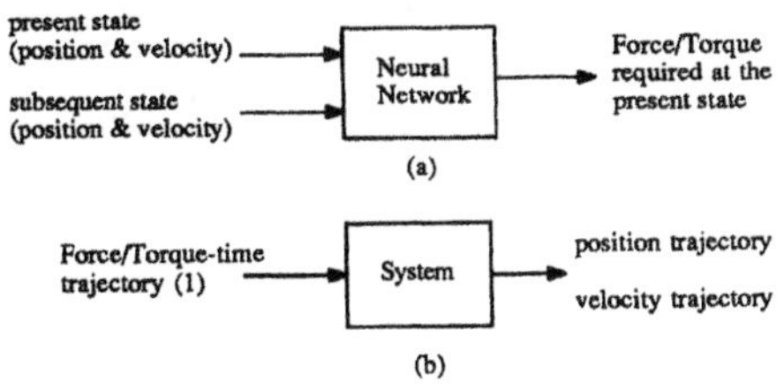

Fig. 5. Training the neural network.

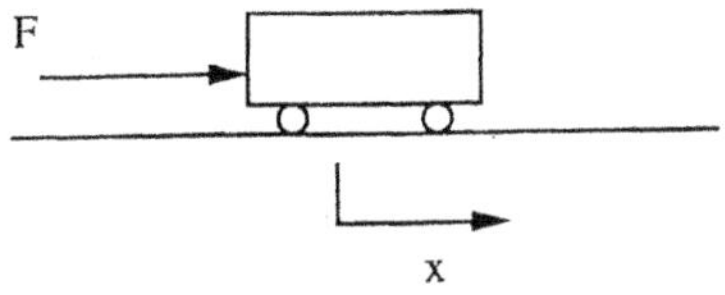

Fig. 6. Linear problem: cart on a rail.

Thirty random trajectories (P's) corresponding to $0 < x_f \leq 1.0$ were used to generate the training data. The standard back-propagation network is used with four inputs (current position and velocity, x_f and final velocity of 0) and one output (force). Two hidden layers with sigmoid functions were employed; the first and second hidden layers had 6 and 8 neurons respectively. The weights in the neural network were updated using the generalised delta rule with learning and momentum coefficients of 0.145 and 0.31 respectively. After training on the order of a few hundred thousand cycles, the neural network controller was able to achieve a typical performance as shown in Fig. 7. The response is similar to a critically damped case with some (maybe) undesirable overshoot. The mass of the cart was changed to simulate plant model changes and the resulting motion is also shown in Fig. 7. The performance against positive and negative disturbances were also simulated. The simulations show that the neural network can provide the entire control function without the need for "conventional" feedback gains or compensators. Furthermore, the controller seems robust towards modeling errors and disturbances.

For controlling nonlinear systems, training the network with force/torque-time profiles is difficult. Such a system was simulated with the cart going over a hill with gravity effects and the performance was unsatisfactory.

NEURAL NETWORK NONLINEAR CONTROLLERS

Training the neural network controller with smooth force/-torque profiles proved difficult for nonlinear systems. Perhaps because not only is the neural network required to plan the path according to the force/torque profiles, but also has to learn the nonlinearities of the system it is controlling. It is therefore logical to hypothesise that the neural network controller will work better if it be decomposed to handle two functions: the first function is to compute the acceleration required to bring the current state to the final state (in effect it does path planning), and the second function computes the required actuating force/torque given the state information and the acceleration required. Two neural networks are therefore employed to serve the two aforementioned functions, as shown in Fig. 8.

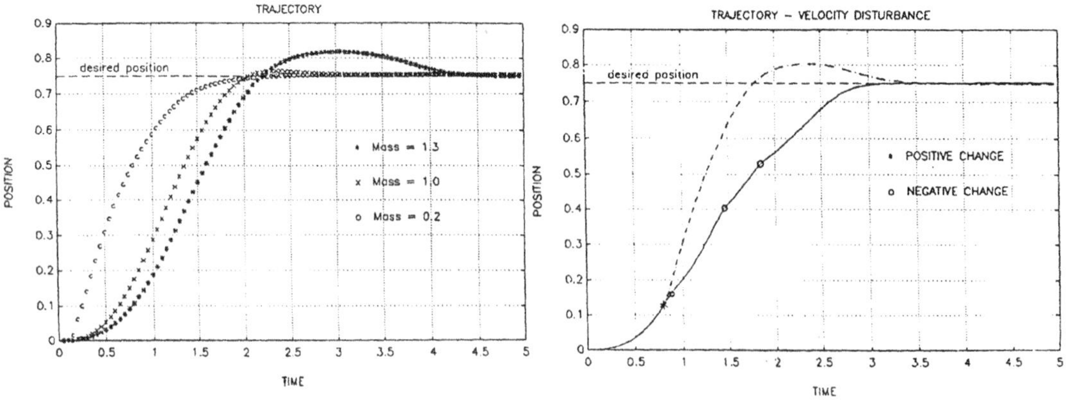

Fig. 7. Results for cart problem.

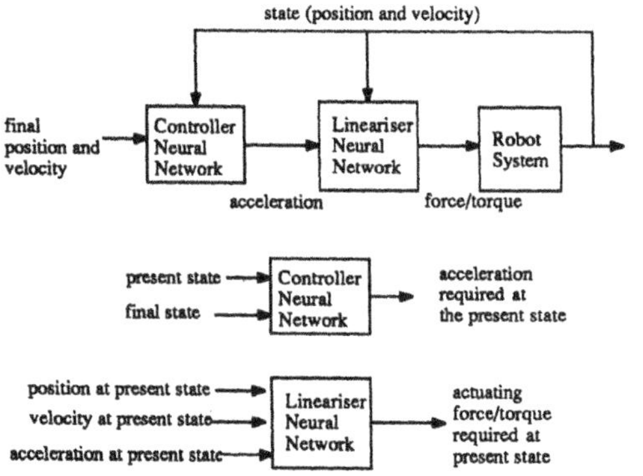

Fig. 8. General-purpose neural-network based motion controller.

The first neural network is referred to as the neural-network-based *controller* and outputs the required acceleration signal given the present and final states. Having the objective of smooth motions, it is desired to have smooth acceleration-time profiles. Equation (1) can also be used to serve this purpose with the force term replaced by acceleration:

$$\text{acceleration} = t^3 - \frac{3P}{2}t^2 + \frac{P^2}{2}t. \qquad (4)$$

The *controller* neural network is then trained in the same manner as in the previous section. Since the output of the controller is acceleration, the robot motion can be computed independent of the plant. The controller can therefore be trained to provide smooth acceleration-time profiles (1) irregardless of the plant it is controlling. The same network can be used for different plants as long as the same acceleration behavior is desired. That is, (3) remains valid as long as the desired acceleration profile is (4).

The second neural network outputs the required actuating force/torque given the required acceleration and present state. The input to this network is then position, velocity and acceleration; and the output is the required force/torque. In effect the network emulates the *inverse dynamics* evaluation and is therefore referred to as the *lineariser* neural network since its implementation inside the control loop linearises the system (Tourassis, 1988). Furthermore, the lineariser neural network has been demonstrated to be able to adapt to plant model changes (e.g., payload variations, unmodeled dynamics) through online training, thus making it suitable for adaptive control (Ang and Andeen, 1990).

Nonlinear System Simulation: One DOF Robot with Gravity

We now consider the nonlinear problem of rotating a link carrying a point mass (Fig. 9) from the initial angular position of zero to a final angular position, starting and ending at zero velocities. This single mass model is suitable as a model of even a lightweight compliant limb with an end mass where the joint is servoed according to the torque as measured at the bending of the limb (Andeen, 1988). For the neural-network-based *controller* that outputs the required acceleration, we utilise the same network for the one mass model with no gravity, but this time the network is trained using acceleration-time profiles of the family (4). No re-training of the network is needed for this simulation since unit mass was used (force = acceleration) for the linear simulation.

The lineariser neural network was used to learn the robot dynamic model:

$$F = mL^2\ddot{\theta} + mgL\cos\theta \qquad (5)$$

where g is the acceleration due to gravity. For purposes of numerical simulations, the parameters of the robot were chosen to be $m = 0.10$ kg and $L = 1$ m.

We use the back-propagation network with two hidden layers of 15 and 5 neurons respectively (from input to output) to learn the robot dynamic model. The two inputs are the joint position and acceleration while the output is the joint torque required.

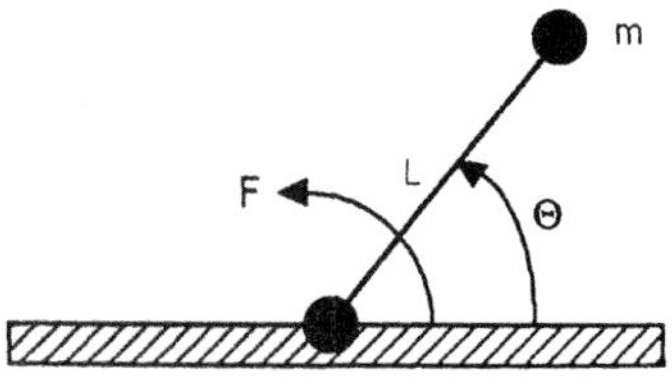

Fig. 9. Nonlinear problem: rotating a link.

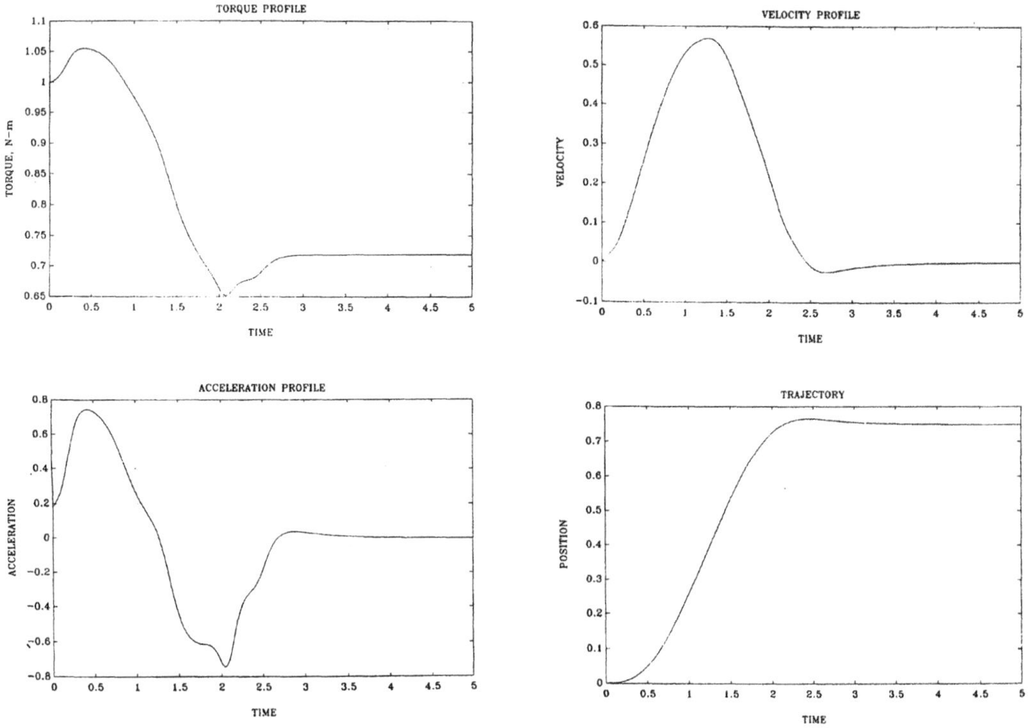

Fig. 10. Results for one-dof robot problem.

The lineariser network is trained using the nominal dynamic parameters ($m = 0.10$ kg and $L = 1$ m) of the robot. The input training data consists of uniformly-distributed random values of joint position and accelerations in the following ranges:

joint position	$0 \leq \theta \leq \pi$ rad
joint acceleration	$-3 \leq \ddot{\theta} \leq 3$ rad/s^2

The output training data consists of the torques computed using (5) from the input training data. The weights in the neural network were updated using the delta rule with learning and momentum coefficients of 0.4 and 0.2 respectively. After training in the order of a million cycles, a worse case accuracy of 0.0015 N·m was achieved for a torque range of ± 1.28 N·m.

We now employ the *controller* and *lineariser* neural networks to accomplish a move from $\theta = 0$ to $\theta = 0.75$ rad according to Fig. 8. Typical controller performance is shown in Fig. 10. The actual acceleration command and torque profiles are also shown in Fig. 10. We note that the acceleration starts from a low value, followed by an increase in magnitude, and then a decrease to zero at the end of the move. This desired behavior is different from "conventional" control algorithms wherein the initial acceleration is typically a very large value since the initial position is in general far from the final position (e.g., computed-torque algorithm). The initial jerks are expected to be minimised with these acceleration profiles. It should be highlighted that the resulting force/torque profile is not exactly (1); the deviation depends upon the nonlinearities of the system.

CONCLUSIONS

The "trainability" of the neural network allows the design of path planners and controllers that have smooth force/-torque-time profiles. Simulation results indicate that neural networks can be used as the controller to select force/torque- and/or acceleration-time profiles according to training to arrive at remote positions in the presence of disturbances. For nonlinear control problems, performance is much better if the overall control function is decomposed into an acceleration controller neural network that outputs the required acceleration, and a lineariser neural network which compensates for nonlinearities. In addition to the advantages of neural networks, speed, sensor fusion, and trainability, the principal advantage is that the force/torque- and/or acceleration-time trajectories can be selected to suit the equipment. This selection ability would be an advantage in controlling lightweight, flexible manipulators.

REFERENCES

Andeen, G.B., and R. Kornbluh (1988) Design of Compliance in Robotics. In *Proceedings of the IEEE International Conference on Robotics and Automation*. pp 276–281.

Ang Jr., M.H., and G.B. Andeen (1990). Adaptive Robot Control Based on a Neural Network Paradigm. In *International Conference on Automation, Robotics, and Computer Vision (ICARCV'90)*. pp. 437–441.

Bayo, E., and B. Paden (1987). On the Trajectory Generation for Flexible Robots. *Journal of Robotic Systems*, Vol. 4, No. 2, 229–235.

Guez, A., and Z.Ahmad (1988). Solution to the Inverse Kinematics Problem in Robotics by Neural Networks. In *Proceedings of the International Joint Conference on Neural Networks*. pp. 617–624.

Miller III W.T., F.H. Glanz, and L.G. Kraft, III (1987). Application of a General Learning Algorithm to the Control of Robotic Manipulators. *International Journal of Robotics Research*, Vol. 6, No. 2, 84–98.

Tourassis, V.D. (1988). Principles and Design of Model-Based Controllers. *International Journal of Control*, Vol. 47, No. 5, 1267–1275.

ADAPTIVE NONLINEAR CONTROL OF A LABORATORY WATER-GAS SHIFT REACTOR

G. T. Wright and T. F. Edgar

*Dept. of Chemical Engineering, The University of Texas at Austin,
Austin, TX 78712, USA*

Abstract. The objective of this research is to demonstrate that good control of a nonlinear system can be achieved over a large operability region when a periodically adapted low-order nonlinear model is used in a nonlinear control strategy. In this paper, we employ a steady-state model and adaptively implement a variation of nonlinear model predictive control (NMPC) on a water-gas shift reactor via simulation. This approach is computationally less demanding, yet retains many of the attractive features of a dynamic model implementation of NMPC. We further discuss the ramifications of nonlinear control in the absence of estimation, and introduce a simple nonlinear parameter estimation scheme to improve performance.

Keywords. Adaptive control; Predictive control; Nonlinear control systems; Optimization.

INTRODUCTION

Typically, processes encountered in chemical engineering are nonlinear. The severity of the nonlinearities associated with a process determines the type of control algorithms which are most suitable for successful control of the process. Traditionally, chemical systems have been approximated by linear second-order-plus-dead-time models and more recently linear convolution models. These simple transfer function models can be used to design model-based control strategies such as Internal Model Control (IMC: Garcia and Morari, 1982), Dynamic Matrix Control (DMC: Cutler and Ramaker, 1980), and Generalized Predictive Control (GPC: Clarke, Mohtadi, and Tuffs, 1987a,1987b). Use of these low-order linear models for control is often adequate when the process nonlinearities are mild, and plant operation is constrained to a small region about a nominal steady-state. However, for highly nonlinear systems such as nonisothermal fixed-bed chemical reactors, this approach may lead to poor control.

Adaptive control compensates for the inadequacies associated with linear control by periodically updating a feedback control law based upon current and past operating conditions. The use of adaptation in this approach results in an inherently nonlinear control strategy. However, parameter estimates usually have little physical significance, and the control system performs best when the parameters of the system vary slowly relative to the rate at which the states of the system change. In practice, this condition cannot be guaranteed. Furthermore, since the parameters are only locally valid, the model often extrapolates poorly to new operating conditions. Because of these deficiencies, there appears to be a need for further investigation of compensation based on nonlinear models, which is the motivation for this paper.

A number of control strategies have been developed for systems with nonlinear process models. Among them are techniques which exactly linearize many important classes of nonlinear systems based upon concepts from differential geometric theory. Henson and Seborg (1990) provide a comprehensive review of this field and related topics. Another class of techniques is Nonlinear Model Predictive Control (NMPC), which is implemented in either of two ways. The first method employs separate algorithms to solve the differential equations and perform the optimization. This sequential solution and optimization strategy has been reported by Asselmeyer (1985), Morshedi (1986), and Economou and co-workers (1986).

The second, more efficient alternative is to use a simultaneous solution and optimization strategy, which is accomplished by discretizing the model differential equations via orthogonal collocation, for instance. The discretized model equations and the model algebraic equations are then included among other constraints in a nonlinear programming problem (NLP) which may be solved by a variety of techniques such as successive quadratic programming (SQP). Biegler (1984) , Cuthrell and Biegler (1987), and Renfro, Morshedi and Asbjornsen (1987) have implemented this strategy to find optimal open-loop manipulated variable trajectories. Patwardhan, Rawlings and Edgar (1990) extended the open-loop algorithms to incorporate feedback, which improved the robustness of the control scheme to modeling errors and disturbances.

The dimensionality of the NLP resulting from the simultaneous solution and optimization strategy may be large. This can make solving them computationally prohibitive for real-time applications. In many cases, however, the dimensionality of the NLP can often be reduced considerably by employing a steady-state model. In this paper we investigate the ramifications of using a steady-state model in an NMPC framework as outlined by Patwardhan and co-workers (1990). We also employ an on-line parameter estimation scheme which is capable of efficiently estimating parameters which appear nonlinearly in low-order nonlinear process models. We apply the methods to a fixed-bed water-gas shift (WGS) reactor simulation.

THE CONTROLLER

For simplicity, we assume that the nonlinear dynamic process is described by the following time-invariant set of differential/algebraic equations:

$$\frac{d\mathbf{x}}{dt} = \mathbf{f}(\mathbf{x}, \mathbf{u}; \mathbf{p}) \tag{1}$$

$$\mathbf{y} = \mathbf{g}(\mathbf{x}, \mathbf{u}; \mathbf{p}) \tag{2}$$

where $\mathbf{y}$ and $\mathbf{u}$ are controlled and manipulated variables respectively, $\mathbf{x}$ is the state variable vector, and $\mathbf{p}$ is a vector of parameters and modeled, unmeasured disturbances.

A steady-state optimization problem for control of this system is given by:

$$\min_{\mathbf{u}} \Phi(\mathbf{y}, \mathbf{x}, \mathbf{u}; \mathbf{p}) \tag{3}$$

subject to:

(i) Model constraints

$$0 = \mathbf{f}(\mathbf{x}_{ss}, \mathbf{u}; \mathbf{p}) \tag{4}$$

$$\mathbf{y}_{ss} = \mathbf{g}(\mathbf{x}_{ss}, \mathbf{u}; \mathbf{p}) \tag{5}$$

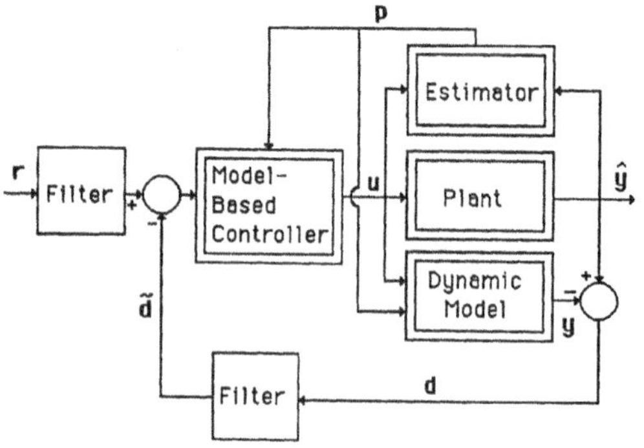

Figure 1: Structure of adaptive model-based controller.

(ii) Bounds on state variables

$$x_l \leq x_{ss} \leq x_u \quad (6)$$

(iii) Bounds on outputs

$$y_l \leq y_{ss} \leq y_u \quad (7)$$

(iv) Bounds on manipulated variables

$$u_l \leq u \leq u_u \quad (8)$$

(v) Bounds on changes in manipulated variables

$$| u - u_{old} | \leq \Delta u_{max} \quad (9)$$

The constrained NLP problem is then solved using SQP.

A block diagram of the nonlinear control structure is shown in Fig. 1. The effect of modeling error is treated as an additive, unmeasured disturbance, and is estimated in a manner similar to DMC (Cutler and Ramaker,1980):

$$\mathbf{d} = \hat{\mathbf{y}} - \mathbf{y} \quad (10)$$

where $\hat{\mathbf{y}}$ is the measured plant output and $\mathbf{y}$ is the predicted model output. This constitutes the feedback portion of the algorithm. If a perfect process model is available, then $\mathbf{d}$ is equal to the additive disturbance in the process output. Of course, this condition can be grossly violated during dynamic transitions if a steady-process model is used to predict the output vector, $\mathbf{y}$. It is, therefore, advantageous to use a dynamic model to calculate the disturbance vector, despite the fact that the optimization is constrained by the steady-state process model. This step is not iterative; it is a simple open-loop simulation and adds little to the computational burden. An exponential filter is also used to further compensate for the oscillatory behavior of $\mathbf{d}$, which could arise due to the absence of dynamic information in the NLP.

The algorithm is constructed as a two-degree of freedom controller. The set-point prefilter block can be used, therefore, to tailor the closed-loop dynamics to achieve the desired response in the servo-problem. The set-point prefilter block is an important component of the algorithm since we do not specifically account for the dynamics of the process in the model-based controller block.

The nonlinear control block consists of the NLP described above. In the absence of constraints except for those of class (i), equations 4 and 5, the objective function is designed to achieve set-point by inverting the plant. In many cases, however, this inversion is not unique as will be demonstrated later in the text. Care must therefore be taken to construct an objective function which leads to the desired process input. In some cases the inverse may not even exist as is the case when a disturbance makes the set-point infeasible. In such instances the plant is inverted in a least squares sense. The following objective functional is used in the simulations presented in this paper:

$$\Phi = (\mathbf{r} - (\tilde{\mathbf{d}}+\mathbf{y}_{ss}))^T\mathbf{Q}_1 (\mathbf{r} - (\tilde{\mathbf{d}}+\mathbf{y}_{ss})) + \mathbf{u}^T\mathbf{Q}_2\mathbf{u} \quad (11)$$

where $\mathbf{r}$ is the vector of set-points; $\tilde{\mathbf{d}}$ is the filtered output disturbance vector, and $(\tilde{\mathbf{d}}+\mathbf{y}_{ss})$ is the corrected model-generated steady-state output.

THE ESTIMATOR

When a fundamental, nonlinear process model is available, it is often desirable to use nonlinear, model-based control. Obviously, the quality of the control is largely a function of the model accuracy. Regardless of the modeling effort, however, plant/model mismatch will always exist. The model deficiencies may be caused by unmeasured disturbances, nonstationary process behavior, or they may be the result of an incomplete model structure. Estimation is, therefore, often a vital component of the model-based control scheme.

Since the estimation problem can be as formidable as the control problem, and since it is often unrealistic to estimate a large number of parameters on-line, a small number of parameters to which the model is highly sensitive should be chosen for estimation. These parameters should preferably represent significant modeling uncertainty. In fact, the parameters to be estimated could be determined by evaluating the parametric sensitivity of the optimal solution of the control problem (Eaton and Rawlings, 1990). Rhinehart and Riggs (1990) have proposed a technique for adjusting one parameter using one measured output according to the following rule:

$$p_t = p_{t-\Delta t} - \alpha \frac{\varepsilon_t}{(\partial\varepsilon/\partial p)_{t-\Delta t}} \quad (12)$$

where

$$\varepsilon_t = \hat{y}_t - y_t \quad (13)$$

$\hat{y}_t$ is the measured output; y_t is the predicted output, and α is a relaxation factor. The sensitivity of the error, ε, to the estimated parameter, p, is evaluated using either a steady-state or a dynamic model. Equations 12 and 13 constitute a single-step Newton iteration in the solution of

$$0 = \hat{y}_t - y_t$$

From a practical perspective this approach is appealing. However, it is limited in the sense that there must be a one-to-one correspondence between parameters to be estimated and measured inputs. That is, extra information is necessarily discarded in the estimation process. Furthermore, in most cases, the largest window size of past data that can be employed is one time step. The natural extension of this idea is to perform one or several complete quasi-Newton iterations in a constrained or unconstrained optimization using one of the many commercially available optimization packages.

If, however, simplicity is a priority, then the following derivation, similar to that presented by Rhinehart and Riggs (1990), permits the unconstrained estimation of nonlinear parameters using a variable window size. This approach is more general, but reduces to their method in some limiting cases. Consider the following cost functional:

$$\min_{\mathbf{p}} J = \frac{1}{2} \sum_{i=1}^{H} \varepsilon_i^T\mathbf{R}\varepsilon_i \quad (14)$$

where

$$\varepsilon_i = y_i - \hat{y}_i \quad (15)$$

and $\mathbf{R}$ is a symmetric, positive-definite weighting matrix. The first-order necessary conditions are given by:

$$\nabla_{\mathbf{p}} J = \sum_{i=1}^{H} \varepsilon_i^T\mathbf{R}\nabla_{\mathbf{p}}\mathbf{y}_i = \mathbf{0}^T \quad (16)$$

To evaluate $\nabla_{\mathbf{p}}\mathbf{y}$, one could employ equations 1 and 2, but this would require that the dynamic sensitivity equations be solved as well. As an alternative, equations 4 and 5 are used, obviating the need to integrate sensitivity equations while retaining much of the nonlinear information. The steady-state parametric sensitivities are given by:

$$\nabla_p y_i = (-\nabla_x g (\nabla_x f)^{-1} \nabla_p f + \nabla_p g)_i \quad (17)$$

It can be shown that the Hessian of the cost functional is given by:

$$\nabla_{pp} J = \sum_{i=1}^{H} \left\{ \nabla_p y_i^T R \nabla_p y_i + \begin{bmatrix} \varepsilon_i^T R \nabla_p (\nabla_p y_i(1)) \\ \vdots \\ \varepsilon_i^T R \nabla_p (\nabla_p y_i(n_p)) \end{bmatrix} \right\} \quad (18)$$

where $\nabla_p y_i(j)$ is the jth column vector of $\nabla_p y_i$, and n_p equals number of parameters. If the Hessian is to be employed in a Newton iteration, it must be positive definite to ensure that a step in a descent direction is taken. If the parameters are sufficiently far from the optimum, there is no guarantee that the Hessian will, in fact, be positive definite. Notice, however, that if the model structure is exact, and if the appropriate parameters are estimated, then at the optimum the second term in the summation of equation (18) will vanish. One might speculate, moreover, that even when the model is inexact, the second term in the summation will be small relative to the first at the optimum provided ε_i is small. We, therefore, neglect this term. What remains is the sum of positive semi-definite matrices. This approximation to the Hessian is therefore, at the very least, always positive semi-definite. To ensure invertibility a constant diagonal matrix, βI, is added to the approximate Hessian (Edgar and Himmelblau, 1988). When this approximation is employed, a quasi-Newton step in the proposed optimization is given by:

$$\Delta p = -\alpha \left[\left\{ \sum_{i=1}^{H} \nabla_p y_i^T R \nabla_p y_i \right\} + \beta I \right]^{-1} \sum_{i=1}^{H} \nabla_p y_i^T R \varepsilon_i$$

If the data horizon, H, is one, then this single-step update is essentially an open-loop, multivariable, normalized MIT-type rule. If, in addition, there is only one measured input, and β is set to zero, this update is equivalent to the update proposed by Rhinehart and Riggs. As a precaution, one should consider limiting the parameter adjustments at each iteration according to the following rule:

$$\Delta p = -\alpha \, \mathrm{sat} \left[\left[\left\{ \sum_{i=1}^{H} \nabla_p y_i^T R \nabla_p y_i \right\} + \beta I \right]^{-1} \left[\sum_{i=1}^{H} \nabla_p y_i^T R \varepsilon_i \right], \gamma \right] \quad (19)$$

where

$$\mathrm{sat}(x, y)_i = \mathrm{sign}(x_i) \min(|x_i|, |y_i|) \quad (20)$$

and γ is a vector containing the maximum step-sizes permitted at each iteration. Absolute limits on the parameters can be set as well.

THE FIXED-BED WATER-GAS SHIFT REACTOR

The water-gas shift (WGS) reaction, shown below, is reversible and mildly exothermic.

$$CO(g) + H_2O \Leftrightarrow CO_2(g) + H_2(g)$$

$$\Delta H_{rxn} = -9.8 \text{ kcal/mole}$$

In typical industrial applications, the dry process feed contains carbon monoxide, hydrogen, small quantities of hydrocarbons, and sulfur impurities. The reaction can be run in either a single adiabatic fixed-bed reactor or in multiple reactors in series when high conversion of carbon monoxide is required. The reactor facility simulated in this study is catalyzed by an iron-oxide WGS catalyst. The dry gas feed components (CO, CO_2, and H_2) are supplied by pressure-regulated cylinders. The flow rate of each component is adjusted via a mass flow controller, one associated with each gas. The steam flow rate is regulated by a computer-controlled metering pump. Total gas flow rates range from 12 to 25 standard liters per minute (SLPM). A WGS reactor model developed by Bell (1990) for a similar system constitutes the real plant in all simulations that follow. This model consists of three pde's representing material and energy balances of the catalyzed bed and the reactor wall, and one ode which constitutes an energy balance on the inlet section of the reactor,which is modeled as a CSTR. For details of the model, Bell (1990) should be consulted. Nominal operating conditions are given in Table 1.

TABLE 1 Nominal Operating Conditions for WGS Reactor.

Pressure	1.37 atm
CO feed flow rate	3.5 SLPM
CO_2 feed flow rate	2.0 SLPM
H_2 feed flow rate	4.0 SLPM
H_2O feed flow rate	5.5 SLPM
Fractional power of inlet heater	0.25
Inlet Temperature	438 K
Effluent CO composition	0.142 wt. %

TABLE 2 Sources of Plant/Model Error

Parameter	Plant	Model
Activation Energy (kcal/mol)	14.0	13.0
Pre-exp constant (10^{-3} mol/s-gm-cat)	0.185	0.165
Spatial Collocation points	5	7

The control objective is to regulate the effluent dry gas composition. One measure of the effluent composition is the CO weight fraction, which is taken to be the controlled variable. The manipulated variable is power to the inlet heater, given as a fraction of the total available power. The primary disturbances to the reactor are fluctuations in the feed temperature and variations in dry gas feed composition and flow rate.

The plant model has several interesting nonlinearities which make linear control unsuitable under many circumstances. We highlight one in particular. Curve A of Fig 2 illustrates how the steady-state CO weight fraction varies with fractional power at a nominal steam flow rate of 5.5 SLPM. The diagram clearly indicates that the steady-state gain varies substantially with power. In fact, for a fractional power setting of approximately 0.27, the reverse reaction begins to dominate the forward reaction and forces the process gain from negative to positive.

RESULTS AND DISCUSSION

The dynamic WGS reactor model is very sensitive to parameter variations in the pre-exponential constant, and the reactor inlet temperature which is assumed to be constant, but could vary dynamically. Therefore, the adaptive algorithm updates these physically significant parameters on-line. In each simulation, relaxation factors of 0.1 and 0.5 were used for the pre-exponential constant and the reactor inlet temperature, respectively. A data horizon of unity was employed. The reactor effluent temperature and composition as well as the temperature at the mid-point of the reactor were used as inputs to the estimator. Plant/model mismatch was introduced as outlined in Table 2. Recall that only two parameters are estimated. Their estimates must, therefore, compensate for all sources of error. This, however, is not strictly possible because some of the errors are structural and others are parametric. One should, therefore, anticipate some variation in the parameter estimates

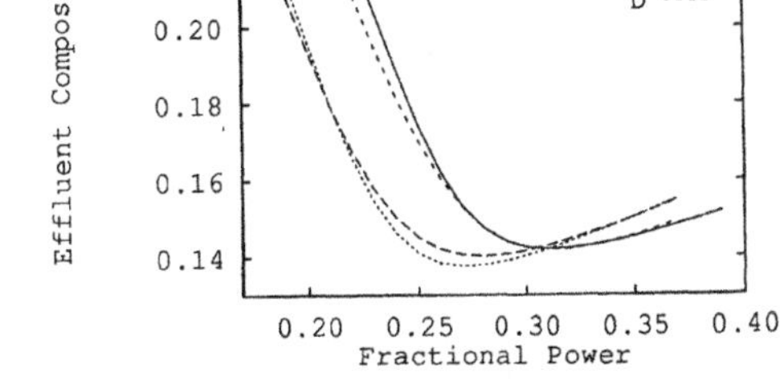

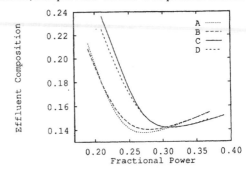

Figure 2: Steady-state relationship of effluent composition to power.

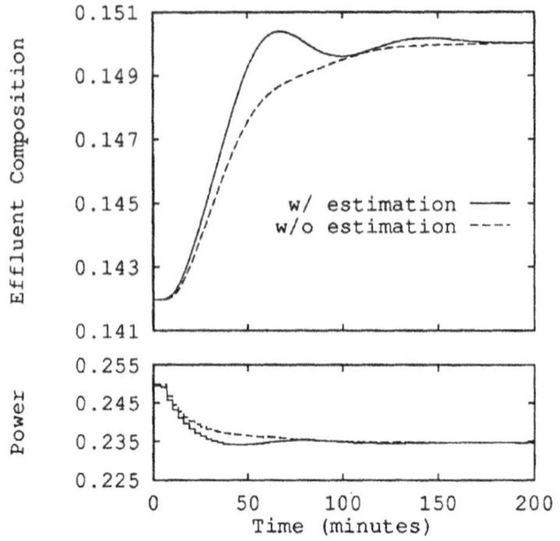

Figure 3: Response for a set-point change to a feasible operating point. (composition set-point = 0.150)

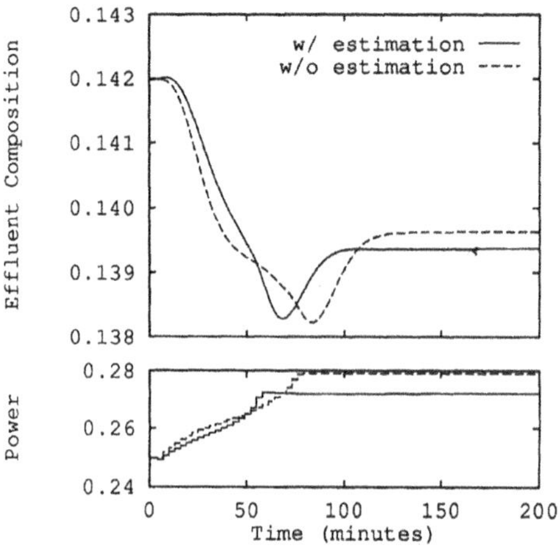

Figure 4: Response for a set-point change to an infeasible operating point. (composition set-point = 0.138, minimum feasible composition = 0.1395)

when any input or disturbance enters the plant. The magnitude of the change in the estimates will vary as a function of the plant's sensitivity to the inputs, and the model's sensitivity to the parameters.

A linear first-order filter with a time constant of 15.6 min. was used as the set-point prefilter. This time constant corresponds to approximately to the dominant time constant of the open-loop response of the reactor to a change in fractional power. Similarly, a first-order filter with a time constant of 15.0 min was used as the disturbance filter in the feedback loop. This value was simply chosen to give a desirable response since little was known about how to choose these values for robustness in nonlinear systems. The sampling time for each experiment was 3 min. The NLP that results from our formulation of the controller consisted of 19 variables and 18 constraints.

Case I: Set-point change to a feasible operating point

Figure 3 illustrates the reactor response when the CO set-point is changed from the nominal value of 0.142 to 0.150. This change moves the plant away from the infeasibility boundary, and should pose little difficulty for a well-tuned linear controller. We have, therefore, depicted only the response of the nonlinear controller with and without estimation. Although a slightly faster, mildly under-damped response is achieved when estimation is employed, estimation does not have a tremendous effect under these circumstances. In fact, in some cases nonlinear control without estimation is slightly better than with estimation. This is not typically the case, but the point needs to be made that estimation does not always guarantee improved performance. Indeed, the parameter variations may lead to mild undesirable process perturbations.

Case II: Set-point change to an infeasible operating point

The minimum effluent CO weight fraction that can be achieved, given the nominal inlet composition and flow rate, is 0.1395 as illustrated by curve A of Fig 2. Figure 4 depicts the reactor response when a CO weight fraction of 0.138 is required of the nonlinear controller. The proposed algorithm, like NMPC (Patwardhan, Rawlings, and Edgar, 1990), recognizes that this set-point is not attainable; it, therefore, moves the effluent composition to either the absolute limit that the plant can achieve or the absolute limit that the model can achieve depending upon the nature of the model error. Notice that the steady-state effluent weight fraction without estimation is not equivalent to that with estimation. This is best explained by considering curves A and B of Fig 2. Recall that curve A represents the steady-state relationship of effluent composition to power, and that it achieves a minimum at approximately 0.27. Curve B represents the model's input/output steady-state relationship. It has a minimum at approximately 0.28. Now consider the term $\mathbf{r} - (\tilde{\mathbf{d}} + \mathbf{y})$ of equation 11. This term could be written as $(\mathbf{r} - \tilde{\mathbf{d}}) - \mathbf{y}$ where $(\mathbf{r} - \tilde{\mathbf{d}})$ could be interpreted as a model set-point generated by feedback. Since the value of the filtered disturbance, $\tilde{\mathbf{d}}$, is such that the model set-point is infeasible at steady-state, the plant moves not to its minimum, but to the model's minimum. The effect of estimation is to reduce plant/model mismatch in the region of interest, or in this case to force the minimum of curve B to coincide with that of curve A. This drives the plant to set-point in a least squares sense. It should also be noted that traditional adaptive control based upon a linear discrete model proved to be most unreliable in this region since the sign of the gain varies.

Case III: Step-change in feed flow rate

The mass velocity of the gas in the reactor is a parameter which drastically affects reactor behavior. Figure 5 shows the response of the reactor to a 10.0% increase in the feed flow rate. Notice first that without estimation the algorithm is incapable of eliminating steady-state offset. In some cases, a disturbance of this nature will make the set-point infeasible, which will naturally lead to offset, but this is not the case here. While the plant still has a feasible set-point, the model representation of the plant does not, much like the previous example. With estimation, however, offset is easily eliminated because the model more accurately represents the process. In fact, Fig 2 illustrates the extent of the model improvement. Curve A represents the nominal plant. Curve C represents the nominal plant with a 10% flow rate increase. Curve B represents the model's steady-state output as a function of power before estimation and curve D represents the model output with estimation using the final estimate values. Figure 6 is a plot of the parameter estimates. It should be noted that good steady-state agreement is obtained despite our not estimating flow rate, which is the source of the disturbance.

CONCLUSIONS

We have presented a new approach for adaptive nonlinear control of a water-gas shift reactor and tested it using simulation. A simple adaptation method has been proposed that appears to compensate for substantial modeling error. We have further developed a natural way to extend this adaptation approach to include constraints on parameters and states. The simulation examples illustrate that under some circumstances adaptation is necessary to eliminate steady-state offset . The necessity of adaptation is a ramification of the feedback method, which can always provide integral action in the linear case provided the controller is designed for asymptotic set-point tracking. Since this approach to control is a limiting case of NMPC, these findings may provide some insight to the NMPC user.

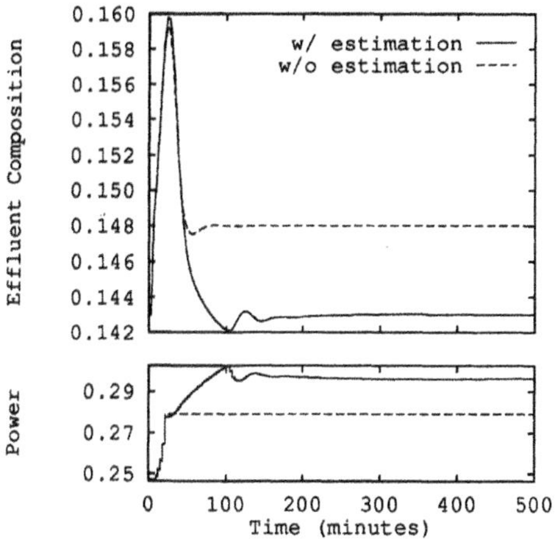

Figure 5: Response to a step-increase in flowrate. (composition set-point = 0.143)

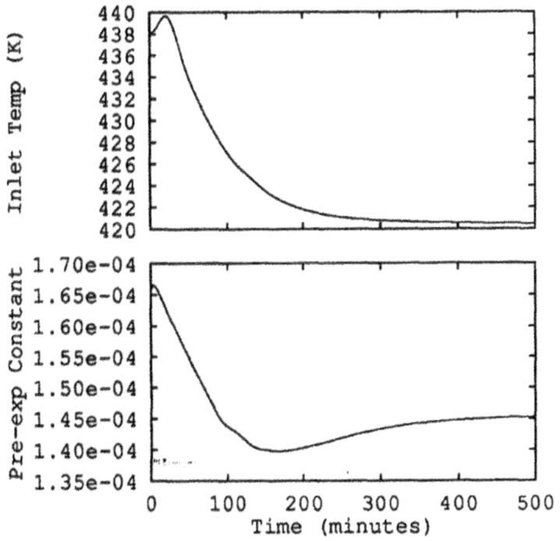

Figure 6: Parameters as a function of time.

REFERENCES

Asselmeyer, B., 1985, Optimal Control for Nonlinear Systems Calculated with Small Computers, *J. Opt. Theory Applic.*, **45**, 533.

Bell, N. H., 1990, Steady-State and Dynamic Modeling of a Fixed-Bed Water-Gas Shift Reactor, Ph.D. Dissertation, The University of Texas at Austin.

Biegler, L., 1984, Solution of Dynamic Optimization Problems by Successive Quadratic Programming and Orthogonal Collocation, *Comput. Chem. Eng.*, **8**, 243.

Clarke, D. W., C. Mohtadi and P. S. Tuffs, 1987a, Generalized Predictive Control -- Part I The Basic Algorithm, *Automatica*, **23**(2), 137-148.

Clarke, D. W., C. Mohtadi and P. S. Tuffs, 1987b, Generalized Predictive Control -- Part II Extensions and Interpretations, *Automatica*, **23**(2), 159-160.

Cuthrell, J. E. and L. T. Biegler, 1987, On the Optimization of Differential/Algebraic Process Systems, *AIChE J*, **33**, 1257.

Cutler, C. R. and B. L. Ramaker, 1980, Dynamic Matrix Control - A Computer Control Algorithm, Joint Automatic Control Conference Proceedings, Paper WP-5B.

Eaton. J. W. and J. B. Rawlings, 1990, Feedback Control of Chemical Processes Using On-line Optimization Techniques, *Comp & Chem Eng*, **14**(4/5),469-479

Edgar, T. F. and D. H. Himmelblau, 1988, *Optimization of Chemical Processes*, Chapter 8, McGraw-Hill, New York, 214

Economou, C. G., M. Morari and B. O. Palsson, 1986, Internal Model Control. 5. Extension to Nonlinear Systems, *Ind. Eng. Chem. Process Des. Dev.*, **25**, 403.

Garcia, C. E. and M. Morari, 1982, Internal Model Control. 1. A Unifying Review and Some New Results, *Ind. Eng. Chem. Process Des. Dev.*, **21**, 308.

Henson, M. A. and D. E. Seborg, 1990, A Critique of Differential Geometric Control Strategies for Process Control, IFAC World Congress, Tallinn.

Morshedi, A. M., 1986, Universal Dynamic Matrix Control, Session VI, Paper No. 2, Chemical Control Conference III, Asilomar, California.

Patwardhan, A. A., J. B. Rawlings and T. F. Edgar, 1990, Nonlinear Model Predictive Control, *Chem. Eng. Commun.*, **87**, 123.

Renfro, J. G., A. M. Morshedi and O. A. Asbjornsen, 1987, Simultaneous Optimization and Solution of Systems Described by Differential/Algebraic Equations, *Comput. Chem. Eng.*, **11**, 503.

Rhinehart, R. R. and J. B. Riggs, 1990, On-Line Dynamic Adaptation of Nonlinear Models 1: Development and Application to Several SISO Processes. Submitted for publication.

A CO-ORDINATED SELF-TUNING EXCITATION AND GOVERNOR CONTROL SCHEME FOR POWER SYSTEMS

C. M. Lim* and T. Hiyama**

*Dept. of Electronic Engineering, Ngee Ann Polytechnic, Singapore 2159
**Dept. of Electrical Engineering and Computer Science, University of Kumamoto, Kumamoto 860, Japan

Abstract: This papers describes a new method of co-ordinating self-tuning excitation and governor-loop stabilisers of a power system for the purpose of enhancing the overall system transient and dynamic stability limits. The effectiveness of the proposed scheme has been evaluated by subjecting the study system to large and small disturbances, and with the system operating over a wide power range. Nonlinear simulation results show that the proposed control scheme is suitable and effective for enhancing the performance of the overall power system.

Keywords: Control applications; parameter estimation; power system control; self-tuning regulators; recursive least squares

INTRODUCTION

It is well-known in the literature that excitation control has become a standard means of enhancing the transient and dynamic stability limits of power systems (Larsen and Swann, 1981; Yu, 1983). In the conventional method of designing an excitation stabiliser or commonly known as a power system stabiliser (PSS), analogue circuits (Keay and South, 1971) are employed. The primary function of a PSS is to produce a control or stabilising signal in order to induce a positive damping torque (Yu, 1983) after the onset of a disturbance. The PSS parameters are usually set to fixed values in order to ensure optimal system performance but for a particular operating point. Consequently, the power system performance is degraded whenever its operating point begins to drift.

In order to overcome the above-mentioned disadvantage of the conventional PSS, the recent trend in designing PSS is towards the utilization of self-tuning control scheme (Cheng, Malik and Hope, 1986; Lim and Hiyama, 1989). The reason being that the design framework of self-tuning control makes it suitable for adjusting the stabiliser parameters in order to compensate for distrubances as well as changes in the system operating conditions. Many different self-tuning control schemes for excitation control have been proposed and, most importantly, encouraging results have been reported (Cheng, Malik and Hope, 1986; Lim and Hiyama, 1990). However, little attention has been given to the co-ordination of self-tuning excitation and governor-loop stabilisers (Ibrahim, Hogg and Sharaf, 1989). Therefore, more research work in this area is needed in order to further demonstrate the potential benefits of self-tuning control for stability enhancement of power systems.

The objectives of this paper are

i. To propose a new method of designing self-tuning excitation and governor-loop stabilisers which are co-ordinated such that the overall power system stability is enhanced;

ii. To illustrate the effectiveness of the proposed control scheme for large and small disturbances, and over a wide range of operating conditions.

POWER SYSTEM UNDER STUDY

The configuration of the power system chosen for study is as shown in Fig. 1. The study system consists of a single-machine and an infinite-bus. The generator is equipped with a faster excitation system and a speed governoring system. It is well-known in the literature that by applying a supplementary signal to the excitation-loop and/or the governor-loop the overall power system stability is enhanced (Yu, 1983).

Here, the objective is to design a self-tuning stabiliser for each loop and to co-ordinate both stabilisers such that the overall system damping of the low-frequency mechanical mode oscillations is enhanced for large and small disturbances.

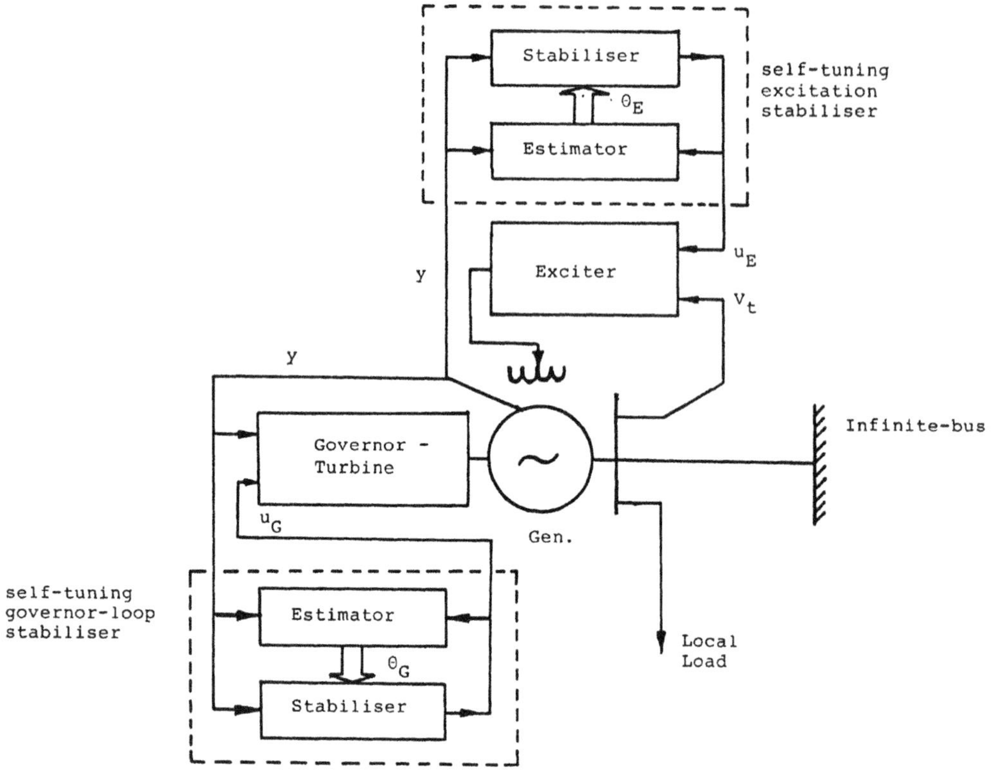

Fig. 1. Study system under co-ordinated self-tuning excitation and governor-loop control.

PROPOSED SELF-TUNING CONTROL SCHEME

Details of the proposed self-tuning excitation and governor-loop stabilisers, and the method of co-ordinating both stabilisers are as follows:

Excitation-Loop Stabiliser Design

In the development of a self-tuning algorithm, a linear discrete-time model is used to describe the dynamics of the study system. For the purpose of designing self-tuning stabiliser for power systems, it has been shown that a 3rd-order discrete-time model is adequate (Lim and Hiyama, 1989, 1990).

Discrete-time model. For the excitation-loop stabiliser design, the discrete-time model is of the form (Lim and Hiyama, 1989)

$$y(k+1) = \sum_{i=1}^{n} a_{Ei}y(k+1-i) + \sum_{i=1}^{n} b_{Ei}u_E(k+1-i) \quad (1)$$

where $y(k)$ and $u_E(k)$ are the system output and excitation control signal at time kT_E; T_E being the sampling period. For a 3rd-order model, n in (1) is set equal to 3 and $(a_{E1}, a_{E2}, \ldots, a_{En}, b_{E1}, b_{E2}, \ldots, b_{En})$ are the model parameters which remain to be estimated.

Performance index. The control objective is to determine $u(k)$ such that $y(k)$ follows a constant reference signal, $y_r(k)$, with zero steady-state error. As such, the internal model of $y_r(k)$ is an integrator and to ensure robust performance the control loop should include this internal model driven by the system tracking error (Lim and Hiyama, 1989). The tracking error, $e_E(k)$, is defined as

$$e_E(k) = y_r(k-1) - y(k) \quad (2)$$

and the integrator output, $v_E(k)$ is given by

$$v_E(k) = y_r(k-1) - y(k) + v_E(k-1) \quad (3)$$

In addition to the above control objective, it is desirable that deviations of the system variables should be minimized. To this end, the following performance index is chosen for the purpose of selecting the optimal excitation control signal

$$I_E = [y(k+1)-y_r(k)]^2 + q'_E[Dy(k+1)]^2 + r_E[u_E(k)]^2 + p_E[v_E(k+1)]^2 \quad (4)$$

where $Dy(.)$ is the first derivative of $y(.)$ and (q'_E, r_E, p_E) are weighting constants.

Optimal excitation control signal. By using the approximation

$$Dy(k+1) = [y(k+1) - y(k)]/T_E \quad (5)$$

the optimal excitation control signal which minimises the above performance index is given by

$$u_E(k) = K_E[(1+p_E)y_r(k) + p_E v_E(k) - (1+p_E+q_E)s_E(k) + q_E y_E(k)] \quad (6)$$

where

$$K_E = b_{E1}/[b_{E1}^2(1 + q_E + p_E) + r_E]$$

$$q_E = q'_E/T_E^2$$

$$s_E(k) = \sum_{i=1}^{n} a_{Ei}y(k+1-i) + \sum_{i=2}^{n} b_{Ei}u_E(k+1-i)$$

Governor-Loop Stabiliser Design

The same power system output, $y(k)$, is chosen as the input to the governor-loop stabiliser.

Discrete-time model. The discrete-time model used in designing the governor-loop stabiliser is of the form

$$y(k+1) = \sum_{i=1}^{n} a_{Gi}y(k+1-i) + \sum_{i=1}^{n} b_{Gi}u_G(k+1-i) \quad (7)$$

where $u_G(k)$ is the governor-loop control signal at time kT_G; T_G being the sampling period. For a 3rd-order model, n in (6) is set equal to 3 and $(a_{G1}, a_{G2}, \ldots, a_{Gn}, b_{G1}, b_{G2}, \ldots, b_{Gn})$ are the model parameters which remain to be estimated.

Performance index. Following the same arguments given for the above excitation-loop stabiliser design, the governor-loop control signal is chosen to minimise the following index

$$I_G = [y(k+1)-y_r(k)]^2 + q'_G[Dy(k+1)]^2 + r_G[u_G(k)]^2 + p_G[v_G(k+1)]^2 \quad (8)$$

The optimal governor-loop control signal is given by

$$u_G(k) = K_G[(1+p_G)y_r(k) + p_G v_G(k) - (1+p_G+q_G)s_G(k) + q_G y_G(k)] \quad (9)$$

where

$$K_G = b_{G1}/[b_{G1}^2(1 + q_G + p_G) + r_G]$$

$$q_G = q'_G/T_G^2$$

$$s_G(k) = \sum_{i=1}^{n} a_{Gi}y(k+1-i) + \sum_{i=2}^{n} b_{Gi}u_G(k+1-i)$$

where (q_G, r_G, q_G) are weighting constants, $v_G(.)$ is the output of an internal model and is given by

$$v_G(k) = y_r(k-1) - y(k) + v_G(k-1) \quad (10)$$

The internal model is an integrator with y_r as its reference.

Self-Tuning Control Algorithm

Each of the above stabilising signal can only be implemented provided the model parameters are known. For a power system, the system model parameters of each loop are unknown and their values vary with the system operating conditions. Here, the recursive-least-squares algorithm is employed to estimate each model parameters at every sampling time and based on the estimated values each stabilising signal is computed, as shown in Fig. 1.

Co-Ordination of Stabilisers

In order to ensure that both the excitation and governor-loop self-tuning stabilisers are well co-ordinated to provide maxmium damping to the power system, the following index is defined

$$J = \sum_{k=0}^{m} y(k)^2 \quad (11)$$

where m is the total number of data points used in the computation of J. The above index can be used to select the best combination of (T_E, T_G) once the weighting constants $(p_E, q_E, r_E, p_G, q_G, r_G)$ have been obtained.

OVERALL SYSTEM DATA

Study System

The performance of the above self-tuning stabilisers has been evaluated through digital simulations. In all the simulations, the generator is represented by a 5th-order nonlinear model, the exciter by a 1st-order model and the governor-turbine by a 4th-order model (Yu, 1983). Furthermore, the output of the exciter as well as the speed of the governor servo gate opening are constrained within physical limits.

Table 1 shows the 3 operating conditions of the power system. Details of the numerical data for the overall system can be found from (Yu, 1983).

TABLE 1 Operating Conditions In Per Unit

op. point	real power	react. power	ter. voltage	remark
OP1	1.100	0.200	1.05	Heavy
OP2	0.095	0.015	1.05	Nominal
OP3	0.700	-0.150	1.05	Light

Self-Tuning Stabilisers

The generator speed deviation was chosen as the input signal to both the excitation and governor-loop stabilisers. The weighting constant parameters (p_E, q_E, r_E) of the self-tuning excitation stabiliser were set to (0.1,0.5,20). The weighting constant parameters (p_G, q_G, r_G) of the self-tuning governor-loop stabiliser were set to (1.5,3,35). The initial model parameter estimates of both stabilisers were chosen to be (2.3,-1.9,0.5,-1.4, 0.4,-0.15) and the covariance matrix of the recursive-least-squares algorithm was reset once every 5 sampling intervals to 700I where I is a 6 by 6 unit matrix.

For both control loops, the output of its internal model and that of its stabiliser were both constrained to ± 0.12 per unit.

The above stabiliser settings were chosen based on experience gained from previous studies (Lim and Hiyama, 1989). It should be noted that the same stabiliser settings were always used in all simulations irrespective of the types of tests and the operating conditions of the study system.

Types of Tests

The following large and small disturbance tests have been simulated to evaluate the performance of the above stabilisers:

T1: A 100 ms short circuit near the infinite-bus;

T2: A - 0.10 per unit step change in input power which is followed by another 0.25 per unit step change 3.5 s later;

T3: A 100 ms short circuit on a line in the transmission network and successful restoration of the faulted line 0.5 s later;

T4: A 0.04 per unit step change in the reference voltage of the excitation system.

The above tests were performed with the study system operating over a wide power range as shown in Table 1.

NUMERICAL RESULTS

In order to co-ordinate the above stabilisers to provide maximum damping to the overall power system, the optimal combination of T_E and T_G needs to be determined. To this end, the power system is subjected to the large disturbance test T1 at the nominal loading condition, i.e., operating point OP2, for different values of T_E and T_G.

The results obtained are evaluated using the performance index J of (11) and are summarised as shown in Tables 2 and 3. From the results shown in Tables 2 and 3, it can be seen that maximum damping is obtained by setting T_E and T_G to 60 ms and 75 ms, respectively. In all the subsequence tests, the above settings for T_E and T_G were always chosen.

TABLE 2 System Performance for $T_E = T_G$

T_E(ms)	45	60	75	90	105
T_G(ms)	45	60	75	90	105
J	283	229	237	197	646

TBALE 3 System Performance for $T_E < T_G$

T_E(ms)	60	60	60	75
T_G(ms)	75	90	105	90
J	171	185	184	238

More test results with the power system being subjected to different disturbances and over a wide range of operation conditions are shown in Figs. 2 to 6.

From the results obtained, it can be seen that the proposed self-tuning control scheme is suitable for enhancing the stability of the study power system for different disturbances and over a wide range of operating conditions.

CONCLUSION

A new method of co-ordinating self-tuning excitation and governor-loop stabilisers for single-machine infinite-bus power systems has been proposed. The effectiveness of the proposed method has been demonstrated.

The proposed method can be readily extended to multimachine power systems using the approach given in Lim and Hiyama, 1990.

REFERENCES

Cheng S.J., Malik O.P. and Hope G.S. (1986). Self-Tuning Stabiliser for a Multimachine Power System. IEE Proc. Pt. C, pp. 176-185.

Ibrahim A. S., Hogg B. W. and Sharaf M. M. (1989). Self-tuning controllers for turbogenerator excitation and governing systems. IEE Proc. Pt. C, pp. 238-251.

Keay F. W. and South W. H. (1971). Design of a System Stabiliser Sensing Frequency Deviation. IEEE Trans. PAS-90, p.707-713.

Larsen E. V. and Swann D. A. (1981). Applying Power System Stabilisers Part I, II and III. Trans. PAS-100, pp.3017 -3041.

Lim C. M. and T. Hiyama (1989). A robust self-tuning power system stabiliser. Presented at 1989 IFAC Sym. on Power Systems and Power Plant Control, Seoul, S. Korea.

Lim C. M. and Hiyama T. (1990). A Self-Tuning Control Scheme for Stability Enhancement of Multimachine Power System. To appear IEE Proc. Pt. C.

Yu Y. N. (1983). Electric Power System Dynamics. Academic Press, New York.

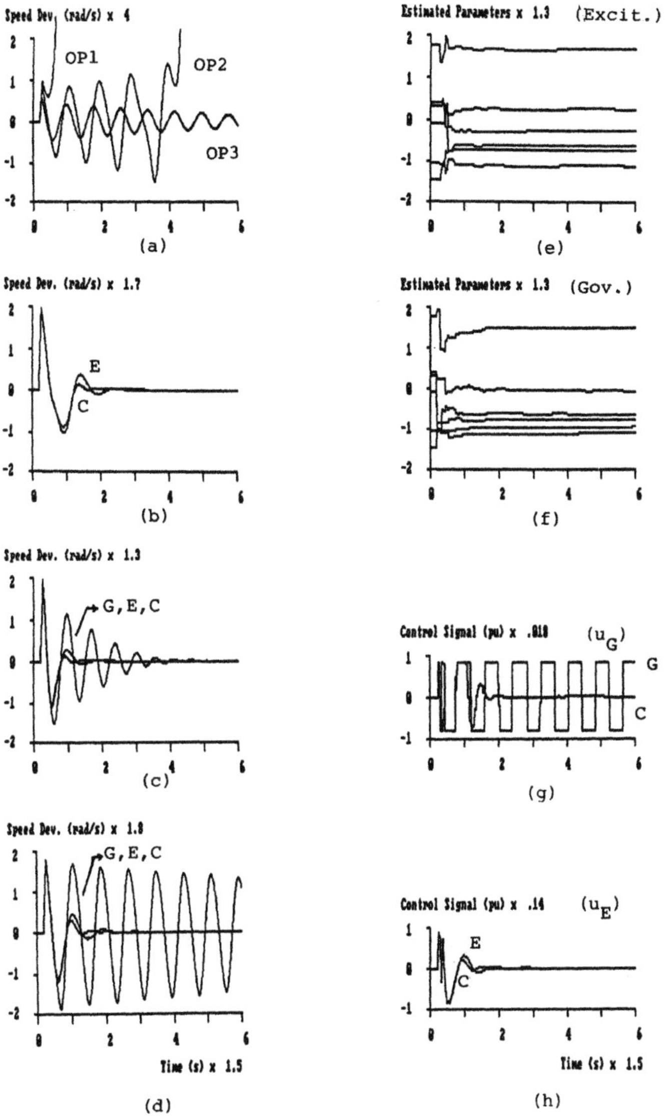

Fig. 2. Nonlinear system responses to test T1 and over a wide power range.

E = self-tuning excitation control
G = self-tuning governor-loop control
C = co-ordinated self-tuning control

OP1, OP2 and OP3 are as defined in Table 1.

(a) open loop control
(b) operating point at OP1
(c) operating point at OP3
(d) - (h) operating point at OP2
(e) - (f) co-ordinated control

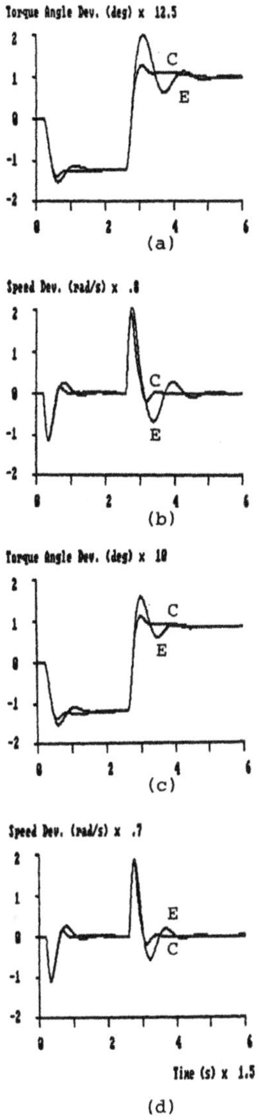

Fig. 3. Nonlinear system responses to test T2 and over a wide power range.

E and C are as defined in Fig. 2.

OP1 and OP2 are as defined in Table 1.

(a) - (b) operating point at OP1
(c) - (d) operating point at OP2

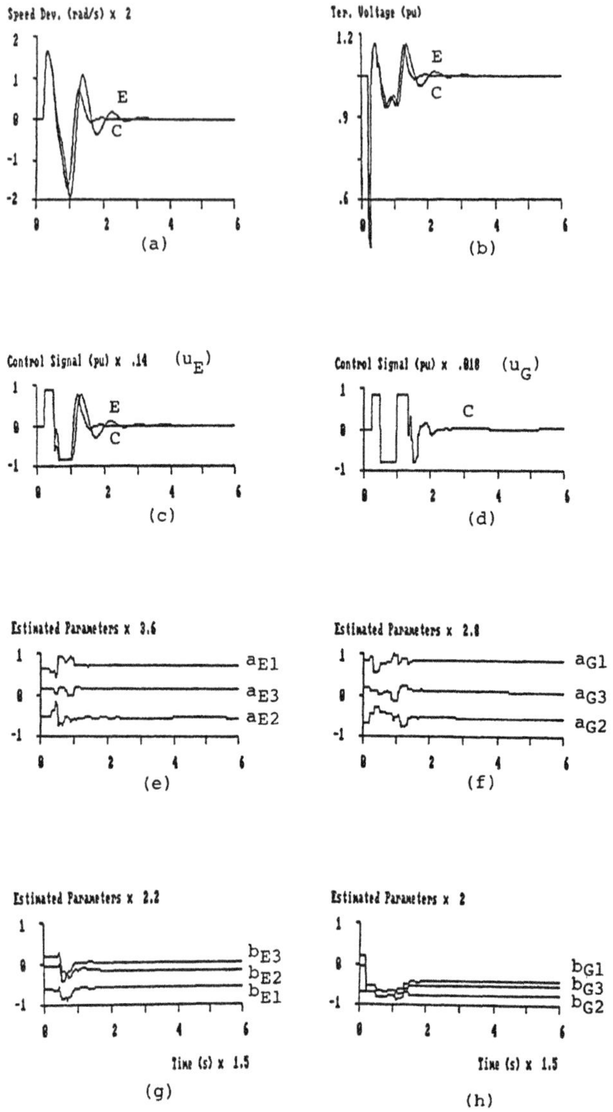

Fig. 4. Nonlinear system responses to test T3 and over a wide power range.

E = self-tuning excitation control
C = co-ordinated self-tuning control

(e),(g) excitation model parameters of co-ordinated control
(f),(h) governor-loop model parameters of co-ordinated control

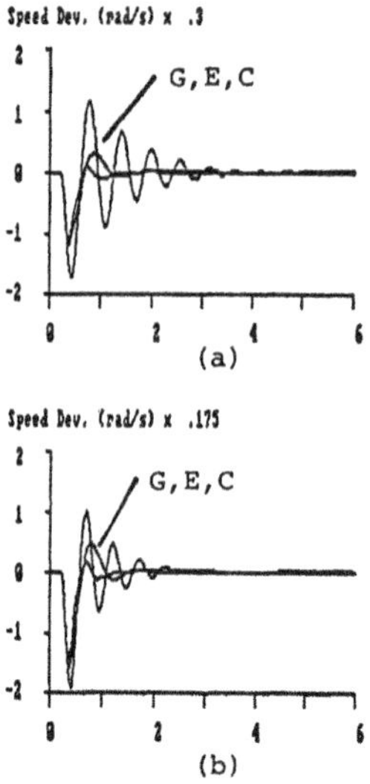

Fig. 5 Nonlinear system responses to test T4.

(a) operating point at OP1
(b) operating point at OP3

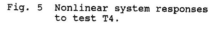

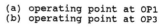

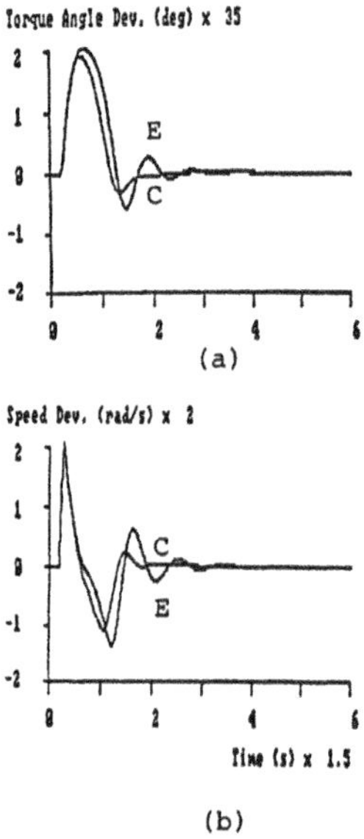

Fig. 6 Nonlinear system responses to test T1. (Transmission network different from that of Figs. 1 - 5)
Real power = 1.2, reactive power = 0.341 and duration of short circuit = 120 ms.

IMPLEMENTATION OF ADAPTIVE CONTROLLERS USING DIGITAL SIGNAL PROCESSOR CHIPS

E. K. Teoh and Y. E. Yee

School of Electrical & Electronic Engineering, Nanyang Technological Institute, Singapore 2263, Singapore

Abstract. The tuning of a classical servo system is often problematical, owing to the presence of system non-linearities, parameter variations and unmeasured disturbances. Adaptive control schemes are of growing interest since the system is self-tuned such that some predefined performance requirement is met, regardless of variations in system parameters. With the recent availability of the powerful intensive-computation based digital signal processor (DSP) chips, adaptive controllers become viable in practice. This paper presents the position control of a DC motor using adaptive-PID and adaptive one-step-ahead control schemes. The experimental results demostrate that the responses of DC motor with adaptive position controllers can adapt to altered process behavior and shows a satisfactory, well damped control performance.

Keywords. Adaptive Control; Real-Time Implementation; DSP Chips; Servomechanism.

INTRODUCTION

In many applications of electromechanical systems, parameters such as inertial and load torque may vary over time. Variation of load torque, manufacturing variations and aging can degrade system performance. For the case of a DC motor with permanent magnets, the direct coupling of the load to the motor axis induces a large sensitivity of the motor behavior to load variations. Therefore, a fixed, linear controller cannot provide an acceptable response under varying load conditions. A PID-based scheme controller, which is auto-tuned and uses a linearizing block and gain scheduling, can be used to improve the above mentioned system performance (Dumont at al,1989). However, the nonlinearities of the system must assume to be known and changes in system dynamics must be predictable. This approach also required much tailoring works. Hence, at this junction, a control system which is capable of keeping track of the variation of the parameters is desirable. This leads to the notion of adaptive control. The advances in microprocessor technology with reduced cost have made it possible to apply adaptive control scheme to electromechanical systems because digital signal processor (DSP) chips reduce cost and development time. In this paper, a digital controller for controlling the position of a DC motor using digital signal processor chips is presented. Two different adaptive control algorithms, namely, adaptive-PID and adaptive one-step-ahead were used. A DSP chip TMS320E15 is chosen for availability of development systems, as well as for its high speed number crunching capability, powerful instruction sets and innovative architecture (Texas Instruments,1987). The motor used is the Baldor "Big MHO" Series M2200 DC permanent magnet servo motor. This motor is the new generation DC motor that delivers a high torque on the motor axis. The high-acceleration torque of new DC motors with permanent magnet permits a direct coupling of the load to the motor axis, avoiding the use of a transmission with its inherent disadvantages (such as backlash and friction) (Bulter at al,1989). The motor has torque ranges of 10 to 50 oz in continuous stall and 65 to 360 oz in peak, with a maximum of 5000 revolution per minute (RPM). Velocity feedback is available from a 1% precision DC tachometer mounted on the motor drive shaft. Position feedback is obtained from an optical encoder linked to the output shaft of the motor. The encoder is a modular 5 volts DC type with TTL output.

On-line testing is performed by linking the DC motor and the position controller to the HP64000 DSP chip emulator. The control schemes were tested using different loads on the motor shaft and with disturbances applied to the motor shaft.

ADAPTIVE POSITION CONTROL OF A DC MOTOR

System model

The transfer function of the permanent DC motor is shown in Figure 1. R_a is the copper resistance, L_m the motor inductance, K_t the torque constant, K_b the back electromotive-force (EMF) constant, J_t the total inertia ($=J_1 + J_m$), f the viscous damping, and M_w coulomb friction. The complete motor transfer function consists of an electrical function $1/(sL_m + R_a)$ and a mechanical function $1/(sJ_t + f)$. Coulomb friction, which is caused by magnetic and mechanical hysteresis, results in a counteracting torque with magnitude M_w and a sign that depends on the sign of the angular velocity of the motor, w_p. This static model of the Coulomb friction reflects only approximately the real friction effects (Walrath, 1984) and is represented by the factor M_w sign (w_p). The Coulomb friction introduces a steady-state error in the transfer from the motor torque, M_w, to w_p (Butler at el, 1989). The DC motor is drived by the Baldor Pulse-Width-Modulated Transistot Servo-drivers TFM Card. The driver consists of pulse-width-modulated amplifiers working in four quadrants operation (Baldor,1988).

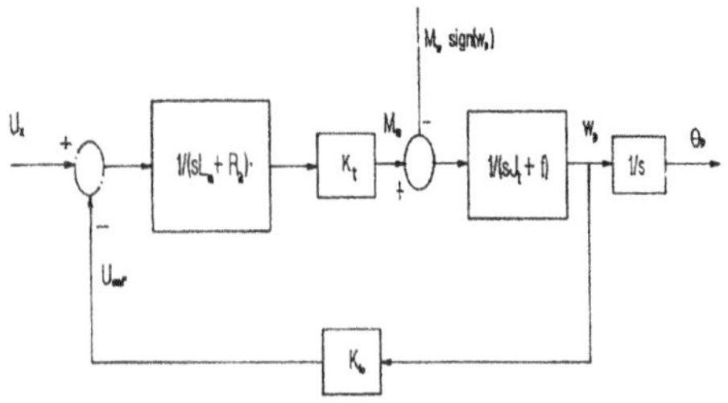

Fig 1. Model of DC motor

<u>Adaptive PID control scheme</u>

Presently, the most commonly used control scheme is the PID scheme. The equation of the discrete PID controller can be stated as below :

$$u(n) = u(n-1)+k_0*e(n)+k_1*e(n-1)+k_2*e(n-2) \qquad (1)$$

where
- $u(n)$ = present output
- $u(n-1)$ = previous output
- $e(n)$ = latest error sample
- $e(n-1)$ = previous error sample
- $e(n-2)$ = oldest error sample
- k = gains constant

Adaptive-PID is an extension of a PID control schemes in real time. The adaptive-PID scheme is achieved by estimating J_1 the load inertia ($J_t = J_1 + J_m$) by means of a recursive least-square method (Goodwin, 1982)(Goodwin and Sin, 1984) and adjusting the values of k_0, k_1 , and k_2 accordingly. Let's look at the corresponding Z-transfer function of equation (1), which is

$$G_R(z) = u(z)/e(z) = Q(z^{-1})/P(z^{-1}) = (k_0+k_1z^{-1}+k_2z^{-2})/(1-z^{-1}) \qquad (2)$$

which leads to

$$P(z^{-1})u(z) = Q(z^{-1})e(z) = Q(z^{-1})[w(z)-y(z)] \qquad (3)$$

where
- $w(z)$ = reference variable
- $y(z)$ = control variable

There are many possible modifications of this basic PID-controller. One possibility is to increase the order in the denominator or in the numerator of $G_R(z^{-1})$. The design of the controller parameters for parameter-adaptive control has to be based on the process parameter estimated. However, for parameter estimation in a closed loop without external pertubations, conditions for the parameters identifiability must be satisfied : the process order m and deadtime d must be known and $m \le 2$ if d=0, $m \le 3$ if d=1,etc (Radke and Isermann, 1987). Experience shows, however, that this condition (for k tends to infinity) is not critical for stochastically disturbed or slowly time-varying processes, as the controller parameters then also vary with time and therefore satisfy the identificability conditions independent of its order (Radke and Isermann, 1987).

<u>Adaptive one-step-ahead control scheme</u>

Adaptive one-step-ahead control scheme is another method to overcome the variation problem faced by PID controller. The adaptive one-step-ahead control scheme predicts the next desired input to a system from a set of parameters. The parameters are estimated using recursive least-square method (Goodwin,1982)(Goodwin and Sin,1984) as in adaptive-PID control scheme.

To demostrate this control scheme, let's refer to a simple equation

$$u_k = (B/A)y^* \qquad (4)$$

where
- u_k = input to a system
- y^* = the desired output
- B, A = parameters to be estimated

If y_{k+1} is the next output, and if u_k is predicted very well from equation (4), then y_{k+1} will converge to y^* . How close y_{k+1} will converge to desired value depends on the accuracy of parameter estimation of A and B. Therefore, the estimation of the load inertia of the motor is very important to get a small final position error.

TMS320E15 DSP CHIP: AN OVERVIEW (TEXAS INSTRUMENTS,1987)

The TMS320E15 uses a modified Harvard-type architecture by having separate program and data buses but at the same time retaining a means of passing information between the two (through the TBLR and TBLW instruction). As a result the TMS320 maintains separate program and data memory in the PROM and RAM respectively. Hence memory data transfer and program instruction fetch can be carried out simultaneously.

The TMS320 stands out from other microprocessor because it has a 32 bit ALU with a direct 16 * 16 multiplier and a 32 bit accumulator which would greatly increase computation speed especially when complex control algorithms are implemented. The TMS320 also has a barrel shifter to allow for quick shift of bits to the left of up to 16-bits. With its modified Harvard architecture, the TMS320E15 operating on a 20 MHz clock has an instruction cycle of only four clock cycles or 200ns. All instructions (except branch instructions, TBLR, TBLW and I/O instructions) take only one instruction cycle to execute. Branch and I/O instructions take two cycles while TBLR and TBLW take three cycles. A comparison with a 8 MHz Intel 8086 will serve to illustrate the vast improvement obtained in using the TMS320E15. A 8086 CALL instruction takes a minimum of 42 clock cycles while a multiply instruction takes between 124 to 139 clock cycles. A similar multiply instruction on the TMS320E15 (MPY) takes only 4 clock cycles.

IMPLEMENTATION OF ADAPTIVE CONTROLLERS FOR A DC MOTOR USING DSP CHIPS

The DSP chip alone cannot achieve the capability to control the position of a DC motor. Certain support circuitry is built and integrated with the DSP chip. A position counter which counts the number of revolution of the motor shaft by decoding the signal of the optical encoder is needed. The DSP chip buffers and decoder chip are needed, so is a digital-to-analog converter for controlling the motor shaft position. The hardware consists of four major builiding blocks which are shown in Figure 2. The heart of the hardware is the TMS320E15 DSP chip which is involved in the direct control and monitoring of the DC motor. Two programs are written in the TMS320E15 Assembly Language, they are the adaptive-PID control program and the adaptive one-step-ahead control program. The flowchart of the adaptive-PID control program is shown in Figure 3. The algorithm used to compute the adaptive-PID output is according to the simple relationships between the motor and the controller parameters (k_o, k_1, k_2 of equation (1)). To illustrate those simple relations, say, if for a certain motor load inertia J2, a certain controller parameters provide a suitable response, a larger value of J, will result in too high an overshoot. A smaller load inertia, however, results in an overdamped, and so non-time-optimal, response. Therefore, controller's parameters are decreased if J, is larger, and are increased if J1 is smaller. The load inertia is estimated using recursive least-square with covariance resetting method (Goodwin, 1982)(Goodwin and Sin, 1984) and adjusting the value of ko, k1 and k2 accordingly.

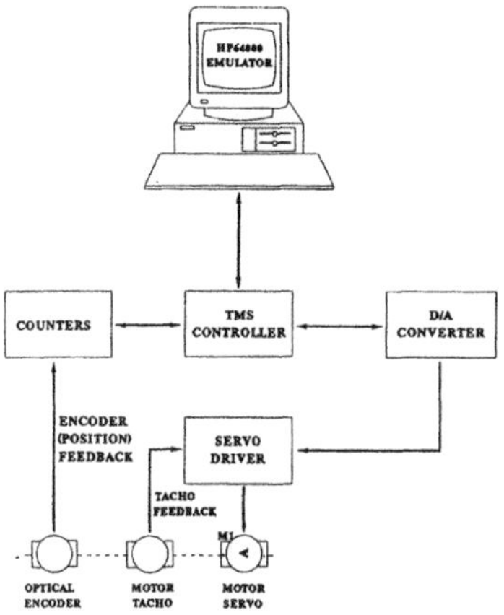

Fig 2. Block diagram of system configuration

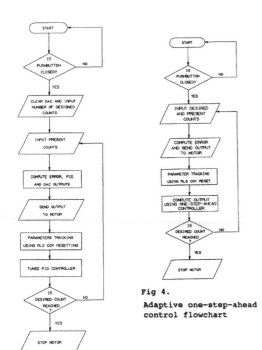

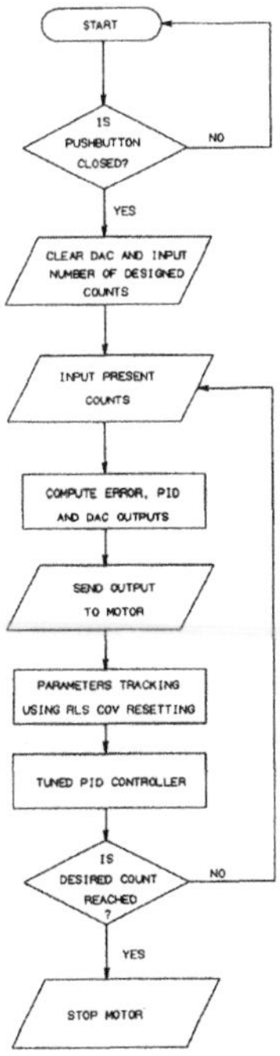

Fig 3. Adaptive-PID program flowchart

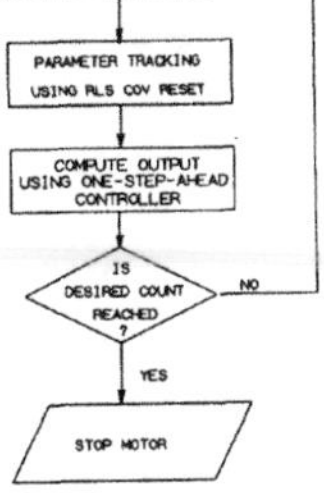

Fig 4.
Adaptive one-step-ahead control flowchart

The adaptive one-step-ahead controller program flowchart is shown in figure 4. The motor load inertia is estimated using recursive least-square with covariance resetting method (Goodwin, 1982)(Goodwin and Sin, 1984). The input to the motor is then computed using one-step-ahead scheme. This is achieved by categorising the inertia load into four catregories: 0 to 1/4 of allowable peak load inertia: 1/4 to 2/4 of allowable peak load inertia; 2/4 to a 3/4 of allowable peak load inertia; and 3/4 to 1 of allowable peak load inertia. Each category has its own controller parameters. If the estimated load inertia falls in category one, the set of controller parameters assigned to this category will be used to compute the next input to the motor.

The instructions MPY, LTD and APAC are extremely useful and are used intensively in the two assembly programs for computation purpose.

ON-LINE TESTING RESULTS AND DISCUSSIONS

The two control schemes were tested using different loads on the motor shaft and with disturbances (brake) applied to the motor shaft. All results shown are for a desired position counts of 1000.

Figure 5 shows the results obtained using adaptive-PID controller with no-load, 500 grammes load and 800 grammes load respectively. All figures show an error of less than ±5 counts. Similarly, figure 6 show the results obtained using adaptive one-step-ahead controller with no-load, 500 grammes load and 800 grames load respectively. All figures also show an error of less than ±5 counts. The results show that both controllers can achieve relatively high accuracy under the presence of inertia variations and disturbances. The time-varying non-linear structure of the dynamic model of multiarticulated robotic manipulators has called for adaptive control methods. The aim of adaptive techniques is to estimate the parameters of the model recursively, and then from estimates of these, the time-varying feedback gain is computed. The hardwares used to control the DC motor can used be to control three out of five axis of a robot manipulators, namely the Base, the shoulder and the Elbow by incorporating two more DACs and driver cards to the present hardware parallelly.

The main concern regarding the real-time application of these schemes to multiple-axis manipulators is the computational complexity involved. The TMS320E15 must continuously monitor the three axis and make necessary changes to reach target position as required by control algorithm.

CONCLUSIONS

The development and implementation of Digital Controllers for position control of a DC motor using the DSP chip, TMS320E15, has provided encouraging results. Real-time position control of the DC motor was investigated by using the controllers with a shaft encoder. Experimental results show that the system responses are good and are as expected theoretically. Adaptive control has become a viable alternative for controlling electromechanical systems. The adaptive control system provides the desired performance throughtout the life of the mechanism. Furthermore, reliability is enchanced since the system meets the expected performance despite of aging of the mechanism.

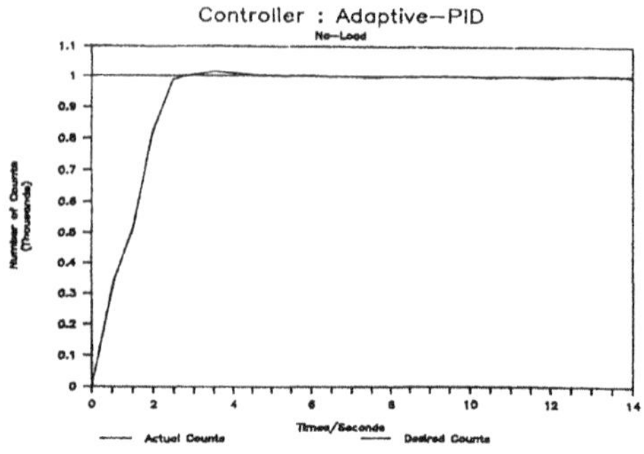

(a) No Load

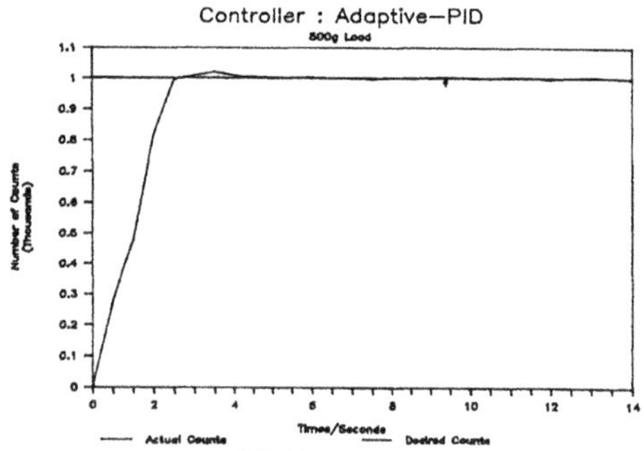

(b) 500g Load

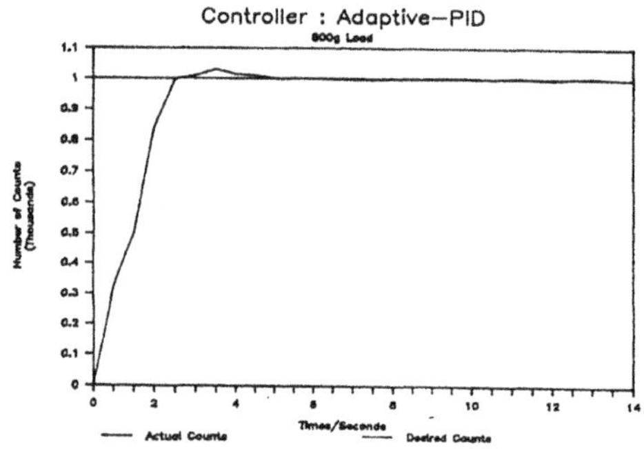

(c) 800g

Fig 5. System response with adaptive-PID controller

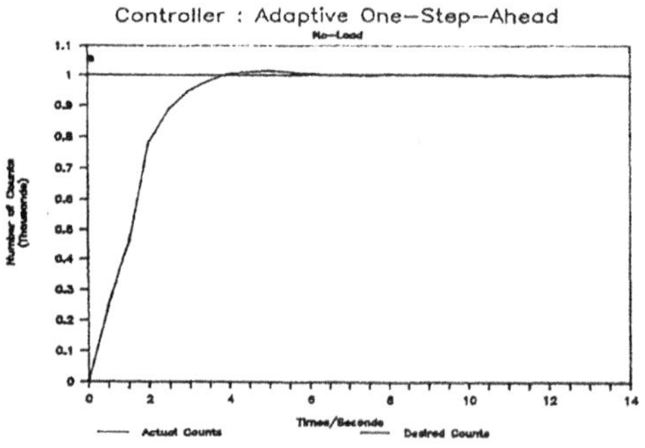

(a) No Load

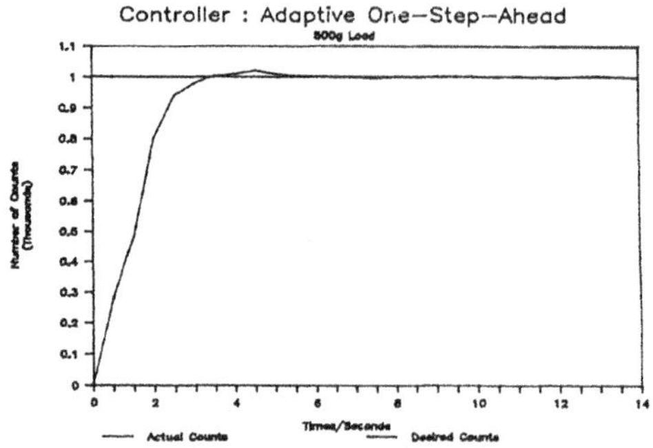

(b) 500g Load

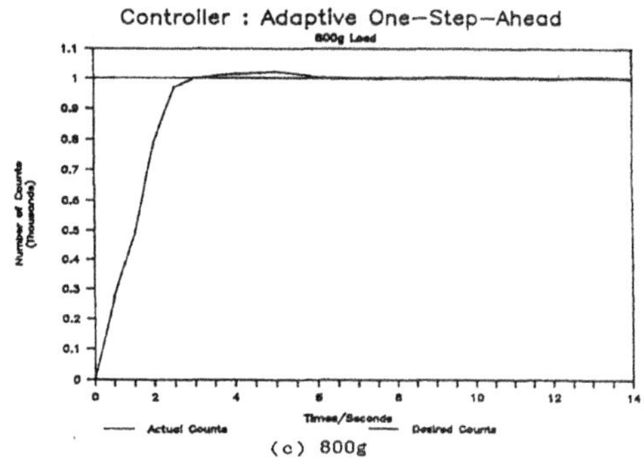

(c) 800g

Fig 6. System response with adaptive one-step-ahead controller

ACKNOWLEDGEMENTS

The authors wish to extend their gratitudes and appreciation to Professor Brian Lee, Dean of School of Electrical and Electronic Engineering, Nanyang Technological Institute and the Applied Research Fund Allocation Committee for providing incentives and financial support for this research project. Thanks are also owed to Song Choo Beng, the Final Year Student at NTI of 1989/90, for his experimental and software development contribution to the research reported here.

REFERENCES

Baldor, (1988). Pulse Width Modulated Transistor Servodriver TFM Instruction Manual.

Butler Hans, G. Handerd and Job Van Amerongen, (1989). Model Reference Adaptive Control of a Direct-drive DC Motor, IEEE Control Systems Magazine, pp 80-84.

Goodwin G. C., (1982). Adaptive Control - A Novel Approach to Control System Design, Second Conference on Control Engineering, Newcastle, 25-27 August.

Goodwin G. C., Kwai Sang Sin, (1984). Adaptive Filtering Prediction and Control, Prentice-hall, New Jersey.

Guy a. Dumont, Juan M. Martin-sanchez and Christos C. Zervos, (1989). <u>Comparison of an Auto-tuned PID Regulator and an Adaptive Preditive Control System on an Industrial Bleach Plant</u>, Automatica, Vol 25, No 1, pp 33-40.

Radke F., R. Isermann, (1987). <u>A Parameter-adaptive PID-controller with Stepwise Parameter Optimization</u>, Automatica, Vol 23, No 4, pp 429-457.

Texas Instruments, (1987). <u>First Generation of TMS320E15 User Guide</u>.

Walrath C.D, (1984). <u>Adaptive Bearing Friction Compensation Based on Recent Knowledge of Dynamic Friction</u>, Automatica , vol 20, pp 717-727.

AUTO-TUNING OF MULTIVARIABLE DECOUPLING CONTROLLERS

C. C. Hang, T. T. Tay and V. U. Vasnani

Dept. of Electrical Engineering, National University of Singapore, Kent Ridge (0511), Singapore

Abstract. Controller parameters for multivariable systems with interacting loops are difficult to tune. However, using feedforward compensators or decouplers, the interaction can be appropriately compensated so that effective non-interacting single-input/single-output processes are obtained. It then allows application of tuning techniques for single-loop PID controllers to the individual loop. This method, though simple in concept, is difficult to implement manually as it is tedious and time consuming. In this paper a method to automate the process of determinig the transfer functions of the decouplers and the tuning of PID parameters is presented. Typical 2-input/2-output plants are used as simulation examples.The controller is implemented on a microcontroller and tested on a portable simulator. Results show significant reduction in interaction and satisfactory tuning performance.

Keywords. Automatic tuning; PID control; Multivariable control systems; Decoupling; Feedforward.

1. INTRODUCTION

Controller parameters for multivariable systems are usually difficult to tune[1]. A typical 2-input/2-output system with interacting loops is shown in Fig 1. For practical reasons it is desirable that PID controllers in single loops, widely used in industries and well known to operators and practising engineers[2], be extended to the design of such multivariable controllers. However the main difficulty in tuning such a loop is that the tuning setting for one loop would upset the tuning of the other loop.

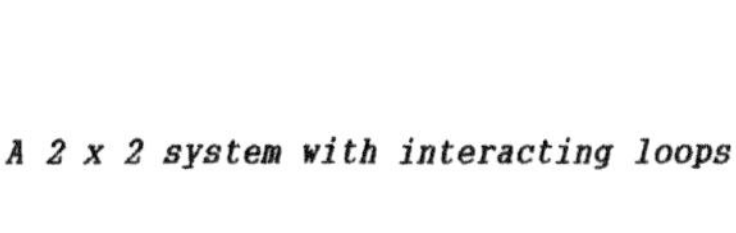

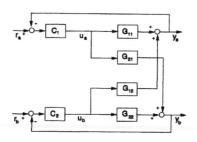

Fig 1: A 2 x 2 system with interacting loops

A simple and intuitively appealing scheme to the above problem is to insert two interaction compensators (or decouplers) much like feedforward controllers, to cancel out or reduce the effect of the interacting loops (Fig 2). In this way two single-input/single-output loops are obtained for which well known PID controller design methodology can be applied. This scheme, though simple in concept, is difficult to implement manually as it is tedious and time consuming.

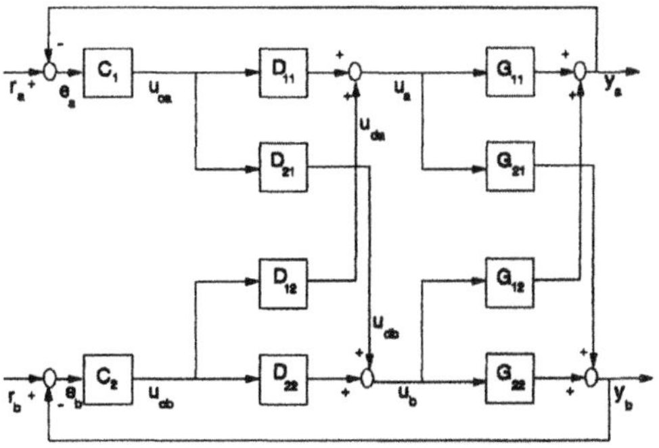

Fig.2: A 2x2 system with interacting loops with decouplers to cancel the effect of the interaction loops.

There are three distinct steps in the implementation of the above scheme. Firstly the transfer functions of both the forward and the interacting loops have to be identified. Secondly the transfer functions identified in step one are used to design the decouplers. Thirdly, the decoupled loops, free of interaction are then tuned using conventional PID tuning rules.

In this paper we present our experience in the automation of the three steps resulting in an auto-tuning multivariable controller. The design and implementation of this controller on a microcontroller based system is also discussed.

2. STRUCTURE OF AUTO-TUNING MULTIVARIABLE DECOUPLING CONTROLLER

In this section, the three steps for designing the multivariable controller are outlined in detail.

2.1 Identification of plant using correlation techniques

In this section, we consider a stable double-input/double- output plant:

$$\begin{bmatrix} y_a \\ y_b \end{bmatrix} = G \begin{bmatrix} u_a \\ u_b \end{bmatrix}$$

$$G = \begin{bmatrix} G_{11} & G_{12} \\ G_{21} & G_{22} \end{bmatrix}$$

With the plant in open loop, a N-sample PRBS is first sent into u_a with u_b held at a constant value and the output y_a, y_b are logged. The set of data collected can be used to identify G_{11}, G_{21}. Subsequently another N-sample PRBS is sent into u_b with u_a held constant. Similarly the values of y_a, y_b (effect due to u_b) are logged. This enables the identification of G_{22}, G_{12} using correlation technique.

PRBS has the property that its auto-correlation function approximates an impulse[3], Fig 3 and can be written as:

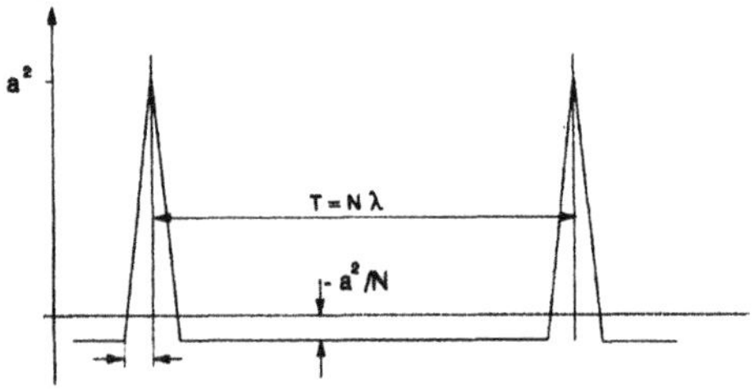

Fig 3: Autocorrelation Function of PRBS

$$\phi_{uu}(k-j) = \sum_{n=-\infty}^{\infty} a^2\lambda\ (1 + 1/N)\ \delta(k-j-nT) - \frac{a^2}{N} \qquad (2.11)$$

where N is the duration of the PRBS signal in number of sampling periods
a is the amplitude of the PRBS signal
λ is the sampling interval
and $T = N\lambda$

Consider a system G(s) with discrete impulse response given by $g(i)_{i=1,2..}$, and a corresponding set of input(u) and output(y) experimental data. Cross-correlation of the input and output data gives:

$$\phi_{uy}(k) = \frac{1}{N}\sum_{i=1}^{N} u(i)y(i+k) \qquad (2.12)$$

Assume that the output y(p) is perturbed by a noise term as follows:

$$y(p) = y_1(p) + n(p) \qquad (2.13)$$

where

$$y_1(p) = \sum_{j=1}^{\infty} g(j)u(p-j)$$ where g(j) is the discrete impulse response

n(p) as the noise term.

Substituting equation 2.13 into 2.12, gives

$$\phi_{uy}(k) = \sum_{j=1}^{\infty} g(j)\ \phi_{uu}(k-j)\lambda + \phi_{un}(k) \qquad (2.14)$$

where $\phi_{un}(k)$ is the cross-correlation between the input signal u(t) and noise n(t) which corrupts the output signal. If the input and noise signal are uncorrelated and n(t) has zero mean then $\phi_{un}(k)=0$. In practice, $\phi_{un}(k) \rightarrow 0$ as the noise term averages out to zero with a long correlation time.

With the length of correlation period (T) chosen to be greater than the settling time of the impulse response, the autocorrelation function for PRBS (equation 2.11) is substituted in 2.14, and after algebraic manipulation[3], the discrete impulse function at time $t=k\lambda$ is obtained as:

$$g(k) = \frac{1}{ma^2\lambda(1 + 1/N)} \left(\phi_{uy}(k) + \phi_{uy}(N)\right) \qquad (2.15)$$

where $m = \frac{1}{2}$ when $k = 0$, otherwise $m = 1$.

The term $\phi_{uy}(N)$ is used to compensate for the negative DC bias in the impulse response due to the negative DC bias in the autocorrelation function of PRBS (see Fig 3).

Using (2.15), the step response of the plant can be reconstructed as follows:

$$y(k) = \sum_{i=0}^{k} g(i)\lambda \qquad (2.16)$$

A first order and dead time model can then be easily extracted from the reconstructed step response as shown in Fig 4 below:

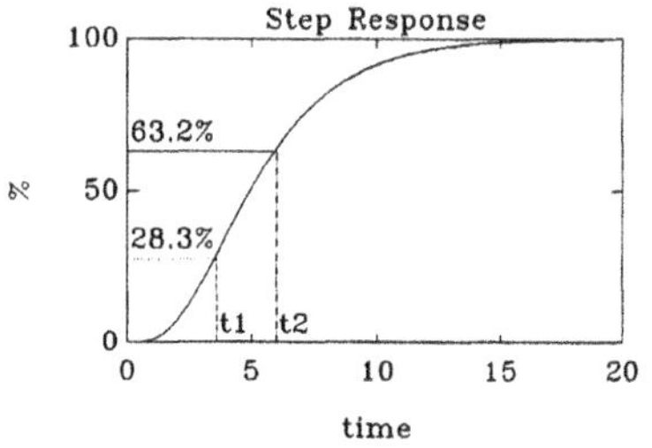

Fig 4: Extracting First Order and Dead Time parameters from Step Response

The extraction of the first order parameters is taken from [4], which proposes that the value of dead time and time constant be selected such that the model and actual responses coincide at two points in the region of high rate of change. The two points recommended correspond to 28.3% and 63.2% of the set point response. Taking t_1 and t_2 respectively as the time at which these points occur, the time constant (T) and dead time (d) are obtained as follows:

$$T = 1.5(t_2 - t_1);$$

$$d = t_2 - T; \qquad (2.17)$$

2.2 Determination of Decoupler Equations

In this subsection, we consider augmenting the double-input/double-output systems of the previous subsection to obtain two decoupled single-input/single-output loops.

In the two variable interacting system (Fig 2), we can write,

$$y_a(s) = G_{11}(s)MV_a(s) + G_{12}(s)MV_b(s)$$
$$y_b(s) = G_{22}(s)MV_b(s) + G_{21}(s)MV_a(s) \qquad (2.21)$$

The decoupler inputs (U) and its output ie. the manipulated variables(MV) are related by the following equations.

$$MV_a(s) = D_{11}(s)u_a(s) + D_{12}(s)u_b(s)$$
$$MV_b(s) = D_{21}(s)u_a(s) + D_{22}(s)u_b(s) \qquad (2.22)$$

For easy implementation of the decouplers, we make

$$D_{11}(s) = D_{22}(s) = 1 \qquad (2.23)$$

Substituting equations 2.22,2.23 into 2.21 the following equations are obtained

$$y_a(s) = (G_{11}+ G_{12}D_{21})u_a + (G_{11}D_{12}+ G_{12})u_b$$
$$y_b(s) = (G_{22}D_{21}+ G_{21})u_a + (G_{22}+ G_{21}D_{12})u_a \qquad (2.24)$$

For complete decoupling, y_a is *only* affected by u_a and y_b *only* by u_b. As such four equations can be obtained from 2.24:

$$G_{11}+ G_{12}D_{21} = H_1$$
$$G_{11}D_{12}+ G_{12} = 0$$
$$G_{21}+ G_{22}D_{21} = 0$$
$$G_{21}D_{12}+ G_{22} = H_2 \qquad (2.25)$$

where H_1 and H_2 are defined as:

$$y_a = H_1u_a$$
$$y_b = H_2u_b \qquad (2.26)$$

Equations 2.25 and 2.26 are then solved to obtain the decoupler equations.

$$D_{12}(s) = -\frac{G_{12}(s)}{G_{11}(s)}$$

$$D_{21}(s) = -\frac{G_{21}(s)}{G_{22}(s)} \qquad (2.27)$$

With the plant model identified, D_{12} and D_{21} can be obtained as in (2.27). The decouplers are implemented digitally using the Tustin approximation..

2.3 Determination of Optimal PID Parameters

Once the decouplers are in place, the transfer function H_1 and H_2 as in (2.26) can be obtained as follows:

$$H_1(s) = G_{11}(s) - \frac{G_{12}(s)G_{21}(s)}{G_{22}(s)}$$

$$H_2(s) = G_{22}(s) - \frac{G_{12}(s)G_{21}(s)}{G_{11}(s)} \qquad (2.28)$$

Substituting $s=j\omega$, a search strategy can then be employed to locate the frequency ω_u for which the phase of each of the decoupled system H_1 and H_2 is 180°.

The search strategy involves an initial guess followed by an iterative search using the Newton-Raphson method to converge to the ultimate frequency ω_u from which the ultimate period t_u can be easily obtained. The ultimate gain is calculated at that frequency. Refined Ziegler-Nichols tuning rules[5] then provide the appropriate PID tuning settings from the ultimate gain and ultimate frequency. These refined rules sets the appropriate set point weighting factor, β, and adjustments to the integral time based on the normalised process gain, κ. The normalised process gain is the product of the ultimate gain and the process steady-state gain. In short the rules are:

for $2.25 < \kappa < 15$;

$$\beta = \frac{15 - \kappa}{15 + \kappa} \qquad (2.29)$$

for $1.5 < \kappa < 2.25$;

$$\beta = \frac{8}{17}\left(\frac{4}{9}\kappa + 1\right)$$

$$\mu = \frac{4}{9}\kappa \qquad (2.30)$$

where μ is the multipying factor in the adjustment of the integral time, T_i,such that:

$$T_i = 0.5\mu t_u \qquad (2.31)$$

The control law is thus modified to:

$$u_c = K_p\left[(\beta r - y) + \frac{1}{T_i}\int e\,dt - T_d\frac{dy}{dt}\right] \qquad (2.32)$$

where K_p is the proportional gain
r is the setpoint
T_d is the derivative time
y is the process output
$e = (r-y)$

2.4 Advantage and limitation of this structure

This procedure is relatively simple and the principle underlying the procedure and its implementation is easily understood. The whole process of collecting data for identifcation and calculation of the decouplers and controller tuning parameters can be fully automated with this new controller structure. Re-tuning can be easily done periodically to cater for drifting of plant parameters. This scheme would work for any open loop stable system. The above scheme, fully automated with features of bumpless transfers, is implemented on an Intel 8096 based microcontroller board. The board can be now used as a stand alone controller allowing ease of portability and implementation.

3.0 ISSUES IN IMPLEMENTATION

3.1 Bumpless Transfer

Once the decoupler transfer functions are obtained, they must be initialized to ensure a bumpless transfer when the loop is closed and the decouplers are put into operation.

Referring to Fig 2, at steady state the values of u_{ca}, u_{cb}, u_{da}, u_{db}, u_a, u_b, y_a, y_b, are at some constant steady values, us_{da}, us_{db}, us_a, us_b, ys_a, ys_b, respectively. Before the loop is closed, only the values of us_a, us_b, ys_a,ys_b are known. From the equations of the decouplers (2.22 & 2.23), the steady state gains, g_{12} and g_{21}, of D21 and D12 respectively can be determined. Hence the following equations can be obtained

$$us_{ca} + us_{db} = us_a$$

$$us_{cb} + us_{da} = us_b$$

and

$$us_{da} = g_{12}us_{cb} \quad , \quad us_{db} = g_{21}us_{ca} \tag{3.11}$$

These equations are solved to obtain us_{ca}, us_{cb}, us_{da},and us_{db} . With the values of us_{ca} and us_{cb}, the integrator in the controller can be initialized for bumpless transfer. These values also enable the decouplers to be initialized. For example D12 has the following typical structure:

$$D12 = \frac{u_{da}}{u_{cb}} = q^{-d}\frac{b_0 + b_1q_1^{-1} + b_2q^{-2}}{a_0 + a_1q^{-1} + a_2q^{-2}} \tag{3.12}$$

i.e.

$$u_{da} = \{\, b_0u_{cb}(k-d) + b_1u_{cb}(k-d-1) + b_2u_{cb}(k-d-2) - a_1u_{da}(k-1) - a_2u_{da}(k-2) \,\}/\, a_0 \tag{3.13}$$

Hence at steady state,

$$u_{cb}(k-d) = u_{cb}(k-d-1) = u_{cb}(k-d-1) = us_{cb}$$

and

$$u_{da}(k) = u_{da}(k-1) = u_{da}(k-2) = us_{da} \tag{3.14}$$

As such the decouplers are initialized and put into operation by closing the loops.

3.2 PID Controller Implemented

The PID controller structure implemented includes features of anti-reset windup and set-point weighting. Anti-reset windup is a necessary feature in cases of actuator saturation as it reduces the build up of the control variable on saturation.

Set-point weighting is included as a feature and its valueis governed by the refined Ziegler-Nichols tuning rules[5].

In addition, bumpless transfer is carried out when the controller is placed in operation by using the values of us_{ca} and us_{ca} calculated in (3.11) above to initialize the integrator in the controller structure.

3.3 Implementation of Controller on a Microcontroller

The Intel 8096 microcontroller board termed as the Universal Controller 96 (UC96) has been designed and built in-house as a general purpose board. It exploits the on-chip features of the 8096 such as 10-bit analog to digital converters, counters/timers, serial port, I/O ports to achieve low chip count and consequently realiability. The card has 12-bit digital to analog converters, as well as 8 K of RAM and 8 K of ROM allowing for programming of a range of control applications. In addition it has a RS232 interface allowing for serial communication with a host computer.

3.4 Issues in Software Development

The Development tools include a *C* language compiler - C96, developed by Intel, which allows for code to be written in high level language and then compiled into object code. The code is then executed on the 8096. The 8096 assembly language -ASM96 consists of a large instruction set including integer divide and multiplication instructions. A floating point library is also provided for real number computations.

Memory buffers of fixed size are allocated to hold process input and output values. These locations are fixed and are then used in the *C* environment for the various calculations.

The mathematical library functions such as inverse tangent function, required in the search for the phase crossover, is however absent from this compiler. This was then done by means of table lookup and interpolation.

The main points that had to be noted in the development are the memory size constraint as well as speed of execution. As such whenever possible, the same allocated data area is re-used again. An example would be in the calculation of the impulse response, the resulting impulse response data points are stored in the same location as the output points were stored.

3.5 Simulation on a Portable Simulator

The algorithm is tested on a portable simulator which is essentially the same microcontroller-based board configured to act as a digital simulator functioning at a sampling interval of 0.1s[2]. It has a wide variety of dynamics including a 2-input/2-output system with interacting loops.

Simulation was done mainly on plant model from [6], however to speed up the simulation, the time constants of the plant were reduced by a factor of 10. Hence the result are still valid upto a scale factor of 10.

The plant model is illustrated as shwon in Fig 5.

Simulation results for

$$G_{11} = G_{22} = \frac{1}{(1 + s)} \quad , \quad G_{12} = G_{21} = \frac{0.5}{(1 + 2s)}$$

with $L_1 = L_2 = 0.5$s and $L_1 = L_2 = 1.0$s are shown in Fig 6 and Fig 7 respectively. In Fig 7, PRBS is sent through u_a with u_b held constant from t=0s to

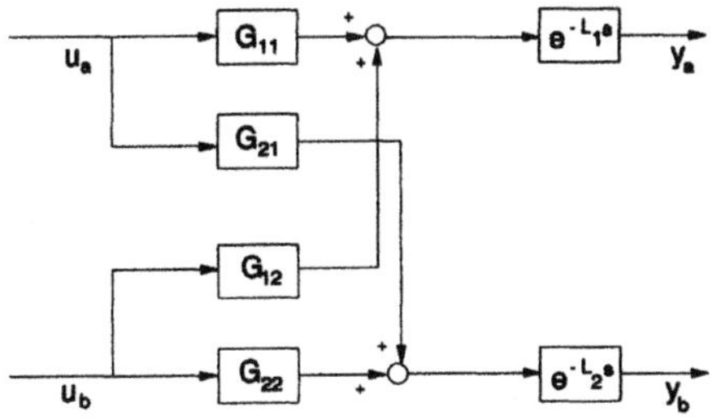

Fig 5 Block diagram of process with interactions

t=37s. From *t*=37s to *t*=46s computation takes place for identifying the processes G_{11} and G_{21}. At *t*=46s to *t*=83s, PRBS is sent out through u_b with u_a held constant. Identification of G_{22}, G_{12} computations for tuning and bumpless transfer take place from *t*=83s to *t*=93s. At *t*= 93s, both loops are closed, a setpoint change on loop *a* to 60% while loop *b* is held at 45%. This is followed by a setpoint change to 60% in loop *b* at *t*=123s, and a setpoint change to 45% in loop *a* again at *t*=143s. Similar processes occur for the second simulation whoose results are shown in Fig 7 and for the rest of the simulations done.

Results for

$$G_{11} = G_{22} = \frac{1}{(1 + 0.5s)^2}, \quad G_{12} = G_{21} = \frac{0.5}{(1 + 2s)}$$

with $L_1 = L_2 = 0.5s$ and $L_1 = L_2 = 1.0s$ are shown in Fig 8 and Fig 9 respectively.

The results are as expected, showing a significant reduction in interaction as a result of decoupling. Tuning results are satisfactory in view of the relative large dead time with respect of the dominant time constant of the decoupled loops. This is achieved with the refined Ziegler-Nichols tuning rules[5].

4.0 CONCLUSION

The results show that the process of determining the appropriate feedforward compensators or decouplers can be automated using a correlation technique to obtain the plant models. With the decoupled single input/single

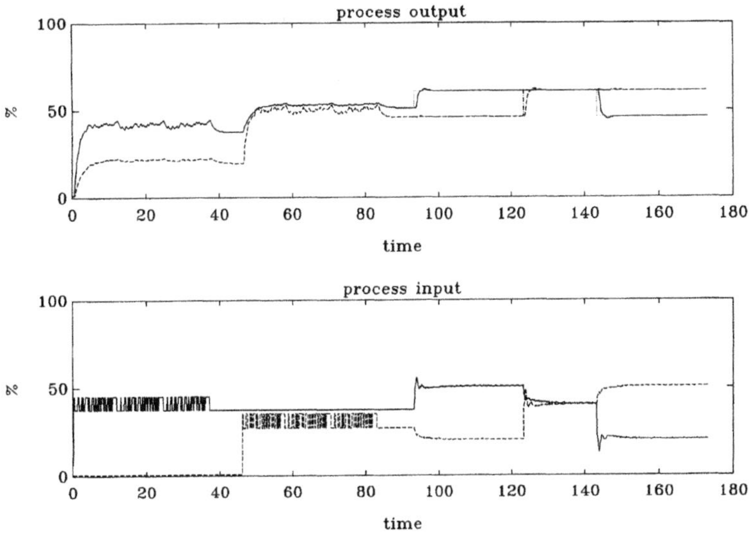

Fig 6: Simulation plots for

$$G_{11} = G_{22} = \frac{1}{(1 + s)}, \quad G_{12} = G_{21} = \frac{0.5}{(1 + 2s)}, \quad L_1 = L_2 = 0.5s,$$

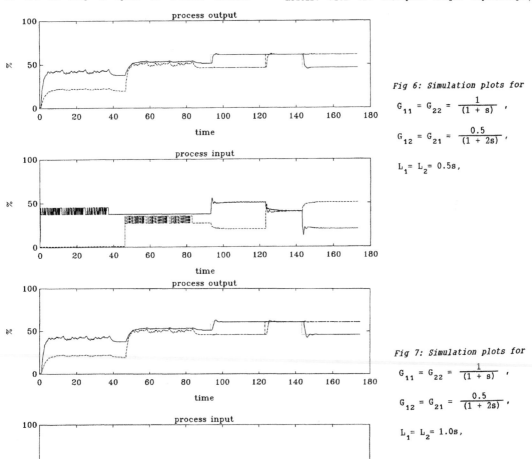

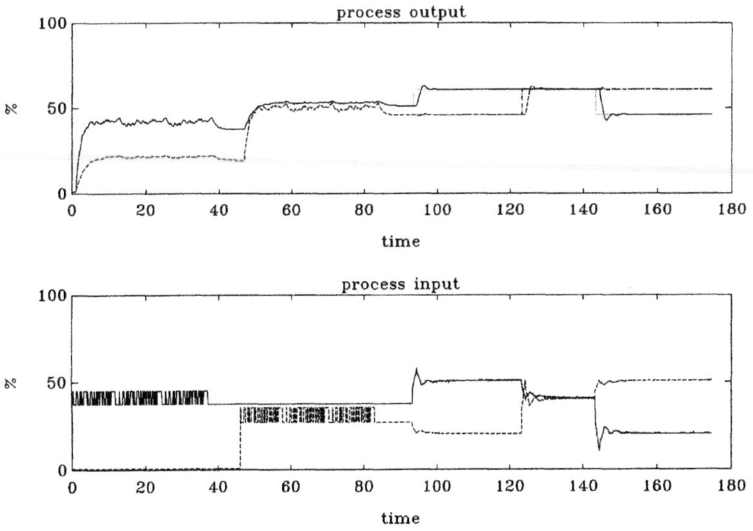

Fig 7: Simulation plots for

$$G_{11} = G_{22} = \frac{1}{(1 + s)}, \quad G_{12} = G_{21} = \frac{0.5}{(1 + 2s)}, \quad L_1 = L_2 = 1.0s,$$

output loops, automatic tuning of PID parameters can be carried out satisfactorily.It is hoped that the availability of auto-tuners for multivariable processes will stimulate more routine appliations of multivariable control in practice.

References

[1] Ochiai S.,D.O. Roark, "A case for multivariable Control". Instrument Technology, Sept 1987, pp67-72.

[2] Hang C.C., K.K. Sin, "Training Simulators For Advanced Process Control",InstrumentAsia Conf, S'pore 1989.

[3] Davies W.D.T., System Identification For Self Adaptive Control 1970 John Wiley.

[4] Smith C.A., A.B. Armando, Principles and Practice of Automatic Process Control, 1985 John Wiley pp 219-220.

[5] Hang C.C., K.J. Astrom, W.K. Ho, "Refinements Of The Ziegler-Nichols Tuning Formula, National University of Singapore, Department of Electrical Engineering, Control Group Technical Report CI-90-1,Apr 1990

[6] Yokogawa,'Simulated Results of Multivariable Control System', Applied Engineering Data, Jan 15 1985.

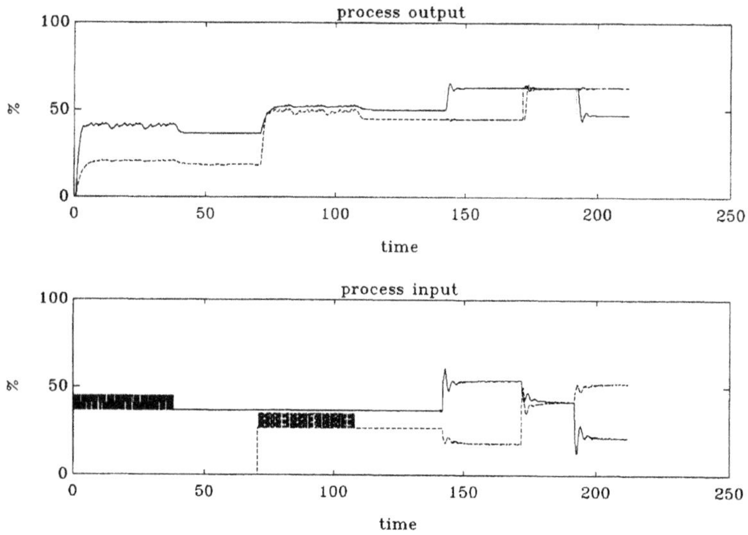

Fig 8: Simulation plots for

$$G_{11} = G_{22} = \frac{1}{(1 + 0.5s)^2},$$

$$G_{12} = G_{21} = \frac{0.5}{(1 + 2s)},$$

$$L_1 = L_2 = 0.5s,$$

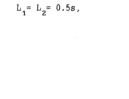

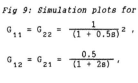

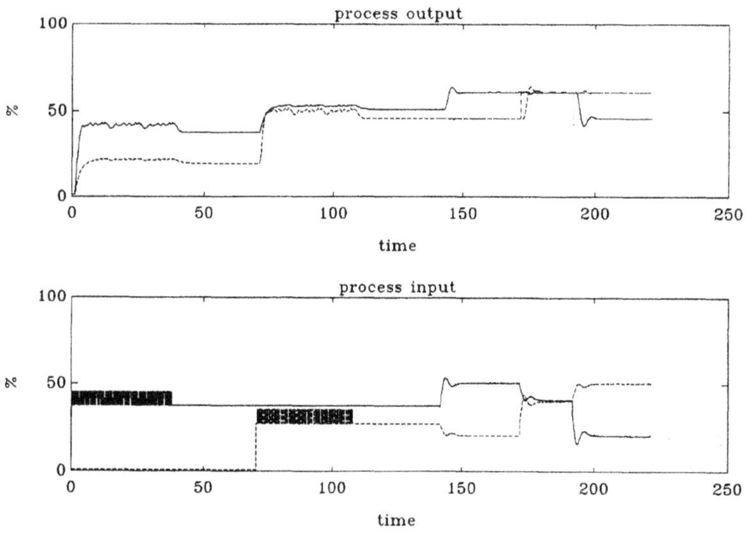

Fig 9: Simulation plots for

$$G_{11} = G_{22} = \frac{1}{(1 + 0.5s)^2},$$

$$G_{12} = G_{21} = \frac{0.5}{(1 + 2s)},$$

$$L_1 = L_2 = 1.0s,$$

ISSUES IN THE DESIGN OF AN ADAPTIVE GPC FOR SYSTEMS REQUIRING FAST SAMPLING

T. H. Lee and E. K. Koh

Dept. of Electrical Engineering, National University of Singapore, Kent Ridge (0511), Singapore

Abstract: The Generalised Predictive Controller (GPC) has been demonstrated in many implementation experiments to be a suitable candidate for adaptive control of a wide range of systems. In this paper, we investigate the application of the adaptive GPC for systems requiring fast sampling. Realisations in the shift and delta operator forms are discussed, and an appropriate delta operator re-design of the adaptive GPC is proposed.

1 Introduction

Fast servos are a feature of many present day control systems including two-axis pedestal systems, stabilised platforms and free gyro stabilised mirror systems. In many of these, the performance obtained with fixed time-invariant controllers are not satisfactory as the system parameters change under operational conditions, for example, with different payloads or with different flywheel speeds in systems with some passive in-built stabilisation. For such systems, the use of adaptive controllers for both automatic tuning as well as fully on-line adaptive purposes are attractive. A feature of such systems is that they require fast sampling both in the absolute sense (typically less than 20 milliseconds) and the relative sense (sampling frequency higher than ten times the bandwidth of the overall system) to achieve acceptable performance. In this paper, we investigate the application of the adaptive GPC [1],[2] to these systems.

First, we consider the performance of the conventional adaptive GPC, which is designed using a shift operator formulation, when it is applied to systems requiring fast sampling when moderate noise is present. Simulation results show that poor performance can result under these circumstances.

We next considered a delta operator realisation of the conventional adaptive GPC in order to investigate if such an approach might help to solve the problem encountered. This is motivated in particular by the very promising results reported in [3] where Middleton and Goodwin considered an alternative discrete-time operator, the delta operator δ, and showed that it had superior numerical properties which are especially important when the digital control system under consideration has a requirement that the sampling frequency be in excess of ten times the bandwidth of the controller. Issues involved in realising the adaptive GPC (designed in the shift operator) using delta operators are discussed.

As it turned out, simply using delta operators to implement controllers designed using a shift operator formulation was not necessarily the best thing to do; there are no clear guidelines and the procedure was fraught with potential for dangerous slip-ups. Based on these observations, a complete delta operator re-design (in contrast to simply a delta operator realisation of a shift operator design) of the adaptive GPC was considered instead. The re-design is presented in the paper, and simulation results indicate that it provides better performance under circumstances of moderate noise.

2 The GPC — conventional shift operator design

2.1 Application of the shift operator adaptive GPC under situations of moderate noise

The GPC designed in shift operator formulation considers the discrete-time plant

$$A(q^{-1})y(t) = B(q^{-1})u(t-1) + \ell$$

where $A(q^{-1})$ and $B(q^{-1})$ are the pole and zero polynomials respectively of the plant, and ℓ models the effect of unmesurable step load disturbances. An equivalent description is

$$A(q^{-1})\Delta y(t) = B(q^{-1})\Delta u(t-1)$$

Using either of the above descriptions of the plant, there are various algorithms that can be used to obtain the estimates $\hat{A}(q^{-1})$ and $\hat{B}(q^{-1})$. In a system which requires fast sampling, the coefficients of the $B(q^{-1})$ polynomial are typically several orders of magnitude smaller than the coefficients of the $A(q^{-1})$ polynomial. This poses numerical problems in real time parameter estimation of $\hat{A}(q^{-1})$ and $\hat{B}(q^{-1})$, especially in situations of moderate noise. In such cases, the estimated parameters may, for example, be accurate within a tolerance of ± 0.01. While this is acceptable for the $\hat{A}(q^{-1})$ estimates, it is totally unacceptable for the $\hat{B}(q^{-1})$ estimates which may be off by as much as 100 per cent! This situation is shown in Figure 1 which is a simulation of the application of the shift operator adaptive GPC to a plant under situations of moderate noise.

2.2 Delta operator implementation of a shift operator design of the adaptive GPC

In view of the previous situation discussed, it seemed logical to consider a delta operator implementation of the GPC (which was designed using a shift operator formulation). In our investigations, we had proceeded with this approach and found that it was not at all a viable option.

As an experiment, we had provided some control engineers with the design and algorithm [1], [2] of the GPC and also the guidelines on using the delta operator for implementation [3]. In all cases, they came back with an implementation coding that used the operator δ^{-1} as a substitution for q^{-1}, firmly believing that the improved properties claimed in [3] would be achieved in their implementation. Naturally, simulation experiments using this coding ran into numerical problems as the basic operator used was δ^{-1} which was a marginally unstable basic operator sensitive to initial conditions.

To some extent, the above is not unexpected as the procedure to translate adaptive controllers designed in the shift operator to a form suitable for implementation in the delta operator is not as straightforward as one might be led to think. To illustrate this, consider the simpler case of translating the d-step ahead controller, designed in the shift-operator, for coding in the delta operator. This is less complicated than the GPC; however it shares a few key common features and will serve as an example to illustrate the complications involved. The design of the d-step

[1] Author to whom all correspondence should be addressed.

ahead controller considers the plant

$$A(q^{-1})y(t) = q^{-d}B(q^{-1})u(t) \tag{1}$$

and the prediction identity

$$1 = A(q^{-1})E(q^{-1}) + q^{-d}F(q^{-1}) \tag{2}$$

Using (1) and (2), the plant can be written in predictive form

$$y(t) = F(q^{-1})y(t-d) + E(q^{-1})B(q^{-1})u(t-d) \tag{3}$$

In the indirect version of the adaptive controller (which is closest to the GPC design as the adaptive GPC is also an indirect adaptive controller), all three expressions are needed for the implementation of the controller. Thus, to translate the indirect adaptive controller into a form suitable for coding and implementation using delta operators, all three expressions must be cast into causal and realisable forms involving delta operators.

One appropriate procedure for converting the above controller into a delta operator form suitable for implementation is as follows: First multiply (1) by q^n, where n is the maximal order of the backward shift in (1). This results in

$$A(q)y(t) = B(q)u(t)$$

This can then be converted to delta operator form

$$\bar{A}(\delta)y(t) = \bar{B}(\delta)u(t)$$

as there exist appropriate $\bar{A}(\delta)$ and $\bar{B}(\delta)$ corresponding to $A(q)$ and $B(q)$. However, this is still not quite in a form suitable to be used in indirect adaptive control for one has to convert it into an equivalent expression involving realisable signals. Thus, the expression should be be shifted n time steps backward to obtain

$$\bar{A}(\delta)y(t-n) = \bar{B}(\delta)u(t-n) \tag{4}$$

This is then the equivalent to (1) involving realisable signals which can be used for parameter estimation.

Next we shall have to write (2) in delta operator form. Thus consider

$$q^n q^{d-1} = q^n q^{d-1}\left\{A(q^{-1})E(q^{-1}) + q^{-d}F(q^{-1})\right\}$$

and the proper delta operator coding of (2) is

$$(1+\delta h)^n(1+\delta h)^{d-1} = \bar{A}(\delta)\bar{E}(\delta) + \bar{F}(\delta)$$

This is then the design identity which has to be solved at each sampling interval for $\bar{E}$ and $\bar{F}$.

Finally, to obtain the proper δ coding of (3), operate on (3) by $q^{n-1}q^d$ to obtain

$$q^{n-1}q^d y(t) = \bar{F}(\delta)y(t) + \bar{E}(\delta)\bar{B}(\delta)u(t)$$

or

$$y(t+d) = \left\{q^{-(n-1)}\bar{F}(\delta)\right\}y(t) + \left\{q^{-(n-1)}\bar{E}(\delta)\bar{B}(\delta)\right\}u(t) \tag{6}$$

The expressions (4), (5) and (6) comprise the delta operator implementations of (1), (2) and (3). It is quite obvious that the procedure is complicated even in this simple case, and the natural conclusion is that a re-design is more appropriate.

3 Delta operator re-design of the adaptive GPC

In the previous section, it has been observed that the conventional adaptive GPC can yield poor performance under conditions of moderate noise in the case of systems requiring fast sampling. It has also been observed that an attempt to simply provide a delta operator realisation of the conventional adaptive GPC involves fairly complicated procedures and is fraught with the potential for dangerous slip-ups. In this section, we consider instead the alternative approach of providing a delta operator re-design of the adaptive GPC.

3.1 Control law

Consider a discrete-time plant described by

$$A_p(\delta)y(t) = B_p(\delta)u(t) + \ell_p \tag{7}$$

where δ is the delta operator $\delta \equiv \frac{q-1}{h}$ and ℓ_p is a constant unmeasurable load disturbance. The plant given in (7) thus is assumed to have a strictly proper transfer function from the input $u(t)$ to the output $y(t)$. However any (but obviously not all) of the coefficients of B_p may take on zero values. The implication of posing the problem in this framework is that the controller that we develop below will not require exact knowledge of the relative degree of the plant (7) but simply an upper bound.

It is also assumed that an upper bound n ($n \geq n_p$) is known for the order of the plant, so that for the development of the control algorithm, the plant is modelled as

$$A(\delta)y(t) = B(\delta)u(t) + \ell \tag{8}$$

with $\deg(A) = n$ and $\deg(B) = n-1$. In (8), A and B are polynomials that differ from A_p and B_p respectively by an arbitrary stable polynomial of degree $n - n_p$ and ℓ differs from ℓ_p by a constant.

For a large class of controllers, some kind of a predictive structure is typically used as a basis for control law design [4]. Therefore, consider the "prediction" identity

$$(\delta + a)^j T(\delta) = A(\delta)\delta E_j(\delta) + F_j(\delta)$$

where $a > 0$ and $T(\delta)$ is a stable monic polynomial with $\deg(T) = \deg(A) = n$. Since $T(\delta)$ is a stable polynomial, this leads to the predictive form

$$(\delta + a)^j y(t) = G_j(\delta)\delta u^f(t) + F_j(\delta)y^f(t)$$

where $G_j(\delta) \equiv E_j(\delta)B(\delta)$, $T(\delta)y^f(t) = y(t)$ and $T(\delta)\delta u^f(t) = \delta u(t)$. For $j = 1, 2, ...N$, we can write the bank of N predictors compactly in vector form as

$$Y = GU + F$$

where

$$\begin{aligned} Y &= [(\delta+a)y(t), ..., (\delta+a)^N y(t)]^T \\ U &= [\delta u(t), ..., \delta^{N-1}(\delta u(t))]^T \\ F &= [f_1(t), f_2(t), ..., f_N(t)]^T \end{aligned}$$

Let

$$W = [(\delta+a)y_m(t), ..., (\delta+a)^N y_m(t)]^T$$

with

$$A_m(\delta)y_m(t) = a_N^* r(t)$$

where $r(t)$ is the reference signal, and $A_m(\delta)$ is a stable polynomial given by

$$A_m(\delta) = \delta^N + a_1^*\delta^{N-1} + ... + a_N^*$$

The choice of the elements of the W vector is important. These may be considered as the appropriate signals that the outputs of the bank of N predictors are expected to match. Here is also where the delta operator re-design differs considerably from the shift operator design of [1].

To derive the control law, consider the criterion function

$$\begin{aligned} J &= (Y - W)^T(Y - W) + U^T\Lambda U \\ &= (GU + F - W)^T(GU + F - W) + U^T\Lambda U \end{aligned}$$

with $\Lambda = \text{diag}(\lambda_j)$. The criterion function is minimised by

$$U = (G^TG + \Lambda)^{-1}G^T(W - F)$$

The control law is thus

$$\delta u(t) = g^T(W - F) \tag{10}$$

where g^T is the first row of $(G^TG + \Lambda)^{-1}G^T$. Note the inclusion of integral action in the control law.

3.2 Parameter estimation

For adaptive control, we will need to adaptively update the G_j and F_j polynomials. This can be done by first estimating the parameters of A and B. By appropriate reorganisation of the available signals it can be shown that

$$y(t) = (T_2(\delta) - \bar{A}(\delta))y^{f_2}(t) + B(\delta)\bar{u}^{f_2}(t)$$

This clearly gives us a suitable estimation model in the linear-in-the-parameters form:

$$y(t) = \psi(t)^T\theta$$

where

$$\theta = [(t_1 - a_1^*), ..., (t_{n+1} - a_{n+1}^*), b_0, ..., b_{n-1}]^T$$

The parameters can then be estimated using any one of a number of suitable methods [4].

Simulation results for this delta operator re-design of the adaptive GPC are shown in Figures 2 through 5. They indicate that improved performance in noise is achieved, and that the re-design retains the desirable properties of the conventional adaptive GPC in being applicable to non-minimum phase and overparameterised systems.

4 Conclusion

In this paper, we have considered the application of the adaptive GPC to systems requiring fast sampling. For this class of systems, the conventional adaptive GPC design using a shift operator formulation can yield poor performance under conditions of moderate noise. The use of delta operators to simply provide a realisation of the controller (designed in shift operator formulation) was complicated and fraught with potential problems. A delta operator re-design was shown to be a much better approach and this yielded improved performance under circumstances of noise while still maintaining the desirable properties of the adaptive GPC.

References

[1] Clarke, D. W., C. Mohtadi and P. S. Tuffs (1987a), "Generalized Predictive Control—Part I. The Basic Algorithm," *Automatica*, Vol. 23, No. 2, pp. 137–148.

[2] Clarke, D. W., C. Mohtadi and P. S. Tuffs (1987b), "Generalized Predictive Control—Part II. Extensions and Interpretations," *Automatica*, Vol. 23, No. 2, pp. 149–160.

[3] R. H. Middleton and G. C. Goodwin (1986), "Improved Finite Word Length Characteristics in Digital Control Using Delta Operators," *IEEE Trans. Automatic Control*, Vol. AC-31, pp. 1015–1021.

[4] Astrom, K. J. and B. Wittenmark (1988), "Adaptive Control," *Addison Wesley Book Company*, Massachussets, U. S. A.

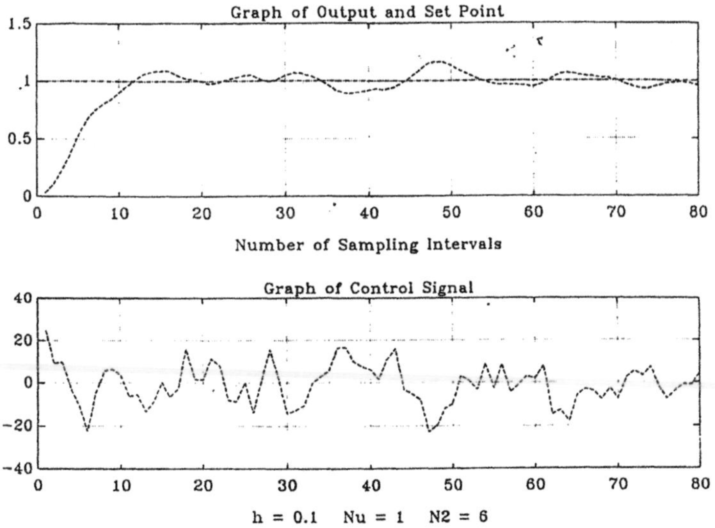

Figure 1 : shift GPC : N2 = 6, Nu = 1, k = 2 with System Noise

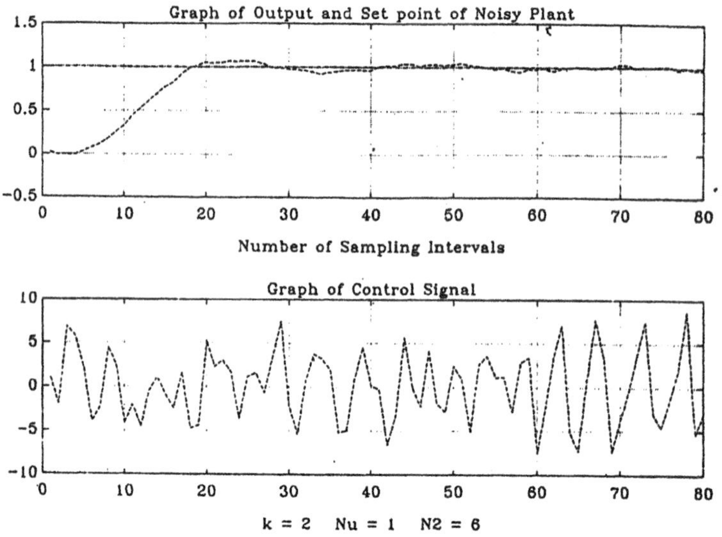

Figure 2 : δ-GPC : N2 = 6, Nu = 1, k = 2 with System Noise

Plant	Transfer Function	Response Curves
non-minimum phase	$\frac{-0.2s + 1}{s(s^2+1)}$	Figure 3
open loop unstable	$\frac{.0.2s + 1}{s(s^2+1)}$	Figure 4
over-parameterised	$\frac{b(s)}{b(s)(s^2+1)}$	Figure 5

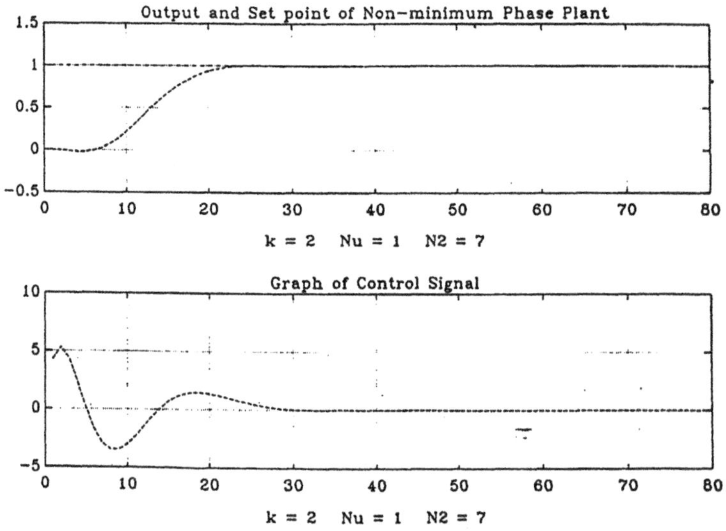

Figure 3 : δ-GPC Controlling a Non-minimum Phase Plant

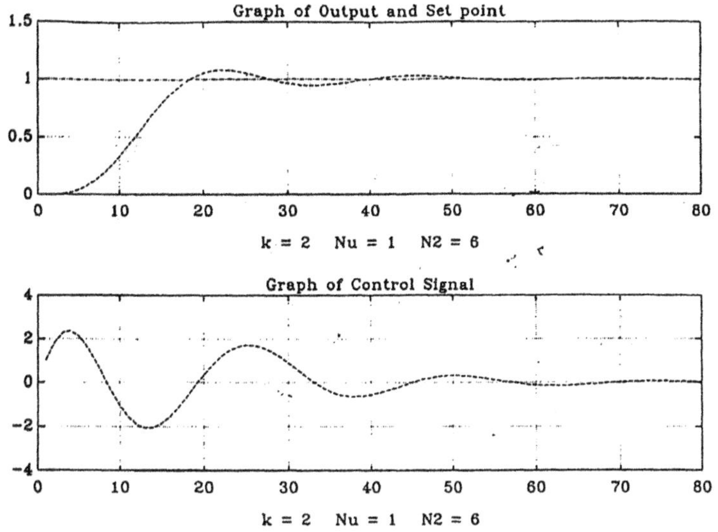

Figure 4 : δ-GPC controlling a Marginally Stable Plant

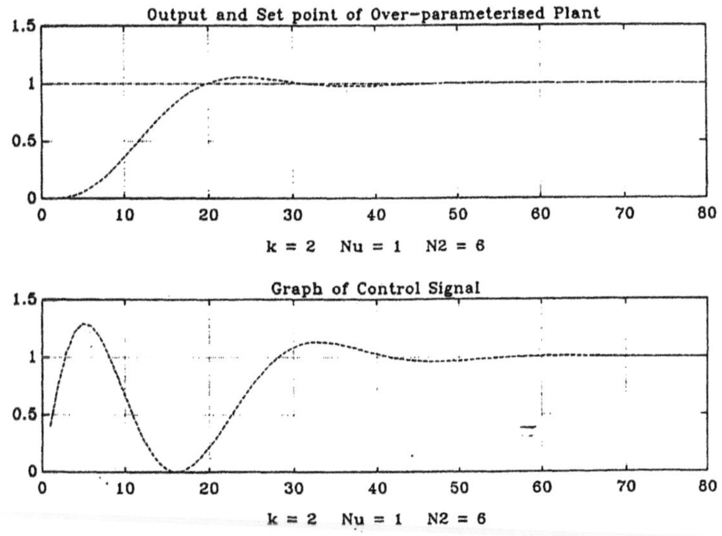

Figure 5 : δ-GPC controlling an Over-parameterised Plant

ADAPTIVE POLE PLACEMENT CONTROL LAW USING THE DELTA OPERATOR

+M. A. Garnero*, G. Thomas**, B. Caron*** and J. F. Bourgeois*

*Electricité de France, Direction des études et recherches, Département A.D.E., B.P 1, 77250 Moret-sur-Loing, France
**Université Lyon I, Laboratoire LAGEP/ECOLE CENTRALE de LYON, Département M.I.S., 36 Avenue Guy de Collongue, 69131 Ecully Cedex, France
***Université de Savoie, Laboratoire LAMII, 41 Avenue de la plaine, B.P. 806 -F 74016 Annecy cedex, France

Abstract.
This paper deals with problems encountered in continuous-time estimation. The process is modelled with the delta operator ($\delta=(q-1)/T_e$ where T_e is the sampling period and q the shift operator). The structure of the discrete-time model with the operator q is supposed to be the same as the corresponding continuous-time model. The parameters of the delta model tend in this case towards those of the continuous-time model when the sampling period tends towards zero. The problems encountered when treating the data of the observation equation are the same as in the continuous-time approach (particularly, the problems of the derivatives). They are discussed here.
A pole-placement control law using the parameters of the delta model is then applied. Two possibilities for the choice of the closed-loop poles will be presented.
Physical experiments on an electrical laboratory furnace are discussed.
Here only the case of processes without time delay will be treated. But a natural robustness of the adaptive control law with respect to eventually non modelled time delays will appear.

Keywords.
Adaptive control; control applications; difference equations; parameter estimation; pole placement.

I] Introduction.

The advantages of the delta modelling are multiple. Numerical problems for estimated parameters are avoided when the sampling period tends towards zero. When modelled with the operator q the process sees its poles tend towards the point (1,0) of the discrete plane, some zeros become unstable when the sampling period becomes small with respect to the closed loop time-constants of the system (ASTROM, K.J. et al.,1984). This disappears with the delta modelling (GOODWIN G.C. et al., 1986; MIDDLETON R.H. et al.,1986).

On the contrary the phase of treatment of data in the observation equation which was obvious in the case of the q operator must be carefully handled as in the continuous-time case. We will present three different methods of rewriting the observation equation. In the first approach, the derivatives of the data are filtered. In the second method, the successive outputs of a cascaded filter with identical stages ("POISSON filter chain" as referred by UNBEHAUEN and RAO (1987,1990)), each element of which having a first order transfer function, are exhibited. In the third method the data are linear combinations of the filtered derivatives of different order of the input/output signals.

+The authors are also with the GRECO CNRS "SARTA".

A pole-placement control law is then applied. It is formulated in the delta operator. It uses the results of the identification procedure to solve the BEZOUT equation. Two choices for the poles of the closed-loop system are proposed. They are based on the frequency content of the signals in the feedback loop.

II] Identification.

The process is supposed to be modelled with the q operator by:

$$(q^2 + a_1 q + a_2)y(k)=(b_1 q + b_0)u(k) \quad (1)$$

Be $\delta =(q - 1)/T_e$. Then (1) becomes:

$$(\delta^2+a'_1\delta+a'_2)y(k)=(b'_1\delta+b'_0)u(k) \quad (2)$$

II.1] First proposal:

The data of equation **(2)** tend towards the derivatives of the input/output signals when the sampling period tends towards zero. **(2)** must be filtered by a low-pass second order filter:

$$x_f(k) = \frac{1}{(1 - \alpha\, q^{-1})^2}\, x(k).$$

Be:

$$x_{f1}(k) = \frac{(1 - \alpha)\, q^{-1}}{(1 - \alpha\, q^{-1})}\, x(k) \text{ and}$$

$$x_{f2}(k) = \frac{(1 - \alpha)^2\, q^{-2}}{(1 - \alpha\, q^{-1})^2}\, x(k).$$

where the signs "f1" and "f2" are the notations for those two filters.

(2) becomes : $v(k) = \Phi^T(k)\, \theta$ **(3)**

with $\Phi^T(k)$ of the form:

$$\left[\frac{-T_e}{(1-\alpha)} (y_{f1}(k) - y_{f2}(k)),\ \frac{-T_e^2}{(1-\alpha)^2} y_{f2}(k) \right.$$

$$\left. \frac{T_e}{(1-\alpha)} (u_{f1}(k) - u_{f2}(k)),\ \frac{T_e^2}{(1-\alpha)^2} u_{f2}(k) \right]$$

and $\theta^T = [a'_1,\ a'_2,\ b'_1,\ b'_0]$.

We can then apply a standard estimation algorithm to equation **(3)** to extract the parameter vector.

II.2] Second proposal.

This proposal has been inspired by the works of UNBEHAUEN and RAO (1987, 1990) from the POISSON Moment Functionals. The data of the observation vector are not the filtered derivatives of the input/output data but the filtered input/output data.

Equation **(2)** is equivalent to equation **(4)** with:

$$y(k) + a''_1 \frac{\alpha}{(\delta + \alpha)} y(k) + a''_2 \frac{\alpha^2}{(\delta + \alpha)^2} y(k)$$

$$= b''_1 \frac{\alpha}{(\delta + \alpha)} u(k) + b''_0 \frac{\alpha^2}{(\delta + \alpha)^2} u(k) \quad \textbf{(4)}$$

of the form $y(k) = \Phi^T(k)\, \theta$ with $\Phi^T(k)$:

$$\left[-\frac{\alpha}{(\delta + \alpha)} y(k),\ -\frac{\alpha^2}{(\delta + \alpha)^2} y(k), \right.$$

$$\left. \frac{\alpha}{(\delta + \alpha)} u(k),\ \frac{\alpha^2}{(\delta + \alpha)^2} u(k) \right]$$

and $\theta^T = [a''_1,\ a''_2,\ b''_1,\ b''_0]$.

The data in $\Phi^T(k)$ are well conditionned because:
- there is no problem of different magnitudes between the derivatives of different order as can occur in the first proposal.
- the data are not correlated between one another as in the discrete-time case with the q operator when the sampling period is small. This is shown on figure 1. The data are the outputs of the chain of filters $\alpha/(\delta+\alpha)$ with the input or output of the process as inputs. This is shown on figure 2.

II.3] Third proposal.

This proposal is similar to the first. But here linear combinations of the derivatives of different order are operated in order to eliminate the great disproportion between the magnitude of the derivatives of different order. The same behaviour of the data in response to a step input is observed for the first and third proposals. The independence between the data of the observation equation is so kept. But the disproportion in magnitude is reduced with this third method. A better well conditionned observation equation is then manipulated.

Equation **(2)** can be shown to be equivalent to:

$$(\delta+\alpha)^2\, y(k) + a''_1 \alpha (\delta+\alpha)\, y(k) + a''_2 \alpha^2\, y(k)$$

$$= b''_1\, \alpha\, (\delta+\alpha)\, u(k) + b''_0\, \alpha^2\, u(k) \quad \textbf{(5)}$$

The data $\{(\delta+\alpha)^i\, y\}_{i=0,1,2}$ and $\{(\delta+\alpha)^i\, u\}_{i=0,1}$ must be filtered when T_e tends towards zero as in the first proposal. If we use the following low-pass filter:

$\frac{f_c^2}{(\delta + f_c)^2}$ we obtain: $v(k) = \Phi^T(k)\, \theta$ **(6)**

with $\Phi^T(k)$ of the form :

$$\left[\frac{-f_c^2}{\alpha} \frac{(\delta + \alpha)}{(\delta + f_c)^2} y(k),\ \frac{-f_c^2}{(\delta + f_c)^2} y(k), \right.$$

$$\left. \frac{(\delta + \alpha)}{(\delta + f_c)^2} \frac{f_c^2}{\alpha} u(k),\ \frac{f_c^2}{(\delta + f_c)^2} u(k) \right]$$

and $\theta^T = [a''_1,\ a''_2,\ b''_1,\ b''_0]$.

Remark: - In these three methods, the identification is performed every L steps with L>1 and the command is sent every step.

III] Pole-placement control law.

III.1] Principle.

The controller structure is shown on figure 3. For the process represented in equation **(2)**: $A'(\delta)\ y(k) = B'(\delta)\ u(k)$ we have:

$$S(\delta) = \delta + \sigma_0$$
$$R(\delta) = r_1\ \delta + r_0$$

The transfer function of the model reference is B_r/A_r.

The observer polynomial is noted $P_0(\delta)$ and is of degree 1: $P_0(\delta) = \delta + p_0$

The zero of the process is not cancelled and we have:
$B_r(\delta)=B'(\delta)\ B_{r1}(\delta)$ and $T(\delta)=B_{r1}(\delta)\ P_0(\delta)$.

The BEZOUT equation is:
$A'(\delta)\ S(\delta) + B'(\delta)\ R(\delta) = A_r(\delta)\ P_0(\delta)$

III.2] Choice of the closed loop poles.

III.2.1] High-frequency gain method.

The high frequency gain of the feedback loop is:

$$G_{HF} = \left[\frac{R(\delta)}{S(\delta)}\right]_{\delta=-2/T_e} = \left[\frac{(r_0+r_1\delta)}{(\sigma_0+\delta)}\right]_{\delta=-2/T_e}$$

$$G_{HF} = \frac{(r_0 - 2r_1/T_e)}{(\sigma_0 - 2/T_e)}$$

and a linear relation is deduced between the coefficients of the controller and the value of G_{HF}. This relation is added

to that resulting from the BEZOUT equation. These equations are then solved with a least-squares method. The unknowns are the controller parameters and a pole of the closed loop polynomial. The other poles are taken into account in the value of the high-frequency gain (in the controller parameters r_0, r_1 and σ_0). The calculated closed loop pole makes a compromise between the dynamics of the other closed loop poles and the value attributed by the user to G_{HF}.

G_{HF} represents the attenuation of the high frequencies by the feedback loop.

<u>Remark:</u> - The same multiple pole can be taken for the model reference and the observer polynomials.

III.2.2] Lead-phase controller.

The analogy between a lead-phase controller and the pole placement controller is made here. We mathematically have the following configuration shown on figure 4.

$D(\delta)=R(\delta)/S(\delta)$ corresponds to a classical lead phase controller.

$$D(\delta) = \frac{r_0 + r_1\delta}{\sigma_0 + \delta} = A\left[\frac{(\delta + 1/V)}{(\delta + \mu/V)}\right]$$

and $A = r_1$; $V = r_1/r_0$; $\mu = \sigma_0 r_1/r_0$.

μ determines the amplification of the high frequency content of the error signal e (see figure 5). It measures the difference between the high-frequency gain and the static gain. μ must not be too high. If so, the poles of the closed loop must be slower: the user demands too much to the closed loop. On the contrary, if $\mu<1$, $D(\delta)$ is no longer a lead-phase controller; the poles of the reference model and the observer polynomials must then be more rapid. Hence the algorithm:

1) The poles of the model reference and the observer polynomials are initia--lized with a great value.
2) Calculus of $\mu = \sigma_0 r_1/r_0$.
3)-If $\mu>10$ and if the mean of the prediction error is below a pre-determined threshold then the poles of the closed loop are slowed down. - If $\mu<1$ then the poles of the closed loop are accelerated.

IV] Physical experiments.

They have been lead on an electrical laboratory monozone furnace. These experiments concern the behaviour of adaptive pole-placement control law with the delta operator, together with the three identification methods presented (figures 6 to 11). The temperature, set point and command are drawn on the same diagram. The scale of the temperature stretches from 0°C to 1000°C and this of the command from 0 to 100%. The sampling period is equal to 2 seconds in all the experiments. These experiments are then compared with the behaviour of a P.I.D. controller with filtered derivative action (figures 12 and 13). This P.I.D. has been tuned for the set-point 800°C, with the ZIEGLER-NICHOLS rules. The performances are better in the case of adaptive control than with P.I.D. control: more rapid reaction to disturbances, a better following of the variations of the set-point and less overshoot.

The Nyquist diagrams of the open loop of the experiment performed in figure 7 are shown in figures 14 and 15. Eventual non modelled delays of respectively 10 s. and 20 s. are added to the open loop transfer function.

V] Conclusion.

We have shown that a continuous-time approach of identification and control with the delta operator was possible. The three presented adaptive methods have proved their great ability to follow variations of set-points and react to disturbances. Moreover thanks to the delta modelling, the sensitivity of the BEZOUT equation to variations of model parameters is reduced. A simple pole-placement control law can so be used together with more sophisticated identification methods. Finally two methods for the choice of the closed loop poles have been proposed. They have proved to be correct in laboratory experiments.

References

ASTROM, K.J., P. HAGANDER and J. STERNBY (1984). Zeros for sampled Systems. AUTOMATICA.,vol.20, n°1, 31-39.

GOODWIN, G.C., R.L. LEAL, D.Q. MAYNE and R.H. MIDDLETON (1986). Rapprochement between continuous and discrete model reference adaptive control. AUTOMATICA, Vol.22, n°2, 199-207.

MIDDLETON, R.H. and G.C. GOODWIN (1986). Improved finite word length characteristics in digital control using delta operators. I.E.E.E. Transactions on Automatic Control., Vol.AC-31, n°11, 1015-1021.

UNBEHAUEN, H. and G.P. RAO (1987). Identification of continuous systems. ELSEVIER SCIENCE PUBLISHERS, pp 167-236.

UNBEHAUEN, H. and G.P. RAO (1990). Continuous-time Approaches to system Identification - A survey. AUTOMATICA, Vol.26, n°1, 23-35.

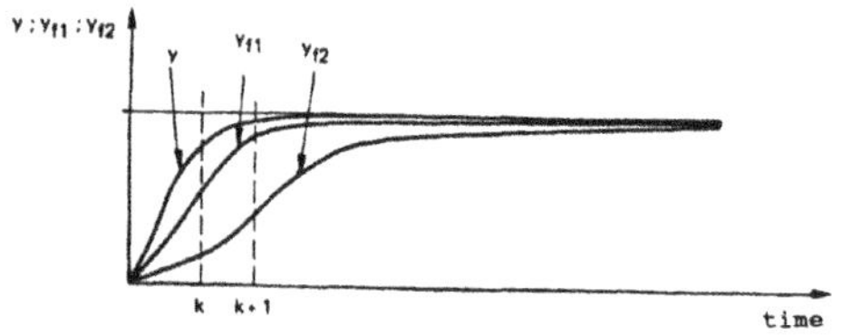

Fig. 1. Evolution of the data y, y_{f1}, y_{f2}.

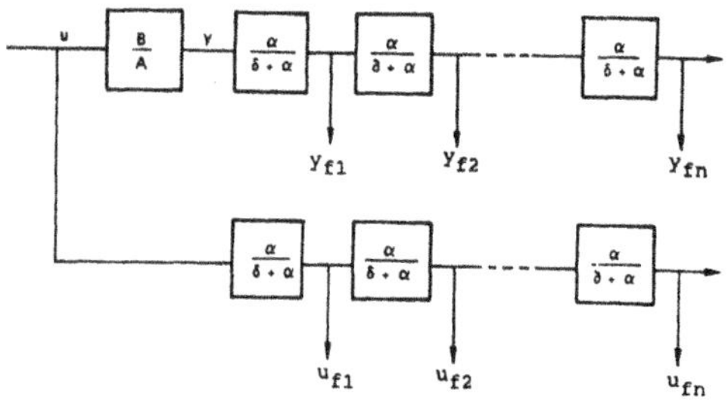

Fig. 2.

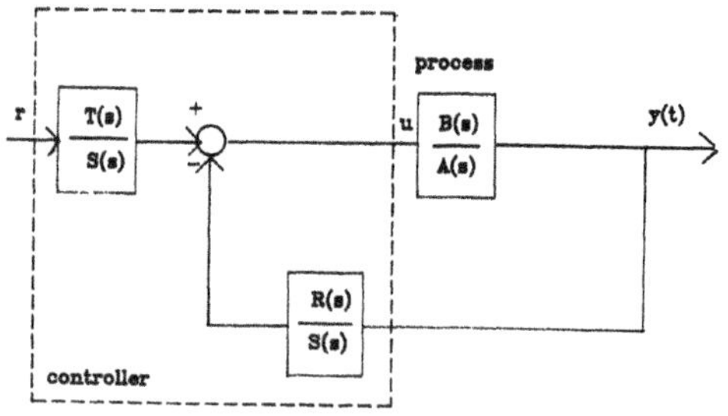

Fig. 3. Controller structure.

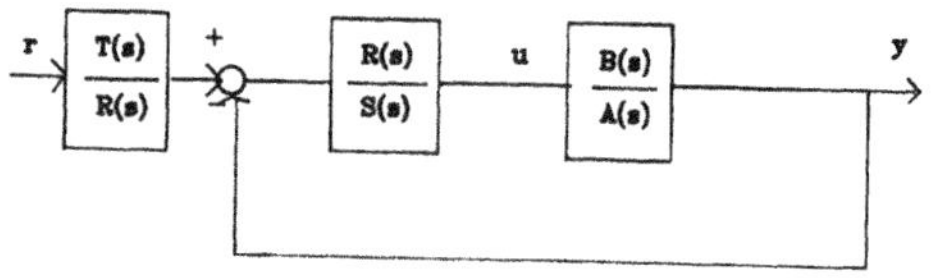

Fig. 4. Modified structure of the pole placement controller.

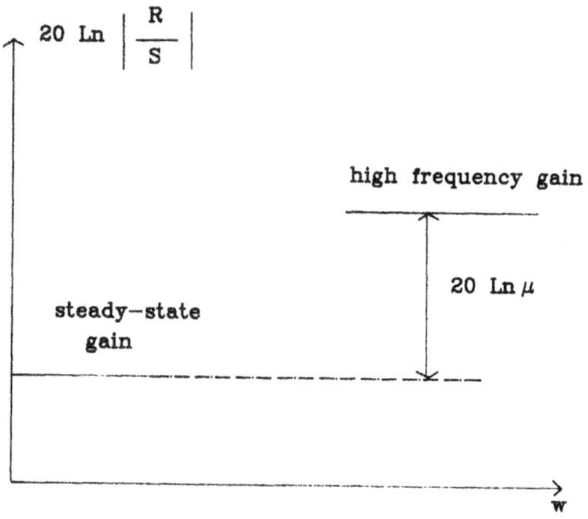

Fig. 5. Analogy with lead-phase controllers.

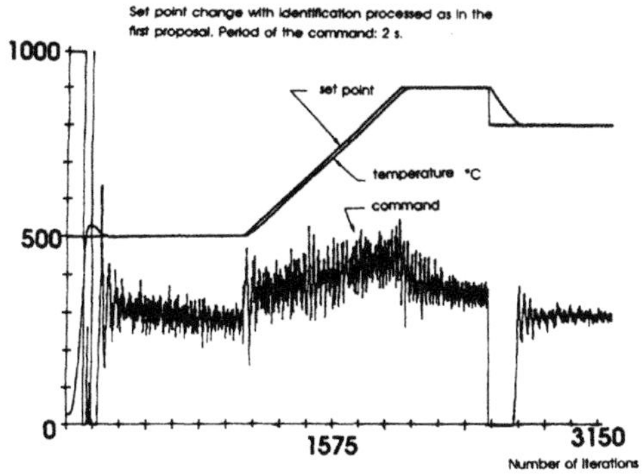

Fig. 6. Set point change with identification processed as in the first proposal. Period of the command: 2 s.

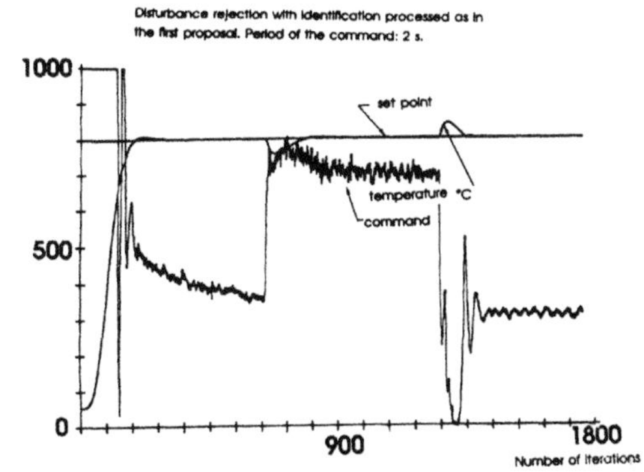

Fig. 7. Disturbance rejection with identification processed as in the first proposal. Period of the command: 2 s.

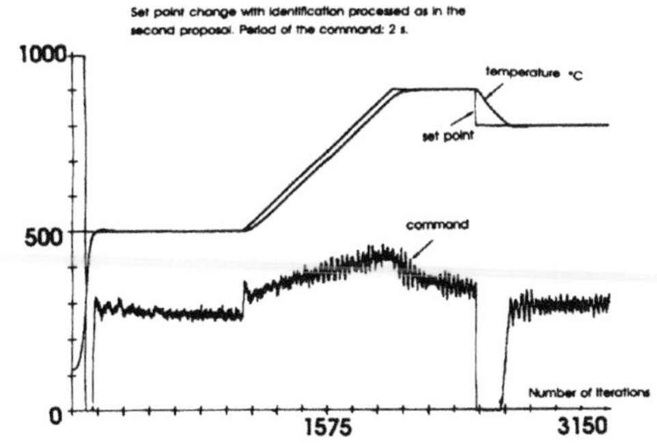

Fig. 8. Set point change with identification processed as in the second proposal. Period of the command: 2 s.

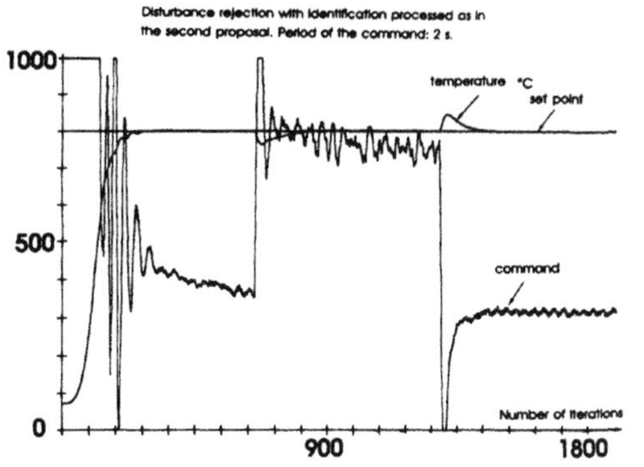

Fig. 9. Disturbance rejection with identification processed as in the second proposal. Period of the command: 2 s.

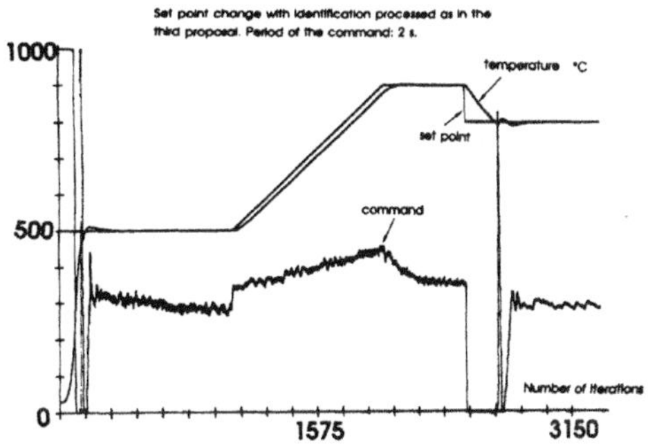

Fig. 10. Set point change with identification processed as in the third proposal. Period of the command: 2 s.

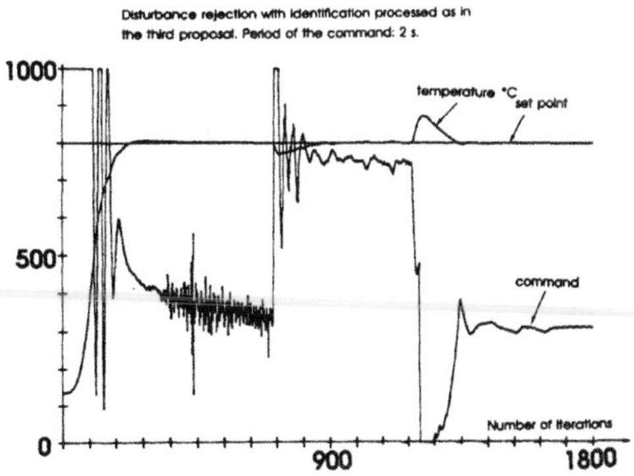

Fig. 11. Disturbance rejection with identification processed as in the third proposal. Period of the command: 2 s.

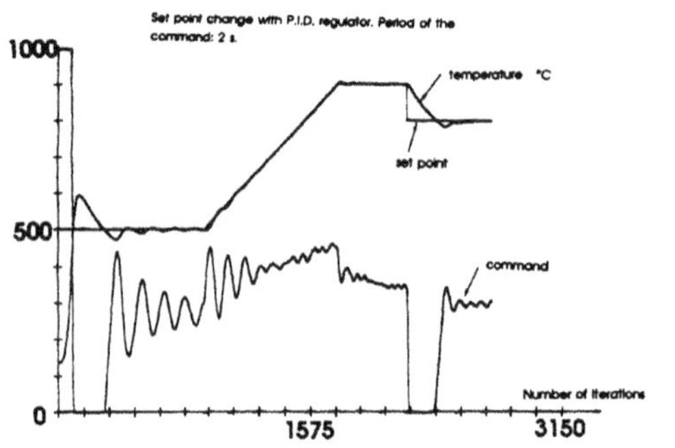

Fig. 12. Set point change with P.I.D. controller. Period of the command: 2 s.

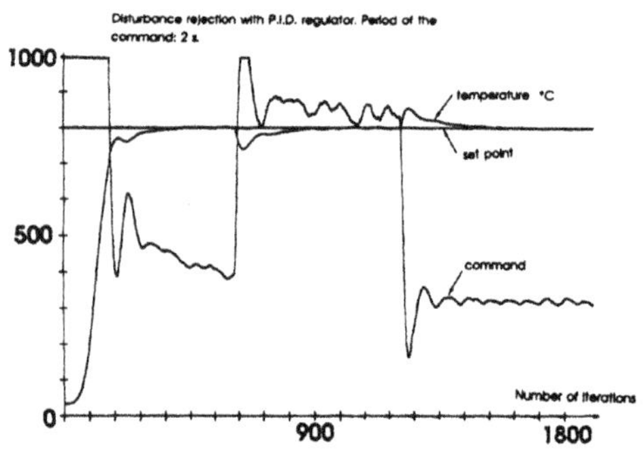

Fig. 13. Disturbance rejection with P.I.D. controller. Period of the command: 2 s.

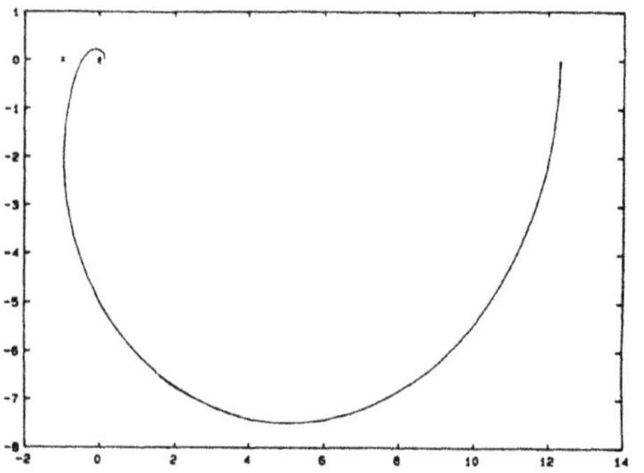

Fig. 14. Robustness with respect to an eventually non modelled delay of 10 s. The parameters of the controller are those of the experiment in fig. 7.

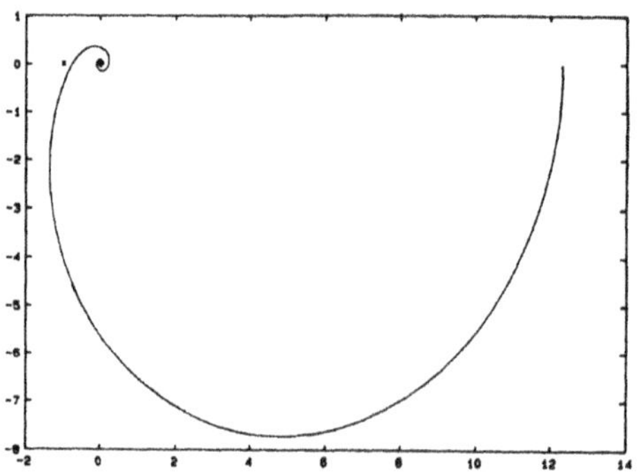

Fig. 15. Robustness with respect to an eventually non modelled delay of 20 s. The parameters of the controller are those of the experiment in fig. 7.

A NON MINIMAL MODEL FOR SELF TUNING CONTROL

K. W. Lim and S. Ginting

School of Electrical Engineering and Computer Science, University of New South Wales, P O Box 1, Kensington, NSW 2033, Australia

Abstract. This paper examines the use of a non-minimal model for self tuning control. This model arises naturally when the plant measurements are sampled at a rate different from that at which plant actuation signals are applied. We consider the case when the measurement sampling period is longer than the control sampling period. An iterative algorithm is proposed for the online conversion of the non-minimal model to the corresponding minimal model. A predictive formulation is used to obtain measurement estimates at the control sampling rate. The combination of model conversion and predictive controller is shown to work in simulation.

Keywords. Multi-rate control, Non-minimal model, Predictive Control, Self-tuning control.

INTRODUCTION

In basic digital control theory (Astrom 1984), it is usually assumed that measurement samples are synchronous with and available at the same sampling rate as control actuation signals. There is early theoretical work dealing with the multi-rate case (Kalman 1959) and recently there has been a resurgence of interest in the applications of multi-rate control systems (Hang 1989, Zhang 1989, Berg 1988). Many of these papers originate from the flight control problem, where the plants are often reasonably well modelled.

In process control problems, there are at least two reasons for looking to multi-rate solutions. Firstly there has been an increased use of feedback control in loops using "quality" measurements as the feedback variable. Some of these "quality" measurements are automated. Many would consist of a periodic laboratory sample. In either case, these measurements are often sampled at a rate relatively slower than that at which control signals are computed.

Secondly, there is an increased awareness (Shinskey 1988) of the need to consider loop interaction for higher performance. In a modern distributed control environment, it is likely that interacting loops may be sampled at different frequencies and would thus require a multi-rate model representation.

In this paper, we develop a self-tuning version of a multi-rate controller. We consider the case where the measurement samples are sampled at longer time intervals than the control signals. For simplicity we describe this controller for a plant modelled by a second order linear transfer function. Section 2 describes the algorithm. Section 3 provides a simple analysis of the non-adaptive version of this controller while Section 4 shows preliminary simulation results.

THE ALGORITHM

Consider the plant described by a second order model:

$$\ddot{y}(t)+\alpha_1\dot{y}(t)+\alpha_2 y(t) = \beta u(t)+d(t) \qquad (1)$$

where y(t) is the measurement signal at time t, u(t) is the control signal, d(t) is the disturbance. α_1, α_2 and β are assumed to be constant coefficients. If this plant were sampled synchronously at a constant sampling period h, then the corresponding difference equation model would be:

$$y(n)+a_1y(n-1)+a_2y(n-2)=b_1u(n-1)+b_2u(n-2)+d(n) \qquad (2)$$

where a_1, a_2, b_1, b_2 : equivalent discrete model parameters of Eqn.1 when sampling period is h and d(n) is a nonmeasurable disturbance.

We assume that the measurement sample rate is related to the control sample rate by an integer multiple ie the control signal is delivered at period h while the measurements are available every M*h seconds. It is simple to show that the corresponding multi-rate model, eg. for M = 2 is

$$y(n)+\mathit{a}_1y(n-2)+\mathit{a}_2y(n-4)= \mathit{b}_1u(n-1)+\mathit{b}_2u(n-2)+\mathit{b}_3u(n-3)+\mathit{b}_4u(n-4)+d(n)$$

where $\mathit{a}_1= 2a_2 - a_1^2$ $\quad \mathit{b}_1=b_1$ $\quad \mathit{b}_3=a_2b_1-a_1b_2$

$\mathit{a}_2= a^2_2$ $\quad \mathit{b}_2=b_2-a_1b_1$ $\quad \mathit{b}_4=a_2b_2$ (3)

Note that Eqn.3 describes a non minimal discrete time model for the plant of Eqn.1 (Lu 1989).

A self tuning controller is developed for this model as follows (Fig. 1) :

Step 1

A simple least squares algorithm is used to obtain estimates for the parameters of the model Eqn 3. This algorithm is run every M*h seconds (Lu 1989).

Step 2

A model conversion process translates the parameters of the non-minimal model to those of the corresponding minimal model (Eqn.2). This translation process works every M*h seconds.

Step 3

The minimal model is used to develop a certainty equivalent controller. In this paper, an algorithm akin to the Generalised Predictive Controller (GPC) is used (Clarke 1987, 1989). This controller runs every h seconds. Since measurements are only available every M*h seconds, a predictive formulation is used for the intermediate time samples.

MODEL CONVERSION

We propose a procedure for converting the model parameters estimated in step 1 for the nonminimal model to those of the equivalent minimal model as in Eqn.2. An algebraic approach would yield a number of redundant equations which are nonlinear in the parameters. In the presence of errors it is postulated that an optimization technique may provide a better solution than an arbitrary choice of solution equation. In this paper an iterative approach due to Gauss (Beightler 1979) and some of its variants will be explored. The solution shows good performance in simulation.

Parameter optimization to obtain the minimal model of Eqn.3 is done as follows.

Define $\theta=[a_1,a_2,b_1,b_2]^T$ parameters of Eqn.2 and

$\underline{\theta}=[\underline{a}_1,\underline{a}_2,\underline{b}_1,\underline{b}_2]^T$ the first approximation of the minimal model parameters of Eqn.2.

$\phi=[\mathit{a}_1,\mathit{a}_2,\mathit{b}_1,\mathit{b}_2,\mathit{b}_3,\mathit{b}_4]^T$ parameters of Eqn.3 ,

$\underline{\phi}=[\underline{\mathit{a}}_1,\underline{\mathit{a}}_2,\underline{\mathit{b}}_1,\underline{\mathit{b}}_2,\underline{\mathit{b}}_3,\underline{\mathit{b}}_4]^T$ is their first approximation calculated from $\underline{\theta}$ using the parameter relation in Eqn.3.

Now the error between identified nonminimal model parameters ϕ and approximation of minimal model parameters $\underline{\theta}$ is given by

$$\varepsilon = \phi - \underline{\phi} = \phi - f(\underline{\theta})$$

Minimize ε in the least squares sense. We consider ψ in the following equation

$$\psi = \varepsilon^T \varepsilon$$

and its first derivative ψ' which is

$$\psi' = 2\gamma\,\underline{\theta} = 0$$

where γ is Jacobian of ϕ

The solution using Gauss method (Beightler 1979) is to iterate

$$\Delta\underline{\theta} = -(\gamma\gamma^T)^{-1}\gamma\psi'$$

until ψ' reaches a small value. In practice a limited number of iterations is adequate for convergence.

DEVELOPMENT OF NON ADAPTIVE MULTI-RATE CONTROLLER

We develop a controller for the plant in the multi-rate model of Eqn.3 using an adaptation of GPC algorithm (Clarke 1987, 1989). The GPC formulation uses the CARIMA model of the plant. For 2nd order plant model in Eqn.2 The corresponding CARIMA model is

$$\mathbf{A}(q^{-1})y(n) = \mathbf{B}(q^{-1})\Delta u(n-1) + \xi(n) \qquad (4)$$

where

$$\Delta = 1 - q^{-1}$$

$$\mathbf{B}(q^{-1}) = b_1q^{-1} + b_2q^{-2}$$

$$\mathbf{A}(q^{-1}) = (1+a_1q^{-1}+a_2q^{-2})\Delta = 1+\mathit{a}_1q^{-1}+\mathit{a}_2q^{-2}+\mathit{a}_3q^{-3}$$

$$\mathit{a}_1 = a_1 - 1;\; \mathit{a}_2 = a_2 - a_1;\; \mathit{a}_3 = -a_2$$

$$d(n) = \xi(n)/\Delta.$$

As $\xi(n)$ is nonmeasurable disturbance the predictive output of the plant can be represented as

$$\begin{bmatrix} y(n+1) \\ y(n+2) \\ \vdots \\ \vdots \\ y(n+N) \end{bmatrix} = \mathbf{g} \begin{bmatrix} \Delta u(n) \\ \Delta u(n+1) \\ \vdots \\ \vdots \\ \Delta u(n+NU-1) \end{bmatrix} + \mathbf{f} \begin{bmatrix} u(n-1) \\ y(n) \\ y(n-1) \\ y(n-2) \end{bmatrix} \qquad (5)$$

where N = prediction horizon
NU = control horizon
$\mathbf{g}$ = lower triangular matrix

Matrices $\mathbf{g}$ and $\mathbf{f}$ are calculated uniquely from plant dynamic parameters $\mathbf{A}$ and $\mathbf{B}$.
The present and future control signal vector $[\Delta u(n)\ \Delta u(n+1) \ldots \Delta u(n+NU-1)]^T$ is calculated through minimisation of the quadratic cost function J_{GPC} as shown in Eqn.6

$$J_{GPC} = \sum_{N_1}^{N_2} e^2(n+j) + \sum_{1}^{NU} \lambda_j \Delta u^2(n+j-1) \qquad (6)$$

where N_1 and N_2 are the costing horizon, λ_j is control weighting, $e(n+j)=w(n+j)-y(n+j)$, $w(n+j)$ is future set point. The present and future control vector which minimises J_{GPC} is given by

$$\mathbf{U} = (\mathbf{g}^T\mathbf{g} + \lambda\mathbf{I})^{-1}\mathbf{g}^T(\mathbf{w} - \mathbf{f}') \qquad (7)$$

where $\mathbf{f}' = \mathbf{f}\,[\Delta u(n-1)\ y(n)\ y(n-1)\ y(n-2)]^T$

$\mathbf{U} = [\Delta u(n)\ \Delta u(n+1) \ldots \Delta u(n+NU-1)]^T$

$\mathbf{w} = [w(n)\ w(n+1) \ldots w(n+N)\ ]^T$

We use the GPC formulation to predict the plant output between output sampling instant. This implies that at each output measurement sampling instant we compute a vector of predicted plant outputs and vector of future control signals between output sampling. For a ratio of 3 input to 1 output samples the predicted output and future control vector taking set point $\mathbf{w}(.)=0$ is shown in Eqn.8 and Eqn.9 respectively. Without loss of generality we consider a specific case NU=M=3.

$$\begin{bmatrix} \mathit{y}(n+1) \\ \mathit{y}(n+2) \\ \mathit{y}(n+3) \end{bmatrix} = \mathbf{g} \begin{bmatrix} \Delta u(n) \\ \Delta u(n+1) \\ \Delta u(n+2) \end{bmatrix} + \mathbf{f} \begin{bmatrix} \Delta u(n-1) \\ y(n) \\ \mathit{y}(n-1 \\ \mathit{y}(n-2) \end{bmatrix} \qquad (8)$$

$$\begin{bmatrix} \Delta u(n) \\ \Delta u(n+1) \\ \Delta u(n+2) \end{bmatrix} = -\,\mathbf{G} \begin{bmatrix} u(n-1) \\ y(n) \\ \mathit{y}(n-1) \\ \mathit{y}(n-2) \end{bmatrix} \qquad (9)$$

where y(.) measured output, $\mathit{y}(.)$ are predicted measurements and $\mathbf{G} = (\mathbf{g}^T\mathbf{g} + \lambda\mathbf{I})^{-1}\mathbf{g}^T$

For analysis, rewrite Eqn.8 and Eqn.9 to Eqn.10 and Eqn.11 respectively.

$$\begin{bmatrix} \mathit{y}(n+1) \\ \mathit{y}(n+2) \\ \mathit{y}(n+3) \end{bmatrix} = \mathscr{g} \begin{bmatrix} \Delta u(n-1) \\ \Delta u(n) \\ \Delta u(n+1) \\ \Delta u(n+2) \end{bmatrix} + \mathscr{f} \begin{bmatrix} y(n) \\ \mathit{y}(n-1) \\ \mathit{y}(n-2) \end{bmatrix} \qquad (10)$$

$$\begin{bmatrix} \Delta u(n-1) \\ \Delta u(n) \\ \Delta u(n+1) \\ \Delta u(n+2) \end{bmatrix} = \mathscr{G} \begin{bmatrix} y(n) \\ \mathit{y}(n-1) \\ \mathit{y}(n-2) \end{bmatrix} \qquad (11)$$

where

$$\mathscr{g} = \left[\begin{array}{c|c} f_{11} & \\ f_{21} & \mathbf{g} \\ f_{32} & \end{array}\right] \qquad \mathscr{f} = \begin{bmatrix} f_{12} & f_{13} & f_{14} \\ f_{22} & f_{23} & f_{24} \\ f_{32} & f_{33} & f_{34} \end{bmatrix}$$

$$\mathscr{G} = -\begin{bmatrix} G_{11} & 1 & 0 & 0 \\ G_{21} & 0 & 1 & 0 \\ G_{31} & 0 & 0 & 1 \end{bmatrix}^{-1} \begin{bmatrix} G_{12} & G_{13} & G_{14} \\ G_{22} & G_{23} & G_{24} \\ G_{32} & G_{33} & G_{34} \end{bmatrix}$$

By substitution of Eqn.11 to Eqn.10 output prediction can be written as Eqn.12.

$$\begin{bmatrix} \mathscr{y}(n+1) \\ \mathscr{y}(n+2) \\ \mathscr{y}(n+3) \end{bmatrix} = [\, -\mathscr{g}\mathscr{G} + \mathscr{f} \,] \begin{bmatrix} y(n) \\ \mathscr{y}(n-1) \\ \mathscr{y}(n-2) \end{bmatrix} \qquad (12)$$

The control vector across the output sampling interval can be obtained from Eqn.11.

$$\begin{bmatrix} \Delta u(n) \\ \Delta u(n+1) \\ \Delta u(n+2) \end{bmatrix} = \underline{\mathscr{G}} \begin{bmatrix} y(n) \\ \mathscr{y}(n-1) \\ \mathscr{y}(n-2) \end{bmatrix} \qquad (13)$$

State model of the plant for a ratio of 3 input to 1 output samples can be written as shown in Eqn.14.

$$X(n+3) = \Phi^3 X(n) + [\Phi^2\Gamma \mid \Phi\Gamma \mid \Gamma] \begin{bmatrix} \Delta u(n) \\ \Delta u(n+1) \\ \Delta u(n+2) \end{bmatrix} \qquad (14)$$

Then defining a new time index such that k+i=n+i*3 ;(i=0,1...), the multi-rate sampled plant can be represented by

$$X(k+1) = \underline{\Phi}\, X(k) + \underline{\Gamma}\Delta u(k)$$

$$y(k) = \underline{C}\, X(k) \qquad (15)$$

where Φ, Γ, and C are state matrices of the plant transfer function Eqn.4.

$$\Phi = \begin{bmatrix} -a_1 & -a_2 & -a_3 \\ 1 & 0 & 0 \\ 0 & 1 & 0 \end{bmatrix} \quad ; \quad \Gamma = [1 \;\; 0 \;\; 0]^T \quad ; \quad C = [b_1 \;\; b_2 \;\; 0]$$

$$\underline{\Phi} = \Phi^3$$

$$\underline{\Gamma} = [\Phi^2\Gamma \mid \Phi\Gamma \mid \Gamma]$$

$$\underline{C} = \begin{bmatrix} C \\ \hline 0\;0\;0 \\ 0\;0\;0 \end{bmatrix}$$

The output estimation and control vector represented in k time index become :

$$\mathcal{Y}(k) = [\, \mathscr{y}(n+1) \; \mathscr{y}(n+2) \; \mathscr{y}(n+3)]^T$$

$$\Delta U(k) = [\, \Delta u(n) \; \Delta u(n+1) \; \Delta u(n+2)]^T$$

and the plant output

$$Y(k) = [\, y(n) \;\; 0 \;\; 0\,]^T \qquad (16)$$

The output prediction and control vector equation in Eqn.12 and Eqn.13 respectively in k time index is shown in Eqn.17 and Eqn.18.

$$\mathcal{Y}(k) = [-\mathscr{g}\underline{\mathscr{G}} + \mathscr{f}][T_1 \mid T_2] \begin{bmatrix} Y(k) \\ \mathcal{Y}(k-1) \end{bmatrix} \qquad (17)$$

and

$$\Delta U(k) = -\,\underline{\mathscr{G}}[T_1 \mid T_2] \begin{bmatrix} Y(k) \\ \mathcal{Y}(k-1) \end{bmatrix} \qquad (18)$$

where

$$T_1 = \begin{bmatrix} 1 & 0 & 0 \\ 0 & 0 & 0 \\ 0 & 0 & 0 \end{bmatrix} ; \quad T_2 = \begin{bmatrix} 0 & 0 & 0 \\ 0 & 1 & 0 \\ 1 & 0 & 0 \end{bmatrix}$$

The close loop state equation can be written by putting together Eqn.15, Eqn.17, and Eqn.18 as shown in Eqn.19

$$\begin{bmatrix} X(k+1) \\ \mathcal{Y}(k) \end{bmatrix} = \begin{bmatrix} \Phi - \underline{\Gamma}\,\underline{\mathscr{G}}\,T_1\underline{C} & -\underline{\Gamma}\,\underline{\mathscr{G}}\,T_2 \\ (-\mathscr{g}\mathscr{G} + \mathscr{f})T_1\underline{C} & (-\mathscr{g}\mathscr{G} + \mathscr{f})T_2 \end{bmatrix} \begin{bmatrix} X(k) \\ \mathcal{Y}(k-1) \end{bmatrix} \qquad (19)$$

Analysis shows that this state equation can have unstable eigenvalues even though the corresponding GPC for a single rate plant has stable closed loop poles. This is due to the essentially open loop estimation of plant output between output samples. To overcome this problem a simple error correction scheme is used as described in the following section.

Correction of output estimates

The algorithm described in Eqn.19 has no correction in the predicted measurements. $\mathcal{Y}(k-1)$ consists of $[\mathscr{y}(n-2)\; \mathscr{y}(n-1)\; \mathscr{y}(n)]$ while **Y**(k) consists of [y(n) 0 0] because y(n+1) and y(n+2) cannot be measured. At time k the past measurement predictions can be corrected using the error between measured output y(n) and estimated output $\mathscr{y}(n)$ as in the following :

$\mathcal{Y}(k-1)$ become

$$\mathcal{Y}(k-1) + K[T_3 \mid T_4] \begin{bmatrix} \mathbf{Y}(k) \\ \mathcal{Y}(k-1) \end{bmatrix}$$

or

$$(\mathbf{I} + \mathbf{K}T_4)\,\mathcal{Y}(k-1) + \mathbf{K}T_3\mathbf{Y}(k) \qquad (20)$$

where

$$\mathbf{K} = \begin{bmatrix} k_3 & 0 & 0 \\ k_2 & 0 & 0 \\ k_1 & 0 & 0 \end{bmatrix} \quad T_3 = \begin{bmatrix} 1 & 0 & 0 \\ 0 & 0 & 0 \\ 0 & 0 & 0 \end{bmatrix} \quad T_4 = \begin{bmatrix} 0 & 0 & -1 \\ 0 & 0 & 0 \\ 0 & 0 & 0 \end{bmatrix}$$

K: Correction gain for past measurement prediction.

Substituting Eqn.20 to (19), the closed loop state equation with backward correction of output estimates can be written as shown in Eqn.21.

$$\begin{bmatrix} X(k+1) \\ \mathbf{Y}(k) \end{bmatrix} = \begin{bmatrix} \underline{\Phi} - \underline{\Gamma}\,\underline{\mathscr{G}}(T_1 + T_2\mathbf{K}T_3)\underline{C} & -\underline{\Gamma}\,\underline{\mathscr{G}}\,T_2(\mathbf{I} + \mathbf{K}T_3) \\ (-\mathscr{g}\mathscr{G} + \mathscr{f})(T_1 + T_2\mathbf{K}T_3)\underline{C} & (-\mathscr{g}\mathscr{G} + \mathscr{f})T_2(\mathbf{I} + \mathbf{K}T_4) \end{bmatrix} \begin{bmatrix} X(k) \\ \mathcal{Y}(k-1) \end{bmatrix} \qquad (21)$$

SIMULATION RESULTS

In this simulation study we choose a poorly damped second order plant with transfer function as shown in Eqn.22 (Hang 1989-b). The plant discrete transfer function and corresponding parameter values when sampled at $h_s = 0.5$ and $h_l = 2.0$ second are shown in Eqn.23.

$$G(s) = \frac{0.1296}{s^2 + 0.216\,s + 0.1296} \qquad (22)$$

$$G(q^{-1}) = \frac{b_1 q^{-1} + b_2 q^{-2}}{1 + a_1 q^{-1} + a_2 q^{-2}} \qquad (23)$$

where for sampling time $h_s = 0.5$ second

$a_1 = -\,1.8670 \qquad b_1 = 0.0156$

$a_2 = \;\; 0.8976 \qquad b_2 = 0.0150$

and for $h_l = 2.0$ second

$a_1 = -\,1.2641 \qquad b_1 = 0.2163$

$a_2 = \;\; 0.6492 \qquad b_2 = 0.1868$

The response of the plant with non adaptive GPC and non adaptive multi-rate control to step setpoint and step load changes are studied. For comparison the response of single rate GPC with sampling time h_s and h_l are shown in Fig.2.a and b respectively. Fig.3 shows the response of the plant with non-adaptive multi-rate controller with input and output sampling time h_s and h_l (ie. M=4) respectively. The control horizon, NU=3, prediction horizon N=10 are used for all simulations. The non-adaptive simulations demonstrate that a faster sampling rate leads to improved dynamic response. The multi-rate response of Fig 3 (used when fast measurements are not available) falls somewhere between those of Figs 2a and 2b.

Fig.4 shows the response and the convergence of the estimated parameters of adaptive single rate GPC with sampling time h_l. Fig.5 shows the response of the multi-rate adaptive control with M=4 (input and output sampling time h_s and h_l) and the convergence of estimated nonminimal model to minimal model parameters of the plant. Comparing Figs 4 and 5, there is an improvement in output response, at the price of more control activity. The parameter estimates in the non-minimal case take a longer time to converge, which accounts for poorer transient behaviour, especially in the first 50 seconds.

CONCLUSIONS

Simulation results show that the algorithm which has been developed is feasible. The problem of choosing an appropriate correction factor for predicted measurement remains. Extension of this algorithm to a multivariable case needs further work

REFERENCES.

Astrom, K.J., and B.Wittenmark (1984). *Computer Controlled Systems : Theory and Design*. Prentice Hall International, Englewood Cliffs, New Jersey.

Beightler, C.S., D.T. Phillips, and D.J. Wilde (1979). *Foundation of Optimization*. Prentice Hall.

Berg, M.C., N. Amit, and J.D. Powel (1988). Multi-rate Digital Control Design. *IEEE Trans. Aut. Control*, **AC-33**, pp.1139-1150.

Clarke, D.W., C. Mochtadi, and P.S.Tuffs (1987). Generalized Predictive Control. Parts 1 and 2. *Automatica*, **23**, 137-160

Clarke, D.W., and C. Mochtadi (1989). Properties of Generalized Predictive Control. *Automatica*, **25**, 859-875.

Hang, C.C., K.W. Lim, and B.W. Chong (1989 a). Dual Rate Adaptive Digital Smith Predictor. *Automatica*, **25**, 1-16.

Hang, C.C., Y.S. Cai, K.W. Lim (1989 b). A Dual-Rate Approach to Auto-Tuning of Pole-Placement Controller. Report CI-89-4 (AUG 1989). Dept. of Electrical Engineering, National University of Singapore.

Kalman, R.E., and J.E. Bertram (1959). A Unified Approach to the Theory of Sampling Systems. *J. Franklin Inst.*, **267**, pp.405-436.

Lu, W., D.G. Fisher (1989). Least Square Estimation with Multi-rate Sampling. *IEEE Trans. Aut. Control*, **AC-34**, pp.669-672.

Shinskey, F.G. (1988). *Process Control Systems, Application, Design and Tuning*. McGraw-Hill, New York, 3rd ed.

Zhang, C., R.H. Middleton, and R.J. Evans (1989). An Algorithm for Multi-rate Sampling Adaptive Control. *IEEE Trans. Aut. Control*, **AC-34**, pp.792-795.

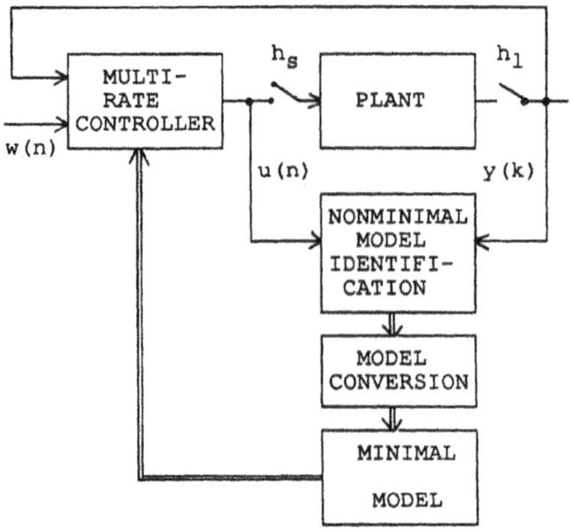

Fig.1 Block diagram of non minimal self tuning control.

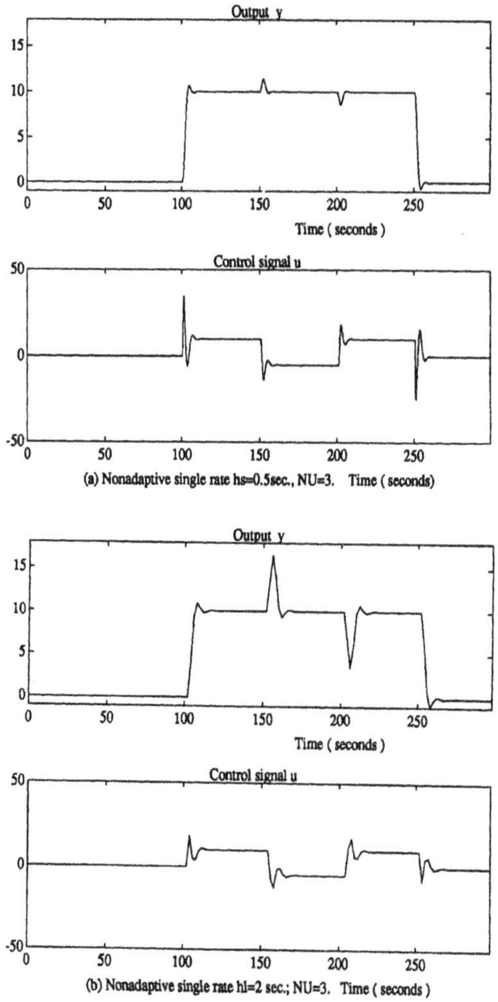

Fig.2 The response of non adaptive single rate GPC.
a. Sampling time h_s= 0.5 second.
b. Sampling time h_l= 2.0 second.

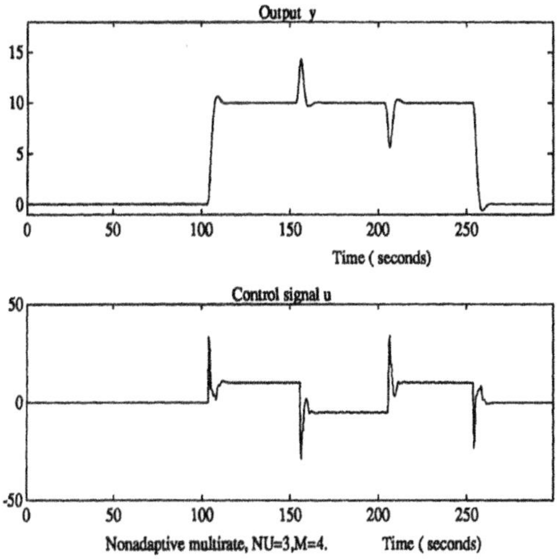

Fig.3 The response of non adaptive multi-rate control

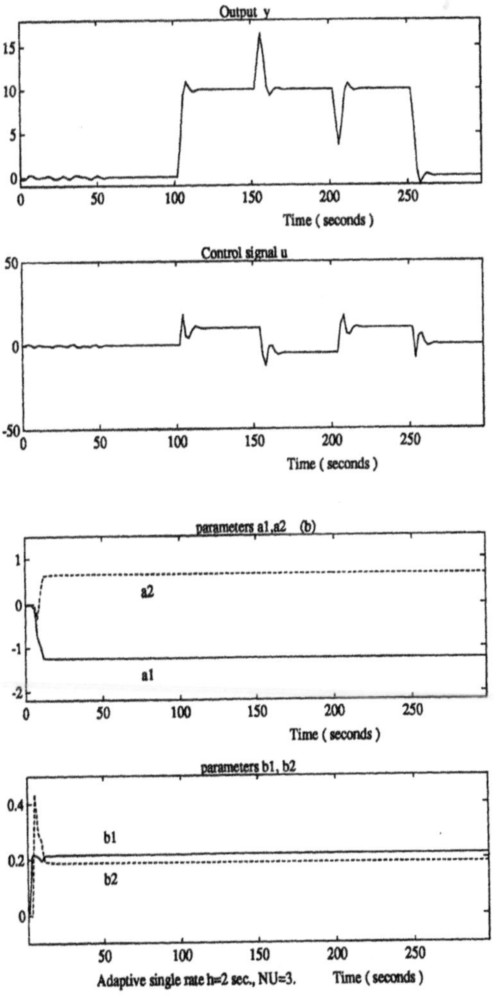

Fig. 4 The response of adaptive single rate GPC h_1= 2 second.

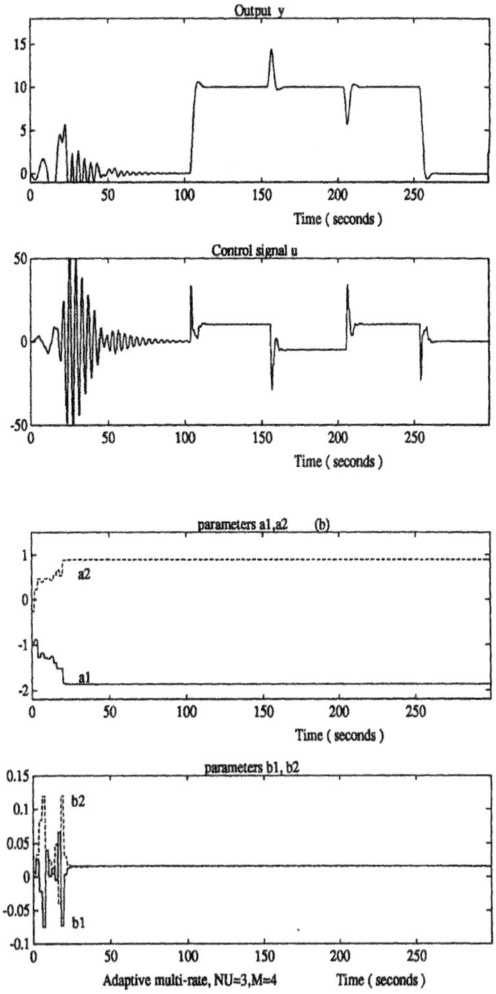

Fig.5 The response of adaptive multi-rate control.

A LEARNING MODIFIED GENERALIZED PREDICTIVE CONTROLLER*

Ning-Shou Xu, Zhang-Lei Wu and Li-Ping Chen

Dept. of Automatic Control, Beijing Polytechnic University, Eastern Suburb, Beijing 100022, PRC

Abstract. This paper is focused on the problem of further improving the robustness of Modified Generalized Predictive Control (MGPC) so as to cope with much larger variations in type of the deterministic disturbance or the dynamics of a controlled plant. A new self-learning scheme --- Learning Modified Generalized Predictive Controller (LMGPC) is proposed. The controller is realized by adding to the structure of MGPC system a learning loop containing 2 sub-loops: The one is for on-line optimization of the optimal moving average filter parameters, and the another for on-line optimization of the MGPC parameters; both are performed using reinforcement algorithm. The functions and operation principle of each sub-loop are explained in detail. Numerical simulations and experimental test have shown the feasibility and good performance of the LMGPC proposed.

Keywords. Learning systems; predictive control; parameter optimization.

INTRODUCTION

As is well known, in the area of selftuning control the Generalized Predictive Control (GPC) proposed by Clarke et al. (1984) has drawn increasing attention for its excellent robustness with respect to the variations of the model-order or time-delay of the controlled plant and also to the stepwise deterministic disturbance which may lead to the offset problem. Keeping the same robustness of GPC, Xu et al. (1988a) presented a modified version of GPC (MGPC) by introducing the following 3 modifications into the basic framework of GPC:
1)The Diophantine equation recursion for the multi-step ahead predictions of the plant output is simply replaced by the plant model recursion, which is much easier for implementing via assembler language in microcomputer and requires much less computational effort; 2)The theoretically approximate parameter values and the tending-to-one theorem for parameter sum of the sampled model of the plant, in the case of relatively small sampling period, are applied in setting the initial parameter estimates so as to improve the quality of the RLS estimation (Xu, 1989). Thus, a satisfactory response to the step set-point change can be gained even during start-up of the control; 3)Employing an optimal moving average filter (Xu, 1988b), the effect of the stepwise or sinusoidal deterministic disturbance may be suppressed without excessive amplification of the random noise.

The MGPC may be briefly reviewed as follows
The plant model is given by

$$A(z^{-1})y(k)=B(z^{-1})u(k)+C(z^{-1})e(k)+g(k) \quad (1)$$

where
$$\left.\begin{aligned} A(z^{-1})&=1+a_1z^{-1}+\cdots+a_nz^{-n}\\ B(z^{-1})&=b_1z^{-1}+\cdots+b_mz^{-m},\ m=n+d_{max}\\ C(z^{-1})&=1+c_1z^{-1}+\cdots+c_nz^{-n}\end{aligned}\right\} \quad (2)$$

* The Project Supported by National Natural Science Foundation of China.

d_{max} is the maximum delay expected, $y(k)$ the plant output, $u(k)$ the plant input, $e(k)$ a zero-mean white noise, $g(k)$ the deterministic disturbance satisfying the following internal model

$$\Psi(z^{-1})g(k)=0 \quad (3)$$

where $\Psi(z^{-1})=1+\Psi_1z^{-1}+\cdots+\Psi_\mu z^{-\mu}$

$$=\begin{cases}1-z^{-1}, & \text{if } g(k) \text{ is stepwise}\\ 1+\Psi_1z^{-1}+z^{-2}, & \text{if } g(k) \text{ is sinusoidal}\\ (1-z^{-1})(1+\Psi_1'z^{-1}+z^{-2}), & \text{if } g(k) \text{ consists of both stepwise and sinusoidal disturbances}\end{cases} \quad (4)$$

The control performance is in the form of

$$J=E\left\{\sum_{j=1}^{N}w_j[y(k+j)-y_r(k+j)]^2+\lambda K_d^2\sum_{j=1}^{M}u_f^2(k+j-1)\right\} \quad (5)$$

where N is the prediction horizon, M(≤N) the control horizon, $y(k+j)$ and $y_r(k+j)$ the future values of the plant output and reference signal at time instant k+j respectively, $w_j>0$ and $\lambda>0$ weighting factors, K_d the d. c. gain of the plant, and $u_f(k)=f(z^{-1})u(k)$ the plant input filtered by the following OMA filter

$$\begin{cases}f(z^{-1})=\Psi(z^{-1})f'(z^{-1})=1+f_1z^{-1}+\cdots+f_{\mu+\eta}z^{-\mu-\eta}\\ f'(z^{-1})=1+f_1'z^{-1}+\cdots+f_\eta'z^{-\eta}\end{cases} \quad (6)$$

where the coefficients of the auxiliary polynomial $f'(z^{-1})$ with order η(≤2 usually) theoretically depend upon those of $\Psi(z^{-1})$ and $C(z^{-1})$ (Xu, 1988).

The minimum variance (MV) predictions of the plant output filtered by $f(z^{-1})$ is

$$\hat{y}_f(k+j)=\begin{cases}f(z^{-1})y(k+j), & j\le 0\\ [-\hat{y}_f(k+j-1),\cdots,-\hat{y}_f(k+j-n)\,\vdots\,u_f(k+j-1),\cdots,u_f(k+j-m)]\,\underline{\theta}, & j>0\end{cases} \quad (7)$$

where $\underline{\theta}=[a_1,\cdots,a_n\,\vdots\,b_1,\cdots,b_m]^\tau$ (8)

may be on-line estimated via RLS method.

Furthermore, assuming all the current and future filtered controls $u_f(k+j)$, $j\geq 0$, to be zero, the MV predictions of the "free filtered output" may be obtained as follows

$$\hat{y}_f'(k+j)=\begin{cases} f(z^{-1})y(k+j), & j\leq 0 \\ [-\hat{y}_f'(k+j-1),\cdots,-\hat{y}_f'(k+j-n)\,\vdots\,\underbrace{0,\cdots,0}_{j},u_f(k-1),\cdots,u_f(k+j-m)]\,\underline{\theta} & 0<j<m \\ [-y_f'(k+j-1),\cdots,-y_f'(k+j-n)\,\vdots\,0,\cdots\cdots,0]\,\underline{\theta}, & j\geq m \end{cases} \quad (9)$$

Then, the relationship between $\hat{y}_f(k+j)$ and $\hat{y}_f'(k+j)$ is

$$\hat{y}_f(k+j)=\hat{y}_f'(k+j)+r_1u_f(k+j-1)+\cdots+r_ju_f(k) \quad (10)$$

where

$$r_j=\begin{cases} b_1, & j=1 \\ b_j-r_{j-1}a_1-\cdots-r_1a_{j-1}, & 1<j\leq n \\ b_j-r_{j-1}a_1-\cdots-r_{j-n}a_n, & n<j\leq m \\ -r_{j-1}a_1-\cdots-r_{j-n}a_n, & j>m \end{cases} \quad (11)$$

Moreover, if we define

$$\hat{y}''(k+j)=\begin{cases} y(k+j), & j\leq 0 \\ \hat{y}_f'(k+j)-f_1\hat{y}''(k+j-1)-\cdots-f_{\mu+\eta}\hat{y}''(k+j-\mu-\eta), & j>0 \end{cases} \quad (12)$$

then the relationship between $\hat{y}''(k+j)$ and the prediction of $y(k+j)$ defined by

$$\hat{y}(k+j)=\begin{cases} y(k+j), & j\leq 0 \\ [-\hat{y}(k+j-1),\cdots,-\hat{y}(k+j-n)\,\vdots\,u(k+j-1),\cdots,u(k+j-m)]\,\underline{\theta}, & j>0 \end{cases} \quad (13)$$

may be expressed as

$$\hat{y}(k+j)=\hat{y}''(k+j)+\hat{r}_1u_f(k+j-1)+\cdots+\hat{r}_ju_f(k) \quad (14)$$

where

$$\hat{r}_j=\begin{cases} r_1=b_1, & j=1 \\ r_j-f_1\hat{r}_{j-1}-\cdots-f_{j-1}\hat{r}_1, & 1<j\leq\mu+\eta+1 \\ r_j-f_1\hat{r}_{j-1}-\cdots-f_{\mu+\eta}\hat{r}_{j-\mu-\eta}, & j>\mu+\eta+1 \end{cases} \quad (15)$$

In the case of constant reference signal y_r, in order to moderate the control and also to simplify the computation, the blunting factor $0<\sigma\leq 1$ may be introduced as follows

$$u_f(k+j)=\sigma^j u_f(k) \qquad j\geq 1 \quad (16)$$

Thus, Eq. (14) reduces to

$$\hat{y}(k+j)=\hat{y}''(k+j)+\hat{r}_j'u_f(k) \quad (17)$$

with $\hat{r}_j'=\hat{r}_j+\sigma\hat{r}_{j-1}+\cdots+\sigma^{j-1}\hat{r}_1$ (18)

Minimizing J with respect to $u_f(k)$ gives

$$u_f(k)=\frac{\sum_{j=1}^{N} r_j'w_j\left[y_r(k+j)-\hat{y}''(k+j)\right]}{\lambda'+\sum_{j=1}^{N}(r_j')^2w_j} \quad (19)$$

with $\lambda'=\lambda(1+\sigma^2+\cdots+\sigma^{2M})K_d^2$ (20)

where $K_d=(b_1+\cdots+b_m)/(1+a_1+\cdots+a_n)$ (21)

and σ may be chosen as

$$\sigma=1-1.5/M \quad (22)$$

Finally, the current control is

$$u(k)=u_f(k)/f(z^{-1}) = u_f(k)-f_1u(k-1)-\cdots-f_{\mu+\eta}u(k-\mu-\eta) \quad (23)$$

Although the MGPC keeps all the good robustness of GPC, many design parameters (such as N, M, λ and f_1') remain to be artificially predetermined, mainly according to the designer's experience. When these parameters are properly chosen under certain conditions, the MGPC can give excellent control performance. If, however, the plant dynamics or the environment disturbance changes drasticly, then the control system may deviate from its optimal operating state when all these parameters rest unchanged.

To cope with still larger variations in the disturbance or plant dynamics, a new self-learning scheme --- Learning Modified Generalized Predictive Controller (LMGPC) is proposed later. The scheme is enlightened by Zoubeidi (1983), in which the reinforcement learning is introduced into the model reference adaptive control system. In a similar way of on-line learning, the LMGPC can adapt the parameters of OMA filter and MGPC to their optimal values in different cases so as to maintain frequently good operating state of the control system.

OVERALL STRUCTURE AND OPERATION PRINCIPLE OF LMGPC

As shown in Fig. 1, the LMGPC is realized by adding to the structure of MGPC system a learning loop containing 2 sub-loops: The one is for on-line optimization of OMA filter parameter f_1', and the another for that of MGPC parameters N, M and λ. In normal cases the MGPC system itself can cope with a plant with linear dynamic model whose order or time-delay varies within a certain range. To overcome the stepwise disturbance that occurs very often in practice, the MGPC is designed such that the OMA filter takes the form of $(1-z^{-1})f'(z^{-1})$, where $f'(z^{-1})=1+f_1'z^{-1}$, i.e. $\eta=1$, for the sake of simplicity. To minimize the amplification of the random noise, the theoretically optimal value of f_1' may be evaluated by using the method in Xu (1988b), with the coefficient $\hat{c}_1$ in the noise polynomial $C(z^{-1})=1+\hat{c}_1z^{-1}$ being roughly preset. (Notice that here $\hat{c}_1\neq 0$ implies not only the system noise but also the error due to the bias in the estimation of a_is and b_is).

The operation principle of the whole learning system loop may briefly be described as follows: Whenever a large error $e_r=y-y_r$ between the plant output y and the desired output y_r occurs and lasts a given time duration N_1 so that

$$I_{p1}=\frac{1}{N_1}\sum_{k=1}^{N_1}|e_r(k)|>\varepsilon_1 \quad (24)$$

the learning system loop is then triggered. After that, a pattern vector

$$VP_1=\left[\hat{P}_{Y(0)},\ \hat{P}_{Y(1)},\cdots,\ \hat{P}_{Y(L)}\right]^T \quad (25)$$

is extracted from the estimated covariance function

$$\hat{P}_{Y(l)}=\frac{1}{N_p}\sum_{k=l+1}^{N_p}[y(k-l)-\hat{\bar{y}}][y(k)-\hat{\bar{y}}] \qquad l=0,1,\cdots,L(\geq 20) \quad (26)$$

where

$$\hat{\bar{y}}=\frac{1}{N_p}\sum_{k=1}^{N_p}y(k) \quad (27)$$

is the estimated mean value of y(k).

If VP_1 illustrates distinct periodic undulation (as in Fig. 2a), then there must appear a sinusoidal disturbance in g(k), So the reinforcement learning unit 1 for on-line optimizing the parameter f_1' in the OMA filter $f(z^{-1})=\psi(z^{-1})(1+f_1'z^{-1})$ should be started with the mark $\beta=1$ for introducing the estimation of ψ_1 (implemented in the MGPC system) into

$$\psi(z^{-1})=(1-z^{-1})(1+\hat{\psi}_1z^{-1}+z^{-2}) \quad (28)$$

The learning unit 1 will put the new $f(z^{-1})$ thus optimized into use in the MGPC system

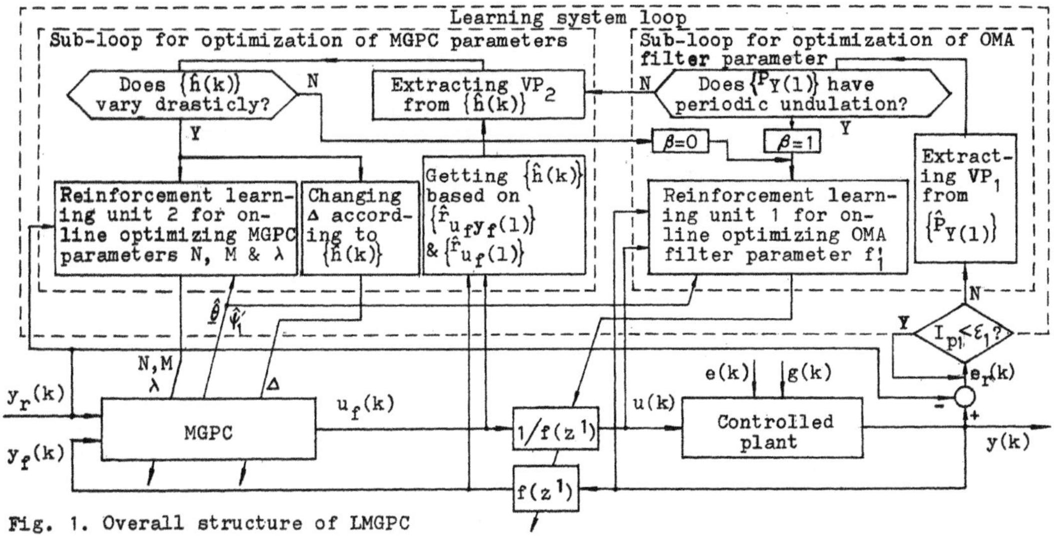

Fig. 1. Overall structure of LMGPC

so as to suppress the sinusoidal disturbance.

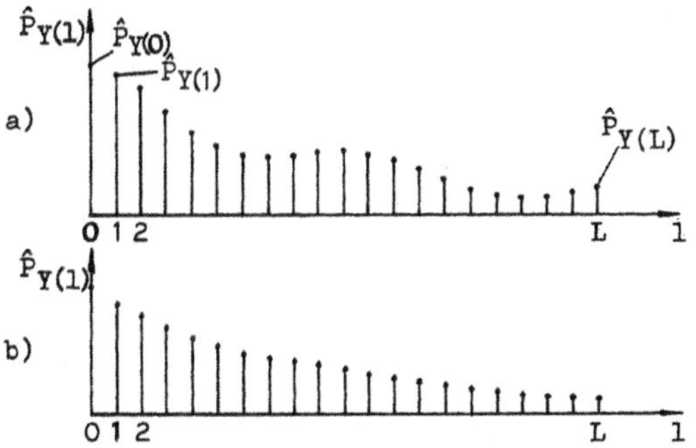

Fig. 2. The covariance function $\hat{P}_{Y(l)}$ of y

If VP_1 shows no distinct undulation (as in Fig. 2b), then it appears that the error e_r may be due to a large variation in the plant dynamics. So the next step is to generate the pattern vector VP_2 via the correlation method as follows

$$\begin{bmatrix}\hat{h}(0)\\ \hat{h}(1)\\ \vdots\\ \hat{h}(L)\end{bmatrix}=\begin{bmatrix}\hat{r}_{u_f}(0) & \hat{r}_{u_f}(1) & \cdots & \hat{r}_{u_f}(L)\\ & \ddots & & \vdots\\ & & \ddots & \hat{r}_{u_f}(1)\\ (\text{sym-metric}) & & & \hat{r}_{u_f}(0)\end{bmatrix}^{-1}\begin{bmatrix}\hat{r}_{u_f y_f}(0)\\ \hat{r}_{u_f y_f}(1)\\ \vdots\\ \hat{r}_{u_f y_f}(L)\end{bmatrix} \quad (29a)$$

or

$$VP_2 = \hat{R}_{u_f}^{-1} \quad \hat{\underline{r}}_{u_f y_f} \quad (29b)$$

where

$$\hat{r}_{u_f}(l)=\frac{1}{N_r}\sum_{j=l+1}^{N_r} u_f(j-l)u_f(j) \quad (30)$$

$$\hat{r}_{u_f y_f}(l)=\frac{1}{N_r}\sum_{j=l+1}^{N_r} u_f(j-l)y_f(j) \quad (31)$$

may be on-line estimated recursively as

$$\hat{r}_{u_f}(l,k)=\hat{r}_{u_f}(l,k-1)+[u_f(k-l)u_f(k)-\hat{r}_{u_f}(l,k-1)]/k \quad (32)$$

$$\hat{r}_{u_f y_f}(l,k)=\hat{r}_{u_f y_f}(l,k-1)+[u_f(k-l)u_f(k)-\hat{r}_{u_f y_f}(l,k-1)]/k \quad (33)$$

If the newly obtained VP_2 differs greatly from the original one, then it is needful to judge at first whether the discrete settling time K_s (as shown in Fig. 3) is reasonable or not using the criterion:

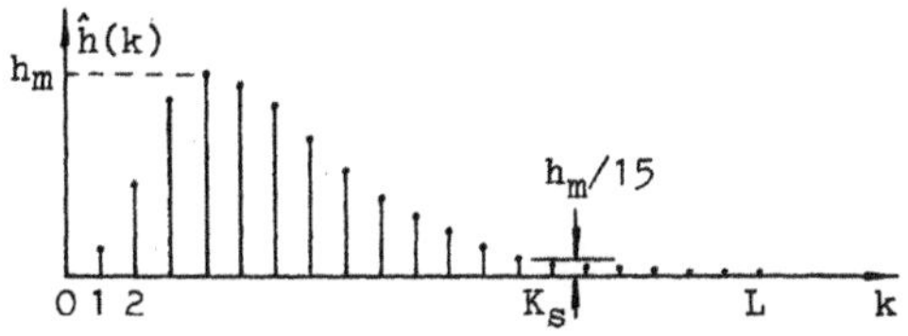

Fig. 3. The plant impulse response $\hat{h}(k)$

$$K_s = 10\text{---}15$$

The sampling period Δ should be gradually increased if K_s is too large, and decreased if too small.

At the same time, the reinforcement learning unit 2 for on-line optimizing MGPC parameters is started. The unit 2 will put the new N, M and λ thus optimized into use in the MGPC algorithm.

Again, if the newly obtained VP_2 does not differ greatly from the original one, then it implies that the error e_r is merely due to the improper presetting of $\hat{c}_1$ which leads to unsuited f_1'. So it is meaningful to renew the value of f_1' by setting β=0 and then starting the reinforcement learning unit 1 still with the normal OMA filter $\bar{f}(z^{-1})=(1-z^{-1})(1+f_1'z^{-1})$.

LEARNING PROCESS OF ON-LINE OPTIMIZATION OF THE OMA FILTER PARAMETER f_1'

As shown in Fig. 4, the main input and output of the reinforcement learning unit 1 are the pattern vector VP_1^q, which corresponds to the current control situation characterized by $\hat{P}_{Y(l)}$, and the OMA filter $f(z^{-1})=\Psi(z^{-1})f'(z^{-1})$ with optimized f_1' in $f'(z^{-1})=1+f_1'z^{-1}$, respectively.

Suppose here f_{1j}', $j=1,2,\cdots,\bar{r}_1$, form an exhaustive classes of response, and VP_1^q, $q=1,2,\cdots,\bar{s}_1$, characterize $\bar{s}_1$ possible different control situations. Within the qth situation, the response probability of the j-th response class is denoted by P_{1j}^q.

If VP_1^q, and hence the current control situation, has never occured before, then the learning unit 1 has to start to accumulate new experience on the optimized f_1' in this situation, with VP_1^q being stored in the memory and still the original value of f_1' being temporarily used in the tentative OMA filter $f'(z^{-1})=1+f_{1i}'z^{-1}$.

The on-line optimization of f_1' is performed as follows. Comparing the filtered output

$$\bar{y}_f(k)=\bar{f}(z^{-1})y(k)$$
$$=\psi(z^{-1})\bar{f}'(z^{-1})y(k) \quad (34)$$

with the estimation of $\bar{y}_f(k)$ based on the identified disturbance-free plant model

$$\hat{\bar{y}}_f(k)=[-\hat{\bar{y}}_f(k-1),\cdots,-\hat{\bar{y}}_f(k-n)\,|\,\bar{u}_f(k-1),\cdots,\bar{u}_f(k-m)]\underline{\hat{\theta}} \quad (35)$$

with
$$\bar{u}_f(k)=\bar{f}(z^{-1})u(k)$$
$$=\psi(z^{-1})\bar{f}'(z^{-1})u(k) \quad (36)$$

gives the filtered output error

$$\bar{e}_f(k)=\bar{y}_f(k)-\hat{\bar{y}}_f(k) \quad (37)$$

The optimization process for f_1' via Hooke-Jeeves method continues untill the following condition

$$I_{p2}=\frac{1}{N_2}\sum_{k=1}^{N_2}|\bar{e}_f(k)|=\min \quad (38)$$

is satisfied. Before the optimized $f(z^{-1})$ thus obtained is posted into the MGPC system, the response probability P_{1i}^q of the optimized f_{1i}' is reinforced and the remaining P_{1j}^q, $j\neq i$, are penalized by using the following reinforcement algorithm (Saridis, 1977):

$$P_{vj}^q(k_q)=\begin{cases} P_{vi}^q(k_q-1)+m_p[1-P_{vi}^q(k_q-1)], & j=i \\ P_{vj}^q(k_q-1)(1-m_p), & j\neq i \end{cases} \quad (39)$$

where k_q is an instant of control being in the qth control situation, $0<m_p<1$ the probability reinforcement factor, and v=1 in the case of VP_1.

Again, if VP_1^q has previously occured, then according to the Bayessian estimation theorem, the value of $f_{1\hat{i}}'$ corresponding to the maximum response probability $P_{1\hat{i}}^q$ among all P_{1j}^q, $j=1,2,\cdots,\bar{r}_1$, is chosen to be directly used in the OMA filter $f(z^{-1})$.

In addition, depending on the mark $\beta=0$ or 1, the internal model polynomial $\psi(z^{-1})$ of g(k) takes its form of either $(1-z^{-1})$ or $(1-z^{-1})(1+\psi_1'z^{-1}+z^{-2})$ respectively so as to match the two different cases where the sinusoidal disturbance appears or not.

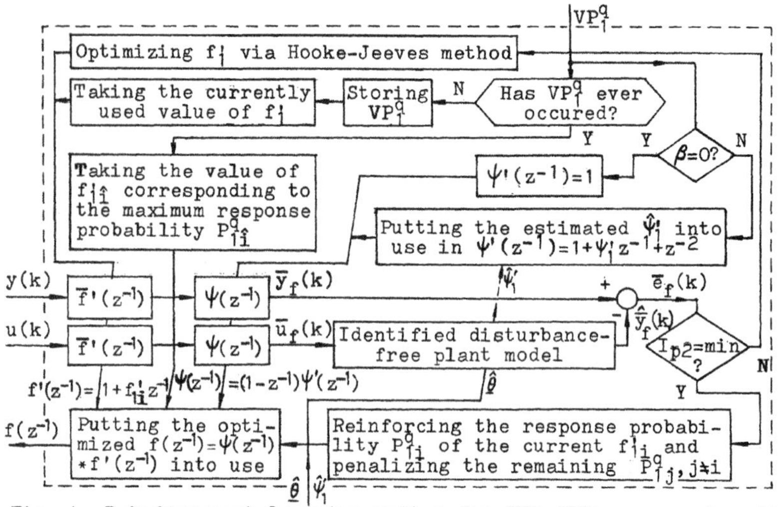

Fig. 4. Reinforcement learning unit 1 for OMA filter parameter f_1'

LEARNING PROCESS FOR ON-LINE OPTIMIZATION OF THE MGPC PARAMETERS N, M & λ

As shown in Fig. 5, the main input and output of the reinforcement learning unit 2 are the pattern vector VP_2^q, which corresponds to the current control situation characterized by $\{\hat{h}(k)\}$, and the combination of the optimized design parameters N, M and λ in the MGPC algorithm, respectively.

Suppose here the combinations $\omega_j=(N_{jn}, M_{jm}, \lambda_{j\lambda})$, $j=1,2,\cdots,\bar{r}_n x\bar{r}_m x\bar{r}_\lambda$ ($j_n=1,\cdots,\bar{r}_n$; $j_m=1,\cdots,\bar{r}_m$; $j_\lambda=1,\cdots,\bar{r}_\lambda$), form an exhaustive class of response, and VP_2^q, $q=1,2,\cdots,\bar{s}_2$, characterize $\bar{s}_2$ possible different control situations. Within the qth situation, the response probability of the jth response class ω_j is denoted by P_{2j}^q.

If VP_2^q, and hence the current control situation, has never occured before, then the learning unit 2 has to start to accumulate new experience on the optimized combination of N, M and λ in this situation, with VP_2^q being stored in the memory and still the original values of N, M and λ used in the MGPC system being temporarily used in the tentative MGPC algorithm.

The on-line optimization of N, M and λ is performed as follows. The tentative closed-loop system (shown in Fig. 5) consists of the tentative MGPC algorithm and the identified disturbance-free plant model defined by $\underline{\hat{\theta}}$. So its output is

$$\hat{y}_f(k)=[-\hat{y}_f(k-1),\cdots,-\hat{y}_f(k-n)\,|\,\hat{u}_f(k-1),\cdots,\hat{u}_f(k-m)]\,\underline{\hat{\theta}} \quad (40)$$

Comparing

$$\hat{y}(k)=\hat{y}_f(k)/f(z^{-1})$$
$$=\hat{y}_f(k)-f_1\hat{y}(k-1)-\cdots-f_{\mu+\eta}\hat{y}(k-\mu-\eta) \quad (41)$$

with the desired output $y_r(k)$ gives the tentative output error

$$\hat{e}_r(k)=\hat{y}(k)-y_r(k) \quad (42)$$

The optimization process for N, M and λ via Hooke-Jeeves method continues untill the following condition

$$I_{p3}=\frac{1}{N_3}\sum_{k=1}^{N_3}|\hat{e}_r(k)|<\varepsilon_3 \quad (43)$$

is satisfied. Before the optimized N, M and λ thus obtained are posted into the MGPC system, the response probability P_{2i}^q of the optimized N, M and λ is reinforced and the remaining P_{2j}^q, $j\neq i$, are penalized by using the same reinforcement algorithm as Eq.(39) but with v=2 in this case of VP_2.

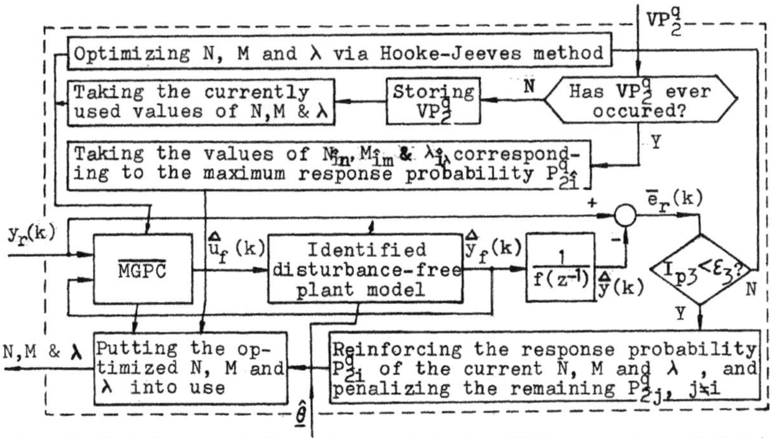

Fig. 5. Reinforcement learning unit 2 for MGPC parameters N,M & λ

Again, if VP^q_2 has previously occured, then the combination $\omega_{\hat{i}}=(N_{\hat{i}n},M_{\hat{i}m},\lambda_{\hat{i}\lambda})$ corresponding to the maximum response probability $P^q_{2\hat{i}}$ among all P^q_{2j}, $j=1,2,\cdots,\bar{r}_n x\bar{r}_m x\bar{r}_\lambda$, is chosen to be directly used in the MGPC system, according to the Bayessian estimation theorem.

NUMERICAL SIMULATIONS AND EXPERIMENTAL TESTS

An extensive set of numerical simulations and experimental tests was made for showing the excellent robustness of the LMGPC to very large variation in type of the external disturbance or the plant dynamics. Hereafter is only a part of their results.

Numerical simulation 1 This simulation was for showing the robustness to environmental disturbance. The simulated plant has the following continuous-time model

$$H(s)=1/(s+1)(s+2)=0.5/(s^2+1.5s+0.5) \quad (44)$$

Taking sampling period $\Delta=1''$ and considering both noise and disturbance, the discrete-time stochastic model of the plant became

$$(1-0.9774z^{-1}+0.2231z^{-2})y(k)$$
$$=(0.1548z^{-1}+0.0939z^{-2})u(k)$$
$$+(1+0.2z^{-1})e(k)+g(k) \quad (45)$$

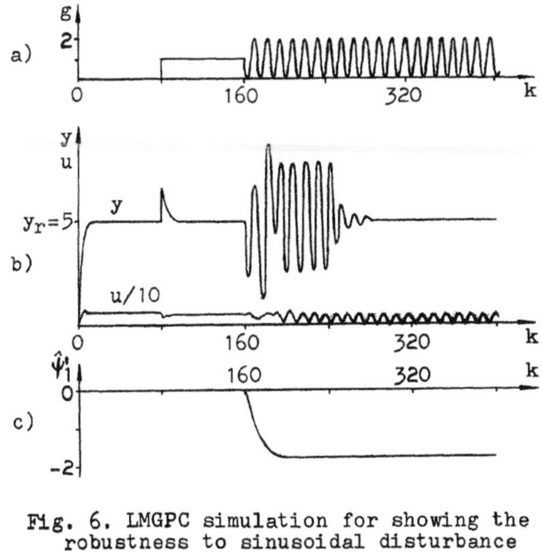

Fig. 6. LMGPC simulation for showing the robustness to sinusoidal disturbance

where $\sigma_e^2=0.01$ and $g(k)$ consists of

stepwise $g_1(k) = \begin{cases} 0, & k\leq 80 \\ 1, & k>80 \end{cases}$ (46)

sinusoidal $g_2(k) = \begin{cases} 0, & k\leq 160 \\ \sin[2\pi(\frac{k}{12})], & k>160 \end{cases}$ (47)

The curve of $g(k)=g_1(k)+g_2(k)$ is shown in Fig. 6a.

The MGPC parameters were set to N=8, M=4(hence $\sigma=1-1.5/M=0.63$), $\lambda=1$, $w_j=1$, $\varepsilon_1=0.06$, $N_1=40$, the model orders in RLS estimation for $\hat{\theta}$ are $\hat{n}=\hat{m}=2$. The exhuastive response classes for the OMA filter parameter f_1' are chosen to be $\{-0.8,-0.6,-0.4,-0.2,0,0.2,0.4,0.6,0.8\}$ and the initial value is $f_1'=0$.

The simulation result of the LMGPC for showing the robustness to the variation in type of the deterministic disturbance $g(k)$ is shown in Fig. 6b. With the quick convergence of the plant parameter estimations $\hat{a}_1$, $\hat{a}_2$, $\hat{b}_1$, $\hat{b}_2$ to their true values from the theoretically approximate values as initial estimates respectively, the MGPC smoothly started up without any overshoot. After k=80 appeared the stepwise disturbance $g_1(k)$ at first, which was easily overcome by the original MGPC itself without offset.

Then, after k=160 appeared again the sinusoidal disturbance $g_2(k)$ with a period of 12, which caused a severe undulation on the curve of y(k), and meanwhile the RLS estimation of ψ_1 was started as shown in Fig. 6c. After k=240 the learning system loop was triggered with $f_1'=0$. The estimated $\hat{\psi}_1$, which was close to the true value $\psi_1=-2\cos(2\pi/12)=-1.732$ after k=183, was then put into use in the $\psi(z^{-1})=(1-z^{-1})(1+\psi_1'z^{-1}+z^{-2})$. Up to k=240 an optimized OMA filter parameter $f_1'=0.6$, which was close to the theoretical value 0.72 in this case, was gained and immediately put into use. Thus, a satisfactory result in overcoming the effect of $g_2(k)$ completely was obtained after k=280.

Numerical simulation 2 This simulation was for showing the robustness to large variation in the plant model order n and d.c.gain K_d. The simulated plant took one of the following two models:

I) The same model as Eq.(45) with g(k)=0 and $K_d=1$;

II) $(1-1.5788z^{-1}+1.2599z^{-2}-0.8583z^{-3}+0.2725z^{-4})y(k)$
$=(0.0305z^{-1}+0.3052z^{-2}+0.2364z^{-3}+0.0185z^{-4})u(k)$
$+(1+0.2z^{-1})e(k)$, $K_d=6.2$ (48)

For the sake of simplicity, $\hat{n}=\hat{m}=2$, $w_j=1$ and even $\lambda=1$ were all fixed throughout, and the exhuastive response classes of the combinations $\omega_j=(N_{jn},M_{jm})$, $j=1,2,\cdots,12$, were chosen such that $N_{jn}\in\{4,8,12,16\}$ and $M_{jm}\in\{2,4,6\}$. The initial values of N and M were set to 12 and 6 respectively.

As shown in Fig. 7, after the plant changed its model from I to II at k=80, i.e. n from 2 to 4 and K_d from 1 to 6.2, the plant output y(k) varied drasticly at first due to the large variations in n and K_d, and the

Fig. 7. LMGPC simulation for showing the robustness to large order & d.c. gain variations

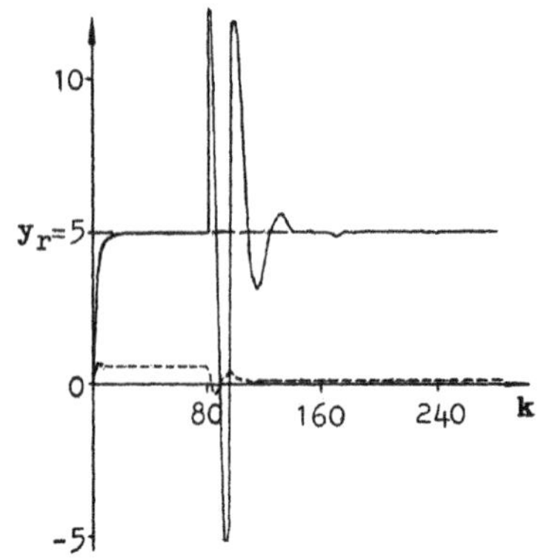

learning system loop was then triggered. After on-line optimizing N and M via the reinforcement learning algorithm, an optimized set of N=8 and M=4 was posted into the MGPC algorithm at k=110, and y(k) soon came back to its original steady value y_r=5.

While, when the plant changed its model from I to a model with n=3 and K_d=1.2, the MGPC itself still coped with this dynamics variation without triggering the learning loop, showing its own good robustness under certain condition.

Experimental tests This set of tests was for showing the feasibility of LMGPC and its robustness to a large variation in the plant dynamics. These tests were performed by the same LMGPC algorithm as in the Simulation 2, implemented via TURBO PASCAL language in a STD-BUS microcomputer system and experienced on a laboratory model process. The model process consists of 3 serially **coupled water vessels (Fig. 8)**. The inlet

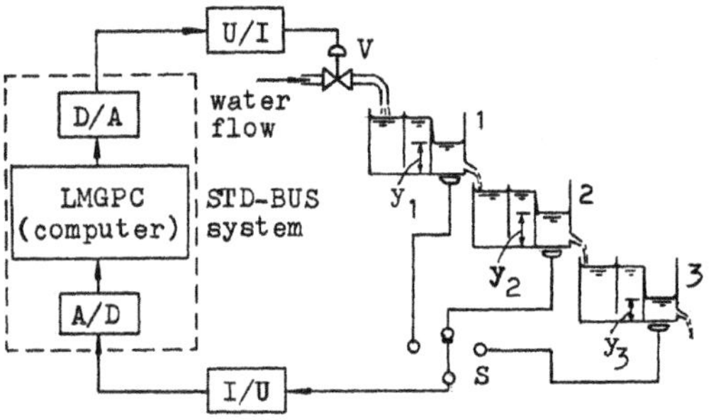

Fig. 8. Laboratory model process

flow rate of the first vessel is controlled by the opening u of the electrical valve V and the water levels y_1, y_2 and y_3 in these vessels are all measured by DP-cells. The dynamic relation between u and y_i is

$$H_i(s)=Y_i(s)/U(s)=\sum_{j=1}^{i}\frac{K_j}{T_js+1}, \quad i=1,2 \text{ and } 3$$

The dynamics of the controlled process can be changed abruptly by the switch S. For the sake of simplicity, throughout the tests n=m=2, $w_j=\lambda=1$, $f_j^1=0$ and only the sub-loop for MGPC parameters was used.

Fig. 9 shows a test run illustrating that the LMGPC can on-line optimize the MGPC parameters N, M and Δ, if they are improperly preset. The process output was kept to be y_2, and the control started with N=8, M=4 and Δ=8". At t=5'44" the learning function came into use, and Δ was adjusted to 16". 9' later, the learning process ended up with the optimized N=12 and M=2, and the control performance did become better.

Fig. 10 shows another test run illustrating that the LMGPC can cope with a very large model order variation. The process control started with a less proper setting: N=8, M=4 and Δ=8". At t=2'40", the process output was switched from y_1 to y_3, i.e. n from 1 to 3. Then the learning system loop was triggered at t=5'12", and Δ was adjusted to 16". 7'30" later, the learning process ended up with the optimized N=12 and M=4, and the control performance become very well.

REFERENCES

Clarke, D.W., C. Mohtadi and P.S. Tuffs (1984). Generalized predictive control. O.U.E. Report, 1555/84 and 1557/84.

Saridis, G.N. (1977). Self-Organizing Control of Stochastic Systems. Marcel Dekker, New York. Chap. 8, pp.319-332.

Xu Ning-Shou. (1988a). Modified generalized predictive control. 8th IFAC/IFORS symposium on Identification and System parameter Estimation, Beijing, China.

Xu Ning-Shou. (1988b). Optimal Moving Average filter with applications to self-tuning control against deterministic disturbances. Journal of Beijing Polytechnic University, 14.

Xu Ning-Shou. (1989). Approximate evaluation of the sampled model parameters for a linear continuous-time system with relatively small sampling period. Int. J. of Systems Sci., 20, 811-838.

Zoubeidi. M. (1983). A Learning Adaptive Controller. Ph. D. Thesis. University of California, Berkeley.

Fig. 9. **A test run showing the ability of LMGPC in on-line optimizing N, M and Δ**

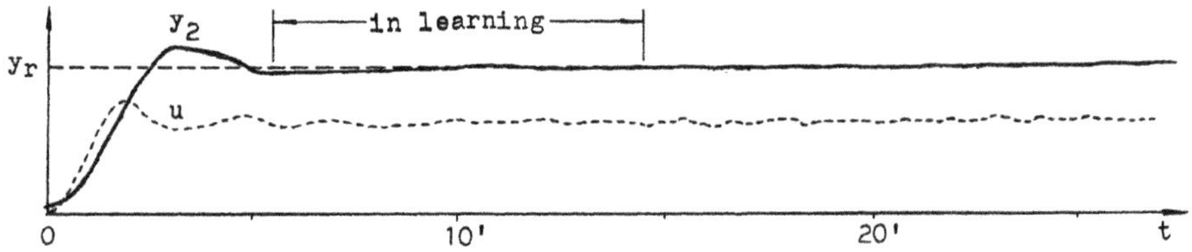

Fig. 10. A test run showing the robustness of LMGPC to large order variation

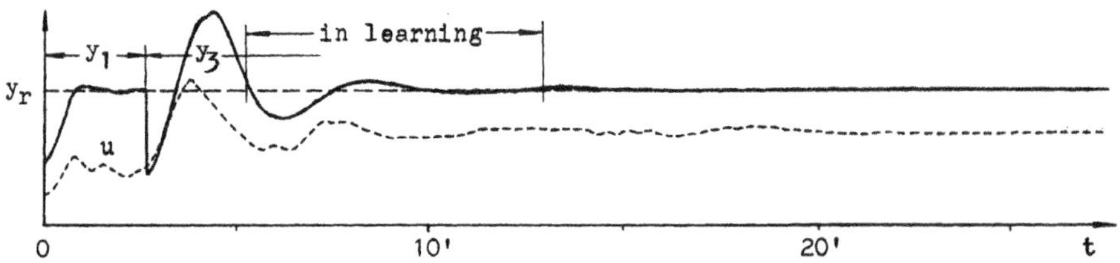

INTELLIGENT ADAPTIVE CONTROL OF MODE-SWITCH PROCESSES

R. A. Hilhorst*, J. van Amerongen*, P. Löhnberg* and H. J. A. F. Tulleken**

**Control, Systems and Computer Engineering Group, Dept. of Electrical Engineering, University of Twente, and Mechatronics Research Centre Twente, P. O. Box 217, 7500 AE Enschede, The Netherlands*
***Process Systems Group, Dept. of Mathematics & Systems Engineering, Koninklijke/Shell-Laboratorium (Shell Research B.V.), P.O. Box 3003, 1003 AA Amsterdam, The Netherlands*

Abstract. The intelligence of controllers has increased over the decades. However, the number of applications of adaptive controllers is still restricted, due to practical limits of the implemented continuous adaptation. For processes which operate only in a limited number of modes (so called mode-switch processes), constant adaptation is not needed or desired. In this paper an intelligent extension of adaptive control will be presented, in which process behaviour can be stored in a memory, retrieved from it and evaluated for each mode of operation. This intelligent memory concept leads to an adaptive control structure which, after a learning phase, quickly adjusts the controller parameters based on retrieval of old information, without the need to relearn every time. This approach has been tested on a simulation model of an assembly robot, but it is directly applicable to many processes in the (petro)chemical industry.

Keywords. Adaptive control; Supervisory control; Automatic tuning; Mode-switch Processes; Bumpless transfer.

1 Introduction

Over the decades, the intelligence of controllers - starting from PID control to more complex controllers like adaptive controllers - has increased. This increase of intelligence is due to the demand for more advanced control. Although the current adaptive controllers have the intelligence to adjust the controller parameters the number of applications is still restricted compared to less smart PID-like control applications (Åström and Wittenmark, 1989). The reason for this is fourfold. Firstly, in many cases the time needed for adaptation is too long. For instance, if a robot picks up a payload of unknown mass, the time needed for adaptation may be just as long as, or longer than, the time needed to transport the payload. Secondly, in order to adapt to changes in the process or in the control criterion, the current adaptive controllers have to introduce deliberate additional excitation to the process. In environments like the process industry, this is not allowed most of the time as it introduces loss of product (quality). Thirdly, most operators want to know exactly what is going on in their process. As the commercially available adaptive control equipment, although advanced, is often less transparent, it is not always easily accepted. Fourthly, continuous adaptation is not always necessary.

In practice one frequently encounters processes of which the dynamics are fairly similar in one mode of operation, but are different in another. Furthermore these processes frequently return in a previous mode of operation. Such processes are encountered in process industry and robotics. For instance, a chemical reactor in which the production of one of the chemical compounds has to be optimized according to the market demand shows this mode-switch behaviour. The same applies for a robot which has to transport payloads with a limited number of different masses.

A new solution to the problems described above is to exploit the mode-switch behaviour of processes. For this purpose it is attractive to extend the current adaptive controllers with an intelligent facility by which the process behaviour in subsequent modes can be recognized, stored and retrieved from memory. These functions should be coordinated by a supervisor. This approach has the advantage that no identification of the process is needed when it returns to an earlier visited recognized mode of operation or when the control criterion has changed. A performance monitor can take care of restarting the adaptation whenever necessary.

In this paper the general structure of a supervisor with an intelligent memory will be described, with the main emphasis on the retrieval of stored information. It is organized as follows. In section 2 the modeling of mode-switch processes is presented. Definitions of a 'mode' and 'mode switch' are presented. In section 3 the structure of the supervisor is discussed. Furthermore, an overview of the functions to be executed by the supervisor is given. In section 4 a detailed discussion about retrieval is presented. In section 5 an application to a robot is shown. Finally in section 6 conclusions are drawn.

2 Modeling of mode-switch processes

Processes, the dynamics of which are fairly similar in one mode of operation but different in another, are obviously non-linear. Such processes can often be modeled quite well by a non-linear model:

$$\dot{x} = f(x, u, \theta) \qquad (1)$$
$$y = g(x, u, \theta)$$

where x is a vector of plant states, u is the vector of control inputs, θ is a vector of process parameters, and f and g are non-linear, time-invariant functions. If a description of the process in the form of equation (1) is known, then a non-linear controller could be designed such that the required control performance can be met. However, the design of robust non-linear controllers is still in its infancy. In addition, in many cases (1) is not available, either because of the cost of producing such a model, or because of the large complexity and uncertainties in the process. Moreover, a complete description often is not needed, as in many cases processes are operated only for small deviations around nominal operating points in a small number of modes. In these cases it may be interesting to have a number of descriptions of the process, each valid in one mode of operation. In order to derive such a description, a space of operation Ω should be defined. This is spanned by all variables which influence the dynamics of the process. As most process variables are constrained, in practice this space will be bounded. The operation vector ω points to a particular realization in this space. Typically, the operation vector will contain information of the nominal state vector $\bar{x}$, input $\bar{u}$ and process parameter $\bar{\theta}$:

$$\omega = (\bar{x}, \bar{u}, \bar{\theta}) \qquad (2)$$

It is assumed that around an operation vector ω, a linearized description for the local dynamics can be obtained, which can be written as:

$$\begin{aligned} \dot{x} &= f'(x, u, \omega) \\ y &= g'(x, u, \omega) \end{aligned} \qquad (3)$$

Such a linear description can be obtained analytically if (1) is available or otherwise by identification of the process around the operation vector ω. E.g. suitable identification techniques will yield the required linear model of the process:

$$\begin{aligned} f'(x, u, \omega) &= f(\bar{x}, \bar{u}, \bar{\theta}) + A_\omega(x-\bar{x}) + B_\omega(u-\bar{u}) \\ g'(x, u, \omega) &= g(\bar{x}, \bar{u}, \bar{\theta}) + C_\omega(x-\bar{x}) + D_\omega(u-\bar{u}) \end{aligned} \qquad (4)$$

where

$$A_\omega = \left.\frac{\partial f}{\partial x}\right|_\omega, \quad B_\omega = \left.\frac{\partial f}{\partial u}\right|_\omega, \quad C_\omega = \left.\frac{\partial g}{\partial x}\right|_\omega, \quad D_\omega = \left.\frac{\partial g}{\partial u}\right|_\omega$$

By the use of identification, linear descriptions of the process can be obtained for various nominal operation vectors. In order to distinguish between the various operation vectors, they will be denoted by ω_i, $i \in \mathbb{N}$. The process model obtained for the i-th nominal operation vector ω_i will be referred to as P^i. The current nominal operation vector will be denoted by ω and the associated linear model for the currently encountered process dynamics by P_ω. In order to obtain a restricted number of models describing the process dynamics adequately over the whole operating space, a metric is defined which indicates the dynamic distance between two models. This metric, which will be discussed in more detail in section 4, is denoted by:

$$d(P^i, P^j) \geq 0 \qquad (5)$$

When $d(P^i, P_\omega) \leq \varepsilon_i$ (ε_i is a real positive number chosen by the designer), the operating vectors ω_i and ω corresponding to the models P^i and P_ω are defined as being in the same mode of operation, which is indicated by Ω_i, where $i \in \mathbb{N}$. The tolerances ε_i and the number N of operation vectors ω_i, $i = 1, \ldots, N$, should be chosen such that the union of all modes of operation will cover the whole space of operation Ω and that a good description of the process is obtained over the whole space of operation Ω. On the other hand the modes of operation should preferably not overlap each other. If overlap occurs, the operation vector ω corresponding with the model P_ω will be classified to mode Ω_i if the following holds:

$$d(P_\omega, P^i) \leq d(P_\omega, P^j), \text{ for all } j \in [1,..,N] \qquad (6)$$

An example given in Fig. 1a is illustrated by Fig. 1b, which shows the result of applying (6) to it.

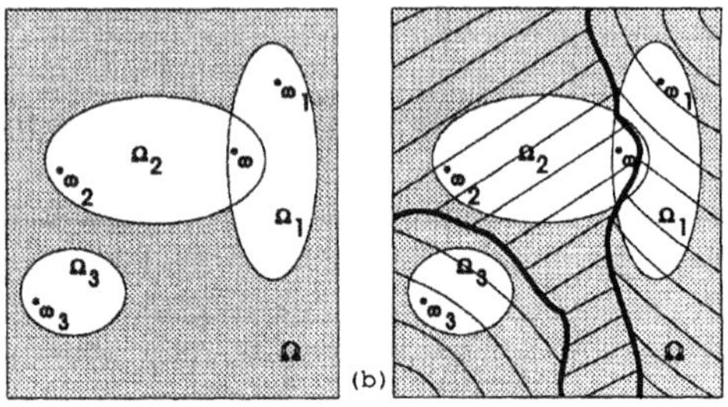

Fig. 1 The space of operation before (a) and after (b) the application of (6).

A mode switch from mode Ω_i to mode Ω_j ($j \neq i$) occurs at that moment later in time where (6) becomes false, i.e.

$$d(P_\omega, P^j) < d(P_\omega, P^i) \quad \text{(switching rule)} \qquad (7)$$

Switches between modes can be divided into three categories: abrupt, fast and slow relative to the dominant time constant in the control loop. In the case of a robot, these changes are due to for instance saturation (abrupt), Coriolis and Centrifugal forces (fast), and aging (slow).

3 Supervisor

In the previous section it was shown that it may be advantageous to use models of the process dynamics only at a limited, representative number of points of operation. In order to use a controller based on one of these models, a memory must be present in which the obtained information for each mode of operation can be stored and from which it can be retrieved. To meet the requirements for industrial usage, such a system should be sufficiently transparent so that operators are willing to work with it. Also a performance monitor should be present to indicate whether the current control performance is acceptable or not. If the current control performance is not acceptable, either a switch to another model in the memory should take place, or an additional identification should be considered in order to add an extra model. A new controller should be designed and installed. These requirements should lead to a system in which conventional adaptation is only used at the times that it is really necessary. In order to meet the stated requirements, an intelligent supervisory structure as shown in Fig. 2 is proposed.

In Fig. 2 the Supervisor supervises the monitoring, retrieval, storage and maintenance of the information in the memory, and the starting of identification and of controller design. The identification module can consist of several process identification methods. The monitor is a device which determines whether the control performance is adequate or not.

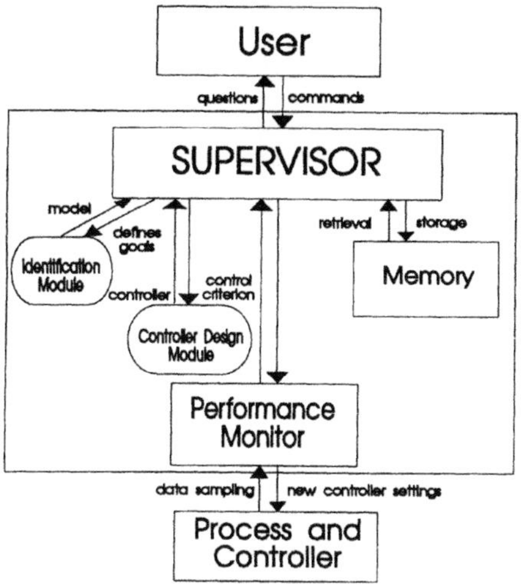

Fig. 2 Supervisor with memory

The supervisor communicates with the user. This supervisor has two strategies: an automatic and a manual one. In the manual strategy, operators of the process have the opportunity to overrule the decisions or proposals made by the memory or to choose other control criteria. The resulting control behaviour at the various modes of operations can be inconsistent, since operators do not all react in the same way when a particular control situation occurs. Such inconsistent control sometimes results in unwanted disturbances of the process. Therefore the automatic strategy is advantageous.

The information obtained during the identification phase is stored in the memory. This information consists of the operation vector ω and/or the controller and/or the model. Depending on the method of retrieval, on the constancy of the control criterion and on the time needed for control design, the controller and/or the model will be stored.

4 Retrieval

A model should be retrieved from memory when it is recognized that according to (6), this model better resembles the current process dynamics than the currently selected model. This comparison requires a sensible dynamic distance function such that the model space $\mathbb{M}$ is a metric space and obeys some additional practical requirements. The model space $\mathbb{M}$ is a metric space (Korn and Korn, 1968) if and only if for each pair of models P^i, $P^j \in \mathbb{M}$ (i.e. $\omega_i, \omega_j \in \Omega$) it is possible to define a real number $d(P^i, P^j)$ such that for all P^i, P^j, $P^k \in \mathbb{M}$ (i.e. $\omega_i, \omega_j, \omega_k \in \Omega$) according to (6):

$$d(P^i, P^i) = 0, \tag{8a}$$

$$d(P^i, P^j) \le d(P^i, P^k) + d(P^j, P^k) \tag{8b}$$

This definition implies $d(P^i, P^j) \ge 0$ and $d(P^i, P^j) = d(P^j, P^i)$. To be able to track mode switches, it must be possible to evaluate the dynamic distance function in a given time interval. In the following a number of possible distance functions will be discussed.

4.1 Euclidean norm between operation vectors

An obvious choice for the dynamic distance function is the Euclidean distance between two operation vectors in the space of operation Ω:

$$d_1(P^i, P^j) := \|\omega_i - \omega_j\| \tag{9}$$

Obviously, the Euclidean norm satisfies the requirement for a metric. The advantage of this norm is the simplicity of evaluation, resulting in an evaluation which can be performed very quickly. Therefore this distance function is very popular in gain scheduling. A disadvantage of this norm is that it requires all elements in the operation vector to be measurable. Furthermore, the distance between two operation vectors is not a measure for the difference between the related dynamics.

4.2 Average output error

Another choice for the dynamic distance function is the L_2 weighted error between the outputs y^i and y^j of processes P^i and P^j respectively in a selected time interval $[t, t+T_s]$:

$$d_2(P^i, P^j) := \int_t^{t+T_s} (y^i(\tau) - y^j(\tau))^2 d\tau \tag{10}$$

When an appropriate observation period T_s is selected, (10) will be a good indication whether the dynamics of those two processes resemble each other. The method requires that the processes have the same initial conditions and the same input during this interval. The average output error can be obtained by placing the several candidate models in memory in a series-parallel structure (see for instance Landau, 1979) with the process. The series-parallel structure will be illustrated for a sampled linear process:

$$\underline{x}_\omega(k+1) = A_\omega \underline{x}_\omega(k) + B_\omega \underline{u}(k) \quad \text{(process)} \tag{11a}$$

$$\underline{x}_{\omega_i}(k+1) = A_{\omega_i} \underline{x}_\omega(k) + B_{\omega_i} \underline{u}(k) \quad (i^{th}\ \text{model}) \tag{11b}$$

Equation (11) shows that this structure compares the actual (linearized) process with the nominal (linearized) process by resetting the model states at the sampling instants.

The advantage of this distance function d_2 (10) compared with the former d_1 (9) is that for d_2 not all elements in the operation vector have to be known. A disadvantage of d_2 is that for d_2 a decision may take more time than for d_1 because of the required observation period T_s (see (10)). The selection of T_s is a compromise between the speed of tracking mode switches and the sensitivity to noise. Retrieval on the basis of d_2 is used in this paper.

5 Application (Robot control)

In order to test the applicability of the supervisory structure where the retrieval is based on the series-parallel structure, the retrieval of a model from those stored in the memory has been studied using a simulation model of a pending two-link assembly robot (Gras and Nijmeijer 1989). This robot is described in the appendix.

The objective of control is to have negligible overshoot, a zero-position error, a fast settling time, and decoupling between the two links. To satisfy these goals, a multi-variable controller was selected. This controller was designed by the partial model matching (PMM) method (Takamatsu and co-workers, 1985), which is an extension of the pole-placement philosophy.

To derive a multi-variable controller by the partial model matching approach, the linearized process model should be written in the form of a transfer function. In the case of the two-link robot, s-transformation of the linearized robot description given in the appendix yields the transfer function relating U(s) to Y(s)

$$H(s) = [- A_0 - A_1 s + Is^2]^{-1} B_0 \tag{12}$$

where:

$Y(s) = (\Phi_1(s), \Phi_2(s))^T$ (angles of link 1 and 2)

$U(s) = (U_1(s), U_2(s))^T$ (torques on motor 1 and 2)

A_0, A_1, B_0 according to the appendix

The linearized robot model together with the PMM controller is depicted in Fig. 3.

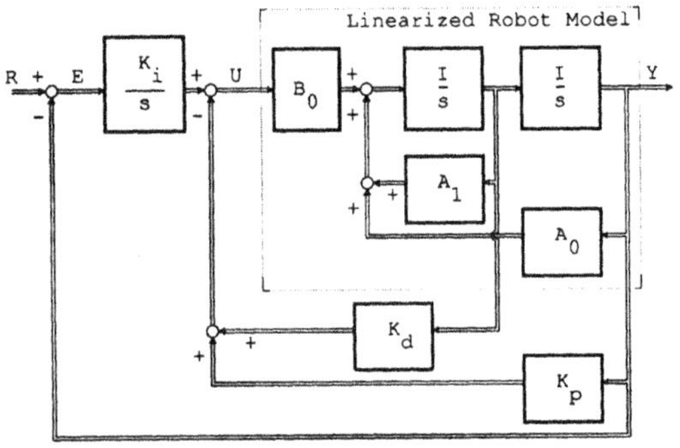

Fig. 3 Linearized robot model with PMM controller

In Fig. 3, K_i denotes the integral gain, K_d the derivative feedback gain, and K_p the proportional feedback gain. The transfer function matrix $M_p(s)$ of the closed-loop system (relating R(s) to Y(s)) is described by:

$$M_p(s) = \left[I + sK_i^{-1}(K_p - H_0) + s^2 K_i^{-1}(K_d - H_1) + s^3 K_i^{-1} B_0^{-1}\right]^{-1} \tag{13}$$

where

$H_i = B_0^{-1} A_i$ $(i = 0,1)$ if B_0 is non-singular.

Choosing a 'reference model' $M_r(s)$ with all non-diagonal coefficients zero leads to a decoupled system:

$$M_r(s) = \left[I + M_1 s + M_2 s^2 + M_3 s^3\right]^{-1} \tag{14}$$

The reference model and the model of the controlled process should have the same transfer functions. Thus the following equations have to be obeyed:

$$\begin{aligned} K_i^{-1}(K_p - H_0) &= M_1 \\ K_i^{-1}(K_d - H_1) &= M_2 \\ K_i^{-1} B_0^{-1} &= M_3 \end{aligned} \tag{15}$$

The poles of the reference model are chosen such that no overshoot occurs and that the settling time is about 5 seconds. These requirements lead to the following closed-loop poles:

$$s_{1,2} = -3.18 \pm 0.348j, \quad s_3 = -2.65 \tag{16}$$

and to the following reference model:

$$M_r(s) = \frac{I}{(1 + s + 0.333s^2 + 0.037s^3)} \tag{17}$$

The sampling rate was taken sufficiently fast, i.e. 50 Hz. As the time needed for controller design was not considered to be important, only the models together with the operation vector were stored in memory. Those models were stored that correspond to the robot states shown in Fig. 4a. The motivation of this rather ad hoc choice is that the robot has to regulate around these states as a part of the assembly cycle shown in Fig. 4b.

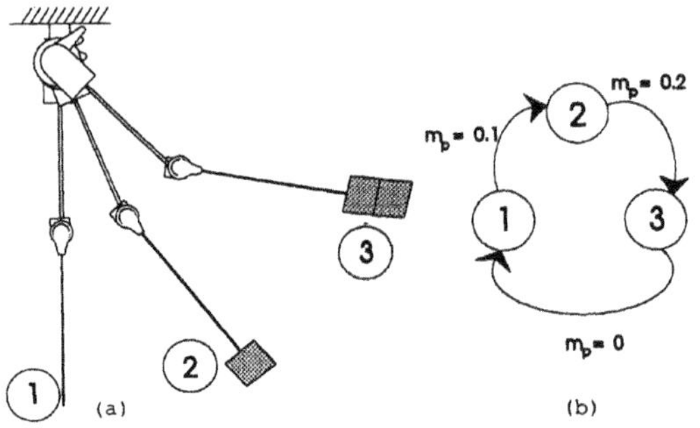

Fig. 4. (a) Robot states (Kruise, 1990). (b) Assembly cycle.

The resulting operation vectors belonging to these states are shown in table I.

TABLE 1 Operation vectors relating to Fig. 4.

	$\bar{\varphi}_1$	$\bar{\varphi}_2$	$\bar{\dot{\varphi}}_1$	$\bar{\dot{\varphi}}_2$	$\bar{\theta}_1 = m_p$	$\bar{u}_1$	$\bar{u}_2$
ω_1	0	0	0	0	0	0	0
ω_2	0.4	0.3	0	0	0.1	14.23	6.44
ω_3	0.8	0.6	0	0	0.2	24.20	9.85

If a switch is made from one model to another, then the feedback gains alter to maintain the desired control behaviour. If no precaution is taken, the control signal will contain bumps at the moments of a mode switch. These bumps can result in an undesired control behaviour. Therefore a bumpless transfer algorithm was added to remove the bumps from the control signal. Should the robot control be based on a PMM controller, a simple algorithm can be designed for removing the bumps in the controller signal by resetting the integrator terms (Åström and Wittenmark, 1990) such that the control signal remains continuous. In this case it would imply that

$$u(t) = u^+(t) \tag{18}$$

where u(t) and $u^+(t)$ are the current and new (model based) control signal respectively. After a mode switch, the control signal $u^+(t)$ (see Fig. 3) is equal to

$$u^+(t) = K_i^+ I(t) - K_p^+ y(t) - K_d^+ \dot{y}(t) \tag{19}$$

where:

$I(t) = \int_0^t e(t)\, d\tau$,

K_i^+, K_p^+, K_d^+ denote the new feedback gains, and e(t) the position error according to Fig. 3. As the position error e and velocity $\dot{y}$ are known at time t, bumpless transfer is realized by resetting the integrator states according to

$$I^{+}(t) = K_i^{+^{-1}} (u(t) + K_p^{+} y(t) + K_d^{+} \dot{y}(t)) \qquad (20)$$

In order to test the applicability of the retrieval method, a simulation was made in which the robot had to carry different payloads of different masses. In this simulation the robot moves through the assembly cycle shown in Fig 4b. At the beginning of the cycle the robot has to pick up a payload of unknown mass (m_3 = 0.1 kg) and transport it to an intermediate cell (φ_1, φ_2) = (0.4, 0.3 rad). In the intermediate cell an additional load (m_4 = 0.1 kg) is picked up, after which the heavier payload has to be transported to an end gate (φ_1, φ_2) = (0.8, 0.6 rad). Finally the robot has to return unloaded to its initial position ($\varphi_1 = \varphi_2$ = 0 rad) to recommence the assembly cycle with the same type of payloads. The initial position and the position of the intermediate cell and the end gate correspond to positions 1, 2 and 3 in Fig. 4a respectively. In (10), an observation period of T_s = 0.1 (s) was selected, which corresponds to 5 samples. In Fig. 5, the responses are shown during one assembly cycle for the memory containing model P^1 only. To compare the simulations of Fig. 5 with the case where switching among models occurs, the simulation was repeated for the memory based on the operating vectors in Table 1. This is shown in Fig. 6. To test the effect of the linearization errors, the designed controller was applied to the linear models resulting in each operating point. The results are shown in figures 5, 6, 7, and 8. In the figures the encircled numbers refer to the assembly states shown in Fig. 4.

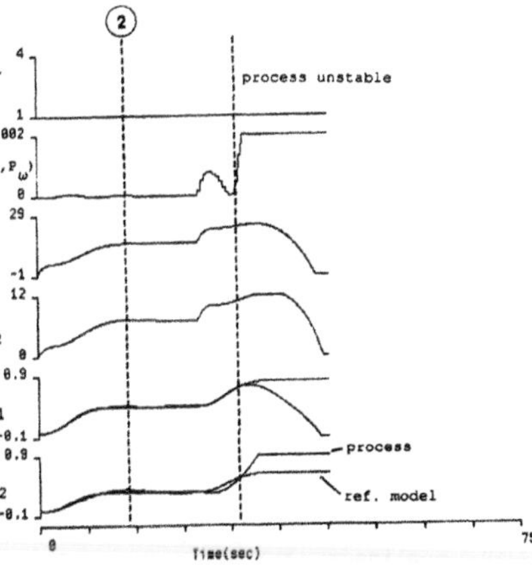

Fig. 5 Simulation with only model P^1 in memory

P^i = selected model (i = 1, 2, 3)

$d(P^i, P_\omega)$ = dynamic distance between P^i and P_ω

u_1, u_2 = torques on motor 1 and 2

φ_1, φ_2 = angles of link 1 and 2

Figure 5 shows that with only model P^1 in the memory, the robot cannot be stabilized over the range of setpoints. Figure 6 shows that the robot stabilizes around each set point and that the responses of the first link almost coincide with those of the linearized model. The angular response of the second link contains overshoot when the robot moves from (φ_1, φ_2) = (0.4, 0.3 rad) to (φ_1, φ_2) = (0.8, 0.6 rad). This result was preceded by an increasing distance between the process and the selected model. When a fourth model at (φ_1, φ_2) = (0.6, 0.45 rad) and payload m_p = 0.2 kg is added to the memory, the described overshoot in φ_2 vanishes as is shown in Fig. 7.

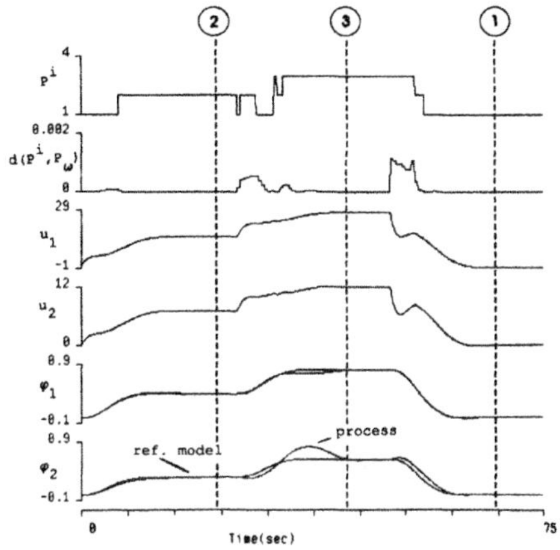

Fig. 6 Simulation with the memory based on Table 1

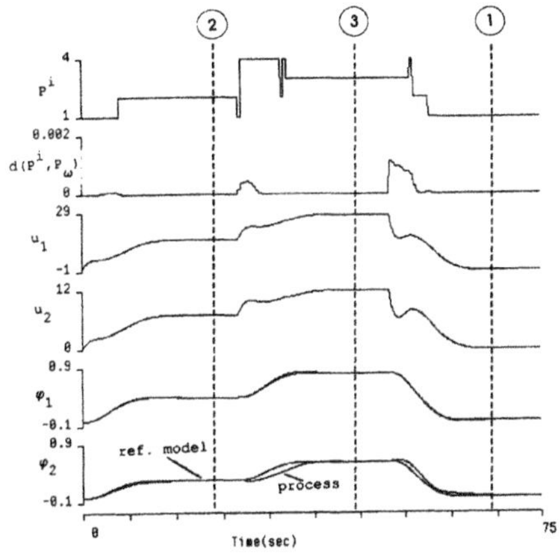

Fig. 7 Simulation with the extra model in the memory.

Figures 5 and 7 show that during the transition phases from one assembly state to another, the choice for the most appropriate model is not always obvious. For instance, when the robot moves from assembly state 2 to assembly state 3, at the beginning model P^1 appears more appropriate than the models corresponding to the assembly states 2 and 3 which would be expected intuitively. At the moments of (almost) steady state, the most appropriate models P^i at assembly state i (i = 1, 2, 3) are selected.

To illustrate the importance of a bumpless transfer, the simulation was repeated without the bumpless transfer algorithm. The result is shown in Fig. 8.

Figure 8 shows that without a bumpless transfer, the control signals contain bumps at each instant of a mode switch. Beside the deteriorated control behaviour, the process of retrieving the best model is heavily disturbed. In some cases, even limit cycles may occur.

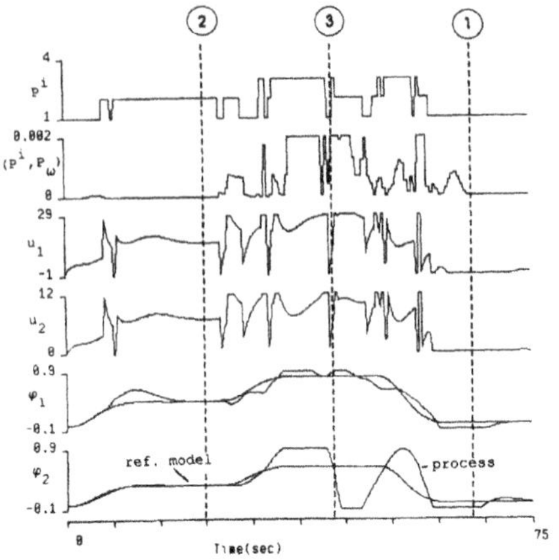

Fig. 8 No bumpless mode switching

6 Conclusions

The presented extension of adaptive control with a supervisor can fulfill the demand for more advanced control without constant adaptation. Simulations of a supervisor applied to a non-linear two-link robot using bumpless transfer show that when appropriate models in the memory are available, the retrieval of models based on a dynamic distance function improves the control compared to non-adaptive control (e.g. only one model in the memory). In the presented case with 4 models in memory, the difference between the response of the (linear) reference model and of a robot is negligibly small.

It must be noted that the retrieval in the presented case is a rather difficult retrieval problem, as the robot moves rapidly through the different modes of operation and thus remains only for a short period of time in one mode of operation. Many processes will remain longer in a mode of operation, so that the problem of retrieval will be less difficult.

This extension of adaptive control has the advantage that the time needed for identification is reduced by using old information obtained from previous adaptation phases. Secondly, no deliberate excitation of the process is needed when a process returns in the same mode of operation. Furthermore, the structure of the supervisor is transparent, so that it is expected to be acceptable in the process industry.

Appendix Robot model

The description of the pending frictionless rigid two-link robot arm (double pendulum), is described by Gras and Nijmeijer (1989). The robot arm is depicted in Fig. A1.

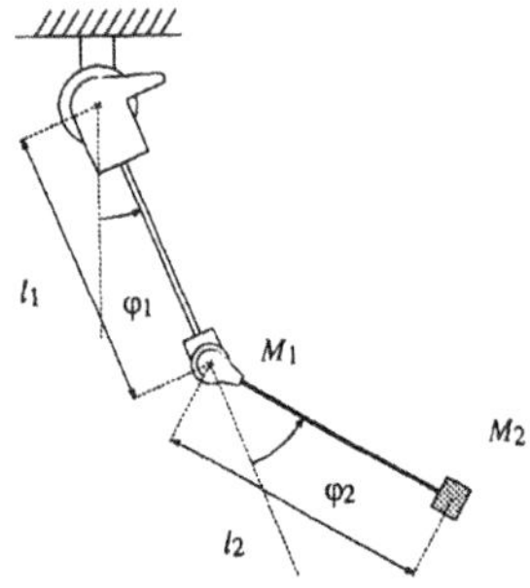

Fig. A1 Two link robot arm (Kruise, 1990)

The robot arm is controlled by applying torques u_1 and u_2 at the joints. The outputs are the angles φ_1 and φ_2 defined in Fig. A1. The dynamics of such a configuration can be given as

$$\frac{d^2}{dt^2}\begin{bmatrix}\varphi_1\\ \varphi_2\end{bmatrix} = M(\varphi)^{-1}\left[-P(\varphi,\dot{\varphi})-Q(\varphi)+\begin{bmatrix}u_1\\ u_2\end{bmatrix}\right] \qquad \text{(A1)}$$

with

$$M(\varphi) = \begin{bmatrix}\alpha + \gamma + 2\beta\cos\varphi_2 & \gamma + \beta\cos\varphi_2\\ \gamma + \beta\cos\varphi_2 & \gamma\end{bmatrix} \qquad \text{(A2a)}$$

$$P(\varphi,\dot{\varphi}) = \begin{bmatrix}-\beta\sin\varphi_2\,(2\dot{\varphi}_1\dot{\varphi}_2+\dot{\varphi}_2^2)\\ \beta\dot{\varphi}_2^2\sin\varphi_2\end{bmatrix} \qquad \text{(A2b)}$$

$$Q(\varphi) = \begin{bmatrix}\delta\sin\varphi_1 + \varepsilon\sin(\varphi_1+\varphi_2)\\ \varepsilon\sin(\varphi_1+\varphi_2)\end{bmatrix} \qquad \text{(A2c)}$$

$\alpha=(m_1+m_2)l_1^2$ (kg/m^2), $\beta=m_2 l_1 l_2$ (kg/m^2), $\gamma=m_2 l_2^2$ (kg/m^2), $\delta=(m_1+m_2)gl_1$ (kg/m^2/s^2), and $\varepsilon=m_2 g l_2$ (kg/m^2/s^2).

The lengths and masses are normalized as $m_1 = m_2 = l_1 = l_2 = 1$. The gravity acceleration g is set equal to 10. If a payload with mass m_p is attached to the robot, then the mass of the second link m_2 increases by m_p. The dynamics in a mode of operation can be determined by exact linearization of (A1) around the operation vector $\omega = (\bar{\varphi}_1,\bar{\varphi}_2,\dot{\bar{\varphi}}_1,\dot{\bar{\varphi}}_2,m_p,\bar{u}_1,\bar{u}_2)^T$. This yields the system:

$$\frac{d}{dt}\begin{bmatrix}\varphi\\ \dot{\varphi}\end{bmatrix} = \begin{bmatrix}0 & I\\ A_0 & A_1\end{bmatrix}\begin{bmatrix}\varphi\\ \dot{\varphi}\end{bmatrix} + \begin{bmatrix}0\\ B_0\end{bmatrix}\begin{bmatrix}u_1\\ u_2\end{bmatrix} \qquad \text{(A3a)}$$

$$y = \begin{bmatrix}\varphi^T & \dot{\varphi}^T\end{bmatrix}^T \qquad \text{(A3b)}$$

where A_0, A_1, and B_0 are the derivatives of (A1) to φ, $\dot{\varphi}$ and u respectively at ω.

Literature

Åström, K.J. and B. Wittenmark (1989). Adaptive Control, Prentice Hall, New York

Åström, K.J. and B. Wittenmark (1990). Computer Controlled Systems, Prentice Hall, New York

Gras, L.C.J.M. and H. Nijmeijer (1989). Decoupling in nonlinear systems: From linearity to nonlinearity, IEE Proceedings, Vol. 136, Pt.D, March, pp. 53-62

Korn, G.A. and T.M. Korn (1968). Mathematical Handbook for Scientists and Engineers, McGraw-Hill, New York

Kruise, L. (1990). Modeling and control of a flexible beam and robot arm, Ph.D. Thesis, University of Twente, Enschede

Landau, Y.D. (1979). Adaptive Control, The model reference approach, Marcel Dekker, New York

Takamatsu, T., I. Hashimoto, Y. Togari, and Y. Hashimoto (1986). Non-interacting control system design of a distillation column by the partial model matching, Proceedings of the IFAC conference on Control of Distillation Columns and Chemical Reactors, Pergamon, Oxford, pp. 237-242

MODEL REFERENCE ADAPTIVE CONTROL OF A CLASS OF NONLINEAR SYSTEMS

F. Ohkawa* and M. Tomizuka**

**Dept. of Control Engineering, Kyushu Institute of Technology, Tobata, Kitayushu 804, Japan*
***Dept. of Mechanical Engineering, University of California, Berkeley, California 94720, USA*

Abstract. A variety of adaptive control methods have been proposed to deal with uncertainties in the control object. Model reference adaptive control (MRAC) is one of the most popular adaptive control methods. While many physical systems exhibit nonlinear characteristics, most of existing MRAC design methods assume that the control object is linear. In this paper, MRAC design method for a class of nonlinear systems is proposed. This class includes systems which are cascade combinations of a static nonlinear part and linear dynamic part. Nonlinearities are limited to saturation, deadzone and those which are composed from saturation and deadzone. The design procedure is developed for each of the two subclasses: nonlinear systems with saturation and those with deadzone. The stability of MRAC for nonlinear systems with saturation is assured by modifying the reference input (desired output) signal when it is judged that the input saturation prohibits the system output from reaching the desired output level. It is shown that an input matching method is suited for dealing with systems with deadzone. The idea of modifying the reference input signal and input matching are combined to deal with nonlinearities which possess both deadzone and saturation characteristics. The effectiveness of the proposed MRAC approach is demonstrated by computer simulation.

Keywords. Model reference adaptive control; Nonlinear systems; Saturation; Deadzone.

INTRODUCTION

A large number of design methods of adaptive control have been proposed to deal with uncertainties of systems. Especially, the MRAC (Model Reference Adaptive Control) is regarded one of the most popular adaptive control methods. However, in most cases, these methods have been considered for time-invariant and linear systems. Therefore, there is no assurance of stability if these methods are applied to nonlinear systems. In fact, there are only a few analysis and design of adaptive controllers for nonlinear systems(Kung and Womack,1984a,1984b; Abramovitch and Franklin,1986; Abramovitch, Kosut and Franklin,1990).

In this paper, the design of the discrete time MRAC is considered for nonlinear systems which are cascade combinations of a static nonlinear block and linear dynamic block. Nonlinear element considered here is the saturation and deadzone. In order to simplify the design procedure, the nonlinear element is decomposed to a saturation element and a deadzone element. We will consider the design of MRAC for systems with saturation and the design for systems with deadzone independently. For systems with saturation, it is difficult to guarantee convergence by the conventional adaptive control method. However, the proposed method shows that asymptotic convergence is assured by adaptively modifying the reference input in the region where the input amplitude is constrained. For system with deadzone, by introducing the new adjustable gain according to the deadzone width, the effect of the deadzone can be compensated and asymptotic stability is assured by the adaptive control algorithm based on input matching .

The MRAC algorithms proposed in this paper have relatively simple structures and guarantee the asymptotic convergence of the error between the reference input and the system output. Digital simulation results will be shown to demonstrate that uncertain nonlinear systems can be controlled effectively by the proposed adaptive control algorithms.

STATEMENT OF THE PROBLEM

In this paper, we consider adaptive control of a class of nonlinear systems shown in Fig.1. Notice that the system nonlinearity is static and is characterized by deadzone and saturation. For convenience, we will discuss the design of MRAC to handle saturation(Fig.3) and the design to handle deadzone(Fig.4) separately.

Linear part

The linear dynamic block is described by the transfer function

$$G(z^{-1})=B(z^{-1})/A(z^{-1}) \qquad (1)$$

where

$$A(z^{-1})=1-a_1z^{-1}-\cdots-a_nz^{-n}$$
$$B(z^{-1})=b_1z^{-1}+\cdots+b_mz^{-m} \qquad b_1\neq 0$$

n and m are assumed to be known but a_i's and b_i's are unknown. Furthermore, $B(z^{-1})$ is assumed to be stable: i.e. characteristic roots are all inside the unit circle of z-plane. Note that the input output dynamics of the linear portion can be described by the difference equation

$$A(z^{-1})y(k)=B(z^{-1})m(k) \qquad (2)$$

where z^{-1} should be interpreted as a one step delay operator.

Saturation

The saturation element is characterized by the widths of the linear region δ_1 and δ_2, the slopes in the linear region, k_n and k_m, and the saturation limit values, M_1 and M_2 as shown in Fig.3. k_n, k_m, M_1 and M_2 are unknown, but δ_1 and δ_2 are assumed to be known. u(k) and y(k) are both measurable, but m(k) is not.

There have been several methods proposed for identification of unknown parameters of this kind of systems(e.g. Yoshimura and Tomizuka,1981). However, adaptive control of the present class of nonlinear plants has not been understood as much as identification(Abramovitch and Franklin,1986). One possible MRAC approach is proposed here.

Under the present problem formulation, the easiest approach is to define m(k) by

$$m(k)=\begin{cases} M_1 & \text{for } \delta_1\leq u(k) \\ k_nu(k) & \text{for } 0\leq u(k)<\delta_1 \\ k_mu(k) & \text{for } \delta_2<u(k)<0 \\ M_2 & \text{for } u(k)\leq\delta_2 \end{cases} \qquad (3)$$

Namely, the relation between m(k) and u(k) becomes

$$m(k)=(k_n+(k_m-k_n)q(k))(l(k)u(k)+p(k)) \tag{4}$$

$$q(k)=\begin{cases}0 & \text{for} \quad 0\leq u(k)\\ 1 & \text{for} \quad u(k)<0\end{cases}$$

$$p(k)=\begin{cases}\delta_1 & \text{for} \quad \delta_1<u(k)\\ 0 & \text{for} \quad \delta_2\leq u(k)\leq \delta_1\\ \delta_2 & \text{for} \quad u(k)<\delta_2\end{cases}$$

$$l(k)=\begin{cases}1 & \text{for} \quad \delta_2\leq u(k)\leq \delta_1\\ 0 & \text{for} \quad \delta_1<u(k)\ \&\ u(k)<\delta_2\end{cases}$$

Deadzone

For the deadzone element shown in Fig.4, the following relation between the input u(k) and the output m(k) holds.

$$m(k)=\begin{cases}0 & \text{for} \quad |u(k)|\leq\varepsilon\\ k_n(u(k)-\varepsilon) & \text{for} \quad u(k)>\varepsilon\\ k_n(u(k)+\varepsilon) & \text{for} \quad u(k)<-\varepsilon\end{cases} \tag{5}$$

If the deadzone width ε is known, linearization of the relation between input and output can be easily achieved by compensating the input as shown in Fig.5. Namely, we add the compensating signal $\varepsilon\,\mathrm{sgn}\,[u(k)]$ to the input u(k) as follows.

$$u'(k)=u(k)+\varepsilon\,\mathrm{sgn}\,[u(k)] \tag{6}$$

$$\mathrm{sgn}\,[u(k)]=\begin{cases}1 & u(k)>0\\ -1 & u(k)<0\\ 0 & u(k)=0\end{cases}$$

Then, the relationship between u(k) and m(k) becomes

$$m(k)=k_n u(k) \tag{7}$$

If ε is unknown, Eq.(6) is replaced by

$$u'(k)=u(k)+s(k)p(k) \tag{8}$$

$$p(k)=\mathrm{sgn}\,[u(k)]$$

where s(k) is the estimate of the deadzone width ε.

In this case, the relation between u(k) and m(k) becomes

$$m(k)=\begin{cases}0 & \text{for} \quad |u'(k)|<\varepsilon\\ k_n u(k)+k_n(s(k)-\varepsilon) & \text{for} \quad u'(k)\geq\varepsilon\\ k_n u(k)-k_n(s(k)-\varepsilon) & \text{for} \quad u'(k)\leq-\varepsilon\end{cases} \tag{9}$$

Now, in order to simplify the design procedure, assume that the sign of u(k) is identical with that of u'(k). This assumption can be avoided by introducing a new variable like p(k).

Then, for the case $|u'(k)|\geq\varepsilon$, we obtain

$$m(k)=k_n(u(k)+(s(k)-\varepsilon)p(k)) \tag{10}$$

Therefore, when s(k) becomes ε, Eq.(10) becomes identical with Eq.(7). It should be noted that the Eq.(10) applies only for $|u'(k)|\geq\varepsilon$. To use Eq.(10) for all possible values of u(k), the parameter k_n must be set zero for $|u'(k)|<\varepsilon$. Then, the parameter k_n is not constant and adaptive algorithms for time invariant systems can not be applied. This problem will be overcome by the input matching MRAC method.

ADAPTIVE CONTROL UNDER THE PRESENCE OF INPUT SATURATION

From Eq.(4), the difference equation (2) is

$$\begin{aligned}y(k+1)&=\sum_{i=1}^{n}a_i y(k+1-i)\\ &+\sum_{j=1}^{m}b_j(k_n+(k_m-k_n)q(k+1-j))(l(k+1-j)u(k+1-j)+p(k+1-j))\\ &=\sum_{i=1}^{n}a_i y(k+1-i)\\ &+\sum_{j=1}^{m}b_j(k_n+(k_m-k_n)q(k+1-j))v(k+1-j)\\ &=\sum_{i=1}^{n}a_i y(k+1-i)+\sum_{j=1}^{m}b_j^* v(k+1-j)\\ &+\sum_{j=1}^{m}c_j q(k+1-j)v(k+1-j)\end{aligned} \tag{11}$$

where

$$v(k)=l(k+1-j)u(k+1-j)+p(k+1-j)$$

$$b_j^*=b_j k_n$$

$$c_j=b_j(k_m-k_n)$$

Letting $y_d(k)$ denote the reference input, we would like to select u(k) to achieve

$$y_d(k+1)=\sum_{i=1}^{n}a_i y(k+1-i)+\sum_{j=1}^{m}b_j^* v(k+1-j)+\sum_{j=1}^{m}c_j q(k+1-j)v(k+1-j) \tag{12}$$

This equation can be solved for v(k), and if the solution v(k) satisfies $\delta_2\leq v(k)\leq\delta_1$, we can achieve the goal by letting u(k)=v(k) and p(k)=0. If the inequality is not satisfied, it is not possible to achieve $y(k+1)=y_d(k+1)$. In such cases, we let v(k)=p(k) and apply this v(k) at time k. Equation (12) can be used for determination of v(k) only if the parameters a_i's, $b_j^*$and c_j have known values. If not, they must be replaced by adjustable gains. That is, v(k), must be determined from

$$y_d(k+1)=\sum_{i=1}^{n}f_i(k)y(k+1-i)+\sum_{j=1}^{m}g_j(k)v(k+1-j)+\sum_{j=1}^{m}h_j(k)q(k+1-j)v(k+1-j) \tag{13}$$

where $f_i(k)$, $g_j(k)$ and $h_j(k)$ are the adjustable gains.

Using Eq.(11) and Eq.(13), we have the following output error equation.

$$\begin{aligned}e(k+1)&=y_d(k+1)-y(k+1)\\ &=\sum_{i=1}^{n}(f_i(k)-a_i)y(k+1-i)+\sum_{j=1}^{m}(g_j(k)-b_j^*)v(k+1-j)\\ &+\sum_{i=1}^{m}(h_i(k)-c_i)q(k+1-j)v(k+1-j)\\ &=\Phi(k)W(k)\end{aligned} \tag{14}$$

$$\Phi(k)=[f_1(k)-a_1\cdots f_n(k)-a_n,\ g_1(k)-b_1^*\cdots g_m(k)-b_m^*,\ h_1(k)-c_1\cdots h_m(k)-c_m]$$

$$W(k)=[y(k)\cdots y(k+1-n), v(k)\cdots v(k+1-m),\ q(k)v(k)\cdots q(k+1-m)v(k+1-m)]$$

Thus, it is assured that e(k) converges to zero and $\Phi(k)$ converges to a constant vector as k tends to infinity by various adaptive algorithms(Landau,1979).

The adaptive control input v(k) is determined from Eq.(13). However, when the input amplitude is constrained, v(k) from Eq.(13) is not necessarily possible to apply.

Now, define T(k+1) as

$$T(k+1)=y_d(k+1)-\sum_{i=1}^{n}f_i(k)y(k+1-i)-\sum_{j=2}^{m}g_j(k)v(k+1-j)-\sum_{j=2}^{m}h_j(k)v(k+1-j)p(k+1-j) \tag{15}$$

Then, Eq.(13) can be rewritten as

$$T(k+1)=(g_1(k)+h_1(k)q(k))v(k) \tag{16}$$

The right hand side of Eq.(16) is

$$(g_1(k)+h_1(k)q(k))u(k) \quad \text{for} \quad \delta_2\leq u(k)\leq\delta_1 \tag{17}$$

$$g_1(k)\,\delta_1 \quad \text{for} \quad \delta_1<u(k) \tag{18}$$

$$(g_1(k)+h_1(k)q(k))\,\delta_2 \quad \text{for} \quad u(k)<\delta_2 \tag{19}$$

From Eq.(17), u(k) from Eq.(13) can be applied if the input u(k) satisfies the following condition

$$T(k+1)=(g_1(k)+h_1(k)q(k))u(k) \quad \delta_2\leq u(k)\leq\delta_1 \tag{20}$$

For the case of Eq.(18) or Eq.(19),

$$T(k+1)=g_1(k)\,\delta_1 \tag{21}$$

or

$$T(k+1)=(g_1(k)+h_1(k)q(k))\,\delta_2 \tag{22}$$

Eq.(13) is not satisfied in general: i.e. for

$$T(k+1)/(g_1(k)+h_1(k)q(k))=T(k+1)/g_1(k)>\delta_1 \tag{23}$$

or

$$T(k+1)/(g_1(k)+h_1(k)q(k))<\delta_2 \tag{24}$$

Eq.(13) is not satisfied. In this case, the convergence of the output error cannot be verified.

Therefore, we propose a method which overcomes the convergence problem by modifying the reference input. Now, modify the reference input $y_d(k+1)$ to x(k+1) which is represented by the following equation.

$$x(k+1)=\sum_{i=1}^{n}f_i(k)y(k+1-i)+\sum_{j=2}^{m}g_j(k)v(k+1-j)+\sum_{j=2}^{m}h_j(k)v(k+1-j)p(k+1-j)+(g_1(k)+h_1(k)q(k))p(k) \tag{25}$$

Then, in this case, from Eq.(11) and (25), the output error equation becomes

$$\begin{aligned}e^*(k+1)&=x(k+1)-y(k+1)\\ &=\sum_{i=1}^{n}(f_i(k)-a_i)y(k+1-i)+\sum_{j=2}^{m}(g_j(k)-b_j^*)v(k+1-j)\\ &+\sum_{j=2}^{m}(h_i(k)-c_i)q(k+1-j)v(k+1-j)\\ &+(g_1(k)+h_1(k)q(k))p(k)-b_1^*v(k)-c_1q(k)v(k)\\ &=\Phi(k)W(k)+\mu(k)\end{aligned} \tag{26}$$

where

$$\mu(k)=(g_1(k)+h_1(k)q(k))(p(k)-v(k))$$

In this case, we set

$$v(k)=p(k)$$

Therefore, $\mu(k)$ becomes zero and we have

$$e(k+1)=\Phi(k)W(k) \quad \text{for} \quad \delta_2\leq u(k)\leq\delta_1 \tag{27}$$

$$e^*(k+1)=\Phi(k)W(k) \quad \text{for} \quad \delta_1<u(k) \text{ or } u(k)<\delta_2$$

Thus, under the condition of boundedness of x(k), the convergence of $\Phi(k)W(k)$ can be assured by the use of the same adaptive algorithm with the selection of e(k+1) or $e^*(k+1)$ as the adaptive error signal according to the condition of $u(k)$. Therefore, the convergence of $e(k)$ to zero is verified under the conditions of $\delta_2 \leq u(k) \leq \delta_1$. The boundedness of $x(k)$ can be assured along the same manner which is presented by Abramovitch and Franklin(1986). It should be noted that the input $u(k)$ is not within the linear region between δ_1 and δ_2, and that the output does not necessarily track the reference input.

ADAPTIVE CONTROL UNDER THE PRESENCE OF DEADZONE

From Eqs. (4) and (10), we obtain the relation between the input and the output of the compensated system :

$$y(k+1) = \sum_{i=1}^{n} a_i y(k+1-i) + \sum_{j=1}^{m} b_j^*(u(k+1-j) + (s(k+1-j) - \varepsilon)p(k+1-j)) \quad (28)$$

where

$$b_j^* = k_n b_j$$

Eq.(28) is solved for $u(k)$ to yield

$$u(k) = \sum_{i=0}^{n} \alpha_i y(k+1-i) + \sum_{j=2}^{m} \beta_j(u(k+1-j) + s(k+1-j)p(k+1-j)) + \sum_{j=2}^{m} \gamma_j p(k+1-j) + (\varepsilon - s(k))p(k) \quad (29)$$

where

$$\alpha_i = \begin{cases} 1/(b_1 k^*) & i=0 \\ -a_j/(b_1 k^*) & i=1 \cdots n \end{cases}$$

$$\beta_j = -b^*_j/b_1^* \qquad j=2 \cdots m$$

$$\gamma_j = b_j^* \varepsilon / b_1^* \qquad j=2 \cdots m$$

Recall that for $|u'(k)| < \varepsilon$, i.e. $m(k)=0$, Eq.(10) is not appropriate. This case will be discussed later.

Eq.(29) suggest that when the plant parameters are unknown the prediction of $u(k-1)$ can be obtained from

$$u_e(k-1) = \sum_{i=0}^{n} f_i(k-1)y(k-i) + \sum_{j=2}^{m} g_j(k-1)(u(k-j) + l(k-j)p(k-j)) + \sum_{j=2}^{m} h_j(k-1)p(k-j) \quad (30)$$

where $f_i(k)$, $g_j(k)$ and $h_j(k)$ are adjustable gains that correspond to unknown parameters α_i, β_j and γ_j respectively.

Now, define the input error as

$$\eta(k) = u_e(k) - u(k) \quad (31)$$

Using Eq.(29) and (30), we have

$$\begin{aligned} \eta(k-1) = & \sum_{i=0}^{n} (f_i(k-1) - \alpha_i)y(k-i) \\ & + \sum_{j=2}^{m} (g_j(k-1) - \beta_j)(u(k-j) + q(k-j)p(k-j)) \\ & + \sum_{j=2}^{m} (h_j(k-1) - \gamma_j)p(k-j) + (s(k-1) - \varepsilon)p(k-1) \\ = & \Phi(k-1)W(k) \end{aligned} \quad (32)$$

where

$$\Phi(k) = [\phi_1(k)\ \phi_2(k)\ \phi_3(k)\ \phi_4(k)]$$
$$W(k) = [w_1(k)\ w_2(k)\ w_3(k)\ w_4(k)]^T$$
$$\phi_1(k) = [f_0(k) - \alpha_0 \ \cdots\cdots \ f(k)_n - \alpha_n]$$
$$\phi_2(k) = [g_2(k) - \beta_2 \ \cdots\cdots \ g(k)_m - \beta_m]$$
$$\phi_3(k) = [h_2(k) - \gamma_2 \ \cdots\cdots \ h(k)_m - \gamma_m]$$
$$\phi_4(k) = [s(k) - \varepsilon]$$
$$w_1(k) = [y(k) \ \cdots\cdots\cdots \ y(k-n)]$$
$$w_2(k) = [u(k-2) + q(k-2)p(k-2) \ \cdots \ u(k-m) + q(k-m)p(k-m)]$$
$$w_3(k) = [p(k-2) \ \cdots\cdots\cdots \ p(k-m)]$$
$$w_4(k) = p(k-1)$$

Input error equation (32) is in the identical form as the output error equation (14). Thus, it is assured that $\eta(k)$ converges zero and $\Phi(k)$ converges a bounded constant vector as k tends to infinity by various adaptive algorithms under the condition of boundedness of $W(k)$.

Now, we define the input $u(k)$ by

$$u(k) = f_0(k)y_d(k+1) + \sum_{i=1}^{n} f_i(k)y(k+1-i) + \sum_{j=2}^{m} g_j(k)(u(k+1-j) + s(k+1-j)p(k+1-j)) + \sum_{j=2}^{m} h_j(k)p(k+1-j) \quad (33)$$

where $y_d(k)$ is the reference input.

Using Eqs.(30) and (33), we have

$$\eta(k) = -f_0(k)e(k+1) \quad (34)$$

where $e(k)$ is the output error. Since $\eta(k) \to 0$ and $f_0(k) \to$ Const.(bounded), the convergence of $e(k)$ to zero is ensured. As mentioned above, $m(k)$ becomes zero for $|u'(k)| < \varepsilon$. Then, Eq.(10) is not satisfied and Eq.(31) becomes

$$0 = \sum_{i=0}^{n} \alpha_i y(k+1-i) + \sum_{j=2}^{m} \beta_j(u(k+1-j) + s(k+1-j)p(k+1-j)) + \sum_{j=2}^{m} \gamma_j p(k+1-j) \quad (35)$$

Therefore, if we use Eq.(30), it is necessary to adapt the adjustable gains as $u'(k)$ tends to zero though $u(k)$ is not zero. Then, the error equation (34) is not appropriate in this case.

$|u'(k)| < \varepsilon$ is caused when $s(k) \neq \varepsilon$. In this case, the following relation holds.

$$m(k) = k_n(u(k) + (s(k) - \varepsilon)p(k)) = 0 \quad (36)$$

Then, we have

$$u(k) = -(s(k) - \varepsilon)p(k) \quad (37)$$

Namely, the input error $\eta(k)$ can be considered as

$$u'(K) - u(k) = (q(k) - \varepsilon)p(k) \quad (38)$$

Therefore, it is not necessary to adapt the adaptive algorithm as $u'(k)$ tends zero under the condition $s(k) \neq \varepsilon$. Then, the problem mentioned above is avoided. Consequently, the convergence of the input error and the output error can be always verified by the use of the error equation (32) even in the case of $|u'(k)| < \varepsilon$.

Now, the boundedness of $W(k)$ can be assured along the same manner which has been presented for the linear system(Nakamura and Suzuki,1982).

The design of MRAC for nonlinear systems with deadzone and different slopes in the positive and negative linear regions is an important problem to be solved.

For the systems with saturation and deadzone, we can easily combine the two design methods proposed in this paper. In such case, MRAC system is designed for nonlinear block shown in Fig.6 and designed by substituting $x(k)$ for $y_d(k)$ in Eq.(33) in accordance with the results of the preceding section. Namely, when the input amplitude is constrained, we define the input $u(k)$ by

$$u(k) = f_0(k)x(k+1) + \sum_{i=1}^{n} f_i(k)y(k+1-i) + \sum_{j=2}^{m} g_j(k)(u(k+1-j) + s(k+1-j)p(k+1-j)) + \sum_{j=2}^{m} h_j(k)p(k+1-j) \quad (39)$$

where $x(k)$ is presented by the following equation.

$$x(k+1) = (\delta p(k) - \sum_{i=1}^{n} f_i(k)y(k+1-i) - \sum_{j=2}^{m} g_j(k)(u(k+1-j) + s(k+1-j)p(k+1-j)) - \sum_{j=2}^{m} h_j(k)p(k+1-j))/f_0(k) \quad (40)$$

DIGITAL SIMULATION

To validate the theoretical results, digital computer simulations were performed.

Saturation

The simulation is carried out for an unknown plant given by

$$G(z^{-1}) = (z^{-1} + 0.8z^{-2})/((1-z^{-1})(1-0.5z^{-1}))$$
$$k_n = 1.5 \quad k_m = 1 \quad \delta_1 = 1 \quad \delta_2 = 1.2$$

Initial conditions of the adjustable gains are

$$f_1(0) = g_1(0) = 1$$

Furthermore, other initial conditions were set to zero.

Fig.7 shows the simulation result for the conventional MRAC system without modifying the reference input. Fig.8 shows the result for the MRAC proposed here. The parameter adaptation algorithm used here is a weighted Least-Square algorithm. The plant output by conventional adaptive control does not converge to the reference input. But, as shown in Fig.8, it is confirmed that the plant output can be controlled well by the proposed adaptive algorithm which modifies the reference input. As shown in Fig.8, at an early stage of the adaptive control process, the reference input is not modified in appropriate directions. The reference input is modified based on the adaptive gains in Eq.(25). Thus, at the initial stage, the parameter errors have not sufficiently converged. An appropriate and optimal modification of the reference sequence is not guaranteed. This point requires further study.

Deadzone

Digital simulations were performed for a two axes positioning system, an X-Y motion control table. The dynamics of each axis is described by

$$G_x(z^{-1})=(1.3z^{-1}+z^{-2})/(1-0.7z^{-1}+0.12z^{-2})$$

and

$$G_y(z^{-1})=(1.25z^{-1}+0.8z^{-2})/(1-0.5z^{-1}+0.06z^{-2})$$

The deadzone element for x-axis is characterized by $\varepsilon_x=0.2$ and $k_{nx}=1$ and that for y-axis is characterized by $\varepsilon_y=0.1$ and $k_{ny}=1$. The reference inputs for x- and y-axis are $y_{dx}(k)$ and $y_{dy}(k)$ respectively.

$$y_{dx}(k)=\sin(0.1k)$$
$$y_{dy}(k)=\cos(0.1k)$$

Notice that this pair of the reference inputs come from a unit circle on the X-Y plane. Fig.9 shows the simulation result of adaptive control without taking into account the deadzone. The result under the proposed adaptive control is shown in Fig.10. It can be seen from these figures that the proposed method improves tracking performace significantly.

Saturation and deadzone

To illustrate and evaluate the effectiveness of the proposed method numerically, the following system with saturation and deadzone is considered.

$$G(z^{-1})=(z^{-1}+0.8z^{-2})/((1-0.7z^{-1})(1-0.8z^{-1}))$$

and

$$k_n=k_m=1 \qquad \delta_1=\delta_2=1 \qquad \varepsilon=0.4$$

The simulation result shown in Fig.11 shows that the proposed MRAC system performs well.

CONCLUSION

We have proposed MRAC schemes for single-input single-output discrete time nonlinear system characterized by a linear dynamics preceeded by a nonlinear saturation block and/or dreadzone block. For saturation, the convergence is guaranteed by adaptively modifying the reference input in the region of saturation. On the other hand, it is shown that the deadzone is compensated by introducing another adjustable gain in a minor feedforward loop and applying the input matching method. Digital simulation results demonstrated the validity of the proposed adaptive algorithms.

In this paper, the algorithms were developed for systems with one step delay. The algorithms can be extended to handle systems with delay steps larger than one.

REFERENCES

Abramovitch, D.Y., R.L.Kosut and G.F.Franklin(1986). Adaptive control with saturation inputs. Proc. 25th Conf. Decision and Control.

Abramovitch, D.Y. and G.F.Franklin(1990). On the stability of adaptive polepacement controllers with a saturating actuator. IEEE Trans. Automatic Control. AC-35. 303.

Kung, M.C. and B.F.Womack(1984a). Discrete time adaptive control of linear dynamic with a two-segment piecewise-linear asymetric nonlinearity. IEEE Trans. Automatic Control. AC-29. 170.

Kung, M.C. and B.F.Womack(1984b). Discrete time adaptive control of linear system with preload nonlinearity. Automatica. 20. 477.

Landau, I.D.(1979). Adaptive Control-The model reference approach. Marcel Dekker.

Nakamura, T. and T.Suzuki(1982). Design method for indirect discrete MRACS based on plant-input representation. Trans. of the SICE. 18. 131 (in Japanese).

Yoshimura, T. and M.Tomizuka(1981). Application of model reference adaptive techniques to a class of nonlinear systems. ASME J. Dynamic Systems, Measurement and Control. 102. 158.

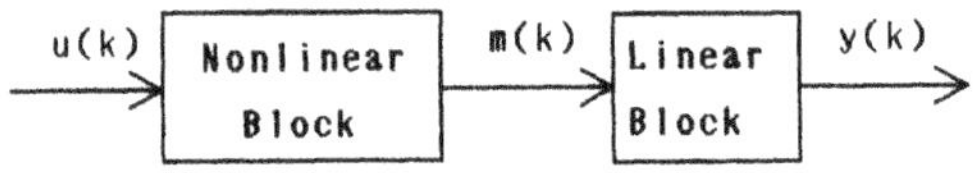

Fig.1 Nonlinear system

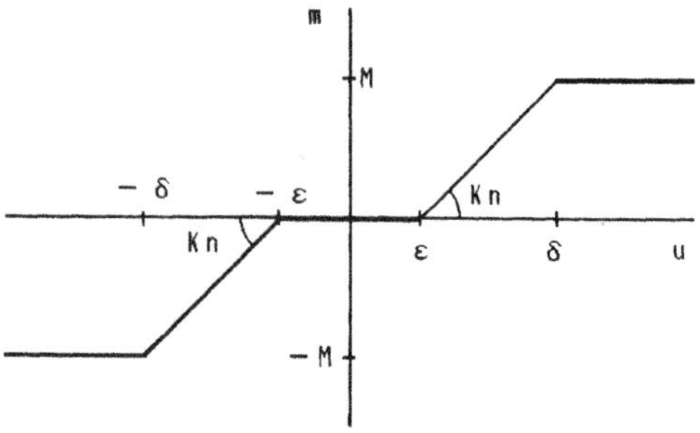

Fig.2 Nonlinear block

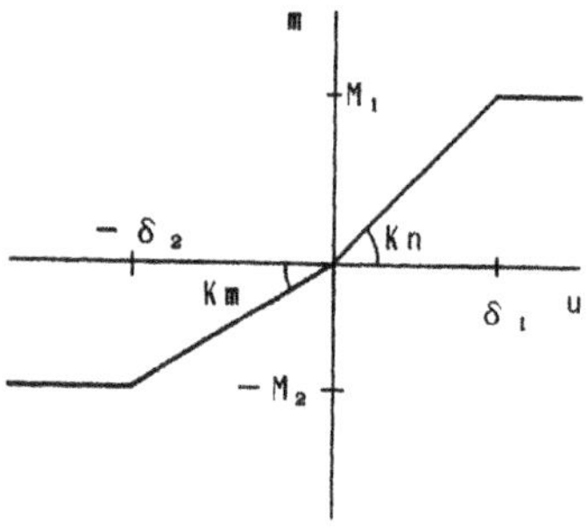

Fig.3 Saturation

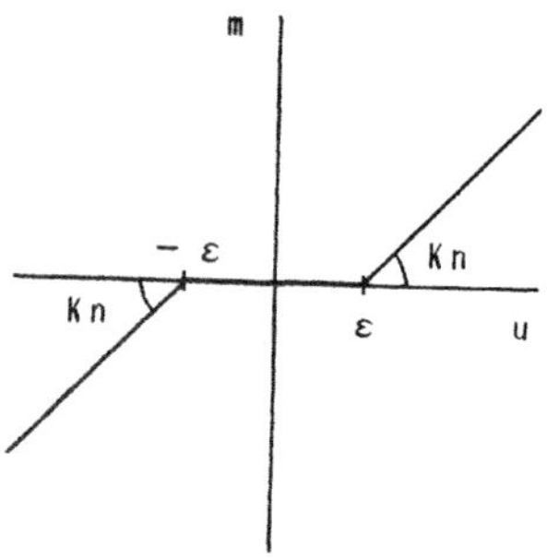

Fig.4 Deadzone

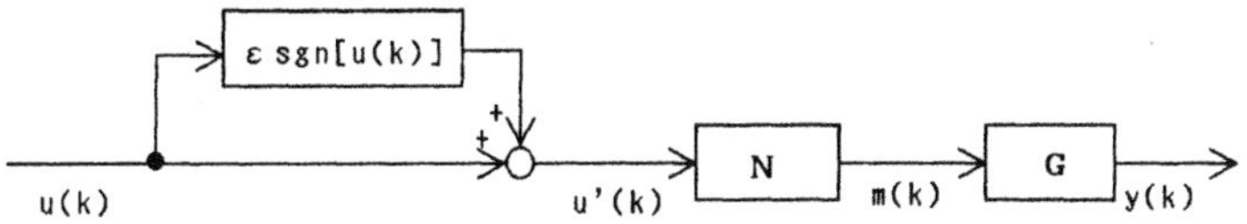

Fig.5 Feedforward compensation of deadzone

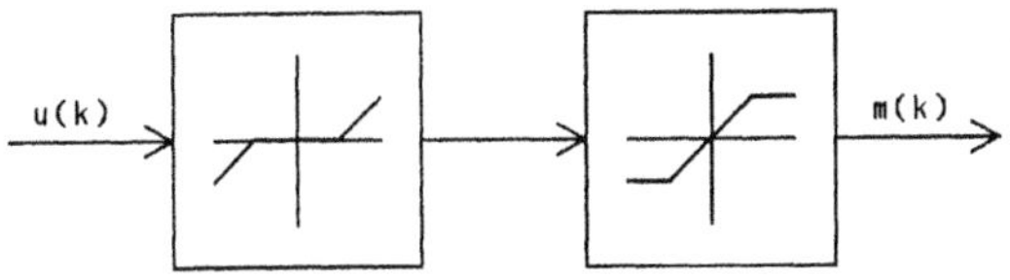

Fig.6 Nonlinear block

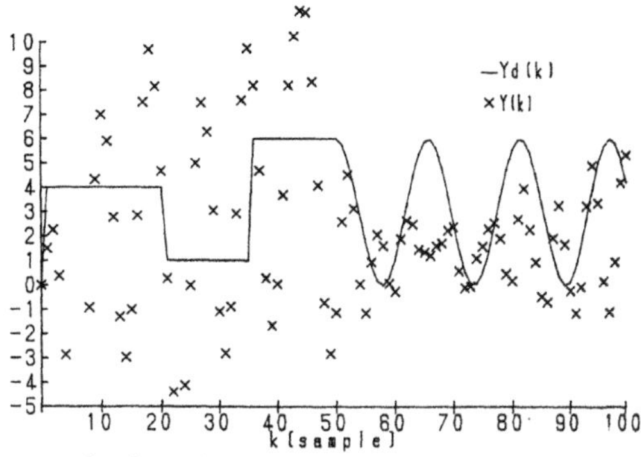

Fig.7 Simulation result of the conventional adaptive control method

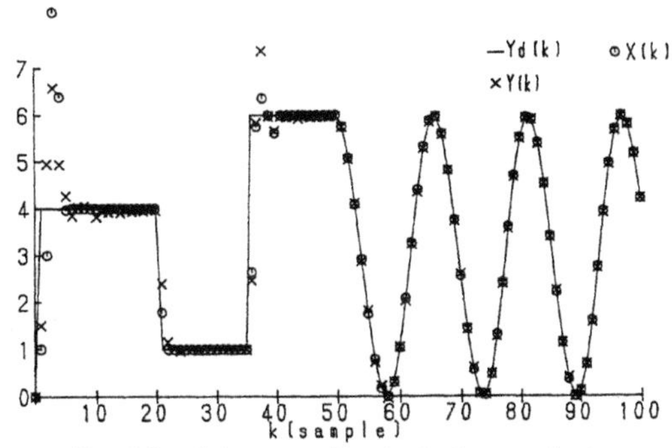

Fig.8 Simulation result of adaptive control

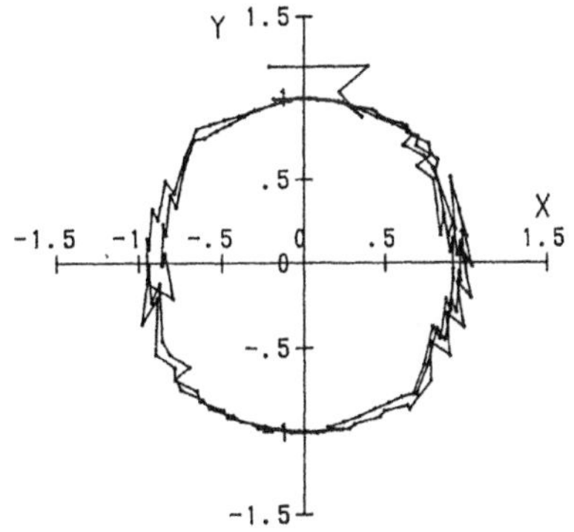

Fig.9 Simulation result of adaptive control without the adaptation of deadzone

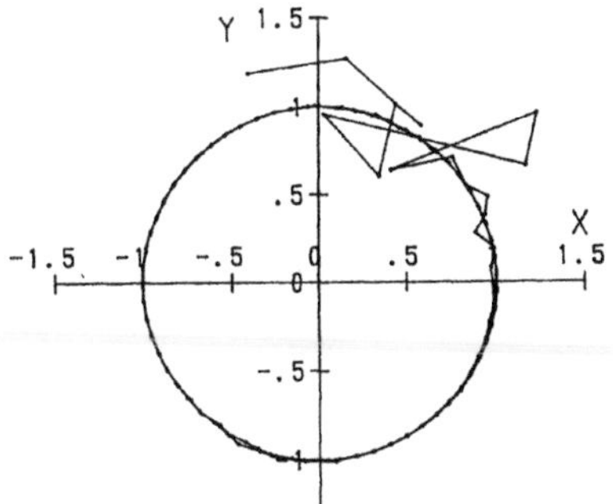

Fig.10 Simulation result of adaptive control

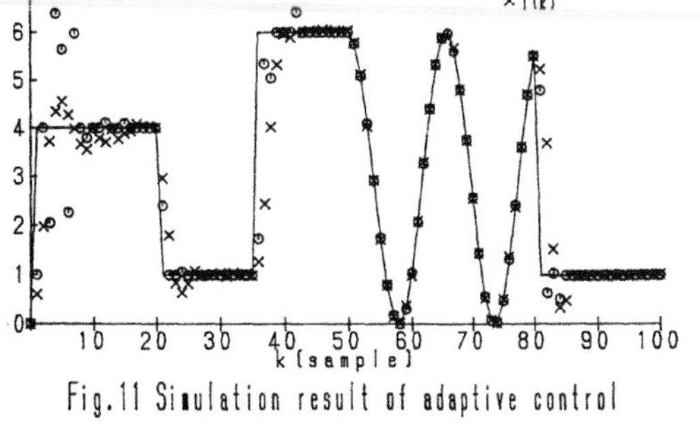

Fig.11 Simulation result of adaptive control

ADAPTIVE FREQUENCY RESPONSE COMPENSATION: EXPERIMENTS ON A HEAT EXCHANGER

Yu Tang and J. Job Flores

Division de Estudios de Posgrado, FI
Universidad Nacional de Mexico, P. O. Box 70-256, 04510 Mexico D.F., Mexico

ABSTRACT

In this paper, the strategy of adaptive frequency response compensation is used to control a pilot plant heat exchanger. We first extend the results of [1] to a set of arbitrary frequencies, describing a procedure for designing adaptive controllers to meet a frequency domain specification given in terms of a desired open loop frequency response at a set of arbitrary frequencies. Then, this method is shown to work well in the experiments of controlling a heat exchanger.

I. INTRODUCTION

Heat exchangers are an essential ingredient in a wide range of industrial applications. The overwhelming majority of heat exchangers used in the industry operates in the closed loop with standard proportional plus integral (PI) controllers. Tuning of the PI constants is a considerably difficult task in view of the complicated dynamic behavior of the system.

In this paper, the strategy of adaptive frequency response compensation is used to control a pilot plant heat exchanger. We first extend the results of [1] to a set of arbitrary frequencies, describing a procedure for designing adaptive controllers to meet a frequency domain specification given in terms of a desired open loop frequency response at a set of arbitrary frequencies. Then, this method is shown to work well in the experiments of controlling a heat exchanger.

The idea of adaptive frequency response compensation is carried out by using a bank of adaptive band-compensators, each of which is intended to compensate the plant frequency response in its corresponding band. The way of compensation above can be viewed as an extension to the conventional methods in which only the plant frequency response at some critical frequency (*e.g.*, gain-crossover or phase-crossover frequency) is compensated. This approach is close in spirit to the quantitative feedback design of [2] and is related with the idea of using some points of the Nyquist curve of the plant to tune a controller used in [3].

The paper is organized as follows: in Section II, we first give a procedure for designing non-adaptive controllers to meet a frequency domain specification and an adaptive frequency response identification algorithm, then we obtain the adaptive frequency response compensation scheme by combining the non-adaptive controller procedure with the identification algorithm. In Section III, after briefly describing the heat exchanger, we present some experimental results. Finally, we give concluding remarks in Section IV.

II. DESIGN PROCEDURE

2.1. Non-adaptive Controller Design

Plant

Consider a single-input single-output linear time-invariant (LTI) discrete-time plant

$$y_t = G_p(q)u_t + d_t \qquad (2.1)$$

where u_t, y_t are plant input and output, respectively, d_t represents the effect of disturbances reflected at the output, and $G_p(q)$ is the plant pulse transfer function. The plant is assumed to be stable, *i.e.*, every bounded input will produces bounded output.

Design Specification

The design specification is given in terms of a desired *open loop* frequency response $G_o(e^{jw})$ in a set of arbitrary frequencies $\Omega := \{w_k\}_{k=0}^{L}$, where L is an integer, and $w_0 = 0$, $0 < w_k < \pi/T_s$ for $k = 1, 2, \ldots, L$, with T_s the sampling time. The choice of the desired open loop frequency response is compromised between performance and robustness of the closed loop system [4,5]. Typically, $|G_o(e^{jw})|$ is large in low frequencies for good performance, and small in high frequencies for robustness.

Controller

We will consider the standard unit feedback controller configuration (Fig. 2.1). Let $G_c(q)$ denote a controller such that

$$G_p(e^{jw})G_c(e^{jw}) = G_o(e^{jw}), \quad \forall w \in \Omega. \qquad (2.2)$$

It is clear that for each $w \in \Omega$, (2.2) defines two algebraic equations that, given the knowledge of the plant Nyquist curve and the desired open loop frequency response at Ω, *i.e.*

$$G_p(e^{jw_k}) := A(k) + jB(k), \qquad (2.3)$$

$$G_o(e^{jw_k}) := A^*(k) + jB^*(k), \qquad (2.4)$$

for $k = 0, 1, \ldots L$, can be solved to get the $L+1$ points of the controller frequency response

$$G_c(e^{jw_k}) := C(k) + jD(k) \qquad (2.5)$$

via the relation

$$\begin{bmatrix} C(k) \\ D(k) \end{bmatrix} = \frac{1}{A^2(k) + B^2(k)} \begin{bmatrix} A(k) & B(k) \\ -B(k) & A(k) \end{bmatrix} \begin{bmatrix} A^*(k) \\ B^*(k) \end{bmatrix}. \qquad (2.6)$$

In (2.6) we have assumed that the plant frequency response is different from zero at the frequency set Ω. If the plant frequency response is zero at some point, then it can not be compensated at that point. Note that $B(0)=B^*(0)=D(0)=0$, since the plant pulse function has coefficients all real.

Controller Implementation Using Lagrange Filter

To get a controller satisfying the frequency response matching equality (2.2), we propose a band-wise implementation. That is, first the error signal is passed through a bank of bandpass filters to split the signal into its spectral components. Then the outcoming signals are fed into a bank of compensators connected in parallel, each of which is intended to compensate the plant frequency response at its corresponding band.

This idea is realized by using Lagrange filter. Lagrange filter is well known in the field of signal processing [6], which is designed, given a desired frequency response in a set of frequencies, by making use of Lagrange interpolation formula. This leads to [5]

$$G_c(q)=\sum_{k=0}^{L}(c(k)+d(k)q^{-1})H_k(q), \qquad (2.7)$$

where $H_k(q)$ are bandpass filters with the center frequency w_k

$$H_k(q)=\begin{cases}\dfrac{1}{1-\mu q^{-1}}H_c(q), & k=0,\\[2mm] \dfrac{1}{1-2\mu\cos(w_k)q^{-1}+\mu^2q^{-2}}H_c(q), & k=1,2,\ldots,L,\end{cases} \qquad (2.8)$$

with comb filter $H_c(q)$

$$H_c(q)=(1-\mu q^{-1})\prod_{k=1}^{L}(1-2\mu\cos(w_k)q^{-1}+\mu^2q^{-2}), \qquad (2.9)$$

and μ the stability factor marginally smaller than the unit in order to avoid pole-zero cancellation in (2.8) along the unit circle, guaranteeing the stability of the filters. The controller coefficients $c(k)$ and $d(k)$ are calculated, given the frequency response of the controller at Ω (see (2.6)), by

$$c(0)=\frac{1}{\alpha(0)}C(0), \quad d(0)=0; \qquad (2.10a)$$

$$\begin{bmatrix}c(k)\\ d(k)\end{bmatrix}=2\begin{bmatrix}\alpha_k & \beta_k\\ \mu[\beta_k\sin(w_k)-\alpha_k\cos(w_k)] & -\mu[\beta_k\cos(w_k)+\alpha_k\sin(w_k)]\end{bmatrix}\begin{bmatrix}C(k)\\ D(k)\end{bmatrix}$$

$$\text{for } k=1,2,\ldots L. \qquad (2.10b)$$

where $\alpha(k)$ and $\beta(k)$, are constants given by

$$\alpha(0):=2^L\prod_{m=1}^{L}[1-\cos(w_m)], \quad \beta(0):=0; \qquad (2.11a)$$

$$\alpha(k)=2^L\cos(\tfrac{2L+1}{2}w_k)[\cos(\tfrac{3}{2}w_k)-\cos(\tfrac{1}{2}w_k)]\cdot\prod_{\substack{m=1\\ m\neq k}}^{L}[\cos(w_k)-\cos(w_m)],$$

$$\beta(k)=-2^L\sin(\tfrac{2L+1}{2}w_k)[\cos(\tfrac{3}{2}w_k)-\cos(\tfrac{1}{2}w_k)]\cdot\prod_{\substack{m=1\\ m\neq k}}^{L}[\cos(w_k)-\cos(w_m)],$$

$$\text{for } k=1,\ldots L. \qquad (2.11b)$$

The overall controller implemented with Lagrange filter is thus given by

$$u_t=\sum_{k=0}^{L}u_t(k), \qquad (2.12)$$

$$u_t(k)=c(k)e_t(k)+d(k)e_{t-1}(k), \qquad (2.13)$$

$$e_t(k)=H_k(q)e_t, \qquad (2.14)$$

$$e_t=r_t-y_t. \qquad (2.15)$$

To end this section let us summarize the design procedure as follows

Non-Adaptive Controller Design Procedure

Design parameters:

$L+1$ = number of frequencies

$\Omega=\{w_k\}_{k=0}^{L}$, $w_0=0$, $0<w_k<\pi/T_s$ for $k=1,2,\ldots,L$

σ = Low bound on $|G_p(e^{jw_k})|$

μ = stability factor of the filters

Data:

Plant frequency response

$$G_p(e^{jw_k}):=A(k)+jB(k), \quad k=0,1,\ldots,L$$

Frequency domain specification:

Desired open loop frequency response

$$G_o(e^{jw_k}):=A^*(k)+jB^*(k), \quad k=0,1,\ldots,L$$

Step 1: If $A^2(k)+B^2(k)\geq\sigma$, calculate the controller frequency response at Ω using (2.6). Otherwise, set $C(k)=D(k)=0$.

Step 2: Calculate the controller coefficients using (2.10).

Step 3: Implement the controller via (2.12)-(2.15) and (2.8)-(2.9).

2.2. Adaptive Identifier

The procedure above needs the knowledge of plant frequency response at the set Ω, *i.e.*, $\{A(k),B(k)\}_{k=0}^{L}$. In adaptive control, this is obtained by on-line identification. Among many existing algorithms [7,8,9], we will use the identification scheme of [10] to identify the plant frequency response at the set Ω. Let

$$\theta_t(k):=[a_t(k) \quad b_t(k)]^T$$

which is updated by

$$\theta_{t+1}(k)=\theta_t(k)+\gamma(k)\frac{\phi_t(k)\varepsilon_t(k)}{1+\phi_t^T(k)\phi_t(k)}, \qquad (2.16)$$

for $k=0,1,\ldots,L$, with initial conditions $\theta_0(k)$. In (2.16) the following signals are needed

regressor:

$$\phi_t(k):=[u_t(k) \quad u_{t-1}(k)]^T, \qquad (2.17)$$

$$u_t(k)=H_k(q)u_t, \qquad (2.18)$$

identification error:

$$\varepsilon_t(k):=y_t(k)-\hat{y}_t(k), \qquad (2.19)$$

$$y_t(k)=H'_k(q)y_t, \qquad (2.20)$$

$$\hat{y}_t(k)=\phi_t^T(k)\theta_t(k), \qquad (2.21)$$

where $H'_k(q)$ is a normalized bandpass filter (*i.e.*, its frequency response at the center frequency is one) defined by

$$H'_k(q)=f_k(z)H_k(q), \quad k=0,1,\dots,L, \qquad (2.22)$$

with $f_k(q)$ being normalizing factors

$$f_k(q)=\begin{cases} 1, & \text{for } k=0, \\ 2\alpha(k)-2\mu[\alpha(k)cos(w_k)-\beta(k)sin(w_k)]q^{-1}, & \text{for } k=1,,\dots,L. \end{cases} \qquad (2.23)$$

The use of the normalized bandpass filters in (2.20) is to minimize the effect of mainlobe error of the bandpass filters in (2.19) [10]. With $\{a_t(k),\ b_t(k)\}$ given by (2.16), the plant frequency response estimate at the set Ω is computed by

$$A_t(0)=a_t(0), \quad B_t(0)=0, \qquad (2.24a)$$

$$\begin{bmatrix} A_t(k) \\ B_t(k) \end{bmatrix} = \frac{1/2}{\alpha^2(k)+\beta^2(k)} \begin{bmatrix} \alpha(k)+\beta(k)ctg(w_k) & \beta(k)csc(w_k) \\ \beta(k)-\alpha(k)ctg(w_k) & -\alpha(k)csc(w_k) \end{bmatrix} \begin{bmatrix} a_t(k) \\ b_t(k) \end{bmatrix}, \quad \text{for } k=1,2,\dots L. \qquad (2.24b)$$

This adaptive identification scheme is summarized as follows.

Adaptive Frequency Response Identification Algorithm

Design parameters:

$\gamma(k)$ = adaptation gain, $k=0,1,\dots,L$

Input:

Plant input and output: $u_t,\ y_t$

Output

Plant frequency response estimate at time *t*: $\{A(k),\ B(k)\}_{k=0}^{L}$

Step 1: Update the parameter $\theta_t(k)$ using (2.16)

Step 2: Calculate the plant frequency response estimate at time *t* $\{A_t(k),\ B_t(k)\}$ via (2.24).

2.3. Adaptive Control

We now consider the following adaptive control problem: given a unknown LTI stable plant, and a desired open loop frequency response at a set of frequencies, design an adaptive controller such that the open loop frequency response matches asymptotically the desired one at the given set of frequencies.

We approach the problem by combining the control design procedure with the adaptive frequency response identification algorithm above. In practice, a bound on the plant frequency response is often available. This knowledge is readily incorporated in the design of the adaptive controller as shown in the following adaptive control algorithm.

Adaptive Control Algorithm

Prior Information:

$$A_{min}(k)\le A(k)\le A_{max}(k),$$

$$B_{min}(k)\le B(k)\le B_{max}(k), \quad \text{for } k=0,1,\dots,L$$

For each sampling time do

Step 1: Identify the plant frequency response at the set Ω $\{A_t(k),B_t(k)\}_{k=0}^{L}$ using Adaptive Frequency Response Identification Algorithm.

Step 2: Set

$$A(k)=\begin{cases} A_{max}(k) & \text{if } A_t(k)>A_{max}(k) \\ A_{min}(k) & \text{if } A_t(k)<A_{min}(k) \\ A_t(k) & \text{otherwise} \end{cases}$$

$$B(k)=\begin{cases} B_{max}(k) & \text{if } B_t(k)>B_{max}(k) \\ B_{min}(k) & \text{if } B_t(k)<B_{min}(k) \\ B_t(k) & \text{otherwise} \end{cases}$$

Step 3: Update the controller coefficients using Step 1-3 of Non-Adaptive Controller Design Procedure.

III. EXPERIMENTAL RESULTS

The above adaptive control strategy was used to control a pilot plant heat exchanger, which consists functionally of [11]: (1) a blower, which operates at a constant rate and circulates an airstream along a polypropylene tube, (2) a shutter at the fan inlet, which manually controls the volume of airflow and is changed in one of the experiments to modify the process dynamics, (3) an actuator, which is a thyristor-fed heating grid that feeds heat into the airstream, and (4) a sensor, which is a thermistor that measures the air temperature and gives a voltage signal (Fig.3.1).

The heat exchanger exhibits several non-linearities due to the actuator and the sensor manifesting in the steady-state gain in different ranges of operation. Fig.3.2 shows the estimated steady-state gain $A(0)$ at several operating ranges when the plant was excited by a random signal uniformly distributed between $[0,1]$ plus a DC component.

In all experiments the sampling time was chosen $T_s=0.25$ *sec.*, which gives a Nyquist frequency equal to *12.56* rad/sec. The frequency domain specification was given in terms of the frequency response of a desired *closed loop* transfer function

$$G_d(q)=\frac{0.02q^{-1}}{1-0.92q^{-1}} \qquad (3.1)$$

at $L+1=4$ frequencies $\Omega=\{0,1,5,10\}$, which corresponds to a first order system with the time constant equal to *3* sec., and steady-state gain equal to *0.25*. The desired open loop frequency response was obtained as

$$G_o(e^{jw_k})=A^*(k)+jB^*(k)=\frac{G_d(e^{jw_k})}{1-G_d(e^{jw_k})},\quad w_k=0,1,5,10.$$

The low bound on the plant frequency response σ was set equal to *0.01*. Others parameters used in the experiments are listed in Table 3.1.

The purpose of the first experiment is to show the behavior of the adaptive control system when the plant dynamic is changed. In this experiment, the heat exchanger operated at the range $25\text{-}35C^o$. During the operation, the plant dynamics was changed by moving the shutter from *20* to *40* degrees and later back to *20* degrees. Fig. 3.3a shows the plant closed loop response and the desired response $(G_d(q)r_t)$ to a slowly varying reference signal $r_t=12+3\sin(0.0628t)$. The plant frequency response at the set Ω is shown in Fig. 3.3b.

The next experiment (Fig.3.4a) shows the plant output when operating at several ranges. Note that the plant frequency response estimate at the set Ω (Fig.3.4b) moves from one convergence values to others according to set point changes.

V. CONCLUDING REMARKS

A scheme of adaptive frequency response compensation at a set of arbitrary frequencies has been presented, which was shown to work well with a pilot plant heat exchanger. The heat exchanger and many industrial processes can be approximated by first-order dynamics with a time delay. For this class of plants, simulation and experimental results have been shown that this adaptive frequency response compensation scheme works well with a few ($L=0$ to 3) frequencies. Also, several theoretical results of adaptive frequency response compensation are available in [1,12].

REFERENCES

[1] Y. Tang and R. Ortega, "Adaptive tuning to frequency response specifications", *Proc. 11th World Congr., IFAC,* Tallinn, 1990, also submitted to *Automatica*

[2] I. M. Horowitz and M. Sidi, "Synthesis of feedback systems with large plant ignorance for prescribed time-domain tolerances," *Int. J. Control*, Vol.16, pp287-309, 1972

[3] K. J. Astrom and T. Hagglund, "Automatic tuning of simple regulators with specifications on phase and amplitude margins," *Automatica,* Vol. 20, pp645-651, 1984

[4] J. C. Doyle and G. Stein, "Multivariable feedback design: concepts for a classic/modern synthesis," *IEEE Trans. Automt. Contr.*, Vol. 26, pp4-16, 1981

[5] Tang, Y. and FLores J.J., "Digital Control synthesis based on frequency response compensation," *Proc. IV Latin american Contr. Congr.*, Puebla, Mexico, 1990

[6] L. R. Rabiner and B. Gold, *Theory and Application of Digital Signal Processing*, Prentice-Hall, 1975

[7] L. Ljung and K. Glover, "Frequency domain versus time domain methods in systems identification," *Automatica,* Vol. 17, pp71-76, 1981

[8] P. J. Parker and R. R. Bitmead, "Adaptive frequency response identification," *Proc. 26th CDC,* Los Angeles, 1987

[9] R.L. Kosut, "Adaptive control via parameter set estimation," *Int. J. Adaptive Contr. Signal Processing,* Vol. 2, pp371-399, 1988

[10] Y. Tang and J. F. Capistran, "Adaptive frequency response identification using the Lagrange Filter," *Proc. American Control Confr.* San Diego, 1990, pp1899-1900, also *Internal Rep.* UNAM, 1989

[11] Feedback Instruments, Ltd., *Process Trainer PT326*, Instructional Manual D326

[12] Y. Tang, "A frequency domain approach to robust adaptive control", *Proc. American Control Conference,* Pittsburgh, Pennsylvania, pp2285-90, 1989

k	freq w_k	adap.gain $\gamma(k)$	plant prior knowledge			
			Amin(k)	*Amax(k)*	*Bmin(k)*	*Mmax(k)*
0	*0.0*	*0.1*	*0.5*	*2.5*	——	——
1	*1.0*	*0.1*	*-1.0*	*5.0*	*-1.0*	*5.0*
2	*5.0*	*0.1*	*-1.0*	*5.0*	*-1.0*	*5.0*
3	*10.0*	*0.1*	*-1.0*	*5.0*	*-1.0*	*5.0*

Table 3.1.

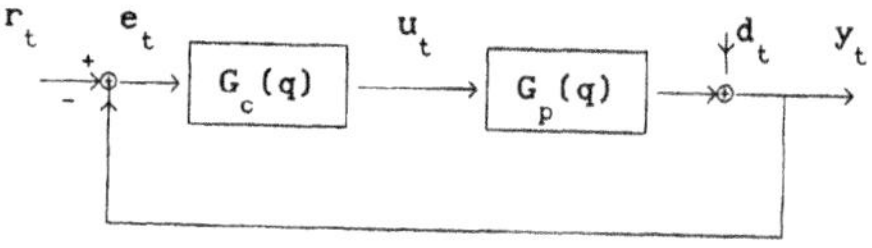

Fig. 2.1 Feedback control system

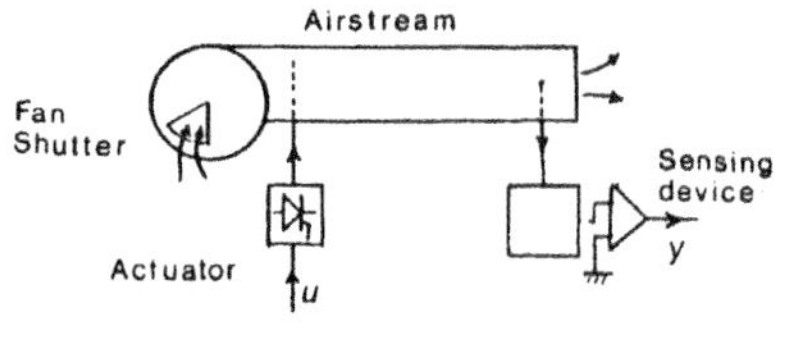

Fig. 3.1. Airstream heating system

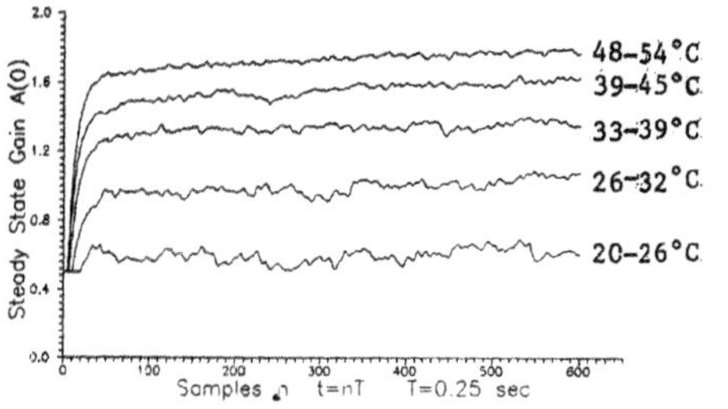

Fig. 3.2. Estimated plant steady-state gain

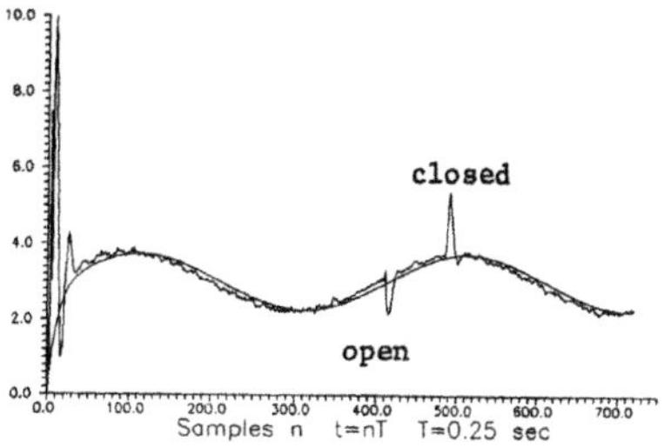

Fig. 3.3a. The plant closed loop response and the desired response $(G_d(q)r_t)$

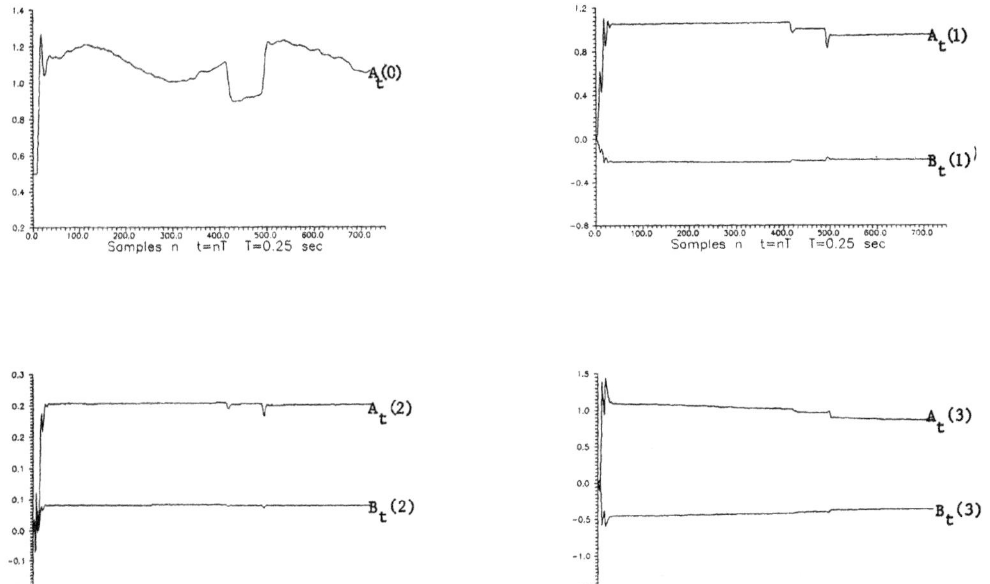

Fig. 3.3b. The plant frequency response estimate at the set Ω

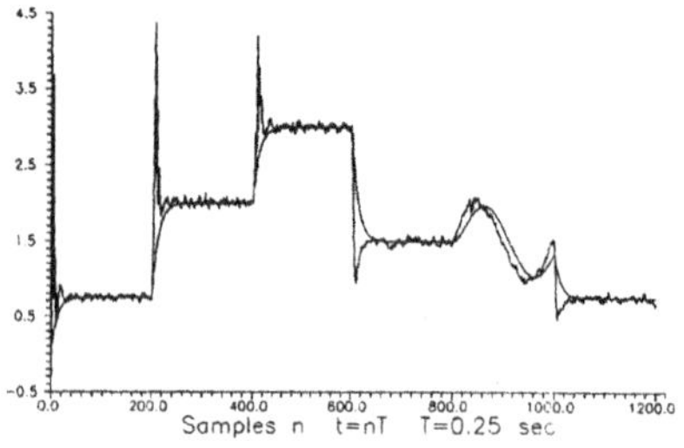

Fig. 3.4a. Desired closed loop response and the plant closed loop response

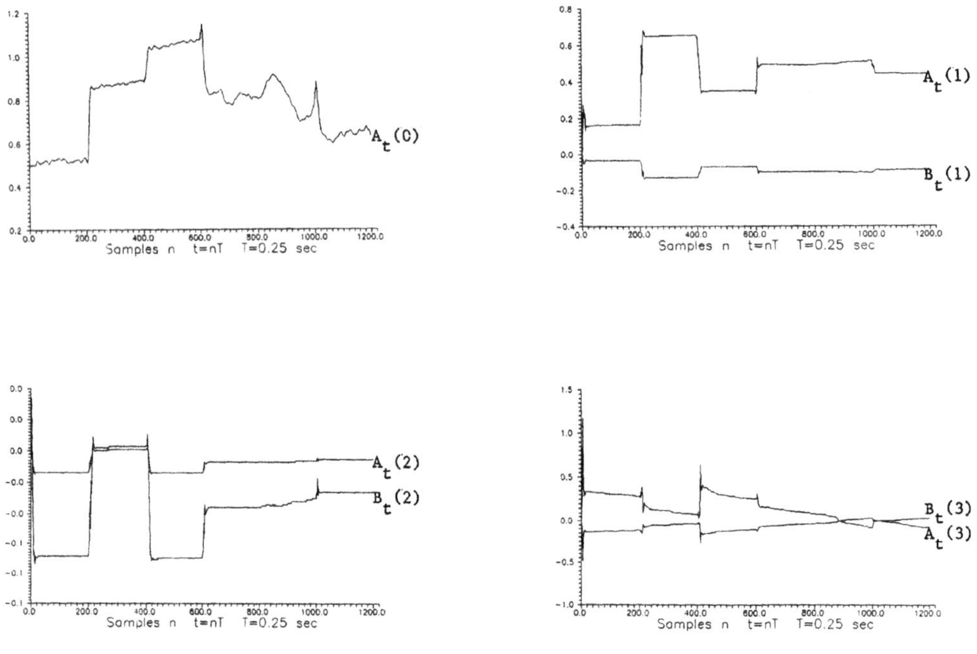

Fig.3.4b. The plant frequency response estimate at the set Ω

THE APPLICATION OF MULTIVARIABLE ADAPTIVE CONTROL TO AN INDUSTRIAL RUN-OF-MINE MILLING PROCESS

G. Metzner* and I. M. MacLeod**

**Measurement and Control Division, Mintek, Private Bag X3015, Randburg 2125, South Africa*
***Dept. of Electrical Engineering, University of the Witwatersrand, 1 Jan Smuts Avenue, Johannesburg, South Africa*

Abstract. The application of multivariable adaptive control to a run-of-mine milling circuit is described. The process is not only highly interactive, but is also open-loop unstable, and has long time delays associated with it. A generalized minimum-variance pole-placement multivariable adaptive controller was implemented and tested on a linearized model of the process, and is currently being tested on an industrial circuit. The paper presents the algorithm used, and the extensions necessary for an industrial application. The results of the simulation demonstrate that the controller can adjust to substantial changes in the process dynamics.

Keywords. Adaptive control; grinding; industrial control; multivariable control systems; parameter estimation; time-varying systems.

INTRODUCTION

The field of adaptive control has over the past two decades matured from a field of largely theoretical developments to the point where algorithms are increasingly robust, and industrial applications are starting to emerge in increasing numbers (Seborg, Edgar and Shah, 1986). An offshoot of this technology has been a number of auto-tuning PID controllers, which are now marketed by most of the major control equipment vendors. There are also several single-variable adaptive controllers commercially available (Astrom and Wittenmark, 1989). Nonetheless, given the potential benefits of adaptive control techniques, the number of practical applications of adaptive control have not been as large as might have been expected. In the authors' view, this is due to the numerous pitfalls that await the would-be user of adaptive control. Only with a thorough understanding of the underlying theory and its implications can the user be assured of a successful and robust adaptive control application.

In the case of multivariable adaptive control, the increased order of the problem has made the proof of global stability without excessive constraints extremely difficult. In addition, most of the proposed algorithms have been extensions of single-variable algorithms, with their associated restrictions. Hence, for a practical application of multivariable adaptive control, it is still necessary to take special precautions to safeguard the stable operation of the controller, and substantial expertise is required in the tuning of the controller parameters.

A multivariable self-tuning pole-placement controller was implemented on a PDP11 process-control computer, and is currently being tested on a full scale industrial run-of-mine (ROM) milling circuit at the reduction plant of the Deelkraal Gold Mining Co., South Africa.

This paper describes some of the experiences gathered by the authors in the process of developing this multivariable adaptive controller. Section 1 describes the ROM milling process, and the rationale for using adaptive control in this application. Section 2 deals with the multivariable adaptive control algorithm and some of the factors necessary for a robust industrial implementation, while Section 4 describes a simulation of the ROM process under multivariable adaptive control.

THE RUN-OF-MINE MILLING PROCESS

The comminution of gold-bearing ore is commonly achieved by crushing, where the ore is broken up from large rocks into particles smaller than about 2cm in diameter. These particles are further reduced by rod-, ball- or pebble mills, to a particle size less than 150µm. The gold-bearing particles are then small enough for most of the gold to be extracted by leaching techniques.

In an effort to save on capital and running costs, traditional comminution processes, which consisted of crushers in series with a rod- or ball mill and several pebble mills, are being replaced by a single ROM mill (Ruhmer, 1985). Comminution of ore in such a mill is achieved in two ways:
(i) crushing and fracturing of rocks in collisions, and
(ii) abrasion through friction within the mill charge.
The crushing mechanism predominates among the larger particles (rocks), while abrasion is the dominant mechanism among the smaller particles. In the majority of ROM mills the required throughput cannot be achieved when operation is purely autogenous. The addition of grinding media such as steel balls is therefore necessary.

Figure 1 shows a diagram of a ROM milling circuit. The circuit elements are the ROM mill; a primary sump with a variable-speed pump, which regulates the flowrate of slurry to the primary hydrocyclone; and a secondary sump, again with a variable-speed pump regulating the flowrate to the secondary hydrocyclone. Solids are fed into the mill via a variable speed belt, and water is added at the inlet to the mill, into the primary sump, and into the secondary sump. The dilution flowrates are monitored, and can be regulated via local control loops.

Owing to the difficulty of relating physical circuit dimensions and ore characteristics to the model parameters, accurate dynamic models of the ROM milling process are difficult to obtain. As a result, most of the current ROM milling-control schemes are based on empirical models. Automatic control generally consists of single-loop PI control of the flowrates and dilutions. The open-loop-unstable elements of the circuit, such as sump levels and the mill load, are controlled by cascading PI controllers with the sump water additions and solids feedrate respectively. However, the process dynamics are highly interactive, and hence the application of a multivariable control scheme is highly desirable (Hulbert, 1989).

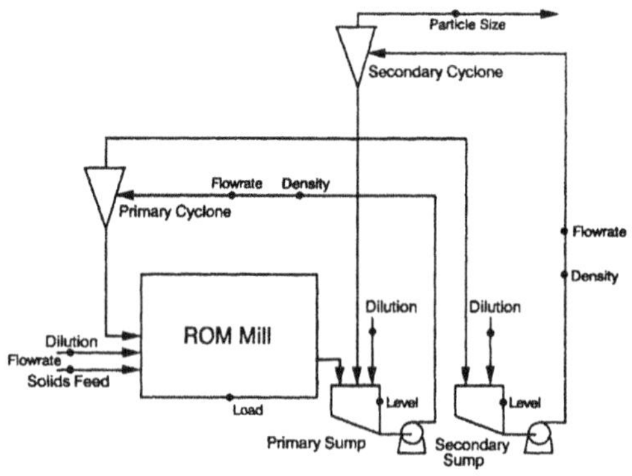

Fig. 1. Diagram of the ROM milling circuit

Hulbert and co-workers (1987) successfully implemented multivariable control on an industrial ROM milling circuit using an inverse Nyquist array controller design. This method, however, does not take into account the time-varying dynamics of the process. Herbst, Hales, and Pate (1988) implemented a control scheme based on an expert system, coupled with a Kalman filter to estimate the hardness of the ore. However, that application does not adequately address the dynamic interactions of the multivariable system.

Several unobservable factors influence the performance of the process. These are the size distribution of the ore fed to the mill, the hardness or friability of the ore, and several other factors such as the state of wear of the mill liners and hydrocyclone classifier spigots, as well as the amount of steel balls added to the mill charge. Hence the dynamics of the process change gradually over a period of weeks and months, as the spigots, mill liners, and lifter bars are worn down and replaced. The size distribution and hardness of the ore may, in contrast, change from shift to shift, as ore is drawn from different parts of the silo, or when different orebodies are mined.

Since these disturbances to the process dynamics are unobservable and may change on a continuous basis, a multivariable adaptive controller offers clear benefits in the control of the ROM milling process.

In an evaluation of the ROM milling process, it was found that the control algorithm must have the following characteristics.

(a) the algorithm must be *multivariable*, since the process is highly interactive. The resulting control system is a 5x5 system with the variables.[1]

Inputs

- feedrate of solids (Solids feed)
- primary sump dilution (Pdil)
- flowrate to the primary cyclone (PCF)
- secondary sump dilution (Sdil)
- flowrate to the secondary cyclone (SCF)

Outputs

- particle size of the product stream (PSM)
- mill load (Load)
- primary sump level (PSL)
- secondary sump level (SSL)
- primary cyclone feed density (PCFD).

(b) the algorithm must be able to control an *open-loop unstable* process, owing to the integrator-type responses of the sump levels and mill load.

(c) the algorithm must be insensitive to *non-minimum phase* responses, such as may result from fractional time delays. It has been found that a fractional time delay may result in a numerator polynomial zero outside the unit circle (Seborg, Edgar and Shah, 1986).

[1] The abbreviations of the variables, as used in the description of the simulation results, are given in parentheses.

(d) the algorithm must be able to control a process that is *nonlinear*, since the operation of the cyclones, as well as of the mill, is non-linear.

(e) the algorithm must be able to control a process in which the *dynamics change with time*, owing to changing conditions in the mill, as the cyclone spigots are changed, or as the feed composition changes.

(f) the algorithm must be robust to measurement or process *noise* and gradual *drift* of operating conditions.

(g) Finally, the algorithm must be able to control a process that has *long time delays* relative to some of its time constants, and where the time delays differ for the different input-output pairs, and may vary with time.

THE ALGORITHM

A recursive extended least-squares (RELS) algorithm with UD-factorization of the covariance matrix was used to estimate the models for each of the outputs with respect to the inputs to the process. A variable forgetting factor (Ydstie, Kershenbaum and Sargent, 1985) was used in combination with an estimation deadzone. The offset term in the model equation was eliminated by high-pass filtering of the measurements.

A multivariable self-tuning pole-placement controller based on the control scheme proposed by McDermott and Mellichamp (1984,1986) was used, with modifications and extensions as necessary. For instance, the algorithm was modified to result in an incremental control algorithm with inherent integral action.

Take the n-input, n-output process model

$$Ay(t) = z^{-k_{ij}}Bu(t) + Ce(t) + d. \tag{1}$$

The model assumed for the control algorithm is[2]

$$A\Delta y(t) = z^{-k_{ij}}B\Delta u(t) + C\Delta e(t), \tag{2}$$

where y, u, and e are n-dimensional vectors of the output, input, and estimation error respectively, and d is a vector of steady-state offsets. Without loss of generality, A and C are monic diagonal matrix polynomials, and B is a full matrix polynomial, all of dimension $n \times n$. The orders of the matrix polynomials are n_a, n_b, and n_c respectively, and Δ is the difference operator, equal to $1 - z^{-1}$.

Note that the orders of the sub-matrix polynomials do not necessarily have to be the same. Hence, if a matrix polynomial is given by:

$$B = B_0 + B_1 z^{-1} + B_2 z^{-2} + B_3 z^{-3} + \ldots + B_{n_b} z^{-n_b},$$

then the order of the first row of B may be different to that of the ith row, and n_b is given by:

$$n_b = \begin{bmatrix} n_{b_1} \\ n_{b_2} \\ \cdot \\ \cdot \\ \cdot \\ n_{b_n} \end{bmatrix}.$$

Convention

So that the form of the ensuing mathematical equations could be simplified, the following convention was adopted. In the case of a multivariable system with different time delays for the various input-output interactions, the delays are given by a matrix of shift operators:

[2] Inherent integral action can be achieved in the controller by assuming the incremental form of the model and a suitable choice of the controller cost function. In addition, the need to estimate the steady-state offset d is eliminated.

$$
z^{-k_{ij}} = \begin{bmatrix} z^{-k_{11}} & z^{-k_{12}} & \dots & z^{-k_{1n}} \\ z^{-k_{21}} & z^{-k_{22}} & \dots & z^{-k_{2n}} \\ \cdot & \cdot & \cdot & \cdot \\ \cdot & \cdot & \cdot & \cdot \\ z^{-k_{n1}} & z^{-k_{n2}} & \dots & z^{-k_{nn}} \end{bmatrix}, \tag{3}
$$

where k_{ij} is the number of sampling intervals equal to the time delay from the ith input to the jth output. However, in the model expression

$Ay(t) = z^{-k_{ij}}Bu(t) + Ce(t) + d$,

the multiplication of $z^{-k_{ij}}B$ is not a matrix multiplication, but rather a scalar multiplication of the (i,j) element of $z^{-k_{ij}}$ and the (i,j) element of the matrix B.

The notation $z^{-k_{ii}}$ denotes a similar matrix of the diagonal elements, but one that operates on the entire row, so that

$$
z^{-k_{ii}} = \begin{bmatrix} z^{-k_{11}} & z^{-k_{11}} & \dots & z^{-k_{11}} \\ z^{-k_{22}} & z^{-k_{22}} & \dots & z^{-k_{22}} \\ \cdot & \cdot & \cdot & \cdot \\ \cdot & \cdot & \cdot & \cdot \\ z^{-k_{nn}} & z^{-k_{nn}} & \dots & z^{-k_{nn}} \end{bmatrix}.
$$

Derivation of the Control Law

The following cost function is defined:

$$\Phi(t + k_{ii}) = Py(t + k_{ii}) + z^{-(k_{ij} - k_{ii})}Q\Delta u(t) - Rr(t), \tag{4}$$

where Φ is a n-dimensional vector denoting the cost function, P and R are $n \times n$ diagonal polynomial matrices, and Q is a $n \times n$ full polynomial matrix. The order of the matrix polynomials is n_p, n_r, and n_q respectively. r is the n-dimensional vector of setpoints.

Let

$$PC = \Delta AE + z^{-k_{ii}}F. \tag{5}$$

The matrix polynomials E and F are diagonal, and of order

$n_{e_i} = k_{ii} - 1$

$$
n_{f_i} = \begin{cases} n_{a_i} + n_{c_i} - k_{ii} + 1 & if \quad k_{ii} \le n_{c_i} + 1 \\ n_{a_i} & if \quad k_{ii} > n_{c_i} + 1 \end{cases} \tag{6}
$$

Pre-multiplying Eq. (2) by E and substituting from Eq. (5) yields

$$PCy(t) = z^{-k_{ij}}EB\Delta u(t) + z^{-k_{ii}}Fy(t) + EC\Delta e(t). \tag{7}$$

$z^{-k_{ij}}CQ\Delta u(t) - z^{-k_{ii}}CRr(t)$ is added on both sides of Eq. (7), and the resulting expression is simplified to

$$C[Py(t) + z^{-k_{ij}}Q\Delta u(t) - z^{-k_{ii}}Rr(t)] = z^{-k_{ij}}(EB + CQ)\Delta u(t)$$

$$+ z^{-k_{ii}}Fy(t) - z^{-k_{ii}}CRr(t) + EC\Delta e(t). \tag{8}$$

The term in square brackets corresponds to the cost function in Eq. (4), except for a shift of k_{ii}. By shifting each row i in Eq. (8) by k_{ii}, we can write:

$$\Phi(t + k_{ii}) = C^{-1}[z^{-k_{ij}-k_{ii}}(EB + CQ)\Delta u(t) + Fy(t) - CRr(t)]$$

$$+ E\Delta e(t + k_{ii}), \tag{9}$$

and the control law minimizing this cost function is therefore

$$z^{-(k_{ij} - k_{ii})}S\Delta u(t) = -Fy(t) + CRr(t), \tag{10}$$

where $S = EB + CQ$. S may be broken up into its diagonal and off-diagonal parts $S = S_D + S_{UL}$. It may be noted that since the matrix polynomials E and C are diagonal,

$$S_D = EB_D + CQ_D$$

$$S_{UL} = EB_{UL} + CQ_{UL}. \tag{11}$$

The control equation may then be modified as follows:

$$S_D\Delta u(t) = -Fy(t) + CRr(t) - z^{-(k_{ij} - k_{ii})}S_{UL}\Delta u(t). \tag{12}$$

Also, since it is not desirable to predict future values of the input, k_{ij} must be greater or equal to k_{ii}. This means that the problem must of necessity be set up in such a way that the smallest time delay in each row lies on the diagonal in Eq. (1).

The closed-loop equation may be found from the process model in Eq. (1) and the control law in Eq. (12):

$$[B_DP + Q_D\Delta A]y(t + k_{ii}) = B_DRr(t) + S_D\Delta e(t + k_{ii})$$

$$+ z^{-(k_{ij} - k_{ii})}C^{-1}[S_DB_{UL} - B_DS_{UL}]\Delta u(t). \tag{13}$$

The following points are apparent from the closed-loop equation.

(a) The delay in the closed loop response is k_{ii}, which is the smallest possible time delay.

(b) The closed-loop characteristic equation is given by $B_DP + Q_D\Delta A$. It is therefore possible to determine the values of P and Q_D from the pole placement equation:

$$B_DP + Q_D\Delta A = T, \tag{14}$$

where T is the diagonal matrix polynomial corresponding to the desired closed-loop poles, and may be specified by the user. In order to give a unique solution to Eq. (14), it is necessary that the orders of the polynomial equations correspond, so that $n_b + n_p = n_q + 1 + n_a = n_t$. The orders of P and Q_D were therefore chosen such that

$$n_p = n_a + 1 \quad , \qquad n_{q_D} = n_b,$$

$$n_t = n_a + n_b + 1. \tag{15}$$

(c) The term containing Δu in Eq. (13) represents the disturbance in the closed-loop system due to a lack of decoupling of the multivariable system. For true multivariable decoupling, it is therefore necessary that this matrix polynomial becomes zero:

$$S_DB_{UL} - B_DS_{UL} = 0. \tag{16}$$

This leads us on to the derivation of the system-decoupling equations. Substitution of the expression for S_D and S_{UL} from Eq. (11), and manipulation of the resulting expression, yields

$$Q_DB_{UL} - B_DQ_{UL} = 0. \tag{17}$$

B_D and B_{UL} have been estimated, and Q_D has been calculated in the pole-placement equation (14). We therefore need to determine the coefficients of Q_{UL} to satisfy Eq. (17). However, the equation is under-parameterized, and hence does not necessarily have a solution. We must therefore find the coefficients of Q_{UL} that minimize the equation:

$$\min_{Q_{UL}} |Q_DB_{UL} - B_DQ_{UL}|. \tag{18}$$

This can be done by setting up a least-squares minimization for each of the $n \times (n-1)$ independent polynomial equations in the coefficients of Q_{UL}.

(d) Rearrangement of the closed-loop equation (13), under the assumption that the multivariable decoupling is perfect, yields:

$$y(t + k_{ii}) = \frac{B_DR}{B_DP + Q_D\Delta A}r(t) + \frac{S_D}{B_DP + Q_D\Delta A}\Delta e(t + k_{ii}).$$

This expression shows that any offsets in the disturbance $e(t)$ will be eliminated, since the Δ term becomes zero at steady state. Integral action is therefore inherent in the algorithm. In addition, it may be seen that for unit steady-state gain, we require that $R = P$. R is commonly chosen as a constant matrix $R = P(1)$.

Implementation Considerations

Sampling rate. Time constants of the process range from approximately 50 to 1000 seconds for the various interactions. With these large differences in time constants it was necessary to compromise on the sampling rate, in order to strike a balance between the estimation of the slow and the faster dynamics. In this case a sampling rate of 10 seconds was chosen.

Time Delays. In addition, the time delays in the system are substantial (up to 45 sampling intervals), so they cannot be estimated with sufficient accuracy by a polynomial such as a Pade approximation. Nor is it feasible to extend the model-numerator polynomial to include the time delay, since this would result in a prohibitively large number of model coefficients to be estimated. Therefore the control algorithm that is used has to formulate the time delays explicitly.

Reduction of the number of estimation coefficients. Even in the case of single-input single-output estimation, a large number of model parameters leads to slow adaptation of the estimated model. In the case of multivariable estimation, the number of parameters increases with the square of the number of inputs and outputs, resulting in a substantial increase in the number of coefficients to be estimated. It is therefore desirable, if not essential, to minimize the number of model coefficients. Where possible, *a priori* process knowledge should be used to decrease the number of model coefficients. One such method is by explicit declaration of the time delays, as described above.

Further reductions may be achieved by setting elements of the multivariable transfer function matrix to zero, if the interactions are known to be nil. Furthermore, in the case of known integrator-type responses, the denominator coefficient was explicitly set to unity, and not included amongst the estimated coefficients. In the case of the 5x5 ROM milling application, this resulted in a reduction of the number of model parameters from 95 to 35.

Filtering. The ROM milling process is by its nature a very noisy process, resulting in problems in the estimation of the model parameters. It was therefore essential to apply both analog and digital filtering to the signals before they were processed in the adaptive controller. It was also necessary to high-pass-filter the signals, and to use an incremental control algorithm in order to deal with the non-zero and slowly varying 'steady state' operating conditions of the process.

In order to improve the filtering characteristics, a 1 second sampling rate was chosen. Analog anti-aliasing filters with time constants of 8 seconds were used prior to sampling. The 1-second sampled signals were then digitally filtered to the frequency range of interest. Note that each of the input signals must be filtered to the same passband as the output signal in that model, otherwise the filter becomes part of the estimated model. A first-order high-pass filter was used to eliminate steady state offsets, as well as low frequency drift, while a third-order low-pass filter was used to eliminate the high-frequency noise, and also acted as a digital anti-aliasing filter for the 10-second sampled signals. Note also that the anti-aliasing filters (and particularly those of higher order) introduce noticeable phase lag (Astrom and Wittenmark, 1989). This is of no consequence in the estimator, in which both the input and output signals are filtered identically. In the controller however, this lag should be approximated by additional time delays in the process model.

Excitation signal. In a noisy process such as the ROM milling process, care should be taken that the information content of the measured signals is larger than the noise content. The noise level is minimized by selective filtering as described above. In addition, the information content may be boosted by perturbing the process. This is achieved by adding a pseudo-random binary sequence (PRBS) to the controller outputs. In the case of multivariable control, an identical but delayed PRBS must be applied to all outputs in order to ensure that the perturbation signals are uncorrelated (Davies, 1970). The magnitude of this delay should be made longer than the sum of the longest time delay in the system and the order of the highest-order model in the estimation. The size of the PRBS was made proportional to the trace of the covariance matrix, so that adequate excitation is available when the model parameters are uncertain.

Estimation deadzone. An estimation deadzone may also be used to reduce the effect of process and measurement noise on the estimator. The use of a variable deadband to increase robustness to higher order unmodelled dynamics (Kreisselmeier, 1986) was evaluated, and was found to be of negligible benefit in the present application, where the high levels of process and measurement noise tend to swamp the effects of higher-order dynamics.

Numerical problems. A UD-factorisation algorithm was used in the recursive calculation of the covariance matrix so as to avoid numerical instability. This was particularly necessary in the case under consideration, since the PDP11 computer used in the implementation of the controller did not offer double-precision calculations. In addition, a check was kept on the solutions of all sets of linear equations. An iterative procedure was used to improve the accuracy of the solution if it was found to be inaccurate.

Both the solution to the RELS algorithm used in this implementation and the solution of the control-synthesis equations were found to exhibit numerical instability, owing to the differences in the order of magnitude of the measured variables. For example, the particle size of the product ranges from 70 to 90% passing 75μm, while the density of the primary cyclone feed ranges between 1000 and 2000kg/m^3. This is a difference of a factor of 50, which is amplified further in the calculations. All the measured variables were therefore normalized with respect to their measurement range, and this was found to eliminate the numerical problem.

Saturation of measurements. The measurement range of the product particle-size analyser is confined to a fairly narrow range, and therefore this measurement frequently saturates. The non-linearity introduced in this way adversely affects the model parameters. The estimator therefore needs to be disabled while saturation of a measurement occurs.

Saturation of controls. A 'safety jacket' design was also mandatory in order to deal with the particular peculiarities of the process, and also with the problems that tend to crop up in a practical application. For instance, the process is open-loop unstable, and hence the saturation of control actions can lead to loss of control. The controller was therefore modified so that it recalculates the controls when one or more of the control actions saturate.

RESULTS OF THE SIMULATION

Before it was implemented on the plant, the controller was developed and tested using simulations based on a non-linear heuristic model of the process. However, this model was found to be highly inaccurate in several respects, and did not reflect the long time delays inherent in the process. Further simulations were therefore conducted, at first with a three-input three-output, and then a five-input five-output empirical linear model of the ROM milling process. The results presented here are based on the simulation with the five-input five-output model of the process. Random disturbances in the process were simulated as pseudo-gaussian noise on the process output measurements, with standard deviations of 0.5, 1.5, 0.2, 0.2, and 5.0 respectively.

From the empirical models of the process, it was established that the system is dominantly first-order with time delays. The order of the model denominator or A-polynomial was therefore chosen as one, giving the added advantage that pole-zero cancellation in the model is avoided. The numerator or B-polynomials were chosen as second order, so as to compensate for different time constants in the different input-output combinations. Higher order polynomials were found to give negligible improvement in performance. For the known integrator-type transfer functions, the numerator polynomial was set to zero order, while the denominator pole was set to unity. The time delays for each input-output pair were set to values established from plant tests. These values remain fixed, and are not updated in the estimator.

The process was sampled at a one-second update rate, and band-pass filtered to cut-off frequencies of 0.0357Hz and 0.00025Hz. These signals were then down-sampled to the 10-second sampling rate of the adaptive controller. The closed-loop pole specification was chosen as second order for all five loops, with both pole locations set at 0.8, 0.9, 0.8, 0.8, and 0.8 for the five loops. Estimation deadzones were used in the estimator, and the variable forgetting factors were kept fairly insensitive, since the high level of process noise would otherwise result in large fluctuations in the model parameters. A random walk and a PRBS of small amplitude were also incorporated into the controller.

Discussion

The simulation results are shown in Fig. 2 and Fig. 3. Fig. 2 shows the five process outputs with their setpoints, and the five corresponding control actions. The multivariable adaptive controller can be seen to control the process effectively, as each of the outputs follows its setpoint. At $t=2500$ seconds, the process model changes, simulating the addition of steel balls to the mill charge, as well as a hydrocyclone spigot change.

The addition of steel balls results in a sudden increase in the mill load, as well as an increased breakage rate. At the same time, the simulated spigot change from 100mm to 80mm diameter results in a large decrease of the circulating flowrate. This in turn results in changes in the time constants of the system of up to 27%. The time delays also change, owing to the decreased flowrate, increasing over a range of several seconds in some interactions, and up to more than 100 seconds in one case.

The discontinuity introduced by the sudden load change, the initial model mismatch, and the changes in time delays, all result in the condition of transient upset shown after the process change at $t=2500$. Fig. 3 shows the model parameters, and the trace of the corresponding covariance matrices, for the same period of time.[3] The covariance matrices can be seen to increase at $t=2500$, owing to a random walk that was instituted when the model mismatch was detected. The parameters in outputs 3 and 4 shown in Figs. 3c and 3d (primary and secondary sump levels), change very little, as would be expected, since these dynamics are not affected by the process changes. The parameter in output 2, shown in Fig. 3b (mill load), changes slightly as the gain of the integrator-type model decreases owing to the increased breakage rate of ore in the mill. The dominant effects can be seen in the model parameters of outputs 1 and 5 shown in Figs. 3a and 3e (product particle size, PSM, and primary cyclone feed density, PCFD). Here, there are large changes in the gains, time constants, and time delays of the process model, as is shown by the changes of the corresponding model parameters at and after the process change at $t=2500$. It may be noted that even before the process change at $t=2500$, the changes in the model parameters are due to the high levels of noise. This could not be eliminated altogether, despite the use of filtering, estimation deadzones, and insensitive forgetting factors.

[3] The stepwise changes in the process parameters are due to the effect of the estimation deadzone, which prevents estimation when the estimation error is small.

After the process change and the re-estimation of the model parameters, the adaptive controller can again be seen to follow the setpoint changes satisfactorily, despite the significant changes in the gains, time constants and time delays of the process.

CONCLUSION

An algorithm for the multivariable adaptive control of a run-of-mine milling circuit has been developed. Even though the number of process interactions (5-input 5-output) is relatively large, and the associated time delays are long, the number of parameters in the estimated model was kept manageable. This was achieved by the use of *a priori* knowledge of the process interactions, and by explicitly specifying the time delays. The effect of process noise was minimized by the use of selective filtering, applying an estimation deadzone, and perturbing the process with a PRBS. By means of simulation it was shown that effective multivariable adaptive control of a ROM milling process is possible, even under severe changes in operating conditions.

The control system described in this paper is currently being commissioned at a South African gold-reduction plant.

ACKNOWLEDGEMENT

This paper is published by permission of Mintek and Gold Fields of South Africa.

REFERENCES

Astrom K.J., and B. Wittenmark (1989). *Adaptive Control.* Addison Wesley, London.

Davies, W.D.T. (1970). *System Identification for Self-Adaptive Control.* Wiley, London. Chap. 3.

Herbst J.A., L.B. Hales, and W.T. Pate (1988). Model based control of SAG mills using real-time expert systems. *CEE Course: Optimisation of Comminution*, University of the Witwatersrand, Johannesburg.

Hulbert D.G., I.K. Craig, M.L. Coetzee, and D. Tudor (1987). Multivariable control of a run-of-mine milling circuit. *SAIMM Colloquium on Milling*, Mintek, Randburg.

Hulbert D.G. (1989). The state of the art in the control of milling circuits. *Preprints of the 6th IFAC Symp. On Automation in Mining, Mineral and Metal Processing*, Buenos Aires.

Kreisselmeier G. (1986). A robust indirect adaptive control approach. *Int. J. Control, 43*, 161-175.

McDermott P.E., and D.A. Mellichamp (1984). A decoupling pole-placement self-tuning controller for a class of multivariable processes. *Proc. 9th IFAC World Congress*, Budapest.

McDermott P.E., D.A. Mellichamp, and R.G. Rinker (1986). Pole-placement self-tuning control of a fixed-bed autothermal reactor. *AIChE Journal, 32,* 1004-1024.

Ruhmer W.T. (1985). A techno-economic evaluation of five routes for the comminution of gold ores in South Africa. *Report No. M182*. Mintek, Randburg.

Seborg, D.E., T.F. Edgar, and S.L. Shah (1986). Adaptive control strategies for process control: a survey. *AIChE Journal, 32,* 881-913.

Ydstie, B.E., L.S. Kershenbaum, and R.W.H. Sargent (1985). Theory and application of an extended horizon self-tuning controller. *AIChE Journal, 31,* 1771-1780.

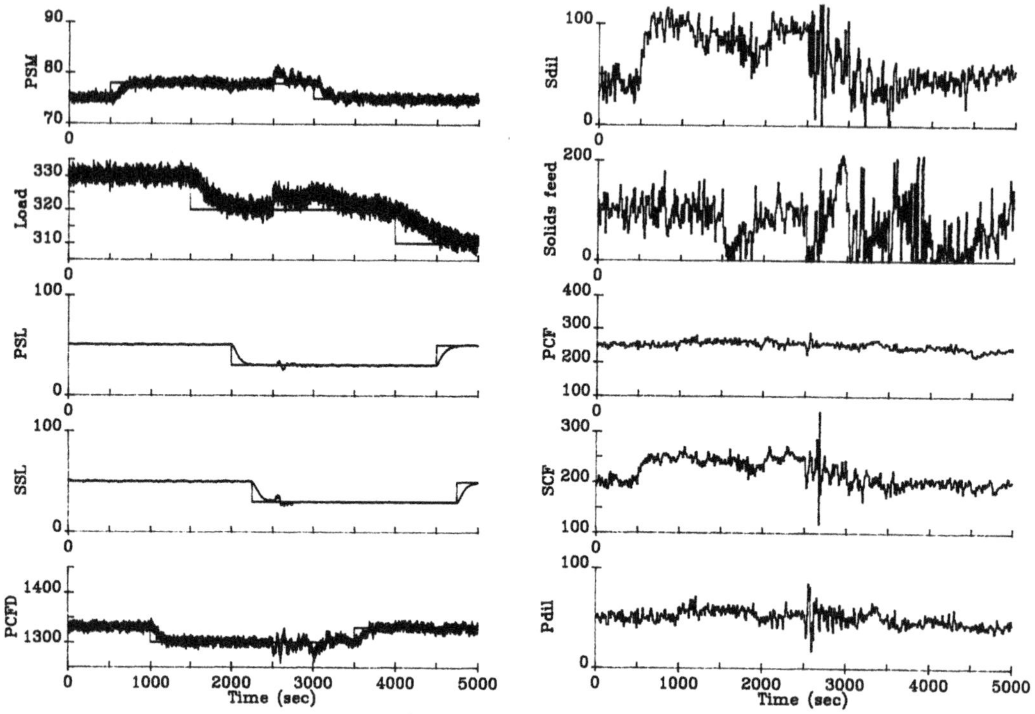

Fig. 2. Response of the ROM model to step changes in setpoints and dynamics.

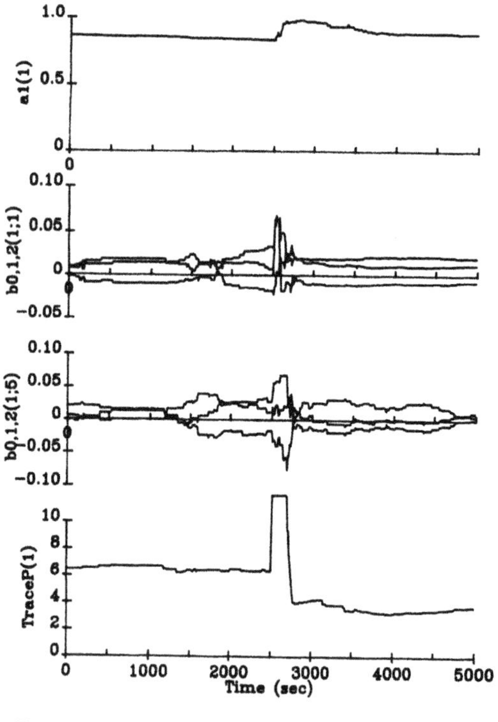

Fig. 3a. Behaviour of loop 1 model parameters for Fig. 2.

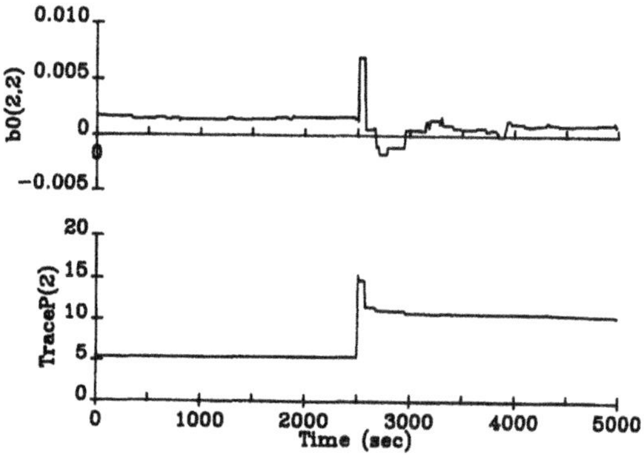

Fig. 3b. Behaviour of loop 2 model parameters for Fig. 2.

The naming convention adopted for the model parameters is as follows. *a* and *b* correspond to the denominator and numerator coefficients respectively. In the case of *b0,1,2(1;5)*, the *b0*, *b1* and *b2* coefficients are given for the model of input-5 on output-1. Similarly, *b0(3;3,5)*, refers to the *b0* coefficients for the model of input-3 on output-3, and input-5 on output-3.

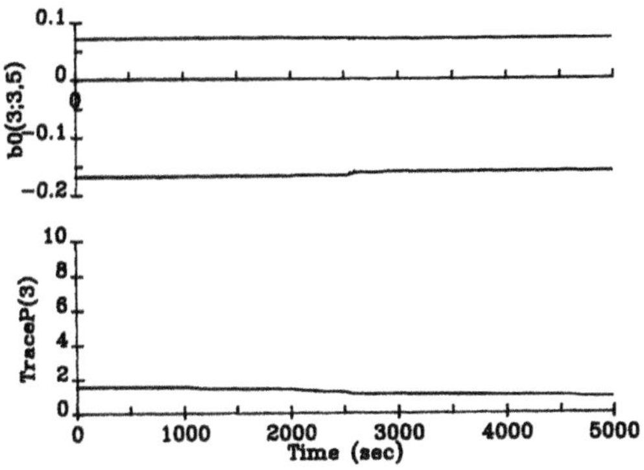

Fig. 3c. Behaviour of loop 3 model parameters for Fig. 2.

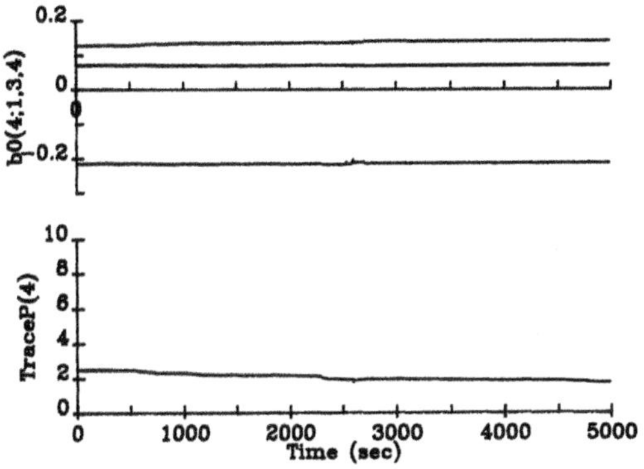

Fig. 3d. Behaviour of loop 4 model parameters for Fig. 2.

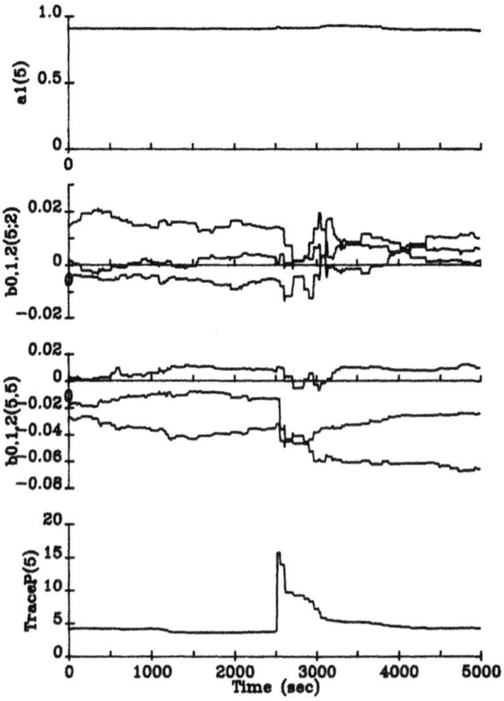

Fig. 3e. Behaviour of loop 5 model parameters for Fig. 2.

INDUSTRIAL DEVELOPMENTS IN INTELLIGENT AND ADAPTIVE CONTROL

S. Yamamoto

Application Engineering Dept., Yokogawa Electric Corporation, 25-1 Nishi Shinjuku, 1-Chome, Shinjuku-ku, Tokyo, Japan

Abstract. With the progress of computer technology, new control methods are being developed. The adaptive control made rapid progress in theory, and various applications for practical use are reported in industrial fields. Various types of products, especially PID type self-tuning controllers are commercially available today.

Recently, applications of fuzzy control are being given attention. The fuzzy control is one of effective methods to automate operators' operations by use of a computer. There are several examples such as the cross-machine directional basis weight control of paper machines and the temperature control of distillation columns.

Key words. Intelligent control, adaptive control, self-tuning control, fuzzy control, predictive control, learning control

1. Introduction

The greatest part of control system still uses PID control in the field of the process control. It has a long history and has many advantages as follows ;

(1) In the functional aspect, it has well balanced actions to reduce offset and to raise response.
(2) From the engineering point of view, it does not require exact process model.
(3) Physical meaning of controller parameters are clear.
(4) Controller parameters can be easily adjusted after installation.

These functions are quite important when developing a new control method.

In spite of the above advantages there are some cases where standard PID does not suffice ;

(1) The process has a long dead time.
(2) The process has a strong non-linearity.
(3) Strong interactions between controlled variables exist.
(4) Process characteristics are not constant.
(5) Frequent load changes occur.
(6) Innegligible process noise exists.

Advanced control methods are developed to overcome these difficulties. However, the introduction of new functions complicates the parameter tuning works and diminishes the above engineering advantages of the PID control. For instance, the feedforward control, Smith's controller and decoupling control require process models and complicates the tuning work.

Thus, the advent of control systems with easy engineering while having high degree of control functions are desired. It can be said that the intelligent control is the one which responds to those demands.

2. PID-type Self-tuning Control

There are various types of adaptive control ;

(1) Open-loop Adaptive Control
(2) PID-type Self-tuning Control
(3) Self-tuning Regulator (STR)
(4) Model Reference Adaptive Control System (MRACS)

This chapter discusses the PID-type self-tuning control (STC).

The ultimate sensitivity method[1], limit cycle method[2] and such were developed in early 1970s. Since the pervasion of micro-processors, the complicated calculation became easy. Accordingly, the method using area method[3], the method based on the modern control theory[4] and the like have appeared as commercial products. In addition, the expert method[5), 6), 7)] which replaces the human tuning work with a computer has appeared in

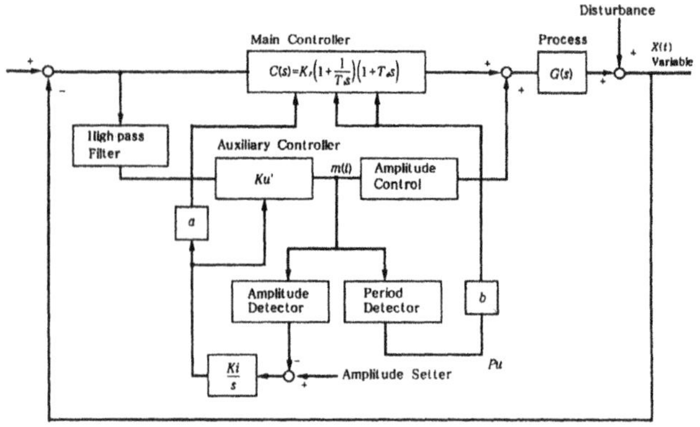

Figure 1 STC by Ultimate Sensitivity Method

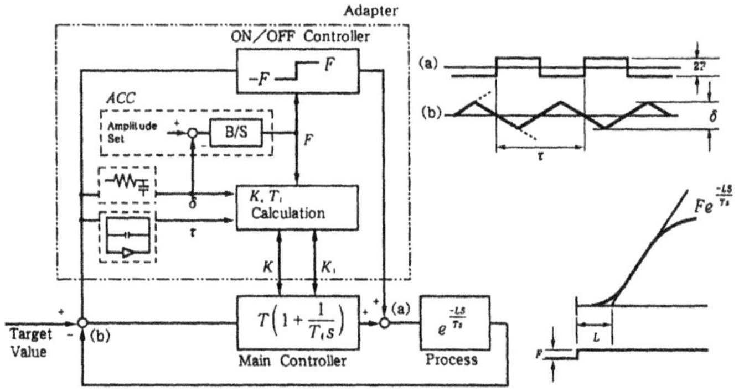

Figure 2 STC by Limit Cycle Method

the market. Nowadays, the ones based on those methods with various added improvement are introduced in the marketplace. Several of these types reported in Japan are surveyed in the following sections.

2. 1 STC by Ultimate Sensitivity Method[1)]

This STC automatically executes Zeigler-Nichols' ultimate sensitivity method. As shown in Figure 1, this STC is constructed by connecting an auxiliary controller (proportional action only) to a main controller (PID controller) in parallel. In the figure, the ultimate oscillation with small amplitude will be generated when the proportional gain of the auxiliary controller is gradually increased. The gain when the oscillation is begun K_u (ultimate gain) is expressed by following equation ;

$$K_u = K_u' + K_p \qquad (1)$$

where,

K_u' : the gain of the auxiliary controller
K_p : the gain of the main controller

Since Ziegler-NicholS' method specifies the gain of a controller as the ultimate gain multiplied by a coefficient, the gain of the main controller at normal controlling condition ($K_u' = 0$) is expressed as ;

$$K_p = \alpha K_u \qquad (2)$$

where,

α : coefficient

By substituting the K_u in equation (2) into equation (1), the relationship between K_u' and K_p at the ultimate oscillation condition is expressed as ;

$$K_p = \frac{\alpha K_u'}{(1-\alpha)} \qquad (3)$$

Therefore, the proportional gain of the main controller K_p is obtained from the K_u' by multiplying,

$$\alpha' = \frac{\alpha}{(1-\alpha)} \qquad (4)$$

The integral action time and derivative action time of the main controller can be obtained directly from the period of the ultimate oscillation by multiplying respective constant.

As for the method applying Ziegler-Nichols' method to auto-tuning, there also is a method proposed by Astrom[8)]. This method can be utilized to obtain approximate process characteristics as the pre-stage to using other STCs.

2. 2 STC by Limit Cycle Method[2)]

This STC utilizes Ziegler-Nichols' limit cycle method. As shown in Figure 2, this STC also is constructed by connecting an auxiliary controller (on/off controller) to a main controller (PI controller) in parallel.

This STC expresses a process in the form of e^{-Ls}/Ts (process model). At first, a limit cycle with small amplitude is generated by the auxiliary controller. Next, the dead time L and time constant T of the process is estimated from the amplitude and period of the limit cycle, by following equations.

$$T=\frac{\tau}{2\delta}\left(F+\frac{k\delta}{\pi}\right) \qquad (5)$$

$$L=\frac{\tau}{4}\left(1-\frac{\frac{k\delta}{\pi^3}}{F+\frac{k\delta}{\pi}}\cdot\frac{\tau}{Ti}\right) \qquad (6)$$

where,

τ : the period of the limit cycle

δ : the peak-to-peak amplitude of the limit cycle

F : the amplitude of the auxiliary controller

k : the proportional gain of the main controller

T_i : the integral gain of the main controller

According to the values of L and T thus obtained, the PI parameters of the main controller are tuned by Ziegler-Nichols' step response method. Although this method is proposed fairly long years ago, there is a product realizing STC by similar concept[10].

2. 3 STC by Least Square Method[4]

There is an example of which STC is applied to the two degrees of freedom PID controller. Figure 3 shows the system configuration of this method. This method is similar to the standard STC configuration.

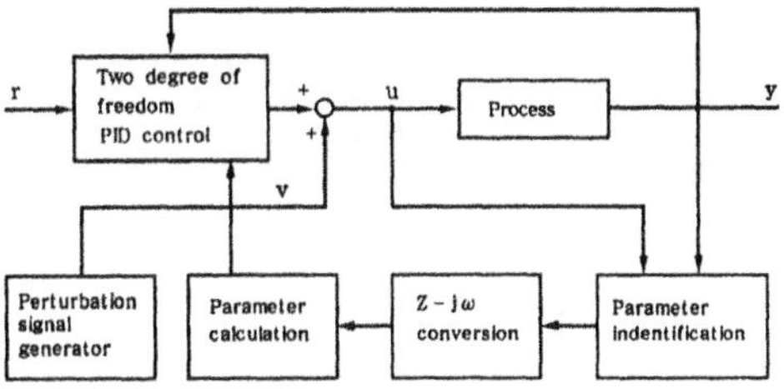

Figure 3 Two Degrees of Freedom PID Auto-Tuning Controller

This STC expresses the process characteristics by an auto-regressive moving average (ARMA) model. By applying a M-series signal to the process as its identifying signal, the parameters of the ARMA model are estimated using least square method. Obtained ARMA model is then converted to a pulse transfer function, and further converted into an s-domain transfer function. The PID parameters contain four reference models (Bessel, Binomial, Butterworth and ITAE minimum), and are obtained by the model matching method.

2. 4 STC by Expert Method

There are two types of STC using expert system ; to correct PID parameters according to the response wave patterns, and using the fuzzy control. An example of latter type is introduced here. Figure 4 shows the system configuration.

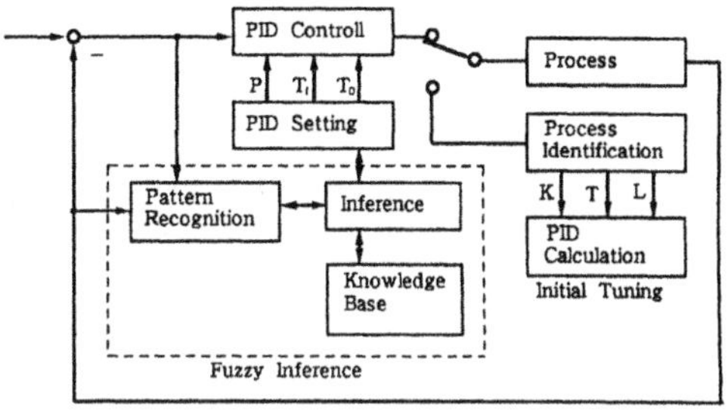

Figure 4 Expert Auto-Tuning System

The area surrounded by dotted lines executes the fuzzy inference. Overshoot amount (OV), oscillation damping ratio (DP), settling time (QR) and the like obtained from the wave form analyzer are used as characteristics values. Control rules are given as "If (overshoot is large, damping ratio is large and settling time is large) then (increase P, increase T_I and decrease T_D)." Table 1 shows the examples of the rules.

Table 1 Control Rules

No.	IF Block			THEN Block		
	OV	DP	QR	K_{PB}	K_i	K_d
1	PB	PB	—	PB	PB	NB
2	PB	PM	PB	PB	NB	ZO
3	PB	PM	ZO	ZO	PB	PB
4	PB	PM	NB	NB	PB	PB
5	PB	ZO	PB	PB	ZO	ZO

$$QR = \frac{\text{Response Time (This time)}}{\text{Response Time (Previous time)}}$$

where,

PB : Positive Big

PM : Positive Medium

ZO : Zero

NM : Negative Medium

NB : Negative Big

2. 5 STC Using Model Fitting Tuning[12]

We have developed an intelligent STC implemented in a single loop controller.

(1) Identification of process characteristics

The STC observes process input and output as shown in Figure 5. STC detects process abnormality, eliminates noises in measured process value and compensates operation level. Then, if a change in process dynamics are detected, a process model is obtained by a model fitting method using the process value PV and the control output MV.

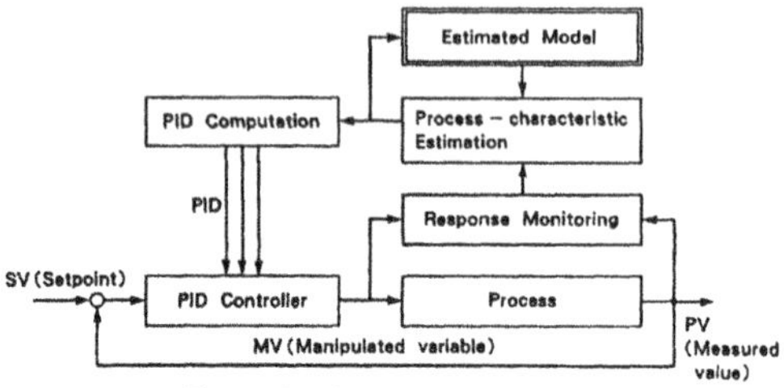

Figure 5 Self-Tuning Function

(2) PID parameter tuning

STC calculates optimum PID parameters as a function of the identified process characteristics, desired response curves and controller types.

3. Open-loop Adaptive Control

In some processes, parameters affecting the process characteristics (auxiliary parameters) can be measured. In these cases, the effect on the controllability can be cancelled out by adjusting the controller parameters according to measured auxiliary parameters.

The gain-scheduling control is an open-loop adaptive control, and works as feedforward-like function. Therefore, it does not correct control parameters after checking the result of adjustment. Nevertheless, since it has an advantage of making the parameters quickly tracking the change of process characteristics, it is largely used in practice. Figure 6. shows the system configuration of the gain-scheduling control. In the gain-scheduling control, parameters to be changed are not only the gain but also includes the terms of dynamic characteristics.

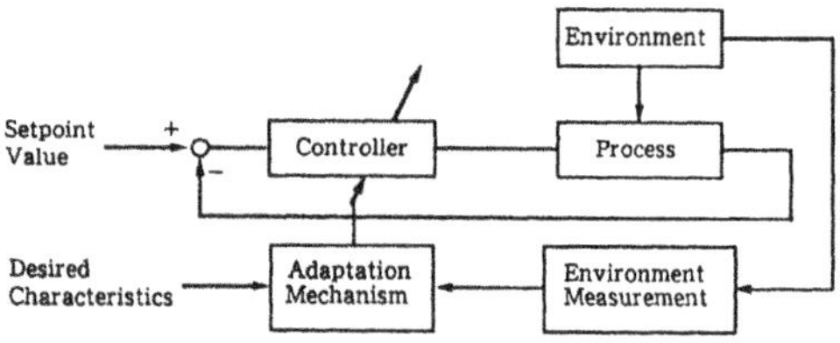

Figure 6 Gain-Scheduling Control

4. Fuzzy Control

4. 1 The Reasoning of Fuzzy Control

In the fuzzy control, empirically obtained control rules are expressed in the form of "IF…THEN…" using the production system. Thus, the feature of the fuzzy control is that control rules are expressed by language. Figure 7 shows an example comparing the fuzzy control method with PID control.

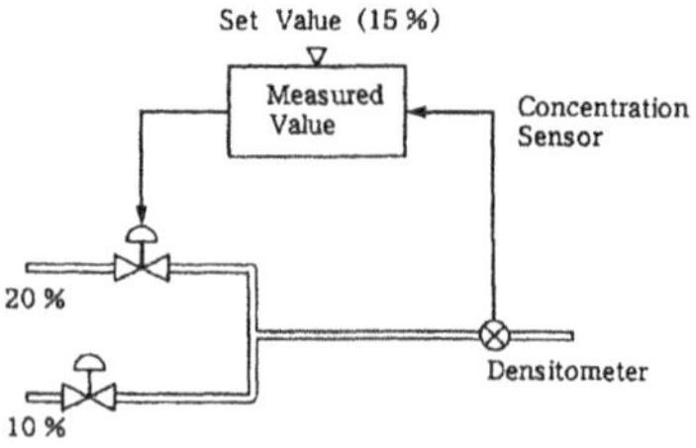

Figure 7 Fuzzy Control Method

[PID control]

Control output $= K_p\ (K_i E + \Delta E)$

E = Setpoint (SV) − Measured value (PV)

$\Delta E = E_n - E_{n-1}$

[Fuzzy control]

Rule 1 : IF deviation is positive big and load is big THEN control valve opening is big.

Rule 2 : IF deviation is positive big and load is small THEN control valve opening is small.

There are several kinds of fuzzy control reasoning.

4. 2 Realization of Fuzzy Controller

Although fuzzy controllers may be realized by dedicated hardware, it is also easy to construct the fuzzy reasoning using a general-purpose computer.

Generally, since process control does not require high-speed, controllers programmed into a general-purpose process control computer can be used satisfactorily.

Following functions are to be required when packaging the controller for process use.

(1) Online data processing
(2) Recipe management
(3) Displays while under operation Trend display
    Reasoning progress display
(4) Engineering
    Control rules alteration
    Membership function setting and modification
    Simulation function
    Control rules check
    Closed-loop controllability confirmation

4. 3 Fuzzy Control Application

Fuzzy control is first applied to actual process by Mamdani in 1974. Then, a period of vacancy elapsed. However, from the middle of 1980's, many applications have emerged.

This control method is favored especially in Japan, and many applications are being reported. Table 2 shows application examples of fuzzy control in Japan. A few of the applications are described in the following section.

Table 2 Examples of Fuzzy Control Application

Industries	Process	Process Features
Petroleum Chemical	• Ethylene Plant Production Control • Dsitillation Tower Bottom Temperature Control • Powder & Fluids Batch Automatic Weighing	• Each cracking furnace has differnt characteristics. • Complicated system ; Cannot cope with a mathematical model. • A great deal of disturbances. • Weighing conditions will change.
Iron & Steel	• Sintering Furnace Return Ore Blending • Converter Gas Supply Control • Blast Furnace Control • Air-heating Furnace Temperature Control • Hot/Cold Rolling mill Thickness Control Form Control	• A great deal of disturbance variation factors. • Long dead-time. • Intermittent and irregular gas generation. • Too much and inaccurate information for opretors.
Power Plant	• Power Plant Start-up Control • Nuclear Power Plant Recirculation System Pump speed Control • Dam Sluice Gate Control	
Ceramics	• Cement Kiln Sintering Control • Cement Mill Feed Quantity Control • Alumina Kiln	• Too small number of measurement points available. • Complicated reaction. • Long retention time. • Each mill has different characteristics.
Water Treatment	• Purification Plant Pre-chlorination Control • Rain Water Treatment Pump Coordinating Control • Water-works Drain Valve Control	• A great deal of disturbance factors. • Long dead-time. • Mathematical modeling is difficult. • Diverse situations. • Control evaluation standards change by situations.
Refuse Disposal	• Garbage Burning Furnace Combustion Measurement and Control	• Combustion modeling is difficult. • Variable measurement is difficult.
Food	• Glutamic Acid • Crystalization Pan Operation Control	• Mathematical modeling is impossible • Complicated reaction. • Measurement is impossible. • Ambiguous causal relations.
Pulp & Paper	• Paper Machine Cross Directional Control	
Electronic Parts Manufacturing	• Crystal Growing Process Temperature Control • Glass Fusion Furnace Temperature Control • Semiconductor Production Setepper Position Control • Clean Room Air Conditioning • LED Production • Ion Injection Control	• Thermal inertia will disturb the control. • Time delay exists in temperature measurement.
Machinery	• Robot Control	
Instruments	• Auto-tuning Controller • Temperature Controller Overshoot Prevention	
Transportation	• Automatic Train Opration • Automatic Crane Operation • Elevator Direct Numerical Control • Anti-skid Braking • Automatic Transmission Control • Shield Method Tunnel Drilling	
Commercial	• Hot-water Supply	
Commodity	• Air conditioner • Laundry Machine • Cleaner • Camera • Vedeo Movie Camera	

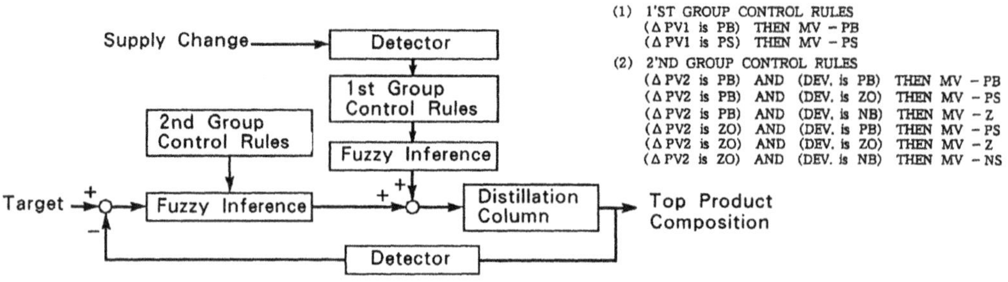

Figure 8 Example of Control Rules

4. 4 Fuzzy Control on Distillation Tower[17)]

The tower is used for the distillation of propylene in an ethylene plant. The propylene is extracted from the tower top, and propane from the bottom. A process gas-chromatograph is installed on the column to monitor and control the composition of top product by manipulating reflux quantity.

There are two groups of control rules. As for the first group, feedfoward empirical control rules which monitor the feed quantity variation are selected. Feedback empirical control rules which monitor the deviation from target composition and the rate of deviation variation are selected for the second group. Each fuzzy reasoning part determines respective reflux quantity which are then added to become the process manipulation value.

Figure 8 shows the block diagram of the control system and examples of the control rules. Fuzzy variable consists of five variables, PB, PS, ZO, NS and NB. Besides, weighted averaging method using point values is used for the reasoning :

$$U = \frac{e_1 U_1 + e_2 U_2}{e_1 + e_2} \qquad (13)$$

where,

e_1, e_2 : Grade
U_1, U_2 : Manipulated variable

Figure 9 shows the result of fuzzy control application . The figure compares the fuzzy control with conventional manual operation in case of tower feed rate change. The maximum composition variation of 0.5% at manual operation is improved to 0.12% by applying the fuzzy control.

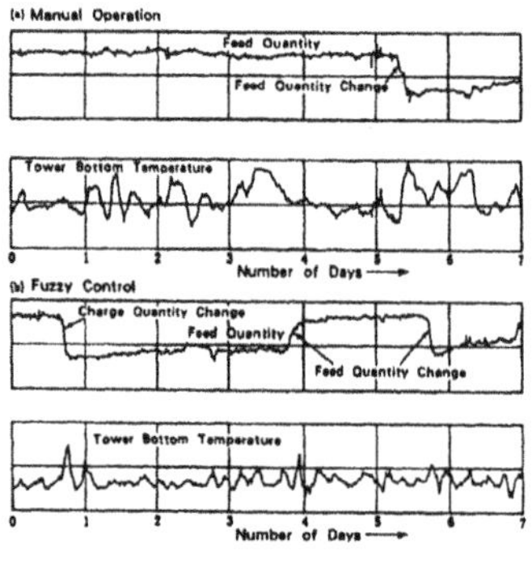

Figure 9 Example of Controllability Comparison

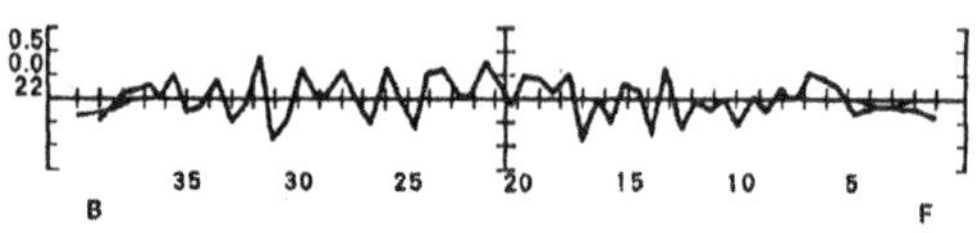

Figure 10 Saw-Tooth Pattern in a Bone Dry Profile

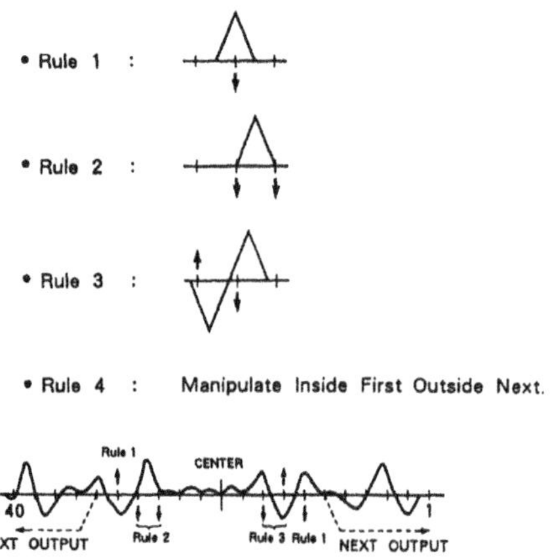

Figure 11 Bone Dry Profile Fuzzy Control Strategy

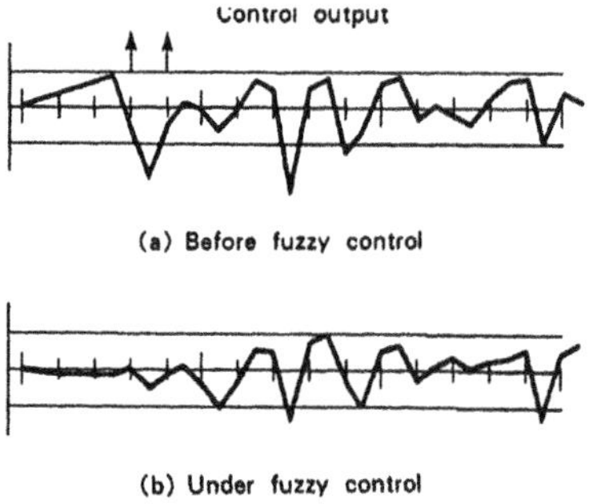

Figure 12 Bone Dry Profile Fuzzy Control Result

4. 5 Profile Control of Paper Machine

It is very important, in the paper machine, to regulate the cross directional bone dry variation and make uniform basis weight paper. It is, however, fairly difficult in practice. In a worst case, variation of saw-tooth form may appear

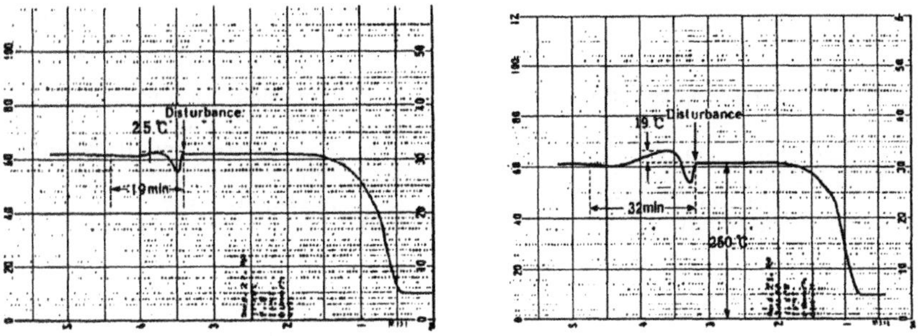

(a) Control Result by UT35 「Super」 (b) Control Result of 2-degree of Freedom Control

Figure 13 Field Test Example

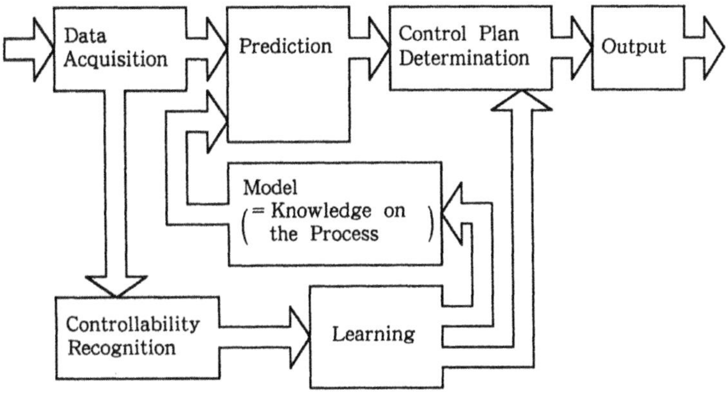

Figure 14 Intelligent Control

between between each slice as shown in Figure 10.

To overcome this problem we have made a trial of applying the fuzzy control. As for control rules, expertise of skilled operators are implemented as shown in Figure 11, and obtained good results as shown in Figure 12.

4. 6 Overshoot Restriction in Temperature Raising Control[18]

There is a case, in temperature control, that it is required to change the temperature without causing overshoot. An example using fuzzy control is described herein.

To restrict the overshoot, an internal target setpoint (SSP), which is lower than the true set point TSP, is set in advance. At the time when PV exceeds the SSP, the SSP is corrected toward the TSP so that the PV converges to the TSP smoothly. The fuzzy control is applied to this corrective action. Besides, the PV settles to the TSP without exceeding the SSP.

FIgure 13 shows the simulation examples of step response, constant value control with slope setting and response to external disturbance. Aresult of a field test shows the improvement in controllability.

4. 7 On Application of Fuzzy Control

Today, the application of fuzzy control is progressing vigorously. Numerous success cases are reported, and their effectivity is being advertised. However, it seems that what part of the fuzzy control structure is effect or from where the advantage is derived compared to conventional PID control and modern control theory, are not well discussed.

The feature of the fuzzy control is that it provides the means for realizing control rules expressed by language, by introducing the mumbership function. As the result, the fuzzy control realizes the following control.

(1) Enables non-linear control action.
(2) Enables the control action in consideration of many input variables.

When looking into the control rules carefully, they often comprise non-linear gain PID controls. Besides, the system comprises a multi-input system, by writing various conditions in the rules.

In addition, the fuzzy control has following advantages, from the engineering point of view.

(1) The contents of control are easy to understand since the control rules are expressed by production rule.
(2) It is not necessary to describe all the cases likely to occur, because the membership functions overlap one another so that

they interpolate the gaps among the rules.
(3) It is not necessary to study the control theory thoroughly.

For above reasons, the fuzzy control seems widely used.

On the other hand, the fuzzy control involves following disadvantages.
(1) There is no theoretical backing for stability and controllability.
(2) The procedure for determining the membership function is not established.
(3) It is difficult to handle dynamics.

There are several conditions necessary for successful introduction of fuzzy control.
(1) Control rules should be expressed by language.
(2) The variables described in control rules should be available as system input signal.
(3) Enables intermittent operation. For instance, the process shoud be able to operate manually by an operator.

5. Conclusion

Adaptive Control, and fuzzy control are discussed as examples of intelligent control, mainly from the industrial point of view.

However, these methods only realize a certain class of the intelligent contol.

As for the indispensable conditions of the intelligent control, as discussed in the introduction, it seems that following functions should be provided when realizing as a controller.

Permits imperfect model (make it to a perfect model by itself through learning)
Enables the recognition of external conditions.
Provides model identification function.
Provides prediction functions.
Enables non-linear-type operation.

Finally, an intelligent controller combining above functions is proposed in Figure 14.

6. Reference

1) T. Kitamori, Study on adaptive control system based on the ultimate sensitivity method, Proceedings of SICE8 Japanese), Vol. 15 No.4 pp549/555 (1971)
2) A. Sumi, Analog simulation of an adaptive controller using a limit cycle, Analog Technique (Japanese), Vol. 12 No.8 (1972)
3) Nishikawa et al. PID parameters autotuning in FUJI MICREX
4) K. Nagakawa (toshiba), An Auto Tuning PID Controller, ISA 1982 pp1353/1362
5) T. Kraus (the Foxboro Company) SELF TUNING CONTROL : AN EXPERT SYSTEM APPROACH,AUTOMATIC CONTROL SYSTEMS, ISA 1984 Houston
6) K. Tachibana et al. Auto-Tuning One-Loop Controller Using Fuzzy Logic to PID Parameter Tuning, Keisou (Japanese), Vol. 31 No.5 1988 pp11/15
7) K. Nomoto et al. The Recursive Fuzzy Reasoning and Its Application to an Auto-Tuning Controller, Proc. International Workshop on Fuzzy System Application, pp81/82, 1988
8) K. Astrom and B. Wittenmark, On self tuning regulator, Automatica Vol. 9 pp185/199 (1973)
9) E. H. Bristol et al.(The Foxboro Co.), Adaptive Process Control by Pattern Recognition, Instrument and Control System, March, 1970 pp101/106
10) M. Iwasaki (Yamatake Honeywell) TDCS3000SSC Self-tuning controller, Keisoku Gijutsu (Japanese) '86.9 pp35/41
11) ASEA, NOVATUNE Catalog
12) H. Takatsu, Intelligent Self-Tuning PID Controller, IFAC, ITAC91 15 – 17Jan. 1991, Singapore
13) N. Yoshitani, Strip temperature control for a heating furnace of continuous annealing and processing line (CAPL) (Japanese), proceeding of SICE
14) Michael J Piovoso et al. Self-Tuning pH Control, A difficult Problem, An Effective Solution, Intech, May 1985 pp45/49
15) M. J. Piovoso, SELF-TUNING CONTROL OF pH, ISA 1984, pp705/723
16) C. Kiparissides et al. Adaptive control of a polymer reactor, IFAC PRP, Belgium 1980, pp325/334
17) Y. Hanakuma et al. Design and Application of Self adjustment Fuzzy Controller to a distillation column. Kagaku Kogaku (Japanese) Vol.16. No.4 (1990), pp667
18) Y.Yasuda et al. Development of a controller Having an Overshoot Suppression, Yokogawa Technical Report Vol. 33 No.4

DESIGN AND IMPLEMENTATION OF AN ADAPTIVE CONTROLLER FOR A HYDRAULIC TEST ROBOT

F. Conrad, P. H. Sorensen, E. Trostmann and J. Zhou

Control Engineering Institute, Technical University of Denmark, DK-2800 Lyngby, Denmark

Abstract

A fast digitally controlled hydraulic test robot has been developed, designed and implemented in the hydraulic laboratory at the Control Engineering Institute, the Technical University of Denmark. The purpose of the test robot is to carry out research activities in hydraulic control system design, digital control and adaptive control and on-line systemidentification. The paper describes the test robot, design and implementation of an adaptive controller. An adaptive geometrical compensation control scheme is proposed for hydraulic robot manipulators.

Keywords: adaptive control, robot control, hydraulic robot, servoactuators, hydraulic control, signal processors, digital control, on-line identification.

INTRODUCTION

Robotics and fluid power control are important basic research and industrial areas within the general area of flexible automation and intelligent machine tools. In view of the need for faster robots, including mobile robots and hydraulic machines such as heavy load manipulators, cranes, earth-moving machines and vibration test facilities, a hydraulic test robot with two links digitally controlled by two linear hydraulic servo actuators has been developed and installed in the hydraulics laboratory of the Control Engineering Institute at the Technical University of Denmark. The digital control system is implemented on an AT&T signal processor.

The use of digital controllers for the control of hydraulic systems is proliferating more and more to areas where the microprocessor is not normally present. It is for instance becoming increasingly interesting to use microprocessor control in earth moving machines, cranes, hydraulically powered lorries as well as hydraulic robots. The general problem in all these machines is the closed loop servo control of a multivariable system using digital control.

The purpose of the robot is to use it in research activities within robot system design, control design, digital control, adaptive control, implementation, system identification and evaluation of multivariable digital controllers for hydraulic systems such as fast robots, cranes, earth moving machines, multi-axis test platforms and special hydraulically powered machines. The installation of a test robot manipulator is part of a research project for the study of MULTIvariable Digital Oil-hydraulic Servo systems called MULTIDOS.

The aim of the MULTIDOS project is mainly to carry out the following activities:

* To develop, design and implement a fast hydraulic test robot.
* Development of design rules for hydraulic robots and other digitally controlled hydraulic machinery.
* Research, design and implementation of hydraulic servo actuators.
* Research, design and implementation of multivariable digital controllers for robots and machines.
* Systems identification and adaptive control of hydraulic manipulators.

Results from the MULTIDOS project are presented in this paper.

THE DESIGNED FAST HYDRAULIC TEST ROBOT MANIPULATOR

The multi-axes design problem

In general, the design problem concerns how to design a multi-axes hydraulically powered mechanical system with its multi servo actuator system and the digital control system itself as an integrated machine system. A schematic block diagram is shown in fig.1.

The aim has been to design and implement a very fast test robot with high performance linear hydraulic servo actuators and components with optimal properties in order to eliminate the unwanted effects of friction (stick-slip), backlash and dead bands. By using linear actuators the robot arms can be directly controlled without gears which are necessary in robot systems involving rotational motors.

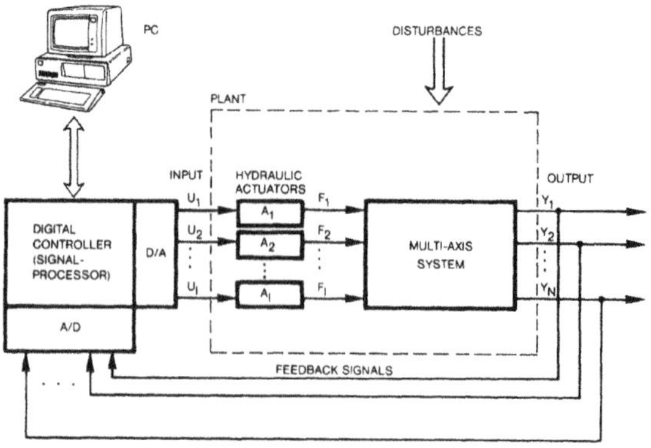

Fig. 1. Schematic block diagram of a multivariable hydraulic system.

For economical reasons, it was decided as a first step to design a two-link test robot system with rigid arms digitally controlled by two high performance linear hydraulic servo actuators. The test robot should be prepared for a later extension with additional axes. The main requirement for the design was that the test robot should be able to move a payload of maximum 50 kg path controlled as fast as possible in the specified work space within the limitation of the existing hydraulic power supply which is 84 l/min at 210 bars. The hydraulic actuators should have hydrostatic bearings to eliminate friction (stick-slip) and equipped with two-stage high frequency response servo valve. The digital controller should be designed and implemented with a very fast processor (signal processor) in order to obtain minimum computational time. The target is to study and identify the physical limitations of implementing a hydraulic digitally controlled robot system with respect to the mechanical subsystem, the hydraulic actuators, the digital controller circuits and the control algorithms itself.

Design concept and mechanical structure

In the MULTIDOS project various mechanical structures for a rigid hydraulic robot controlled by two linear actuators have been studied. This work has been continued. The aim has been to find promising design concepts and mechanical structures in order to solve the design problem described above. The design of the two arms should be shaped with respect to minimizing the moving masses of the arms without loss of the necessary stiffness of the arms and the base-unit fixed to the floor. Further, modelling, design, simulation and analysis applying CAD/CAE- and CADCS-software have resulted in the implemented hydraulic robot manipulator shown in fig. 2. For a description of the systems engineering methods involved in the design we refer to (Conrad et.al., 1990c)

For economical reasons, the arms have been designed as welded steel parts build up by modifications of standard steel profiles. Particular attention has been focused on the bearing design to eliminate backlash and minimize friction (stick-slip).

The base-unit is also built up as a rigid welded steel construction. It is bolted onto a heavy cast iron machine platform placed in the floor of the hydraulic laboratory.

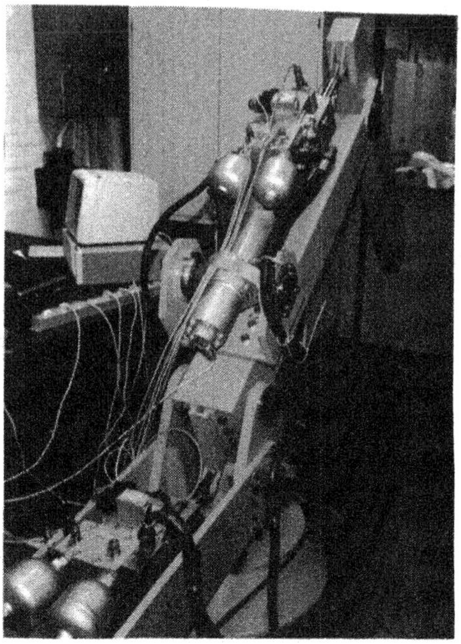

Fig. 2. The implemented TUD hydraulic test robot manipulator.

The two hydraulic servo actuators have been manufactured by the German company Mannesmann Rexroth GmbH. The hydraulic cylinders can be controlled by super high servo valves (MOOG, Rexroth and others). The actuators are equipped with a hydraulic accumulator in the supply port and an other one in the return port to cope with the dynamic flow demands.

The test robot is equipped with transducers for measuring: positions, velocities and accelerations of the moving piston shafts and the pressures in the hydraulic chambers. Further, the TCP (tool centre point) can be equipped with transducers.

As mentioned above, today the pressure and flow are limited by the existing hydraulic power station which can deliver a maximum flow rate of 84 l/min at a pressure up to 210 bars. Its is already planned to install a new more powerful supply unit in order to extend the available pressure and flow range. The robot manipulator system is described in detail in (Conrad et.al., 1990a).

The Dynamic Model of the Robot Manipulator.

The robot arm dynamics can be described by a Lagrange-Euler equation of motion

$$D(q)\ddot{q} + H(q,\dot{q}) + G(q) = \tau \qquad (1)$$

where q, $\dot{q}$, and $\ddot{q}$ are vectors signifying the joint positions, velocities, and accelerations, respectively, $D(q)$ is an acceleration-related matrix, $H(q,\dot{q})$ is a Coriolis and centrifugal force vector, $G(q)$ is a gravitational force vector, and τ is an applied torque vector generated by the actuators.

The actuator equations consist of a servo valve dynamic equation, a cylinder continuity equation, a static valve equation, and a piston force equation. Since the two actuators, one for each link, are equal, it is assumed that they have the same static and dynamic characteristic. For the sake of brevity, we shall drop the subscribes for the variables which can be applied on one of the two actuators without confusion in sequentia.

The dynamic response of the servo valves can approximately be described by the transfer function:

$$\frac{x_v}{e_g}=\frac{k_v}{s^2/\omega_s^2+2\xi_s s/\omega_s+1} \qquad (2)$$

where x_v is the displacement of the spool valve from null, e_g is the signal voltage input to the amplifier, k_v is the amplifier gain, and $\omega_s(=190\,Hz)$ and $\xi_s(=0.63)$ are the natural frequency and damping ratio of the servo valve, respectively, which are estimated by means of the measured frequency response of the servo valve.

The static equation of the servo valve can approximately be described by the nonlinear flow equation:

$$Q_f=\begin{cases} c_v w x_v\sqrt{(P_s-P_f)/\rho} & for \quad x_v>0 \\ c_v w x_v\sqrt{(P_s+P_f)/\rho} & for \quad x_v<0 \end{cases} \qquad (3)$$

where

- Q_f flow across the piston,
- c_v discharge coefficient of the valve orifices,
- w area gradient of the spool valve,
- P_s supply pressure,
- P_f pressure drop across the piston, and
- ρ mass density of oil.

Assuming that the piston is centred and that the cylinder has two symmetrical volume chambers, the continuity equation of the cylinder is:

$$Q_f=A_t\cdot\dot{x}+\frac{V_t}{4\beta}\cdot\dot{P}_f+c_{sl}\cdot P_f \qquad (4)$$

where:

- V_t total volume of the cylinder chambers,
- c_{sl} total leakage coefficient of the piston,
- β effective bulk modulus of the system,
- A_t end area of the piston, and
- $\dot{x}$ piston velocity.

The force equations of the two pistons are:

$$A_t\cdot P_{1f}=m_t\cdot\ddot{x}_1+B_t\cdot\dot{x}_1+m_t\cdot g\cdot\sin(w_1-\gamma)+F_1$$

$$A_t\cdot P_{2f}=m_t\cdot\ddot{x}_2+B_t\cdot\dot{x}_2+m_t\cdot g\cdot\sin(\theta_1-w_2+\psi)+F_2 \qquad (5)$$

where the subscripts 1 and 2 denote link 1 and 2, respectively, m_t is the total mass of the piston, B_t is the viscous damping coefficient and F_1 and F_2 are the actuator drive forces.

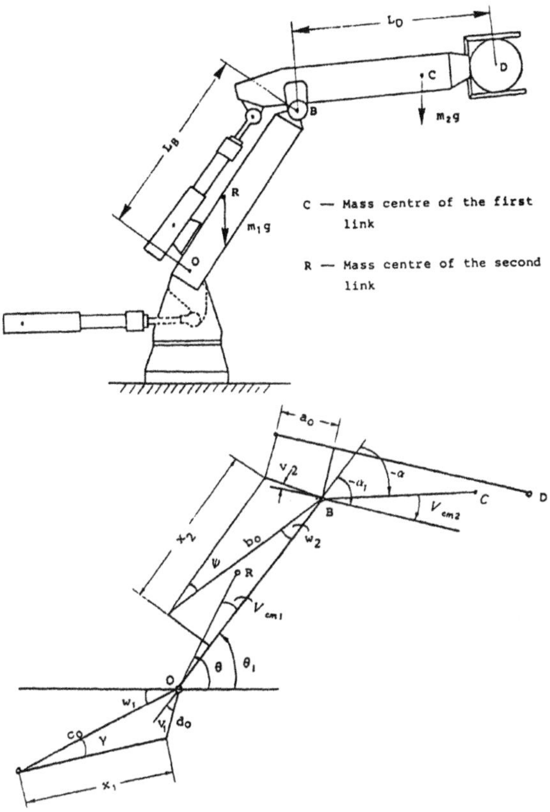

Fig. 3. The link geometry of the test robot.

The joint torques and the actuator drive forces are related by the following equations:

$$\tau_1=|d_0\cdot\sin(\gamma)|\cdot F_1 \qquad \tau_2=|b_0\cdot\sin(\psi)|\cdot F_2 \qquad (6)$$

where

$$\gamma=Arctg\left(\frac{d_0\cdot\sin(\pi+w_1+V_{cm1}-v_1-\theta)}{d_0\cdot\cos(\pi+w_1+V_{cm1}-v_1-\theta)+c_0}\right)$$

$$\psi=Arctg\left(\frac{a_0\cdot\sin(\pi+w_2+V_{cm2}+v_2+\alpha)}{a_0\cdot\cos(\pi+w_2+V_{cm2}-v_2+\alpha)+b_0}\right)$$

where the geometric parameters are defined on Fig. 3.

DESIGN AND IMPLEMENTATION OF THE CONTROL SYSTEM

It was early decided to design and implement a fast flexible digital control system which would be suited for experimental research of digital controller design and test of various algorithms and which also could meet the user requirements of programming in a convenient high level language (C, Modula 2 or others).

In fig. 4 is shown schematically the designed digital control system and its connection via interfaces to the hydraulic servo actuator system. The key digital component is the signal processor, which performs very fast floating point operations and which as

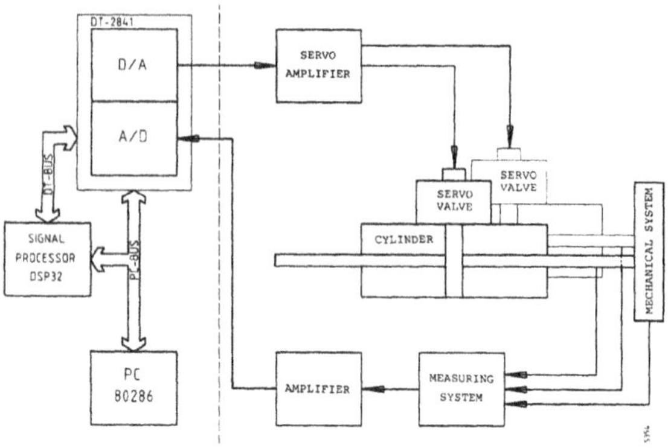

Fig. 4. Schematic layout of the hydraulic robot control system.

required can be programmed conveniently in a high level language. For a detailed description of the signal processor system refer to (Conrad et. al., 1990b).

The signal conversion is required to have a resolution of at least 12 bit ADC and DAC. This is considered sufficient. For the feedback of as many dynamic variables as possible the ADC has 16 channels and there are 2 channels DAC for the two control signals. A later extension of the robot system to more than two links would therefore require additional ADC/DAC capability. The data conversion equipment must have direct communication with the signal processor for fast data transfer.

A schematic block diagram for the ADC/DAC system is given i fig. 5.

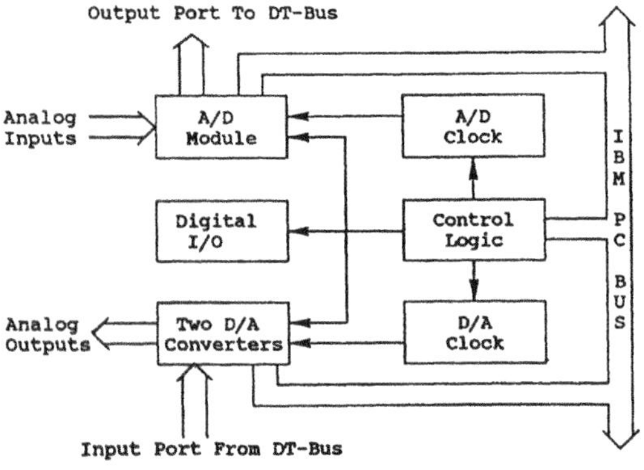

Fig. 5. ADC/DAC system layout.

Program development is performed on a PC-based system, where source code written in C or in Assembler is compiled and linked to generate executable code. This code in turn is downloaded to the Signal processor after having undergone testing and verification in a processor simulator. The program development routine is depicted in fig. 6.

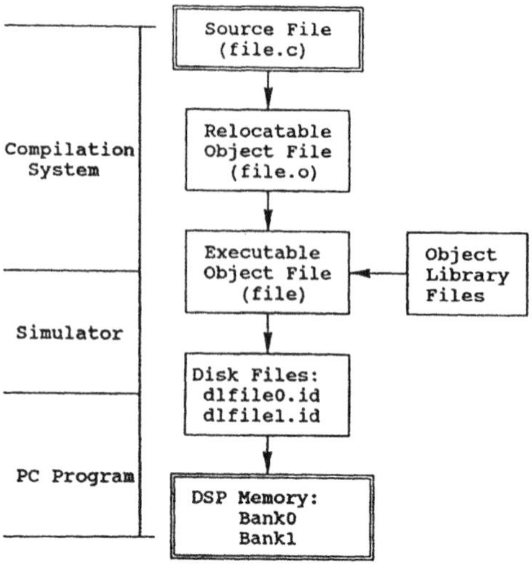

Fig. 6. Flow chart for program development.

Communication between the ADC/DAC system and the signal processor at the moment takes place in the way that the measured and converted feedback signal are transferred via the fast DT-bus shown in fig. 4. The control signal is transferred via the slower PC-bus. Program execution is controlled by interrupts induced on the AD-converter by the PC. In the case of a state-space type controller data flow and control logic is shown in fig. 7.

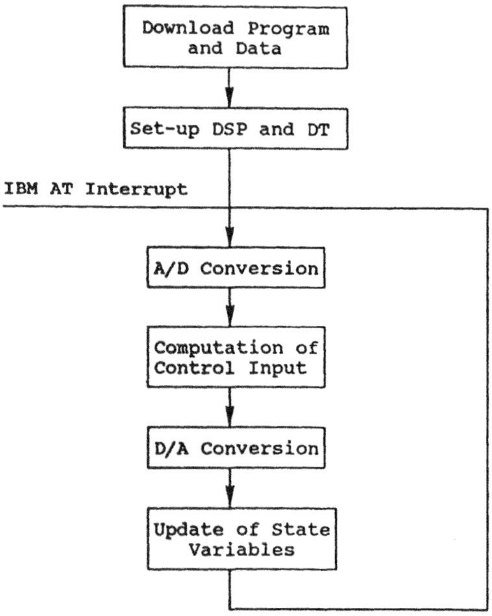

Fig. 7. Data flow chart for a state space based controller.

The arm is equipped with a measuring- and instrumentation system including devices for the measurement of various dynamic variables, devices for the supply of the valve control currents, and a security system for safety. There is the possibility for the measurement of piston position, essential for closed loop position control. This is achieved through an LVDT position transducer integrated in the two designed cylinders. There is at the moment no possibility for measuring tool-centre-point position directly. The system will later be equipped with velocity transducers. Furthermore, the cylinder chamber absolute pressures can be measured with four strain gauge based pressure transducers. The acceleration of each piston is measured with two capacity type accelerometers. The main characteristics of the transducer system is shown in table I.

Table I. The TUD Test Robot Measuring System.

Variable	Transducer	Range	Linearity
Position	Messotron	0-650 mm	±0.5 %
Accelera-tion	Zetra	0-150 g	1.0 %
Pressure	HBM PK4A	0-650 bar	0.5 %

The instrumentation system incorporates a set of servo valve amplifiers providing the current control input to the servo valves with additional current limiting functions and dither application. These are standard manufacturer supplied current amplifiers.

For safety reasons the arm is surrounded by a fence connected to the security system which automatically shuts down oil flow in the case that the perimeter fence door is opened or if an emergency button is pressed.

The spectrum of transducers chosen allow for the investigation of a large variety of control schemes. These can range in complexity from simple single axis PID-controllers to the multivariable and adaptive controller with varying numbers of feedback signals from single axis position feedback to full state feedback controllers. It is one of the aims of the MULTIDOS project to compare the performance and feasibility of these various control schemes.

ADAPTIVE CONTROLLER DESIGN AND RESULTS

The hydraulic test robot can be described by a nonlinear, time-varying state-space equation

$$\dot{X}(t) = f(X(t), u(t)) \tag{7a}$$

$$y(t) = g(X(t)) \tag{7b}$$

where X(t) is an $n \times 1$ dimension state vector.

The dynamics of the test robot consists of actuator dynamics and arm dynamics. The dynamic equations for each actuator includes a servo valve dynamic equation, a cylinder continuity equation, a static valve equation, and a piston force equation. A second order transfer function with an eigenfrequency of 190 Hz can be used to describe approximately the servo valve dynamic response. Assuming stiff links and joints, the system dynamics is then described by a 10'th order nonlinear state-space equation. For each of the two links the 10 state variables include joint position, joint velocity, pressure drop across the piston, and two states describing spool displacement. It is assumed that the pistons are symmetric with motion around the centre position.

By linearizing the nonlinear state-space equation, an approximated linear, time-invariant state space equation is obtained. Because computer control is considered, a discrete-time state-space equation is given:

$$X(k+1) = AX(k) + Bu(k) \tag{8a}$$

$$y(k) = CX(k) \tag{8b}$$

where y(k) is a joint output vector.

An adaptive geometrical compensation scheme was proposed by (Zhou et.al., 1990) for the hydraulic test robot. This control scheme consists of feedforward geometric compensation, computed from the desired robot joint velocity, and an adaptive feedback controller. See fig. 8.

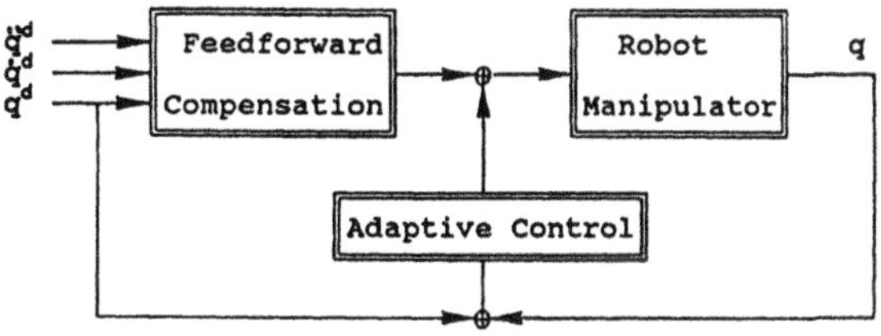

Fig. 8. The adaptive control scheme with feedforward compensation.

Feedforward compensation involves the calculation of a control signal on the basis of the desired joint position, velocity and acceleration. Since a calculation of the complete dynamic equations of the system is too time consuming, an approximate control signal is calculated. As shown in fig. 9 for each link the total valve flow can be divided into two parts the displacement flow and the compressibility and leakage flow. In the case of the test robot it turns out that the dominant flow is the displacement flow, which is a function of piston velocity only.

Therefore, for gross feedforward compensation the control signal may be computed from the desired velocity, neglecting pressure dependence on valve flow and piston motion. This leads to a feedforward compensation scheme which is a purely geometric relation between flow and velocity, hence the name geometric compensation:

$$u^d(t) = \begin{bmatrix} u_1^d \\ u_2^d \end{bmatrix} = \begin{bmatrix} k_{1g} \cdot A_t \cdot \dot{x}_1^d(t) \\ k_{2g} \cdot A_t \cdot \dot{x}_2^d(t) \end{bmatrix} \tag{9}$$

where k_{ig} (i=1, 2) are constant gains specified by the program and $\dot{x}_i^d$ (i=1, 2) are the desired piston velocities. The computations necessary for calculating the above control signals are three multiplications for each joint only.

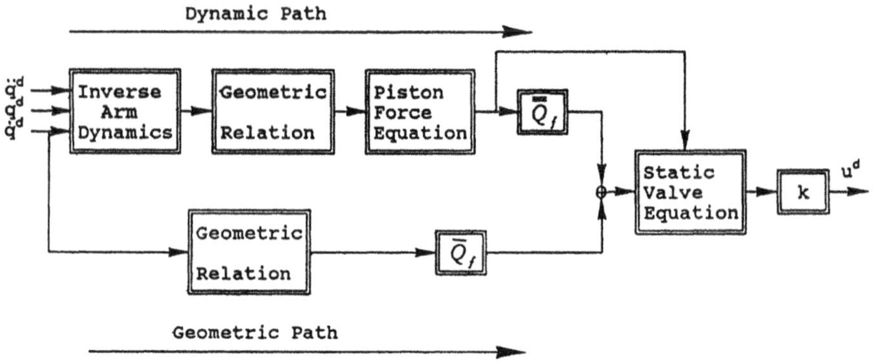

Fig. 9. The signal flow diagram for computing the feedforward compensation.

To secure sufficient dynamic response and for stationary error compensation an **adaptive feedback controller** is implemented. For a detailed account of the adaptive controller we refer to (Zhou, 1989). The controller is an independent joint controller for each link separately.

We assume that the perturbation of each joint can be modelled separately by the model:

$$A(z^{-1})\delta q(k) = B(z^{-1})\delta u(k-1) + \frac{1}{D(z^{-1})}e(k) \quad (10)$$

$$A(z^{-1}) = 1 + a_1 z^{-1} + a_2 z^{-2}$$

$$B(z^{-1}) = b_0 + b_1 z^{-1}$$

$$D(z^{-1}) = 1 + d_1 z^{-1} + d_2 z^{-2}$$

$$\delta q(k) = q^d(k) - q(k)$$

where k denotes the sampling instant. Based on the linear model, the generalized minimum variance controller, can be used to compute the feedback control. The cost function is:

$$J = E\{(\delta q(k+1))^2 + \mu(\delta u(k))^2\} \quad (11)$$

where $E(\cdot)$ is the expectation operator conditional on data up to time k; and μ is the control weight. The minimisation of the cost function gives:

$$\delta u(k) = -\frac{G\delta q(k)}{\hat{D}\hat{B} + \mu/\hat{b}_0} \quad (12)$$

$$G = z(1 - \hat{A}\hat{D}).$$

where "$\hat{}$" denotes the estimates of the unknown polynomials A, B and D by using the recursive generalized least squares algorithm.

Simulation results indicate that the geometric compensation technique together with the proposed adaptive controller gives an overall controller with good static and dynamic performance. In fig. 10 is shown a comparison of simulations with complete and geometric compensation. The maximum angular position error is less than 5% over the whole range from minimum to maximum piston position. The desired motion is a simultaneous motion of both links with piece-wise constant acceleration. It is seen that the control signal in this motion saturates, but still the controller manages good dynamic performance.

The mathematical operations for computing the adaptive feedback control input consist of three parts: identification and controller design and updating.

In the identification part there are 6 model parameters to be estimated. The computational expense for the identification of the perturbation model is 9 divisions and 129 multi.+add. operations.

The calculations for updating the control parameters and adjusting the control input include 2 divisions and 29 multi.+add. operations.

Including the identification calculations, the calculations for computing the feedback control input require a total of 11 divisions and 158 multi.+add. operations for one joint. Assuming that one division takes 20 instruction cycles, all calculations require approximately 378 multi.+add. operations for one joint.

Adding the computational expense of the feedforward nominal control, it requires about 762 multi.+add. operations for two joints. The DSP can perform one multi.+add. operation of the form: a = b + c*d, in one instruction. A multiplication operation or an addition operation can be considered as special case of the multi.+add. operation.

Assuming that two memory fetches and one memory store are required for each operation and one memory fetch and one memory store take one instruction, respectively, all calculations required for computing the control inputs of the two joints require about 3048 instructions. Thus the proposed control scheme can be performed in about 0.5 $m$sec..

CONCLUSIONS.

As an important result of the MULTIDOS project a very fast two-link hydraulic digitally controlled test robot manipulator with a maximum payload of 50 kg and a maximum speed of about 3.5 m/s in the tool point has

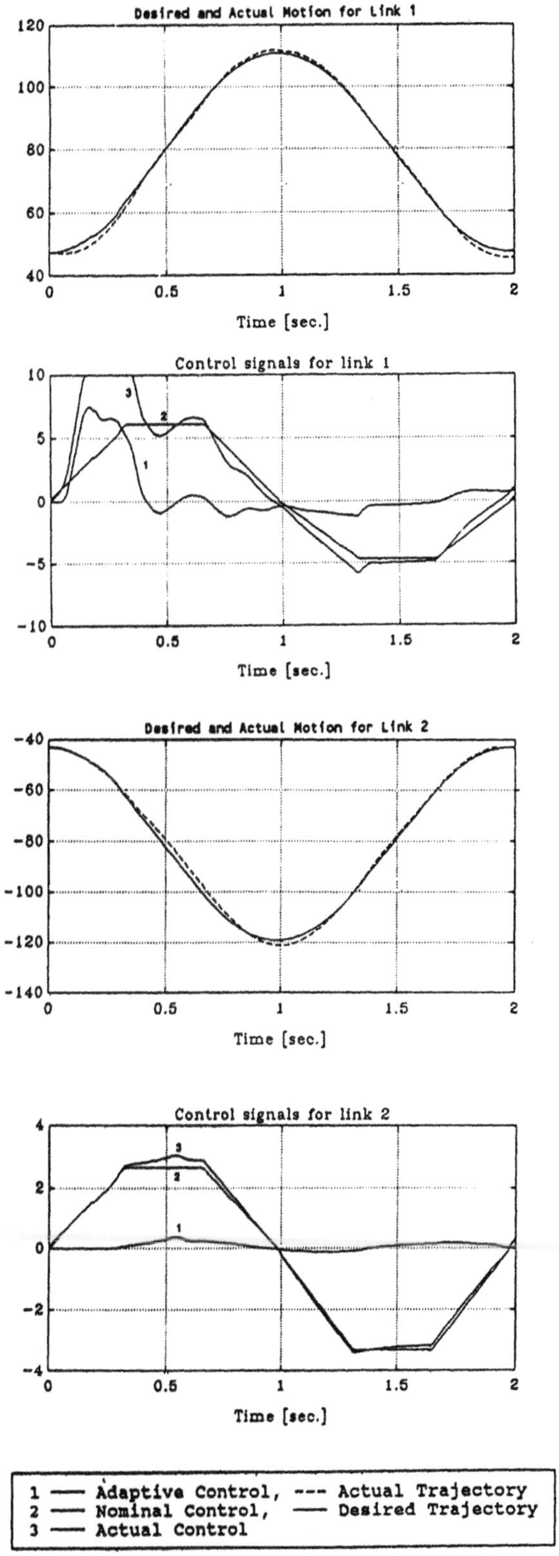

Fig. 10. The simulation results.

been developed and implemented. The robot is used for further design and evaluation of hydraulic actuators, multivariable and adaptive control schemes and for the implementation of digital algorithms on a digital signal processor.

The obtained simulation results using an adaptive perturbation control approach may be regarded as promising. Therefore this will be investigated experimentally in the future research work.

REFERENCES.

Conrad,F., P.H.Sørensen, E.Trostmann, J.J.Zhou (1990a). Design and Digital Control of a Fast Hydraulic Test Robot Manipulator. Proc. of the 9th BHRA International Conference on Fluid Power, Cambridge, England, 25-27 April.

Conrad,F., P.H.Sørensen, E.Trostmann, J.J.Zhou (1990b). Signal Processors for the Control of a Multivariable Hydraulic Robot System. Proc. of the 3th Bath International Fluid Power Workshop, Bath, England, 13-15 September.

Conrad,F., P.H.Sørensen, E.Trostmann and J.J.Zhou (1990c). CAD-Methods for the Design and Implementation of a Fast Digital Controlled Hydraulic Test Robot Manipulator. Proc. of 21'st International Symposium on Industrial Robots, 24-25 October, Copenhagen.

Zhou,J. (1989). Adaptive Control and Application to Hydraulic Systems. Ph.D.-thesis. Control Engineering Institute, Technical University of Denmark, Lyngby.

Zhou,J., F.Conrad, and P.H.Sørensen (1990). An Adaptive Geometric Compensation Control Scheme for Hydraulic Manipulators. Proc. of IEEE International Workshop on INTELLIGENT MOTION CONTROL, 20-22 August, Bogazici University, Istanbul, Turkey.

IMPLEMENTATION OF IMPEDANCE CONTROL USING SLIDING MODE THEORY

Z. Lu and A. A. Goldenberg

Robotics and Automation Laboratory, Dept., of Mechanical Engineering, University of Toronto, Toronto, Ontario, Canada, M5T 1A4

Abstract. This paper presents an approach to the implementation of impedance control using sliding mode control theory. Sliding surface is determined by the form of targeted impedance. Based on the dynamic model in the Cartesian coordinate, a sliding mode controller is designed. The achievement of targeted impedance and preservation of stability in the presence of model uncertainties and external disturbances are the key issues in the design. Under the sliding mode, the system trajectories are forced to reach and to remain on the sliding surface so that the desired targeted impedance is achieved. The controller has been experimentally verified on a RAL direct-drive arm. The arm has two degree of freedom with direct drive parallel five-bar link. The primary results indicate that the approach taken is valid. The results of the experiment are presented in this paper.

Keywords. Variable structure control; impedance control; force control; bang-bang control; robust control; robots.

INTRODUCTION

There are two approaches to compliant motion control of manipulators; force control, and impedance control. In force control, forces are commanded along those directions constrained by the environment and positions are commanded along other directions in which the manipulator is free to move. In impedance control, the relationship (called targeted impedance) between the position and the force of a robot on each constrained direction is controlled in order to obtain a desired dynamic response when the robot is in contact with an environment.

The present paper is concerned with the impedance control of robotic arms. The specific objective of this paper is to develop a robust control scheme for the implementation of the targeted impedance. The achievement of the targeted impedance and the preservation of the stability in the presence of bounded model uncertainties and external disturbance are of main concern in the design of the control scheme.

In the robotic literature, various forms of impedance control has been implemented (Hogan, 1985; Kazerooni, 1986; Whitney, 1987). Most of them can be divided into two groups: one, basically a PD position controller, does not use the dynamic model; the other one uses the exact model to decouple the robot into a set of linear systems. Since the robot impedance depends on its configuration, it is impossible for a PD position controller with a constant gain to implement the desired impedance in its whole working space (Anderson and Spong, 1987). As to the second one, an exact model is needed in the controller to compensate the nonlinear part of the model, so that the desired impedance can be achieved. In practice, however, obtaining the exact model of a robot is difficult. In addition, since there will be more chances of collision between the robot and the environment resulting from the uncertainty in the constrained work space, the effect of external disturbances on the robot is more severe. Therefore, a chosen impedance controller should be robust to the uncertainties in the system parameters and exogenous disturbances. The goal of this paper is to develop such a robust impedance controller to satisfy the above requirements.

In this paper, a sliding mode impedance controller has been developed. The impetus to develop a sliding mode impedance controller is its distinct stability in the presence of uncertainties. When the system is in the sliding mode, the trajectories of a robot system are forced to reach and be maintained on the sliding surface, which is determined by the targeted impedance, so that the targeted impedance is realized without need of an exact model of the robot. The problem of chattering in the sliding mode control is alleviated by using an integral and proportional function in the control inputs. The merits of this control scheme are: (1) the high-frequency components in the control input are reduced and the performance in the vicinity of the sliding surface is improved; (2) the robustness of stability on the sliding surface is ensured.

In the following, we first present the theoretical background of the design of the sliding mode impedance controller. Then we give a description on the experimental setup and a discussion on the experimental results. At last a brief conclusion is given.

DESIGN OF CONTROLLER

Robot manipulator tasks are usually specified in the Cartesian space. In the context of impedance control, the target impedance of a robot is always expressed in the Cartesian space. Hence an effective and direct approach is to handle the trajectory planning, manipulator modeling, and controller design directly in the Cartesian space.

Dynamic Model of Robot

Consider a non-redundant rigid manipulator with n joints, we represent the position and orientation of its end-effector by n independent parameters x_i, $1 \le i \le n$ in a fixed Cartesian space. Then the dynamics of the manipulator, in the absence of friction and disturbances, can be expressed as

$$H(X)\ddot{X} + C(X,\dot{X})\dot{X} + G(X) = F + F_e, \qquad (1)$$

where $H(X)$ is a $n \times n$ symmetric positive definite inertia matrix, $C(X,\dot{X})$ is a $n \times n$ Coriolis centripetal matrix and $G(X)$ is a $n \times 1$ vector of gravity forces, F is a $n \times 1$ vector of applied forces (or torques), F_e is a $n \times 1$ vector of external force acting on the end-effector of the robot.

Targeted Impedance of Manipulator

Impedance control does not attempt to track the motion or force trajectories directly but rather attempts to regulate the relationship between the position and force called the desired mechanical impedance or targeted impedance. The general form of targeted impedance including mass, stiffness and damping terms, can be expressed as

$$M_m\ddot{X} + C_m(\dot{X} - \dot{X}_c) + K_m(X - X_c) = F_e, \qquad (2)$$

where M_m is a $n \times n$ positive definite inertial matrix, C_m and K_m are the $n \times n$ semi definite damping matrix and stiffness matrix respectively. X is the actual position of the robot , X_c is the command position of the robot and F_e is the interaction force acting on the end-effector of the robot.

In this paper, we assume that the targeted impedance has been determined, and we will focus on the achievement of the targeted impedance given in (2).

Sliding Surface

For a n-dimensional state-space system with m-dimensional control inputs, the sliding surface $\sigma(X) = 0$ is defined as [DeCarlo and Zak, 1988]:

$$\sigma(X) = [\sigma_1(X), \cdots, \sigma_m(X)]^T = 0$$

where $\sigma_i(X) = 0$ is the ith switching surface, X is the state vector. From this equation we know that the sliding surface $\sigma(X) = 0$ is a $(n - m)$-dimensional manifold in the system state space. The sliding surface is designed such that the system response restricted to $\sigma(X) = 0$ has a desired dynamic behavior. Since equation (2) expresses the desired dynamic response of robot, it determines the form of sliding surface. But the system of a robot in (1) is a $2n$-dimensional system $(\dot{X}, X)$ with a n-dimensional control input F; the sliding surface is a n-dimensional manifold, which is not equal to the dimension of the targeted impedance in (2); it requires the sliding surface to be a $2n$-dimensional manifold.

In order to implement the targeted impedance of (2), we use the results in [Chan, 1989] to integrate equation (2), that leads to

$$M_m\dot{X} + C_m(X - X_c) + K_m \int (X - X_c)\,dt = \int F_e\,dt, \qquad (3)$$

then we define the sliding surface $\sigma(X) = 0$ as

$$\sigma(X) = \dot{X} + M_m^{-1}C_m(X - X_c) + M_m^{-1}K_m \int (X - X_c)\,dt - M_m^{-1}\int F_e\,dt = 0, \qquad (4)$$

and call $\sigma(X)$ as the impedance measure error. When $\sigma(X)$ goes to zero, that means the state trajectories of the system reach the sliding surface $\sigma(X) = 0$, the dynamic behavior of the robot is determined by equation (4), which also satisfies the targeted impedance (2). Actually, the integration of equation (2) introduces a new n-dimensional state vector $Z = \int X dt$ or $\dot{Z} = X$ in the sliding surface (4); the $2n$-dimensional robot system incorporating with this n-dimensional state vector Z, forms a new $3n$-dimensional system $(\ddot{Z} = \dot{X}, \dot{Z} = X, Z)$, then the sliding surface is $2n$-dimensional, which is identified with the dimension of target impedance. When the targeted impedance only involves damping and stiffness terms, such as

$$C_m(\dot{X} - \dot{X}_c) + K_m(X - X_c) = F_e, \qquad (5)$$

since this equation is n-dimensional, there is no need to introduce a new state vector $Z = \int X dt$. The sliding surface can be directly defined as

$$\sigma(X) = (\dot{X} - \dot{X}_c) + C_m^{-1}K_m(X - X_c) - C_m^{-1}F_e = 0. \qquad (6)$$

Controller Design

The problem to implement the targeted impedance, as seen from the above discussion, is equivalent to finding a control law so that the impedance measure error $\sigma(X)$ will go to zero (see (4) and (2)). In other words, the control law assures the existence of a sliding mode (that means the trajectories of system are forced to reach and be maintained on the sliding surface) in the system.

The sufficient conditions for the existence of sliding mode can be briefly expressed as: there exists a positive definite function $V(t,\sigma,X)$ with respect to σ, i.e., (i) $V(t,X,0) = 0$; for all $t \in R^+$ and $X \in R^n$; and (ii) $V(t,\sigma,X) > 0$; for all $\sigma \neq 0$ and $t \in R^+$, $X \in R^n$, and the total time derivative of V for the system (1) satisfies (iii): $\dot{V} < 0$; for all $\sigma \neq 0$. The more strict mathematical description and proof of above conditions are given in [Utkin, 1978].

From above arguments, the design of the controller can be formulated as: first to choose a proper function that satisfies the conditions (i) and (ii), and then to find a control law that makes the derivative of this function satisfy condition (iii).

Control law for the exact dynamic model of robot. We first consider the case; there is no uncertainty in the model of the robot and no disturbance. The control law is

$$F = H\ddot{X}_s + C\dot{X}_s + G(X) - F_e - K_c sgn(\sigma) \qquad (7)$$

where

$$\dot{X}_s = -M_m^{-1}C_m(X - X_c) - M_m^{-1}K_m \int (X - X_c)\,dt + M_m^{-1}\int F_e\,dt,$$

and K_c is a $n \times n$ diagonal matrix with $k_{ci} > 0$, $sgn(\sigma)$ is a $n \times 1$ vector. F_e is the interaction force between the robot and environment and is measured by force sensor.

Proof. We choose the function $V(\sigma,X)$ as

$$V(\sigma,X) = 0.5\sigma^T(X)H(X)\sigma(X), \qquad (8)$$

since the inertial matrix $H(X)$ of a robot is positive definite; the function $V(\sigma, X)$ satisfies the conditions (i) and (ii). Now we prove that its derivative is negative. The differentiation of $V(\sigma,X)$ leads to

$$\dot{V}(\sigma,X) = \sigma^T H\dot{\sigma} + .5\sigma^T \dot{H}\sigma = \sigma^T(H\dot{\sigma} + C\sigma),$$

here the property of skew-symmetry of $(\dot{H} - 2C)$ is used. Substitution of equation (3) and (4) into $\dot{V}$, we get

$$\dot{V}(X,\sigma) = \sigma^T(F + F_e - H\ddot{X}_s - C\dot{X}_s - G(X)). \quad (9)$$

Combining equation (7) and (9) leads to

$$\dot{V}(X, \sigma) = \sigma^T K_c sgn(\sigma) = -\sum_{i=1}^{i=n} k_{ci}\,|\sigma_i| < 0, \quad \sigma \neq 0.$$

The conditions for the existence of the sliding mode are satisfied, thus the trajectories of the system will be restricted on the sliding surface $\sigma(X) = 0$; according to (4) or (6), the targeted impedance is realized by this control scheme. This control law is derived based on the exact model of a robot. In the following we extend our discussion to the robust implementation of impedance control.

Control law for non-exact dynamic model of robot. In most cases, the exact model of robot is not obtained so easily. An estimated model is only used in a practical controller. The control law for a non-exact model of robot is,

$$F = \hat{H}\ddot{X}_s + \hat{C}\dot{X}_s + \hat{G}(X) - F_e - K_v sgn(\sigma), \quad (10)$$

where $\hat{H}(X)$, $\hat{C}(X,\dot{X})$ and $\hat{G}(X)$ refer to the modeled values of matrices $H(X)$, $C(X,\dot{X})$ and $G(X)$ respectively, K_v is

$$K_v = \delta h\,|\ddot{x}_s|_2 + \delta c\,|\dot{x}_s|_2 + \delta g + \alpha \qquad \alpha > 0, \qquad (11)$$

here, δh, δc and δg are the bounds of uncertainty in $H(X)$, $C(X)$ and $G(X)$ respectively. These bounds satisfy the following conditions,

$$|(\hat{H} - H)v|_2 \leq \delta h\,|v|_2;$$

$$|(\hat{C} - C)v|_2 \leq \delta c\,|v|_2;$$

$$|(\hat{G} - G)|_2 \leq \delta g, \qquad (12)$$

where v is any vector with n components, and $|\cdot|_2$ is Euclidean vector norm. In the following, we will show that the condition for the existence of a sliding mode is satisfied by the control law (10).

Proof. Substituting equation (10) into (9) and using the relation of $(X\ Y) \leq |X|_2 |Y|_2$ leads to

$$\dot{V}(X, \sigma) = \sigma^T[(\hat{H} - H)\ddot{X}_s + (\hat{C} - C)\dot{X}_s$$
$$+ (\hat{G} - G) - K_v\, sgn(\sigma)],$$
$$\leq |\sigma|_2(|(\hat{H} - H)\ddot{X}_s|_2 + |(\hat{C} - C)\dot{X}_s|_2$$
$$+ |\hat{G} - G|) - K_v\,|\sigma|_1,$$

where $|\cdot|_1$ is the l_1 norm of a vector. Using the above conditions of uncertainty bounds, we have

$$\dot{V} \leq |\sigma|_2(\delta h\,|\ddot{X}_s|_2 + \delta c\,|\dot{X}_s|_2 + \delta g) - K_v\,|\sigma|_1$$
$$\leq (\delta h\,|\ddot{X}_s|_2 + \delta c\,|\dot{X}_s|_2 + \delta g - K_v)\,|\sigma|_1,$$

Substituting (11) into above equation, we get

$$\dot{V}(X,\sigma) \leq -\alpha\,|\sigma|_1 < 0; \qquad |\sigma| \neq 0.$$

The relation of $|\cdot|_1 \geq |\cdot|_2$ is used in above proof.

The target impedance is realized without the requirement of an exact model in the control scheme; the derivative of $V(\sigma, X)$ is negative, and the conditions of existing a sliding mode are satisfied.

Continuous control law. Discontinuous control laws, such as those given in (7) and (10) will give rise to chattering in the control input along the sliding surface. Chattering is undesirable; it represents a high-frequency signal component in the control input, which may excite unmodeled high-frequency dynamics. In order to avoid this problem, it is necessary to modify the above control laws by introducing a continuous function to approximate the discontinuous function in the neighborhood (boundary layer) of the sliding surface. Several continuous functions to replace the discontinuous sign function can been found in [DeCarlo, 1988]. One of method is to replace $sgn(\sigma)$ with a saturation function $sat(\frac{\sigma}{\varepsilon})$, as suggested in the paper by [Slotine, 1985]. The function $sat(w)$ is defined by

$$sat(w) = \begin{cases} 1; & w > 1, \\ w; & -1 < w < 1, \\ -1; & w < -1. \end{cases} \qquad (13)$$

With the replacement of $sgn(\sigma)$ by $sat(\frac{\sigma}{\varepsilon})$ in (7) or (10), the trajectories are directed toward the switching planes only for $|\sigma| > \varepsilon$, but when $|\sigma| < \varepsilon$, there is no such guarantee; $\dot{V}(X, \sigma)$ is no longer ensured negative. So that the dynamics of the state trajectory inside the $|\sigma| < \varepsilon$ is only an approximation to the desired dynamics on the sliding surface. Here, we introduce an integral and proportional function $F(\sigma)$ to replace the switching function $sgn(\sigma)$. The form of $F(\sigma)$ is

$$F(\sigma) = \begin{cases} sgn(\sigma); & |\sigma| > \varepsilon, \\ K_I\int_{t_i}^{t}\sigma dt + K_p\sigma; & |\sigma| \leq \varepsilon. \end{cases} \qquad (14)$$

where K_I and K_p are $n \times n$ diagonal constant matrices; $k_{pi} = \frac{1}{\varepsilon}$. When $|\sigma| > \varepsilon$, the value of function (14) is equal to that of (13); the system trajectories are restricted within a ε-neighborhood of the sliding surface. On the other hand, when the system trajectories $(t = t_i)$ come into the ε-neighborhood of the sliding surface, the action of the term, $\int_{t_i}^{t}\sigma dt$, will continuously confine the trajectories of the system to the sliding surface; $\sigma = 0$, so that the dynamics inside the $|\sigma| < \varepsilon$ is improved to close to the sliding surface no matter how large a ε we choose; usually a large ε is helpful to reduce the chattering in the control input. Experimental results confirm the effectiveness of such continuous approximations.

The block diagram for the whole control system is shown in Fig. 1. It can be roughly divided into three parts: the controller, the robot, and the environment. In the controller, the impedance measure error σ is computed from equation (4) or (6), and the sliding mode mechanism assures the error σ to be zero under the model uncertainties and disturbance.

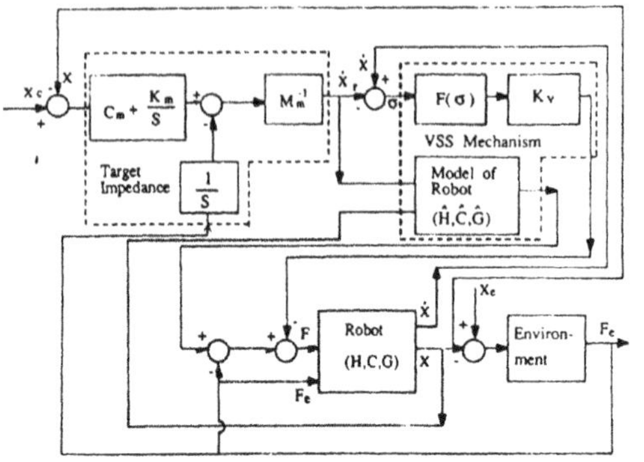

Fig. 1. The structure of the controller.

DESCRIPTION OF EXPERIMENT

Before giving the experimental results we briefly describe here the experiment setup, the dynamic model of the robot, and the design of the impedance controller.

Equipment

The experimental system consists of a two degree-of-freedom arm which is actuated by two dc motors with amplifiers and optical encoders for joint angle measurements, a force sensor to measure the arm tip forces, a planar vertical wall that acts as a constrained surface and a COMPAQ 386/25 personal computer for controlling the arm.

The schematic structure of the experimental arm is shown in Fig. 2. The arm is one of the two direct-drive robot arms developed at the Robotics and Automation Laboratory of the University of Toronto. This arm was designed to be used to evaluate some existing as well as the newly developed control schemes. The work space of the arm is in a horizontal plane. Since the arm is not dynamically mass-balanced, it presents full coupling effects.

The two motors DMA 1050 and DMB 1030 by Yokogawa Inc. providing the torque for joint 1 and joint 2 respectively, are mounted on a rigid supporting frame at the arm base. The maximum torques of the motors are 50 N.m and 30 N.m. The two motors are controlled by power drivers SD1050A-1 and SD1030B-1 respectively.

The joint position is measured by incremental optical encoder integrated in each motor, with a very high resolution of 1,024,000 pos/rev for joint 1 and 655,360 pos/rev for joint 2. The joint velocity is indirectly obtained from the encoders to take the advantages of their high resolution.

The forces acting on the end-effector of the arm are measured by a force sensor JR3(S10/S200) located on the end of link 2. The maximum force in x and y direction the sensor can measure is 100 lbs force. The end-effector consists of a bar and a rolling ball. The bar is mounted on the force sensor and stretches out in x direction. On the tip of the bar there is a roller, the purpose to assemble a roller on the tip is to reduce the friction force when the tip is contacting and moving along the surface of the wall.

The wall is actually an aluminum plane. The wall and arm are mounted on the same table. The surface of the wall is perpendicular to x direction.

Position, velocity and force measurements are sent to the COMPAQ 386/25 for torque computation. The control program was written in the C language. A sampling period of 15 ms. is chosen.

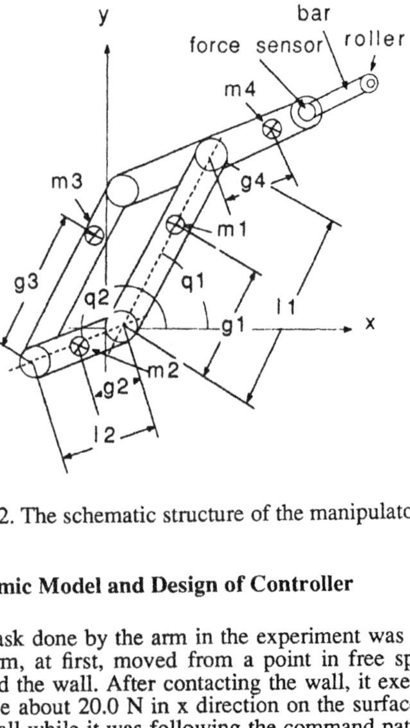

Fig. 2. The schematic structure of the manipulator.

Dynamic Model and Design of Controller

The task done by the arm in the experiment was that the arm, at first, moved from a point in free space toward the wall. After contacting the wall, it exerted a force about 20.0 N in x direction on the surface of the wall while it was following the command path in y direction along the surface of the wall.

Dynamic model. The dynamic model of the manipulator in joint space can be derived from Lagrange's equations as

$$h_{11}\ddot{q}_1 + h_{12}\ddot{q}_2 + c_{12}\dot{q}_2 = \tau_1,$$
$$h_{21}\ddot{q}_1 + h_{22}\ddot{q}_2 + c_{21}\dot{q}_1 = \tau_2,$$

where

$$h_{11} = I_1 + m_1 g_1^2 + I_3 + m_3 g_3^2 + m_4 l_1^2,$$
$$h_{21} = h_{12} = (m_3 l_2 g_3 - m_4 l_1 g_4)\cos(q_2 - q_1),$$
$$h_{22} = I_4 + m_4 g_4^2 + I_2 + m_2 g_2^2 + m_3 l_2^2,$$
$$c_{12} = -(m_3 l_2 g_3 - m_4 l_1 g_4)\sin(q_2 - q_1)\dot{q}_2,$$
$$c_{21} = (m_3 l_2 g_3 - m_4 l_1 g_4)\sin(q_2 - q_1)\dot{q}_1.$$

The values of above parameters are give in the working document (Karunakar and co-workers 1989). The dynamic model in task space can be easily obtained by using the relationship between the task space model and joint space model.

Targeted Impedance. Based on the requirement of the task, the form of targeted impedance is chosen as

$$c_{11}(\dot{x}-\dot{x}_c)+k_{11}(x-x_c)=f_x,$$

$$c_{22}(\dot{y}-\dot{y}_c)+k_{22}(y-y_c)=0,$$

where $c_{11}=c_{22}=50\,N.sec./m$, $k_{11}=k_{22}=500\,N./m$. The values of cii and k_{ii} were chosen to maintain the system stable. Actually, from above equations, the arm was impedance controlled in x direction to avoid arising a large force when the arm contacted the wall, and position controlled in y direction in order to follow the command path closely.

Sliding surface. The sliding surface is determined by above targeted impedance as

$$c_{11}(\dot{x}-\dot{x}_c)+k_{11}(x-x_c)-f_x=0,$$

$$c_{22}(\dot{y}-\dot{y}_c)+k_{22}(y-y_c)=0.$$

Impedance Measure Error. The impedance measure error is

$$\sigma_x=(\dot{x}-\dot{x}_c)+c_{11}^{-1}k_{11}(x-x_c)-c_{11}^{-1}f_x,$$

$$\sigma_y=(\dot{y}-\dot{y}_c)+c_{22}^{-1}k_{22}(y-y_c).$$

EXPERIMENTAL RESULTS

The purpose of this experiment is to demonstrate the performance characteristics of the control system under the effect of practical factors such as unmodeled dynamics, various noises, and disturbances, which are usually neglected in the theoretical analysis.

Command trajectory. The command trajectory are fifth-order polynomials interpolated in task space between starting point $\mathbf{p}_s=[0.47m\ 0.28m]^T$ and ending point $\mathbf{p}_e=[0.57m\ 0.33m]^T$ with a maximum speed of 0.1 m/s and zero velocities and accelerations at the starting and ending points. The command trajectories of the arm tip position on x and y axes are shown by dotted line in Figures 3 and 4.

Actual trajectory. The actual trajectories on x and y axes are shown by solid line in Figures 3 and 4. Before the end-effector contacted the wall, the arm were following the command trajectories in free space for about $t=2.6\,sec.$. At the moment of collision between the arm and wall, the motion of the arm was forced to stop in x direction by the wall, but was continue to follow the command trajectory in y direction along the surface of the wall.

Interaction force. The interaction forces in x and y directions are shown by solid and dotted line respectively in Fig. 5. These forces before the contact were zero. At the moment of the collision, a large collision force in x direction was caused to stop the motion of the arm in a very shot time. The value of the collision force is proportional to the speed of the arm approaching the wall. In the experiment the approaching speed is about 0.1 (m/s). After the collision, the interaction force f_x was back to a small value. The interaction force f_y is contributed by the friction force between the end-effector and the surface of the wall. The value of this force was small in the experiment as a roller located on the tip of the arm (Fig. 2).

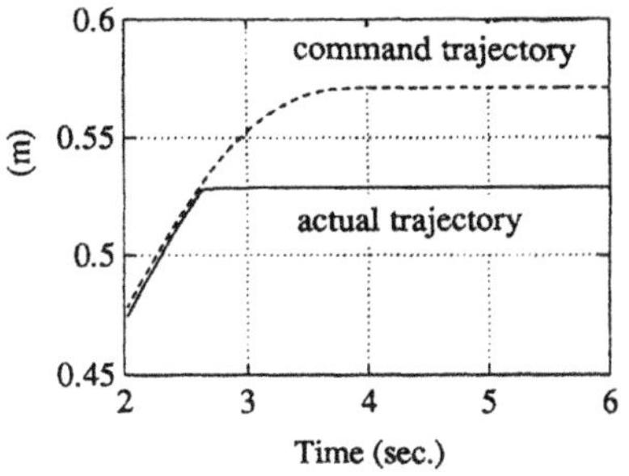

Fig. 3. The trajectories in x direction.

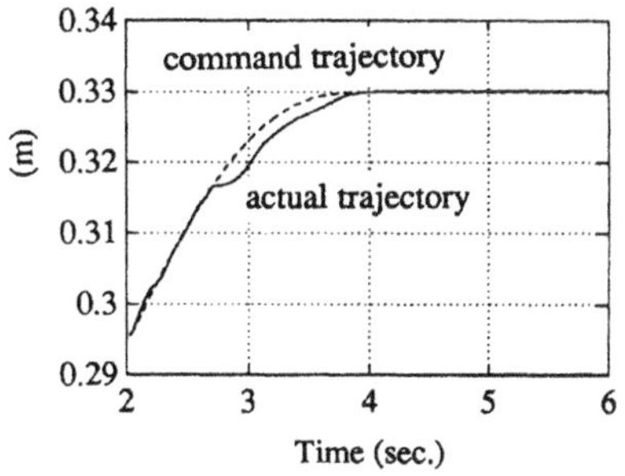

Fig. 4. The trajectories in y direction.

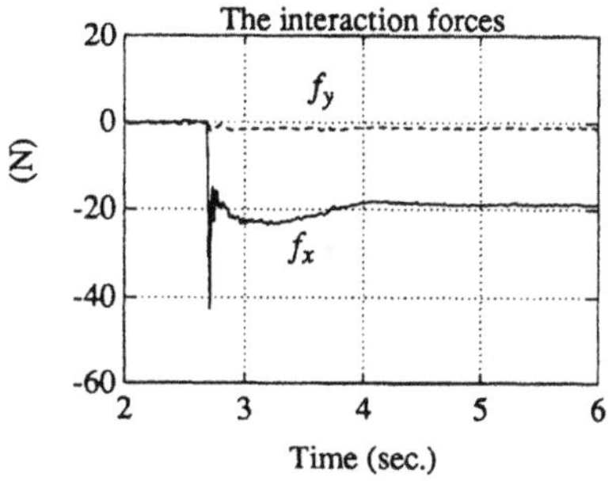

Fig. 5. The interaction forces f_x and f_y.

Impedance measure error. The impedance measure errors on x and y axes are plotted by solid and dotted line respectively in Fig. 6. At first, when the arm was in free motion, both of them were small. At the moment of collision there was a big jump in σ_x but soon back to 0.25 (m/s). About 1 sec. later, its value returned to 0.02 (m/s). Since the force caused by collision had little component in y direction, it affected the motion in y direction very small. The value of σ_y was small (less than 0.05). There are various factors that produce the errors in the system. Particularly in this experiment, the friction force and the noise in force measurement are believed to contribute a significant portion of the tracking errors. The Coulomb frictions at the motor shafts are about $4\sim 7N.m$ at motor 1 and $2\sim 4N.m$ at motor 2. It is difficult to compensate for these completely. An overcompensation for the friction was seen to result in vibration of the arm. The noise in force measurement was about $0.8N$. It brought about tracking errors when robot was in motion.

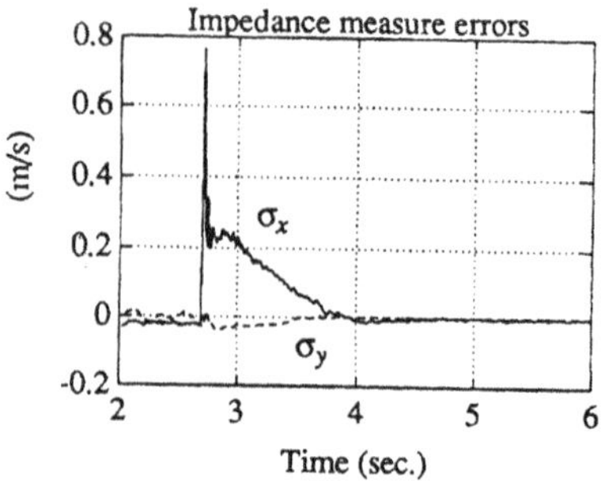

Fig. 6. The impedance measure errors σ_x and σ_y.

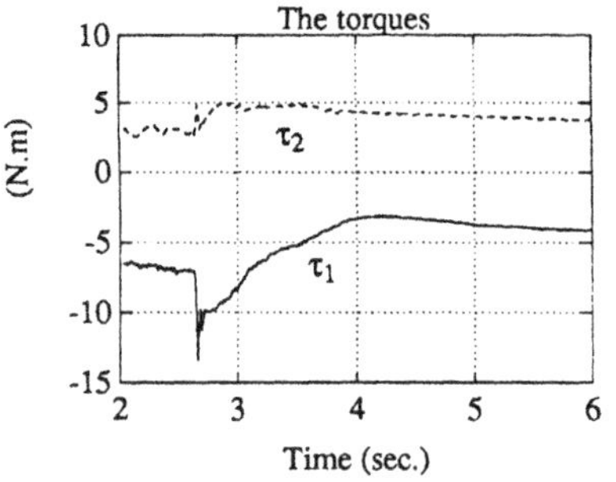

Fig. 7. The control torques τ_1 and τ_2.

Control input. The control inputs τ_1 and τ_2 are shown in Fig. 7. Since a continuous function was used in the controller, there was almost no chattering in τ_1 and τ_2.

CONCLUSION

This paper discusses the problem of implementing the impedance control by using sliding model theory. An impedance controller is developed. Its robustness property is guaranteed by the sliding mode control. The sliding surface is determined by the targeted impedance. The state trajectories of the system are forced to reach and remain on the sliding surface by sliding mode control in the face of model uncertainty and disturbances. A continuous function used in the controller is seen to eliminate the chattering and to maintain the performance in the vicinity of the sliding surface. The experimental results show the validity of the proposed controller for use in a practical situation.

ACKNOWLEDGMENT

The authors are grateful to Jacek Wiercienski and Ron Baker for looking after the design and the fabrication of the arm, to Pawel Kuzan for working on the electronic hardware connection and software interface and to Chris Symczyk for writting control subroutings to make the system ready for experimentation.

REFERENCES

Anderson, R. J., Spong, M. W., 1987, "Hybrid Impedance Control of Robotic Manipulators". *Proc. 1987 IEEE Int. Conf. on Robotics and Automation.pp.1073-1080.*

Chan, S. P. and Gao, W. B., 1989, "Variable Structure Model-Reaching Control Strategy for Robot Manipulators", *Proc. 1989 IEEE Int. Conf. on Robotics Automat.*, pp. 1504-1508.

DeCarlo, R. A. and Zak, S. H., 1988, "Variable Structure Control of Nonlinear Multivariable Systems: A Tutorial", *Proc. IEEE* Vol. 76, NO. 3, pp. 212-232.

Goldenberg, A. A., 1988, "Implementation of Force and Impedance Control in Robot Manipulators", *Proc. 1988 IEEE Int. Conf. on Robotics Automat.*, pp. 1626-1632.

Hogan, N., "Impedance Control: An Approach to Manipulation: Part I- Theory", "Part II - Implementation", "Part III- Applications", *ASME J. of Dynamic Systems, Measurement and Control,* Nov., 107, pp. 1-24.

Karunakar, S., and co-workers, 1989, "RAL Direct-Drive Arms Project Progress Report on Phase-3", *RAL Working Document,* No. DD-4/89.

Kazerooni, H., Sheridan, T.B., Houpt, P.K., 1986, "Robust Compliant Motion for Manipulators: Part I - The Fundamental Concepts of Compliant Motion", "Part II - Design Method", *IEEE Journal of Robotics and Automation,* Vol. RA-2, No. 2, pp. 83-105.

Slotine, J. J. E. and Li, W., 1988, "Adaptive Manipulator Control: A Case Study", *IEEE Trans. on Automation Control,* Vol. 33, No. 11, pp. 995-1003.

Utkin, V.I., 1978, "Sliding Modes and Their Application in Variable Structured System", Moscow, *Soviet Union: MIR Publishers* .

Whitney, D. E., 1987, "Historical Perspective and State of the Art in Robot Force Control", *Int. J. of Robotics Research,* Vol. 6. No. 1, Spring, pp. 3-14.

ORTHONORMAL FUNCTIONS IN IDENTIFICATION AND ADAPTIVE CONTROL

G. A. Dumont, Ye Fu and A.-L. Elshafei

Dept. of Electrical Engineering, c/o Pulp and Paper Centre, 2385 East Mall, University of British Columbia, Vancouver, BC, Canada V6T 1W5

Abstract. This paper discusses the use of orthonormal functions to represent the dynamics of processes, both for identification and for adaptive control. Orthonormal functions that form a complete base in $L_2[0,\infty)$ are appropriate for describing open-loop stable systems. We discuss the properties of Laguerre-based estimates, the influence of unmodelled dynamics, and the choice of the optimal time scale. Finally, we present some adaptive predictive control schemes based on this representation.

INTRODUCTION

Although much work in the identification and adaptive control fields using the ARMAX representation has been done over the past twenty years or so, there has recently been an increased interest in alternatives, possibly more flexible representations, such as impulse or step response models. In this paper, we look at the use of series of orthonormal series representations.

Interest in orthonormal series representations of signals goes back to the classical work of Wiener and Leeduring the 1930's on network synthesis using Laguerre functions. This classical work is best summarized in the book by Lee (1960). In the 1950's and 1960's, many papers dealing with the use of laguerre functions for representing transients signals appeared, e.g. Ward (1954), Head (1956), Clowes (1965). The problem of choosing the time scale was further considered by Parks (1971). During the 1970's and the 1980's, Laguerre functions were primarily studied from the point of view of data compression. Exceptions are the use of digital Laguerre functions for synthesis of digital filters by King and Paraskevopoulos (1971), and the papers by Rusev (1978), and Nurges and Jaaksoo (1982) and Nurges (1987).

In the early 1980's, our own work in adaptive process control, and in particular on the problem of varying dead time led us to look at the classical work on Laguerre functions. This led to the development of an auto-tuning scheme for PID controllers (Zervos et al., 1985; Dumont et al., 1985; Zervos et al., 1988), and later to the development of novel adaptive control schemes (Dumont and Zervos, 1986; Zervos and Dumont, 1988a-c). Those schemes have since been successfully tested on various industrial plants, see for instance Dumont et al. (1990). The relative success of this approach has recently prompted other researchers to take a closer look at the theoretical properties of identification of stable systems by Laguerre networks, Wahlberg (1989), Makila (1990).

This paper presents some of the background behind our work and extends some of our previous results. We emphasize the advantages of using Laguerre series representations for modelling stable dynamical systems, with discussions on the influence of unmodelled dynamics, choice of the Laguerre filter time scale, and uncertainty quantification. Laguerre-based adaptive predictive control laws are then presented. Finally, we discuss our current industrial eperience with the various schemes we have so far developed.

LAGUERRE FUNCTIONS

Hilbert Spaces

In this paper, we limit ourselves to signals in the Lebesgue space $L_2[0,\infty)$of square-integrable functions $y(t)$:

$$\int_0^\infty y^2(t)dt < \infty \quad (1)$$

The impulse response of an asymptotically stable system lies in L_2. Such a system is said to be L_2 stable, and in this paper we limit ourselves to such systems. Defining the inner product as

$$\langle x, y\rangle = \int_0^\infty x(t)y(t)dt \quad (2)$$

we can define an orthonormal sequence $\{\psi_i\}$ if it satisfies:

$$\langle \psi_i, \psi_j\rangle = \delta_{i,j} \quad (3)$$

where $\delta_{i,j}$ is the Dirac delta-function. In additon, the set $\{\psi_i\}$ is said to be complete if every $y \in L_2[0,\infty)$ can be expanded into the series:

$$y = \sum_{i=1}^{\infty} \langle y, \psi_i \rangle \psi_i \tag{4}$$

then, for any real $\epsilon > 0$, there exists a positive integer N such that

$$\left\| y - \sum_{i=1}^{N} \langle y, \psi_i \rangle \psi_i \right\| < \epsilon \tag{5}$$

An inner product space with the property of completeness is called a Hilbert space. Then, we can state the classical Riesz-Fisher theorem, Zemanian (1968):.

Theorem

Let $\{\psi_i\}$ be a complete orthonormal set as specified above, and let $\{c_i\}$ be a sequence of real numbers such that $\sum_{i=0}^{\infty} |c_i|^2$ converges. Then, there exists a unique $y \in L_2[0,\infty)$ such that $c_i = \langle y, \psi_i \rangle$. Consequently,

$$y = \sum_{i=1}^{\infty} c_i \psi_i \tag{6}$$

in the sense of convergence in $L_2[0,\infty)$.

The above means that the impulse response of any stable system can be represented with arbitrary accuracy by a series of orthonormal functions, i.e. for any function $y \in L_2[0,\infty)$, and a positive ϵ, then there exists $N > 0$ such that

$$\int_0^{\infty} \left| y(t) - \sum_{i=1}^{N} c_i l_i(t) \right|^2 dt < \epsilon \tag{7}$$

In this paper, we use Laguerre filters as orthonormal basis. We made that choice for two reasons. Our field of applications is process control, meaning that we primarily have to deal with generally well damped systems presenting long and varying time delays. In this context, Laguerre filters because of their similaity with Pade approximants are efficient. Laguerre filters are also attractive from an implementation point of view.

Continuous Laguerre Functions

In the continuous time domain, the Laguerre fonctions are defined as:

$$l_i(t) = \sqrt{2p}\frac{\exp(pt)}{(i-1)!}\frac{d^{i-1}}{dt^{i-1}}\exp(-2pt) \tag{8}$$

In the Laplace transform domain, the Laguerre filters are defined as:

$$L_{i(s)} = \sqrt{2p}\frac{(s-p)^{i-1}}{(s+p)^i} \quad , i = 1, \cdots, N \tag{9}$$

These Laguerre filters are readily implemented by a simple ladder network. To use those functions, one must discretize them. With fast sampling, one could simply use the δ-operator (Middleton and Goodwin, 1989) which can be shown to preserve orthonormality. Alternatively, by discretizing each block, a discrete-time state-space representation is obtained (Zervos and Dumont, 1988):

$$\begin{aligned} \underline{l}(t+1) &= A\underline{l}(t) + \underline{b}u(t) \\ y(t) &= \underline{c}^T\ \underline{l}(t) \end{aligned} \tag{10}$$

In this representation, A is a lower triangular $N \times N$ matrix, and $\underline{b}$ an $N \times 1$ vector. The vector $\underline{c}$ contains the Laguerre gains needed to represent th signal $y(t)$. it is also known as the Laguerre spectrum of the system.

Discrete Laguerre Functions

The transfer function of discrete Laguerre functions can be described as:

$$L_k(q^{-1}) = \frac{\sqrt{1-a^2}}{q-a}\left(\frac{1-aq}{q-a}\right)^{k-1} \tag{11}$$

where in terms of Z-transform, ($q^{-1} = z^{-1}$). This set is also known as the set of Meizner polynomials, see King and Paraskevopoulos (1971). It can be proved orthonormal and complete in L_2. Here, a discrete signal $y(n)$ ($n = 1, 2, \cdots$) is said to be in L_2 iff:

$$\sum_{n=0}^{\infty} y^2(n) < \infty \tag{12}$$

IDENTIFICATION

Correlation Method

From the Riesz-Fisher theorem, we know that

$$c_i = < y, l_i > = \int_0^{\infty} y(t) l_i(t) dt \tag{13}$$

This is however only true when the outputs of the Laguerre filters are orthonormal, i.e. when the input is zero-mean white noise with unit variance. Then, the cross-correlation between the plant output and the output of the ith Laguerre filter gives the ith Laguerre gain. This is the classical method for obtaining the Laguerre gains. In practice, if we have n data points obtained with sampling interval T, this suggests the estimate:

$$\hat{c}_i = \frac{1}{n}\sum_{k=1}^{n} y(kT) l_i(kT) \tag{14}$$

If the process output is corrupted by a stationary random process $w(.)$ independent of u, i.e.

$$y(t) = \sum_{1}^{\infty} c_i l_i(t) + w(t) \quad (15)$$

Then, given u, the expected value of $\hat{c}_i$ over the random variable w is:

$$\begin{aligned} E[\hat{c}_i|u(\cdot)] &= \frac{1}{n}\sum_{j=1}^{\infty} c_j \sum_{k=1}^{n} l_j(kT) l_i(kT) \\ &+ \frac{1}{n}\sum_{k=1}^{n} l_j(kT) E(w) \end{aligned} \quad (16)$$

It is easy to see that it will be equal to c_i only if u is a white noise process, and if n is large. Thus, this method is not really suitable to exploit the results from a single, finite-duration experiment. In practice, we shall use least-squares identification.

Least-Squares

As seen before, any stable plant can be represented exactly by an infinite Laguerre series. In practice, we shall obviously use a truncated series. Let the actual plant be described by the following discrete representation:

$$y(t) = G_0(q)u(t) + G_u(q)u(t) + w(t) \quad (17)$$

where q denotes the usual forward shift operator. G_0 and G_u respectively denote the nominal model and the unmodelled dynamics, and are given by:

$$\begin{aligned} G_0(q) &= \sum_{i=1}^{N} c_i L_i(q) \\ G_u(q) &= \sum_{i=N+1}^{\infty} c_i L_i(q) \end{aligned} \quad (18)$$

For all practical purposes, the unmodelled dynamics can also be represented by a large, but truncated Laguerre series. Indeed, as seen before, given the finite word length of computers, the actual plant can be exactly represented within the accuracy of the system by a large, but truncated series. Introducing the following notation,

$$\begin{aligned} \underline{\phi}^T(k) &= [l_1(kT), \cdots, l_N(kT)] \\ \underline{\phi}_u^T(k) &= [l_1(kT), \cdots, l_N(kT)] \\ \Phi^T &= [\underline{\phi}(T) \cdots, \underline{\phi}(nT)] \\ \Phi_u^T &= \left[\underline{\phi}_u(T) \cdots, \underline{\phi}_u(nT)\right] \\ \underline{Y}^T &= [y(T), \cdots, y(nT)] \\ \underline{W}^T &= [w(T), \cdots, w(nT)] \end{aligned} \quad (19)$$

we can write

$$\underline{Y} = \Phi^T \underline{\theta}_0 + \Phi_u^T \underline{\theta}_u + \underline{W} \quad (20)$$

The least-squares estimate $\hat{\underline{\theta}}$ of $\underline{\theta}_0$ is then given by

$$\hat{\underline{\theta}} = \left[\Phi\Phi^T\right]^{-1}\Phi\underline{Y} \quad (21)$$

Taking the expectation yields:

$$\begin{aligned} E\left[\hat{\underline{\theta}}\right] &= \underline{\theta}_0 + E\left\{\left[\Phi\Phi^T\right]^{-1}\Phi\Phi_u\theta_u\right\} \\ &+ E\left\{\left[\Phi\Phi^T\right]^{-1}\Phi\underline{W}\right\} \end{aligned} \quad (22)$$

If the input is white, because of the orthonormality of the Laguerre filters, the second term on the rhs of the above equation will be zero. Furthermore, if $w(\cdot)$ is zero mean, the third term will vanish. Thus, contrary to the ARMAX representation, here we can get unbiased estimates of the nominal plant in the presence of unmodelled dynamics and coloured noise. Thus,

$$E\left(\hat{G}\right) = G_0 \quad (23)$$

The above property, together with the fact that the regressor does not depend on the process output presents several advantages when trying to quantify the uncertainty of the estimated model as in Goodwin and Salgado (1989). Rewriting Eq. (17) as

$$y(t) = G_0(q) + \eta(t) \quad (24)$$

where $\eta(t)$ denotes the modelling error, and from Goodwin and Salgado (1989), we can write

$$\begin{aligned} &\left|\hat{G} - G\right| = \\ &|G_u|^2 + \left|\hat{G} - G_0\right|^2 - 2\Re\left[\left(\hat{G} - G_0\right)G_u^*\right] \end{aligned} \quad (25)$$

where

$$\begin{aligned} \hat{G} - G_0 &= v^T P \frac{1}{T}\int_0^T \phi(t)\eta(t)dt \\ v^T &= [l_1(s)\cdots l_N(s)]|_{s=j\omega} \end{aligned} \quad (26)$$

Here, v^T is known a-priori and does not need to be approximated as in the transfer function approach. If a white noise input is used, then the above expression is very small, or zero, and only the a-priori knowldege on unmodelled dynamics will then be left in Eq. (25). However, this will not be true for non-white input, or in closed-loop operation. Wahlberg (1989) further studies the error bounds, consistency and statistical properties of this estimate. In particular it is shown that the mapping $(1 + ae^{i\omega})/(e^{i\omega} + a)$ improves the condition number of the covariance matrix, and that because of the implicit assumption that the system is of low-pass nature, the asymptotic covariance of the estimate will decrease for high frequencies. Makila (1990) studies both Hankel norm and L^∞ approximations of stable systems by Laguerre filters, and in particular derives achievable L^∞ error norms for delay systems.

Choice of Time Scale

Obviously, the choice of the time scale of the Laguerre filters is affecting the quality of the approximation as well as the rate of convergence of the laguerre spectrum. This problem has been looked at in the past, in particular in Clowes (1965), Parks (1971) and Clement (1982) for continuous Laguerre functions, and by Nurges (1987) for discrete Laguerre functions. Here, we extend the work of Parks (1971) to discrete Laguerre functions and systems with time delay. Given a stable discrete transfer function

$$G(q^{-1}) = \frac{B(q^{-1})q^{-d}}{A(q^{-1})} \tag{27}$$

where d is the unknown delay, we can express it by using Laguerre series as:

$$G(q^{-1}) = \sum_{k=1}^{\infty} c_k L_k(q^{-1}) \tag{28}$$

where L_k are the previously defined discrete Laguerre functions. and c_k are the Laguerre coefficients. The following theorem gives the Laguerre time scale a which minimizes the index:

$$J = \sum_{k=1}^{\infty} k c_k^2 \tag{29}$$

The minimization of this performance index forces rapid convergence of the Laguerre spectrum and gives an analytical solution for the optimal a. For a proof of the theorem, see Fu and Dumont (1990).

Theorem

Let $h(n)$ be the impulse response of the discrete plant, and as in Parks (1971) define M_1, M_2 as:

$$M_1 = \frac{1}{\|h\|^2}\sum_{n=0}^{\infty} n h^2(n)$$
$$M_2 = \frac{1}{\|h\|^2}\sum_{n=0}^{\infty} n[\Delta h(n)]^2 \tag{30}$$

M_1, M_2 are constants for a given discrete plant. The optimum Laguerre time scale which minimizes the index J in Eq. (26) is given by

$$a_{opt} = \frac{2M_1 - 1 - M_2}{2M_2 - 1 + \sqrt{4M_1M_2 - M_2^2 - 2M_2}} \tag{31}$$

and the optimal value of the index is:

$$J_{opt} = \frac{1 + \sqrt{4M_1M_2 - M_2^2 - 2M_2}}{2} \tag{32}$$

As the delay increases, M_1, M_2 will increase and the optimum time scale will generally decrease. To better see this, let M_1^d, and M_2^d be the values of M_1, and M_2 with delay d. We can then write

$$M_1^d = M_1^1 + (d-1)$$
$$M_2^d = M_2^d + 2(d-1)(1-M_3) \tag{33}$$

where

$$M_3 = \frac{1}{\|h\|^2}\sum_{n=0}^{\infty} h(n)h(n+1) \tag{34}$$

Example: Given a continuous plant

$$G(s) = \frac{p}{s+p} \tag{35}$$

and a sampling interval T, the discrete transfer function of the plant is

$$G(q^{-1}) = \frac{(1-e^{-pT})q^{-1}}{1-e^{-pT}q^{-1}} \tag{36}$$

It can be shown that if the delay is $d = 1$, the optimum Laguerre time scale is $a_{opt} = e^{-pT}$ and $J_{opt} = 1$. This results is as expected. When the delay is large, a_{opt} will decrease and J_{opt} will increase. Fig. 1 shows the values of the cost function J vs the Laguerre time scale a for different delays d. Here, the discrete plant pole is $p = 0.6$.The minimum cost function value is also shown by the line across those index functions. For this example, the value of M_3 is 0.6. According to Fu and Dumont(1990), the optimum Laguerre time scale when the delay goes to infinity is

$$\lim_{d\to\infty} a = \frac{M_3}{1+\sqrt{1-M_3^2}} = \frac{1}{3} \tag{37}$$

This is clearly verified by Fig, 1.

For a second order stable continuous transfer function, the optimum Laguerre time scale depends on the location of the poles and zeros, and the sampling frequency. The optimum Laguerre time scale will be in [-1,1] and if the second order system is poorly damped, a_{opt} will very likely be negative. See Fu and Dumont (1990) for further discussions.

CONTROL

Extended-Horizon Control

In Zervos and Dumont (1988), a simple predictive control law was developed in order to build an adaptive controller based on the above representation. In an adaptive mode, the Laguerre spectrum gain vector is estimated e.g. via least-squares. Proofs of global stability of the resulting scheme are given in Zervos and Dumont (1988). In Dumont et al (1990), results of an industrial test and a discussion of the robustness of that adaptive control scheme are presented. The representation of the unmodelled dynamics by a Laguerre series is equivalent to assuming that the unmodelled dynamics impulse response is bounded by a decaying exponential, a common assumption. A non-adaptive robustness result show that the closed-loop system is stable if the difference between the actual system and the estimated one is SPR. Because of the flexibilty of the representation by Laguerre filters, it is easy to increase the complexity of the model until satisfactory control is achieved within the desired bandwidth. Another possibilty that remains to be explored is to exploit the convergence pattern of the Laguerre gains to estimate uncertainty bounds.

Generalized Predictive Control

Consider a j-step ahead predictor, the future output is

$$\hat{y}(t+j) = \underline{c}^T \underline{l}(t+j) \tag{38}$$

where

$$\begin{aligned} \underline{l}(t+j) &= A^j \underline{l}(t) + A^{j-1}\underline{b}u(t) + A^{j-2}\underline{b}u(t) \\ &+ \cdots + \underline{b}u(t+j-1) \end{aligned} \tag{39}$$

The above predictor is utilized in the performance index

$$J = \sum_{j=1}^{n_2} (\hat{y}(t+j) - y_{sp}(t+j))^2 + \beta \Delta u^2(t) \tag{40}$$

The control law which minimizes the above performance index is

$$\begin{aligned} &u(t) = \\ &\epsilon\Big[m(y(t) - \hat{y}(t) - y_{sp}(t)) + \underline{k}^T \underline{l}(t) + \beta u(t-1)\Big] \end{aligned} \tag{41}$$

where

$$\begin{aligned} \underline{k}^T &= -\sum_{i=0}^{n_2-1} s_i \underline{c}^T A^{i+1} \\ \epsilon &= \frac{1}{\underline{s}^T \underline{s} + \beta} \\ m &= \sum_{i=0}^{n_2-1} s_i \end{aligned} \tag{42}$$

$\underline{s}^T = [s_0 s_1 \cdots s_{n_2-1}]$, s_i is the ith step-response coefficient and $y_{sp}$is the set-point.

The above control law corresponds to GPC with a one-step control horizon, a prediction horizon of n_2 steps and a control weighting factor β. Assuming an exact plant-model match, the closed-loop stability can be assessed by the following theorem.

Theorem

Let the system be described by Eq. (10) and controlled using the control law in Eq. (37). Assume that the open-loop system is stable, $\beta \geq 0$, and ϵ is sufficiently small. Then, the closed—loop system is stable.

In an adaptive scheme, it is not true that $y(t) = \hat{y}(t)$. The closed —loop system is

$$\begin{aligned} \begin{bmatrix} \underline{l}(t+1) \\ u(t+1) \end{bmatrix} &= \begin{bmatrix} A & \underline{b} \\ \epsilon \underline{k}^T A & \epsilon \underline{k}^T \underline{b} + \epsilon\beta \end{bmatrix} \begin{bmatrix} \underline{l}(t) \\ u(t) \end{bmatrix} \\ &+ \begin{bmatrix} \underline{0} \\ m\epsilon \end{bmatrix} (y(t) - \hat{y}(t) - y_{sp}(t)) \end{aligned} \tag{43}$$

The above theorem together with Theorem 2.1 in Payne (1987) can be used to prove the global stability of the proposed predictive controller in the adaptive scheme.

Theorem

Let the system be described by Eq. (10), then, provided that the least-squares algorithm is used to find $\hat{c}$ such that $\dim(\underline{\hat{c}}) = dim(\underline{c})$ and that the stability conditions stated in the previous theorem are satisfied, then the closed-loop system is globally stable.

For further discussion, see Elshafei et al. (1991).

CONCLUSIONS

This paper has discussed the use of Laguerre series representation in identification and control. Several advantages can be derived when using such models, in particular on systems dominated by long and varying time delays. The Laguerre representation has been shown to have attractive properties in the presence of unmodelled dynamics. A way of choosing the Laguerre time scale that provides rapid convergence of the Laguerre spectrum has been discussed. Adaptive control schemes based on this representation have also been briefly discussed. Our industrial experience with such schemes has so far been very promising. Indeed several applications are currently being developed, both in North-America and Europe, and a commercial implementation may soon be available.

ACKNOWLEDGMENTS

This research was partially supported by the Natural Sciences and Engineering Research Council of Canada under Grants No. A-5960 and No.

IRC0100044 and by the Pulp and Paper Research Institute of Canada.

LITERATURE CITED

Clement, P.R. (1982). Laguerre functions in signal analysis and parameter identification. *J. Franklin Inst.*, **313**, No. 2, 85–95.

Clowes, G.J. (1965). Choice of the time-scaling factor for linear system approximation using orthonormal Laguerre functions. *IEEE Trans. Autom. Control*, **AC–10**, 487–489.

Dumont, G.A., and C.C. Zervos (1986). Adaptive control using Laguerre functions. *IFAC Workshop on Adaptive Control and Signal Proc.*, Lund, Sweden.

Dumont, G.A., C.C. Zervos, and P.R. Belanger (1985). Automatic tuning of industrial PID controllers. *1985 American Control Conf.*, Boston, MA.

Dumont, G.A., C.C. Zervos, and G. Pageau (1990). Laguerre-based adaptive control of pH in an industrial bleach plant extraction stage. *Automatica*, **26**, 781–787.

Fu, Y., and G.A. Dumont (1990). Choice of time scale for discrete Laguerre approximations. *Submitted for publication.*

Elshafei, A-L, G.A. Dumont, and A. Elnaggar (1991). Perturbation analysis of GPC with one-step control horizon. *Automatica*, accepted for publication.

Goodwin, G.C., and M.E. Salgado (1989). A stochastic embedding approach for quantifying uncertainty in the estimation of restricted complexity models. *Int. J. Adapt. Control and Signal Proc.*, **3**, 357–374.

Head, J.W. (1956). Approximation to transients by means of Laguerre series. *Proc. Camb. Phil. Soc.*, 640–651.

Lee, Y.W. (1960). *Statistical Theory of Communication.* Wiley, New York.

King, R.E., and P.M. Paraskevopoulos (1971). Digital Laguerre filters. *Circuit Theory and Appl.*, **5**, 81–91.

Makila, P.M. (1990). Approximation of stable systems by Laguerre filters. *Automatica*, **26**, 333–345.

Nurges, U. (1987). Laguerre models in problems of approxmation and identification of discrete systems. *Avtomatika i Telemekhanika*, No. 3, 88–96.

Nurges, U., and U. Jaaksoo (1982). Laguerre state equations of multivariable discrete time systems. *Proc. 8th IFAC World Congress*, Kyoto, Japan.

Parks, T.W. (1971). Choice of time scale in Laguerre approximations using signal measurements. *IEEE Trans. Autom. Control*, **AC–16**, 511–513.

Payne, A.N. (1987). Stability result with application to adaptive control.*Int. J. Control*, **46**, 249–262.

Rusev, P.K. (1978). On the representation of analytic functions by Laguerre series. *Dokl. Akad. Nauk SSSR*, **240**,

Wahlberg, B. (1989). System identification using high-order models, revisited. *28th IEEE Conf. on Decision and Control*, Tampa, Flda.

Ward, E.E. (1954). The calculation of transients in dynamical systems.*Proc. Camb. Phil. Soc.*, 49–59.

Zemanian, A. H. (1968). *Generalized Integral Transformations.* 1987 Republication, Dover, New York.

Zervos, C.C. (1988). Adaptive control based on orthonormal series representation. *Ph. D. Dissertation*, University of British Columbia, Vancouver, Canada.

Zervos, C.C., P.R. Belanger, and G.A. Dumont (1985). On PID controller tuning using orthonormal series identification. *IFAC Workshop on Adaptive Control of Chemical Processes*, Frankfurt-am-Main, FRG.

Zervos, C.C., P.R. Belanger, and G.A. Dumont (1988). Controller tuning using orthonormal series identification. *Automatica*, **24**, 165–175.

Zervos, C.C., and G.A. Dumont, (1988a). Deterministic adaptive control based on Laguerre series representation. *Int. J. Control*, **48**, 2333–2359.

Zervos, C.C., and G.A. Dumont, (1988b). Laguerre functions in stochastic self-tuning control. *1988 IFAC Workshop on Robust Adaptive Control*, Newcastle, Australia.

Zervos, C.C., and G.A. Dumont, (1988c). Multivariable self-tuning control based on Laguerre series representation. *Proc. of Int. Workshop on Adaptive Strategies for Industrial Use,*, Banff, Canada.

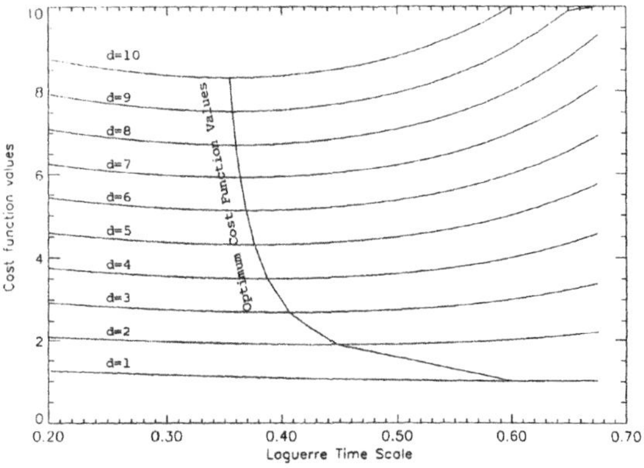

Cost Function Values for different Delay Cases

Fig.1

ADAPTIVE CONTROL FOR A ROBOT ARM AND A ROBOT DRIVE SYSTEM WITH ELASTICITY

T. Suehiro*, F. Ohkawa, K. Kurosu**, and T. Yamashita****

**Electronics Division, Mechanics and Electronics Research Institute, Fukuoka Industrial Technology Center, 3-6-1 Norimatsu, Yahatanishi-ku, Kitakyusyu-shi, 807 Japan*
***Dept. of Control Engineering, Kyushu Institute of Technology, 1-1 Sensui-cho Tobata-ku, Kitakyushu-shi, 804 Japan*

Abstract. First, this paper presents an application of a Model Reference Adaptive Control System (MRACS) to the position control, and the force feedback control of a flexible robot arm. Second, this paper investigates the application of an MRACS to the speed control, and the position control of a DC servo system, and a Direct-Drive(DD) servo system with a flexible transmission element. The flexible robot arm is simply modelled as a flexible cantilever beam. The servo systems with a flexible transmission element are treated as a servo mechanism with a one degree of freedom oscillatory load. Using discrete-time mathematical models as a lumped parameter system, the discrete MRACS is designed for each discrete model. Experimental results have shown that the control performance is good and the MRACS is effective.

KeyWords. Model Reference Adaptive Control; Flexible Robot Arm; Flexible Transmission Element; Position Control; Force Control; Speed Control; d.c. motors; d.d. motors.

INTRODUCTION

It is always necessary to develop techniques which can control systems precisely and with high performance. In manipulators and servo mechanisms, the dynamic parameters, for example, the friction of a reduction gear and the inertia moment of loads, are time-varying and unknown in advance. So it is difficult to control those systems accurately by conventional control methods. Many researchers have been studying effective control methods for these systems. Model Reference Adaptive Control System(MRACS) is one of the control methods which can effectively control systems with such unknown dynamic characteristics (Ohkawa,1982; Tomizuka,1983; Goodwin,1984).

Furthermore, recently it is required to develop light robot arms and robot drive systems which can handle a heavy payload and operate at a high speed to increase ability and economize operating energy. This is a difficult problem since light robot arms and joints tend to vibrate at operation due to the lack of rigidity. On the other hand, a flexible(light) arm has the advantage that it is good for force feedback control, as it is easy to sense the force applied to the tip due to its elasticity(compliance), unlike rigid arms. Thus the flexible arm can be used as a robot arm which can easily perform intelligent tasks such as assembly, polishing and so on.

Direct-Drive(DD) servo systems have been developed recently to solve such problems as friction and backlash. However, manipulators are apt to vibrate, because of their structure like a cantilever beam. Moreover, parameters such as load inertia are variable. Therefore, it is important to consider the vibration problem when making the mechanical structures lighter and smaller for both DC and DD servo systems.

Many papers have discussed the analysis and synthesis of the flexible systems described above. Flexible arms have been treated as a distributed parameter system in most of the papers. These methods are complicated in construction and control strategy(Book,1983; Fukuda,1984; Sakawa, 1985). In most industrial robots and servo mechanisms, only the end of the arm or the state for loads is controlled. Thus, it is important and practical to develop simple methods, such as a lumped parameter system. However, only a few simple methods applying a discrete MRACS have ever been presented for the control of the flexible systems.

First, this paper describes an MRACS design for the position control and the force feedback control of a flexible robot arm. The mathematical formulations of the flexible beam are modelled as a lumped parameter system from the viewpoint of a simple structure and the adaptive control. The models are discretized through a sampling period in order to use a microcomputer as a controller. The discrete MRACS is designed for the discrete models. The experimental studies for the position control were carried out using a CCD camera and an image processor to measure the position of the tip of the arm. In the experiment of force feedback control, a load cell was used to measure the force acting on the end of the arm. A dc servo motor was used as an actuator. Next, an MRACS for the speed control, and the position control of a DC servo system, and a DD servo system with a flexible transmission element is designed. The speed and the position of the load were measured with a rotary encoder in the experiment. The MRACS with the relatively simple structure and algorithm can be designed. Experimental results have shown the high control performance and the effectiveness of the constructed MRACS.

POSITION CONTROL OF A FLEXIBLE ARM

Mathematical Model

We consider a simple flexible arm as shown in Fig.1. The flexible arm is modelled as a flexible cantilever beam with a concentrated mass at the tip simulating the payload. The beam is driven by an actuator at the base and revolves about a joint. It vibrates only in bending but not in torsion. The higher order vibrations are small and are neglected in the model. Thus, only the first mode of the arm dynamics are treated. By letting the controlled variable be only the position of the tip of the arm, a discrete model as a lumped parameter system is used to construct a simple adaptive control system.

θ (rotation angle of the arm) and α (deflection angle of the arm) are selected as generalized coordinates. The analytical dynamical model for the arm can be obtained by applying the Lagrange equations.

$$\ddot{\theta} = -\frac{D_{\theta}}{J}\dot{\theta} + \frac{KL^{2}}{2J}\sin 2\alpha + \frac{\tau}{J} \quad (1)$$

$$\ddot{\alpha} = \frac{D_\theta}{J}\dot{\theta} - \frac{KL^2}{2}\left(\frac{1}{J} + \frac{1}{ML^2+I}\right)\sin 2\alpha - \frac{\tau}{J} \tag{2}$$

where the parameters are:

L :length of the arm I :moment of inertia about payload
J :moment of inertia along axis M :mass of the payload
K :elastic modulus of the arm D_θ :viscosity coefficient
τ :applied torque to the arm

By defining the controlled variable(output) as

$$y = \theta + \alpha \tag{3}$$

Then, using Eq.(1), (2) and (3) yields

$$\begin{aligned} \ddot{y} &= \ddot{\theta} + \ddot{\alpha} \\ &= -\frac{1}{2}\frac{KL^2}{ML^2+I}\sin 2\alpha \\ &\fallingdotseq -\omega_n^2\alpha \\ &= -\omega_n^2(y-\theta) \end{aligned} \tag{4}$$

$$\omega_n^2 = \frac{KL^2}{ML^2+I}$$

From Eq.(4), the transfer function from $\theta(t)$ to $y(t)$ is

$$G(s) = \frac{Y(s)}{\theta(s)} = \frac{\omega_n^2}{s^2 + 2\zeta\omega_n s + \omega_n^2} \tag{5}$$

where the damping term ζ is added to model the air friction, the viscosity of the arm, and so on.

The time constant of the dc servo motor is small enough to be ignored. The transfer function of the motor can be expressed approximately by

$$F(s) = k_1/s \tag{6}$$

The pulse transfer function from the input (applied voltage to the motor) $u(k)$ to the output $y(k)$ becomes

$$\begin{aligned} W(z) &= Y(z)/U(z) \\ &= Z\left\{\frac{1-e^{-Ts}}{s}\cdot F(s)\cdot k_2\cdot G(s)\right\} \\ &= \frac{c_1z^2 + c_2z + c_3}{z^3 - d_1z^2 - d_2z - d_3} \end{aligned} \tag{7}$$

where
k :sampling index T :sampling time
k_2:reduction ratio of the gear
c_i and d_i ($i=1\sim3$) are constants related to T, k_1, k_2, ζ, and ω_n.

Adaptive Control

The MRACS is constructed to compensate for variations and uncertainties of parameters. The MRACS for the system described by the following equation is designed(Ohkawa, 1982).

$$y(k+1) = \sum_{j=1}^{n}\{a_j y(k+1-j) + b_j u(k+1-d-j)\} \tag{8}$$

where a_j, b_j are unknown parameters and the d represents the time delay. The output error is defined as

$$e(k) = x_m(k) - y(k) \tag{9}$$

where $x_m(k)$ is the reference response (reference model output). Now, the adaptive control input $u(k)$ is given by

$$u(k) = \left\{x_m(k+d+1) - \sum_{j=1}^{n} f_j(k)\, x_m(k+d+1-j) - \sum_{j=2}^{n} g_j(k)\, u(k+1-j) - q(k)\right\} / g_1(k) \tag{10}$$

where $f_j(k)$, $g_j(k)$ are adaptive gains, and $q(k)$ is an auxiliary input signal.
The adaptive algorithm to assure $\lim_{k\to\infty} e(k) \to 0$ is

$$\Phi(k+1) = \Phi(k) - \frac{\rho\,\varepsilon(k+1)W(k)^t}{\gamma + W(k)^tW(K)} \tag{11}$$

where

$$0 < \rho < 2 \qquad \gamma > 0$$

$$\begin{aligned} \Phi(k) = [\, & f_1(k) - a_1, \cdots\cdots, f_n(k) - a_n, \\ & g_1(k) - b_1, \cdots\cdots, g_n(k) - b_n] \end{aligned} \tag{12}$$

$$\begin{aligned} \varepsilon(k+1) = {} & e(k+1) - \sum_{j=1}^{n} f_j(k-d)\, e(k+1-j) \\ & - \delta(k) - q(k-d) \end{aligned} \tag{13}$$

$$\delta(k) = \{\Phi(k-d) - \Phi(k)\}\, W(k) \tag{14}$$

$$\begin{aligned} W(k) = [\, & y(k), \cdots\cdots, y(k+1-n), \\ & u(k-d), \cdots\cdots, u(k+1-d-n)]^t \end{aligned} \tag{15}$$

The auxiliary input signal is

$$q(k) = -\sum_{j=1}^{n} h_j(k)\, e(k+1-j) - \sum_{j=1}^{n} h_{n+j}(k)\, q(k-j) \tag{16}$$

where $h_j(k)$ are known variable gains related to $f_j(k)$.

Experiment

Figure 2 shows the configuration of the experimental system. The arm is made of a stainless steel plate, whose length 750 mm, width is 30 mm, and thickness is 2.5 mm. The arm has a mass(0.5 kg, 0.75 kg) to simulate a payload at the end. The arm is driven by an actuator through a harmonic gear (reduction ratio is 80:1). The actuator is a dc servo motor(rated torque is 0.314 N·m) which is controlled by a microcomputer through a servopack. A CCD camera and an image processor are used to measure the position at the tip of the arm. Adaptive algorithm is calculated by a NEC PC-9801vm microcomputer.

Figure 3 shows the open loop time response of the flexible arm. The approximate values of ζ and ω_n are 0.0034 and 12.6, respectively. Thus, it can be found that the vibration of the tip does not converge easily on account of the arm's flexibility. Experiment was carried out under the following conditions:

$n=3 \quad d=2 \quad \rho=\gamma=1 \quad T=0.16$ sec
$f_1(0) = f_2(0) = g_2(0) = g_3(0) = 0$
$f_3(0) = 1 \quad g_1(0) = -5$

where, taking into account the computation time of the computer and the response time of the sensors, a time delay is introduced. The remaining initial conditions were zero. Figure 4 shows one of the adaptive control results. It can be seen from this figure that the tip of the arm tracks to the reference response very well. Experiments under several different conditions have also shown the similar results.

FORCE FEEDBACK CONTROL OF FLEXIBLE ARM

Mathematical Model

Figure 5 shows a simple model of a force feedback control system. In this section, we deal with the problem of controlling the force generated at the tip when the flexible beam as shown in Fig.1 presses against a fixed object. Assuming that the elasticity of the flexible arm is represented by the mechanical system(compliant system) shown in Fig.6. The following transfer function is derived on condition that the armature inductance of a dc motor is small enough to be negligible.

$$G(s) = Y(s)/U(s) = \frac{\mu_1}{\lambda_1' s^2 + \lambda_2' s + \lambda_3'} = \frac{K}{s^2 + 2\zeta\omega_n s + \omega_n^2} \tag{17}$$

where

$\lambda_1' = R_a m/(K_T N)$
$\lambda_2' = (R_a c/K_T + K_m)/N$
$\lambda_3' = R_a k_s/(K_T N)$
$\zeta^2 = (c R_a + K_m K_T)^2/(4 k_s m R_a^2)$
$\omega_n^2 = k_s/m \quad K = 2 E I K_T N/(m R_a L^2)$
$y(=F)$:force acting on the tip of the arm
u :input voltage $\quad R_a$:armature resistance of the motor
K_m :induced voltage constant $\quad K_T$:torque constant
N :reduction ratio of the gear
m :coefficient of mass and inertia
c :coefficient of viscosity $\quad k_s$:coefficient of elasticity
EI :modulus of the flexural stiffness of the arm
L :length of the arm

The pulse transfer function is given by

$$W(z) = Y(z)/U(z) = Z\left\{\frac{1-e^{-Ts}}{s} \cdot G(s)\right\} = \frac{c_1 z + c_2}{z^2 - d_1 z - d_2} \tag{18}$$

where c_i and $d_i (i=1,2)$ are constants related to T, ζ, ω_n and K.

Experiment

The experiments were carried out for two cases where the position of the acting point is (Case 1) constant, and (Case 2) variable as shown in Fig.7 and 8. The drive system and the controller are shown in Fig.3. The arm is made of a stainless steel plate which is 800 mm long, 30 mm wide, and 2 mm thick. A load cell is used to measure the force acting on the tip of the beam. The conditions of the experiment are as follows:

$n = 2 \quad d = 2 \quad \rho = \gamma = 1$
Case 1 : $f_1(0) = 0.8 \quad f_2(0) = 0.2$
$g_1(0) = -0.5 \quad g_2(0) = 0 \quad T = 0.18$ sec
Case 2 : $f_1(0) = 0.8 \quad f_2(0) = 0.185$
$g_1(0) = -0.1 \quad g_2(0) = 0.1 \quad T = 0.12$ sec

The adaptive control results are shown in Fig.9 and 10. It can be seen from these figures that the output tracks the reference response very well in both case 1 and case 2.

DC SERVO SYSTEM

Mathematical Model

Here, the model of the DC servo system with a flexible transmission element is simplified as a servo mechanism with an oscillatory load of one degree of freedom as shown in Fig.11. Under the assumption that the armature inductance of the dc motor can be neglected, the state equation and the output equation for this model can be written by

$$\dot{x} = A_C x + B_C u \tag{19}$$

$$y = C_C x \tag{20}$$

where

$$A_C = \begin{bmatrix} a_{11} & a_{12} & a_{13} & a_{14} \\ 1 & 0 & 0 & 0 \\ a_{31} & a_{32} & a_{33} & a_{34} \\ 0 & 0 & 1 & 0 \end{bmatrix} \qquad B_C = \begin{bmatrix} b_{11} \\ 0 \\ 0 \\ 0 \end{bmatrix}$$

$x = [\dot{\theta}_M \ \ \theta_M \ \ \dot{\theta}_L \ \ \theta_L]^t$

u :input voltage $\quad C_C$:observation matrix $\in R^{1\times 4}$
$a_{11} = -\frac{1}{J_M}\left(\frac{K_T \eta K_E}{R_E} + \frac{D_S \eta}{R_G^2} + D_M\right)$
$a_{12} = -K_S \eta/(J_M R_G^2) \quad a_{13} = D_S \eta/(J_M R_G)$
$a_{14} = K_S \eta/(J_M R_G) \quad a_{31} = D_S/(J_L R_G)$
$a_{32} = K_S/(J_L R_G) \quad a_{33} = -(D_L + D_S)/J_L$
$a_{34} = -K_S/J_L \quad b_{11} = K_T \eta K_A/(J_M R_E)$
$R_E = R_A + K_A K_{AF}$
θ_M :rotation angle of the motor
$\dot{\theta}_M$:rotation speed of the motor
θ_L :rotation angle of the load
$\dot{\theta}_L$:rotation speed of the load
J_M :moment of inertia of the motor
D_M :coefficient of viscous friction of the motor
K_S :spring constant
D_S :coefficient of viscous damping of the spring element
J_L :moment of inertia of the load
D_L :coefficient of viscous friction of the load
R_G :reduction ratio of the gear
η :efficiency modulus of the gear
R_A :armature resistance of the motor
K_T :torque constant
K_E :induced voltage constant
K_A :gain of the current amplifier
K_{AF}:gain of the current feedback

Discretizing Eq.(19) and (20) by using the difference method, the following equations are obtained.

$$x(k+1) = A x(k) + B u(k) \tag{21}$$

$$y(k) = C x(k) \tag{22}$$

where

$A = I + T A_C \quad B = T B_C \quad C = C_C$
$x(k) = [\dot{\theta}_M(k) \ \ \theta_M(k) \ \ \dot{\theta}_L(k) \ \ \theta_L(k)]^t$
I :unity matrix $\in R^{4\times 4}$
k :sampling index $\quad T$:sampling time

By letting $C = [0 \ \ 0 \ \ 1 \ \ 0]$ and $[0 \ \ 0 \ \ 0 \ \ 1]$, the pulse transfer functions for $\dot{\theta}_L(k)$ and $\theta_L(k)$ become

$$W_S(z) = Y(z)/U(z) = \frac{d_1 z + d_2}{z^3 - c_1 z^2 - c_2 z - c_3} \tag{23}$$

$$W_P(z) = Y(z)/U(z) = \frac{d_1' z + d_2'}{z^4 - c_1' z^3 - c_2' z^2 - c_3' z - c_4'} \tag{24}$$

where $c_i (i=1 \sim 3)$, $d_j (j=1,2)$, $c_p' (p=1 \sim 4)$ and $d_q' (q=1,2)$ are constants related to $a_{st} (s=1,3,\ t=1 \sim 4)$, b_{11}, T.

Experiment

The set-up of the experimental system is shown in Fig.12. The actuator is a dc servo motor(rated torque is 0.314 N•m) which is controlled by a microcomputer through a servo (current) amplifier. The load is driven by the motor through a harmonic gear(reduction ratio is 80:1) and a torsional spring. The speed and the position of the load is measured with a rotary encoder(12bit). A microcomputer NEC PC-9801vm is used to execute the adaptive algorithm and calculate the control input. Three springs are used in the experiments, and the characteristics are given in Table 1. The loads are made variable by using attachable and removal disks as shown in Fig.13. Figure 14 shows the the open loop time response of the system. It can be found from these figures that the characteristics of the response depend on the conditions of springs and loads. Figure 15 shows the response of load speed under classical PI control. Adaptive control experiments were carried out under the following conditions:

(I)Speed control

$n=3 \quad d=3 \quad \rho=\gamma=1 \quad T=0.1\ \text{sec}$
$f_1(0)=1.5 \quad f_2(0)=0.5 \quad f_3(0)=0$
$g_1(0)=1.5 \quad g_2(0)=-1 \quad g_3(0)=-1$

(II)Position control
$n=4 \quad d=4 \quad \rho=\gamma=1 \quad T=0.1\ \text{sec}$
$f_1(0)=1 \quad f_2(0)=f_3(0)=f_4(0)=0$
$g_1(0)=-1 \quad g_2(0)=g_3(0)=g_4(0)=0$

Figure 16 and 17 show the adaptive control results of the speed and the position, respectively. The followings can be observed from Fig.15, 16 and 17: changing the conditions of the spring and the loads in Fig.15 from (a) to (b), the output can't track the reference under PI control. However, under adaptive control, the output follows the reference of the model very well even under different conditions.

DD SERVO SYSTEM

Mathematical Model

Here, we consider a DD servo system with a flexible transmission element as in Fig.11, however, since the system is DD, there is no reduction element. The transfer functions for the speed and the position are obtained respectively as

$$G_S(s) = Y(s)/U(s) = \frac{\mu_1 s + \mu_2}{s^2 + \lambda_1 s + \lambda_2} \tag{25}$$

$$G_P(s) = Y(s)/U(s) = \frac{\mu_1 s + \mu_2}{s\,(s^2 + \lambda_1 s + \lambda_2)} \tag{26}$$

Since the time constant of the motor is very small, the relation between u and $\dot{\theta}_M$ can be approximated as follow

$$\dot{\theta}_M = K_1 u \tag{27}$$

where

K_1 :gain constant
$\lambda_1 = (D_L + D_S)/J_L \qquad \lambda_2 = K_S/J_L$
$\mu_1 = K_1 D_S/J_L \qquad \mu_2 = K_1 K_S/J_L$

Then, the transfer functions are

$$W_S(z) = Y(z)/U(z) = Z\left\{\frac{1-e^{-Ts}}{s}\cdot G_S(s)\right\} = \frac{\xi_1 z + \xi_2}{z^2 + \nu_1 z + \nu_2} \tag{28}$$

$$W_P(z) = Y(z)/U(z) = \frac{\xi_1' z^2 + \xi_2' z + \xi_3'}{z^3 - \nu_1' z^2 - \nu_2' z - \nu_3'} \tag{29}$$

where $\nu_i(i=1,2)$, $\xi_j(j=1,2)$, $\nu_p'(p=1\sim3)$ and $\xi_q'(q=1\sim3)$ are constants related to $\lambda_s(s=1,2)$, $\mu_t(t=1,2)$, T.

Experiment

The structure of the experimental system is similar to Fig.12 except that the dc motor and harmonic gear in the figure are exchanged for a dd motor(rated torque 34.32 N・m). The conditions of the experiment are as follows:

(I)Speed control
$n=3 \quad d=1 \quad \rho=\gamma=1 \quad T=0.05\ \text{sec}$
$f_1(0)=1.5 \quad f_2(0)=-2 \quad g_1(0)=2 \quad g_2(0)=g_3(0)=0$

(II)Position control
$n=4 \quad d=2 \quad \rho=\gamma=1 \quad T=0.1\ \text{sec}$
$f_1(0)=1 \quad f_2(0)=f_3(0)=0$
$g_1(0)=1 \quad g_2(0)=-0.5 \quad g_3(0)=g_4(0)=g_5(0)=0$

Using the springs and loads shown in the preceding section in this experiment, the open loop responses similar to Fig.14 were obtained. Figure 18 and 19 show the adaptive control results of the speed and the position, respectively. It can be seen that even if we change the conditions for the springs and the loads, the speed and the position outputs track to the reference very well.

CONCLUSION

The discrete MRACS has been applied to the control of a flexible arm and servo systems with a flexible transmission element. By using discrete-time single-input single-output model, an adaptive control system with a simple algorithm and structure can be designed, and good control response can be obtained despite low sampling rates. Furthermore, though it is difficult to control those flexible systems well, the use of the MRACS makes it possible to maintain high performance over a wide range of dynamical variation and uncertainty. Thus MRACS is an alternative control method for lighter and smaller robot and servo systems.

REFERENCES

Book, W.J. and Majette, M.C (1983). Controller design for flexible, distributed parameter and frequency domain techniques. Trans. ASME. J. Dyn. Syst. Meas. Control. 105. pp245-254.

Fukuda, T. and Kuribayashi, Y (1984). Precise positioning and vibration control of flexible robotic arms with consideration of joint elasticity. Proc. of IECON'84. pp410-415.

Goodwin, G.C. and K.S.Sin (1984). Adaptive filtering prediction and control. Prentice-Hall.

Ohkawa, F. and Yonezawa, Y (1982). A model reference adaptive control system for discrete multivariable systems with time delay. Int. J. of Control. 36-6. pp925-934.

Sakawa, Y., Matsuno, F and Fukushima, S (1985). Modeling and feedback control of a flexible arm. J. Robotic Systems. 2-4. pp453-472.

Tomizuka, M. and Horwitz, R (1983). Model reference adaptive control of mechanical manipulators. IFAC Adaptive Systems in Control and Signal Processing. pp27-32.

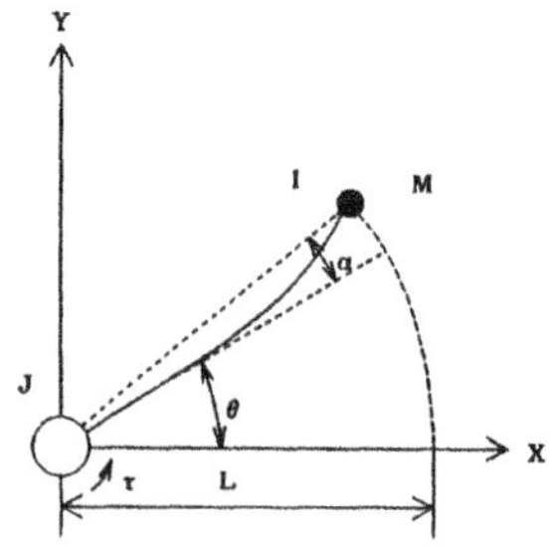

Fig.1. A flexible arm

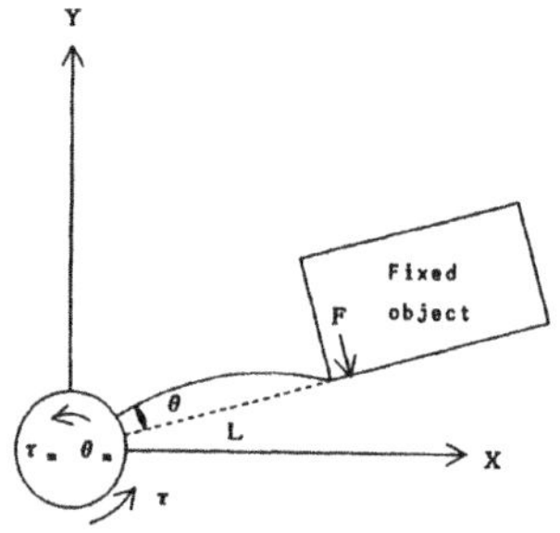

Fig.5. Simple model of a force feedback control system for a flexible arm

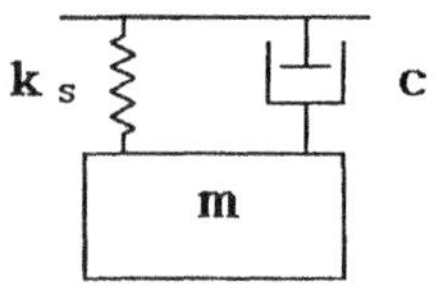

Fig.6. A compliant mechanism

TABLE 1 Characteristics of the Springs

Spring	Length [mm]	Outside diameter [mm]	Diameter of wire [mm]	Compressive spring constant [N/mm] (Specified value)	Torsional spring constant [N/rad] (Calculated value)	Natural frequency [rad/sec] (Experimental value)	Material
Spring 1	100	40	3	2.24	114.44	21.98	SWP-A
Spring 2	120	40	3	1.79	91.89	20.35	
Spring 3	97	26	2	1.14	24.61	11.18	

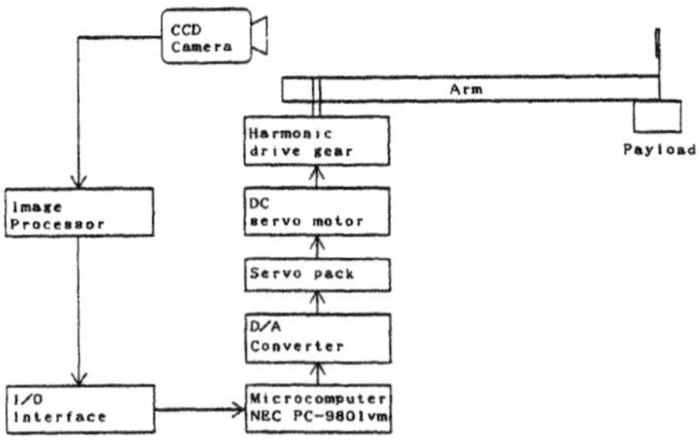

Fig.2. Experimental system

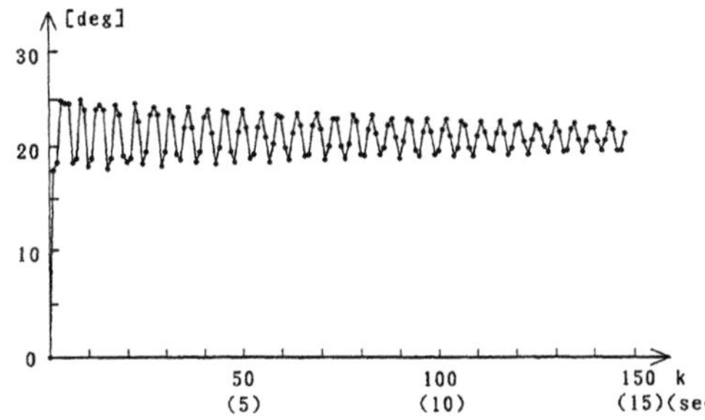

Fig.3. Open loop response of the arm (u=3 V)

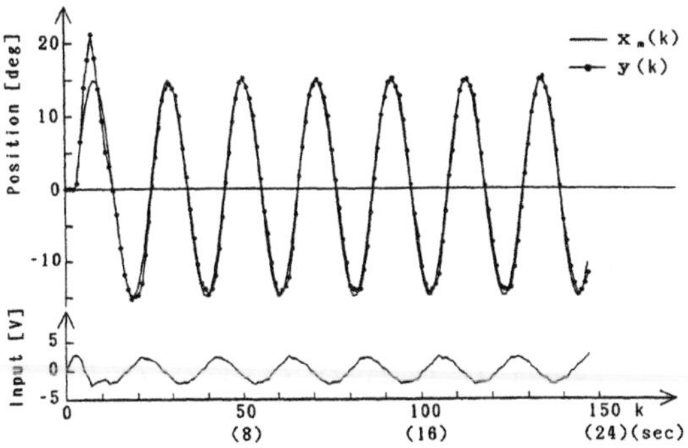

Fig.4. Adaptive control response (M=0.75 kg)

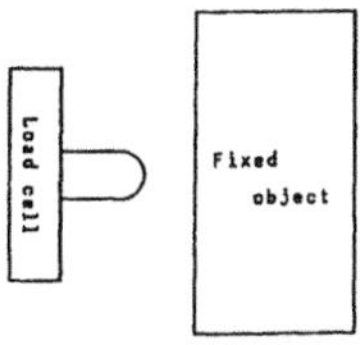

Fig.7. The case that the position of the acting point is constant (Case 1)

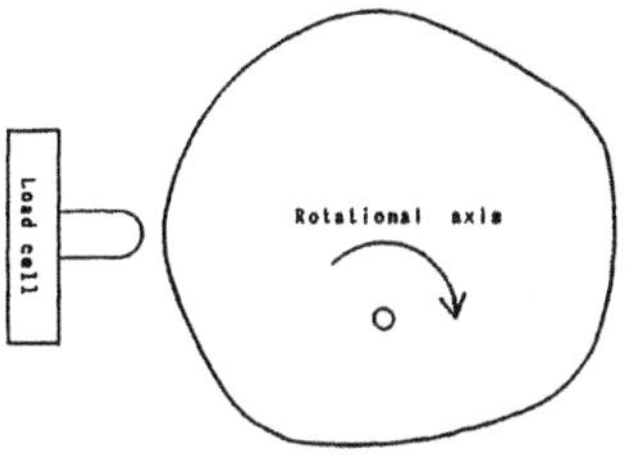

Fig.8. The case that the position of the acting point is variable (Case 2)

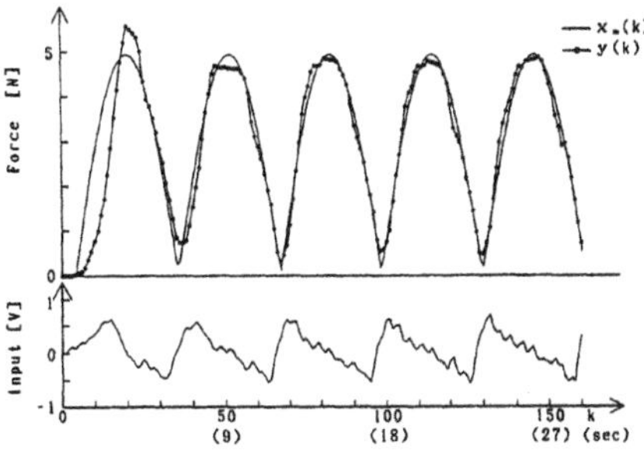

Fig.9. Adaptive control response (Case 1)

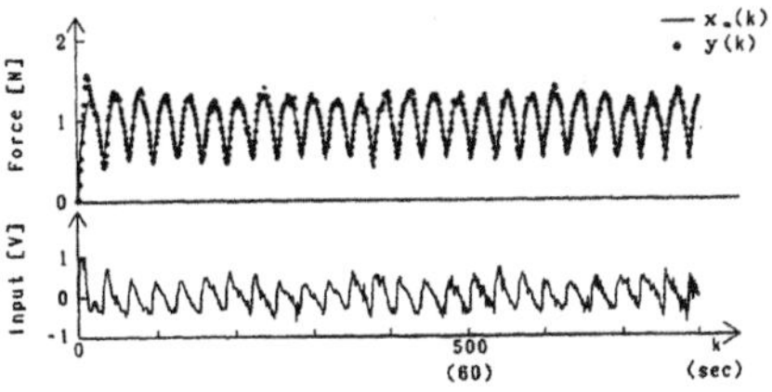

Fig.10. Adaptive control response (Case 2)

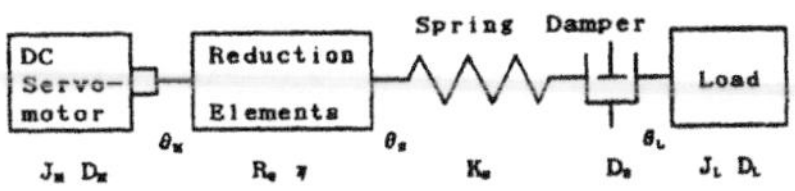

Fig.11. A simple model of a DC servo system with a flexible transmission element

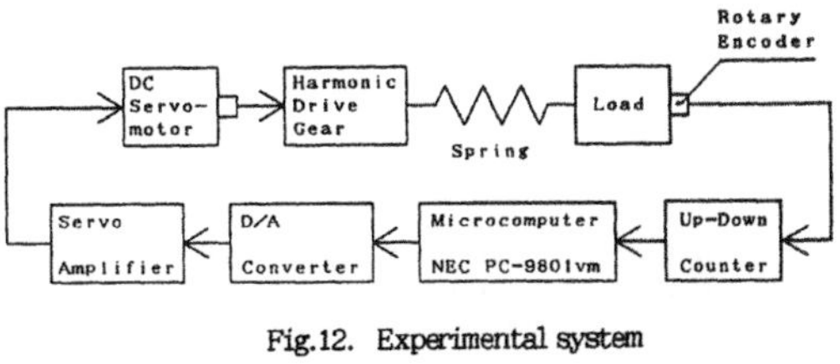

Fig.12. Experimental system

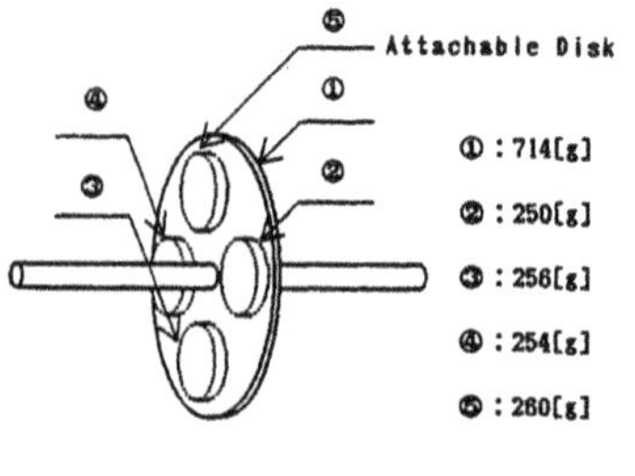

Fig.13. Load

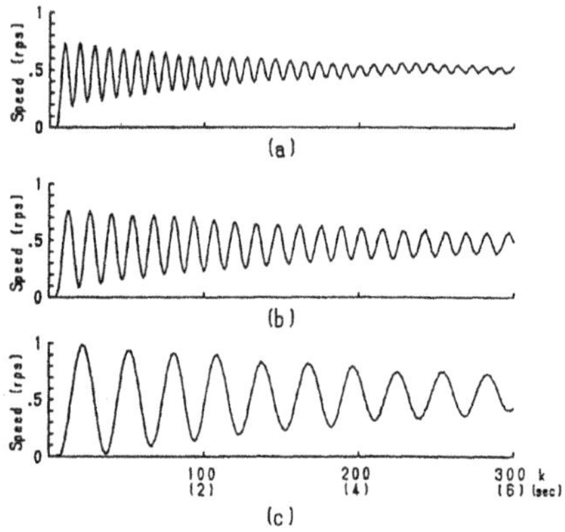

(a) Spring 1, Load ①
(b) Spring 2, Load ①+②+④
(c) Spring 3, Load ①+②+③+④+⑤
Fig.14. Open loop response (u=1.5 V)

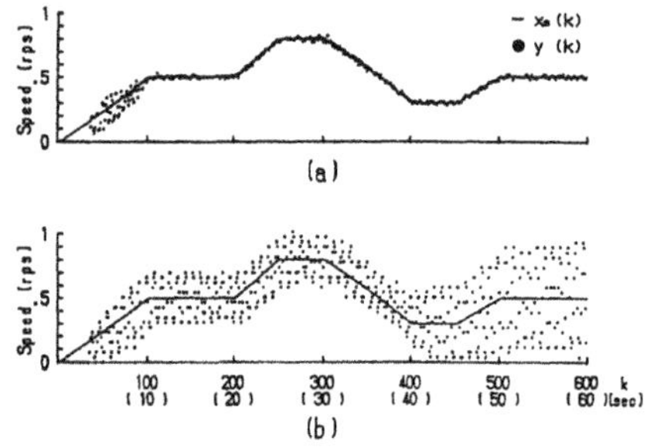

(a) Spring 3, Load ①
(b) Spring 3, Load ①+②+③+④+⑤
Fig.15. PI control response (Speed control)

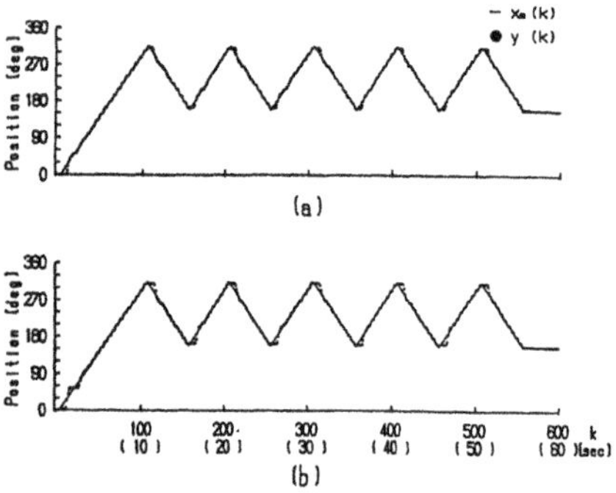

(a) Spring 2, Load ①+②+③+④+⑤
(b) Spring 3, Load ①
Fig.17. Adaptive control response (Position control)

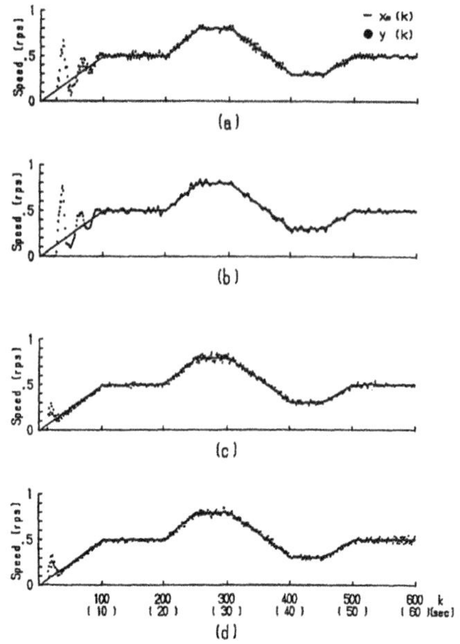

(a) Spring 3, Load ①
(b) Spring 3, Load ①+②+③+④+⑤
(c) Spring 1, Load ①+②+④
(d) Spring 2, Load ①
Fig.16. Adaptive control response (Speed control)

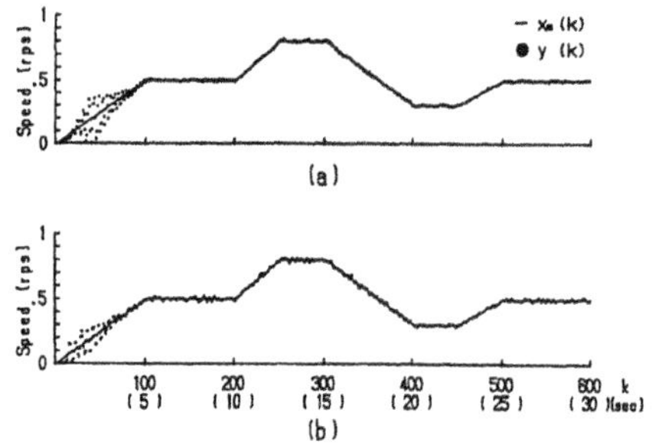

(a) Spring 2, Load ①+②+③+④+⑤
(b) Spring 3, Load ①
Fig.18. Adaptive control response (Speed control)

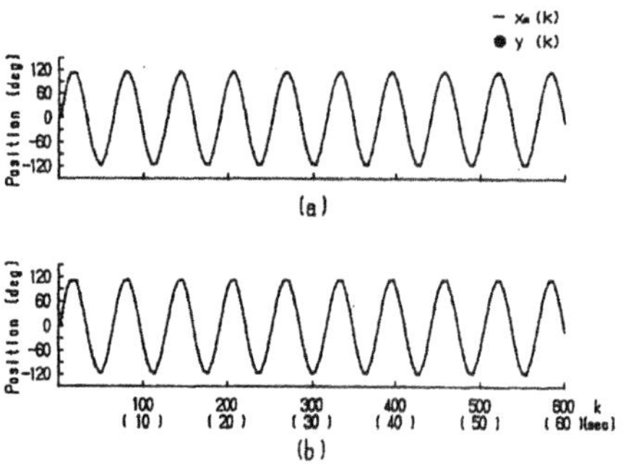

(a) Spring 1, Load ①+②+③+④+⑤
(b) Spring 2, Load ①+②+④
Fig.19 Adaptive control response (Position control)

SELF-TUNING CONTROL WITH FUZZY RULE-BASED SUPERVISION FOR HVAC APPLICATIONS

K. V. Ling, A. L. Dexter, G. Geng and P. Haves

Dept. of Engineering Science, University of Oxford, Parks Road, Oxford OX1 3PJ, UK

Abstract. This paper examines the development of fuzzy rule-based supervisors for a self-tuning controller based on the well-known Generalised Predictive Control (GPC) algorithm. First, to compensate for the input non-linearity, a fuzzy rule-based gain scheduling scheme is proposed that makes use of qualitative prior knowledge about changes in the plant gain over the entire operating region. Next, a simple fuzzy rule-based supervisor is proposed that adjusts the tuning knobs of the controller according to expert opinion based on qualitative descriptions of application dependent performance criteria, so that the performance of the control loop will improve gradually. The performance of the combinations of fuzzy rule-based supervisor and self-tuning controller is evaluated using a detailed component-based simulator developed to test building energy management systems.

Keywords. Air conditioning; Control application; Expert systems; Fuzzy control; Supervisory control.

INTRODUCTION

There is considerable interest in applying self-tuning control techniques to Heating, Ventilating and Air-Conditioning (HVAC) plants; particularly if the control algorithms are suitable for implementation on a low-cost distributed microcomputer system. One of the major difficulties in applying self-tuning or adaptive control in this area is that HVAC plants are highly non-linear and subject to severe disturbances. The controller must function over wide operating range to compensate for the changes in weather conditions and internal heating loads from day to night and from season to season.

Parameter estimation will be sensitive to the time varying gain resulting from the non-linear behaviour of the actuators and valves, and to the large disturbances that can occur during the occupancy period. Outside of the occupancy period, the heating or cooling loads acting on the plant are relatively small and it is usually only possible to estimate the parameters of a model at an operating point that is very different from that occurring during occupancy.

The expert control approach is followed here. First, to compensate for the input non-linearity, a fuzzy rule-based gain scheduling scheme is proposed that makes use of qualitative prior knowledge about changes in the plant gain over the entire operating region. Next, a simple fuzzy rule-based supervisor is proposed that adjusts the tuning knobs of the controller according to expert opinion based on qualitative descriptions of application dependent performance criteria [4], so that the performance of the control loop will improve gradually.

The fuzzy rule-based approach to dealing with uncertainty in control systems [11] has some apparent advantages when implementing "jacketing rules". Unlike most of the expert system based supervisors, only a small amount of memory is used and the decisions can be calculated in the time scales necessary for real-time applications, so that an embedded microprocessor of modest computer power of the type found in the outstations of a building energy management system can be employed in the implementation. The fuzzy rule-based approach [13] also makes it easy for an experienced operator to express the strategy or protocol for controlling a HVAC plant using linguistics variables.

FUZZY RULE-BASED GAIN SCHEDULING

A major practical difficulty with applying self-tuning and adaptive controllers is that amplitude-dependent nonlinearities appear to the estimator as a rapidly varying system gain. One possible way of dealing with such problem is to use a gain scheduling method to compensate for the non-linearity. The gain scheduling mechanism can be seen as a piece-wise approximation to perfect compensation of the non-linearity. The disadvantage of this method is that the scheduling regions and the compensation gains have to be properly chosen. To smooth the effect of gain scheduling on the controller, qualitative descriptions of the plant gain are defined and the rules that select the gains according to the scheduled regions are made fuzzy. [5]

For example, an air handler consists of a cooling coil and a heating coil, each controlled by varying the water flow rate by means of a three-port diverting valve. In many instances, the valve gain varies inversely with flow. Variable pressure drop also reduces the valve's rangeability. Therefore the effectiveness of the coil depends on the amount of valve opening and must be compensated. A gain schedule can be set up that uses the amount of valve opening as the input to the schedule.

Three sets are used to describe the position of the actuator:

```
CLOSE        is between  0% and  30%
HALF OPEN               20% and  80%
OPEN                    70% and 100%
```

The process gain is described by the following five singletons:

```
MUCH SMALLER  is  20% of nominal gain
SMALLER           50%
NOMINAL          100%
LARGER           200%
MUCH LARGER      500%
```

The rules that implement the fuzzy gain schedule are of the form:

```
if the coiling coil actuator is HALF OPEN
     then the gain is LARGER
```

The gain schedule modifies the output of the incremental controller before applying it to the process. The gain is estimated based on the previous values of the control signal applied to the process, using the qualitative knowledge expressed in the fuzzy rule base decribed earlier. One reason for modifying the increments of the control signals instead of the full value one is that the controller is operating on a linearised model about some operating point.

FUZZY RULE-BASED ADJUSTMENT OF GPC TUNING KNOBS

Previous attempts at applying Generalized Predictive Control [1] to controlling the temperature of the air in a zone of a building [3] have indicated that reliable control is only possible when a relatively long sampling interval is used, with the tuning knobs set to their default values for mean level control. The resulting control action was significantly de-tuned and the control performance was comparable only to that of a well-tuned PI controller.

Unfortunately the internal heating and cooling loads in the zone of many buildings have significant high frequency components and the sampling period of the controller must be shortened if these disturbances are to be rejected [9].

Estimating the parameters of discrete-time models becomes difficult when the sampling rate is high in comparison with the bandwidth of the system. Furthermore, with a short sampling interval, the controller becomes increasingly sensitive to unmodelled dynamics so that an observer polynomial must be used to increase robustness; and a very large prediction horizon is required to cover the significant part of the plant response; resulting in a much de-tuned controller. Although the properties of the various GPC tuning knobs have been discussed extensively in the literature [2], their precise effect on performance and robustness will depend on the nature of the system under control.

One possible approach to dealing with the problems of using a short sampling interval for the controller is to use dual-rate sampling [6], in which a model is first identified using a lower sampling frequency, and then transformed to allow the controller to operate at a higher sampling frequency. The approach described here is to configure GPC as a de-tuned controller, and use a rule-based supervisor to adjust the tuning knobs to achieve an appropriate balance between performance and robustness. The initial configuration can be based on the previous experience gained from using GPC at a longer sampling interval, with the tuning knobs settings amended to suit the higher sampling rate.

The particular tuning knobs that are considered in this work are the upper limit of the prediction horizon and the bandwidth of the observer polynomial. The prediction horizon, N2, is the active tuning parameter in the output horizon configuration of GPC. In general, the speed of the response varies with N2; approaching that of minimum variance control as N2 decreases, and that of the more conservative mean-level control as N2 becomes large [10]. The observer polynomial may be interpreted as a design polynomial that adjusts the robustness of the controller to unmodelled dynamics [12]. A suitable balance between desired performance and robustness/stability of the control loop can be achieved by adjusting both the prediction horizon and the bandwidth of the observer polynomial.

For example, in the case of zone temperature control, the supervisor could assess the performance of the controller according to the comfort level that it maintains in the zone under control. The comfort level is measured by estimating the percentage of people that would be dissatisfied (or PPD) if they were working in the zone. The average value of the PPD over the occupancy period is used as the index of performance. The total distance that the valve travels during the occupancy period is used as the measure of control activity and hence stability [4].

The supervisor adjusts the tuning knobs each day at the end of the occupancy period so that the performance of the control loop will improve gradually. It is assumed that initially the control is stable and the performance is acceptable.

The Supervisory Rules

Since expert rules are used in the supervisor, the number of rules must be kept small if we are to ensure that the set of rules is consistent and complete. The use of fuzzy inferencing allows a small number of sets to be used to describe the variables and reduces the number of rules needed to define the operation of the supervisor.

Definition of the Fuzzy Sets

Accepted values were used as a guide to the ranges for the three sets used to describe dissatisfaction. Thus :

dissatisfaction is described as

```
LOW          if PPD is between   5% and  30%
ACCEPTABLE                      15% and  40%
HIGH                            30% and 100%
```

The numerical value of the valve travel was normalised to the maximum possible travel during the occupancy period and the following ranges were selected for the three sets used to describe stability :

stability is described as

```
GOOD      if NVT is between  0% and   5%
MARGINAL                     4% and  10%
POOR                         5% and 100%
```

```
where NVT stands for Normalised Valve Travel
```

The ranges for the two sets describing the changes to be made to the prediction horizon were chosen to be certain percentages of the largest value that would be used for the horizon. The value of N2 for mean level control is approximately (5T+Td)/Ts, where T, Td and Ts are the dominant time constant, time delay of the process and sampling time respectively.

In applications of this type, the dominant time constant is of order 10 minutes, the time delay of order 5 minutes, and the sampling interval 1 minute; suggesting a value for N2 of 50 for mean level control. Thus :

```
An INCREASE  in N2 is between
    0 and  10% of N2 for mean level control

 A REDUCTION in N2 is between
    0 and -10% of N2 for mean level control
```

Changes to N2 are rounded down and the new value of N2 is limited to lie in the range 10 to 50.

The ranges for the two sets that describe the changes to the t1 parameter, which is the observer pole, are chosen as follows

```
An INCREASE  in t1 is between 0 and  0.1
   REDUCTION                  0 and -0.1
```

The new value of t1 is limited to lie in the range 0.5 to 0.95

Overlapping triangular membership functions are defined for each of the sets so that there will be no discontinuities in the values generated by the rules.

Definition of the Rules

The following supervisory rules were used to adjust the two tuning knobs:

```
The N2 parameter is increased
      if the dissatisfaction is LOW or ACCEPTABLE
      and the stability is POOR

The N2 parameter is reduced
      if the dissatisfaction is HIGH
   or if the dissatisfaction is ACCEPTABLE and
         the stability is GOOD

otherwise
     the N2 parameter is maintained at its present
     value


The t1 parameter is increased
      if the dissatisfaction is LOW and
         the stability is MARGINAL or POOR
   or if the dissatisfaction is ACCEPTABLE and
         the stability is MARGINAL or POOR
   or if the dissatisfaction is HIGH but
         the stability is also POOR

The t1 parameter is reduced
      if the dissatisfaction is ACCEPTABLE or HIGH
         and the stability is GOOD

otherwise
     the t1 parameter is maintained at its present
     value
```

The main idea underlying the development of the rules is that to improve stability, the change in t1 must be positive, whereas to improve dissatisfaction, the change in N2 must be negative. However, since the tuning knobs affect both performance and stability, the rules take into account the current levels of dissatisfaction and stability in choosing the size of the changes. Whenever a conflict of interests occurs, the rules will adjust the tuning knobs to ensure that the stability is good.

TEST RESULTS

Two control loops were used to evaluate the fuzzy rule-based approach described earlier. The supply air temperature control loop is fast, has different time delays for heating and cooling, and is very non-linear. This is used to test the fuzzy gain scheduling scheme. The zone air temperature control loop is also subject to the non-linearities of the air-handling plant but is slower and has a longer time delay. The fuzzy rule-based supervisor is used on this loop.

A detailed, component-based, dynamic simulation of the building shell, HVAC plant, and controls was used to evaluate the controller. The simulation program HVACSIM+, which was developed originally at the U.S. National Institute of Standards and Technology, has been used in the evaluation of building control schemes [7].

In the simulation used here, the heating plant is deliberately undersized; the response of the cooling coil includes the flow-rate dependent time-constant for the coils, rate limits for the control valve actuators and hysteresis in the actuator linkages. The room model has a significant time delay of a magnitude that might be encountered in a large internal space. Other dynamic effects such as heat capacity of the walls, furniture and air in the room, and the time constant of the temperature sensor are also included. Figure 1 is a diagram of the simulated building and plant showing each of the control loops.

The test period consists of 24 hours in which the heating loads is offset during the occupancy period (9-17h) by solar and internal gains. Temperature profiles simulating the effect of occupancy, solar gains and lighting are included to serve as disturbances to the process.

Application 1: Supply Temperature Control

Figure 2 compares the performance of the supply-air temperature control with and without gain scheduling compensation. The non-linearity of the process can be seen from the differences in changes in the control signals to achieve the same amount of change in supply temperature. As the cooling coil had a larger gain than the heating coil, control became oscillatory when the operating point shifted to cooling. Such qualitative process knowledge was used to set up a fuzzy gain schedule to compensate for the variation in process gain.

Application 2: Zone Temperature Control

Figure 3 illustrates the effects of the disturbances on the air temperature in the zone and the corresponding control signal. Solid lines were the results of using GPC at 60sec sampling interval. The behaviour of controller when using a 240sec sampling interval (dashed lines) is also shown for comparison.

Table 1 shows the effects of the different settings of the tuning knobs on the criteria which forms the qualitative basis for the rule base in the tuning supervisor. In all cases, the controller uses a 60sec sampling interval. In general, the dissatisfaction level decreases with decreasing N2, with corresponding increase in distance valve travel. With slower observer pole, the dissatisfaction level increases, together with less distance in valve travel.

Table 1: Effects of the tuning knobs on the performance and stability criteria

	Observer Pole		
	0.5	0.8	0.95
N2			
10	SI=395.59 PI=12.47	SI=27.96 PI=15.45	SI=10.03 PI=18.14
24	SI=17.57 PI=16.71	SI=12.77 PI=20.91	SI=5.86 PI=37.22
50	SI=15.75 PI=18.34	SI=11.81 PI=23.37	SI=6.07 PI=42.31

where PI: performance index, PPD
SI: stability index, distance valve travel

Figure 4 shows the adjustments to the tuning knobs made by the supervisor and the resulting changes in dissatisfaction and stability over a period of ten days.

CONCLUSIONS

AI and expert systems have been recognised as vehicles to incorporate experience from the control system designers and on a supervisory level of automatic control. Fuzzy control seems to be an interesting alternative to perform these tasks automatically. A simple fuzzy rule-based supervisor could be considered as a crude form of extremum controller and converges slowly to the "optimal" settings of tuning knobs. Recent work [11] suggests that on-line generation of the rules may be possible. The appropriateness of the tuning knobs selected and the behaviour of other sets of supervisory rules that use more detailed qualitative measures will be examined. Further tests are now being performed on an experimental air-handling unit.

Fuzzy gain scheduling provides a means of incorporating the uncertain prior knowledge about the process and improves stability and/or performance of the controller. The use of such form of prior knowledge to aid the parameter estimator to improve model estimation is currently being investigated.

Acknowledgements – K.V. Ling is supported by Trend Control Systems Limited. The work described here forms part of the research programme investigating industrial applications of self-tuning control.

References

[1] Clarke,D.W., Mohtadi,C. and Tuffs,P.S. (1987). Generalized predictive control. Parts 1 and 2. Automatica, Vol.23, p.137-160.

[2] Clarke,D.W. and Mohtadi,C. (1989). Properties of generalized predictive control, Automatica, Vol.25, p.859-875.

[3] Dexter,A.L. and Haves,P. (1989). A robust self-tuning controller for HVAC applications, ASHRAE Trans. V.95, Pt.2., 1989

[4] Dexter,A.L. and Trewhella,D.W. (1990) Building control systems: a fuzzy rule-based approach to performance assessment. Building Serv. Eng. Res. Technol., Vol.11, No.4. to be published.

[5] Geng,G. (1990). Control of a Non-linear Process. DPhil Thesis. In preparation.

[6] Hang,C.C., Lim,K.W. and Chong,B.W. (1989) A dual-rate adaptive digital smith predictor. Automatica, Vol.25, p.1-16.

[7] Haves,P. (1987). The application of simulation to the evaluation of building energy control systems. Proc. UKSC '87 Conference, U.K. Simulation Council, Bangor, U.K., September.

[8] John,R.W. and Dexter,A.L. (1989). Intelligent Controls for building services, Building Serv. Eng. Res. Technol., Vol.10, No.4.

[9] Lim,K.W. and Ling,K.V. (1989). Generalized predictive control of a heat exchanger, IEEE Control Systems Magazine, Oct 1989. p.9-12.

[10] McIntosh,A.R., Shah,S.L. and Fisher,D.G. (1989). Selection of tuning parameters for adaptive GPC. ACC 1989, 1846-1851.

[11] Pedrycz, (1989). Fuzzy Control and Fuzzy Systems, Wiley.

[12] Robinson,B.D. and Clarke,D.W. (1990). Robustness effects of a prefilter in generalized predictive control. Proc. IEE Pt.D, to be published.

[13] Van Naute Lemke,H.R. and Wang,D. (1985) Fuzzy PID supervisor. Proc. IEEE conf. on decision and control, p.602-608.

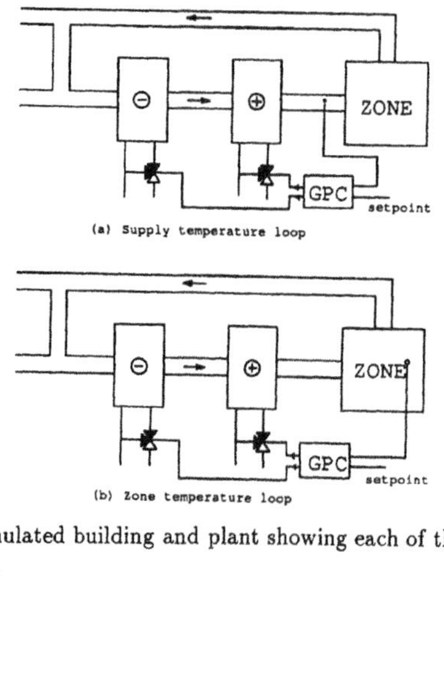

Figure 1: Simulated building and plant showing each of the control loops.

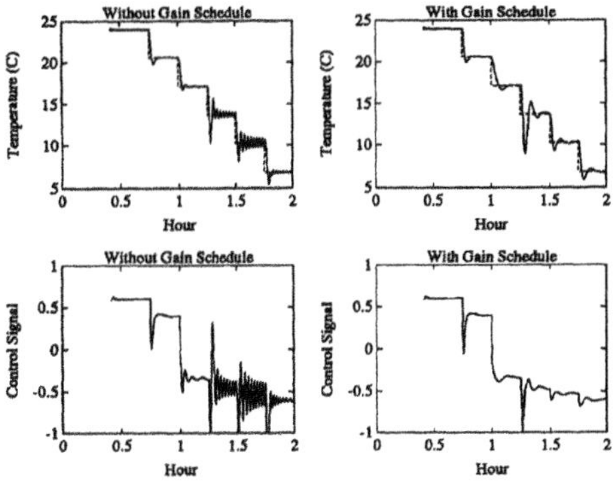

Figure 2: Effect of fuzzy gain schedule.

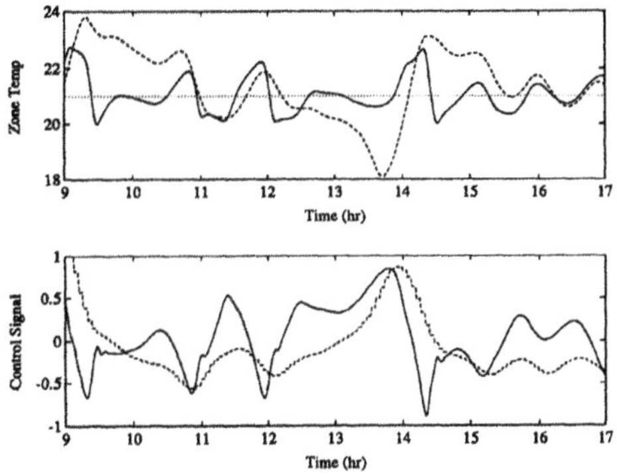

Figure 3: Typical variations in zone temperature over the occupancy period.

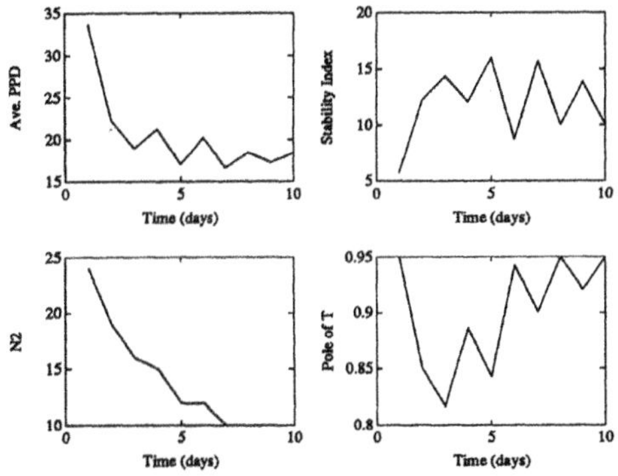

Figure 4: Adjusments to tuning knobs made by the supervisor and the resulting changes in dissatisfaction and stability over of ten days.

AN EXPERT SYSTEM FOR THE MULTIVARIABLE CONTROLLER DESIGN

J. Lieslehto, J. T. Tanttu and H. N. Koivo

Tampere University of Technology, P.O. Box 527, SF-33101 Tampere, Finland

Abstract. In this paper a software package for the design of centralized and decentralized multivariable controllers is represented. The software package consists of numerical calculation programs and an expert system. The paper first describes the different subtasks of the multivariable controller design. Next a short descriptions of the design and analysis methods included in the software package is given. The description of the implementation of the expert system finishes the paper.

Keywords. Expert systems; multivariable control systems; PID control

INTRODUCTION

The task of designing controllers for multivariable systems involves two major design steps. First a designer has to select the control structure and then design and tune the controllers. Several numerical methods are available for these tasks and usually many of them are needed during the design process. The best possible results are only achieved if the designer is able to choose the right methods and use them in a correct way. That is why he must have a considerable amount of knowledge about different design and analysis methods. The expert system techniques have made it possible to make this knowledge a part of the control design software.

In this paper a software package for the design of centralized and decentralized multivariable controllers is represented. The software package consists of numerical calculation software and an expert system. The numerical methods offer several functions for the interaction analysis and control structure selection as well as for the tuning and analysis of MIMO PID controllers. A continuous-time transfer function matrix model of the process is needed as initial data. The expert system monitors the actions taken by the user. Based on the information gathered during the design process it is able to assist the user in problem situations. The software package has been developed on a Macintosh IIfx computer. The numerical methods are coded using Matlab (Moler and others, 1987). The knowledge base of the expert system is build using Nexpert Object (Neuron Data, 1989) and the user interface is implemented as a HyperCard stack (Apple Computer, 1989).

CONTROLLER DESIGN

The design process is divided into four different subtasks as described in Figure 1. The calculation of interaction matrices and interaction measures are the two subtasks of the control structure selection. After the control structure selection controllers are tuned and analysed. These two steps are often repeated in an iterative manner.

Interaction matrices help the designer to find possible candidates for the control structure of a decentralized controller. These matrices contain information about the interactions between inputs and outputs.

Based on the information from the interaction matrices the designer selects one or more candidates for the control structure. Interaction measures are then used to estimate the performance deterioration of decentralized controllers compared to a full centralized controller. If the interaction measures indicate possible problems then the designer has to look at the interaction matrices again and try to find another control structure. Otherwise he is ready to start controller tuning.

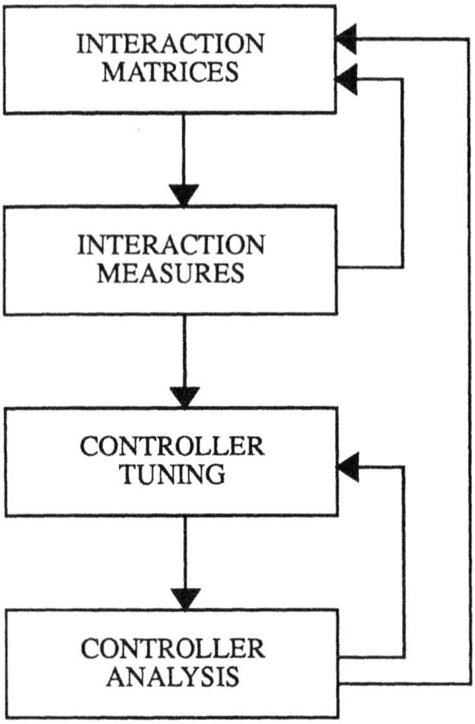

Figure 1: The subtasks of the controller design.

There are several methods for the tuning of multivariable PID controllers. These methods usually give some rough tuning matrices and the user can find the final tuning matrices by multiplying the rough tuning matrices with fine tuning parameters. The tuning is typically an iterative process. Based on the results from different analysis methods the values of fine tuning parameters are changed and the controllers are retuned.

The last step is to analyse the performance and robustness of the designed controller. This is usually done using both simulations and frequency-domain methods. If the analysis indicates any problems the designer has to change the fine tuning parameters or try some other tuning methods. Of course, he can also change the control structure.

INTERACTION MATRICES

The control structure should be selected in such way that the inputs and outputs that strongly interact with each other are connected with a feedback. The number of possible control structures rapidly increases as the number of inputs and outputs increases. Several methods proposed in the literature aim to produce an interaction matrix which describes interactions between the inputs and outputs of a MIMO process. These methods help a designer to find possible control structures. The software package includes the classical relative gain array method proposed by Bristol (1966), the dynamic relative gain array method proposed by Tung and Edgar (1981) and the singular value analysis method by Lau, Alvarez and Jensen (1985). A new interaction matrix is also introduced. This interaction matrix is based on the scaling of the input and output variables.

Bristol's relative gain array is probably the most well-known interaction matrix. Each element in a relative gain array is the ratio of two gains representing first the process gain in an isolated loop and, second, the apparent process gain in the same loop when all other control loops are closed. The ratio of these gains defines the relative gain array with elements

$$\mu_{ij} = \phi_{ij}\gamma_{ji} \tag{1}$$

where ϕ_{ij} is an element in the steady-state gain matrix $G(0)$ and γ_{ji} is an element in the inverse of the steady-state gain matrix $G^{-1}(0)$. Bristol suggested that the proper input-output pair for single loop control is the one having the largest positive μ_{ij} value.

The relative gain analysis is a steady state analysis and does not explicitly include dynamic effects. Several extensions including also the dynamic effects have been proposed. Out of them the software package includes the dynamic RGA proposed by Tung and Edgar. The elements of the dynamic RGA are calculated as follows

$$\alpha_{ij} = g_{ij}(s)\gamma_{ji} \tag{2}$$

where $g_{ij}(s)$ is an element in the process transfer function matrix $G(s)$ and γ_{ji} is an element in the inverse of the steady-state gain matrix $G^{-1}(0)$. Both the RGA and the dynamic RGA can be calculated only if the number of inputs equals to the number of outputs.

The third method in the software package is proposed by Lau, Alvarez and Jensen. This method is based on singular value analysis and it is applicable also for process models with unequal numbers of inputs and outputs. Interactions are analysed using the singular value spectral representation of a $m \times n$ open loop matrix transfer function $G(s)$ whose rank is q. This representation is

$$G(s) = \sum_{i=1}^{q} \sigma_i(s) W_i(s), \tag{3}$$

where σ_i are the singular values of G and W_i are the dyadic expansion matrices defined in terms of singular vectors as follows

$$W_i(s) = z_i(s) v_i^H(s). \tag{4}$$

Through this expansion the system transfer function is expressed as a linear combination of the q nodal contributions. Each of the nodal terms consists of a scaling factor σ_i and a rotational transformation W_i. Lau, Alvarez and Jensen suggested that the maximum entries in the rotational matrices define the input-output pairings for single loop control.

The fourth method is based on the scaling of input and output variables. Although large gain between an input and an output indicates strong interaction, the process gain matrix can't be directly used for interaction analysis, because it depends on the scaling of input and output variables. Anyway, by rescaling the input and output variables we can find a scaled gain matrix whose elements are directly comparable with each other. In the scaled gain matrix the sum of the absolute values of the elements of each row and column is one. SISO control loops should be selected in such way that the scaled gains between the corresponding inputs and outputs are as large as possible.

INTERACTION MEASURES

Using interaction matrices we are able to find possible candidates for the control structure. Interaction measures are used to find out whether these control structures can be used or not. Interaction measures give the user information about the stability of the decentralized control and the loss in performance caused by these control structures. The interaction measures included in the software package are the generalized diagonal dominance criteria (Mees, 1981; Limebeer, 1982) and the structured singular value intercation measure (Grosdidier and Morari, 1986). The software also includes functions to test whether a system is decentralized integral controllable or not.

Morari and Zafiriou (1989) have described the general ideas behind the interaction measures included in the software package. A controller

$$C = diag\{C_1, C_2, \ldots, C_n\} \tag{5}$$

is to be designed for the system

$$\tilde{G} = diag\{G_{11}, G_{22}, \ldots, G_{nn}\} \tag{6}$$

such that the block diagonal closed-loop system with the transfer matrix

$$\tilde{H} = \tilde{G}C(I + \tilde{G}C)^{-1} \tag{7}$$

is stable. Both interaction measures express the constraints imposed on the choice of the closed-loop transfer matrix $\tilde{H}$ for the block diagonal system, which guarantee that the full closed-loop system

$$H = GC(I + GC)^{-1} \tag{8}$$

is stable. These constraints are functions of the relative error matrix

$$L_H = (G - \tilde{G})\tilde{G}^{-1}. \tag{9}$$

The generalized diagonal dominance criteria states that the closed loop system H is stable if

$$|\tilde{h}_j(i\omega)| < \rho^{-1}(|L_H(i\omega)|) \quad \forall j, \omega \tag{10}$$

where ρ denotes the spectral radius. The SSV interaction measure gives the following constraint for the stability of the closed loop system H

$$\bar{\sigma}(\tilde{H}_j(i\omega)) < \mu^{-1}(L_H(i\omega)) \quad \forall j, \omega \tag{11}$$

where μ denotes the structural singular value. In both cases it is assumed that $G(s)$ and $\tilde{G}(s)$ have the same RHP poles and that $\tilde{H}(s)$ is stable.

A plant G is decentralized integral controllable (DIC) if there exists a diagonal controller CKs^{-1} (where K is a diagonal gain matrix with positive entries) with integral action such that the closed loop system is stable and the gain of any subset of loops can be reduced to $K_\epsilon = diag\{k_i\epsilon_i\}, 0 \leq \epsilon_i \leq 1$ without affecting the closed loop stability. This means that any subset of loops can be detuned or taken out of service while maintaining the stability of the rest of the system. The software package includes functions that calculate the following conditions that are used to test whether a system is DIC.

$$det(G^+(0)) > 0 \tag{12}$$

$$Re\{\lambda_i(G^+(0))\} \geq 0 \quad \forall i \tag{13}$$

$$Re\{\lambda_i(L_H(0))\} \geq -1 \quad \forall i \tag{14}$$

$$RGA : \lambda_{ii}(G(0)) > 0 \quad \forall i \tag{15}$$

$G^+(0)$ is derived from the staedy state gain matrix $G(0)$ by multiplying each column with $+1$ or -1 such that all diagonal elements are positive, L_H is the relative error matrix and λ_i is an eigenvalue. The last condition tests the diagonal elements of the relative gain array. All conditions are necessary but not sufficient.

CONTROLLER TUNING

The software package includes three algorithms for the tuning of MIMO PID controllers. The first one was proposed by Davison (1976). This method was later modified by Penttinen and Koivo (1980). The third method is based on the ideas of internal model control presented by Rivera, Morari and Skogestad (1986). A version of the IMC method has been developed that enables the user to use the method for the tuning of multivariable PID controllers. All three tuning methods allow the design engineer to retune the controller using fine tuning parameters.

Davison proposed the following tuning matrices for a PI-controller

$$K_p = \delta G^{-1}(0) \qquad (16)$$
$$K_i = \epsilon G^{-1}(0). \qquad (17)$$

The inverse of the steady state gain matrix $G^{-1}(0)$ is the rough tuning matrix and δ and ϵ are the fine tuning parameters.

If there exists a state space representation of the transfer function matrix $G(s)$

$$\dot{x} = Ax + Bu \qquad (18)$$
$$y = Cx \qquad (19)$$

then the transfer function matrix has the series expansion

$$G(s) = \frac{CB}{s} + \frac{CAB}{s} + \frac{CA^2B}{s} + \dots \qquad (20)$$

Penttinen and Koivo used the first term of the series expansion as the rough tuning matrix of the P-part of the controller

$$K_p = \delta(CB)^{-1}. \qquad (21)$$

The selection aims to decouple the system at high frequencies. The I-part of the controller is selected as in Davison's method.

Rivera, Morari and Skogestad have proposed a SISO PID controller tuning method based on the Internal Model Control (IMC) principle. The

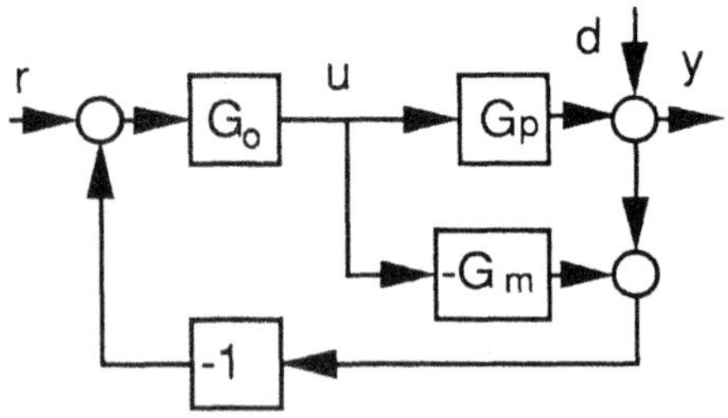

Figure 2: Internal Model Control.

structure of an IMC controller is depicted in the Figure 2. The IMC design process is divided into two steps. First the process model G_m is factored

$$G_m = G_+ G_- \qquad (22)$$

so that G_+ contains all the time delays and unstable zeroes of the model; consequently G_-^{-1} is stable and causal. However it may be improper. So the IMC controller is defined by

$$G_o = G_-^{-1} F \qquad (23)$$

where F is such a low-pass filter that G_o is proper or if derivative action is allowed has a zero excess at most 1. Rivera, Morari and Skogestad suggest that a good choice for the filter is

$$F = \frac{1}{(1+\epsilon s)^n} \qquad (24)$$

The time constant ϵ of the filter is the fine tuning parameter of the IMC method. The classic feedback controller G_c is related to the IMC controller via the transformation

$$G_c = \frac{G_o}{1 - G_m G_o}. \qquad (25)$$

For several low order transfer functions this IMC design procedure leads to PI and PID controllers.

The following heuristic method has been used in the software package to calculate the tuning matrices of a MIMO PID controller. First a SISO PID controller is designed for each element $g_{ij}(s)$ of the transfer function matrix $G(s)$ using the IMC method as described above. As a result we get the tuning parameters $k_{p_{ij}}, k_{i_{ij}}, k_{d_{ij}}$ of a PID controller separately for each element $g_{ij}(s)$ of the process model. Then the tuning matrix for the P-part of the MIMO controller is calculated as follows

$$K_p = \begin{bmatrix} 1/k_{p_{11}} & 1/k_{p_{12}} & \cdots & 1/k_{p_{1n}} \\ 1/k_{p_{21}} & 1/k_{p_{22}} & \cdots & 1/k_{p_{2n}} \\ \vdots & \vdots & \ddots & \vdots \\ 1/k_{p_{n1}} & 1/k_{p_{n2}} & \cdots & 1/k_{p_{nn}} \end{bmatrix}^{-1} \quad (26)$$

The I-part and D-part of the controller are calculated in a similar way. Each SISO controller can have a different fine tuning parameter, although usually good results are achieved when all fine tuning parameters have the same value.

CONTROLLER ANALYSIS

The software package includes several functions for the analysis of the performance and robustness of the designed control system. Both frequency domain measures and time domain simulations can be used.

In the frequency domain the user can analyse the nominal performance of the designed controller by computing the sensitivity and complementary sensitivity functions. The software package also includes functions that measure robust stability. Both unstructured and structured uncertainty models can be used. An uncertainty can be described either as a multiplicative input uncertainty, multiplicative output uncertainty, additive uncertainty or inverse multiplicative output uncertainty. Also functions for the computation of robust performance measures are available.

The user can also simulate the behaviour of the closed loop system. The software package has an easy-to-use function for the simulation of multivariable delayed processes. The simulation program allows the user to choose different reference signals, controller noises and measurement noises.

EXPERT SYSTEM

The software package includes approximately 80 MATLAB functions. Around thirty of them are so called intelligent functions. After running such a function the user is able load the calculated results into the knowledge base of the expert system. This way the expert system gathers information about the design process. Using this information the expert system is able to help the user in possible problem situations.

The structure of the software package is depicted

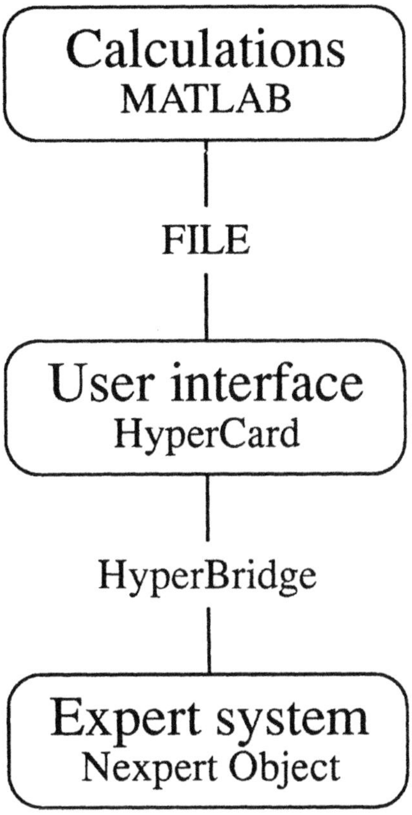

Figure 3: The structure of the software package.

in Figure 3. The numerical methods are used by the designer just like any other MATLAB toolbox. When an intelligent function is used its results are written into a file. If the user wants, he can load these results into the knowledge base of the expert system. A HyperCard program reads the results from the file and makes the necessary changes to the knowledge base. Data can be transferred between HyperCard and Nexpert Object using a program called HyperBridge. From the HyperCard user interface the designer can start the inference engine of the expert system. The expert system analyses the design session so far and draws conclusions. As a result of this analysis the user will be shown HyperCards containing helpful information.

One of the cards of the user interface stack is shown in Figure 4. This card contains information about the input and output parameters of a simulation function. At the bottom of the card there are seven buttons that are used to control the expert system and the user interface. Clicking the button "Matlab" makes the Matlab window active. "Restart" empties the

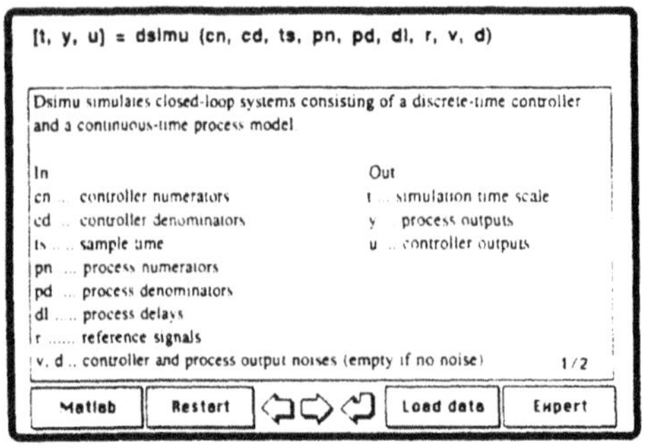

Figure 4: A user interface card.

knowledge base and starts a new session. "Load data" runs the program that reads data written by intelligent functions from a file and makes appropriate changes into the knowledge base. The button "Expert" is used to start the inference engine of the expert system. The three arrow buttons are used to find the previous and next cards in the stack and to return to the previously shown cards. Some words in the text of a card are mouse-sensitive. Those words are highlighted when the cursor is over them. The user can get further information about these subjects by clicking the mouse when the word is highlighted.

CONCLUSIONS

In this paper a software package for the design of centralized and decentralized multivariable controllers is represented. The software package consists of numerical calculation software and an expert system. The numerical methods offer several functions for the interaction analysis and control structure selection as well as for the tuning and analysis of MIMO PID controllers. During the design process the expert system is continuously observing the actions taken by the user. Based on the gathered information the expert system is able assist the user in possible problem situations.

REFERENCES

Apple Computer (1989). *HyperCard User's Guide.* Apple Computer Inc., Cupertino, CA.

Bristol (1966). On a new measure of interaction for multivariable process control. *IEEE Trans.on Autom.Control,11.* 133-134.

Davison (1976). Multivariable tuning regulators: the feedforward and robust control of general servo-mechanism problem. *IEEE Trans.Aut.Control,21.* 35-47.

Grosdidier, Morari (1986). Interaction measures for systems under decentralized control. *Automatica,22.* 309-319.

Lau, Alvarez, Jensen (1985). Synthesis of control structures by singular value analysis: dynamic measures of sensitivity and interaction. *AIChE Journal,31.* 427-439.

Limebeer (1982). The application of generalized diagonal dominance to linear system stability theory. *Int.J.Control,36.* 185-212.

Mees (1982). Achieving diagonal dominance. *Systems and Control Letters,1.* 155-158.

Moler, Herskovitz, Little, Bangert (1987). *Matlab for Macintosh computers: User's guide.* MathWorks Inc., Sherborn, MA.

Morari, Zafiriou (1989). *Robust process control.* Prentice-Hall, Englewood Cliffs.

Neuron Data (1989). *Nexpert Object.* Neuron Data Inc., Palo Alto, CA.

Penttinen, Koivo (1980). Multivariable tuning regulators for unknown systems. *Automatica,16.* 393-398.

Rivera, Morari, Skogestad (1986). Internal model control. 4. PID controller design. *Ind.Eng.Chem.Process Des.Dev.,25* 252-265.

Tung, Edgar (1981). Analysis of control-output interactions in dynamic systems. *AIChE Journal,27.* 690-693.

EXPERT SELF-TUNING PI(D) CONTROLLER

R. Devanathan

*School of Electrical and Electronic Engineering,
Nanyang Technological Insitute, Nanyang Avenue, Singapore 2263*

Abstract. The development of a self-tuning PI(D) controller is discussed in this paper. The controller is based on an expert system approach using the pattern recognition technique. The unique features of the controller developed are: (i) the incorporation of knowledge relating to a simple identification technique for the process characteristic parameter through observation of two closed loop load responses of the process under normal PI(D) control action and (ii) selection of P, I and D settings based on minimum integrated error (IE) criterion. The paper dicusses new theoretical results on the parametrisation of P, I, and D settings for minimum IE criterion in terms of process characteristic ratio, the technique used for the process identification, and the microprocessor-based hardware and software implementation of the self-tuning controller together with the results of tests conducted on the developed controller using a laboratory type pilot process.

Keywords. PID controller; PI controller; integrated error; regulator; self-tuning controller.

INTRODUCTION

Different types of self-tuning controllers have been proposed and applied in practice (Warwick, 1988, Harris and Billings, 1981). Self-tuning proportional, integral and derivative (PID) controllers form an important subclass of these. Three types of adaptive PID controllers have been proposed. Firstly, the classical methods of Ziegler and Nichols (1942) are incorporated into an automated tuning phase which is followed by control using fixed PID parameters. In a method due to Astrom and Hagglund (1984), limit cycle oscillations are enforced on the system to be controlled by a relay in order to obtain both angular frequency and critical gain values. The second method is one in which the PID parameters are varied slowly on-line so as to account for plant variations. An example of the second technique is the Foxboro EXACT system (Higham, 1986). The controller witnesses process response patterns such as overshoot peaks and uses tuning rules based on the heuristic approach developed by the control engineers over the years. The method used does not assume any process model. The third method using a recursive parameter estimation technique directly feeds new plant parameters to the PID controller which is then based on the latest estimates available (Ortega and Kelly, 1984, Wittenmark, 1979).

In this paper, a self-tuning PID controller is described which somewhat falls into the second type mentioned above. However, the controller developed assumes a first order dead time model for the process and incorporates knowledge based on optimum controller settings for minimum integrated error (IE) criterion consistent with a specified damping. Minimum IE criterion lends itself as a simple and yet an effective criterion for optimum control system peformance (McMillan, 1983). Also, a simple relationship exists between IE and the PID settings which need to be adjusted optimally. The main objection to IE criterion is that IE is zero for perfect oscillation and thus by itself does not constitute stability. But then, if the control system damping is maintained otherwise, the above objection can be overruled (Shinskey, 1979, 1988). To summarise the rest of the paper, the following section introduces the analytical framework to be used followed by new theoretical results on the parametrisation of P, I, and D settings for minimum IE criterion. The application of the results to a simple process identification and a subsequent self-tuning technique is described next followed by a description of the hardware and the software aspects of the development of the self-tuning controller. The test results obtained using a laboratory pilot process is described next followed by conclusions.

PRELIMINARIES

A process, under PID control, can be

described by the following two equations corresponding to the phase (lag) angle and gain respectively around the feedback loop at the loop period (T_o).

$$\tan^{-1}(2\pi T_1/T_o) + 2\pi T_d/T_o = \pi + \theta \qquad (1)$$

$$(100/P)\sqrt{1+\tan^2\theta}\; K_p \Big/ \left\{\sqrt{1+(2\pi T_1/T_o)^2}\right\} = X \qquad (2)$$

where K_p is the process dc gain, P is the proportional band ,X is the loop gain and θ (negative for lagging angle) is the controller phase angle which is given in terms of the integral (I) and the derivative (D) settings at the loop period T_o as

$$\tan\theta = (2\pi D)/T_o - 1/\{2\pi(I/T_o)\} \qquad (3)$$

The loop gain X represents control system damping.For quarter amplitude damping, for example, X=0.5.Putting

$$x = I/T_o \qquad (4)$$

Eq.(3) can be written as

$$\tan\theta = (2\pi x)/k - 1/(2\pi x) \qquad (5)$$

where k is the ratio of the integral to the derivative action as in

$$k = I/D \qquad (6)$$

MINIMUM IE CRITERION

Define normalised parameters

$$R = T_d/T_1 \qquad (7)$$

$$N = T_o/T_d \qquad (8)$$

For an arbitrary value of R in the range $0<R<\infty$, put

$$2\pi/N = \pi - \pi y/2 + \theta, \quad 1 > y > 0 \qquad (9)$$

Using Eq.(7)-(9), Eq.(1) and (2) become

$$2\pi/N = \pi - \pi y/2 + \theta = R\tan(\pi y/2) \qquad (10)$$

$$P = (100K_p/X)\cos(\pi y/2)/\cos\theta \qquad (11)$$

Using Eq.(4) and (8),I can be expressed as

$$I = x N T_d$$

$$= 2\pi x T_d \,/\, (\pi - \pi y/2 + \theta) \qquad (12)$$

The IE for a unit load disturbance can be shown to be directly proportional to the product PI of the settings(Shinskey,1988). Using Eq.(11) and (12), we get

$$PI = (200K_p T_d/X) \cdot \{(x\cos\pi y/2)/\{(\pi-\pi y/2+\theta)\cos\theta\} \qquad (13)$$

The following result can now be stated.

Theorem 1

For the control system as defined by Eq.(1)-(3), the minimum IE criterion for a load disturbance is satisfied at the controller phase angle θ given by

$$\{\pi x\sec\theta/(\cos\theta+\pi x\sin\theta)\} + \tan\theta =$$

$$\{1+\sin^2(\pi y/2)\}/\{\pi - \pi y/2 + \theta + (\sin\pi y)/2\} \quad \ldots \qquad (14)$$

for a given value of k where y is as defined in Eq.(9).Using k as a variable, IE can be further minimized by choosing k = 4 , corresponding to maximum derivative action in an interacting type of PID controller, and the corresponding controller phase angle θ is given by

$$\tan\theta + \sec\theta =$$

$$\{1+\sin^2(\pi y/2)\}/\{\pi - \pi y/2 + \theta + (\sin\pi y)/2\} \quad \ldots \qquad (15)$$

Proof:- See Appendix-A

Corollary 1

Consider the control system defined by the equations (1)-(3) under PI control.The minimum IE criterion is satisfied at the controller phase angle θ given by

$$-\cot\theta =$$

$$\{1+\sin^2\pi y/2\}/\{\pi - \pi y/2 + \theta + (\sin\pi y)/2\} \qquad (16)$$

Proof:-By putting D = 0 in Eq.(3), Eq.(14) can be simplified to Eq.(16) directly.

OPTIMUM CHARACTERISTICS

Combining Eq.(14) and (5) , it is possible to express x (= I/ To) as a function of R corresponding to the minimum IE condition and for a given value of k. Fig.1 shows such a plot for k=4 both for PI and PID control.Also, the loop natural period (T_u) corresponding to the case when the controller angle $\theta = 0$, can be obtained, in terms of T_d, for a given R, from Eq. (17) below.

$$2\pi/N_u = \pi - \pi y_u/2 = R\tan(\pi y_u/2) \qquad (17)$$

where

$$N_u = T_u/T_d \qquad (18)$$

For the same R, obtain from Fig. 1, x (or equivalently θ using Eq.(5)).Using θ and R, find the loop period (T_o) from Eq.(10) and (5).One can thus transform the optimum curves of Fig.1 into those of Fig.2 (by multiplying by the factor T_o/T_u as a function of R) where the normalized ratio (I/T_u) is plotted against the process characteristic ratio R both for PI and PID control.The result of Fig.1 has been verified for the particular case of the capacity dominant process (R -->0) as provided by Shinskey (1979).Also , for the case of the PI controller, combining Fig.1 and 2, one can obtain the plot of Fig.3

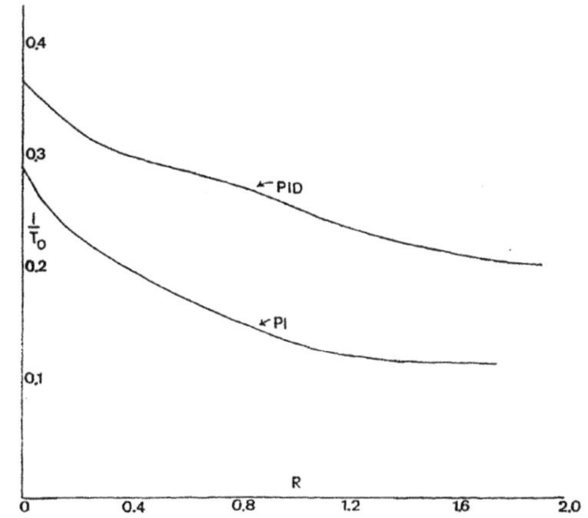

Figure 1. I/T_o vs.R, PID (k= 4),PI Control

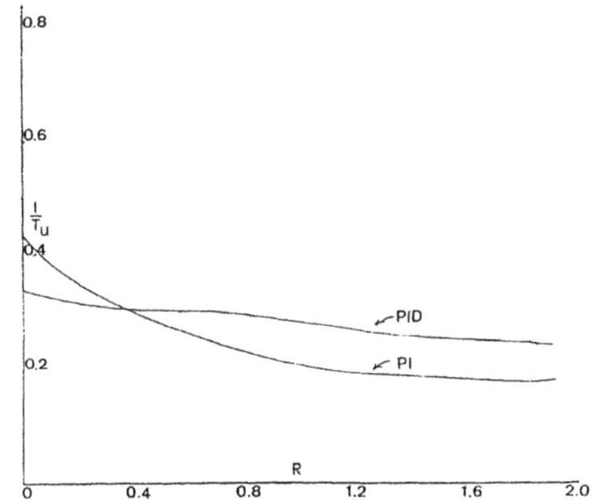

Figure 2. I/T_u vs.R, PID (k=4), PI Control

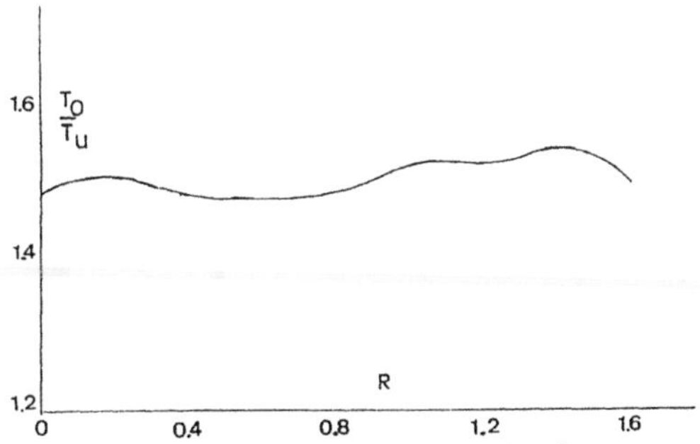

Figure 3. T_o/T_u vs.R ,PI, Control

where the normalised ratio T_o/T_u is nearly constant at the value of 1.5 for all R. Shinskey (1988) has indicated heuristically that the optimum setting for the PI controller is such that the above ratio is 1.35 for all R.The optimum loci corresponding to Fig.1 and 2 can also be obtained in a similar way for the case when k > 4.

PROCESS IDENTIFICATION

To be able to use the optimum results of Fig.2 , one needs to know the value of R for a given process.The procedure to determine R should be such that the disturbance to the process is minimum. Theorem 2 below provides a simple technique to determine R from two closed loop load responses under PI(D) control such as in Fig.4.

Theorem 2

Let T_{o1} and T_{o2} be the loop periods observed from damped responses as in Fig.4 corresponding to two known but different controller angles θ_1 and θ_2 respectively with the unknown process under PI(D) control.Then

$$R = (\pi - \psi + \theta_1)/\tan\psi \tag{19}$$

where ψ is given by

$$\tan(\delta + \alpha\psi) = \alpha \tan\psi \tag{20}$$

where

$$\delta = \pi + \theta_2 - \alpha(\pi + \theta_1) \tag{21}$$

$$\alpha = T_{o1}/T_{o2} \tag{22}$$

Further, T_u is given by

$$T_u = T_{o1}(\tan\psi / \tan\phi) \tag{23}$$

where ϕ is given by

$$(\pi - \phi)/\tan\phi = R \tag{24}$$

Proof:- see Appendix- B.

SELF-TUNING CONTROLLER

Hardware

The hardware configuration for the self-tuning controller is as given in Fig.5.The hardware can be divided into the following modules:(i) microprocessor and memory (ii)process measurement input (iii) control output (iv) control parameter entry (v) digital display (vi)alarm system

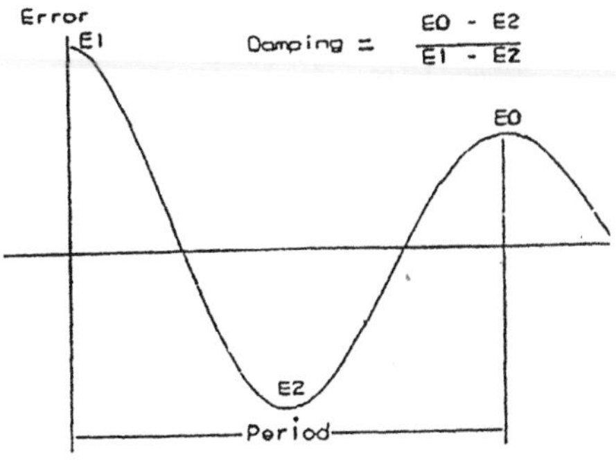

Figure 4. Control System Load Response

and (vii) RS 232 interface to the host where the expert system is residing.All the modules are supported by both the hardware and the software.The microprocessor and memory modules perform the function of PID control, data manipulation and control, and data storage.The process measurement input module acquires the measurement from the field transducer.The control output module outputs the analog manipulating signals to the process.The control data entry module provides a means by which the control parameters are entered into the system.The digital display module displays the essential system parameters. The alarm module caters to the necessary alarm functions for the safe operation of the control system.The RS 232 interface module provides the data and the communication link between the PID controller and the personal computer (PC/XT) where the expert system is running.

Controller Software

The software for the 8086-based PID controller is written in assembly language for the following functions:(i)general routines (ii)PID computation (iii) keypad scanning (iv) display (v)alarm (vi)data conversion (vii) initialisation and (viii) input/ output for process control.The controller card provides the conventional PID and PI actions.A communication driver is required for communication between the expert system and the microprocessor-based controller.Based on low level predicates written in the C language , the programme sets up polled transmission.The received data in the asynchronous communication adaptor will result in an interrupt and the bytes will be placed in a circular input queue.Similarly, characters to be sent will also be queued up at the output buffer.For the communication,software handshake is used.

Expert System Software

The expert system consists of the following: (i) knowledge base of domain facts and heuristics associated with the problem (ii) inference procedure for utilising the knowledge base and (iii) working memory .Turbo-Prolog is the language used for the implementation of the expert system.It is a declarative language designed primarily for symbolic data processing and is based on first-order predicate calculus.The main programme called 'foxcom7' consists of the expert programme, the serial communication driver and a timer programme.A flowchart for the expert system is as shown in Fig.6.The clauses relate to the facts and the rules concerning the application.In the PID controller, the recognition of response of the process is of primary concern.The rules are provided as to how to classify the response in terms of over damped and oscillatory responses.Rules are also required for the following:(i) to identify the direction of load response (ii) to calculate damping and period from the response (iii) to compute the parameters ψ , R and T_u as per the look-up table provided based on the solution of Eq.(19), (20),(23) and (24) (iv) to compute the I and D settings as per a look-up table provided based on the optimum curves of Fig.2 (v) to tune the P setting only up or down by a fixed proportion as per the observed and the required damping (vi) to update the dynamic database as the history of tuning develops and (vii) to properly sequence the different operations in the expert system.

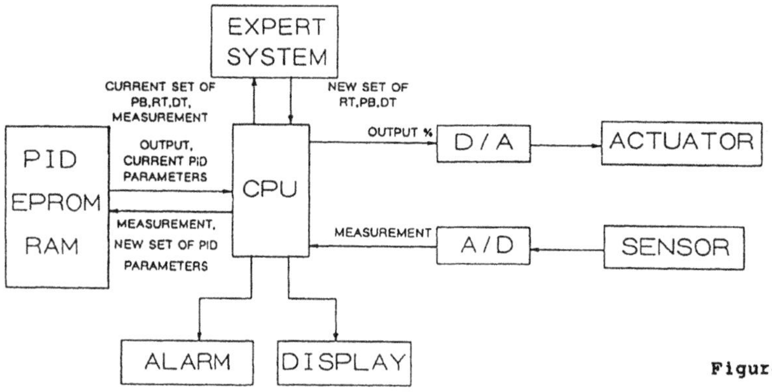

Figure 5. Expert PID Controller

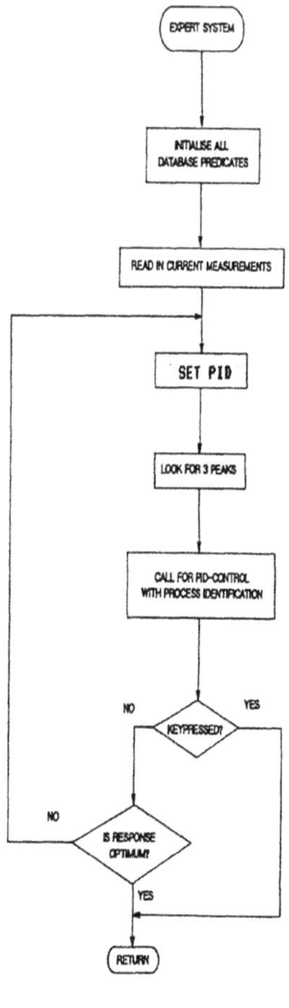

Figure 6. Flowchart for Expert PID Controller

EXPERIMENTAL RESULTS

Fig.7 shows a liquid pressure process in which water flows through two control valves V_1 and V_2.The pressure control loop is closed around the pressure transmitter and the valve V_2 using the microprocessor-based PID controller.The transmitter and the valve are connected to the controller via an analog Foxboro Spec-200 nest. The valve V_1 is kept open by a fixed amount using a Spec-200 display station and is a convenient point to introduce disturbance into the control system.Table 1 shows the results of the experimental trials conducted.The second and the third columns show the P,I and D settings used and the corresponding loop period obtained during the identification phase.The fourth column shows the corresponding controller angle computed in each case.The fifth column shows the value of R as calculated using Eq.(19).The last but one column shows the natural period T_u obtained. The optimum I, D settings are shown in the last column of Table 1.After setting the I and D values , the proportional band is adjusted by the expert system by trial and error as per the damping required.The proportional band used is also shown in the last column of Table 1.Fig.8 shows a typical load response used to identify the process as in Table 1.It was required to make an independent verification for the value of R given in Table 1. Some substantial delay was expected in the microprocessor-based controller incorporating a first order digital filter of time constant 0.01 min on the pressure measurement and using a polling mechanism in its software implementation.An ultimate gain test on the closed loop system resulted in the cycling as shown in Fig.9 which provided a value of T_u=6.9 s which is close to that

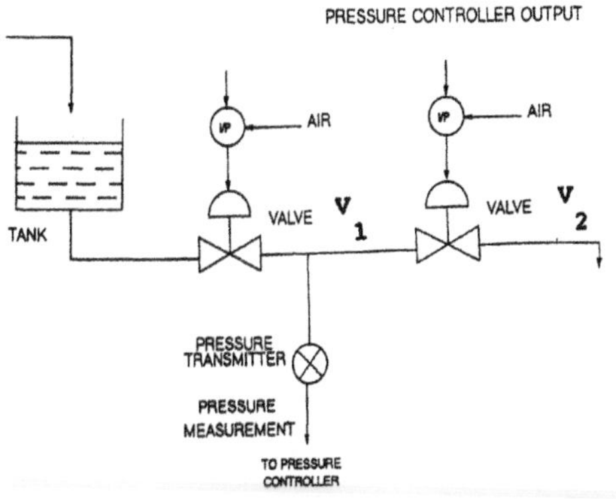

Figure 7. Liquid Pressure Pilot Process

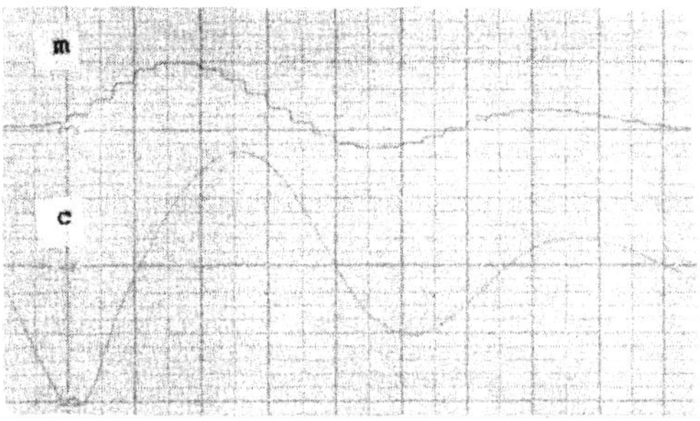

Figure 8. Closed Loop Load Response

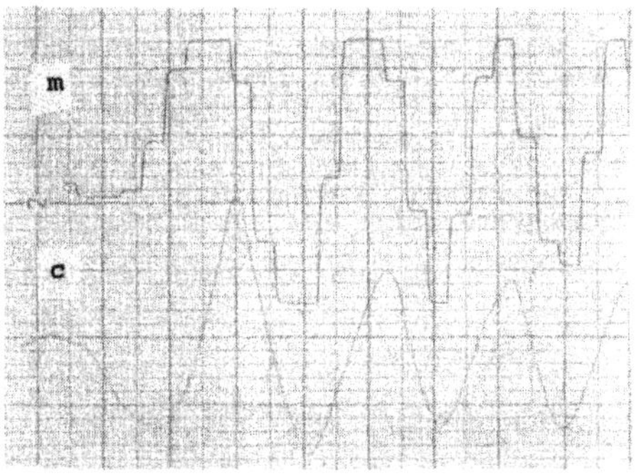

Figure 9. Cycling with Ultimate Period

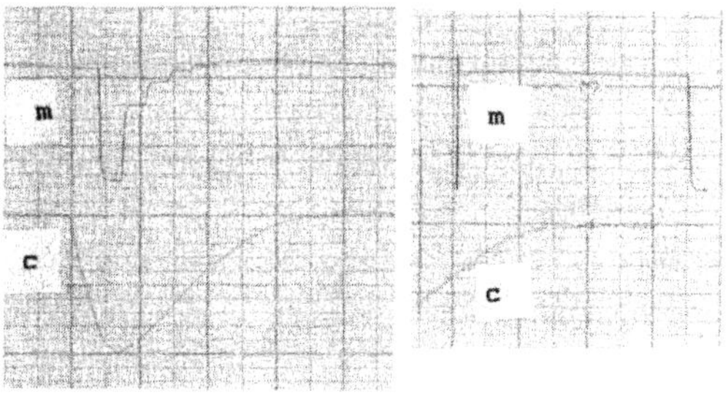

(i) Load Response (ii) Setpoint Response

Figure 10.Tuned Controller Response

Table 1 Results of Experimental Trials

	Identification							Optimum Settings		
No.	P	I m	D m	T_o s	θ rad	R	T_u s	P	I m	D m
1	100	0.01	0	14.4	-1.315	-	-	-	-	-
2	30	0.5	0	7.5	-0.04	1.22	7.5	100	0.04	0.01
3	35	0.1	0	8.4	-0.219	1.37	7.87	100	0.04	0.01

given in Table 1,thus verifying the value of R indirectly.Finally, a typical load response and a set point response obtained with the tuned controller are shown in Fig. 10.

CONCLUSIONS

A simple technique using natural damped closed loop responses has been shown to yield the ultimate period T_u within a resonable error. T_u can then be used to obtain the optimum I and D settings based on the minimum IE criterion while P is adjusted for the required damping.A self-tuning controller based on this principle has been implemented.The results obtained through experimental trials on the controller are encouraging.

ACKNOWLEDGEMENT

The author would like to thank Prof. Brian Lee , Dean, School of EEE, Nanyang Technological Institute,Singapore, for the the provision of facilities towards the research reported in this paper.He also would like to acknowledge the help provided by Mr.Toh Kar Ann in setting up the microprocesor hardware and software for the controller.

REFERENCES

Astrom,K.J.and Hagglund,T.(1984).Automatic tuning of simple reactors with specifications on phase and amplitude margins.Automatica,20,5,645-651.

Harris C.J. and Billings, S.A.(1981)Self-tuning and Adaptive Control: Theory and Applications: Peter Penguins:IEE, London.

Higham,E.J.(1986). Industrial Digital Control Systems, Ch. 15, Peter Penguins: London.

Ortega,R. and Kelly,R.(1984)"PID self-tuners:some theoretical and practical aspects",IEEE Trans. Ind.Elec.,IE-31,4,332-338.

McMillan,G.K.(1983)Tuning and Control Loop Performance,Instrument Society of America Monograph: Research Triangle Park, North Carolina.

Shinskey,F.G.(1979)Process control systems. McGraw-Hill: New York, 86-98.

Shinskey,F.G.(1988)Process control systems, McGraw-Hill: New York,121-122.

Warwick,K. Implementation of self-tuning controllers,Peter Penguins:IEE London.

Wittenmark,B.(1979).Self-tuning PID controllers based on Pole Placement.Lund Institute of Technology, Report: No. TFRT- 7179.

Ziegler,J.G. and Nichols, N.B., (1942) "Optimum settings for automatic controllers",Trans. ASME, 64, Nov.759-768.

Appendix-A

Using Eq.(10),Eq.(13) is written as

$$PI = (200K_p \pi R T_d / X).$$

$$(x\sin \pi y/2)/\{(\cos\theta)(\pi-\pi y/2+\theta)^2\} \quad (A.1)$$

Differentiate Eq.(A.1) with respect to θ while keeping k constant to yield

$$\{(\cos\theta)(\pi-\pi y/2+\theta)^2\}.$$

$$\{(dx/d\theta)\sin \pi y/2+(\pi x/2)(\cos \pi y/2)(dy/d\theta)\}$$

$$=x(\sin \pi y/2)[-(\sin\theta)(\pi-\pi y/2+\theta)^2$$

$$+2(\cos\theta)(\pi-\pi y/2+\theta)(1-(\pi/2)(dy/d\theta))] \quad (A.2)$$

Differentiating Eq.(10) with respect to θ

$$dy/d\theta = 2\cos^2(\pi y/2)/\pi(R+\cos^2\pi y/2) \quad (A.3)$$

Differentiating Eq.(5) with respect to θ

$$dx/d\theta = \pi x^2 \sec^2\theta /(1+\pi x\tan\theta) \quad (A.4)$$

Substituting Eq.(A.3) and (A.4) into Eq.(A.2) and simplifying yields Eq.(14) as required.From Eq.(A.1) it follows that the product PI reaches a minimum with respect to k when x is a minimum with respect to k which corresponds to the minimum value of k, viz., k=4 (see Eq.(5)).Putting k=4 into Eq.(14) yields Eq.(15).Hence the result.

Appendix-B

Using Eq.(10) for the two observed responses, and simplifying,

$$T_{o1} / T_{o2} = \tan(\pi y_2/2)/\tan(\pi y_1/2) =$$

$$\{\pi-\pi y_2/2+\theta_2\}/\{\pi-\pi y_1/2+\theta_1\} \quad (B.1)$$

Similarly using Eq.(10) and (17),

$$T_u / T_{o1} = \tan(\pi y_1/2)/\tan(\pi y_u/2)$$

$$=\{\pi-\pi y_1/2+\theta_1\}/\{\pi-\pi y_u/2\} \quad (B.2)$$

Eq.(B.1) and (B.2) yield

$$\tan(\delta+\alpha\psi) = \alpha\tan\psi \quad (B.3)$$

where

$$\psi = \pi y_1/2 \quad (B.4)$$

$$\delta = \pi+\theta_2-\alpha(\pi+\theta_1) \quad (B.5)$$

Eq.(B.3) and (B.10) yield Eq.(19).

Also, from Eq.(10), and (17), one can get

$$T_u = T_{o1}(\tan\psi / \tan\phi) \quad (B.7)$$

where ϕ is such that

$$R = (\pi-\phi) / \tan\phi \quad (B.8)$$

INVESTIGATIVE STUDIES FOR A KNOWLEDGE-BASED INTELLIGENT ADAPTIVE GENERALISED PREDICTIVE CONTROLLER

S. Nungam, T. H. Lee

Dept. of Electrical Engineering,
National University of Singapore, Kent Ridge (0511), Singapore

Abstract

This paper presents preliminary results on tuning the Generalised Predictive Controller (GPC) using an expert system technique. The closed-loop poles that result from applying the GPC to a first order plant are investigated. The plants are classified according to the pole-zero locations and the closed-loop behavior is studied in detail when the tuning parameters are varied. The tuning rules can be formulated from this knowledge and are implemented in the expert system to tune the GPC to satisfy the user specifications. Implementation using an expert system shell, NEXPERT, shows that this knowledge-based tuning procedure gives consistently good results

Keywords. Knowledge-based; intelligent machines; expert systems; tuning; GPC.

1. Introduction

In recent work [1],[2], the use of an expert system methodology for intelligent PID controllers have been considered and found to be attractive. The intelligence comes from investigating the PID controller in detail and providing heuristic rules for its tuning. This approach has worked rather well in realising an intelligent controller. However, the range of processes that can be satisfactorily controlled using the PID controller is restricted to those where the normalised dead-time is less than 1.5. For processes where the normalised dead-time is larger than 1.5, some other control strategies should be considered. In order to complement the intelligent PID controller of [1],[2], we consider issues in the design of an intelligent adaptive Generalised Predictive Controller (GPC).

Although the GPC has many advantages [3],[4], the algorithm is fairly complicated and general analysis of its properties is extremely difficult. Some stability results were presented in the original work of the GPC [3],[4],[5] and the related work by Kwon et al [9], but these were mainly concerned with limiting cases. Therefore the results are rather difficult to apply for tuning the GPC in an systematic way in order to satisfy the closed-loop performance which are usually specified in qualitative term , e.g. " as fast as possible", " acceptable control input", "small overshoot",etc. In typical applications, the tuning parameters such as the prediction horizon(N), the control horizon(NU) and the control weighting sequence(λ) are chosen in an ad hoc manner. These problems motivate us to investigate issues in the design of an intelligent adaptive GPC using an expert system framework. The idea is to analyse simple structures which are representative of a class of systems and develop heuristic rules for it. The rules will be implemented in an expert system in order to tune the GPC in an intelligent way.

In this paper, we present the preliminary results of our investigation using GPC for first order plants. Section 2 reviews the GPC algorithm and defines the plant description used in our investigation. Section 3 analyses the effects of varying N and λ on the closed loop pole location. Section 4 discusses the user specifications and formulates the tuning rules. Finally, conclusions are given in Section 5.

2 Overview of the GPC algorithm

In this section we review the GPC algorithm as given in [3]. In the deterministic case, the plant is described by

$$A(q^{-1})\, y(t) = B(q^{-1})\, u(t-1) + d. \qquad (1)$$

where d is a constant disturbance signal [6]. An alternative description of (1) is

$$A(q^{-1})\Delta y(t) = B(q^{-1})\Delta u(t-1) \qquad (2)$$

where $\Delta = 1-q^{-1}$.

This nonminimal description of the plant is useful because it automatically incorporates integral action in the resulting control law.

The strategy for the GPC can be summarized as follows:

(a) At the present time t, it is assumed that the command signal $W = [w(t+1), w(t+2)....w(t+N)]$ over the future horizon is available.

(b) A j-step ahead predictor $y(t+j)$ is derived by solving the prediction identity

$$1 = E_j(q^{-1})A\Delta + q^{-1}F_j(q^{-1}) \qquad (3)$$

where $j = 1,2,...N$.

The predictor can be written in vector form as

$$Y = GU + F$$

$$\text{where } Y = [\, y(t+1), y(t+2), ...y(t+N)\,]^T$$

$$U = [\, \Delta u(t), \Delta u(t+1)......\Delta u(t+N-1)]^T$$

$$F = [\, f(t+1), f(t+2)......., f(t+N)]^T$$

and $f(t+j)$ is the component of $y(t+j)$ composed of signals which are known at time t.
G is a lower-triangular matrix of dimension N x N (see Clarke et al [3] [4])

(c) The control sequence u(t), u(t+1), ...,u(t+N-1) is obtained by minimizing the cost function

$$J = \sum_{j=1}^{N} \left([y(t+j)-w(t+j)]^2 + \lambda[\Delta u(t+j-1)]^2 \right) \quad \lambda \geq 0 \tag{4}$$

where N is the output horizon and λ is the control weighting sequence. N and λ are tuning parameters that have to be specified beforehand.
The control increment vector is calculated from

$$U = (G^T G + \lambda I)^{-1} G^T (W-F) \tag{5}$$

(d) Only the first control input u(t) is applied at the present time and at the next time instance t+1, the overall procedure is repeated. Hence

$$u(t) = u(t-1) + g^T(W-F) \tag{6}$$

where g^T is the first row of $(G^T G + \lambda I)^{-1} G^T$.
For simplicity we define g^T as a row vector of

$$g^T = [\ \bar{g}_1 \ \ \bar{g}_2 \ \ \bar{g}_3 \ \cdots \ \bar{g}_N \ ] \ .$$

3.Closed-loop investigation of the GPC with first order plant

In this section,we analyse the closed-loop behaviour of the GPC with first order plant and indicate how the tuning parameters affect the closed loop response. This is used to help us find the rules for tuning the GPC.

In our preliminary investigation, we consider the first order plant of the form as (1) where

$$A(q^{-1}) = 1 + a\,q^{-1} \quad \text{and} \quad B(q^{-1}) = b0 + b1q^{-1}$$

From (6), the first incremental control input can be written as

$$\Delta u(t) = g^T(W-F) \tag{7}$$

We assume that the future command sequence [w(t+j): j= 1,2 ..] is constant which is equal to the current setpoint w(t).

Hence (7) can be written as

$$\Delta u(t) = [\ \bar{g}_1 \ \bar{g}_2 \ \ldots \bar{g}_N\ ] \begin{bmatrix} w(t) - f(t+1) \\ w(t) - f(t+2) \\ \vdots \quad\quad \vdots \\ w(t) - f(t+3) \end{bmatrix} \tag{8}$$

For this particular plant, we have

$$f(t+1) = g_{11}\, q^{-1}\Delta u(t) + F_1 y(t)$$
$$f(t+2) = g_{22}\, q^{-1}\Delta u(t) + F_2 y(t)$$
$$\vdots \qquad\qquad \vdots \qquad\qquad \vdots$$
$$f(t+N) = g_{NN}\, q^{-1}\Delta u(t) + F_N y(t) \ .$$

and (8) can be written as

$$\Delta u(t) = \bar{g}_1 w(t) - \bar{g}_1 g_{11} q^{-1}\Delta u(t) - \bar{g}_1 F_1 y(t) \ +$$
$$\bar{g}_2 w(t) - \bar{g}_2 g_{22} q^{-1}\Delta u(t) - \bar{g}_2 F_2 y(t) \ +$$
$$\vdots \qquad\qquad \vdots \qquad\qquad \vdots$$
$$\bar{g}_N w(t) - \bar{g}_N g_{NN} q^{-1}\Delta u(t) - \bar{g}_N F_N y(t) \ .$$

$$\text{or } \Delta u(t) = S1\, w(t) - S2\, q^{-1}\Delta u(t) - Fs\, y(t) \tag{9}$$

where

$$S1 = \sum_{i=1}^{N} \bar{g}_i \ , \quad S2 = \sum_{i=1}^{N} \bar{g}_i g_{ii} \quad \text{and} \quad Fs = \sum_{i=1}^{N} \bar{g}_i F_i$$

From (1) and (9) we can write closed-loop transfer function as

$$\frac{y(t)}{w(t)} = \frac{S1\, q^{-1} B}{A\Delta + q^{-1}\,(\,S2\, A\Delta + B\, Fs)} \tag{10}$$

Since S1 is a scalar depending on N and λ ,it is clear from (10) that closed-loop zero is the same as the open-loop zero

For simplicity of analysis, we first consider the GPC with Nu = 1 and a first order plant with no zero (b1=0). In this case, the closed-loop characteristic equation can be written as

$$1 + \left[(a-1) + \frac{M(N)}{G(N)+\lambda} \right] q^{-1} + a \left[\frac{G(N)}{G(N)+\lambda} - 1 \right] q^{-2} = 0 \tag{11}$$

where

$$M(N) = (a-2a^2) + (1-a^2)N + (-a+2a^2-a^3)(-a)^N + a^3 a^{2N}$$
$$G(N) = (2a-a^2) + (1-a^2)N - (2a-2a^2)(-a)^N - a^2 a^{2N}$$

and $a \neq -1$.

Details of the analysis is given in [10].

3.1 Closed-loop pole and output horizon N

In this subsection, we investigate the closed loop behaviour of the system when the value of N is varied with λ = 0. In this case, the last term of (11) is zero. The closed loop system has a pole which is fixed at the origin and another pole at

$$p = -a+1- \left[\frac{(a-2a^2)+(1-a^2)N+(-a+2a^2-a^3)(-a)^N + a^3 a^{2N}}{(2a-a^2)+(1-a^2)N-(2a-2a^2)(-a)^N - a^2 a^{2N}} \right] \tag{12}$$

We can see that (12) is a monotonic function in N, provided 'a' is negative which is normally the case for plants in process control.

The following cases may be deduced from (12)

(a) For N=1, the closed-loop pole is at p = 0 for all values of 'a'. The system is equivalent to a stable dead-beat controller.

(b) For open loop stable plants, ($|a| < 1$) as $N \to \infty$, the closed-loop pole monotonically approaches the open loop pole .

(c) For open loop unstable plants ($|a| > 1$), the closed-loop pole monotonically approaches the (1,0) point as $N \to \infty$.

Examples of the variation of the closed loop pole for open-loop stable and unstable plants are shown in Figure 1 curve (a) and Figure 2 respectively.

Thus far, we considered only a first order plant (stable and unstable) with no zero. The results are summarised as follows:
Such a plant can be stabilized by the GPC with λ = 0 and N $\geq$ 1. The larger the value of N chosen, the slower is the closed-loop response. However if the first order plant has an open loop zero (b1 $\neq$ 0), the closed loop behaviour is slightly more complicated. In the following paragraphs, we will discuss how the location of open loop zero affects the closed loop pole as N is varied.

Before we go into details, let us define the open loop zero location (β) as

$$\beta = -b1/b0 \ .$$

The closed loop characteristic equation is of the same form as (11) where

$$M(N) = (a-2a^2)-\beta(a^3-a^2+3a-1)+\beta^2(a-a^2-1) +(1-\beta)^2(1-a^2)N+[-\beta(1-a)^3+(\beta^2-a)(1-a)^2](-a)^N + a(\beta+a)^2a^{2N}$$

$$G(N) = 2\beta(a^2-a+1)+(2a-a^2)+\beta^2(2a-1)+(1-\beta)^2(1-a^2)N +[-2\beta(1-a)^2+(2\beta^2-2a)(1-a)](-a)^N-(\beta+a)^2a^{2N}$$

The details of the analysis is also given in [10].

By using the same method as described previously, the results can be concluded as follows.

For $N = 1$ and $\lambda = 0$, the closed-loop transfer function has both a pole and a zero at the location of the zero of the open loop plant. Hence if the zero is in the unit circle, the closed-loop system is equivalent to a stable dead-beat controller.

For a stable plant, the root locus asymptotically approaches the open loop pole as $N \to \infty$. For an unstable plant, the root locus approaches the (1,0)-point as $N \to \infty$.

As N is increased from 1 to infinity, the following conclusions can be drawn.

For stable plants

Case 1. Either the plant has a zero outside the unit circle on the left half of the z-plane, or it has a zero inside the unit circle on left side of the open loop pole. The root locus starts from the open loop zero and asymptotically approaches the open loop pole.

Case 2. The plant has the zero inside the unit circle on the right side of the open loop pole. The root locus asymptotically approaches the open loop pole from the right side as shown in Figure 1 curve (b). In this case, we can see that the closed-loop system can be stabilized by any value of $N > 1$. However the closed-loop response is always slower than the open loop response.

In the following two cases, the plant has a zero outside the unit circle on the right half of the z-plane. These cases are of special interest. Before we investigate further, let us define

$$L_1 = (-1 - L_2/L_3) \quad \text{where}$$

$$L_2 = -\beta(1-a)^3+(\beta^2-a)(1-a)^2+2\beta(1-a)^2-(2\beta^2-2a)(1-a)$$

$$\text{and } L_3 = a(\beta+a)^2+(\beta+a)^2$$

case 3. The plant has the pole-zero configulation such that $L1 > 0$. As shown in Figure 3 curve (a), the root locus starts from the zero, moves upward and then down before turning upwards and approaches the open loop pole. In this case we can see that the closed loop system can be stabilized if N is greater than a certain value. In addition, it can be shown that the closed-loop response will be faster than the open loop response if

$$N > N_0 \quad \text{where}$$

$$N_0 = \text{Log}(L_1)/\text{Log}(-a)$$

case4. The plant has pole zero location such that $L1 < 0$. The root locus approaches the asymptote without crossing it, as shown in Figure 3 curve (b). In this case we can also stabilize the system with N greater than a certain value. However the setpoint response will never be improved.

For unstable plant

case 1. The plant has a pole outside the unit circle. The plant may have a zero either outside the unit circle in the left half plane or inside the unit circle. The root locus starts from the zer and monotonically approaches the (0,1) point. In this case, the closed loop system can be stabilized with $N > 1$.

In the following two cases, the plant has a zero outside the unit circle in the right half plane.

case 2. The plant has a zero on the right side of the open loop pole. The root locus starts from the zero, moves into the unit circle, and then turns back and approaches the unit circle from the left,as shown in Figure 4 curve (a). In this case we can stabilize the system if N is greater than a certain value.

case 3. The plant has a zero on the left side of the open loop pole. The root locus approaches the unit circle from the right without moving into the unit circle, as shown in Figure 4 curve (b). In this case the closed-loop system cannot be stabilized by any value of N.

3.2 Closed loop pole and control weighting λ

One of its advantages of the GPC is that it provides a tuning parameter λ which directly relates to the control input. It is clear from the cost function (4) that if λ increases, the control input will be smaller. In this section we investigate the closed-loop behaviour when λ is varied and N is kept constant.

If $\lambda > 0$, the last term of equation (11) will no longer be zero. Hence the equation has two roots, one moving from the origin and another from the existing pole 'p' which depends on N. As λ is increased to infinity, the roots approach the open loop pole and the (0,1) point on the unit circle. The family of root loci of a stable plant and an unstable plant for various values of N are shown in Figures 5(a) and (b) respectively.

For a stable plant, if N is selected such that the root locus start inside unit circle, we can see that both the breakaway and the break-in points of the root locus are inside the unit circle. Hence the closed-loop system can be stabilized with any value of $\lambda > 0$. For the unstable plant, the break-in point is always outside the unit circle. Therefore λ can only be increased to a certain value beyond which the system becomes unstable.

In order to avoid the system from being too oscillatory, the value of λ should not be increased too much so that the closed loop poles are too far away from the breakaway point.

In summary, for both stable and unstable plants, increasing λ results in lower control input signal. But if we keep increasing λ, the overshoot of the response will be large.

3.3 The closed loop of the plant with dead-time

In the previous two sections, our investigation is based on a plant with no dead-time. The same method can be applied for a plant with dead-time.

It can be shown that the closed-loop poles of the plant

$$y(t) = \frac{B(q^{-1})}{A(q^{-1})} u(t) \text{ with the GPC } N = N_0 \text{ and } \lambda = \lambda_0$$

is exactly the same as the closed loop poles of the plant

$$y(t) = \frac{B(q^{-1})z^{-d}}{A(q^{-1})} u(t) \text{ with the GPC } N = N_0 + d$$
$$\text{and } \lambda = \lambda_0$$

where d is the time delay as a multiple of the sampling period. Using the second formulation, we can see that the dead-time can be taken into account in the tuning procedure very easily. Please see [10] for details.

4. Tuning Rules

In this section, we use the knowledge from the above sections and propose the rules for tuning the GPC to satisfy the user specifications. These rules are formulated in a systematic way so that they can be implemented in the knowledge-base of an expert system.

4.1 User Specification

As mentioned earlier, we use the GPC to complement the PID controller when the process has long dead-time. Hence the GPC has to be tuned in order to achieve the same performance as required for the PID controller. That is, the closed loop must be stable and the response should be fast, with small overshoot and acceptable control input. Users have to specify the maximum overshoot allowed for a step command and the maximum input limit. The expert system will then tune the GPC so that the closed loop response is as fast as possible, while the overshoot as well as the control input do not exceed the limit.

4.2 Rule deduction

In Section 3 we studied the closed-loop behavior when N and λ were varied separately. We now find the rules for adjusting both parameters to satisfy the above requirement.

These rules are implemented in NEXPERT which is an expert system shell on a HP 9000/360 running Unix. The implementation follows the framework which we have developed in [8]. In this framework, NEXPERT performs off-line tuning procedure by using a mathematical model of the plant and then sends the tuning parameters obtained to the GPC to control the plant.

The rules that will be formulated below are used for tuning all the plants mentioned, except for cases 2 and 4 of a stable plant and case 3 of an unstable plant which are currently under further investigations.

Before we deduce the tuning rules, let us define OVmax and INmax as the maximum overshoot and maximum control input limit respectively. These two values are to be specified by the user. The rules can be formulated as follows

Rule(1). IF OVmax and INmax are specified
THEN start the GPC algorithm (use a step command input) with a very small value of λ and a large value of N .

Rule(2) IF the control input is lower than two times of INmax
THEN decrease N until the control input is about two times of INmax

Rule(3) IF the control input is about two times of Inmax or larger than INmax
THEN increase λ until the control input is equal or slightly lower than INmax and the overshoot does not exceed OVmax

Rule(4) IF the overshoot exceeds OVmax as λ is increasing
THEN increase N by one and repeat rule(3)

Rule(5) IF the overshoot and the control input do not exceed OVmax and INmax respectively
THEN the user specification is satisfied with the current value of N and λ,
AND use these values for the GPC to control the plant.

In the first rule, we set λ to a very small value in order to ensure that $(G^TG + \lambda I)$ is invertable. The value of N is initially chosen to be large. In fact, for a first order plant with no zero, the initial value of N can be any value greater than the plant dead-time because the closed loop system is always stable. However, it is chosen to be large so that this rule can also be applied to a plant with a zero. Such a plant requires N to be larger than a certain value in order to guarantee stability.

In the second rule the value of N is adjusted until the control input is higher than the acceptable limit. After that, λ is increased in order to reduce the control input to the limit. From experience, we found that the control input should be about twice the limit before λ is adjusted. In fact, it can be higher this value without any effect on the final tuning result. However the tuning process will take longer time. The tuning results for a stable and unstable plant with no zero are shown in Figure 6 and Figure 7 respectively.

For the nonminimum-phase stable plant case3 and unstable plant case2, the above rules have to be augmented by a new rule. This rule is deduced from curve (a) of Figure 3 and Figure 4. We can see that the value of N should not be less than N_{min} otherwise the closed system will be brought to the unstable region very easily.

In the figures, N_{min} is the value of output horizon that corresponds to minimum value of the root locus. It is rather complicated to obtain N_{min} analytically. In the tuning procedure ,however, we can easily detect N_{min} by checking the control input. It is clear from the root locus that as N is decreased ,the control input becomes larger and reaches its maximum value at $N = N_{min}$. Hence the rule for detecting this point can be deduced as follows.

Rule(6) IF the plant is nonminimum phase
And the control input reaches the maximum value while N is being decreased
THEN stop decreasing N
And execute Rule(3).

This rule can be easily added to the knowledge-base of NEXPERT without affecting the existing rules. This is one of the advantages of representing knowledge in an expert system framework. Figure 8 shows an example of the tuning result for a nonminimum-phase plant

5. Conclusions

In our preliminary investigation, we show that the GPC is suitable for complementing the intelligent PID controller. It provides tuning parameters that can be systematically tuned in order to satisfy the same specifications as required by the PID controller. Moreover, it is applicable to a wide range of processes.

Heuristic knowledge for tuning the GPC is obtained mostly by studying the closed-loop pole locations as the tuning parameters N and λ are varied. We chose the control horizon NU = 1 throughout the study in order to keep the analysis simple. It is shown that NU = 1 is adequate for any first order plant. For higher order plants however, NU > 1 may be preferable.

The preliminary results of our investigation show that the idea of combining heuristic and theoretical knowledge for tuning the GPC is feasible. The same idea can be extended to higher order plants which are more complicated and, consequently, more rules are needed. In such cases, the advantages of implementation in an expert system framework will be fully realized.

References

[1] Astrom, K.J , C.C. Hang and P. Persson (1989),"Towards Intelligent PID Control", Proceeding of IFAC Workshop on AI in real-Time Control.

[2] Hang, C.C. ,K.J. Astrom and W.K.Ho(1989), "Refinement of the Ziegler-Nichols Tuning Formula",National University of Singapore, Department of Electrical Engineering, Control Group Technical Report CI-89-9.

[3] Clarke,D.W.,C.Mohtadi and P.S.Tuffs(1987a), "Generalized Predictive Control-Part I. The Basic Algorithm," Automatica, Vol.23, No.2, pp. 137-148.

[4] Clarke,D.W.,C.Mohtadi and P.S.Tuffs(1987b), "Generalized Predictive Control-Part II. Extension and Interpretations,"Automatica, Vol.23, No.2, pp. 149-160.

[5] Clarke, D.W.,C.Mohtadi (1989)," Properties of Generalized Predictive Control", Automatica, Vol.25,No.6,pp.859-875.

[6] Lee, T.H., W.C. Lai and K.H. Kwek(1989a), "Extended generalised predictive control incorporating feedforward,with application to real-time control of a heat-exchanger," 1990 ACC.

[7] Lee T.H., W.C. Lai and C.C. Hang(1989b), "Multivariable GPC using a feedforward paradigm,with application to real-time control of a coupled electric-drive pilot plant," Dept. of Electrical Engineering, National Univ. of Singapore, Control Group, Technical Report.

[8] Yue, P.K., T.H. Lee and C.C.Hang(1989), "Application of expert system methodologies in real-time process control," Proceeding of IEEE TENCON 1989,pp.421-424.

[9] Wook Hyun Kwon and Dae Gyu Byun (1989), "Receding horizon tracking control as a predictive control and its stability properties", INT.J. CONTROL, VOL.50, NO. 5 1807-1824.

[10] S.Nungam and T.H. Lee (1990) "Investigative studies for a Knowledge -based Intelligent Adapive Generalised Predictive Controller", Dept. of Electrical Engineering, National Univ.of Singapore, Control Group, Technical Report

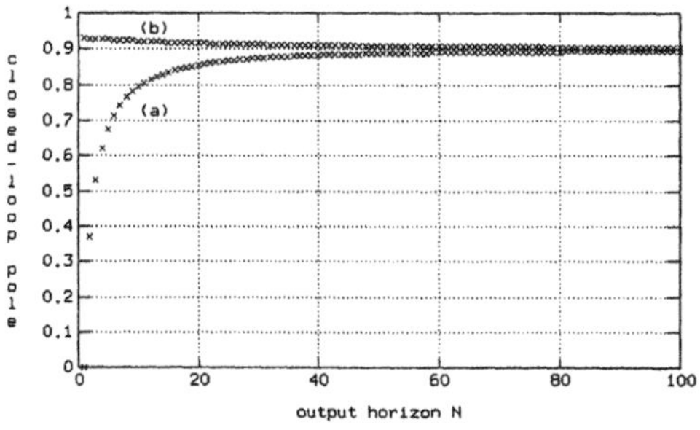

Figure1. The root locus of a stable plant with curve(a): a = -0.9, b0 = 0.1 and b1 = 0. curve(b): a = -0.9, b0 = 1 and b1 = -0.93

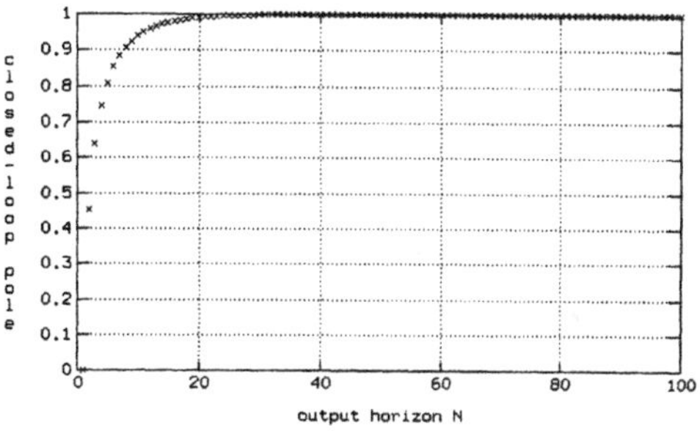

Figure2. The root locus of an unstable plant with a = -1.2, b0 = 1, and b1 = 0

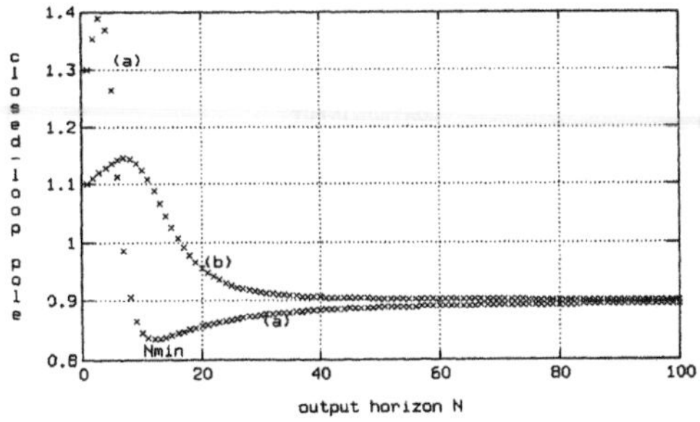

Figure3. The root locus of a stable plant with curve(a): a = -0.9, b0 = 1, b1 = -1.3, (L1>0) curve(b): a = -0.9, b0 = 1, b1 = -1.1, (L1<0)

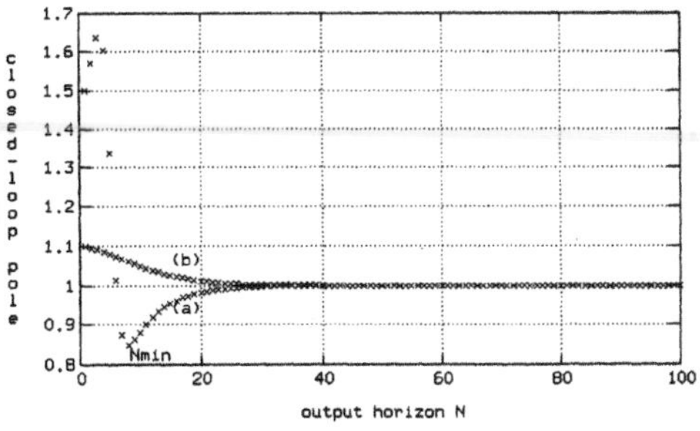

Figure4. The root locus of an unstable plant with curve(a): a = -1.2, b0 = 1,and b1 = -1.5 curve(b): a = -1.2, b0 = 1,and b1 = -1.1

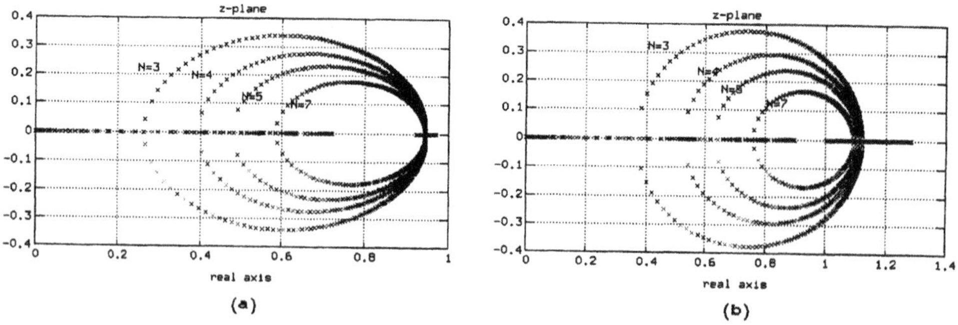

Figure5. The root contours of (a) a stable plant and (b) an unstable plant

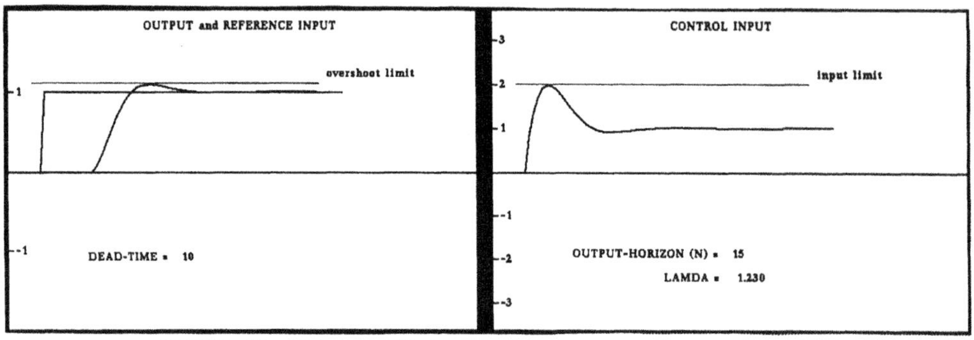

Figure6. The tuning result for a stable plant
with a = -0.9048, b0 = 0.0952, b1=0
maximum overshoot = 10 %
and maximum control input limit = 2 times of a step reference input

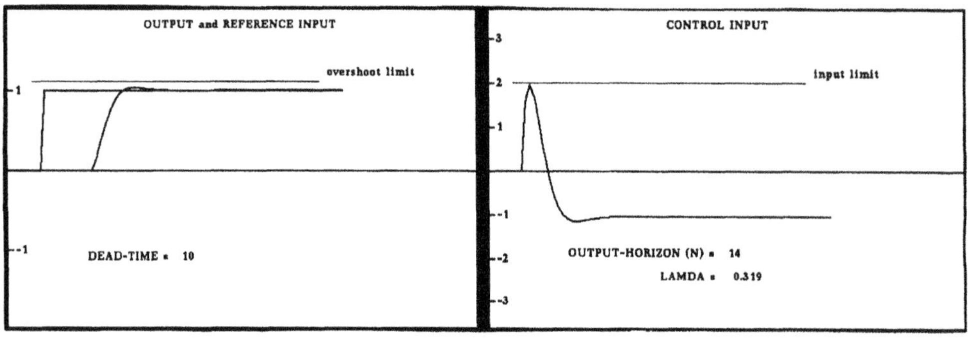

Figure7. The tuning result for an unstable plant
with a = -1.1052, b0 = 0.1, b1 = 0

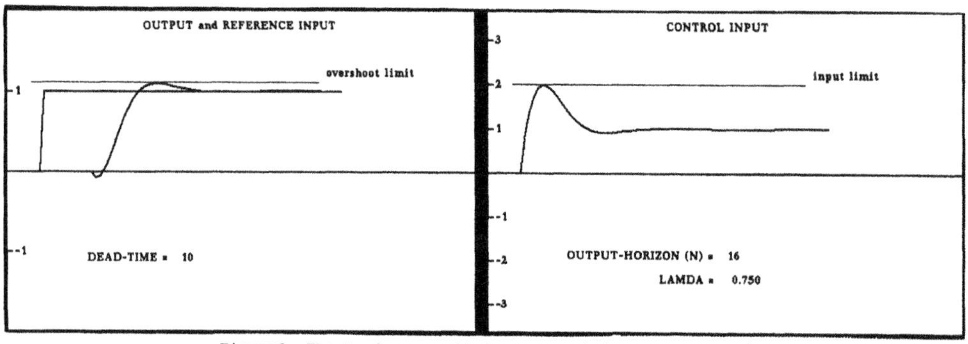

Figure8. The tuning result for a nonminimum-phase plant
with a = -0.9048, b0 = -0.1, b1 = 0.1952

BOUND-BASED WORST-CASE PREDICTIVE CONTROLLER

S. M. Veres and J. P. Norton

School of Electronic and Electrical Engineering,
University of Birmingham, Edgbaston, Birmingham B15 2TT, UK

Abstract. The computation of parameter bounds allows the synthesis of worst-case control, taking account of the uncertainty in the model after each update. The uncertainty of the model is characterized by bounds on the model parameters, which are used in calculating control values. The paper first points out deficiencies of existing self-tuning control methods based on parameter estimation. Then a worst-case controller (WCC) is introduced which properly accounts for model uncertainties. Next parameter uncertainty is linked to control performance in the bounding predictive controller (BPC). Its combined calculation of uncertainty sets and control inputs makes this approach very different from traditional self-tuners; in BPC there is no separation of estimation and control. Simulation examples shows the excellent performance of BPC.

Keywords Self-tuning control, adaptive control, parameter-bounding identification

INTRODUCTION

A number of algorithms for computing bounds on the parameters of a model of a dynamical system have been developed in recent years (Belforte and Milanese 1982, Fogel and Huang 1982, Mo and Norton 1988, Walter and Piet-Lahanier 1987, 1988, Veres and Norton, 1990). The bounds define the set of parameter values giving model-output errors within prescribed bounds, and may be regarded as the result of mapping the uncertainty in the observations into uncertainty in the model. Such bounded-parameter models are a natural basis for designing a controller to meet performance requirements expressed as inequalities in the system output error and control input. Treating the control design problem in this way has the advantages of reflecting classical control-design practice and allowing worst-case design. It also avoids the practical necessity of adopting certainty equivalence; this factor is important in achieving robust adaptive control based on parametric models.

Study of the well known methods of self-tuning control suggests that a better description of parameter uncertainties is much needed for e.g. pole-placement (PP), generalized-minimum-variance (GMV), generalized-predictive-control (GPC) and linear-quadratic-Gaussian (LQG) self-tuners. In the second section we briefly review some deficiencies of traditional self-tuners together with some of the attempts to make them more robust. Later sections show that parameter bounding can contribute more than just a new estimation technique to adaptive control; it makes possible a dual-control effect, balancing estimation against control performance. In the third and fourth sections the worst-case controller (WCC) and the bounding predictive controller (BPC) are presented. It will be seen that no separation principle between estimation and control design is applied in BPC. The last section gives simulation results.

MOTIVATION OF PARAMETER-BOUNDING

Early methods of self-tuning control equipped with good parameter estimators perform well with modest levels of disturbance and slowly changing plant dynamics. Minimum-variance self-tuning control, (Aström and Wittenmark 1973, Clarke and Gawthrop 1975) and its generalization (Clarke, 1984, Clarke et al. 1985) is relatively robust against errors in plant order but sensitive to some errors in dead time. Another approach based on closed-loop pole placement (Wellstead et al. 1979, Aström and Wittenmark 1980) proves insensitive to dead-time variations but sensitive to model overparametrization. Both approaches can be made to cope with non-minimum-phase models. Robustness against high-frequency noise and unmodelled dynamics is improved by the introduction of observer dynamics and disturbance-rejection filtering, and by careful choice of plant model structure and reference model. Generalized pole-placement control (GPP) (Lelic and Zarrop 1986) is designed to improve transient response and overall control performance by applying a multistep cost function. Generalized predictive control (GPC) (Clarke, Mohtadi and Tuffs 1987) aims to achieve a larger degree of robustness. GPC retains good features of GMV and pole placement while improving robustness against plant variations. Proper choice of cost function and incorporation of integral action (by using CARIMA models) can make the method more robust (Clarke and Mohtadi 1989) as found earlier in pole placement and GMV. GPC requires the specification of cost horizons, a control horizon and input weighting as design variables, as well as model orders. Special or limiting cases of GPC are dead-beat, GMV and LQG self-tuners and the methods of Peterka (1984), Ydstie (1984) and De Keyser and Van Cauwenberghe (1981,1985). A detailed analysis of relations of GPC and other methods is given by Clarke and Mohtadi (1989). Robust adaptive control has also been developed by Middleton, Goodwin and Wang (1988, 1989) using relative dead zones. A robust least-squares point estimator is combined with careful design of the control inputs while accounting for unstructured dynamics and structured disturbances.

All these methods are based on point estimators of the model parameters. Good tracking relies on selection of an adequate adaptive estimator, which in turn depends on the a priori assumptions made on possible model changes (see Ljung and Gunnarsson, 1990 for a survey). The commonest approach, recursive-prediction-error estimation with a scalar forgetting factor, can be far from optimal (Benveniste 1987). Optimal parameter tracking can be achieved in suitable circumstances by treating it as state estimation (Norton 1975), but if tracking is to be optimal the covariances describing the statistics of parameter changes (including modelling errors) and observation errors must be known. In practice they may have to be estimated, and may be poor initially. They may also be unable to reflect abrupt changes of the plant. The estimates of parameter uncertainty (error covariances) are then poor. Poor parameter estimates do not always cause problems for self tuning, however. A well known result (Åström and Wittenmark 1973) for MV control with LS estimation is that the self-tuning controller can work well in the steady state even if the model structure is incorrect. Nevertheless there are cases, especially for time-varying plants, where poor parameter estimates can lead to temporary (" bursting") or total loss of stability.

In this paper we take a different approach from point estimation and use direct computation of parameter uncertainty to design a robust worst-case adaptive controller. To set it in context we now discuss in more detail some deficiencies of point estimation in self-tuning controllers.

(1) The phenomenon of temporary loss of stability or sporadic bursts has been analysed by Anderson (1985). Instability can be caused by a drift of parameter values due to a temporary loss of identifiability. Such a loss can result from lack of persistency of excitation in the reference inputs. Practical reference inputs may well be insufficiently exciting; for instance they can be constant for long periods of time. When this happens individual parameters can drift into locations where closed-loop instability occurs, even though some functions of the parameters are well estimated. From this moment the feedback provides large exciting inputs, improving the parameter estimates and making the adaptive system stable again.

In principle there are at least two ways to prevent temporary loss of stability. One method is adding a persistently exciting signal to the control input (Caines and Lafortune 1984, Chen 1984, Chen and Caines 1985, Chen and Guo 1986,1987). The other method is to evaluate the uncertainty of the parameter estimates and apply on-line experiment design to improve the estimates if necessary. This latter approach is advocated in the present paper in the framework of parameter bounding identification and worst-case control design.

(2) Poor estimation of the parameters can cause not only temporary but fatal instability of the closed-loop system. This can happen if the plant is rapidly changing and the parameter estimator is not "tuned" properly. Then the estimation-control feedback cycle constitutes an unstable nonlinear system. For illustration we have picked out two well known methods: pole placement and generalized predictive control. Figs. 1 and 2 show sets of closed-loop poles produced by control using various realisations of the model parameters from a Gaussian distribution with covariance as shown. Each figure shows the poles for the mean parameter values, taken as differing somewhat from the true plant parameters, and separately for a large number of realisations so as to show the additional scatter due to the parameter covariance. The continuous-time system satisfies the equations

$$dx_t^1 = -(9+v_t)x_t^2 - 6x_t^1 + 10u_t^1$$

$$dx_t^2 = x_t^1 + Cu_t^2 , \quad y_t = x_t^2 \qquad (2.1)$$

where u_t^1 is the control input and u_t^2 is a random step signal considered as an unobservable disturbance. Let the sampling time be h. v_t is a sine wave with amplitude 3 and period $12s$. For $C=0$ the discrete-time zero-order-hold equivalent model is

$$y_h(k) + a_1 y_h(k-1) + a_2 y_h(k-2) = b_1 u_h(k-1) + b_2 u_h(k-2) ,$$

$$k=1,2,3,...$$

where $y_h(k) \equiv y(kh)$. For $h=0.6$ the coefficients of the "true" plant are $a_1=-0.3306$, $a_2=0.0273$, $b_1=0.5968$, $b_2=0.1773$. The discrete-time model

$$y_h(k) + a_1 y_h(k-1) + a_2 y_h(k-2) =$$

$$= b_1 u_h(k-1) + b_2 u_h(k-2) + b_3 u_h(k-3) , \quad k=1,2,3,... \qquad (2.3)$$

is estimated by recursive least squares (RLS). On reflection it is clear that the estimation of model (2.3) is well conditioned, by which we mean that the asymptotic information matrix is nonsingular. The estimator could be better adjusted by Kalman filtering or using the method of Chen & Norton (1986).

As an illustration of the dangers of self-tuning control, Fig. 1(b) shows the locations of the closed-loop poles when imprecisely estimated parameter values and the "true" plant defined above are involved in pole-placement control. Fig. 1 (b) shows the locations of poles for 300 independent realisations of the parameter values. For each realisation the pole-placement

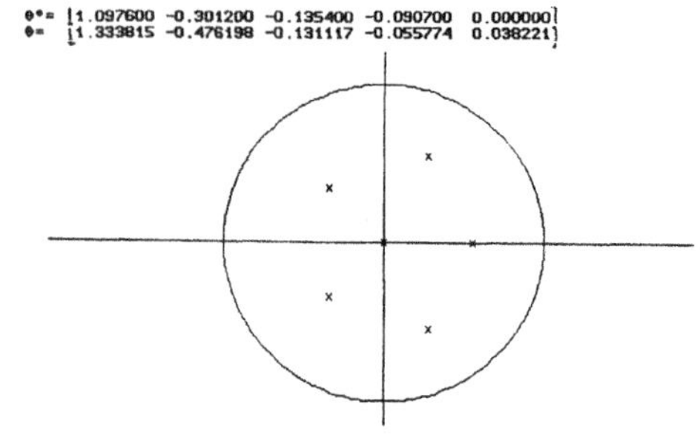

a) for the mean parameter vector

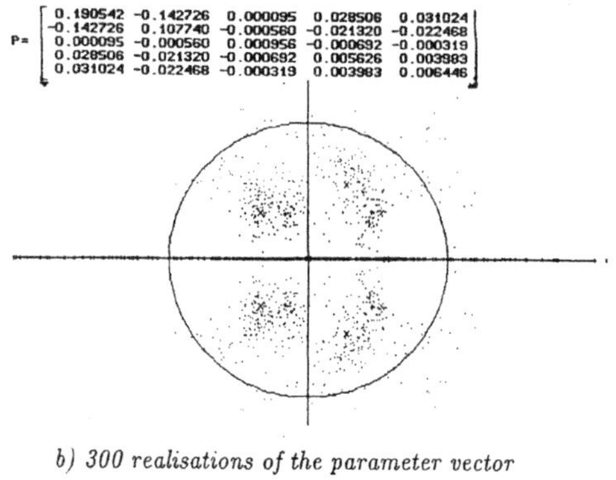

b) 300 realisations of the parameter vector

Fig. 1 *Closed-loop poles for the pole-placement method at a single time instant*

control law was computed and used in combination with the true plant to provide the true closed-loop poles plotted in Fig. 1 (b) . Fig. 2 shows a similar experiment for the GPC predictive controller. These figures are not meant to prove general results; they merely demonstrate the need for concern about the stability of the real plant when a self-tuner is attached to it. In many of the cases simulated the closed-loop poles are close to the unit circle, and for many parameter values with the given covariance the poles are outside the unit circle.

One way to increase reliability of a self-tuning controller is to get more information about parameter uncertainty and only apply control inputs which account for these uncertainties. For this purpose the assessment of parameter uncertainty has to be good. If the noise and parameter changes are white and of known covariance in a linear system, Kalman or nonlinear filtering can be applied to ensure good parameter estimates and good uncertainty evaluation on average. When uncertainty is treated statistically, one can only minimize the probability of the closed-loop system becoming unstable. Parameter-bounding techniques are available now for a large class of models, and provide a deterministic way of ensuring stability.

The aim of the present paper is to discuss new ideas

for this approach. One important feature will be that parameter bounding will be combined with on-line experiment design to avoid large parameter uncertainty regions. The separation of estimation and control is decreased, giving dual control.

$$\theta^* = [1.097600\ \ -0.301200\ \ -0.135400\ \ -0.090700\ \ 0.000000]$$
$$\theta = [1.333815\ \ -0.476198\ \ -0.131117\ \ -0.055774\ \ 0.038221]$$

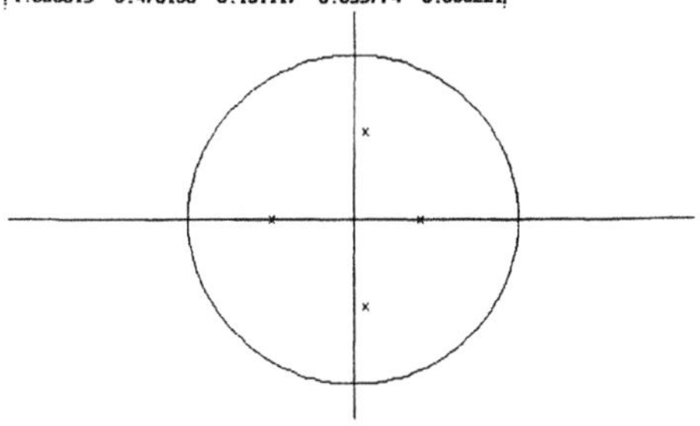

a) for the mean parameter vector

$$P = \begin{bmatrix} 0.190542 & -0.142726 & 0.000095 & 0.028506 & 0.031024 \\ -0.142726 & 0.107740 & -0.000560 & -0.021320 & -0.022468 \\ 0.000095 & -0.000560 & 0.000956 & -0.000692 & -0.000319 \\ 0.028506 & -0.021320 & -0.000692 & 0.005626 & 0.003983 \\ 0.031024 & -0.022468 & -0.000319 & 0.003983 & 0.006446 \end{bmatrix}$$

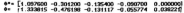

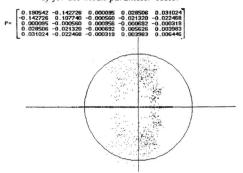

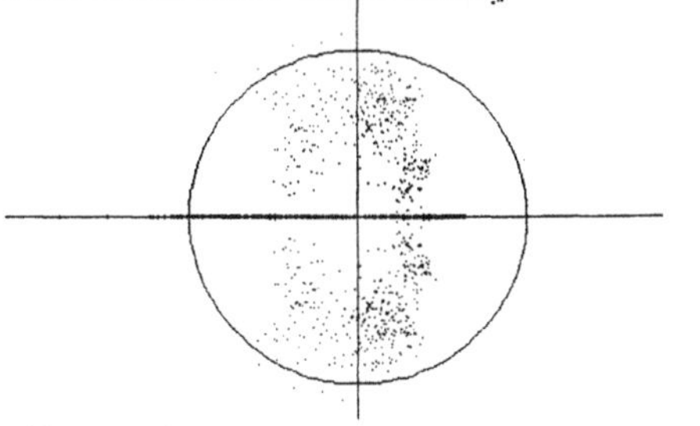

b) 300 realisations of the parameter vector

Fig. 2 *Closed-loop poles for GPC at a single time instant*

WORST CASE CONTROL BY PARAMETER BOUNDING

A new type of adaptive control algorithm is pesented, based on a bounded-noise, linear-in-the-parameters model. Using a polytope parameter-uncertainty region, an optimal control input is computed for a fixed time horizon. The parameter-uncertainty regions are computed from a fixed number of input-output pairs.

Consider the following general linear-in-the-parameters model:

$$y_{t+1} = \sum_{i=1}^{p} a_i f_i(y_{t-i+1}) + \sum_{i=1}^{r} b_i g_i(u_{t-i}) + e_t \qquad (3.1)$$

where a_i and b_i are unknown parameters, f_i and g_i are known, possibly nonlinear, functions and $\{e_t\}$ is a bounded noise sequence satisfying

$$e_t \in [r_t^l, r_t^u] = E_t$$

The output-prediction equation is

$$y_{t+k} = \sum_{i=1}^{p} a_i(k) f_i^k(y_{t-i+1}) + \sum_{i=1}^{r} b_i(k) g_i^k(u_{t-i}) + e_t(k), \ 1 \le k \le N, \qquad (3.2)$$

where

$$r_t^u(k) < e_t(k) < r_t^u(k)$$

A special case of these bounded models is where $f_i^k(y)=y$ and $g_i^k(u)=u$ for each possible i and k. Then

$$y_{t+1} = \sum_{i=1}^{p} a_i y_{t-i+1} + \sum_{i=1}^{r} b_i u_{t-i+1} + e_t, \quad r_t^u < e_t < r_t^u \qquad (3.3)$$

Let $\mathfrak{D}_t$ denote the set of parameter vectors $\theta=(a_1,\dots,a_p,b_1,\dots,b_r)$ satisfying (3.3), which is a convex polytope. From (3.3),

$$y_{t+2} = \sum_{i=1}^{p} a_i(2) y_{t-i+1} + b_1 u_{t+1} + \sum_{i=1}^{r} b_i(2) u_{t-i+1} + e_{t+1} + a_1 e_t \qquad (3.4)$$

$$\begin{aligned} y_{t+3} &= a_1(2)\Big(\sum_{i=1}^{p} a_i y_{t-i+1} + \sum_{i=1}^{r} b_i u_{t-i+1} + e_t\Big) + \\ &+ \sum_{i=1}^{p-1} a_{i+1}(2) y_{t-i+1} + b_1 u_{t+2} + b_1(2) u_{t+1} + \\ &+ \sum_{i=0}^{r-1} b_{i+1}(2) u_{t-i+1} + e_{t+2} + a_1 e_{t+1} \\ &= \sum_{i=1}^{p-1} (a_{i+1}(2) + a_1(2) a_i) y_{t-i+1} + a_p y_{t-p+1} + \\ &+ b_1 u_{t+2} + b_1(2) u_{t+1} + \sum_{i=1}^{r-1} (b_{i+1} + a_1(2) b_i) u_{t-i+1} + b_r u_{t-r+1} + \\ &+ e_{t+2} + a_1 e_{t+1} + a_1(2) e_t = \\ &= \sum_{i=1}^{p} a_i(3) y_{t-i+1} + b_1 u_{t+2} + b_1(2) u_{t+1} + \sum_{i=1}^{r} b_i(3) u_{t-i+1} + \\ &+ e_{t+2} + a_1 e_{t+1} + a_1(2) e_t \end{aligned} \qquad (3.5)$$

etc. for $k=4,\dots,N$.
Denote by $\{y_t^*\}$ the sequence of set points. The control performance criterion function is

$$\mathfrak{C}(t) = \max_{e_{t+i} \in E_{t+i},\, i=1\dots n} \Big\{ \bigvee_{k=d}^{N} |y_{t+k}^* - y_{t+k}| \Big\} \qquad (3.6)$$

where d denotes a lower bound for the dead time between input-output based on physical considerations; if nothing sure is known about d then d can be taken as 1.

Our aim is to find an adaptive controller which minimizes $\mathfrak{C}(t)$ by computing $u_t, u_{t+1}, \dots, u_{t+k-1}$ on the basis of a continuously updated bounded parameter region $\mathfrak{D}_t$ and the past inputs and outputs $u_{t-1}, u_{t-1}, \dots$, and y_t, $y_{t-1}, \dots$. When $\mathfrak{C}(t)$ is optimized for the full variation of $e_{t+i} \in E_{t+i}$ and $\theta \in \mathfrak{D}_t$, then essentially a finite-horizon, worst-case-optimal control is calculated, which guarantees that whatever the modelling error within $\mathfrak{D}_t$, the output will be within a bounded (worst-case) region about the setpoints y_t^* . In the case of modelling by (3.6), the algorithmic problem is the optimization of a multilinear form over a convex polytope. The order of the multilinear form is equal to the time horizon over which the input is optimized. Thus the control inputs $u_t, u_{t+1}, \dots, u_{t+k-1}$ are calculated to minimize

$$\max_{\theta \in \mathfrak{D}_t} \mathfrak{C}(t)$$

i. e.

$$(\tilde{u}_t(t), \dots, \tilde{u}_{t+k-1}(t)) = \arg\min_{(u_t, u_{t+1}, \dots, u_{t+k-1})} \max_{\theta \in \mathfrak{D}_t} \mathfrak{C}(t) \qquad (3.7)$$

As time elapses, the remaining input control sequence will be

$$\hat{u}_t = \tilde{u}_t(t) .$$

Further on the control law defined in (3.7) will be referred to as the *explicit bounding predictive worst-case controller (WCC).* In the *implicit bounding predictive WCC,* the parameter bounds for the predictor equations for different prediction steps will be calculated instead of the parameter-bounding set $\mathfrak{D}_t$. Denote by $\mathfrak{P}_t^1, \dots, \mathfrak{P}_t^k, \dots, \mathfrak{P}_t^n$ the parameter-bounding sets for the coefficients in the predictor equations

$$\begin{aligned} y_{t+k} &= \sum_{i=1}^{p} a_i(k) y_{t-i+1} + \sum_{i=1}^{k-1} b_1(i) u_{t+k-i} + \sum_{i=1}^{r} b_i(k) u_{t-i+1} + \\ &+ \sum_{i=1}^{k-1} a_1(i) e_{t+k-i-1} + e_{t+k-1}, \quad k=1,\dots,n \end{aligned} \qquad (3.8)$$

so that each bounding set $\mathfrak{P}_t^k$ is $(p+r+2k-2)$-dimensional. In our applications $\mathfrak{P}_t^k$ can be a polytope, ellipsoid, orthotope or simplex. Parameter-bounding updating of (3.8) is an errors-in-variables problem because of the occurrence of the error terms e_{t+i}, $i=0,\dots,N-1$. Recursive updating of polytopes or ellipsoids for errors-in-variables problems is discussed by Veres and Norton (1990). Based on the bounds calculated on line for the predictor equations, the control law is

$$(\tilde{u}_t(t), \dots, \tilde{u}_{t+n-1}(t)) = \arg\min_{(u_t, u_{t+1}, \dots, u_{t+n-1})} \max_{k=1\dots n} \tilde{\mathfrak{C}}_k(t) \qquad (3.9)$$

where

$$\bar{\mathcal{E}}_k(t)=\sup_{\theta^k\in\mathcal{P}_k(t),e_{t+i}\in E_{t+i},i=1\ldots n}|y^*_{t+k}-\hat{y}_{t+k}|,$$

and

$$\theta^k=[a_1(k),\ldots,a_p(k),b_1(1),\ldots,b_1(k\text{-}1),\\ ,b_1(k),\ldots,b_r(k),a_1(1),\ldots,a_1(k\text{-}1)]^T$$

$$\phi^k=[y_t,\ldots,y_{t-p+1},u_{t+k-1},\ldots,u_{t-r+1},e_{t+k-2},\ldots,e_t]^T$$

and

$$\hat{y}_{t+k}=[\phi^k]^T\theta^k$$

with the prescribed set point sequence $\{y^*_i\}$. Optimization of (3.9) can be solved without too much computational difficulty.

Next we describe a geometric algorithm for the calculation of the optimal inputs as defined in (3.9) if the parameter uncertainty regions are polytopes. First we establish a lemma for the calculation of $\bar{\mathcal{E}}_k(t)$. Denote the set of vertices for $\mathcal{P}_k(t)$ by $\mathcal{V}_k(t)$ and let $E_t=[-\delta,\delta]$, $t\in\mathbb{Z}$.

Lemma 3.1

$$\bar{\mathcal{E}}_k(t)=\max_{\theta^k\in\mathcal{V}_k}\max_{j=-1,1}\{j[\tilde{\phi}^k(j)]^T\theta^k-jy^*_{t+k}\}$$

where

$$\tilde{\phi}^k(j)=[\bar{\phi}^k_1(j)sign(\theta^k_1),\ldots,\bar{\phi}^k_d(j)sign(\theta^k_d)]^T,\ d=dim(\theta^k),\ j=-1,1.$$

and $\bar{\phi}^k(j)$, $j=-1,1$ are obtained from ϕ^k by replacing all its noise components by δ or $-\delta$:

$$\bar{\phi}^k(1)=[y_t,\ldots,y_{t-p+1},u_{t+k-1},\ldots,u_{t-r+1},\delta,\ldots,\delta]^T$$

and

$$\bar{\phi}^k(-1)=[y_t,\ldots,y_{t-p+1},u_{t+k-1},\ldots,u_{t-r+1},-\delta,\ldots,-\delta]^T.$$

Proof

$$\bar{\mathcal{E}}_k(t)=\sup_{\theta^k\in\mathcal{P}_k(t),e_{t+i}\in E_{t+i},i=1\ldots n}|y^*_{t+k}-\hat{y}_{t+k}|=$$

$$=\sup_{\theta^k\in\mathcal{P}_k(t)}\ \sup_{e_{t+i}\in E_{t+i},i=1\ldots n}\max\{y^*_{t+k}-\hat{y}_{t+k},\ \hat{y}_{t+k}-y^*_{t+k}\}$$

It is easy to see that

$$\sup_{e_{t+i}\in E_{t+i},i=1\ldots n}\{y^*_{t+k}-\hat{y}_{t+k}\}=y^*_{t+k}-\bar{\phi}^k_{-1}(-1)^T\theta^k$$

$$\sup_{e_{t+i}\in E_{t+i},i=1\ldots n}\{\hat{y}_{t+k}-y^*_{t+k}\}=\bar{\phi}^k(1)^T\theta^k-y^*_{t+k}.$$

Since a linear form over a convex polytope can only take its extreme values at the vertices of the polytope, the identities

$$\sup_{\theta^k\in\mathcal{P}_k}\{j\bar{\phi}^k(j)^T\theta^k-jy^*_{t+k}\}=\max_{\theta^k\in\mathcal{V}_k}\{j\bar{\phi}^k(j)^T\theta^k-jy^*_{t+k}\},\ j=-1,1$$

conclude the lemma. □

Assume that M>0 is an upper bound for the absolute values of the inputs. Define $U(n)$ as the domain of future inputs

$$\bar{u}_t=[u_t,\ldots,u_{t+n-1}]^T\in U(n)=\prod_{k=1}^{N}[-M,M].$$

It can be seen by Lemma 3.1 that $\max_{k=1\ldots n}\bar{\mathcal{E}}_k(t)$ is a piecewise linear function over $U(n)$, which can be represented as the maximum of a finite number of linear forms:

$$\max_{k=1\ldots n}\bar{\mathcal{E}}_k(t)=\max_{i\in I}L_i(\bar{u}_t).\qquad(3.10)$$

Each linear form $L_i(\bar{u}_t)$ can also be viewed as a hyperplane H_i in the space of the box $B(n)=U(n)\times[-N,N]\subset\subset R^{n+1}$, where $N>0$ is a sufficiently large number. Thus the space "above" the function $L_i(\bar{u}_t)$ can be calculated by the well known procedure of polytope updating (Mo and Norton, 1988, Walter and Lahanier, 1988). Then the minimum of (3.10) can be found by going through the vertices of this polytope. To be more explicit denote by S_i the halfspace in R^{n+1} above the linear form $L_i(\bar{u}_t)$. Then the polytope

$$B(n)\cap\bigcap_{i\in I}S_i$$

can represent the space above the function

$$\max_{k=1\ldots n}\bar{\mathcal{E}}_k(t),$$

for a sufficiently large choice of $B(n)$. Thus we have shown the following.

Lemma 3.2 The minimization of (3.9) leads to finding the vertex $[(\bar{u}^v_t)^T,C^v]=[u^v_t,\ldots,u^v_{t+n-1},C^v]$ of the polytope

$$B(n)\cap\bigcap_{i\in I}S_i$$

with a minimal last component C^v.

In the on-line control procedure the current input u^v_t is applied to the plant at time t and the rest of the values $(u^v_{t+1},\ldots,u^v_{t+n-1})$ will be lost and recomputed at the next time instant.

As presented above we have applied a prediction horizon $1,\ldots,n$. As in GPC, the prediction horizon could be modified by a lower and upper index n_1 and n_2 by optimizing the criterion (3.9) with a modified

$$\bar{\mathcal{E}}_k(t)=\sup_{\theta^k\in\mathcal{P}_k(t),e_{t+i}\in E_{t+i},i=n_1\ldots n_2}|y^*_{t+k}-\hat{y}_{t+k}|.$$

It is obvious that the solution **for this** optimization problem is a slight modification of that in case of prediction horizon $1,\ldots,n$.

One may have the initial impresssion that the controller presented above is the GPC associated with another, maybe better estimator. The essential difference is that there is now no single point estimate but a set of feasible parameter points, and the control requirement is to perform well for any such feasible parameter values. Thus "uncertainty equivalence" of estimator and controller is guaranteed. To be more precise, the criterion (3.9) selects the future inputs on a minimax basis, minimizing the maximal error over the parameter uncertainty region, thereby accounting for the worst possible parameter values: worst-case design.

Evaluation of closed-loop stability and performance is not possible in the way shown in Figs. 1 and 2 for point estimates with confidence ellipsoids. To study the robust behaviour of WCC a separate technical apparatus is required which needs much more space than the length of this paper allows and will be discussed elsewhere. Instead, we will proceed by discussing the possible incorporation of on-line experiment design, so as to give a dual effect, i.e. a trade off between estimation and control performance.

BOUNDING PREDICTIVE CONTROLLER

Three important issues will be addressed: equation-error bounds and memory specification, a theorem on guaranteed control performance, and adaptation of error bounds. The special case of the above WCC with prediction horizon n=1 is considered [but the following definitions and statements can easily be generalized to $n>1$.]

a) Guaranteed control performance
Denote by $S_t(\delta)$ the parameter set obtained at time t with the equation-error bound δ; $S_t(\delta)$ is a strip bounded by two parallel hyperplanes in the parameter space, corresponding to the extremes of the model output, δ each side of the observed output.

Definition 4.1 A sampled system $\mathcal{I}$ is said to have *B-memory* at least m with equation-error bound $\delta>0$ iff for any t

$$\bigcap_{i=0}^{m} S_{t-i}(\delta) \neq \emptyset$$

To show the meaning of this definition we can reformulate it as follows. Let $P_t^m(\delta) \equiv \bigcap_{i=1}^{m} S_{t-i}(\delta)$. For given $\delta>0$, m is such that at any time t there is a parameter vector $\theta^* \in P_t^m(\delta)$ with equation error

$$|y_t - \phi_t^T \theta^*| < \delta$$

which also means that locally the system can be approximated by the linear model with parameter θ^* and equation-error bound δ. The largest m for which $P_t^m(\delta) \cap S_t \neq \emptyset$ for every t will be called the *B-memory* at error level δ and denoted by $m^*(\delta)$.

Lemma 4.1 If $\mathcal{I}$ has at least memory m at error level $\delta>0$ then the WCC control law has a setpoint-following error bounded by

$$|y_t - w_t| \leq \delta + \min_{u_t} \max_{\theta \in P_t^m(\delta)} | w_t - \phi_t^T \theta^* | ,$$

where w_t is the set-point sequence to be followed.

The proof of Lemma 4.1 is obvious. The importance of this lemma is that it highlights the fact that the loss function

$$L_t^m(\delta) = \min_{u_t} \max_{\theta \in P_t^m(\delta)} | w_t - \phi_t^T \theta^* |$$

has to be kept as small as possible. Obviously,

$$L_t^{m_1}(\delta) \geq L_t^{m_2}(\delta) \ \ if \ \ m_1 \leq m_2 \leq m^*(\delta) \text{ and } .$$

and

$$L_t^m(\delta_1) \geq L_t^m(\delta_2) \ \ if \ \delta_1 \geq \delta_2 \ and \ m \leq m^*(\delta_1), \ m \leq m^*(\delta_2)$$

Thus it follows that for various δ values $L_t^m(\delta)$ takes its minimum value at $m=m^*(\delta)$. In Fig. 4.1 "$\odot$" denotes the

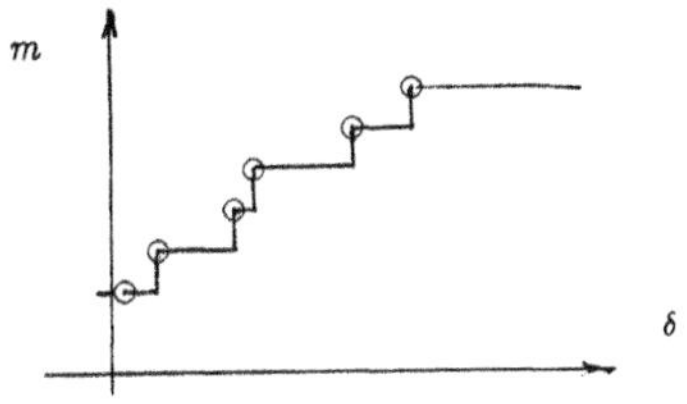

Fig 4.1 *B-memory function m^* of $\delta > 0$*

pairs of (m,δ) where $m=m^*(\delta)$ and $\delta = \inf\{\delta>0 / m^*(\delta) \geq m\}$ hold, i.e. those (m,δ) values where $L_t^m(\delta)$ is expected to take its minimum value. If an upper bound is specified for δ then we only have to select the minimum place of $L_t^m(\delta)$ from a finite number of possibilities. Thus we have arrived at the following simple but useful theorem.

Theorem 4.1 (*Bounding predictive controller*)
If the B-memory function is known for all $\delta>0$, it can guaranteed that the controller

$$u_t = \arg\min_{u_t} \min_{m \geq 1} L_t^m(\delta \leq m^*(\delta))$$

follows the set point with error not worse than

$$|y_t - w_t| \leq \delta + L_t^{\bar{m}}(\delta^*)$$

where $\delta^* = \min\{\delta \ / \ m \leq m^*(\delta)\}$ and $\bar{m} = m^*(\delta^*)$.

The proof is again fairly simple. If we can find good "estimates" for the memory function $m^*(\delta)$ then we might expect good performance from our controller.

b) Assessment of the memory function $m^*(\delta)$.

Assuming that we have no prior information, we rely on past measurements on the system for the evaluation of $m^*(\delta)$. If at some time anything goes wrong with set-point following then it will be immediately reflected in $m^*(\delta)$ and our controller will adapt to the changed system. Note that $m^*(\delta)$ is also a descriptor of the rate of change of the linear approximation to the system.

Let us choose first a discrete set of candidate error bounds $\Delta = \{\delta_1, \delta_2, \ldots , \delta_N\}$. Calculate $m_t^*(\delta)$ *at time* t by the recursion

$$m_t^*(\delta_i) = \min\left\{\bar{m}_t(\delta_i), \ m_t^*(\delta_i)\right\},$$

where

$$\bar{m}_t(\delta_i) = \max\left\{m / \ P_t^{m+1}(\delta_i) \neq \emptyset\right\}.$$

Introducing a "large memory" M as a design parameter,

$$m_t^*(\delta_i, M) = \min\left\{\bar{m}_t(\delta_i), \ \ldots, \ \bar{m}_{t-M}(\delta_i)\right\},$$

can be taken instead of the previous definition of $m_t^*(\delta_i)$.

To summarize, in the controller described above we do not have to specify a constant equation-error bound δ. The uncertainty of the model is actually assessed in conjunction with the control performance, measured here in the form of deviations from the prescribed set-point sequence. The separation principle is not used in this approach, since the system is not first estimated and then controlled on the basis of the estimate. Here parameter bounding is carried out according to an analysis of guaranteed control performance, and *the parameter uncertainty set is strongly influenced by the control requirements* (set points). Certainty equivalence is replaced by full consideration of the uncertainty of modelling, as can be seen from the definition of WCC design.

If the best guaranteed performance $\mu = \min\{\delta_i + L_t^m(\delta_i)\}$ is insufficiently small, on-line experiment design can be applied. It can easily be connected with the WCC controller by allowing for suboptimal minimax solutions of the controller. Again this problem has to be treated elsewhere (Veres and Norton, 1990).

SIMULATIONS

In this section simulations will show the performance of GPC and pole-placement control (based on recursive LS point estimation) and bounding predictive controller (BPC) defined above with prediction horizon $n=1$. General conclusions cannot be drawn from a single example, but these simulations show that BPC has promise. The simulated continuous-time system is that given in (2.1) with a changing parameter $v_t = 3\sin(t/3\pi)$ and a random step disturbance u_t^2 with amplitude 0.5. C=1. The sampling interval was 0.2s. Figs. 5.1 to 5.3 show the

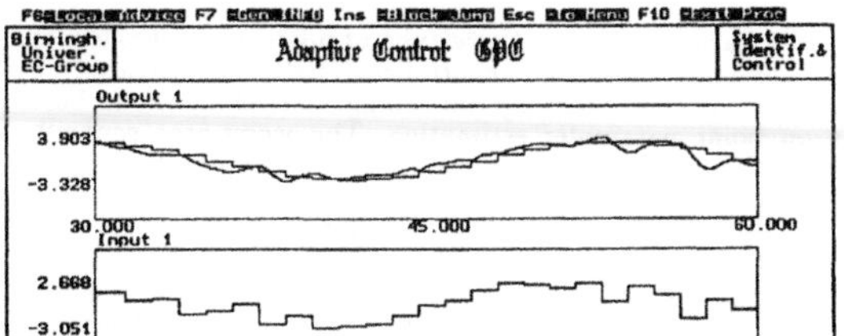

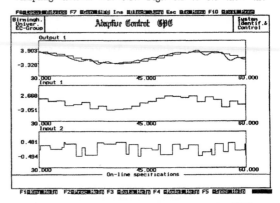

Fig 5.1 *Output, set-point and input sequences for GPC*

outputs, step set-point sequence and input values for GPC, pole placement and BPC, respectively. Although more tuning of the design parameters of GPC and pole placement might improve their perfomance, the present

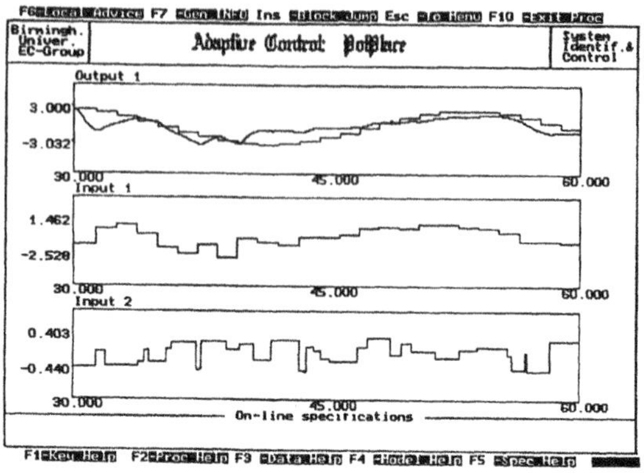

Fig 5.2 *Output, set-point and input sequences for pole placement*

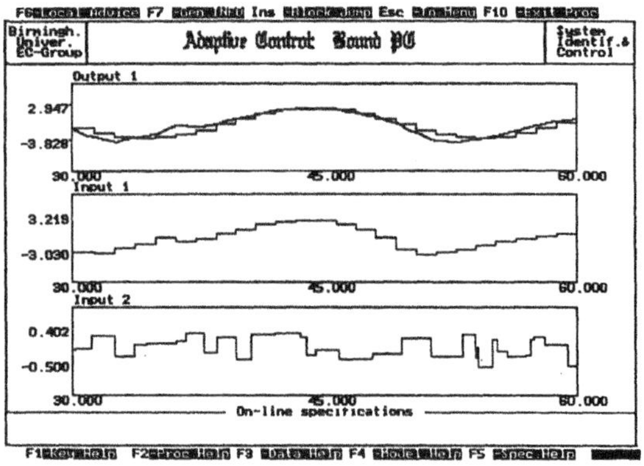

Fig 5.3 *Output, set-point and input sequences for BPC*

examples show better set-point following by BPC. In the preceding sections we gave some explanation of the greater robustness of BPC; there is room for further systematic robustness analysis of BPC.

CONCLUSION

We have considered a new type of adaptive controller, the bounding predictive controller (BPC), motivated by deficiencies of existing self-tuning control methods based on point parameter estimation. The worst-case controller (WCC) has also been introduced, which properly accounts for uncertainties in the model parameters. Parameter uncertainty has been linked to control performance in the BPC, combining calculation of parameter uncertainty regions and control inputs, a quite different approach from traditional self-tuners. In BPC there is no arbitrary and suboptimal separation of estimation and control. The price of a high degree of robustness is, however, a considerably higher computational demand than for traditional methods.

References

Anderson, B.D.O. (1985). Adaptive Systems, Lack of Persistency of Excitation and Bursting Phenomena, *Automatica*, **21**, 247-258.

Åström, K.J. and B. Wittenmark (1973). On self-tuning regulators. *Automatica*, **9**, 185-199.

Åström, K.J. and B. Wittenmark (1989). *Adaptive Control.* Addison-Wesley, Reading, Massachusetts.

Åström, K.J. and B. Wittenmark (1980). Self-tuning controllers based on pole-zero placement. *Proc. IEE*, **127**, 120-130.

Benveniste, A. (1987). Design of adaptive algorithms for the tracking of time varying systems. *Int. J. of Adaptive Control and Sign. Proc.*, 1, 3-29.

Caines, P. E. and Lafortune, S.(1984). Adaptive control with recursive identification for stochastic linear systems, *IEEE Trans. Autom. Control*, **29**, 312.

Chen, H.F. (1984). Recursive system identification and adaptive control by use of the modified least squares algorithm. *SIAM J. Control and Optim.*, **22**, 758.

Chen, H.F. and L. Guo (1986). Optimal stochastic adaptive control with quadratic index. *Int. J. Control*, **43**, 869-881.

Chen, H.F. and L. Guo (1987). Asymptotically optimal adaptive control with consistent parameter estimates. *SIAM J. Control and Optimization*, **25**(3), 559-575.

Chen, M.J. and Norton, J.P. (1987). Estimation technique for tracking rapid parameter changes. *Int. J. of Control*, 45, 1387 -1398.

Clarke, D.W. and P.J. Gawthrop (1975). Self-tuning controller. *Proc. IEE*, **122**, 929-934.

Clarke, D.W. (1984). Self-tuning control of nonminimum-phase systems. *Automatica*, **20**, 501-517.

Clarke, D. W., P.P. Kanjilal and C. Mohtadi (1985). A generalized LQG approach to self-tuning control. *Int. J. Control*, **41**, 1509 -1544.

Clarke, D. W., Mohtadi, C. and P.S. Tuffs (1987) Generalized Predictive Control—Part I. The basic algorithm. *Automatica*, **23**, 137-148.

Clarke, D. W., Mohtadi, C. and P.S. Tuffs (1987) Generalized Predictive Control—Part II. Extentions and interpretations. *Automatica*, **23**, 149-160.

Clarke, D. W., Mohtadi, C. (1989). Properties of Generalized Predictive Control. *Automatica*, 25

De Keyser, R.M.C. and A.R. Cauwenberghe (1981). Self-tuning predictive control. *Journal A*, **22**, 167-174.

Lelić, M.B. and Zarrop, M.B. (1986). A generalized pole-placement self-tuning controller. *Control Systems Centre Report No.* 657. *UMIST, Manchester, U.K.*

Ljung, L. and S. Gunnarsson (1990). Adaptation and tracking in system identification - a survey. *Automatica*, 26, 23-35.

Middleton, R.H., G.C. Goodwin (1988). Adaptive control of time varying linear systems. *IEEE Trans. Aut. Control*, **AC-33**(2), 150-155.

Middleton, R.H., G.C. Goodwin, D.J. Hill and D.Q. Mayne (1988). Design issues in adaptive control. *IEEE Trans. Aut. Control*, **AC-33**(1), 50-58.

Middleton, R.H., Goodwin, G.C. and Y. Wang (1989). On the robustness of adaptive controllers using relative deadzones. *Automatica*, **25**, 889-896.

Mo, S.H. and Norton, J.P. (1988) Recursive parameter-bounding algorithms which compute polytope bounds, *Proc. 12th IMACS World Congr., Paris, July* 1988 ,477-480.

Norton, J.P. (1975). Optimal smoothing in the identification of linear time-varying systems. *Proc. IEE*, 122, 663-668.

Norton, J.P. (1987). Identification and application of Bounded -parameter Models. *Automatica*, 4, yp. 497-507.

Peterka, V. (1984). Predictor-based self-tuning control. *Automatica*,**20**, 39-50.

Veres, S.M. and J.P. Norton (1989). Structure identification of parameter-bounding models by use of noise-structure bounds. *International Journal of Control*, 50, 639-649.

Veres, S.M. and J.P. Norton (1990) Identification of errors in variables models by parameter bounding methods. *Research Memorandum No.* 25, *School of Electronic and Electrical Engineering, University of Birmingham, U.K.*

Walter, E. and Piet-Lahanier, H. (1988) Estimation of parameter bounds from bounded-error data: a survey, *Proc. 12th IMACS World Congr., Paris, July* 1988 ,467-472.

Wellstead, P.E., D. Prager and P. Zanker (1979). Pole-assignment self-tuning regulator. *Proc. IEE*, **126**, 781-787.

Ydstie, B.E. (1984). Extended horizon adaptive control. *Proc. IFAC 9th World Congress*, Budapest, Hungary.

AN EXPERT-ADAPTIVE CONTROLLER

Z. P. Liu* and R. S. Cao**

**Zhejiang Province Electric Power Test and Research Institute, Hangzhou, PRC*
***Dept. of Chemical Engineering, Zhejiang University, Hangzhou, PRC*

Abstract. In this paper, an Expert-Adaptive PID Controller is developed. The Self-Tuning strategies are based on on-line pattern recognition and PID self-tuning knowledge-base. The knowledge-base is constructed with the interior relations between patterns of control curves and PID parameters. Also some important control experiences are built into the knowledge-base. That makes the self-tuning of PID parameters effective and efficient. Meanwhile, a new method of process parameters estimation (Characteristic Vector Approach) is proposed. Its convenience, highly estimation precision and, especially, strong noise resisting make the approach to have many superiorities. With the simplified model estimated by this approach, the pre-tuning of PID is proper and reliable.

Keywords. Adaptive control; Expert systems; Knowledge-base; Parameter estimation; Self-tuning controller;

INTRODUCTION

Adaptive control has been rapidly developed in recent decades. Many methods were proposed based on model identification and parameters estimation (Astrom, 1983; Goodwin, 1977; Kofahl, 1986; Seborg, 1986). Usually, adaptive control is supported by accurate-mathematical model. The more accurate the model is, the better control quality you might get. However, the accurate model for an adaptive algorithm sometimes means complicated calculation that makes very difficult to control a process effectively and successfully.

It is known that a combination of Artificial Intelligence (A I) and Expert System (ES) with control theories and control engineering would bring about good prospects. In many cases, control experts tune the parameters of a controller according to error versus time curves based on their knowledge and experiences, rather than on complicated algorithms. For example, if the curve of a controlled variable is overdamped, then the proportional gain Kc of a controller is increased.

In fact, when a process is controlled, there are interior relations between the shapes of a control curve and the parameters of a controller, which is an important basis for the tuning actions of a control expert. It is also found that each step of tuning of a control expert is ambiguous but directing to the optimum value of PID validly. This kind of tuning method is captivating because of its fast convergency and independence on a process model. A controller developed with such a principle will be intelligent and universal to various controlled process.

In this paper, a new kind of adaptive PID controller—Expert-Adaptive Controller—is presented. The controller searches a set of optimal PID parameters automatically with a built-in knowledge-base. The self-tuning strategies of the controller are based on a comparision of the pattern of error versus time curve with the desired pattern, rather than obtaining a process model. The built-in knowledge-base, which stores the self-tuning strategies and some important control experiences, makes the controller smart and successful in controlling a process, and is distinguished from others based on system identification and parameter estimation. Uncertain problems, such as varing gain Kp, time constant Tp and dead time Dp, etc., can be solved with the controller.

The preliminary PID parameters can be set either by an experienced control engineer or with the pre-tuning program of the controller. The pre-tuning approach, which is called Characteristic Vector Approach proposed here, is reliable and accurate even when the process is much noisy.

SELF-TUNING STRATEGIES BASED ON KNOWLEDGE-BASE

The pattern of a typical control curve could be described as follows (see Fig.1)

$$A1=(P3-P2)/(P1-P2) \tag{1}$$
$$A2=Ti/T \tag{2}$$
$$A3=Td/T \tag{3}$$

where,

Ti; integral time,
Td; derivative time,
P1, P2, P3, and T are shown in Fig.1.

With Ai (i=1 trough 3) the curve as shown in Fig.1 is abstracted and characterized. A control criterion vector V is defined as;

$$V=(A1 \quad A2 \quad A3) \tag{4}$$

Various criteria corresponding to different control requiement can be derived from it. For example,

$$V^{ITAE} = (A1^{ITAE} \quad A2^{ITAE} \quad A3^{ITAE}) \tag{5}$$

infers such a control requiement;

$$\int_0^{\infty} \left| e(t)\cdot t \right| dt = \min \tag{6}$$

Other control criteria such as IAE, ISE, etc., can be obtained just by taking the corresponding criterion vectors, in which the elements have different values.

There are interior relations between criteria vector and controller PID parameters. For a typical process, the relation between proportional gain Kc and V can be involved in the following formula as Ti and Td are constant;

$$\begin{aligned} Kc &= g(V) = \\ &= g(V^{*}) + \left[(A1-A1^{*})\frac{\partial}{\partial A1} + (A2-A2^{*})\frac{\partial}{\partial A2} + (A3-A3^{*})\frac{\partial}{\partial A3}\right] g(V)\Big|_{V=V^{*}} + \frac{1}{2!}\left[(A1-A1^{*})\frac{\partial}{\partial A1} + \right. \\ &\left. + (A2-A2^{*})\frac{\partial}{\partial A2} + (A3-A3^{*})\frac{\partial}{\partial A3}\right]^{2} g(V)\Big|_{V=V^{*}} + \\ &+ \dots + \frac{1}{n!}\left[(A1-A1^{*})\frac{\partial}{\partial A1} + (A2-A2^{*})\frac{\partial}{\partial A2} + \right. \\ &\left. + (A3-A3^{*})\frac{\partial}{\partial A3}\right]^{n} g(V)\Big|_{V=V^{*}} + \dots \end{aligned} \tag{7}$$

So for Ti and Td.

A large number of simulation studies and some important control experiences helped us build the main algorithm for self-tuning PID parameters in the knowledge-base, i.e.,

$$\begin{aligned} Kc(i) &= \left[\left(\sum_{j=0}^{m} Aj^{*}\, Xj^{*}\right) / \left(\sum_{j=0}^{m} Aj\, Xj\right)\right] Kc(i-1) \\ Ti(i) &= \left[\left(\sum_{j=0}^{m} Bj^{*}\, Xj^{*}\right) / \left(\sum_{j=0}^{m} Bj\, Xj\right)\right] Ti(i-1) \\ Td(i) &= Q(\, Ti(i)\,) \end{aligned} \tag{8}$$

where,
  $Xj = f(V)$,
  V; control criterion vector,
  Aj and bj; constants under certain situation and also vary with V,
  *; desired value.
Aj and Bj are very important which contain the interior relations between PID and V.

As can be seen, prior to each tuning the result of previous tuning of PID would be evaluated. If the control criterion vector is very close to the desired one, the PID parameters would not be adjusted, or,

$$\begin{aligned} Kc(i) &= Kc(i-1) \\ Ti(i) &= Ti(i-1) \\ Td(i) &= Td(i-1) \end{aligned} \tag{9}$$

The PID parameters would be adjusted if the control criterion vector deviates from the desired one. As we know, adjustments to any of Kc, Ti or Td will affect the control result as well as the corresponding criterion vectors. The self-tuning algorithm could coordinate the three parameters' changed values. The algorithm will keep running until the criteria vector is within its desired range, or, the optimum PID is achieved.

A large number of simulation studies for various self-regulating processes have shown the self-tuning algorithm has a very satisfactory convergency of PID parameters. Figure 3 and figure 4 show some details about it.

A NEW METHOD OF PROCESS PARAMETERS ESTIMATION FOR PRE-TUING

The Expert-Adaptive Controller calls for a preliminary PID value, which can be set by either an operator or the pre-tuning algorithm automatically. For the later, a simplified process model is got at first with the proposed new method of parameters estimation. This method is able to keep a high estimation accuracy even the process is much noisy. After that, a set of suitable preliminary PID parameters of pre-tuning can be achieved based on the model.

It's reasonable for a self-regulating process to be described approximately as

$$\frac{Kp\, e^{-Dp\, s}}{Tp\, s+1} \tag{10}$$

where,
  Kp, Tp and Dp represents the gain, time constant and dead-time of the process respectively.

A lot of approaches for estimation of process parameters have been proposed. But it's still necessary to develop a new method because of some shortcomings of the previous approaches either in their inconvenience such as complicated calculation or in the gap to practical applications.

A new estimation method for process parameters, which is called Characteristic Vector Approach, is proposed here. For a step response curve yn(t) (cf. Fig.5), the true response curve is y(t)(without distorted by noise or disturbance). A characteristic vector is defined as follows;

$$\begin{aligned} W(k) &= (Q1 \quad Q2 \quad Q3 \quad Q4 \quad Q5) \\ &= \left(\sum_{i=1}^{k} yn(i)\cdot Ts \quad \sum_{i=1}^{k} yn^{2}(i)\cdot Ts+Cn \quad A \quad y(\infty) \quad k\right) \end{aligned} \tag{11}$$

where,
  Ts; sampling period,
  A; amplitude of the test signal(step change)
  yn(i); sampled output of the process at i,
  Cn; correcting factor to noise.

The estimating parameters for the process by eq.(10) are got immediately as Ts is small and k large enough;

$$\begin{bmatrix} \hat{Kp} \\ \hat{Tp} \\ \hat{Dp} \end{bmatrix} = \begin{bmatrix} Q4/Q3 \\ 2(Q1/Q4-Q2/Q4)^2 \\ Q5+2Q2/Q4-3Q1/Q4 \end{bmatrix} \qquad (12)$$

where,

$\hat{}$: estimating value.

The superiority of the approach is its convenience (see eq. (12)) and strong noise resisting achieved by the construction of the characteristic vector. As can be seen, the elements of the vector actually are either determinated, such as Q3 and Q5, or based on area of the curve as Q1, meanwhile, Q1, Q2 and Q4 is obtained statistically. The algorithm also offers a high coincidence between the original process and its simplified model estimated with the approach.

By using the result of eq.(12), the preliminary PID parameters can be accomplished well for various control criteria (Liu,1989).

SIMULATION STUDIES

The effectiveness and superiority of the proceding algorithms are verified also by a large number of simulation tests for various processes.

One of the self-tuning examples is shown as Fig.3. The control system is shown as Fig.2, and

$$Gp(s) = \frac{e^{-s}}{3s+1}$$

$$Gd(s) = \frac{1}{3s+1}$$

$$D(s) = \frac{0.25}{s}$$

The desired control criterion vector is

$V^* = (0.07 \quad 0.4 \quad 0.16)$

An improper preliminary PID values are

Kc=3.8
Ti=1.2
Td=0.4

These poor PID parameters cause a bad control curve C(0), and its poor criterion vector is,

V(0)=(0.87 0.31 0.10)

After twice self-tunings, the criteria vector (V(2)) is very close to the desired one V^*.

A convergence diagram for a process,

$$Gp(s)=\frac{e^{-2s}}{(1.2s+1)^5}$$

is shown as Fig.4. It shows a fast convergency under various initial PID values, even far from the final convergence point.

Generally, a set of optimum PID parameters can be achieved within three steps of self-tuning, even one or two steps in some cases.

Simulations also show the new approach of process parameters estimation has an excellent property. One example is given below.

Suppose the process is,

$$Gp(s) = \frac{e^{-s}}{(s+1)^2 (3s+1)(7s+1)^2}$$

and with noise Nd: $N(0, 0.05^2)$. The actual step response curve yn(t) (as shown in Fig.5) is distorted from the true response curve y(t) because of noise and disturbance signals. By means of Characteristic Vector Approach, the estimated model is

$$\hat{G}(S) = \frac{0.9921 e^{-9.3878s}}{9.8741s+1}$$

the corresponding curve is $\hat{y}(t)$, which is very close to y(t). With the pre-tuning algorithm a set of preliminary PID parameters are achieved for the control criterion of ITAE. Fig.6 shows the pre-tuning result. C(t) is an optimum tuning curve with the process Gp(s) while $\hat{C}(t)$ is a pre-tuning curve based on the estimated model $\hat{G}(s)$. As can be seen, the two curves are very similar.

CONCLUSION

Both self-tuning and pre-tuning algorithm built in the knowledge-base of the controller revealed their effectiveness and superiorities. The self-tuning strategies having a fast **convergence** are not only to a typical process but also to various process that implies the controller is an universal one. In addition, the Characteristic Vector parameters estimating approach is very practical to use for its convenience,high recognition accuracy and strong noise resisting.

ACKNOWLEDGEMENT

The science fundation of the science committee of Zhejiang province and Zhejiang university is a great support for this subject and should be deeply acknowledged.

REFERENCE

Astrom, K.J. (1983). Theory and applications of adaptive control-a survey. Automatica, Vol. 19, No. 5, pp. 471-486.

Goodwin, G.C., R. L. Payne (1977). Dynamic system

identification; experiment design and data analysis, Academic press, New York.
Kofahl, R. (1986). Self-tuning of PID controller based on process parameter estimation. Journal A., 27(3), pp401-417.
Liu, Z. P. (1989). Research on an Intelligent self-tuning controller. Master thesis, Zhejiang University.
Seborg, D.E., et al (1986). Adaptive control strategies for process control: a survey. A.I.Ch.E. journal, pp.881-913.

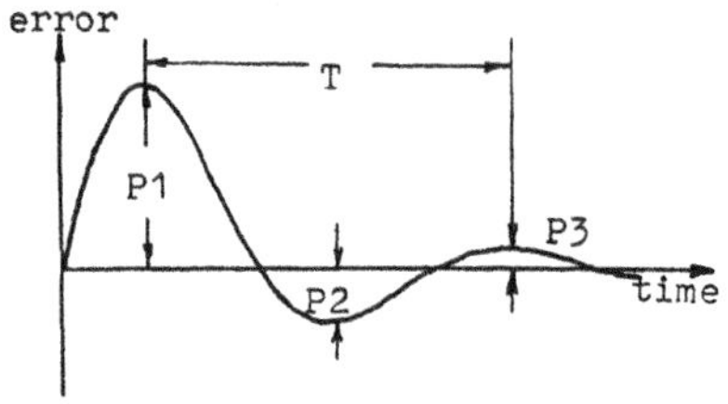

Fig. 1. The Pattern of a Control Curve

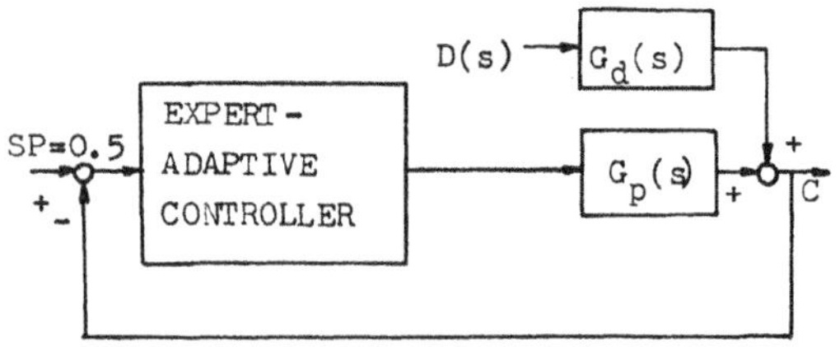

Fig. 2. The Simulated Control System

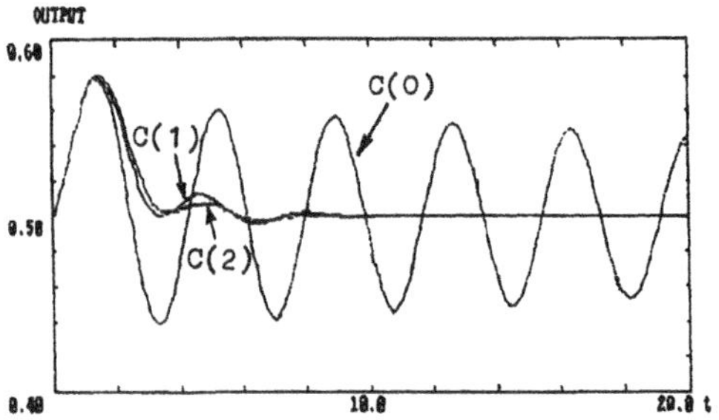

Fig. 3. An Example of Self-tuning

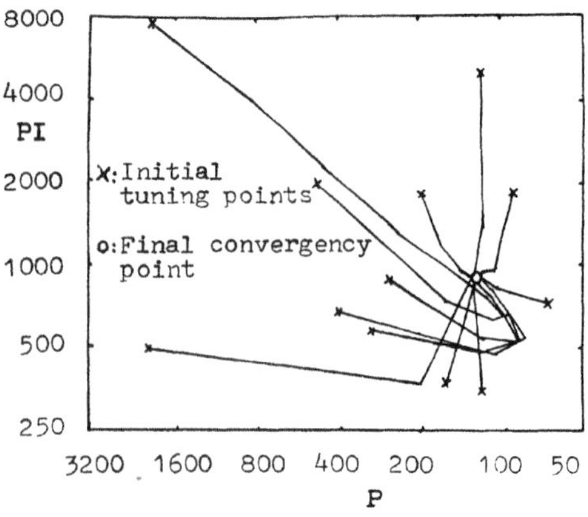

Fig. 4. The Convergence diagram of Self-tuning

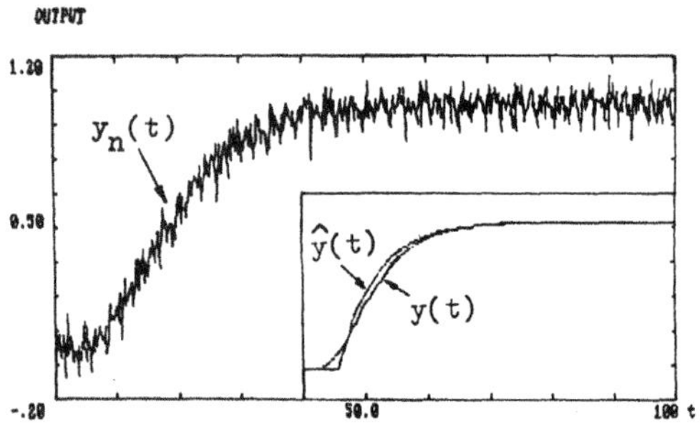

Fig. 5. The Estimation Result with the Characteristic Vector Approach for a noisy process

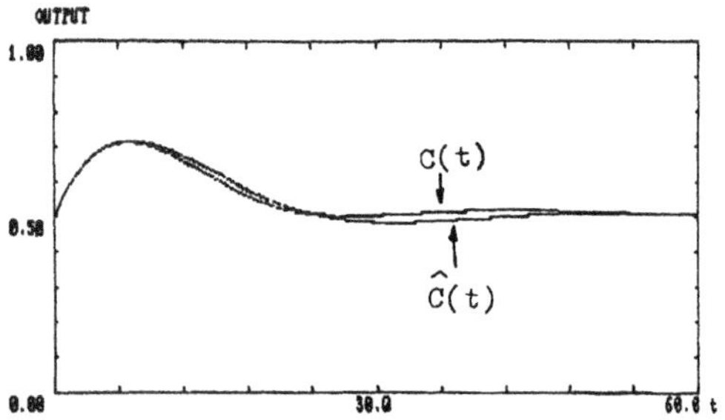

Fig. 6. The Result of Pre-tuning

AN EXTENDED HORIZON SELF TUNING LOAD FREQUENCY CONTROLLER

K. A. Lee and H. Yee

School of Electrical Engineering, University of Sydney, NSW 2006, Australia

Abstract. An explicit extended horizon controller for the frequency and tie line power control of interconnected reheat thermal power systems is described. The design of the controller is based on a multistep cost function whose minimization yields the required control sequences. The control algorithm offers several tuning parameters for performance enhancement and their effects on overall load frequency control responses are studied.

Keywords. Electric power systems; load frequency control; power system control; predictive control; self-tuning control.

INTRODUCTION

Control of frequency and tie-line power flows (LFC) continues to be of interest in the operation of interconnected power systems. The primary objective of LFC is to maintain a balance between power generation and load demand through automatic regulation of the controlled generating units in response to frequency and tie-line power deviations.

In LFC, each control area in the interconnection must operate sufficient generating capacity to balance its generation and power interchange schedule to its load, and also to provide its contribution to frequency regulation. To accomplish this, tie line bias control is used with each control area attempting to adjust its area control error (ACE) to zero. The ACE is given by

$$ACE = \Delta P_t + B\Delta f \tag{1}$$

where ΔP_t and Δf are the deviations in net tie line interchange and frequency respectively. B is the area frequency bias constant. The control objective, therefore, is to design local area controllers to regulate ACE.

Various control techniques have been considered. Schemes based on classical and optimal control theories (Calovic, 1984; Hiyama, 1982; Kothari and Nanda, 1988; Tripathy, Hope and Malik, 1982) were essentially fixed parameter designs and generally failed to maintain optimum quality of performance over a wide range of changing conditions. Control designs employing self tuning techniques have the ability to adapt to changing system characteristics and recent studies have shown their potential (Aly, Abdel-Magid and Wali, 1984; Lee and Yee, 1990a; Lee, Teo and Yee, 1990; Sheirah and Ald-El-Fattah, 1984; Vajk and co-workers, 1985; Yamashita and Miyagi, 1989).

A common feature in the design of the controller has been the use of a single-step optimization method to generate the control law. However, the success of implementation depends on the correct choice of key parameters such as dead-time and, to a lesser extent, model-order. Due to these limitations the design of self tuning control schemes has been focused towards long-range predictive control (LRPC) methods. Of particular interest in the family of LRPC is the generalized predictive control (GPC) algorithm (Clark, Mohtadi and Tuffs, 1987a 1987b). It uses the receding-horizon approach to predict the system output over a horizon greater than the maximum possible delay of the system, making assumptions regarding future control actions. GPC was later enhanced to incorporate pole placement strategy (Lelić and Zarrop, 1987).

This paper describes the design of a simple self tuning control scheme based on a multi-step optimization method to form an extended horizon controller for LFC. The effectiveness of the proposed control algorithm is verified by simulation study on a two-area thermal power system.

CONTROL DESIGN

For the purpose of on-line estimation, each control area is described by a SISO model of the autoregressive moving average form (ARMAX)

$$Ay(t) = Bu(t-1) + C\xi(t) \tag{2}$$

where $y(t)$, $u(t)$ are the output and input respectively, $\xi(t)$ is a $(0, \sigma^2)$ uncorrelated noise sequence and A, B and C are the usual scalar polynomials in the backward shift operator q^{-1}. In this paper, the optimization method used to generate the control law is based on a multi-step cost function given by

$$J = \mathcal{E}\left[\sum_{i=1}^{N_y} \phi_i(t)^2 + \lambda \sum_{i=1}^{N_y} u(t+i-1)^2\right] \tag{3}$$

where $\phi(t)$ is an auxillary output defined as

$$\phi(t) = y(t+i) + (Q_n/Q_d)u(t-1) \tag{4}$$

Q_n, Q_d are finite polynomials in q^{-1}, and N_y, λ are the maximum output horizon and scalar control weighting respectively. The derivation of the control law commences with two polynomial identities to partition the noise and input sequences into past and future values according to

$$\begin{aligned} C &= AF_i + q^{-i}G_i \\ BF_i &= CE_i + q^{-i}T_i \end{aligned} \tag{5}$$

and the optimal control sequences over the range of N_y can be derived as (Lee and Yee, 1990b)

$$\begin{aligned} \mathbf{u} = {} & (\mathbf{E}^T\mathbf{E} + \lambda\mathbf{I})^{-1}\mathbf{E}^T C^{-1}[-\mathbf{G}y(t) - \\ & \mathbf{T}u(t-1)] \end{aligned} \tag{6}$$

where

$$\begin{aligned} \mathbf{T}^T &= [T_1 + C(Q_n/Q_d) \; \ldots \\ &\quad\; T_{N_y} + C(Q_n/Q_d)] \\ \mathbf{u}^T &= [u(t) \; \ldots \; u(t+N_y-1)] \\ \mathbf{G}^T &= [G_1 \; \ldots \; G_{N_y}] \end{aligned} \tag{7}$$

and $\mathbf{E}$ is the $N_y \times N_y$ matrix

$$\mathbf{E} = \begin{bmatrix} e_0 & 0 & \cdots & 0 \\ e_1 & e_0 & \cdots & 0 \\ \vdots & & & \vdots \\ e_{N_y-1} & \cdots & & e_0 \end{bmatrix} \tag{8}$$

By defining $\mathbf{k^T}$ as the first row of $(\mathbf{E}^T\mathbf{E} + \lambda\mathbf{I})^{-1}\mathbf{E}^T$, i.e.,

$$\mathbf{k}^T = [k_1 \; k_2 \ldots \; k_{N_y}]$$

the control law can be obtained as

$$\mathcal{F}u(t) + \mathcal{G}y(t) = 0 \tag{9}$$

where

$$\begin{aligned} \mathcal{F} &= CQ_d + q^{-1}[\mathbf{k}^T Q_d \mathcal{T} + k^* C Q_n] \\ \mathcal{G} &= Q_d \mathbf{k}^T \mathbf{G} \\ k^* &= \sum_{i=1}^{N_y} k_i \\ \mathcal{T} &= [T_1 \; \ldots \; T_{N_y}] \end{aligned} \tag{10}$$

To incorporate pole assignment strategy into the control algorithm, the closed loop system equation is obtained, using (2) and (9), as

$$y(t) = C\mathcal{F}\xi(t)/(A\mathcal{F} + q^{-1}B\mathcal{G}) \tag{11}$$

and the assignment of the closed loop pole set $\mathcal{P}$ can be easily achieved by using (5), (10) and selecting Q_n and Q_d to satisfy

$$\mathcal{A}^* Q_d + q^{-1} \mathcal{B}^* Q_n = \mathcal{P} \tag{12}$$

where

$$\begin{aligned} \mathcal{A}^* &= A + q^{-1}(\mathbf{k}^T \mathbf{M}) \\ \mathcal{B}^* &= Ak^* \end{aligned} \tag{13}$$

and $\mathbf{M}^\mathbf{T} = [M_1 \; \ldots \; M_{N_y}]$ is formed using the polynomial identity

$$B = AE_i + q^{-1} M_i \tag{14}$$

The calculation of the control sequence $\mathbf{u}$ involves the inversion of $(\mathbf{E}^T\mathbf{E} + \lambda\mathbf{I})$. This matrix is singular if the control weighting $\lambda = 0$ and the dead-time > 1, i.e. some leading coefficients of $B(q^{-1})$ are zero and $\mathbf{E}$ is formed using (5). Also, the computational effort involved will be costly for large value of N_y.

These problems can be avoided by adopting a technique, as used in the GPC algorithm, of imposing a control horizon N_u in the control law. It is assumed that the control increment $\Delta u(t+j) = 0$ for $N_u \leq j < N_y$. This reduces the number of optimization variables to N_u by imposing $N_y - N_u$ constraints on $\mathbf{u}$. The control constraints lead to the following substitutions in the control law (9)

$$\mathbf{u} \rightarrow \mathbf{u}^*, \quad \lambda\mathbf{I} \rightarrow \mathbf{\Lambda}, \quad \mathbf{E} \rightarrow \mathbf{E}^*$$

where

$$\mathbf{E}^* = \begin{bmatrix} e_0 & \ldots & 0 & 0 \\ \vdots & & & \\ e_{N_u-1} & & e_1 & e_0 \\ e_{N_u} & & e_2 & \sum_{j=0}^{1} e_j \\ \vdots & & & \\ e_{N_y-1} & & \ldots & \sum_{j=0}^{N_y-N_u} e_j \end{bmatrix}$$

$$\mathbf{\Lambda} = diag(\lambda \ldots \lambda, N_y - N_u + 1) \tag{15}$$

and $\mathbf{E}^*$ and $\mathbf{\Lambda}$ are now $N_y \times N_u$ and $N_u \times N_u$ matrices respectively, and $\mathbf{u}^*$ satisfies the N_u equations obtained by minimizing (3) subject to the control constraints.

APPLICATION TO LFC

A two-area reheat thermal power system model (Lee and Yee, 1990) has been adapted for studying the proposed control scheme. Fig. 1 shows the schematic. A second order stochastic ARMAX model was used and its parameter estimates were used in the recursive solution of polynomial identities throughout the extended costing horizons to generate the control.

The series of step load disturbance superimposed with gaussian noise shown in Fig. 2 was applied to area 1 (referred to as the contingent area hereafter) and area 2 acted as an assisting area. The local area variables, Δf and ΔP_t, were recorded every 0.5 second and accumulated to produce a mean value at each LFC cycle. The non-zero mean component of the load disturbance was accounted for by integrating a simple load disturbance accommodation algorithm (Lee and Yee, 1990; Lee, Teo and Yee, 1990) into the self tuning control law.

Control Performance Criteria

The criteria used for the control performance in LFC is minimum accumulation of time error and minimum inadvertent interchange. Failure of the control area to regulate its ACE will cause frequency deviations which in turn will result in a time error. The effect is cumulative for all control areas. The time error is thus proportional to the total *magnitude* and *duration* of the load/generation imbalances. The inadvertent interchange is the time integral of the difference between the area's net actual interchange and net scheduled interchange, and is mainly caused by the bias response to frequency deviation.

Effect of Tuning Parameters on LFC Responses – Simulation Results

The tuning parameters in the self tuning LFC scheme are the output horizon N_y, control horizon N_u, and the control weighting factor λ. The effect of these parameters on LFC responses is shown in the figures, where the contingent and assisting area are indicated by solid and dotted lines respectively.

Output horizon N_y. N_y, the number of future values of $y(t)$ in the cost function, has a significant effect on the speed of response. The N_u and λ were set to 1 and 0 respectively and N_y was initially assigned a value of 2 (the order of the system model). It was found that $N_y = 10$ gave the best overall response. Larger values resulted in little improvement and caused an increase in control excursions. The results are shown in Fig. 3 and Fig. 4 for $N_y = 10$. ΔP_t and Δf are reduced in both contingent and assisting areas, the settling time is shorter and there is significant reduction in the time error and inadvertent interchange.

Control horizon N_u. As previously mentioned, N_u is chosen mainly to reduce the computational effort. It was found that the larger values of N_u (subject to $N_u < N_y$) generally provide more active and responsive control action at the expense of more computation. Fig. 5 shows the LFC responses obtained with $N_y = 10$, $\lambda = 0$ and $N_u = 4$, and the improvement compared to Fig. 4 is clearly shown. Other values of N_u were studied but it was found that for increases beyond $N_u = 4$ there was no perceptible improvement in the performance.

Control weighting factor λ. Using the same setting as in Fig. 5, a range of λ values was tested but the performances were found to be insensitive to λ. An intuitive explanation is that the use of $N_u = 4$ has already catered for the weighting in the control excursions and the inclusion of λ in the control law thus has negligible effect. To verify this, the value of N_u was reset to 1 as used in Fig. 3 and Fig. 4, and λ was then varied. The responses obtained with values of λ in the range of 0.2–0.5, as shown in Fig. 6, are similar to those shown in Fig. 5.

CONCLUSIONS

A preliminary study of the application of an extended horizon self tuning controller to LFC has been described. This study has shown that the design of a self tuning LFC scheme based on a multi-step cost function minimization has provided flexibility in the use of tuning parameters to improve system response. In particular, the use of a longer output horizon will lead to reduction in load/generation imbalances with minimum accumulated time error and inadvertent interchange, and the use of a control horizon and/or control weighting factor will provide more active control performance and improved response. The optimal choice of the tuning parameters is worth further investigation.

Work is continuing in the use of LRPC techniques with coordinated area corrective control for the time error and inadvertent interchange. Preliminary results obtained are promising.

REFERENCES

Aly, G., Abdel-Magid, Y.L. and Wali, M.A. (1984). Load frequency control of interconnected power system via minimum variance regulator. *Electr. Power Syst. Res.*, 7, 1–11.

Calovic, M. (1984). Automatic generation control : decentralized area-wise optimal solution. *Electr. Power Syst. Res.*, 7, 115–139.

Clark, D.W., Mohtadi, C. and Tuffs, P.S. (1987a). Generalized predictive control–Part I. The basic algorithm. *Automatica*, 23, No. 2, 137–148.

Clark, D.W., Mohtadi, C. and Tuffs, P.S. (1987b). Generalized predictive control–Part II. Extensions and interpretations. *Automatica*, 23, No. 2, 149–160.

Hiyama, T. (1982). Optimisation of discrete-type load-frequency regulators considering generation-rate constraints. *IEE Proc. Pt. C*, 129, No. 6, 285–289.

Kothari, M.L. and Nanda, J. (1988). Application of optimal control strategy to automatic generation control of a hydrothermal system. *IEE Proc. Pt. D*, 135, No. 4, 268–274.

Lee, K.A. and Yee, H. (1990a). Self tuning load frequency controller for interconnected power systems including effects of nonlinearities. *Proc. ACC*, 2100–2105.

Lee, K.A. and Yee, H. (1990b). Long range predictive controller for power system load frequency control. *Tech. Report*, School of Elect. Eng., University of Sydney.

Lee, K.A., Teo, C.Y. and Yee, H. (1990). Self tuning algorithm for automatic generation control in interconnected power system. *Electr. Power Syst. Res.*, accepted for publication.

Lelić, M.A. and Zarrop, M.B. (1987). Generalized pole-placement self-tuning controller. Part 1. Basic algorithm. *Int. J. Control*, 46, No. 2, 547–568.

Sheirah, M.A. and Ald-El-Fattah, M.M. (1984). Improved load frequency self tuning regulator. *Int. J. Control*, 39, No. 1, 143–158.

Tripathy, S.C., Hope, G.S. and Malik, O.P. (1982). Optimisation of load-frequency control parameters for power systems with reheat steam turbines and governor deadband nonlinearity. *IEE Proc. Pt. C*, 129, No. 1, 10–16.

Vajk, I., Vajta, M., Keviczky, L., Haber, R., Hetthéssy, J. and Kovács, K. (1985). Adaptive load frequency control of Hungarian Power System. *Automatica*, 21, No. 2, 129–137.

Yamashita, K. and Miyagi, H. (1989). Load frequency self tuning regulator of interconnected power system with unknown deterministic load disturbances. *Int. J. Control*, 49, No. 5, 1555–1568.

Fig. 1. Self tuning LFC scheme

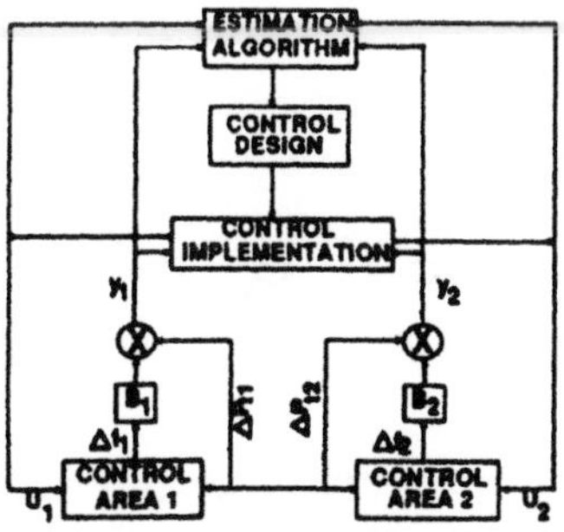

Fig. 2. Load disturbance in the contingent area

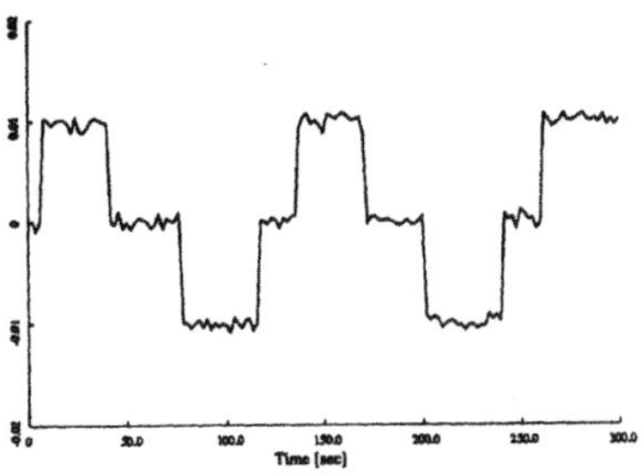

Fig. 3. LFC responses with $N_y = 2$, $N_u = 1$ and $\lambda = 0$

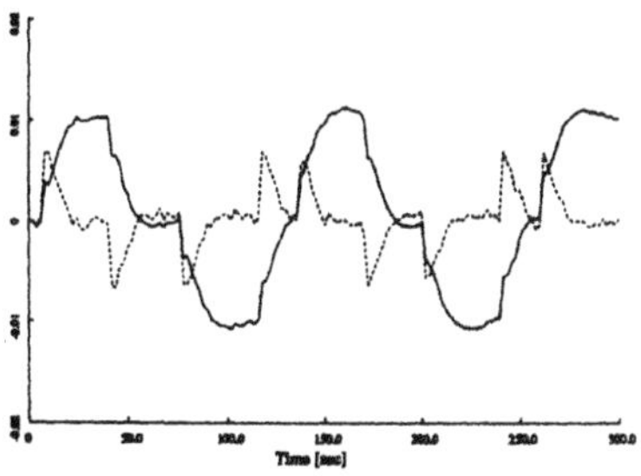

power deviation (puMW)

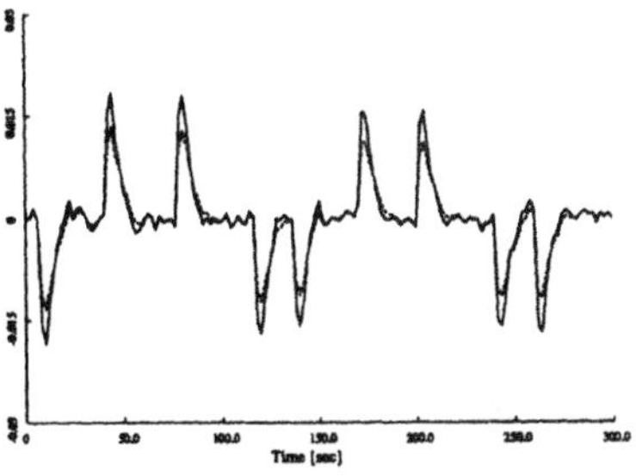

frequency deviation (Hz)

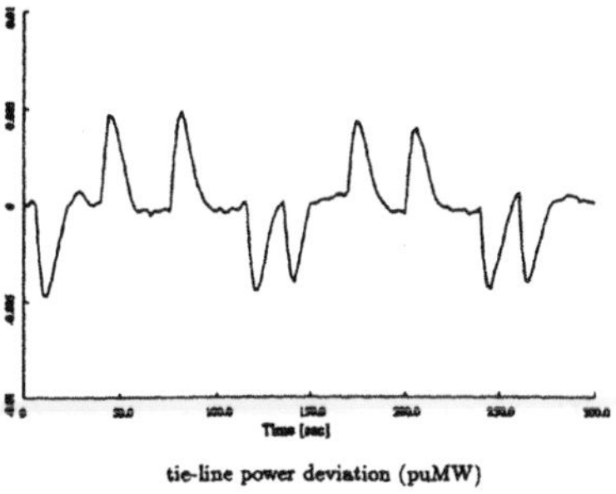

tie-line power deviation (puMW)

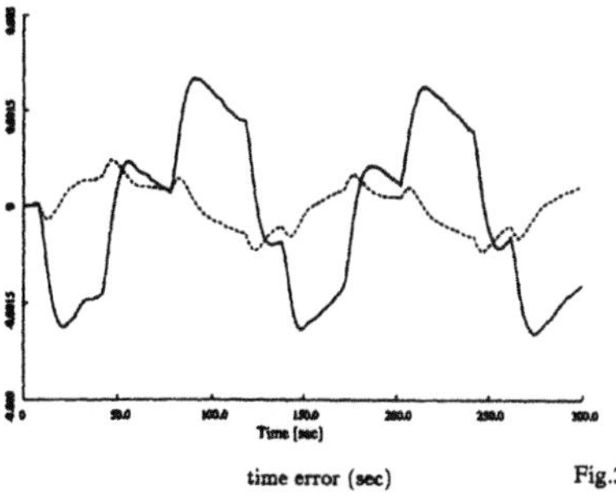

time error (sec)

Fig.3.......

Fig.3 (cont'd)

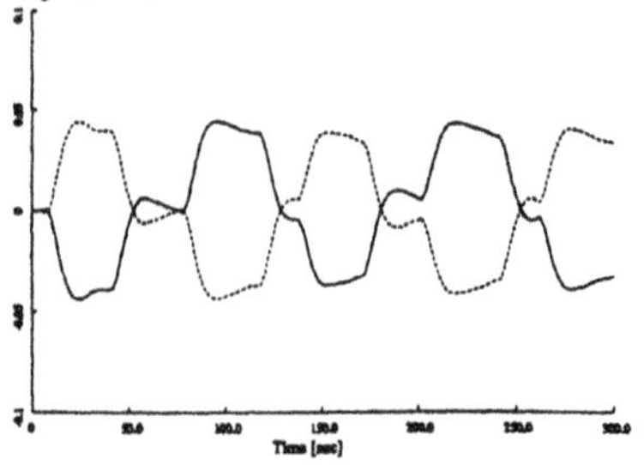

inadvertent interchange (puMW sec)

Fig. 4. LFC responses with $N_y = 10$, $N_u = 1$ and $\lambda = 0$

Fig.4

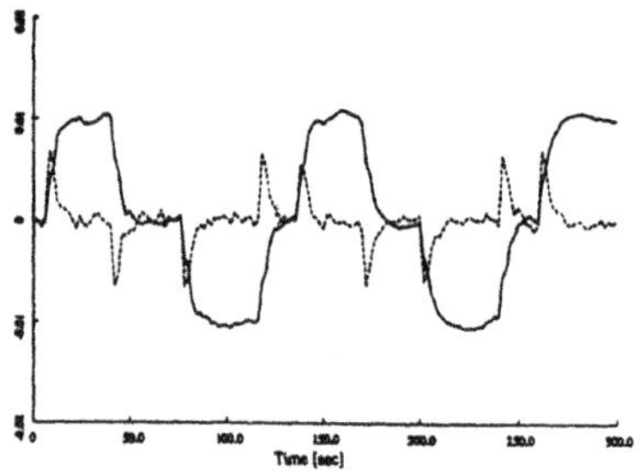

power deviation (puMW)

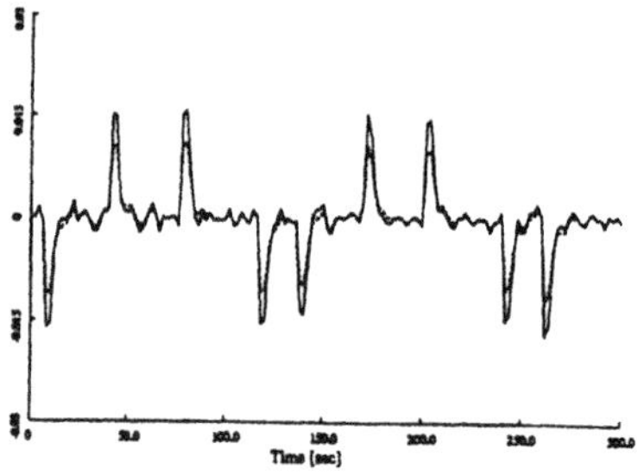

frequency deviation (Hz)

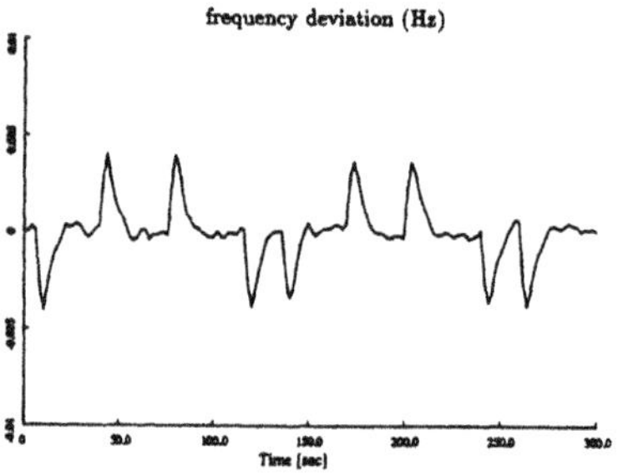

tie-line power deviation (puMW)

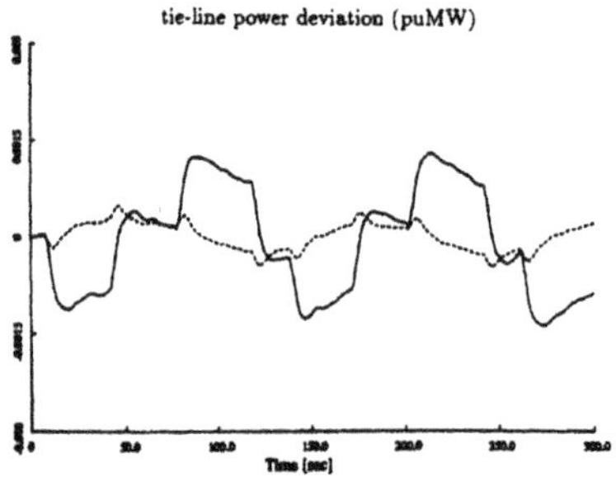

time error (sec)

Fig.4 (cont'd)

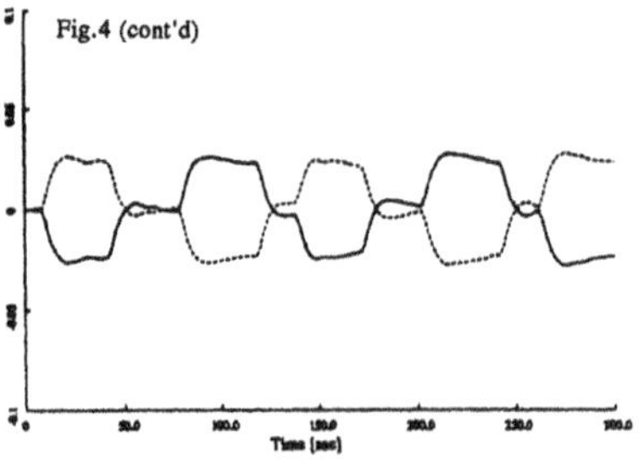

inadvertent interchange (puMW sec)

Fig. 5. LFC responses with $N_y = 10$, $N_u = 4$ and $\lambda = 0$

Fig.5

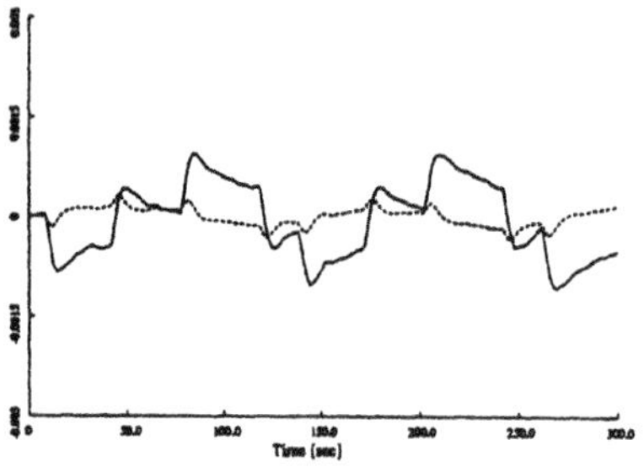

time error (sec)

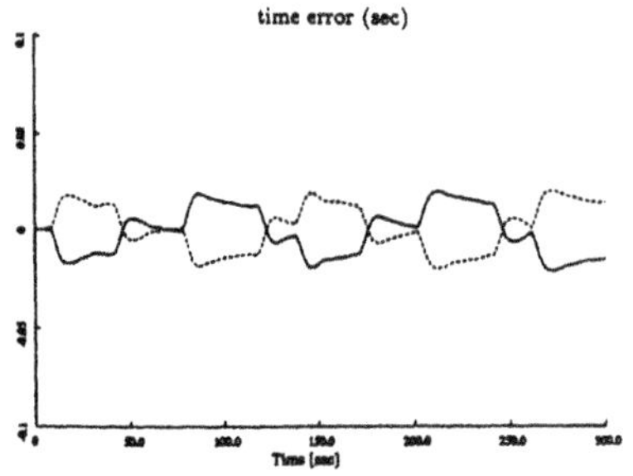

inadvertent interchange (puMW sec)

Fig. 6. LFC responses with $N_y = 10$, $N_u = 1$ and $\lambda = 0.5$

Fig.6

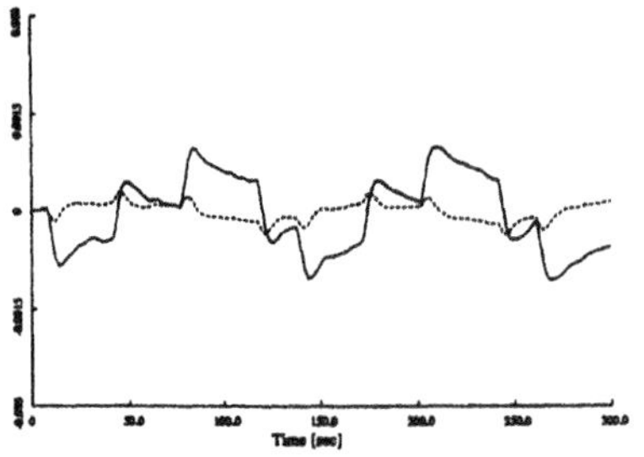

time error (sec)

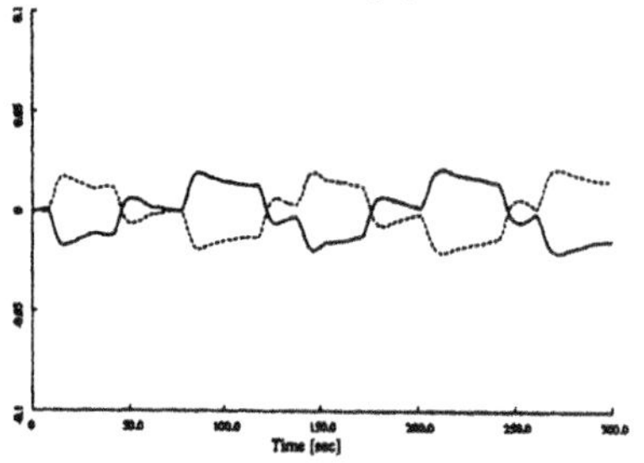

inadvertent interchange (puMW sec)

DESIGN AND IMPLEMENTATION OF A PARALLEL MULTIVARIABLE ADAPTIVE CONTROLLER FOR A FREE GYRO STABILISED MIRROR

T. H. Lee*, A. N. Poo, E. K. Koh*, M. K. Loh** and S. L. Tay*****

**Dept. of Electrical Engineering, National University of Singapore, Kent Ridge (0511), Singapore*
***Dept. of Mechanical and Production Engineering, National University of Singapore, Kent Ridge (0511), Singapore*
****Defence Science Organisation, 20 Science Park Drive, Singapore 0511*

Abstract. Fast close loop response requires sampling intervals in the milli-second range, which is difficult to achieve with uniprocessor computers. Parallel computers are able to achieve faster computation through load sharing. In this paper, we present our results in controlling the gyro stabilised mirror using transputers.

Keywords : digital control, adaptive control, parallel processing, transputer

1. Introduction

This paper considers the design and implementation of the adaptive multivariable MIMO GPC for controlling a free gyro stabilised mirror using a multiprocessor methodology.

The Generalised Predictive Controller (GPC) developed by Clarke et al [1], [2] has been shown to be robust in controlling both SISO and MIMO systems. Here, we investigate its applicability in the control of a free gyro stabilised mirror system.

A schematic diagram of a gyro stabilised mirror platform for which the controller is to be applied is shown in Figure 1. The system consists of the following modules:

(a) a flywheel and its drive;
(b) gimbals that provide two degrees of freedom to the wheel and drives for skewing; and
(c) a mirror that is geared to the gimbal.

Such a gyro stabilised platform is generally an integral element of precision pointing systems. In its normal mode of operation, it is mounted on a mobile platform (aircraft, ships or vehicles) and consequently needs some form of feedback control to maintain caging.

The control of a mirror gyro system is difficult in two aspects. First, there is a strong coupling in the two channels so that input for one channel will significantly affect the other channel. A second difficulty is the fast sampling required. For good controller implementation, such systems should be sampled at the millisecond range. With the speed of micro-computer systems, it would not be possible to complete the control calculations in this time frame. Parallel computers presents a viable solution to reducing computation time.

The use of parallel computers to implement the MIMO GPC also serves as a platform for investigating the use of parallel computers for real time digital control [3].

[1]Author to whom all correspondence should be addressed.

2. Theory

2.1 Gyro Mirror Transfer Function

The mirror gyro system has the following z-domain transfer function when the system is sampled at 0.01 second.

$$A\begin{bmatrix} y_1 \\ y_2 \end{bmatrix} = \begin{bmatrix} B_{11} & B_{12} \\ B_{21} & B_{22} \end{bmatrix}\begin{bmatrix} u_1 \\ u_2 \end{bmatrix} \tag{1}$$

where

$$A = 1 - 1.4516q^{-1} + 0.8418q^{-2} - 1.3132q^{-3} + 0.9230q^{-4}$$

$$B_{11} = 10^{-3}*(0.2542q^{-1} + 0.5716q^{-2} - 0.5748q^{-3} - 0.2430q^{-4})$$

$$B_{12} = 10^{-3}*(0.2415q^{-1} - 0.1973q^{-2} - 0.2475q^{-3} + 0.2047q^{-4})$$

$$B_{21} = 10^{-3}*(0.5427q^{-1} - 0.4795q^{-2} - 0.5427q^{-3} + 0.4807q^{-4})$$

$$B_{22} = 10^{-3}*(0.2385q^{-1} + 0.5362q^{-2} - 0.5394q^{-3} - 0.2280q^{-4})$$

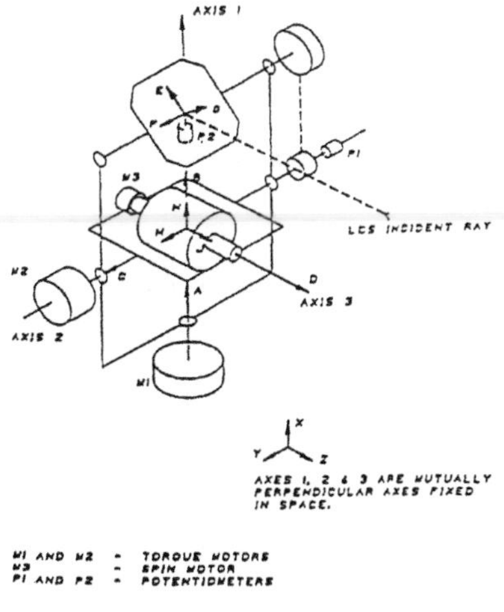

Figure 1: Free gyro stabilised mirror

From the magnitudes of the transfer function coefficients, it can be seen that there is considerable interaction between the two channels.

This transfer function describes the characteristics of the system to be controlled under nominal operating conditions. The system parameters, in fact drift depending on the operating conditions. The controller will employ an on-line parameter estimation technique to continuously adapt to a changing system and the controller will be tuned accordingly.

2.2 MIMO GPC

In this section , we summarise the derivation of the MIMO GPC.

Consider a MIMO linear deterministic process of the form

$$A(q^{-1})\ y = B(q^{-1})\ u + d \qquad (2)$$

where y and u are the process' m x 1 outputs and inputs vectors respectively and d represents the effect of load disturbances in the system. $A(q^{-1})$ and $B(q^{-1})$ are polynomial matrices in the backward shift operator. Equation (1) can also be written as

$$A(q^{-1})\Delta\ y = B(q^{-1})\ \Delta u \qquad (3)$$

where $\Delta = 1 - q^{-1}$. Writing the process transfer function in this form has the added advantage of incorporating integral action in the resultant controller.

Consider the Diophantine equation

$$I = E_j\ A\Delta\ + q^{-j}F_j \qquad (4)$$

where E_j and F_j are unique polynomial matrices of degrees j and n respectively. If a realisation involving A as a diagonal matrix is chosen, E_j and F_j will also be diagonal.

Substituting (4) into (3) and rearranging, we obtain

$$y(t+j) = E_jB\ \ u(t) + F_j\ y(t)$$

$$= G_ju(t+j-1) + G_ju(t-1) + F_jy(t)$$

$$= G_j\ u(t+j-1) + f_j \qquad (5)$$

where $G_ju(t+j-1)$ contains all future terms and f_j represents all known quantities at time t. y(t+j) is a state which we would like to achieve at time (t+j).

To derive a control law, consider forming a bank of parallel predictors of size N by varying the value of j in equation 5 from 1 to N. By matching our bank of parallel predictors with the bank of set point settings w, and adopting a least square error minimisation approach, we arrive at the control law

$$\Delta u(t) = (G^T*G)^{-1}G^T(w-f) \qquad (6)$$

$$u(t) = \Delta u(t) + u(t-1) \qquad (7)$$

where G is a square matrix of all the G_js and f is a vector of the f_js.

Note that N directly affects the closed-loop performance of the system. By increasing N, it is possible to slow down the closed loop response and decrease the magnitude of the control signal. However, increasing N increases the computation load as G is proportionally increased. One way to lighten the computation load is to introduce a variable Nu, where Nu * m is the number of columns in G. While N may be seen as the number of steps required to shift the process output to the set point, Nu may be interpreted as the number of steps within which the control signal is allowed to change. For most systems, setting Nu to 1 is sufficient.

For a more thorough treatment of the GPC and multivariable GPC, please refer to [1] and [2].

3. Parallel Design

3.1 System Design

In essence, parallel processing involves splitting a large task into smaller concurrent tasks. One design method, known as DARTS (Design Approach for Real-Time Systems) [4], outlines a procedure to systematically decompose a task. It starts with identifying the major functions in the task and the data flow through each function. This is followed by breaking each function into simple transforms. Transforms are then grouped into tasks. DARTS provides a set of criteria for deciding whether a transform should be structured as a separate task or grouped with other transforms into one task. Subsequently, task interfaces are defined.

Ideally, each of these tasks will be processed on a processor. In most practical instances, the number of processors is limited and a mapping strategy must be adopted to allocate the concurrent processes to the processors. There are two main groups of mapping strategies, namely the flood-fill strategy and the manual strategy.

3.2 Mapping Strategies

3.2.1 Flood-fill or Processor Farm Mapping

This method is applicable irrespective of the number of parallel processors available. The job is divided into a master task and a number of similar worker tasks. The master task breaks up the job into a number of smaller independent sub-jobs to be performed by the worker tasks. The worker task is a collection of sub-jobs routines. The number of sub-jobs usually greatly exceeds the number of processors available so that manually mapping sub-jobs to processors would be tedious. A worker task may not have to execute all the sub-job routines. Details of the sub-job are sent to the worker task which selects the correct routine to perform the sub-job. Results are sent back to the master task. There is no preference for a worker task over another to perform a sub-job. A worker task is sent a new job once it completes the previous job until all the sub-jobs are completed. The master task combines all the results to arrive at the final result.[5]

The processor farm mapping technique is applicable in situations where it is possible to formulate a master task which will split the job into a number of independent sub-jobs which do not need to communicate with other sub-jobs. The number of sub-jobs should also greatly exceed the number of processors. For example, in the GPC, the Diophantine Equation can be solved using a processor farm technique. This technique enables processors to be added without changing the software. The resultant system is also more robust as failure in one processor would degrade the system but not cause total failure.

3.2.2 Manual Strategy

In situations where the number of sub-jobs is small, a manual allocation strategy could be adopted. The job would first be divided into a number of concurrent sub-jobs. The sub-jobs are then assigned to the processors in one of the following manner.

Heuristic Approach: Sub-jobs are mapped to processors to maximise concurrency. Sub-jobs which execute in parallel are assigned to different processors. Results from a sub-job may be passed to another processor for further processing. Since sub-jobs are of unequal length, a shorter job may have to wait for a longer job before proceeding, thus losing parallelism. The overall performance of the system depends on the longest executing job. Inherently, the number of independent sub-jobs limits the number of processors that can be used.

Parallel Branch Approach: This approach minimises the amount of interprocess communication by maximising parallelism. Sub-jobs are designed to balance the computation load across all the processors. Each task must calculate all the sub-jobs needed to arrive at the result. Efforts are duplicated to minimise communication. This strategy ensures a high level of processor utilisation but does not guarantee shorter execution time since there are duplicated effort.

Hybrid Approach: This is a combination of the first two approaches. A main task is split into smaller tasks using the heuristic approach while critical paths within the smaller tasks are computed using the parallel branch approach.

A comparison of execution timings [6] using the three approaches showed that the hybrid approach gives the best performance. The parallel branch approach maximises processor utilisation but is slowed by the duplicated efforts. The heuristic approach wastes no effort but is limited by the inherent sequentialism and the number of processors that may be used.

4. Implementation Issues

4.1 Limitations on Sampling Interval

In the GPC-type controller where the parameters of the control law are not directly estimated, the control calculations can be divided into three (software) processes. In process 1, the system parameters are identified using the system's input and output signals to generate a transfer function expressed in the $\hat{A}$ and $\hat{B}$ polynomials. Process 2 uses the result from process 1 to formulate the control law parameters. The matrix polynomials E_js and F_js in the Diophantine equation and G and $\bar{G}$ are calculated in this process. Finally, the control law is calculated in process 3.

In a uniprocessor system, the three processes have to be executed in sequence. The sampling interval must be longer than the total time required to compute all three processes sequentially. For a multiprocessor system, the three processes can be executed consecutively so that the sampling interval now only depends on process 3. Processes 1 and 2 may be computed at a slower rate than the sampling rate. During the interval when the new system estimates and

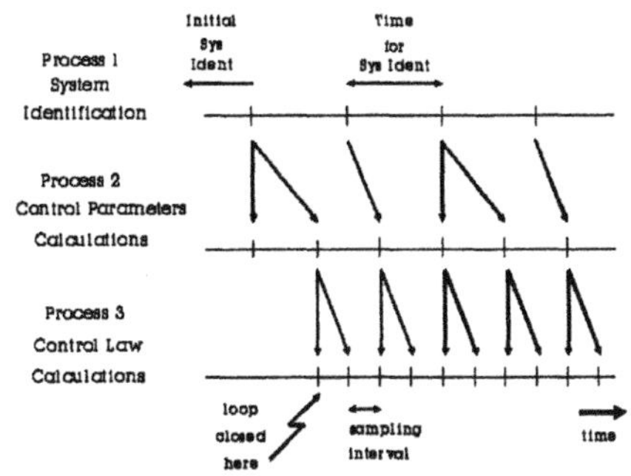

Figure 2 Diagram showing Origins of Parameters Used in Concurrent Processes

control parameters are being computed, calculations in process 3 are based on the most recent set of available parameters. (Please refer to Figure 2 for illustration.) This arrangement will have little adverse effect on controller performance considering the high sampling rate and assuming that the system drifts to be small. Thus by placing processes 1 and 2 and process 3 on different processors, a faster sampling rate is achieved.

4.2 Major Modules in the Adaptive GPC Controller

We have partitioned the adaptive GPC controller into a network of seven communicating modules. A bubble diagram that shows the flow of data between modules is shown in Figure 3.

The Coordinator Module oversees the entire controller operation. Its functions include sequencing critical events such as system output sampling and control signal output, activating suspended processes and maintaining a database of all parameters essential to the controller's operation. The database can only be accessed through the coordinator module to ensure data integrity.

The input/output controller module is responsible for periodically sending control signals to the system and reading system outputs.

The keyboard controller accepts aperiodic inputs from the system operator. Such inputs include changes in set point and controller parameters. Although keyboard inputs are aperiodic, setting a separate module for this function would ensure that no input will be lost.

The system identification module conducts on-line estimation of system parameters using the system's previous input and output signals stored in the database. The results are passed to the control parameter calculation module for further processing.

The control parameter calculations module calculates the matrix polynomials E's, F's and G's that are used in the control law. The matrix inversion in the control law will be calculated in this module and stored in the central database.

The control law calculation module is a time critical module and has to be kept small for fast sampling systems. This module involves weighing the past system input and output signals according to F and $\bar{G}$ and combining with

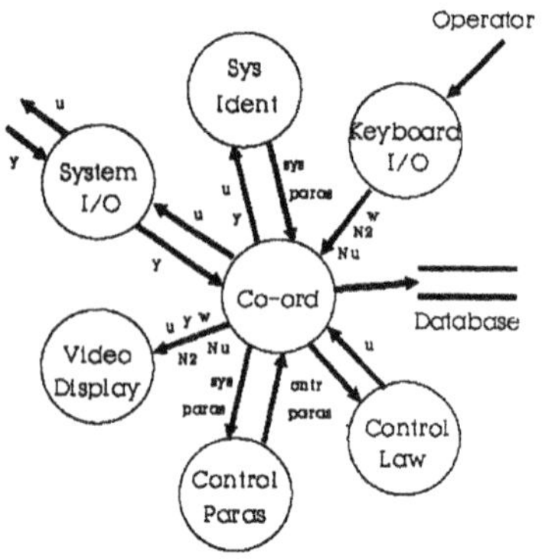

Figure 3. Bubble Diagram

the matrix inversion results from the control parameter module to calculate the control signal.

The video display module shows the state of the system and the parameters values used in the controller.

The topology adopted is a star topology where information are centralised in the coordinator module. The centralised database feeds each module the information required in their computations and is updated by the results. This configuration of having a centralised database reduces sequentialism in the controller as described in section 4.1. However, to reduce the volume of data to be passed between modules, data will not be repeated so that modules, like the control law module, which require past values in the computations will have to set up a local database.

4.3 Mapping Modules onto a Four-Node Transputer

The previous section has outlined the major modules in the MIMO GPC controller. Since the number of modules is small, a manual mapping strategy is adopted. While we can follow the heuristic mapping approach and assign each module to a transputer, it is apparent that the resulting performance will not be optimal. It is obvious that the input/output modules, though important, are less computation intensive while the controller module will be heavily loaded. It is therefore economical to group smaller modules and split bigger ones for better overall controller performance.

The result of grouping/splitting modules is shown in Figure 4. There will be five tasks to be loaded onto a four node transputer. Since one of the tasks will be placed on the host computer, we have an ideal situation where the number of processors available equals the number of concurrent tasks. In most situations more than one task may have to time-share a processor.

The Master Task is the coordinator of the entire controller. It consists of the coordinator module, the input/output module and the keyboard module. Since these modules are not computation intensive, grouping three modules into one task will not affect the performance of the task.

The System Identification Task handles the entire on-line system parameter identification function. Depending on the availability

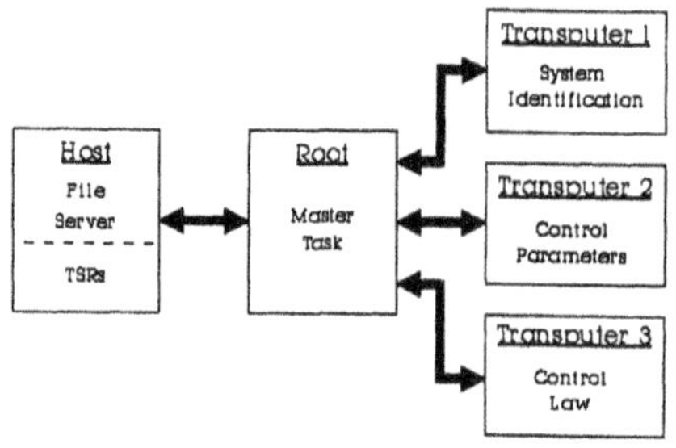

Figure 4. Information Passing between Processors

of processors, this task may be further broken up so that each sub-task handles the identification of parameters of a channel in the system.

The control parameter module and control law module each constitutes a task. Control parameter calculations is the most involved task in the controller as matrix inversion is required. If the size of the matrix to be inverted is very large, it may be beneficial to allocate more resources to this task. The calculations in the control law task is simple but its time critical nature justifies setting it as an independent task.

The display module makes use of the bios on the host computer system to display information on the screen. It would therefore be more appropriate to leave this module on the host system, in the form of terminate and stay resident (TSR) programs that can be activated through a software interrupt.

4.4 Hardware Implementation

The proposed controller will be run on a 20 MHz 80386-based computer with four T800-20MHz processors mounted on a B008 compatible board. Interface to the process will be through a Data Translation DT2821 data acquisition card.

The Master Task must be placed in the root transputer to enable fast access to the peripherals on the host computer. The video modules will stay resident on the host computer. The remaining tasks can be placed on any of the transputers.

Physical links must be present between the root transputer and the other nodes to facilitate communication.

4.5 Software Implementation

The software for the controller is written in Parallel C and standard C. Parallel C is used to code programs that are to be run on the transputers while standard C will be used for the TSR routines.

In the case of the Coordinator Task which is made up of three modules, the software can be written such that each module constitutes a parallel thread. The modules will then be executed in parallel on a time sharing basis.

5. Performance

We present here the results of simulating the controller and also describe the time frame that we expect to work in when the controller is implemented.

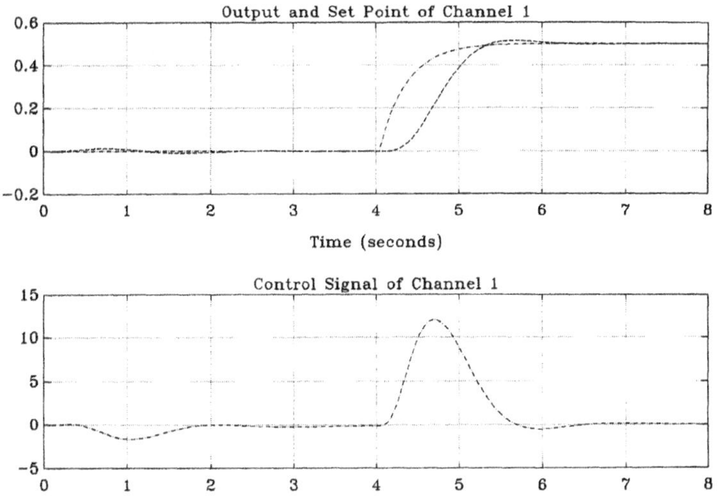

Graph 1

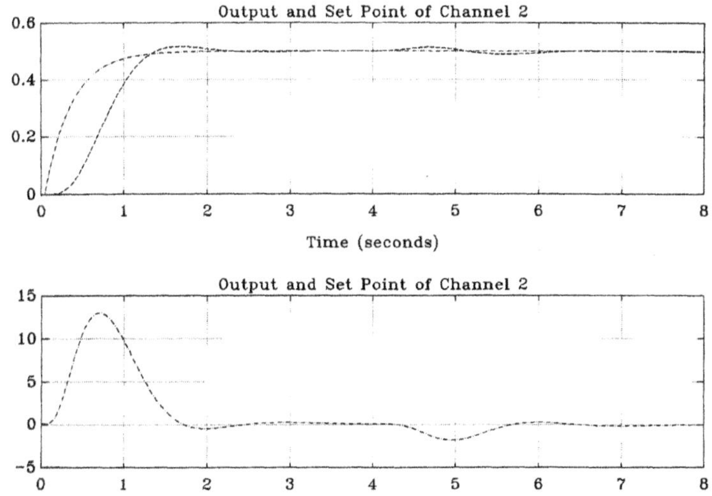

Graph 2

5.1 Simulation Results

The simulated controller was used to control the system described in section 2.1. Performance of the controller with N=20 and Nu=1 are shown in graphs 1 and 2. The system output shows that the controller has minimised the coupling between the two channels.

5.2 Timings

We have implemented the controller on the transputer and have achieved sampling rates of 20 milliseconds. With a conventional uniprocessor (20 MHz 80386 machine), the achievable sampling rate was 0.15 second. This timing includes the time to check for operator input and display the state of the controller.

The timing breakdown for the three processes in the controller are shown in Table 1.

Process	Time (ms)
System Ident	91.7
Control Paras	34.6
Control Law	17.6

Table 1. Timing Breakdown for Processes in Controller

6. Conclusion

In this paper, we have presented our experience in implementing an adaptive controller using parallel computers. We have described in detail the design process of breaking the controller into small tasks and grouping tasks to form concurrent processes. We also showed that the parallel controller was able to achieve sampling rates of about 20 ms, a significant improvement over the uniprocessor controller.

References

[1] D.W. Clarke, C. Mohtadi and P.S. Tuffs, Generalised Predictive Control: Part 1: the basic algorithm, Part 2: extensions and interpretations, Automatica 23 (2)(1987)

[2] S.L. Shah, C. Mohtadi and D.W. Clarke, Multivariable adaptive control without a prior knowledge of the delay matrix, Systems & Control Letters 9 (1987)

[3] P.J. Fleming (ed), Parallel Processing in Control - the Transputer and other architectures, IEE Control Engineering Series 38, Peter Peregrinus Ltd, 1988

[4] H. Gomaa, A Software Design Method for Real-time Systems, Commun-ACM 27, 9 Sept 1984, pg 938-949

[5] 3L Technical Notes No 8, Processor Farms, 3L Ltd, 18 Jan 1990

[6] F. Garcia-Nocetti, H.A. Thompson, M.C.F. De Oliveira, C.M. Jones and Prof P.J. Fleming, Implementation of a Transputer-based Flight Controller, IEE Proceedings~D V137 No 3 May 1990

EXPLICIT PID SELF TUNING CONTROL FOR SYSTEMS WITH UNKNOWN TIME DELAY

A. Besharati Rad* and P. J Gawthrop**

**Singapore Polytechnic, Dept. of Electronics & Communications, 500 Dover Road, Singapore 0513*
***Dept. of Mechanical Engineering, The University, Glasgow G12 8QQ, UK*

ABSTRACT

An explicit self tuning control algorithm is proposed where the underlying control law is the celebrated PID algorithm. The design is based on a continuous time approach. This algorithm can be applied to a class of systems which can be modelled by a second order system with time delay. The parameters of the system including time delay are estimated recursively by a derivative free least squares algorithm. The controller settings can then be selected based on the information obtained from the identification part.

1. Introduction

Although the philosophy of adaptive control has been around for about four decades; it was only during the last fifteen years that the theory could actually be realised in practice. In recent years, the availability of powerful digital computers and the very idea of potentials of adaptive control has motivated considerable amount of work being done on this subject. These works have refined or extended the original theory and brought about new ideas such as self tuning control to the extent that there is now a solid theory backed by a large body of literature on this area of research.

As a result, many new control algorithms have emerged which are superior to traditional PID controllers. However, it seems that the process control industry is a bit cautious if not suspicious about this huge development.

The possible reasons for this attitude include:

♦ PID controllers are regarded as "jack of all trades" in the process control industry.

♦ Developments in classical control and especially tuning PID controllers were largely due to close co-operation between universities and engineers in industry whereas, in general, adaptive control was born in academic world and brought up there. It was considered more an academic topic than a practical solution to difficult actual processes up to nine to ten years ago.

♦ Adaptive control is based on advanced control theory which is not appreciated by many plant operators.

♦ For sometime, there was a mis-understanding that self tuning controllers would substitute the existing PID controllers; this ,of course, can not be justified. Andreiev (Andreiev 1981) makes a conservative guess that in a typical plant, about fifty percent of all process control loops are being run in manual (open loop) rather auto (closed loop) mode. In practice this figure would be much lower; therefore more than fifty percent of all process control loops can be adequately controlled by PID controllers.

Many years of experience have proved that PID controllers are versatile enough to control a wide variety of processes, however even if a nearly optimal set of parameters are selected, the process operating points will certainly change due to many variables acting on the system. In order to return to optimum performance, the controller parameters should be readjusted. That is the selection of the appropriate gain, integral and derivative time constants - so-called Tuning for better performance. This procedure is still very much a manual operation performed by a skilled operator.

In view of the causes mentioned above and since the tuning procedure is mathematically defined, there has been a strong tendency towards automating the tuning process, thus a new chapter in adaptive control has been opened; that is PID adaptive controllers. Names such as PID self tuners, PID auto tuners and intelligent PID controllers are all attributed to a class of adaptive controllers which are essentially PID controllers but tuned automatically.

In this paper, a new adaptive controller is proposed whereby the underlying control law is of PID structure. The design is based on a continuous time approach in contrast to conventional discrete time approach. The proposed controller can be applied to a class of systems which can be modelled by a generic second order system with unknown time delay. The parameters of the system including time delay are estimated recursively by a derivative-free least squares algorithm published recently (Gawthrop et al, 1989). The parameters of the PID controller are then calculated by the formulae suggested by Pemberton (Pemberton, 1972) based on the estimated parameters of the system.

The rest of this paper is organised as follows. In section 2, the work done in this area is reviewed briefly and some of the PID adaptive controllers in process control market are examined. This is followed by introduction of the new algorithm in section 3. In section 4, the proposed algorithm is assessed by some simulations. Section 5 concludes the paper.

2 State of Art

There have been different approaches to the problem of deriving a PID like adaptive controller. However, all of these can be classified under two broad classifications; namely *Model Based (Parameter Estimation)* or *Expert systems (Pattern recognition)*.

The model based approach is a special case of self tuning control where the structure of the controller is pre-fixed to that of the PID algorithm. The parameters of the model of the system are continually updated to match the input output behavior of the actual process. The PID controller is then tuned based on the estimated system parameters.

The so called pattern recognition approach is basically automating the manual process of tuning a PID controller. The theme of most of these "packaged" designs is an intelligent PID controller that can reason logically within some defined context. The process recovery curve (reaction curve) is observered after a disturbance or a set point upset or whenever tuning is required; the appropriate PID adjustments are then made to produce the desired damping and overshoot.

In this section, some of the reported PID adaptive algorithms are annotated. Most of these algorithms are treated exclusively in discrete time domain. There are now more than 70,000 of these adaptive loops in operation (Astrom, 1989); most of which are based on pattern recognition techniques. This confirms that industry has given vote of confidence to this type of PID adaptive controllers.

2.1 Model based PID self tuners

One of the early model based implementations was a model reference PID self tuner by Hawk et al (Hawk, 1982). The system model was recursively updated by an Instrumental variable algorithm. The PID parameters were then selected through interactive communication of operator and controller.

Nishikawa et al (Nishikawa, 1984) proposed an alternative algorithm for auto tuning PID controllers. When tuning is required an intentional disturbance is applied to estimate the process parameters. The authors acknowledge that use of perturbation signal is not desirable. However, they argue that the signal is a small pulse which does not disturb the normal operation of the plant significantly. After estimating the system parameters, the PID parameters are chosen so as to minimise the weighted ISE (Integral of Square Error). This method -like previous method- is essentially a man-machine interactive tuning procedure. The authors have tested this algorithm in real processes and have reported that sufficiently good settings of PID parameters is obtained.

Gawthrop (Gawthrop 1986) introduced a continuous version of PID self tuners. His method is a special case of the continuous time self tuning control (Gawthrop, 1987). He raises the question as to why PID controllers are so common? He then concludes that there should be something about the dynamics of most of processes which make such a control action to be appropriate. He shows that with reasonable assumptions about the dynamics of the process to be controlled, the PID algorithm is automatically derived and there is no need to impose it on the adaptive control structure. These assumptions are suitable modelling of non-zero mean disturbances which lead to integral action and a first or second order system which will give rise to PI or PID algorithm respectively.

Other alternative designs have been pole-placement PID self tuners (Warwick, 1988), (Banyasz, 1982), and (Ortega, 1984). All these approaches share the following: they are based on discrete time self tuning control, and the parameters of the system are identified by recursive least squares estimation. The difference between them is how the PID parameters are updated. Warwick and Ortega adopt a pole placement design control law but fit a PID algorithm to control structure. Banyasz et al derive an explicit formula ensuring prescribed overshoot of the process to update PID parameters.

2.2 Pattern Recognition PID self tuners

Taylor's "Micro-Scan 1300 (Andreiev, 1977) controller was one of the early PID adaptive controllers. The distinct feature of this model was that it was one of the first PID auto-tuners supplied as a "packaged product". Before that, self tuning control was in use in large computer installations based on direct digital control in which the self tuning algorithm was in the form of software resided on a main frame or minicomputer. Basically, this is an adaptive gain controller where the gain is varied according to a preprogrammed gain schedule. When there is no upset to the plant, it performs exactly like an ordinary PID controller tuned for fastest response with low gain. As the error exceeds a preset value, the gain increases dynamically with the error.

Leeds & Northrop's "Electromax V" single loop controller (Andreiev, 1981) was introduced in 1981. It was the first generation of PID adaptive controllers based on pattern recognition. During the tuning, the plant is upset by a signal -as large as the plant can handle- to determine the reaction curve of the process. The controller watches the recovery curve and corrective resettings of the original PID parameters are initiated.

One of the interesting auto tuning techniques is reported by Astrom (Astrom, 1984). The method is essentially based on the Ziegler and Nichols closed loop formula suggested in a classical paper (Ziegler, 1942). In this method, a relay is implemented in parallel with the PID controller. The system actually operates as a relay controller in the tuning mode and as an ordinary PID controller in normal operation. The aim of the relay in the loop is to find critical gain and period which is needed in order to apply the Zeigler and Nichols tuning rules. When the system is under relay control, it drives the system into a limit cycle with frequency equal to that at which the plant phase is -180 degrees. The gain at this frequency is estimated from the limit cycle amplitude. This information is used to calculate the PID parameters. The relay auto-tuner concept has been implemented in products from SattControl and Fisher Controls Inc. The SattControl ECA40 is critically examined by Lee and Newell (Lee, 1988) and is reported to have excellent performance.

Finally, the latest and the most popular model is the Foxboro's "EXACT" controller. Exact stands for EXpert Adaptive Controller Tuning (Kraus, 1984). Foxboro describe this approach to adaptive control as an expert, or artificial intelligence, technique in the sense that it is patterned after an expert control engineer's method of manually tuning a PID controller. As previous models, the controller monitors the closed loop response following a setpoint or load disturbance perturbation and finds the optimum values of PID parameters subject to user predefined damping and overshoot constraints. A second order, lightly damped closed loop model has been selected as the desired system response. Thus a step change in the setpoint or the load is expected to cause error signal responses to match this pattern. If the PID parameters are such that the closed loop response does not resemble the anticipated oscillatory decaying results, calculations are performed and the PID parameters are adjusted accordingly.

The distinctive feature of this controller is that it is adaptive in the sense that it continually checks to confirm if adaptation is necessary whereas in the other controllers the tuning procedure is started manually. The perfomance of this controller is tested in (Nachtigal, 1986). It is concluded that this controller demonstrate very good performance.

The above are not the only PID auto tuners available in the market but they ,rather, indicate the trend in which the process control market has been moving towards adaptive control. From the foregoing discussion, one concludes that although more elaborate predictive self tuners outperform these simplified controllers; the fact is that, at present, the majority of the controllers in use are of the much easier pattern recognition type which do not require a mathematical model of the system under control than those more difficult controllers based on process identification. Astrom and Wittenmark 's recent book (Astrom, 1989) is a very good source for further information on these controllers.

3. Explicit PID self tuing control

In this section, a new algorithm for adaptive control of systems with unknown time delay is presented whereby the underlying control law is the familiar PID algorithm. It is acknowledged ,of course, that the quality of control for systems with time delay is much superior with advanced control techniques rather than PID controllers; however, there are many such systems which can be adequately controlled by PID algorithm and the proposed algorithm can be used for these type of processes. There is no point in providing performance of a higher quality than that which is considered adequate by the user.

Before going further, some preliminaries are needed which define the context in which this technique is derived.

3.1 Mathematical Model of the System

For the purpose of identification and control, one is not interested in high order models, process dynamics of such systems are neither known nor susceptible to accurate measurement; furthermore, few state variables can actually be measured directy for feedback, so higher order models are of limited utility even if the process dynamics are theoretically of high order. Other practical advantages of lower order modelling is outlined in Ashworth (Ashworth, 1982). It was recognised long ago (Oldenbourg, 1948) that most systems can be represented by a second order system with time delay as:

$$\bar{y}(s) = \frac{K\, e^{-sT}}{(\tau_1 s+1)(\tau_2 s+1)}\, \bar{u}(s) \tag{3.1.1}$$

where

y(s) = *Output Signal,*
u(s) = *Input Signal,*
K = *Process Gain,*
T = *Time Delay,*
τ_1 and τ_2 = *Process Time Constants,*

The validity of the above model is confirmed from both theoretical and experimental evidence. Kim and Friedly (Kim, 1974) have tested this model for large staged systems. An extensive bibliograpy of the different chemical processes that have been approximated successfully by this model is given in (Koppel, 1965), (Latour, 1967) and (Bohl, 1976). All agree that the above model effectively approximates most physical systems. A distinct advantage of this model, for the purpose of this study, is that the number of parameters to be identified are at most four.

3.2 PID Algorithm

PID controllers have been in successful use in process plants all over the world for more than 60 years. What is about them that make them so flexible?. One Possible reason would be that PID controllers consist of three fundamental control laws which when combined together act very much like the natural reaction of human beings in interaction with outside world. The proportional action responds to the present situation (middle frequencies), while the integral action sums up the past history (low frequencies) and derivative action predicts into future (high frequencies). This combination is the so-called "common sense" - that is PID controller. The tuning procedure ,therefore, can be interpreted as determination of the individual effect of these three control laws on the process to be controlled.

In Laplace notation, these control laws are written as:

$$\bar{u}(s) = K_p \left[\, 1 + \frac{1}{T_i\, s} + T_d\, s \,\right] \bar{e}(s) \tag{3.2.1}$$

In the above equation,

u(s) = *Controller output,*
e(s) = *Controller input,*
K_p = *Proportional gain,*
T_i = *Integral time constant,*
T_d = *Derivative time constant,*

Equation (3.2.1) is known as Non-Interacting PID algorithm which is not the only representation. Most commercial PID controllers fit under another classification called Interacting PID algorithm which is defined as:

$$\bar{u}(s) = K_p \left[\left(1 + \frac{1}{T_i s}\right)\left(1 + T_d s\right)\right] \bar{e}(s) \tag{3.2.2}$$

The settings of an interacting controller can always be transformed to that of non-interacting by the following equations:

$$(K_p)_n = K_p \left(1 + \frac{T_i}{T_d}\right) \tag{3.2.3}$$

$$(T_i)_n = T_i + T_d \tag{3.2.4}$$

$$(T_d)_n = \frac{T_i\, T_d}{T_i + T_d} \tag{3.2.5}$$

where "n" represents the respective parameters of a non-interacting controller. It must be noted that the settings of a non-interacting controller can be transformed to an interacting one only if $(T_i)_n > 4(T_d)_n$.

3.3 Tuning law for PID parameters

Apart from the many heuristic tuning methods based on either open loop or closed loop system (Ziegler, 1942), (Cohen, 1953), (Coon, 1956) and (Miller, 1967); there have been attempts to give rules for tuning controllers based on the evaluation of some measure of performance (Grinten, 1963), (Haalman, 1965) and (Pemberton, 1972). The proposed adaptive PID algorithm in this paper is based on the results published by Haalman and Pemberton.

In order to derive his tuning formulae, Haalman has made two assumptions the validity of which had been confirmed both in theory (Hazebrock, 1950) and in practice. These two assumptions are quoted here:

(i) Controller and process can be matched in such a way that the open loop transfer function is the same for all combinations.

(ii) A purely integral controller yields good control of a purely dead time process.

With regard to assumption (i) , all the open loop transfer functions of (system + controller) are equivalent. Therefore, if the settings of one of these is found by either simulation or trial and error, the settings of the rest would be implied from this one.

With regard to assumption (ii), a pure time delay process with an integral controller is simulated. The results of several runs and experimental observations indicated that a ***minimum mean square error*** occurs when $T_i = (3/2\ T)$.

The controller settings for different processes suggested by Haalman is given in Table (3.3.1). As it is seen, with these controller adjustments the open loop transfer function of different (system+controller) reduce to that of pure time delay and integral action. It should be noted that the settings for a second order system with time delay in Table (3.3.1) is based on an interacting PID controller. If these settings are converted to a non-interacting PID algorithm, they would give the same controller settings for integral and derivative time constants as suggested by Pemberton (Pemberton, 1972); the gains are different by a factor of third. This is very interesting since Pemberton derived his formulae through a purely theoretical approach. He assumes that the system is modelled by equation (3.1.1) and then defines the control objective to be set point tracking after a delay of T time units, where T is the system time delay. That is:

$$\bar{y}(s) = e^{-sT}\,\bar{w}(s) \qquad \textit{where } \bar{w}(s) = \textit{Set Point}$$

He finally shows that the controller which will meet this control objective is indeed a PID controller.

Table (3.3.2) shows the model together with controller settings as suggested by Pemberton. He reckons that these settings may give higher overshoot than desired in some cases; to overcome this, he suggests the derivative time constant to be quarter of integral time constant. This is called "Mod Ideal PID Settings". These settings have been critically examined and evaluated by (Hang, 1979) and are shown to give better control than other methods such as Ziegler and Nichols.

Tabel 3.3.1 Controller Settings Recommended by Haalman

Controller Transfer Function	*Control Action*	*Process Transfer Function*	*Open Loop T.F.*	K_p	T_i	T_d
e^{-sT}	*I*	$\frac{1}{T_i\,s}$	$\frac{e^{-sT}}{T_i s}$		$\frac{3}{2}T$	
$\frac{e^{-sT}}{\tau_1\,s}$	*P*	K_p	$\frac{e^{-sT}}{(\frac{\tau_1}{K_p})\,s}$	$\frac{2\tau_1}{3T}$		
$\frac{e^{-sT}}{1+\tau_2 s}$	*P.I*	$K_p(1+\frac{1}{T_i s})$	$\frac{e^{-sT}}{(\frac{\tau_2}{K_p})\,s}$	$\frac{2\tau_2}{3T}$	τ_2	
$\frac{e^{-sT}}{\tau_1 s(1+\tau_2 s)}$	*P.D*	$K_p(1+T_d s)$	$\frac{e^{-sT}}{(\frac{\tau_1}{K_p})\,s}$	$\frac{2\tau_1}{3T}$		τ_2
$\frac{K\,e^{-sT}}{(1+\tau_2 s)(1+\tau_1 s)}$	*PID*	$K_p[1+\frac{1}{T_i s})(1+T_d s)]$	$\frac{e^{-sT}}{(\frac{\tau_2}{K_p})\,s}$	$\frac{2\tau_2}{3T\,K}$	τ_2	τ_1

Table 3.3.2 Controller Settings Recommended by Pemberton

Process Transfer Function	$\frac{K\,e^{-sT}}{(1+\tau_1 s)(1+\tau_2 s)}$
Controller Transfer Function	$K_p(1+\frac{1}{T_i s}+T_d s)$

Controller Adjustment	Ideal PID	Mod Ideal PID
K_p	$\frac{2(\tau_1+\tau_2)}{3TK}$	$\frac{2(\tau_1+\tau_2)}{3TK}$
T_i	$\tau_1+\tau_2$	$\tau_1+\tau_2$
T_d	$\frac{\tau_1\,\tau_2}{\tau_1+\tau_2}$	$\frac{T_i}{4}$

3.4 The Proposed Algorithm

We are now in a position to describe the rationale behind the proposed algorithm. The parameters of the system including the unknown time delay which is assumed to fit to a transfer function given by equation (3.3.1) are estimated recursively by a polynomial identification technique recently published (Gawthrop, 1989). The question is: given K, τ_1, τ_2 and T (in our case, their estimates), what are the best values for K_p, T_i and T_d. Best values or optimum values generally depend on the requirements and demands. In this case, the optimum settings are based on minimising the square of the error as given by Haalman.

Therefore, the formulae given in either of Tables (3.3.1) and (3.3.2) can be used to update the parameters of PID controller. Figure (3.4.1) shows the block diagram of the explicit PID adaptive control system arrangement.

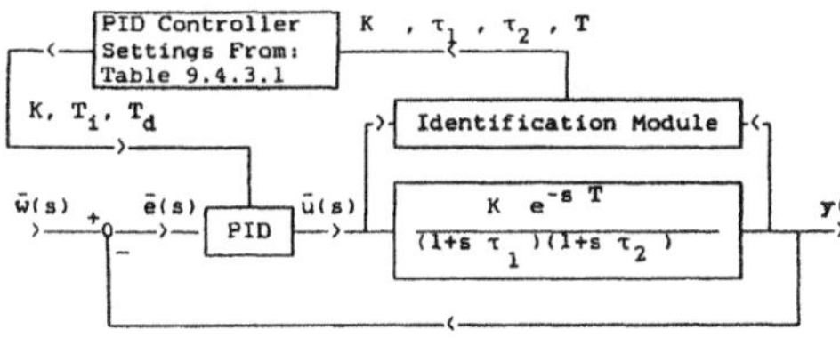

Figure 3.4.1 Adaptive PID Algorithm

3.4.1 Parameter Estimation

The parameters of the system are dentified by a special least square estimation algorithm. The system is modelled as:

$$\bar{y}(s) = e^{-sT}\,\frac{b_1 s^{n-1}+b_2 s^{n-2}+\ldots+b_n}{s^n+a_1 s^{n-1}+a_2 s^{n-2}+\ldots+a_n}\,\bar{u}(s)+\bar{\varepsilon}(s)$$

In the above equation u(s), y(s) and ε(s) are the Laplace transformed input, output and disturbance respectively. The details of the identification algorithm is treated elsewhere (Gawthrop et al, 1989); however, suffice to say that there are six special cases considered each with a different parameter vector q as defined below:

Algorithm 1: Linear least squares

$$\Theta_1 = [\,a_1, \ldots, a_n\,;\, b_1, \ldots, b_n\,]^T$$

Algorithm 2: Time delay identification: exact model

$\Theta_2 = T$ *(a scalar)*

Algorithm 3: Time delay identification: approximate model

$\Theta_2 = T$ *(a scalar)*

Algorithm 4: Identification of combined time delay and time constant

$\Theta_3 = [\,T, a_1, \ldots, a_n\,]^T$

Algorithm 5: Identification of linearised system

$\Theta_4 = [\,T, a_1, \;, a_n\,;\, b_1, \ldots, b_n\,]^T$

Algorithm 6: Decoupled time delay and rational dynamics identification

$\Theta_4 = [\,T, a_1, \;, a_n\,;\, b_1, \ldots, b_n\,]^T$

3.4.2 PID Parameters Update

The PID design parameters are updated at each sample interval based on the information obtained from the identification part.Either Haalman or Pemberton tuning rules will be used for this purpose.

It should be noted that it is also possible to update the PID parameters using the open loop Zeigler and Nichols for those systems that can be modelled by a time delay, a time constant and a gain; i.e:

$$\bar{y}(s) = \frac{K\,e^{-sT}}{\tau\,s+1}\,\bar{u}(s) \qquad (3.4.2.1)$$

For systems that can be represented by the above model, Z-N rules for a PID controller are as follows:

$$K_p = \frac{1.2\tau}{TK} \qquad T_i = 2T \qquad T_d = 0.5T \qquad (3.4.2.2)$$

However, it is shown (Hang, 1979) that Pemberton tuning rules are superior to Z-N open loop and closed loop techniques for both set point and disturbance upsets.

4. Simulations

In this section, the algorithm described in the preceding section is tested by simulated examples which are aimed to show the performance of this method. The algorithm is applied to systems with known time delay as well as unknown time delay. These simulations are programmed in Conic (Kramer, 1984) and implemented on a SUN 3 workstation. The set point signal for the following simulations is a square wave of amplitude one and period of 20 time units for examples 1 and 3 and 40 time units for other examples. The sample interval is 0.1 time units. In the following simulations, the top left window is the input to the system (PID output), the top right window, are the set point and controlled variable, the bottom left window are the estimated system parameters and bottom right are the PID updates.

Example 1

In this example, PID controller is used to control a pure time delay. The system is described as:

$$\bar{y}(s) = e^{-s}\,\bar{u}(s)$$

The assumed system (initial conditions) is:

$$\bar{y}(s) = e^{-2s}\,\bar{u}(s)$$

Time delay is estimated by algorithm 2 of section 3.4.1. Initial Cost, forgetting factor and N the order of truncation are chosen as 2, 0.001 and 0.01 respectively. Figure (4.1) shows the adaptive control of pure time delay with a PID controller.

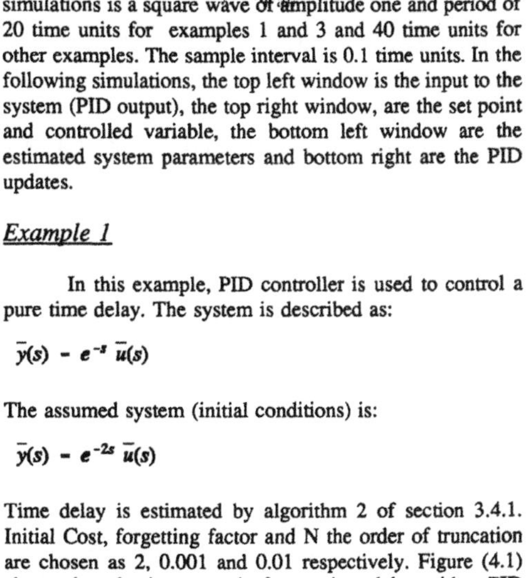

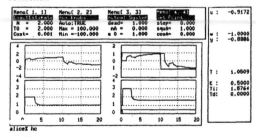

Figure 4.1 Adaptive control of pure time delay system

Example 2

Adaptive control of a first order system with time delay is shown in Figure (4.2). The system is:

$$\bar{y}(s) = \frac{2\,e^{-s}}{s+2}\,\bar{u}(s)$$

The aim of this example is how the algorithm can track the system parameters if they are changed. At time 10, the system is changed to:

$$\bar{y}(s) = \frac{2.4\,e^{-s}}{s+2.4}\,\bar{u}(s)$$

As it is shown the PID parameters adapt themselves to change in system. In this example, the Initial variance and forgetting time are selected to be 10000 and 1000 respectively.

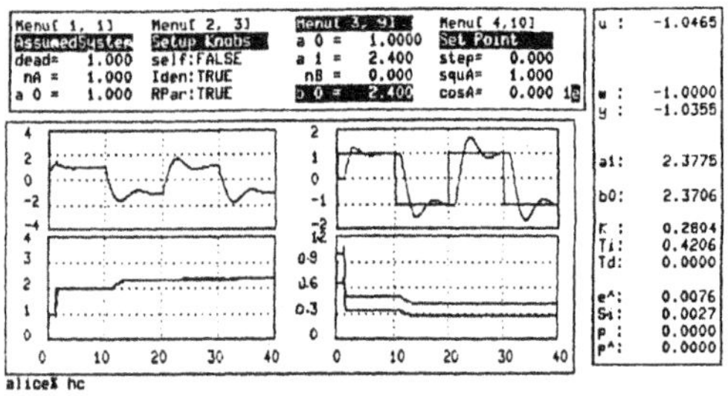

Figure 4.2 Adaptive control of a first order system

Example 3

This example is similar to previous one but for a second order system with known delay. The system is:

$$\bar{y}(s) = \frac{e^{-s}}{(4s+1)(s+1)}\,\bar{u}(s)$$

At time 20, $a_1 = 1.25$ is changed to $a_1 = 1.5$. Figure (4.3) shows the response of this system. The Initial Variance and forgetting time are the same as in preceding example.

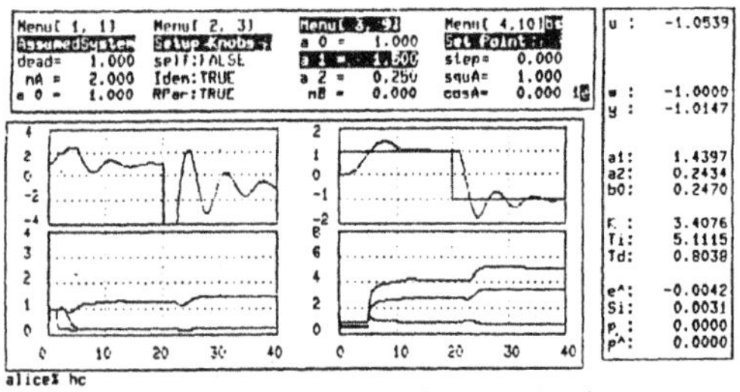

Figure 4.3 Adaptive control of a second order system

Example 4

A first order system with unknown time delay is simulated. The system is modelled as:

$$\bar{y}(s) = \frac{e^{-s}}{2s+1}\,\bar{u}(s)$$

The parameters of this system are estimated by algorithm 4 of section 3.4.1. The PID parameters are updated by both Haalman and Pemberton tuning rules. Other design parameters are selected as:

Assumed initial system:

$$\bar{y}(s) = \frac{e^{-2s}}{3s+1}\,\bar{u}(s)$$

Cost11, cost22, forgetting factor and order of truncation are chosen to be 0.01, 0.01, 0.01 and 4 respectively. Figures (4.4) and (4.5) show the response for both tuning methods.

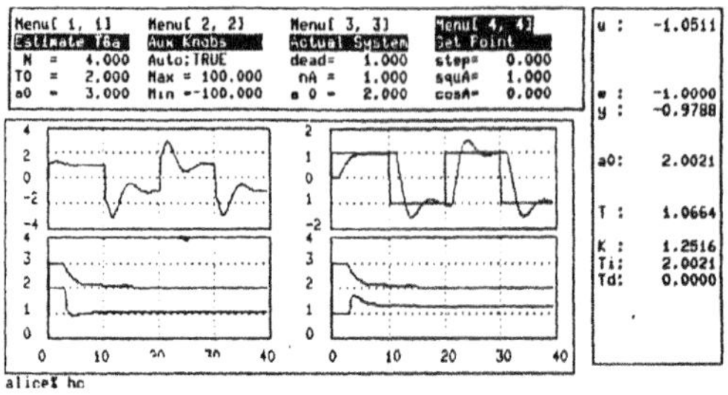

Figure 4.4 Adaptive Tuning By Haalman Method

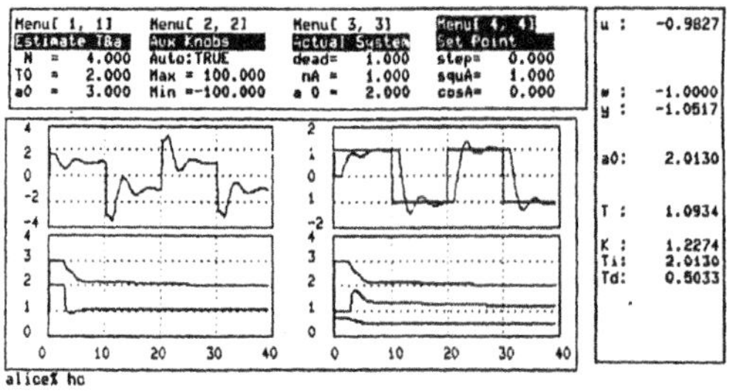

Figure 4.5 Adaptive Tuning By Pemberton Method

Example 5

This last example is the adaptive control of a second order system modelled as:

$$\bar{y}(s) = \frac{e^{-s}}{(4s+1)(s+1)}\,\bar{u}(s)$$

Algorithm 5 of section 3.4.1 is used to identify the time delay and other system parameters. The assumed initial system is:

$$\bar{y}(s) = \frac{e^{-1.5s}}{s^2+s+1}\,\bar{u}(s)$$

Figures (4.6) shows the response of this system under PID control. In this case, Pemberton's mod ideal method is used to update PID parameters.

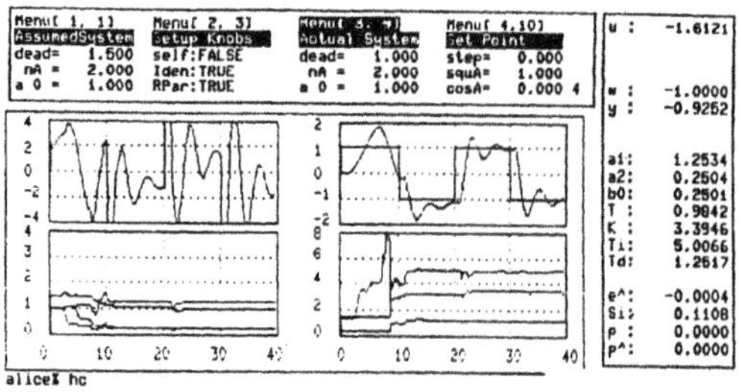

Figure 4.6 Second Order System With Unknown Time Delay

5. Conclusions

PID controllers are revisited and a new explicit PID adaptive algorithm is developed. In deriving the PID parameters, a set of tuning rules are used which have been shown to produce better results than those due to Zeigler and Nichols closed loop criteria. It was suggested that Z-N open loop tuning rule could also be used.

Almost all the algorithms developed for adaptive PID control are based on discrete time setting except Gawthrop's. The proposed algorithm is entirely in continuous time.

A recursive and derivative free least square algorithm is used for identification of time delay and system parameters.

The important feature of this algorithm is its capability to control systems with unknown time delay with a PID algorithm. There are not many algorithms for PID adaptive control that can be applied to this class of systems.

The proposed PID self tuner can also easily be tailored to other identification methods such as the technique proposed by (Besharati Rad, 1990). It is also possible to implement this algorithm in a discrete time setting.

Finally, the inherent limitation of PID algorithm for complex systems is acknowledged, but it is argued that PID algorithm has some valuable history behind it and undoubtedly a rather respectable future ahead. For a typical process plant, it is impossible to design custom-made controllers for each of the hundred loops; it is a must to use a single control structure, it does not seem that any other controller can replace PID controllers in this respect. Thus, it is worth to develop algorithms which enhance the performance of this controller and solve the tuning problem associated with them.

List of References:

Andreiev, N (1977) : *"A process controller that adapts to signal and process conditions"*, Control Engineering, PP. 38-40.

Andreiev, N (1981) : *"A new dimension: A self tuning controller that continually optimizes PID constants"*, Control Engineering, pp. 84-85.

Ashworth, M. J. (1982) : *"Feedback design of systems with significant uncertainty"*, Research studies, Lechworth, Herts, U.K.

Astrom, K.J. and Hagglund, T (1984) : *"Automatic tuning of simple regulators with specifications on phase and amplitude margins"*, Automatica, Vol. 20, pp. 645-652.

Astrom, K.J. and Wittenmark, B (1989) : *"Adaptive control"*, Addison-Wesley publishing company.

Banyasz, C and Keviczky, L (1982) : *"Direct methods for self tuning PID regulators"*, pp. 1395-1400 in Proceedings of the 6th IFAC symposium on Identification & System parameter Estimation, ed. Bekey, G.A. and Saridis, G.N., Washington D.C.

Besharati Rad, A. and Gawthrop, P. J. (1990) : *"Continuous time identification and control of systems with unknown time delay"*, International Conference on Automation, Robotics and Computer Vision, Singapore, 1990.

Bohl, A.H. and McAvoy, T.J. (1976) : *"Linear feedback vs. time optimal control, 1. The servo problem"*, Ind. Eng. Chem. Process. Des. Dev. , Vol 15, (1), pp. 24-29.

Cohen, G.H. and Coon, G.A. (1953) : *"Theoretical consideration of retarded control"*, Trans. ASME.

Coon, G.A. (1956) : *"How to find controller settings from process characteristics"*, Control Engineering, Vol. 3, (5), pp. 66-76.

Gawthrop, P.J. (1986) : *"Self tuning PID controllers: Algorithms and implementation"*, IEEE Trans. AC., Vol. AC-31, (3), pp. 201-209.

Gawthrop, P. J. , Nihtila, T and Besharati Rad, A (1989) : *"Recursive parameter estimation of continuous time systems with unknown time delay"*, Control-Theory and Advanced Technology (C-TAT), Vol. 5, (3), pp. 227-248.

Grinten van der, P.M.E.M. (1963) : *"Finding optimum controller settings"*, Control Engineering.

Haalman, A. (1965) : *"Adjusting controllers for deadtime processes"*, Control Engineering.

Hang, C.C. , Tan, K.K., and Ong, S.L. (1979) : *"A comparative study of controller tuning formulae"*, I.S.A Conference, pp. 467-476.

Hawk, W.M., Hoopes, H.S., and Lewis, R.C. (1982) : *"A self tuning controller"*, in ISA Proceedings, ISBN 0-87665-701-8.

Hazebrock, P and Van der Waerden, B.L. (1950) : *"Theoretical consideration on the optimum adjustment of regulators"*, Trans. ASME, Vol. 72, pp. 309-322.

Kim, C. and Friedly, J.C. (1974) : *"Approximate dynamic modelling of large staged systems"*, I & E C Process Design and Development, Vol. 13, (2), pp. 177-181.

Kopper, L.B. and Aiken, P.M. (1969) : *"A general process controller"*, Ind. & Eng. Chem. Process Des. & Dev., Vol. 8, (2), pp. 174-184.

Kraus, T.W and Myron, T.J. (1984) : *"Self tuning PID controllers uses pattern recognition approach"*, Control Engineering.

Kramer, J., Magee, J., and Sloman, M. (1984) : *"A software architecture for distributed computer control systems"*, Automatica, Vol. 20, (1), pp. 93-102.

Lee, P.L. and Nowell, R.B. ((1988) : *"Evaluation of PID controller"*, Process & Control Engineering , pp. 44-49.

Latour, P.R., Koppel, L.B. and Coughanowr, D.R. (1967) : *"Time optimal control of chemical processes for set point changes"*, I & E C Process Design and Development, Vol. 7, (3), pp. 345-353.

Miller, J.A., Lopez, A.M., Smith, C.L. and Murrill, P.W. (1967) : *"A comparison of controller tuning rules"*, Control Engineering.

Nachtigal, C.L. (1986) : *"Adaptive controller simulated process results: Foxboro EXACT and ASEA Novatune"*, Proceddings of American Control Conf.; Vol. 3, pp. 1434-1439.

Nishikawa, Y., Sannomiya, N, Ohta, T. and Tanaka, H (1984) : *"A method for auto-tuning of PID parameters"*, Automatica, Vol. 20, (3), pp. 312-332.

Oldenbourg, R.C. and Sartorius, H (1984) : *"Dynamics of automatic controls"*, Published by ASME, New York.

Ortega, R. and Kelly, R. (1984) : *"PID Self tuners: Some theoretical and practical aspects"*, IEEE Trans. on I.E., Vol. IE-31, (4), pp. 332-338.

Pemberton, T.J. (1972) : *"PID: The logical control algorithm"*, parts I & II, Control Engineering, Vol. 19, May, July, pp. 66-67, 61-63.

Warwick, K. (1981): *"Simplified algorithms for self tuning control"*, in Implementation of self tuning controllers, ed. By Warwick, K. , pp. 96-124.

Ziegler, J.G. and Nichols, N.B. (1942) : *"Optimum settings for automatic controllers"*, Trans. ASME., pp. 759-768.

LQG ADAPTIVE CONTROL OF PROLONGED NOISE EFFECTS

T. Hesketh and D. J. Clements

*Dept. of Systems and Control, The University of New South Wales,
P.O. Box 1, Kensington, NSW 2033, Australia*

Experiments leading to adaptive control of mould level during continuous steel slab casting are described. The process is non-linear and undermodelled, exhibits non-linearities, and the noise inherent in the process and sensors is non-stationary and significant. An adaptive LQG controller is described which models the noise effects and cancels them, and which produces robust model parameter estimates. Experiments indicate its efficacy, and on-line results are presented.

Introduction

This paper reports investigations into the description and removal of noise effects undertaken in conjunction with experiments concerned with mould level control for continuous steel slab casting.

Superficially, the casting process under investigation is simple. Molten steel is poured from a ladle into a tundish, where a constant head of molten metal is maintained. A slide gate (valve) controls the flow of steel through a nozzle into the mould, which is a heat exchanger jacket, cooled by the flow of water. In the mould, the steel slab solidifies inwards from the surface which is in contact with the cooling jacket. The slab is continuously withdrawn from the bottom of the mould and subjected to further cooling, constrained to form a continuous bar of required size by a complex series of rollers. The slab is later cut into lengths before transportation to other parts of the steel production process. Figure 1 shows the nature of the process.

The quality of the steel slab is affected by the cooling environment, while the rate of production is dependent on being able to maintain a steady level of steel in the mould, and a smooth rate of flow. The level of steel in the mould is affected by several factors: movements of the slide gate which adjust flow rate and produce wave effects; standing waves produced by the flow of steel into the mould; the casting speed (rate of withdrawal of the slab from the mould); electromagnetic stirring; and shaking of the mould to prevent adherence of the slab to the walls.

Any model of the process has to reflect both the dominant integral nature of the process, and its tendency to display oscillatory behaviour due to the wave effects. The oscillations are regarded as being associated with the controller actuation which moves the slide gate and nozzle, and with the impact of random unmeasureable disturbances. These last effects are treated as noise; owing to their nature, they possess a signal bandwidth which is within that of the process model, and so cannot be removed by filtering. Further, random disturbances will have an enduring effect unless measures are taken to cancel the dynamics. Difficulties arise in controlling mould level because of the combination of noise and non-linearity inherent in the process, instruments and actuators.

The Adaptive Controller

An adaptive controller has been configured specifically for the mould level control. The controller is implemented within RTS, which is a real-time control package written for PC's and which has an adaptive control capability (Hesketh, 1988a and b). The package supports on-line identification and control, together with some signal and model analysis, and provides facilities for simple simulation to verify controller behaviour.

The mould level controller is hierarchical, with two cascaded loops. The fast inner loop controls the position of the slide gate valve. The outer mould level control loop provides an actuation which acts as a command reference for the inner loop, the controlled variable being the actual mould level (Table 1).

Inner Loop	
Set Point	Slide Gate Reference
Measurement	Slide Gate Position
Actuation	Slide Gate Control Valve
Outer Loop	
Set Point	Mould Level Reference, s(t)
Measurement	Mould Level, y(t)
Actuation	Slide Gate Reference, u(t)

Table 1: Signals and Loops

The polynomial description of the outer loop process is:

$$\begin{aligned}(1+\tilde{A}(q))y(t) &= B(q)u(t)+E(q)d(t)+\\ &\quad (1+\tilde{C}(q))e(t)+k\end{aligned}$$

The offset and slow drift effects, k, are removed by integral action, resulting from the incorporation of an internal model polynomial $\Delta = 1 - q^{-1}$. The controller is also capable of feedforward action to cancel measured disturbances.

The adaptive controller relies on identification of the model parameters. It has been suggested that robustness requires matched filtering of all signals (Middleton et al, 1988). It is not always clear how the filters should be selected; this problem was also addressed in the experiments (Hesketh, Clements and Williams, 1989). Additional measures include: turning off identification when the information content of the excitation is insufficient (Hesketh and Sandoz, 1987); the use of deadbands in the identifier; and control of the gain of the identifier, where the trace of the covariance matrix is determined according to the requirements of the controller for rapid or slow parameter convergence. Normalisation has also been used with success (Praly, 1986), but no dynamic scaling was undertaken in this study, although all signals were carefuly scaled at the outset to ensure means close to zero, and standard deviations close to 1, reducing numerical difficulties. The adaptive controller required the identification of A_i, B_i, E_i and C_i for the model:

$$\begin{aligned}(1+A_f)(1+A_i)y(t) &= B_fB_i\Delta u(t)+\\ &\quad E_fE_i\Delta d(t)+\\ &\quad (1+C_f)(1+C_i)\hat{e}(t)\\ (1+A_i)y_f(t) &= B_i\Delta u_f(t)+\\ &\quad E_i\Delta d_f(t)+\\ &\quad (1+C_i)\hat{e}(t)\end{aligned}$$

$$\begin{aligned}y_f(t) &= \frac{(1+A_f)}{(1+C_f)}y(t)\\ u_f(t) &= \frac{B_f}{(1+C_f)}u(t)\\ d_f(t) &= \frac{E_f}{(1+C_f)}d(t)\end{aligned}$$

$$\hat{e}(t) = y(t) - \hat{y}(t)$$

The polynomial coefficients are identified using a Sequential Prediction Error algorithm (Goodwin and Sin, 1984). Recognition of the differencing can be exploited to reduce the number of degrees of freedom (Hesketh, Clements and Williams, 1990).

The controller gains are produced by an LQG algorithm (Silverman, 1976; Hesketh and Sandoz, 1987). The algorithm depends on a non-minimal state space representation of the polynomial model. Note the introduction of the set point which is made possible by differencing. The LQ problem is solved by an efficient square root numerical algorithm requiring a series of simple orthogonal transformations. The resulting controller is insensitive to variations of the estimated model parameters. With the introduction of the setpoint, s_t, the state space equations are:

$$x_t = \mathcal{A}x_{t-1} + \mathcal{B}\Delta u_{t-1}$$

with

$$x_t = \begin{bmatrix}\Delta u_{t-1}\\ \Delta u_{t-2}\\ \vdots\\ \hat{y}_t - s_t\\ y_{t-1} - s_t\\ \vdots\\ \Delta d_t\\ \Delta d_{t-1}\\ \vdots\\ \hat{e}_t\\ \hat{e}_{t-1}\\ \vdots\end{bmatrix};\mathcal{B} = \begin{bmatrix}1\\ 0\\ \vdots\\ b_1\\ 0\\ \vdots\\ 0\\ 0\\ \vdots\\ 0\\ 0\\ \vdots\end{bmatrix};$$

$$\mathcal{A} = \begin{bmatrix}0 & 0 & \cdots\\ 1 & 0 & \cdots\\ \cdots & \cdots & \cdots\\ & \mathcal{S} & \\ \cdots & \cdots & \cdots\\ \cdots & 1 & 0\end{bmatrix}$$

$$\begin{aligned} \mathcal{S} &= [\; b_2 \;\ldots\; a_1 \;\ldots\; e_1 \;\ldots\; c_1 \;\ldots\;] \\ y_t &= \mathcal{G}x_t + e_t \\ \mathcal{G} &= [\; 0 \;\ldots\; 1 \;\ldots\; 0 \;\ldots\; 0 \;\ldots\;] \end{aligned}$$

A controller is required which, subject to the constraint of the above state space system, will minimise the cost function:

$$J = \sum_{k=0}^{\infty} y_k^T Q^T Q y_k + Du_k^T R^T R Du_k$$

A numerically suitable algorithm requires appropriate initialisations, and a series of orthogonal transformations (Silverman, 1976).

$$R_0 = R; Q_0 = 0; P_0 = \mathcal{G}Q$$

$$T_k \begin{bmatrix} R_{k-1} & Q_{k-1} \\ P_{k-1}\mathcal{B} & P_{k-1}\mathcal{A} \end{bmatrix} = \begin{bmatrix} R_k & Q_k \\ 0 & P_k \end{bmatrix}$$

Following iteration of this sequence until steady state, matrices R_k and Q_k are obtained, and the feedback gains are calculated as:

$$\begin{aligned} F_k &= -R_k^{-1} Q_k \\ \Delta u(t) &= -F_k x(t) \end{aligned}$$

The principle contribution of this paper is the manner in which the noise dynamics are cancelled by controller action. This is of particular importance for the mould level control problem, which exhibits the noise difficulties already described. In most adaptive schemes that have been applied successfully, the noise is dealt with by filtering for purposes of identification, as above, but the controller design is usually deterministic. Although observer-based designs, with observer gains being derived from the noise model, have been proposed (Shieh et al, 1983), the time varying nature of the noise, and poor identification of the noise model have made such schemes difficult to apply to real plant.

In this paper, both modelling errors and noise effects are cancelled by feedback controller action. This results in a more conservative controller with reduced variability of the measured variable, as will be demonstrated below.

Results

The results of both simulation experiments and on-line closed loop control are presented. The simulation shows most clearly the effectiveness of the noise cancellation, results borne out by later on-line control of the mould level.

The purpose of the simulations was to provide a process with control difficulties similar to those of the true plant, rather than one which exhibited behaviour emulating accurately the true mould level system. A block diagram of the analog simulation is shown in Figure 3. The process is modelled as an integrator, with a time constant of 1 second. All signal ranges are ±10 volts. Gaussian noise, bandlimited to 5 Hz and with a standard deviation of 3.16 volts, is filtered and added to the process input. Also added to the input is a sinusoid of amplitude 0.5 volts and frequency 2 Hz. This has the potential to affect model identification. The integrator output is passed through a non-linearity ($y(t) = 0.1x(t)^2$).

Signals were sampled at 100 millisecond intervals, and the control actuation and model updates were computed at 300 millisecond intervals.

The control loop was closed with the aid of the control program RTS, executing on a PC. Initially, the loop was closed with unity feedback; gradually, the unity feedback signal was replaced by a control signal computed using the feedback gains derived from the identified model. Such a technique is valuable for establishing adaptive control of an unstable process. The actuation is given by:

$$u(t) = \alpha F_1(t) - (1-\alpha) F_k x(t)$$

where $F_1(t)$ is any stabilising control action, $-F_k x(t)$ is the output of the adaptive controller, and α controls the proportion with which the two signals are mixed. As α is decreased, the closed loop model improves, until the adaptive controller takes over completely.

Controller without noise model	
Model	$(1+\sum_{i=1}^{3} a_i q^{-i}) y(t) = (\sum_{i=1}^{5} b_i q^{-i}) u(t)$
Std. Dev. (y(t))	1.6795
Std. Dev. (u(t))	1.9967
Controller with noise model	
Model	$(1+\sum_{i=1}^{3} a_i q^{-i}) y(t) = (\sum_{i=1}^{5} b_i q^{-i}) u(t) + (1+\sum_{i=1}^{5} c_i q^{-i}) e(t)$
Std. Dev. (y(t))	1.2125
Std. Dev. (u(t))	1.4679

Table 2: Simulated Models and Performance Measures

Two controllers were derived for purposes of comparison (Table 2). The controllers differ in that the first uses no noise model, while the second contains the full noise model described earlier. The table gives contrasting performance measures. Figures 4 & 5 show the simulation behaviour. The main points to note are:

- A pair of oscillatory poles is identified, in addition to the integrator. These poles assist in describing the behaviour of the band-limited noise.
- The noise model will always contain a factor $(1 - q^{-1})$. This is allowed for by a factor $(1 - \beta q^{-1})$ in the pre-filter. $1 + A$ and $1 + C$ should be co-prime for purposes of control, requiring $\beta < 1$.
- The variability of the actuation is less when the noise model is used. From this it is surmised that the noise model enables better prediction of future measurement behaviour, reduces the tendency for the controller to react unnecessarily, and results in more conservative action.
- The variability of the measurement decreases.
- Controller performance will have improved as determined by any of a variety of performance measures.

The simulation results are borne out by on-line control experiments. Figure 6 shows the result of applying a controller to the mould level control, of the same structure and order as that used for the simulation. The first part of the controlled sequence shows the performance before the application of RTS, (under control of a commercially available adaptive system), while the second part of the sequence shows the performance with RTS, again indicating the potential for increased conservatism and reduced sensitivity.

Conclusions

Many processes exhibit combinations of noise and non-linearity, requiring cancellation of the enduring noise dynamics. An adaptive controller suitable for such a purpose has been presented. This paper has described simulations which indicate the success of the approach for reducing the effects of noise, and has presented the result of online control indicating that the benefits translate to control of a real process.

Acknowledgements

This work was undertaken with the support and help of poduction and electrical engineering staff at the BHP Slab Caster, Port Kembla, Australia. The particular contribution of Mr. R.D. Williams of the Slab Caster Section who assisted with online experiments is gratefully acknowledged.

References

1. Silverman, L.M., (1976), "Discrete Riccati equations: alternative algorithms, asymptotic properties and systems theory interpretations", in G.T. Leondes (Ed.), Control and Dynamic Systems, Vol. 12, Academic Press, New York.
2. Shieh, L.S., Wang, C.T. and Tsay, Y.T., (1983), "Fast sub-optimal state-space self-tuner for linear stochastic multivariable systems", IEE Proc., 130, Pt. D, 4, 143-154.
3. Goodwin, G.C. and Sin, K.S., (1984), "Adaptive filtering, prediction and control", Prentice Hall, Englewood Cliffs.
4. Praly, L., (1986), "Global stability of a direct adaptive control scheme with respect to a graph topology", In *Adaptive and Learning Systems - Theory and Applications*, ed. K. S. Narendra, New York: Plenum Press.
5. Hesketh, T. and Sandoz, D. J., (1987), 'Application of a multivariable adaptive controller', *ACC Proceedings*, 1301-1306.
6. Middleton, R.H., Goddwin, G.C., Hill D.J. and Mayne, D.Q., (1988), "Design issues in adaptive control", IEEE Trans. on Aut. Contr., 33, 1, 50-58.
7. Hesketh, T. (1988), 'RTS - Realtime System: Theoretical Basis', Technical Report UNSW-DSC-TR-88-03.
8. Hesketh, T. (1988), 'Realtime System for Adaptive Control: User's Manual', Technical Report UNSW-DSC-TR-88-04.
9. Hesketh, T., Clements, D.J. and Williams, R., (1989), 'Experimental Report: Mould Level Control, BHP Slab Caster', Technical Report UNSW-DSC-TR-89-03.
10. Hesketh, T. and Clements, D.J., (1990), 'Adaptive mould level control for continuous steel slab casting', to be submitted.

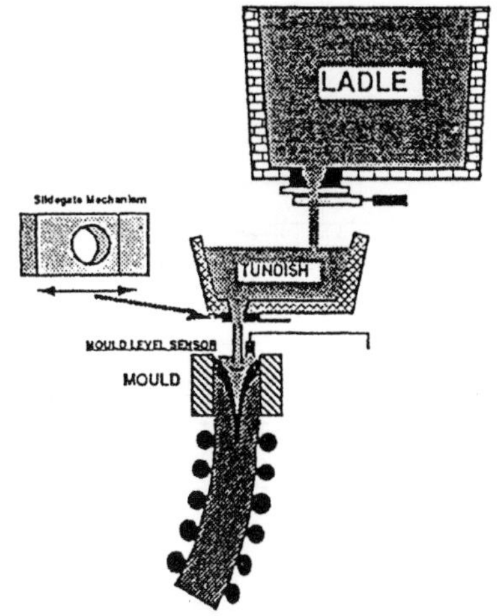

Figure 1: Mould Level Process - Overview

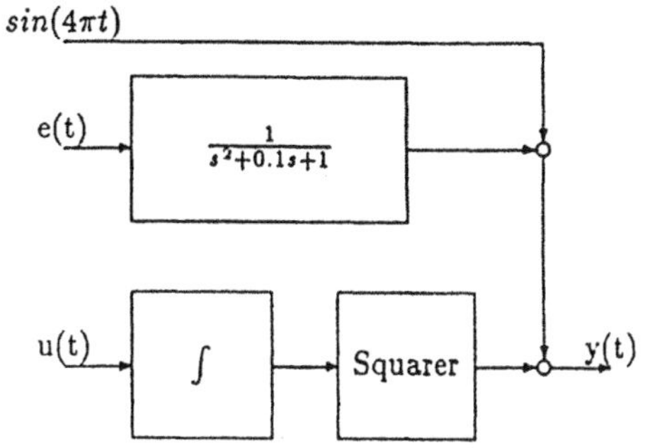

Figure 2: Simulation Block Diagram

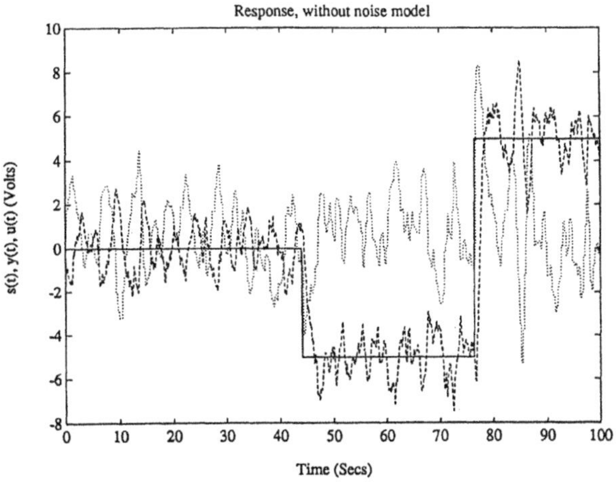

Figure 3: Simulation Result - No Noise Model

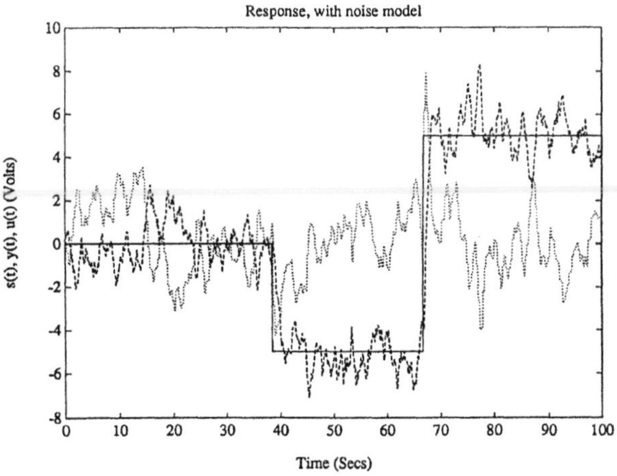

Figure 4: Simulation Result - Noise Model

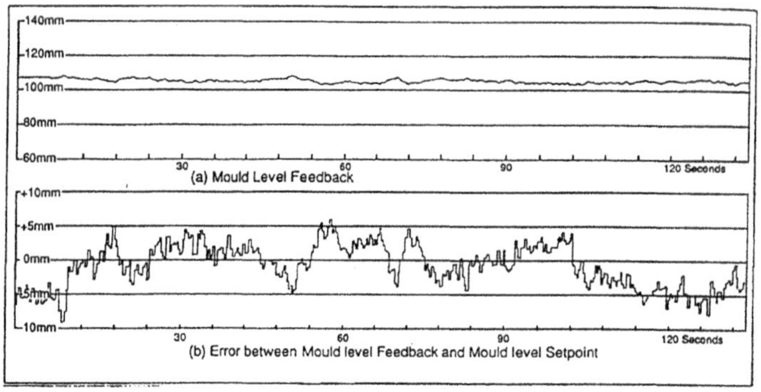

Figure 5: On-Line Control of Mould Level

Performance of Commercial
Self-Tuning Controller

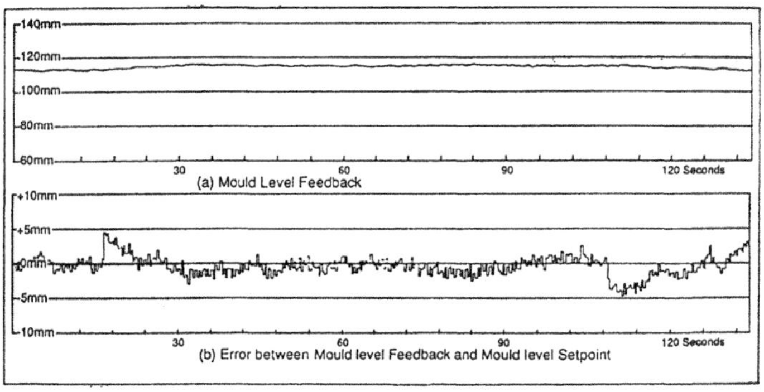

Figure 6: On-Line Control of Mould Level

Performance of RTS

AN EXPERIMENTAL STUDY OF A DUAL-RATE AUTOTUNED POLE-PLACEMENT CONTROLLER

Y. S. Cai and C. C. Hang

Dept. of Electrical Engineering, National University of Singapore, Kent Ridge, Singapore 0511

Abstract

A dual-rate approach to auto-tuning of pole-placement controller allows the use of a large sampling interval for parameter estimation and control during autotuning, and a small sampling interval for normal control. The main merit of the dual-rate autotuned control system is in the simultaneous achievement of robust estimation and reduced computation load during tuning, and improved performance in the regulation of deterministic and stochastic load disturbances. This algorithm has been implemented on an universal controller card based on an Intel 8096 microcontroller. The dual-rate autotuned pole-placement microprocessor-based controller has been applied to a pilot plant consisting of a resonant aerodynamic object. The experimental results show that the proposed controller performs very well.

Keywords. Auto-tuning; pole-placement; dual-rate system.

1. Introduction

The advent of microprocesor technology has spurred the commercialization of automatic tuning (autotuning) methods for PID controllers (Astrom and Hagglund 1984). Advanced control techniques such as pole-placement require the knowledge of process model and are also more sensitive to the variation of model parameters. While many sophisticated adaptive control methods have been proposed in the literature (Allidina and Hughes 1980, Astrom and Wittenmark 1980, Wellstead et al. 1979 and 1981), autotuning and periodic re-tuning are often sufficient in practice. This greatly reduces the complexity in applying advanced control and improves their chance of being accepted by practising engineers.

The majority of auto-tuning methods for the explicit pole-placement controllers published to-date use a single sampling rate in that the controller output and the model parameters are updated at the same frequency. From a modelling point of view, a relatively large sampling interval has been recommended in the literature (Hang et al. 1985, Ljung 1985, Isermann 1981). It is more stable numerically. Computational load for estimation especially when various heuristic logics are used to improve practical robustness will also reduce consequently. However, from a regulation point of view a small sampling interval is preferred by practising engineers as any static load disturbances can be detected and rejected as soon as possible (Hagglund and Astrom 1989, Berg et al. 1988). This is especially important for an underdamped process for which pole-placement is an ideal controller (Hang et al. 1989). It has been shown recently (Berg et al. 1988) that the sampling rate for good load disturbance rejection should be several times larger than that widely accepted for good setpoint response. A higher bandwidth for the closed-loop system can also be specified only if a sufficiently small sampling interval is used.

Both the advantages of a larger sampling interval for parameter estimation and a smaller sampling interval for control can be realized in a dual-rate auto-tuning method first proposed by Hang, Lim and Chong (1989) for a digital Smith Predictor. Its extension to the more complicated case of the pole-placement controller has been reported (Hang, Cai and Lim 1989). It consists of a parameter estimator with a slow sampling rate, a model conversion algorithm to compute the model parameters for a fast sampling rate, a controller parameter updating algorithm, and a controller with a fast sampling rate.

The paper is organized as follows. The dual-rate autotuned pole-placement design is briefly reviewed in section 2. A microcontroller implementation is given in section 3. A performance evaluation of the dual-rate autotuned pole-placement controller using a pilot plant is presented in section 4. Concluding remarks are given in section 5.

2. Dual-rate autotuned pole-placement controller

Consider a single-input/single-output process described by

$$G(q^{-1}) = \frac{q^{-k}\,B(q^{-1})}{A(q^{-1})} = \frac{y(t)}{u(t)} \qquad (1)$$

where q^{-1} is the backward shift operator, A and B are polynomials in q^{-1}, with A being monic and k representing the time delay of the process in number of samples. Variables y and u are the process output and input respectively. The degrees of polynomials $A(q^{-1})$ and $B(q^{-1})$ are written as deg A and deg B. It is assumed that A and B are coprime.

In a pole-placement design, the objective is to find a controller such that the closed-loop transfer function from the command input u_c to the output y is given by

$$G_m(q^{-1}) = \frac{q^{-k} B_m(q^{-1})}{A_m(q^{-1})} \qquad (2)$$

where A_m and B_m are coprime and A_m is monic. The zeros of A_m are assumed to be inside the unit circle and well damped.

A general linear regulator can be described (Astrom and Wittenmark 1984) by

$$R(q^{-1})\ u(t) = T(q^{-1})\ u_c(t) - S(q^{-1})\ y(t) \qquad (3)$$

Combining equations (2) and (3), we obtain

$$B_m = g\ B \qquad (5)$$

and

$$g = \frac{A_m(1)}{\sum_{i=0}^{n_b} b_i} \qquad (6)$$

where n_b is the order of polynomial $B(q^{-1})$ and b_i is the coefficient of polynomial $B(q^{-1})$.

In the standard pole-placement design (Astrom and Wittenmark, 1984), one simply solves the following Diophantine equation

$$A\ R + q^{-k}\ B\ S = A_m\ A_0 \qquad (7)$$

to obtain R and S, with the following choices

$$\deg S = \deg A - 1 \qquad (8)$$

$$\deg R = \max(\deg A_m + \deg A_0 - \deg A,\ \deg B + k - 1) \qquad (9)$$

A_0 is the user-defined observer polynomial (Astrom and Wittenmark, 1980). Finally one computes

$$T = \frac{A_0\ B_m}{B} \qquad (10)$$

During autotuning, the new parameters estimated will be used to compute the new control law as in adaptive control. It is evident that the computation load for single-rate auto-tuning is quite severe as the Diophantine equation (7) has to be solved on-line at every sampling interval in addition to the estimation of process model parameters in the A and B polynomials. This will be grossly aggravated by the choice of a small sampling interval, which is necessary for fast rejection of load disturbances as discussed earlier. The reason is that the delay k will increase rapidly with reduced sampling intervals and, from equation (9), the deg R will increase correspondingly. On the other hand, it is well known (Ljung 1985) that a relatively large sampling interval is essential for robustness of parameter estimation, especially if the dead time is not accurately known. Thus, in the single-rate auto-tuning of a pole-placement controller, it is only practical to use a large sampling interval at the expense of load regulation.

The practical limitations of the single-rate auto-tuning method for pole-placement controller can be overcome by the use of dual sampling rates: a small sampling interval for normal control and a large sampling interval for on-line parameter estimation and control during auto-tuning. The controller computation using the parameter estimation results is, however, not quite straight forward. As the parameters identified by the estimator correspond to a sampling frequency slower than the controlling frequency, it cannot be used directly in the Diophantine equation (7) to tune the controller. There is thus a necessity to convert the identified slow model to a fast model having the same sampling interval as the controller. There is also a necessity to switch the control sampling rate to the same rate as that of the parameter estimation during auto-tuning and this will be elaborated below.

2.1 Model conversion

As proposed in Hang et al (1989), model conversion can be done by the pole-zero mapping technique. The slow model is first converted from z-plane back to s-plane; it is then converted from s-plane back to z-plane, but at a different sampling frequency. In practice, dead time plus first order or second order models are often sufficient for the purpose of on-line modelling, especially when slow sampling is used (Isermann, R., 1981). The derivation for first order system with dead time and possible fractional delay is shown below.

For a slow model sampled at h second:

$$G(q^{-1}) = \frac{b_1\ q^{-1} + b_2\ q^{-2}}{1 - a_1\ q^{-1}}\ q^{-d}$$

$$= \frac{b_1\ (q^{-1} + (b_2/b_1)\ q^{-2})}{1 - e^{\alpha h}\ q^{-1}}\ q^{-d} \qquad (11)$$

For the equivalent fast model sampled at h/N second:

$$G' = \frac{c\ b_1(q^{-1} + \mathrm{sign}(b_2/b_1)(|b_2/b_1|)^{1/N}\ q^{-2})}{1 - e^{\alpha h/N}\ q^{-1}}\ q^{-Nd} \qquad (12)$$

where c = matching constant such that $G(q^{-1})$ and $G'(q^{-1})$ are having the same static gain (i.e. at q = 1). The design of the controller parameters is based on $G'(q^{-1})$ instead of $G(q^{-1})$. The above is valid if the fractional delay is smaller than h/N. Otherwise a more refined formula should be used (Cai et al, 1987). The derivation for second order system with dead time has been given in Hang et al (1989).

2.2 Algorithm for dual-rate auto-tuning

During auto-tuning when the process model is being identified, there are two possible control sampling rates. The first possibility is the continued use of the fast control sampling rates, as in the case of the adaptive Smith Predictor (Hang et al, 1989). However, it has been found that although the poles of the process and the static gain could be correctly identified, the zeros could not be accurately identified as the control signal is changing during the parameter estimation interval (Cai et al 1987). The accuracy of the zeros is not critical to the Smith Predictor but is crucial in the pole-placement design of eqn. (7). It is therefore essential that the second alternative of switching the control also to the same slow sampling rate for estimation be adopted during auto-tuning. This will ensure that the process zeros can be correctly estimated at the expense of increased load regulation variance during auto-tuning.

The complete dual-rate autotuned pole-placement control algorithm with the desired closed-loop poles and observer poles specified by the polynomials A_m and A_0 respectively is given as follows. The subscripts l and s refer to the large and small interval respectively.

Step 1. Switch the control sampling interval to h_l while injecting perturbation signal if the process states are not sufficiently excited. Then estimate the parameters of the following process model by the simple recursive least squares or other suitable method (Isermann 1981). The forgetting factor may be set to 1. The initial value of the covariance matrix is chosen to be large, such as one thousand times the unit matrix.

$$A_l\ y(t_l) = B_l\ u(t_l - k_l) \tag{13}$$

Step 2. Introduce $B_{ml} = g_l B_l$ and choose g_l such that $B_{ml}(1) = A_{ml}(1)$. Then determine the polynomials R_l and S_l such that

$$A_l\ R_l + q^{-k_l}\ B_l\ S_l = A_{ml}\ A_{0l} \tag{14}$$

Step 3. Use the control law

$$R_l\ u(t_l) = T_l\ u_c(t_l) - S_l\ y(t_l) \tag{15}$$

where

$$T_l = g_l\ A_{0l} \tag{16}$$

Steps 1 to 3 are repeated at each large sampling interval h_l until process parameters in polynomial A_l and B_l have converged. Then switch off parameter updating and switch back to normal control at small sampling interval using the following procedure.

Step 4. Map the estimated model to those corresponding to the model at the small sampling interval h_s:

$$y(t_s) = B_s\ u(t_s - k_s) \tag{17}$$

Step 5. Introduce $B_{ms} = g_s B_s$ and choose g_s such that $B_{ms}(1) = A_{ms}(1)$. Then determine the polynomials R_s and S_s such that

$$A_s\ A_s\ R_s + q^{-k_s}\ B_s\ S_s = A_{ms}\ A_{0s} \tag{18}$$

Step 6. Use the control law

$$R_s\ u(t_s) = T_s\ u_c(t_s) - S_s\ y(t_s) \tag{19}$$

where

$$T_s = g_s\ A_{0s} \tag{20}$$

Note that steps 1, 2, 3 are repeated at each large sampling interval h_l; at the end of auto-tuning, steps 4 and 5 are done only once and they can take any amount of time, not necessarily limited to one small or large sampling interval; then step 6 is executed at every small sampling interval h_s as required. Thus a great reduction in computation load during auto-tuning is achieved.

Note that in a dual-rate auto-tuning pole-placement system, we must ensure that the gain cross-over frequency of the transfer function $\frac{S_s(q^{-1})\ B_s(q^{-1})}{R_s(q^{-1})\ A_s(q^{-1})}$ with fast sampling rate is less than π/h_l. In this case, the most important frequency range of the system is still nearby π/h_l. Thus, We can use the dynamic model obtained with large sampling interval for the controller design at the small sampling interval.

If the gain cross-over frequency of the transfer function $\frac{S_s(q^{-1})\ B_s(q^{-1})}{R_s(q^{-1})\ A_s(q^{-1})}$ with fast sampling rate is greater than π/h_l, we should increase controller order, reduce the desired closed-loop bandwidth or increase the small sampling interval to redesign the controller at small sampling interval until the above condition is satisfied.

3. Microcontroller Implementation

The dual-rate autotuned pole-placement controller has been implemented on an universal controller card based on an Intel 8096 microcontroller and applied to the pilot plant consisting of a resonant aerodynamic object. The universal controller is complete with microcontroller, digital I/O, ADC and DAC. Both C96-language and assemblly language can be used in the controller programming. A microcomputer can be linked to universal controller card through serial port using RS-232. During routine operation, the microcomputer keyboard is used as a functional keyboard, providing on-line operator control over changes in setpoint and sampling interval, etc, while the microcomputer CRT display serves as the video monitor for trend logging and other operator interface functions. It is a low-cost controller and development system that is easy to use. It facilitates practical implementation of modern control design during prototype development and for controller commercialization.

4. Experimental Results

The pilot plant as shown in Fig. 1 is one consisting of a resonant aerodynamic object (Schmidtbaue 1986) where the angular orientation of a hinged rectangular plate is controlled by blowing air stream at it with a variable speed fan. The fan is drived by a DC motor with pulse width modulation and the plate rotation angle is measured by a low-friction potentiometer. The pilot plant thus contains rich dynamics for testing pole-placement controller: fan motor time constant; air transportation lag; resonant poles (plate acting as pendulum); high disturbance level (air turbulence). It is also easy to change the dominant time constant and deadtime of the plant making it a versatile pilot plant for studing the effects of parameter changes.

The performance of single-rate and dual-rate autotuned pole-placement controllers will be compared in terms of: (1) deterministic disturbance rejection ; (2) stochastic disturbance rejection, using the above pilot plant. The nominal model of the plant has been identified off-line as:

$$G(s) = \frac{276\, e^{-0.2s}}{(s + 4)(s^2 + 1.5\, s + 40)}$$

It thus has a pair of complex poles with damping factor of 0.12. Its step response is shown in Fig. 2. Note that it is very noisy due to air turbulence.

4.1 Design details

Three auto-tuned controllers will be studied. The first two correspond to single-rate approach with sampling intervals of 0.05 second and 0.2 second, respectively. The third is a dual-rate approach with first auto-tuning (identification and control) at sampling interval of 0.2 second and after parameter convergence the control will be switched back to fast control with sampling interval of 0.05 second.

The desired closed-loop poles are characterized by of a third order continuous-time system denominator of (s + 2)(s + 5)(s + 5). Choices of $\ell = 1$ and $A_0(q^{-1}) = 1$ have been used in the controller design.

4.2 Performance comparison

The responses to static and stochastic load disturbance after auto-tuning are compared in Fig. 3 for the three cases. A static load disturbance was applied at a marked moment and removed at another marked moment. The stochastic disturbance is natural being caused by air turbulence around the aerodynamic object. It is evident that the dual-rate system outperforms the single-rate system in static load regulation and stochastic disturbance rejection. It is comparable with the single-rate system with the small sampling interval. Note that the single-rate system with h_s=0.05 sec is tuned off-line as on-line auto-tuning is not feasible. This is because the computation of both the parameter estimation and control cannot be completed by the single-chip controller within 0.05 sec. Yet another practical observation is that the time taken to tune the controller is larger for the single-rate system with small sampling interval, being 5 second for h_s=0.05 sec and 4 second for h_l = 0.2 sec. The superiority of a dual-rate system is evident.

5. Conclusion

In the dual-rate autotuned pole-placement controller, a large sampling interval is used for parameter estimation and control during auto-tuning, achieving benefits of reduced computation as well as improved estimation robustness. It uses a smaller sampling interval for normal control which has the benefit of improved regulation to load disturbances. The aerodynamic object with lightly damped dynamics and stochastic disturbance has served as a suitable pilot plant to demonstrate the practical significance and achievable performance of the dual-rate autotuned pole-placement.

REFERENCES

[1] Alidina, A. Y., and Hughes, F. M., 'Generalised self-tuning controller with pole assignment,' Proc. IEE, 127, 13-18, (1980).

[2] Astrom, K. J., and Hagglund. T.,'Automatic tuning of simple regulators with specifications on phase and amplitude margins,' Automatica, 20, 645-651, 1984.

[3] Wellstead, P.E., D. Prager, and R. Zanker, 'Pole assignment self-tuning regulator,' Proc. IEE, 126, 781-787 (1979).

[4] Astrom, K.J. and B. Wittenmark, 'Self-tuning controller based on pole-zero placement,' Proc. IEE, 127, 120-130 (1980).

[5] Wellstead, P.E. and S.P. Sanoff, 'Extended self-tuning algorithm,' Int. J.Control, 34, 433-455 (1981).

[6] Hang, C.C., K.W. Lim, and T.T. Tay, 'On practical applications of adaptive control,' Proc. Int. Workshop on Adaptive systems Theory and Applications, Yale University, 1985.

[7] Astrom, K.J. and B. Wittenmark, Computer Controlled Systems: Theory and Design, Prentice-Hall, Englewood Cliffs, New Jersey, 1984.

[8] Isermann, R., Digital Control Systems, Springer-Verlag, 1981.

[9] Ljung, L., 'On the estimation of transfer function,' Automatica, 21, 677-696 (1985).

[10] Hang, C.C., Lim, K.W and B.W. Chong, 'A dual-rate adaptive digital Smith predictor,' Automatica, V01. 25, No. 1, pp. 1-16, 1989.

[11] Berg, M.C., Amit, N., And Powell, J.d., 'Multirate digital control system design,' IEEE Trans, Automatic control, Vol. AC-33, No. 12, 1139, 1988.

[12] Hagglund, T. and Astrom, K.J., 'An industrial adaptive PID controller ,' Proceedings of the IFAC symposium on adaptive systems in control and signal processing, Glasgow, Vol. 1, 293, 1989.

[13] Cai Y.S., C.C. Hang, and K.W. Lim, 'A dual-rate self-tuning pole-placement controller', Report CI-87-5, National University of Singapore (1987).

[14] C.C. Hang, Y.S. Cai and K.W. Lim, 'A dual-rate approach to auto-tuning of pole-placement controller', Report CI-89-4, National University of Singapore (1989).

[15] B. Schmidtbauer, 'A low-cost development tool for microprocessor based controller design,' Proceedings of the IFAC symposium on Low Cost Automation, Valencia, Spain, Vol. 1, 293-298, 1986.

Figure 1. Pilot plant

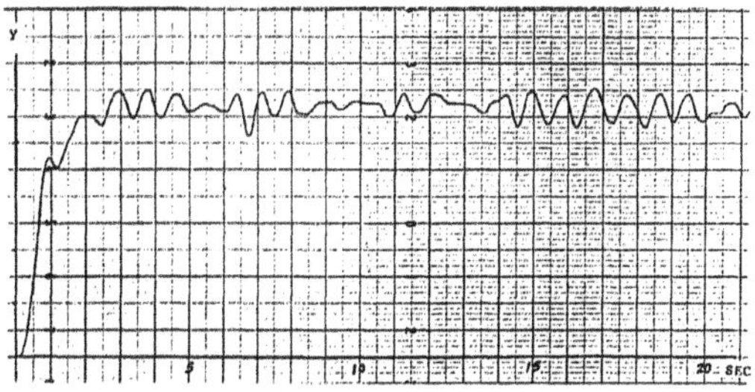

Air turbulence around the aerodynamic object

Figure 2.

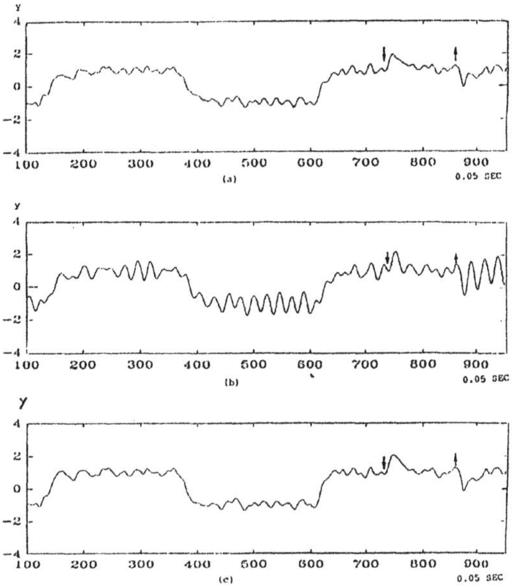

(a) Dual-rate system with $h_l = 0.2$ sec and $h_s = 0.05$ sec.
(b) Single-rate system with $h = 0.2$ sec.
(c) Single-rate system with $h = 0.05$ sec.

Figure 3.

INTELLIGENT MULTIOBJECTIVE OPTIMAL CONTROL

Li-Min Jia and Xi-Di Zhang

Signal and Communication Research Institute,
China Academy of Railway Sciences, Beijing 100081, PRC

Abstract. In the paper, the MLMO principle that More Large and more complex the controlled process is , More Objectives there are to be satisfied to control the process with quality is proposed. And the concept and approach of process partition (PP) is also put forward for modelling the complex process with varying characteristic. To multiobjectively control the process with linguistic uncertainty, the result-related fuzzy multiobjective predictive control approach is proposed and finally the intelligent multiobjective optimal control(IMOP) which is a combination of the process partition(PP), the fuzzy multiobjective predictive control(FMPC) methods and the expert system techniques is put forward and applied to control the train-travelling process with good results.

Keywords. Intelligent control; fuzzy control; multiobjective control; optimal control; expert system; artificial intelligence; process control process partition; automatic train control.

INTRODUCTION

Since the introduction of intelligent control(Fu,1971) and with the development of AI,OR and control theory, intelligent control with its promising features and rapid development has drown more and more attention in the field of control(Meystel and others, 1985,1987). As the intersection of Artificial Intelligence(AI), Control Theory(CT), System Theory(ST), Information theory(IT) and Operational Research(OR),Intelligent Control(IC) with its multidisciplinary property will be the most forceful tool to control the processes of larger & larger scale and increasing complexity. The relation between IC and other disciplines can be expressed as following,

$$IC = AI \cap CT \cap ST \cap IT \cap OR$$

In practice, we often encounter the process which is very complex and affected by numerous environment factors and especially, it undertakes a great change in characteristic during its operation. It is very difficult to describe such a process by a general precise mathmatical model,even if there may be one, it is impossible to reflect its characteristic satisfactorily. For these processes, the conventional control approches are not suitable to be used and ordinary adaptive approches are also not suitable. But, in most cases, man, with his a priori qualitative knowledge and experiences about the controlled process, can control such a kind of processes satisfactorily to meet multiple objectives. The reason for this is that the man controller has the intelligence which the controllers based on conventional control theory do not have, and which includes the ability to deal with uncertainty, to self-adjust the control strategies, to predict the control result, and especially, the multiobjectiveness of control.

As for multiobjective control on complex

large processes, it has not drown enough attention in the field of control. To control a complex process with high quality, there must be some explicit or/and implicit objectives. It seems that More large and more complex the controlled process is, More objectives there are to be satisfied to control it with quality, and we call this the MLMO principle. The multiobjectivization of control is essential for controlling the complex process with high quality.

Modelling the behavior of the process is the essential prerequisite to the realization of control. To model the behavior of complex processes, the concept and approach of the process partition(PP) is proposed, and after that the fuzzy multiobjective predivtive control(FMPC) is also put forward and concequently the intelligent multiobjective optimal control(IMOC) which is the combination of PP, FMPC and the expert system techniques is put forward to realize the automatic control of the complex processes which have varying characteristic and imprecise objectives. Finally, the application of IMOC approach to the train-travelling control process is given to show the feasibility of the concepts and approaches proposed.

PROCESS PARTITION

In practice, there is such kind of processes which are so complex and their characteristics vary so great that their behavior can not be modelled by a single model, but we can partition the process into several sub-processes with different characteristic. Every sub-process, whose characteristic has not great change, can be described by a suitable sub-model with more preciseness. These sub-models may be mathmatically crisp or fuzzy according to specific situations and at any time given, the controlled process can be characterized by only one sub-process. Because of the great characteristic differences, the control algorithms and control objectives of the sub--processes may be also different.

Let the controlled process is P, its behavior is characterized by the movement of the characteristic varible(CV) vector V in the CV space S. So the partition of P is equal to the partition of S. When S is partitioned into n subspaces, S1, S2, ..., Sn, P is partitioned into n corresponding subprocesses P1, P2, ..., Pn.

Every subspace is characterised by the characteristic function $\mu_{S_i}(v)$, i=1,2,...,n. $\mu_{S_i}(v)$ is defined as:

$$\mu_{S_i}(v) = \begin{cases} 1 & v(t) \in S_i \\ 0 & V(t) \notin S_i \end{cases} \quad i=1,2,\cdots,n \qquad (1)$$

This kind of partition is referred as the crisp partition, when $\mu_{S_i}(v)$ takes value in [0.1] then the partition is a fuzzy one in which the subspaces have the non-crisp boundary. Here we concern only the crisp partition.

The partition of S must satisfy the following conditions:

$$\forall t \ \forall v, \quad \exists! i \ \mu_{S_i}(v) \neq 0, \quad i=1,2, \cdots, n \qquad (2)$$

$$\forall t \ \forall v, \ \forall i, j \ \mu_{S_i}(v) \wedge \mu_{S_j}(v) = 0 \quad i \neq j, \ i, j=1,2, \cdots, n \qquad (3)$$

$$\forall t \ \forall v, \ \bigvee_{i=1}^{n} \mu_{S_i}(v) = 1 \qquad (4)$$

(2) means that, at any time t, V belongs to only one subspace, equally say, the process is characterized by only one subprocess. (3) means that the subspaces are disjoint, that is, the subprocesses have crisp differences in characteristic. By (4), we mean that all the subspaces compose the whole CV space and the controlled process is characterized completely by the subprocesses.

Then every subprocess Pi can be described by a suitable model Mi, i=1,2,...,n.

$$Mi: \ U \times Si \rightarrow S \qquad (5)$$

where U is the space of control vector u(t). When V(t) is in Si, then P is at the Pi described by Mi, that is, the controlled process can be expressed by:

$$P: \ v(t) = Mi(\ v(t), u(t)\ ), v(t) \in Si, \quad i=1,2, \cdots, n \qquad (6)$$

The general model of the process is

$$M = \bigcup_{i=1}^{n} Mi \ \text{and} \ P = \bigcup_{i=1}^{n} Pi \qquad (7)$$

and

$$\forall t, \ \exists! i \ M = Mi \ \text{and} \ P = Pi, \ i=1,2, \cdots, n \qquad (8)$$

As the controlled process is partitioned into several subprocesses with the characteristic not changing greatly, we can find more easily the control algorithms to control the subprocesses. Let the control algorithm corresponding to the subprocess Pi is Ai, i=1,2,..., n, then the whole control algorithm is $A = \bigcup_{i=1}^{n} A_i$

$$A_i: U \times S_i \rightarrow U \tag{9}$$

we have also,

$$\forall t. \exists ! i \ u(t) = A_i(v(t), M_i(u(t), v(t))), \quad v(t) \in S_i \quad i=1,2,\cdots,n \tag{10}$$

The partition of process makes it possible to use different kind of models to describe the process with high preciseness, and to orchestrate different control algorithms to control the process with high quality. And Mi,Ai may be mathmatically precise or fuzzy , may be numerical or heuristical.

INTELLIGENT MULTIOBJECTIVE OPTIMAL CONTROLLER

The general description

The intelligent multiobjective optimal controller, as shown in Fig.1 , consists of two levels: the control level and the intelligent coordinating level. The control level is composed of several fuzzy multiobjective predictive controllers(or/and any other possible controllers), every controller controls one subprocess.The intelligent coordinating level is in reality an expert system, its function is to supervise the process and the controllers and to orchestrate the controllers to make the control results optimal.

Fuzzy multiobjective predictive controller

It is well known that, for many complex processes, the control quality of man controller is superior to that of many sophisticated control systems. One reason for this is that the process model on which the control systems are based can not copy the real behavior of the process. Another reason is that these controllers control the process according to only one objective predetermined, but the man controller controls the process by evaluating multiple objectives which, in most cases, are linguistic and imprecise. But these linguistic objectives can be characterized by suitable fuzzy sets(Zadeh,1965) according to the knowledge and experiences of man controllers. The existing fuzzy control approaches (Dubois and other, 1980) is more superior to the conventional control and optimal control approaches in dealing with the uncertain and heuristic knowledge and in control of the processes with some uncertainty. But these fuzzy control methods are not result-related, that is, they can not reflect the real-time evaluation of the control objectives by the man controller. So the fuzzy multiobjective predictive control (FMPC) approach is proposed.

The FMPC goes with four steps:

Step 1: Define possible control alternatives.

Step 2: Predict the future process outputs for every control alternative according to the process model and the control algorithm.

Step 3: Evaluate the degree of satisfactory of the outputs to the control objectives concerned.

Step 4: Select the control which makes the objectives best satisfied as the control output to the process.

That is FMPC selects the control output by predicting the control result and according to whether the results satisfy the objectives. The detailed FMPC algorithm is given as following.

Let the model of the controlled process is:

$$\begin{cases} T: S \times U \rightarrow S \\ O: S \rightarrow Y \end{cases} \tag{11}$$

where S, Y and $U=\{u_1, u_2, \ldots, u_n\}$ are the state space, output space of the process and the control set respectively. Mappings T and O may be precise or fuzzy according to specific situation. Then the process output at time t is:

$$y(t) = O(s(t)) = O \circ T(s(t-1), u(t-1)) \tag{12}$$

where 1 refers to a sample period and $\circ$ is a mapping composite operator. The control objective set $G=\{G1, G2, \cdots, Gk\}$ can be characterized by suitable fuzzy sets, whose membership function is defined as:

$$\mu_{G_j}: Y \rightarrow [0, 1], \quad j = 1, 2, \cdots, k \tag{13}$$

The satisfactory degree of y(t) to objective Gj is

$$\mu_{G_j}(y(t+1))=\mu_{G_j}(\,O\circ T(\,s(t),u(t)\,)), \qquad (14)$$
$$j=1,2,\cdots,k$$

which is also the satisfactory degree of the control u(t) to Gj, when the process state is s(t) at time t. Then for Gj only, the optimal control at time t is:

$$u_{oj}=\{u_{oj}/\mu_{Coj}(u_{oj})=\underset{i}{Max}\,\mu_{Coj}(u_i)\}$$
$$=\{u_{oj}/\underset{i}{Max}\,\mu_{Gj}(y(O\circ T(s(t-1),u(t-1)=u_i))\} \qquad (15)$$
$$j=1,2,\cdots,k$$

where the $\mu_{Coj}(\cdot)$ is the membership function of the induced optimal control set Coj for the objective Gj. When all the objectives $G=\{G1,G2,\ldots,Gn\}$ are considered, we get

$$\mu_G(\hat{y}_i(t+1))=\mu_{\bigcap_j G_j}(\hat{y}_i(t+1))$$
$$=\circledcirc\,\mu_{G_j}(\,O\circ T(s(t),u(t)=u_i)) \qquad (16)$$
$$i=1,2,\cdots,n$$

where $\hat{y}_i(t+1)$ is the predictive value of $y_i(t+1)$ when u(t) is u_i and $\circledcirc$ is the operator corresponding to the conjunctive operation of sets, and usually taken as Min=$\wedge$. Then the optimal control set at time t is:

$$\mu_{Co}(u_i)=\mu_G(\hat{y}(t+1))$$
$$=\bigwedge_j\mu_{G_j}(O\circ T(\,s(t),u(t)=u_i)), \qquad (17)$$
$$i=1,2,\cdots,n$$

and the optimal control at time t is:

$$u_o=\{u_o/\mu_{Co}(u_o)=Max\;\mu_{Co}(u_i)\}$$
$$=\{u_o/\mu_{Co}(u_o)=\bigvee_i\bigwedge_j\mu_{G_j}(\,O\circ T(s(t),u(t)=u_i))\}$$
$$(18)$$

That is, by evaluating the satisfactory degree of the predictive value $\hat{y}_i(t+1)$ when $u(t)=u_i$ to the control objectives $G=\bigcap_{j=1}^{k}G_j$, the FMPC selects the control which has the best control results as the optimal control output to the process at time t.

But in practical applications, the control objectives are of different importance and this difference must be embodied in the control. We use weight to express the different importances of the objectives. The weight set is $W=\{Wj/j=1,2,\cdots,k\}$, then (16) becomes

$$\mu_G(\hat{y}_i(t+1))=\mu_{\bigcap_j(G_j\Delta W_j)}(\,O\circ T(s(t),u(t)=u_i)\,)$$
$$i=1,2,\cdots,n \qquad (19)$$

where Δ is weighting operator. As Wj is irrelevant to the behavior of the process, so (19) becomes

$$\mu_G(\hat{y}_i(t+1))=\bigwedge_j\mu_{(G_j\Delta W_j)}(O\circ T(\,s(t),u(t)=u_i\,))$$
$$=\bigwedge_j\mu_{G_j}(O\circ T(s(t),u(t)=u_i))\Delta W_j \qquad (20)$$
$$i=1,2,\cdots,n$$

by application of (20) to (17) and (18), the optimal control at t is obtained as:

$$u_o=\{u_o/\underset{i}{Max}\,\mu_{Co}(u_i)\}$$
$$=\bigvee_i\bigwedge_j[\mu_{G_j}(O\circ T(\,s(t),u(t)=u_i\,))]\Delta W_j \qquad (21)$$

There are several definitions of Δ, the determination of Δ depends on concrete situations. Here we give some definitions of Δ. When we take arithmetic weighting, Δ is defined as arithmetic product, with the constraint:

$$\sum_{j=1}^{k}W_j=1 \qquad (22)$$

and (20) becomes

$$\mu_G(\hat{y}_i(t+1))=\bigwedge_j\{\mu_{G_j}(O\circ T(\,s(t),u(t)=u_i\,)\}\cdot W_j,$$
$$i=1,2,\cdots,n \qquad (23)$$

When we take the logical pessimistic weighting, Δ is defined as minimum operator $\wedge$ which leads to

$$\mu_G(\hat{y}_i(t+1))=\bigwedge_j\{\mu_{G_j}(\hat{y}_i(t+1))\wedge W_j\} \qquad (24)$$

and if the logical optimistic weighting is used, then Δ is defined as maximum operator $\vee$, and (24) changes to

$$\mu_G(\hat{y}_i(t+1))=\bigwedge_j\{\mu_{G_j}(\hat{y}_i(t+1))\vee W_j\} \qquad (25)$$
$$i=1,2,3,\cdots,n$$

Finally, when we prefer to use exponential weighting, Δ will be defined as exponential operation, and we get

$$\mu_G(\hat{y}_i(t+1))=\bigwedge_j[\mu_{G_j}(\hat{y}_i(t+1))]^{W_j} \qquad (26)$$

The FMPC algorithm given above, in practical usage, can be expressed as n control rules which have the form following

⋮

R_i : IF (u(t) IS u_i ⟶ y(t+1) SATISFY $\bigcap_j$ GJ)
THEN u(t) IS u_i
ELSE (27)
R_{i+1}: IF (u(t) IS u_i ⟶ y(t+1) SATISFY $\bigcap_j$ Gj)
THEN u(t) IS u_{i+1}

where Ri is the i-th rule, which means that if the process output y(t+1) satisfies the objectives $\bigcap_{j=1}^{k} G_j$ when $u(t)=u_i$, then u_i is taken as the control output at time t. From (27) we can see that FMPC is a kind of result— related control(RRC) and the reasoning method adopted by (27) is naturally the result-related reasoning.

Intelligent coordination level

The intelligent coordination level is in reality an expert system, as shown in Fig.1 which is composed of the knowledge base(KB); the data base(DB); the inference engine (IE); the knowledge-updating unit(KU); the man-machine interface(MMI) and the process--controller interface(PCI). The DB consists of the real-time input/output data of the controlled process, various immediate results, the states of the process and the controllers,etc. The function of the DB is to record on-line the behavior of the process and the controllers, to provide the real-time I/O data necessary to the KB and the KU.

Kownledge representation is a key issue in constructing a KB, there are many approaches available (Nilsson, 1980), but whether an approach is used is situation-dependant. In our case, the KB is composed of the fact subbase and the rule subbase. The fact subbase includes the following information: the a priori knowledge and experences about the process, the characteristic of the subprocesses, the control objectives of the subprocesses, the model of the subprocesses and the critical conditions of the transformation from one subprocess to another subprocess. The rule subbase , referring to the coordinating rules, is a set of production rules, which take the form:

IF (situation) THEN (action).

The production rule has been considered very suitable for process control(Astrom and others, 1986). The coordinating rule subbase is the kernel of the intelligent coordination level. The rule has the form:

$$
\begin{array}{ll}
 & \vdots \\
CR_i : & \text{IF } (v(t) \in S_i) \text{ THEN } M \text{ IS } M_i \text{ AND } P \text{ IS } P_i \\
 & \qquad \text{AND } A \text{ IS } A_i \text{ WITH } G_i = \bigcap G_l \\
 & \text{ELSE} \\
 & \vdots
\end{array} \qquad (28)
$$

Where CRi refers to i-th coordinating rule. By the rule we mean that if the characteristic variable of the process is in the subspace Si, then the process is at the subprocess Pi, the behavior of the process is described by the submodel Mi and consequently the control algorithm(controller) used is Ai with the control objectives $G_i = \bigcap_{l=1}^{N_i} G_l$, here, Ni refers to the number of control objectives in Pi.

The KU is actually a set of identification algorithms whose function is to update the models of the subprocesses by using the process data provided by the data base. Every identification algorithm identifies on--line the model of one subprocess. The KU updates, at any time necessary, the models Mi of the subprocesses in the KB, as the result, makes it possible that the Mi can trace the real behavior of the process and the control results can be predicted with accuray and consequently the optimal control is issued.

THE APPLICATION OF IMOC IN TRAIN-TRAVELLING CONTROL

The train travelling process is a very complex process, under different working condition, the control strategies and objectives are also different, so is the characteristic of the process itself. The control quality of the conventional train control methods such as linearized model tracing, restricted speed tracing and other optimal control with saving energy as the optimization objective is inferior to that of an excellent skilled driver. The reason for this is that the control approaches cited above can not deal with the influence of complicated environment factors on the control and ignores some essential travelling objectives such as the travelling comfort, the accuracy of stopping, etc. But the driver can take many factors with uncertain nature into his consideration and deal with various imprecise control objectives in his control.

To control the train travelling with high quality, the control system must be embodied with some kind of intelligence. The IMOC can meet the need.

The partition of train travelling process

The train travelling process(TTP) is partitioned into four subprocesses with different control objectives according to different working conditions. The four subprocesses are the speeding up, constant-speed travelling, speed-adjustment braking and train-stop braking subprocess. The control objectives which must be satiafied are: safety, preciseness of running time, running comfort, energy saving, traceability of the target speed and stopping accuracy. These objectives, although being linguistic, can be characterized by suitable fuzzy sets based on the knowledge and experience of drivers and the experts concerned.

The control set consists of three subsets: the traction control, power braking control and the pneumatic braking control subset. Every subset includes several control alternatives predefined corresponding to the positions of power notch or/and braking notch.

The train intelligent controller has the same structure as that shown in Fig.1. Its control level consists of four FMPCs which control the four subprocesses respectively. The coordination level coordinates the use of the four FMPCs to make the train travelling satisfy the objectives mentioned above optimally.

The knowledge representation in the train IMOC

The knowledge representation is a key issue in constructing the intelligent controller. The knowledge base of the IMOC of train travelling control are mainly composed by a fact subbase and a coordinating subbase. The fact subbase includes the following facts:

1. the knowledge about the line which takes the section as a knowledge unit is represented by the frame structure which includes the knowledge about the section concerned, for example, the length of the section, the grade, the curve, etc.

2. the control objectives which are the reflection of the knowledge and experience of the driver are represented by the membership functions of suitable fuzzy sets.
3. the models of the subprocesses which take the form of mathematic equations.
4. the critical conditions of the transformation between the subprocesses which are represented by first order predicate logic(caculus).

The coordinating rules take the representation approach of the production system with the structure of IF— THEN — ELSE.

The Simulation

The simulation of the application of the proposed IMOC to train travelling control has been done, the train of 3000 tons is drown by an Shaoshan-I electric locomotive. The simulation has been done on a typical line including several sections with different environment conditions. The simulating results are given by Zhang and other (1990) and the results are satisfying and couraging.

CONCLUSIONS

The intelligent control as the intersection of AI,CT,ST,IT and OR is a forceful tool to the control of complex processes with large scale and can deal with not only the mathmatical knowledge but also the heuristical knowledge. The multiobjectivization of the control is an effective means to improve the quality of the control on complex processes. The MLMO priciple that More large and more complex the controlled process is, More objectives there are to be satisfied to control the process with quality might have the universal meaning. The process partition is an effective approach to model the complex processes. The fuzzy multiobjective predictive control which overcomes the result-irrelated property of conventional fuzzy control approaches is especially suitable for controlling the complex processes with uncertainty. As the combination of PP, FMPC and the expert system techniques, the intelligent multiobjective optimal control(IMOC) is proved to be a promising approach to process control. It can in principal be used in situations in which the process control is concerned. The application of IMOC to the control of the train travelling process has proved the feasibility of IMOC.

The further work should be focused on the approaches to partition the process fuzzily and especially the corresponding intelligent control approaches based on such kind of partition.

REFERENCES

Astrom, K.J., J.J. Anton, and K.-E. Arzen (1986). Expert Control. Automatica, 22, 277-286.

Dubois, D. and H. Prade (1980). Fuzzy sets and systems—Theory and Applications. Academic Press, New York.

Fu, K.S. (1971). Learning control systems and Intelligent control systems: An intersection of Artificial Intelligence and Automatic control. IEEE Trans. AC, AC-16, 70-72.

Meystel, A., J.Y.S. Luh (Ed.)(1987). Proc. of IEEE International symposium on Intelligent control, Philadelphia, P.A., U.S.A.

Meystel, A. et al.(Ed.)(1985). Proc. of IEEE Workshop on Intelligent control, Rensselaer Polytechnic Institute, Troy, New York, U.S.A.

Nilsson, N. (1980). Principles of Artificial Intelligence. Toga, Palo Alto, CA.

Zadeh, L.A. (1965). Fuzzy Sets. Information and Control, 8, 338-353.

Zhang, X.-D.,and L.-M. Jia (1990). Intelligent control of train operation. Research Report. China Academy of Railway Science, Beijing, China.

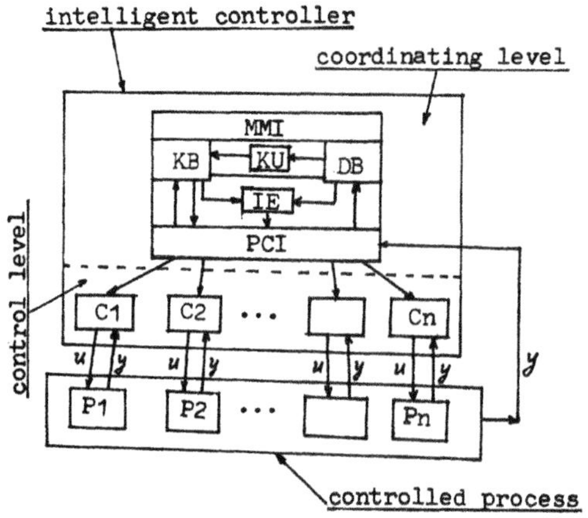

Fig. 1. The general structure of the Intelligent Multiobjective Optimal Controller.

MODEL ACCURACY IN SYSTEM IDENTIFICATION

L. Ljung

Dept. of Electrical Engineering, Linköping University, S-581 83 Linköping, Sweden

Abstract

To assess the quality and accuracy of an identified model is really the most important issue in system identification. We shall here describe several aspects of this problem.

1 Introduction

A model is never a true description of a real process. A model will always contain errors and discrepancies compared to the behaviour of the actual system. The system identification problem is thus not to achieve a "correct" system description, but one that is "sufficiently accurate" for the application in question. A crucial part of this problem is to assess the model accuracy from data.

In this contribution we shall discuss the problem of model accuracy and how to assess it from data. We shall in Section 2 give a brief outline of the system identification machinery, and will then, in Section 3 describe the two fundamentally different, contributions to the model error, the "random error", and the "bias error". Section 4 then deals with the random error while Section 5 is concerned with the bias error.

2 The System Identification Machinery

A typical problem

Here is a typical system identification problem: We observe inputs $u(t)$ and output $y(t)$ from a system $t = 1, \dots, N$. We want to construct a model of the system, and may seek a model of the simple form

$$y(t) + ay(t-1) = b_1u(t-1) + b_2u(t-2) \qquad (1)$$

It thus just remains to determine suitable values of the parameters a_1, b_1 and b_2. This could be done by the well known *least squares* method.

$$\min_{a_1,b_1,b_2} \sum_{t=1}^{N}(y(t) + ay(t-1) - b_1u(t-1) - b_2u(t-2))^2 \quad (2)$$

The minimizing values $\hat{a}^N$, $\hat{b}_1^N$ and $\hat{b}_2^N$ can in this case be easily computed since (2) is a quadratic function. They give the model

$$y(t) + \hat{a}^N y(t-1) = \hat{b}_1^N u(t-1) + \hat{b}_2^N a(t-2) \qquad (3)$$

of the system. This simple system identification problem is a special case of a broad class of model building problems, which we now describe:

Training Sets and Mathematical Models

Here is an archetypical problem in science and human learning:" We are shown a collection of vector pairs $\{[y(t); x(t)] \cdot t = 1, \dots N\}$. Call this "the training set". We are then shown a new value $x(N+1)$ and are asked to name a corresponding value $y(N+1)$." The variable t could be thought of as time, but could be anything. The vectors $y(t)$ and $x(t)$ may take values in any sets (finite sets or subsets of $\Re^n$ or anything else) and the dimension of $x(t)$ could very well depend on t (and could be unbounded). The formulation covers most kinds of classification and model building problems.

How to solve this problem? The mathematical modelling approach is to construct a function $\hat{g}_N(t, x(t))$ based on the "training" set, and to use this function for pairing $y(t)$ to new $x(t)$:

$$\hat{y}(t) = \hat{g}_N(t, x(t)) \qquad (4)$$

Where do we get the function g from? Essentially we have to search for it in a family of functions that is described (parameterized) in terms of a finite number of parameters. These parameters will be denoted by θ. The family of candidate model functions will be called a *model structure*, and we write the function as

$$g(t, \theta, x(t)) \qquad (5)$$

The value $u(t)$ is thus matched against the "candidate" $g(t, \theta, x(t))$:

$$y(t) \sim g(t, \theta, x(t)) \qquad (6)$$

The search for a good model function is then carried out in terms of the parameters θ, and the chosen value $\hat{\theta}_N$ gives us

$$\hat{g}_N(t, x(t)) = g(t, \hat{\theta}_N, x(t)) \qquad (7)$$

The case (1) corresponds to

$$\begin{aligned} \theta &= (a, b_1, b_2) \\ x(t) &= (y(t-1), u(t-1), u(t-2)) \\ g(t, \theta, x(t)) &= -ay(t-1) + b_1u(t-1) + b_2u(t-2) \end{aligned} \qquad (8)$$

In general the function g is a mapping from the set where $x(t)$ takes its values to the space where $y(t)$ takes its values. All kinds of parameterizations are possible, from ones that are tailor-made for the application to general orthogonal functions expansion and neural net structures.

Signals and Dynamical Systems

The general formulation above fits into conventional system identification, which corresponds to particular functions g. We have already shown that the simple (ARX) model (1) fits into the framework.

In general the task to form pairs $[y(t), \varphi(t)]$ is the one-step ahead prediction problem. For example, first order ARMA model of $\{y(t)\}$

$$y(t) + ay(t-1) = e(t) + ce(t-1) \tag{9}$$

is obtained for

$$\theta = (a \quad c); x(t) = (y(0), y(1) \ldots y(t-1))$$

$$g(t, \theta, x(t)) = \sum_{k=0}^{t-1} (c-a)(-c)^{t-k-1} y(k) \tag{10}$$

etc. Fuzzy - or verbal - dynamical models can be obtained if $x(t)$ and $y(t)$ take on values like "the oven is very hot", "the oven is warm", "the water is boiling" and so on. The function g would then be some kind of a table - perhaps implemented in an expert system shell - and its parameters θ would describe the structure of the table.

Fitting Model Structures to Data

The leading principle for choosing θ clearly is to have $g(t, \theta, x(t))$ perform well on the training set, that is to make

$$y(t) \text{ close to } g(t, \theta, x(t)) \; t = 1, \ldots, N \tag{11}$$

This principle applies also to the case where x and y assume non-numeric values, if only "close" can be appropriately defined. Most numeric schemes select $\theta = \hat{\theta}_N$ so that

$$\sum_{t=1}^{N} \| y(t) - g(t, \theta, x(t)) \| \tag{12}$$

is minimized for some norm $\| \cdot \|$ ("norm" should here be taken in a broad sense) or so that $y(t) - g(t, \theta, x(t))$ is uncorrelated with information in $x(t)$.

Typical Asymptotic Properties

A key question is: How good are the estimates obtained by (12)? The typical analysis goes as follows: Suppose that the pairs $[y(t), x(t)]$ really are related by

$$y(t) = g_0(t, x(t)) + v(t) \tag{13}$$

where $\{v(t)\}$ is an as yet undefined sequence

- **Consistency**: Suppose there is a "true" system description available within the model structure. We translate that as for some θ_0, $g(t, \theta_0, x(t)) = g_0(t, x(t))$ and $v(t)$ is white noise. Then $\hat{\theta}_N$ will converge to θ_0 as N increases to infinity, and the difference $\sqrt{N}(\hat{\theta}_N - \theta_0)$ will converge in distribution to a Gaussian random variable (i.e. $\hat{\theta}_N$ tends to θ_0 with the "rate" $\sim 1/\sqrt{N}$)
- **Convergence**: Suppose no "true" description is available in the model structure, but assume that $\{v(t)\}$ in (13) is white noise. Then $\hat{\theta}_N$ will converge to a value θ_* such that $g(t, \theta_*, x(t))$ approximates $g_0(t, x(t))$ as well as possible in the chosen norm in (12). Moreover $\sqrt{N}(\hat{\theta}_N - \theta_*)$ converges in distribution to a Gaussian random variable.
- **Making a sieve finer and finer** An interesting particular case is when the true system is assumed to belong to a very broad class of models, that cannot be parameterized by a finite number of parameters. However this class can be thought of as "the limit" of increasing model structures, that are parameterized by more and more parameters. (Think e.g. of an infinite dimensional system that can be seen as the limit of finite impulse response models as the number of coefficients tends to infinity). Mathematically this can be written as

$$g_0(t, x(t)) \text{belongs to } cl(\cup_{d=1}^{\infty} g_d(t, \theta^d, x(t)) \tag{14}$$

where the vector θ^d contains d parameters. To deal with this case it is customary to employ more and more parameters as more and more data becomes available. That is d becomes a function of $N : d(N)$.

If $\{v(t)\}$ in (13) is white noise and if $d(N)$ is chosen to increase to infinity slowly enough with N, we then have that the model will approach the true system as the number of data tends to infinity. This can be written formally as

$$\hat{g}_{d(N)}(t, \hat{\theta}_N^{d(N)}, x(t)) \longrightarrow g_0(t, x(t)) \text{ as } N \to \infty \tag{15}$$

With this we conclude our brief exposé of the main stream system identification. See the textbooks, e.g. [7] and [9] for the details of the general methods and results. Of course, system identification covers many other topics like how to compute the parameter values that minimize (12) and how to select the data so that they are as informative as possible.

3 Contributions to the Model Error

Model quality is related to what we could call "model stability" (in the statistical sense), i.e. how much the resulting model varies when computed from different data sets. Regardless of the statistical formalities that can be developed around this, it is obvious that one must be suspicious if the resulting model varies considerably with the data set it was computed from. Conversely, we will develop confidence in a model which is returned to us, with small variations, from different measured data sets, under varying experimental conditions, and perhaps using different identification techniques.

Model errors come in two, fundamentally different, shapes. One is those errors that are due to the fact that the measurements and the system are affected by disturbances.

If an experiment is repeated with exactly the same input signal, one does not obtain exactly the same output signal, and thus not exactly the same model for this reason. Such model discrepancies we call *random errors*. They can typically be reduced by making measurements over larger periods.

The other type of model errors stem from deficiencies in the model structure. The model is simply not capable of describing the system, even if fitted to noise free data. With some abuse of terminology, we shall call such errors "*bias-errors*". Bias errors are recognized as variations in the model when it is fitted to data sets, collected under different conditions (even when the record lengths are such that the random error is insignificant). The reason is that the experimental conditions (operating point, input character, feedback etc) enhance different aspects of the system's properties, and the model is fitted to the dominating aspects of the system's behaviour.

We shall in the next two sections discuss the random error, and the bias error, respectively.

4 The Random Error

Consider for simplicity a linear regression model

$$y(t) = \varphi^T(t)\Theta + e(t) \tag{16}$$

It could be, for example, a linear ARX-model

$$A(q)y(t) = B(q)u(t) + e(t). \tag{17}$$

It is well known that the random error, i.e. the *variance* of the results, (least squares) estimate $\hat{\Theta}_N$, based on N observations is

$$E(\hat{\Theta}_N - \Theta_0)(\hat{\Theta}_N - \Theta_0)^T \approx \frac{\lambda}{N}(E\varphi(t)\varphi^T(t))^{-1} \tag{18}$$

in case there is a true description

$$y(t) = \varphi^T(t)\Theta_0 + e(t) \tag{19}$$

available within the structure (16).

Here λ is the variance of the true innovations $\{e(t)\}$. In this case we also have a simple and natural *estimate* of the covariance matrix

$$E(\hat{\Theta}_N - \Theta_0)(\hat{\Theta}_N - \Theta_0)^T \approx \hat{\lambda}_N(\sum_{t=1}^{N}\varphi(t)\varphi^T(t))^{-1} \tag{20}$$

$$\hat{\lambda}_N = \frac{1}{N}\sum_{t=1}^{N}(y(t) - \varphi^T(t)\hat{\Theta}_N)^2 \tag{21}$$

Under assumptions that a true description (eventually) becomes available in the model structure, one can also develop nice asymptotic (in model order) expressions for the variance of the frequency function associated with (17):

$$E\mid \frac{\hat{B}_N(e^{i\omega})}{\hat{A}_N(e^{i\omega})} - \frac{B_0(e^{i\omega})}{A_0(e^{i\omega})}\mid^2 \approx \frac{n}{N}\cdot\frac{\Phi_v(\omega)}{\Phi_u(\omega)} \tag{22}$$

Here n is the order of the model (17), N is the number of data, $\Phi_u(\omega)$ is the spectra of the input $\{u(t)\}$ and $\Phi_v(\omega)$ is the spectra of the measurement noise

$$v(t) = \frac{1}{A_0(q)}e(t)$$

(Subscript "zero" refers to a true system description). See, e.g. [7].

Now, consider the much more realistic case that a true description (19) is **NOT** available. An expression for the random error can still be developed. It goes as follows: Let Θ^* and $\varepsilon(t,\Theta^*)$ be defined by

$$\varepsilon(t,\theta) = y(t) - \varphi^T(t)\Theta$$

$$\Theta^* = \arg\min_{\Theta} E\varepsilon^2(t,\Theta)$$

Then

$$\hat{\Theta}_N \to \Theta^* \text{ as } N \to \infty \tag{23}$$

and

$$E(\hat{\Theta}_N - \Theta^*)(\hat{\Theta}_N - \Theta^*)^T \approx \frac{1}{N}P \tag{24}$$

where

$$P = \bar{R}^{-1}Q\bar{R}^{-1} \tag{25}$$

$$Q = \lim_{N\to\infty}\frac{1}{N}\sum_{t=1}^{N}\sum_{s=1}^{N}E\varphi(t)\varphi^T(s)\cdot\varepsilon(t,\Theta^*)\varepsilon(s,\Theta^*) \tag{26}$$

(See, e.g. [7]).

A very important point now is that in contrast to (18), (20), it is not easy to estimate P in (25) - (26) from data. A first guess, to replace Θ^* in (26) by $\hat{\Theta}_N$ gives $Q = 0$. To use (20) - (21) may give quite unreliable results.

[3] has developed a technique to estimate the actual covariance matrix P in (25) - (26). The technique is based on monitoring a parallel recursive estimation algorithm, and gives quite reliable estimates of the actual random error, also in the case of undermodelling.

5 The Bias Error

It is a fundamental problem to estimate the bias error. The reason is that if we don't know a model structure where the system fits, the system could be "anything", and how can we then, from a finite data record, draw any conclusions about it? In practice one has to introduce assumptions about the system - prior information - and evaluate the resulting model in the light of these.

We shall in this section give some insights into this problem.

Frequency fits

First of all, suppose we assume that the true system indeed is linear but possibly infinite dimensional:

$$y(t) = G_0(q)u(t) + e(t) \tag{27}$$

and that we use a model of simpler structure:

$$y(t) = G(q,\Theta)u(t) + H_*(q)e(t) \tag{28}$$

Then the resulting model estimate $\hat{\Theta}_N$ will converge to Θ^* where

$$\Theta^* = \arg\min \int_{-\pi}^{\pi} \mid G(e^{i\omega},\Theta \mid - G_0(e^{i\omega})\mid^2 \frac{\Phi_u(\omega)}{\mid H_*(e^{i\omega})\mid^2}d\omega \tag{29}$$

We thus obtain that model in the set of candidate models that is closest to the true transfer function in a frequency norm with the weighting function

$$W(\omega) = \frac{\Phi_u(\omega)}{| H_*(e^{i\omega}) |^2} \quad (30)$$

This is however only an implicit characterization of the bias error - we still don't know how *large* the error is.

A direct estimate of the bias error

In the (quite) special case of a FIR model with white, Gaussian input and no measured noise the following result can be shown [8]. Let $P(\omega)$ be the variance of the transfer function estimate (the random error) at frequency ω. It can be computed from the matrix P in (24) - (26) in a straightforward way. Then

$$P(\omega) \approx \frac{n}{N} | G_0(e^{i\omega}) - G(e^{i\omega}, \Theta^*) |^2 \quad (31)$$

The point now is that $P(\omega)$ can be *computed*, using the estimate of P described in the previous section. We can thus, in this case, estimate directly the bias error using (31)!

Estimating Bias Using Priors

A most interesting approach is due to Goodwin et al, see e.g. [1]. The idea is to prove some assumption about the true system, typically like:"The true frequency function is a random function, that deviates from an unknown second (say) order system. The variance function of this deviation is known". Based on such assumptions and on an estimated second (say) order model, one can then estimate the discrepancy between the model and the unknown, true system's frequency function.

Estimating Bias From Residuals

Another approach is to analyse the residuals from a fitted, low order model. They contain information about the total model error that could be recovered by spectral analysis or further modelling. This approach has been pursued by [4].

A family of related methods is obtained by developing a high order model in parallel with the low order one to be used in practice. In principle the exact bias error could then be estimated as the number of data tends to infinity.

Hard error bounds

A special wish from the control designers is to have "hard" error bounds; i.e. certain statements that the true (linear) plant is to be found as a Nyquist plot between two envelopes in the complex plane.

There are recent attempts to develop such bounds directly in the frequency domain, [6] and [2]. It is clear that such bounds can be given only under two assumptions:

- Some prior knowledge of the system is necessary. (After N observations we know absolutely nothing about the impulse response coefficients beyond N otherwise).
- A hard bound on additive disturbances is necessary. (Otherwise we can be fooled by finite date to an arbitrarily large extent).

However, given two assumptions of this kind, there is no problem to develop hard frequency domain bounds using more conventional techniques (nominal model obtained by least squares, the hard bounds by an ellipsoidal set membership algorithm); [5], [11], [10]. It is for the future to formulate prior knowledge that is both sound engineering-wise and suitable for estimation algorithms. Clearly these issues on model mismatch are as important for e.g. failure detection and filter design, as for control design.

6 Conclusions

We have pointed to a number of techniques by which the accuracy of an estimated model can be judged. Generally speaking, all traditional methods are based on the assumption that the true system can be described in the chosen model structure. This is a serious limitation. Although several techniques now exist that can be used to estimate both the bias error and the random error (under the assumption of possible under modelling) the research on this topic is still going on.

References

[1] G.C. Goodwin and M. Salgado. A stochastic embedding approach for quantifying uncertainty in estimation of restricted complexity models. *Int. J. of Adaptive Control and Signal Processing*, 3():333–356, 1989.

[2] A.J. Helmicki, C.A. Jacobson, and C.N. Nett. Identification in H_∞: a robustly convergent non-linear algorithm. In *Proc. American Control Conference*, San Diego, CA, 1990.

[3] H. Hjalmarsson. *On Estimation of Model Quality in System Identification.* Technical Report LIU-TEK-LIC-1990:51, Dept. of Electrical Engineering, Linköping University, 1990.

[4] R.L. Kosut. Adaptive robust control via transfer function uncertainty estimation. In *Proc 1988 American Control Conference*, Atlanta, GA, 1988.

[5] R.L. Kosut, M. Lau, and S. Boyd. Identification of systems with parametric and nonparametric uncertainty. In *Proc. 1990 American Control Conference*, San Diego, 1990.

[6] R.O. Lamaire, L. Valavani, M. Athans, and G. Stein. A frequency domain estimator for use in adaptive control systems. In *Proc. 1987 American Control Conf.*, Minneapolis, MN, 1987.

[7] L. Ljung. *System Identification - Theory for the User.* Prentice-Hall, Englewood Cliffs, N.J., 1987.

[8] L. Ljung. System identification in a noise-free environment. In *Proc. IFAC Symp. Adaptive Control and Signal Proc.*, pages 29–38, 1989.

[9] T. Söderström and P. Stoica. *System Identification.* Prentice-Hall Int., London, 1989.

[10] B. Wahlberg and L. Ljung. On the estimation of transfer function error bound. In *1991 European Control Conf.*, Grenoble, France, 1991.

[11] R.C. Younce and C.E. Rohrs. Identification with non-parametric uncertainty. In *Int. Conf. on Circuits and Systems (ISCAS)*, New Orleans, LA, 1990.

A MICROCONTROLLER-BASED ADAPTIVE POSITION CONTROLLER FOR A D.C. MOTOR

B. C. Lim, T. H. Lee[1] and S. K. Chan

Dept. of Electrical Engineering, National University of Singapore, Kent Ridge (0511), Singapore

Abstract - **In this paper, we present the design and implementation of a microcontroller-based (Intel 8096) adaptive position controller for a d.c. motor. We modified the Liapunov design of [1] to use the framework of adaptive control of partially known systems. This modification also ensures global stability of the overall system but it reduces the number of parameters to be adapted. To ensure robustness of the adaptation, additional features incorporated include user-selectable adaptation dead-zones and bounds on the adapted parameters. To speed up the controller, we wrote our own floating-point library to replace the library provided by the Intel development system for the 8096 chip. This resulted in floating-point operations that are thrice as fast as the original library provided by Intel.**

1. INTRODUCTION

In most servo applications, the controllers used are typically fixed and time-invariant. While this is acceptable in most situations, it is not ideal in cases where the operating conditions (payload, for example) change. In such cases, the controller should really be re-tuned to achieve tighter control. Having an adaptive controller is one solution to this problem as it has the capability to automatically adjust its control gains to track any change in the plant to be controlled.

While having the adaptation option in a controller is desirable, there is the accompanying attribute that the number of computations required are an order of magnitude larger than in a fixed controller. This is a particularly severe problem in servo applications as acceptable sampling times are often of the order of 10 milliseconds and less. Under such situations, it becomes a challenge to design and implement the adaptive controller using inexpensive and easily available microprocessors.

While some may argue that one can always make use of faster and more powerful processors like the Intel's 80386 or Digital Signal Processing chips to achieve the required sampling rate, we have chosen to demonstrate that a simpler and cheaper microcontroller like the Intel's 8096 chip can do the job. Cost of implementation is always an important issue in industrial applications, particularly if the volume of production is large.

[1] Author to whom all correspondence should be addressed

Designing around the 8096 microcontroller requires care in ensuring that the code is as efficient as possible. The adaptive control algorithm used is a modified version of the Liapunov design method [1]. Several improvements for the particular case of position control of a d.c. motor are incorporated in our modification. Firstly, we modify the method of [1] to use the framework of adaptive control of partially known systems. While this modification also ensures global stability of the overall system, the number of adapted parameters is reduced, resulting in a decrease in computations. To ensure robustness of the adaptation, additional features incorporated include user-selectable adaptation dead-zones and bounds on the adapted parameters.

The next innovation we incorporated in our implementation was the re-design of a custom floating point library. This resulted in floating point multiplications and divisions that were thrice as fast as that provided by the floating point library in the Intel development system.

2. Adaptive Controller for a D.C. Motor using the Partially Known System Approach

2.1 Adaptive controller using the Partially Known System Approach

The adaptive control uses a modified version of the Liapunov design method of [1]. The number of adapted parameters is reduced by one by making use of structural information on servo motors.

Consider the transfer function of a d.c. servo motor.

$$\frac{\Theta(s)}{U(s)} = \frac{g}{s+\frac{1}{\tau}} \cdot \frac{1}{s} \qquad -(1)$$

where

$\Theta(s)$: position output of the d.c. motor

$U(s)$: voltage input to the d.c. motor

τ : mechanical time constant of the d.c. motor

The electrical time constant of the motor is ignored as it is negligible compared to the mechanical time constant. Written in the time domain, the equation becomes

$$\ddot{\theta} + a_1\dot{\theta} = gu(t) \qquad -(2)$$

where $a_1 = \frac{1}{\tau}$

In state-space form, we have

$$\begin{bmatrix}\dot{\theta}\\ \omega\end{bmatrix} = \begin{bmatrix}0 & 1\\ 0 & -a1\end{bmatrix}\begin{bmatrix}\theta\\ \omega\end{bmatrix}+\begin{bmatrix}0\\ 1\end{bmatrix}gu(t) \qquad -(3)$$

Let the reference model be given by

$$\ddot{\theta}_m + a_1^*\dot{\theta}_m + a_2^*\theta_m = a_2^*\theta_{ref} \qquad -(4)$$

In state-space form, we have

$$\begin{bmatrix}\dot{\theta}_m\\ \omega_m\end{bmatrix} = \begin{bmatrix}0 & 1\\ -a_2^* & -a_1^*\end{bmatrix}\begin{bmatrix}\theta_m\\ \omega_m\end{bmatrix} + \begin{bmatrix}0\\ a_2^*\end{bmatrix}\theta_{ref} \qquad -(5)$$

Let the control law be given by

$$u(t) = \alpha_1(t)(\theta(t)-\theta_{ref}(t)) + \alpha_2(t)\omega(t) \qquad -(6)$$

Substituting (6) into (3), we have

$$\begin{aligned}\begin{bmatrix}\dot{\theta}\\ \omega\end{bmatrix} &= \begin{bmatrix}0 & 1\\ g\alpha_1 & g\alpha_2 - a_1\end{bmatrix}\begin{bmatrix}\theta\\ \omega\end{bmatrix}+\begin{bmatrix}0\\ -g\alpha_1\end{bmatrix}\theta_{ref}\\ &= \begin{bmatrix}0 & 1\\ -a_2^* & -a_1^*\end{bmatrix}\begin{bmatrix}\theta\\ \omega\end{bmatrix}\\ &+ g\begin{bmatrix}0\\ 1\end{bmatrix}[\tilde{\alpha}_1 \ \ \tilde{\alpha}_2]\begin{bmatrix}\theta\\ \omega\end{bmatrix}\\ &+ \begin{bmatrix}0\\ -g\alpha_1\end{bmatrix}\theta_{ref}\end{aligned} \qquad -(7)$$

where

$$\tilde{\alpha}_1 = \alpha_1 + \frac{a_2^*}{g}$$

$$\tilde{\alpha}_2 = \alpha_2 + \frac{a_1^* - a_1}{g}$$

Let

$$y_p \triangleq \begin{bmatrix}\theta\\ \omega\end{bmatrix}$$

$$A_m \triangleq \begin{bmatrix}0 & 1\\ -a_2^* & -a_1^*\end{bmatrix}$$

$$b \triangleq \begin{bmatrix}0\\ 1\end{bmatrix}$$

$$\tilde{\alpha} \triangleq [\tilde{\alpha}_1 \ \ \tilde{\alpha}_2]^T$$

Then equation (7) may be written as

$$\dot{y}_p = A_m y_p + gb\tilde{\alpha}^T y_p + \begin{bmatrix}0\\ -g\alpha_1\end{bmatrix}\theta_{ref} \qquad -(8)$$

Equation (5) may be written as

$$\dot{y}_m = A_m y_m + \begin{bmatrix}0\\ a_2^*\end{bmatrix}\theta_{ref} \qquad -(9)$$

Let $e \triangleq y_p - y_m$

Subtracting (9) from (8) gives

$$\begin{aligned}\dot{e} &= A_m e + gb\tilde{\alpha}^T y_p + \begin{bmatrix}0\\ -g\alpha_1 - a_1^*\end{bmatrix}\theta_{ref}\\ &= A_m e + gb\tilde{\alpha}^T\begin{bmatrix}\theta-\theta_{ref}\\ \omega\end{bmatrix}\\ &= A_m e + gb\tilde{\alpha}^T X\end{aligned} \qquad -(10)$$

where $X = \begin{bmatrix}\theta-\theta_{ref}\\ \omega\end{bmatrix}$

Consider Liapunov function candidate

$$V(e(t),\tilde{\alpha}(t)) = \frac{1}{2}e^T Pe + \frac{1}{2}\tilde{\alpha}^T\Gamma^{-1}\tilde{\alpha}|g| \qquad -(11)$$

where P and Γ are symmetric positive definite matrices to be defined later.

Differentiating (11), we arrive at

$$\dot{V}(e,\tilde{\alpha}) = \frac{1}{2}(\dot{e}^T Pe + e^T P\dot{e}) + \tilde{\alpha}^T\Gamma^{-1}\dot{\tilde{\alpha}}|g| \qquad -(12)$$

Substituting (10) into (12), we have

$$\dot{V} = \frac{1}{2}e^T(A_m^T P + PA_m)e + g\tilde{\alpha}^T Xe^T Pb + \tilde{\alpha}^T\Gamma^{-1}\dot{\tilde{\alpha}}|g| \qquad -(13)$$

Let $Q = A_m^T P + PA_m$ where Q is 2 x 2 and negative definite.

By choosing

$$\dot{\tilde{\alpha}} = -sign(g)\Gamma Xe^T Pb \qquad -(14)$$

equation (13) becomes

$$\dot{V}(e(t),\tilde{\alpha}(t)) = \frac{1}{2}e^T Qe \le 0 \qquad -(15)$$

Equation (14) defines the adaptive law:

$$\begin{bmatrix}\dot{\tilde{\alpha}}_1\\ \dot{\tilde{\alpha}}_2\end{bmatrix} = -\mathrm{sign}(g)\Gamma\begin{bmatrix}\theta-\theta_{ref}\\ \omega\end{bmatrix}e^T Pb \qquad -(16)$$

Γ and Q are chosen by the designer. P may be calculated from the specified values of Q and A_m

Proof of Asymptotic Stability:

The quadratic form $V(e(t),\tilde{\alpha}(t)) = \frac{1}{2}e^T Pe + \frac{1}{2}\tilde{\alpha}^T\Gamma^{-1}\tilde{\alpha}|g|$ has a derivative that is lesser or equal to zero, i.e.

$$\dot{V}(e(t),\tilde{\alpha}(t)) = \tfrac{1}{2}e^T Qe \le 0$$

This implies that e^2 and $\tilde{\alpha}^2$ are both bounded. Therefore $|e|$ and $|\tilde{\alpha}|$ are also bounded.

Further, $|e|$ is bounded implies that θ and ω are bounded, since θ_m and ω_m are bounded by the choice of a stable reference model.

Next,

$$\int_0^\infty \frac{d}{dt}V(\tau)d\tau = \int_0^\infty \frac{1}{2}e^T Qed\tau$$

$$V(\infty) - V(0) = \int_0^\infty \frac{1}{2}e^T Qed\tau$$

$$\int_0^\infty e^T Qed\tau = 2(V(\infty) - V(0))$$

$\therefore \int_0^\infty e^T Qe \ d\tau$ is bounded.

Hence $|e| \in L^2$

Since e, $\tilde{\alpha}$, θ and ω are all bounded, we can deduce that $\dot{e}$ is bounded from equation (10)

Finally, $\dot{e}$ is bounded and

$$|e| \in L^2 \quad \Rightarrow \quad \lim_{t\to\infty} e = 0$$

$$\therefore \ \theta \rightarrow \theta_m$$

$$\omega \rightarrow \omega_m$$

end of proof

2.2 Advantages of using the Partially Known System Approach

The structural information of the motor transfer function was used to design our adaptive controller. This allowed us to reduce the number of control gains to two instead of three. At the same time, asymptotic stability was maintained. By cutting down the number of parameters to be adapted, the time needed for calculation was significantly reduced. This is particularly important in servo applications because the sampling interval must be small. (Roughly 10 milliseconds or less).

In terms of performance, we observed no significant difference between the tracking performance in the case using All State-Variables Measurable and that using the Partially Known System Approach.

3. Design Guidelines for Digital Implementation

Although global asymptotic stability is assured in the design which is done in the continuous-time, the same cannot be assumed when implemented digitally.

A common requirement is that the sampling frequency must be at least twice that of the highest frequency component in the system (commonly known as the Nyquist frequency). In an adaptive system, we cannot merely consider the frequencies found in the control loop; we need to consider the frequency components in the adaptive loop as well. From our simulations as well as the actual implementation, we propose the following heuristic rule for choosing the sampling time, h:

$$h < \frac{1}{20} \min\left(\tau , \frac{1}{\rho_{max}} \right)$$

where τ and ρ_{max} are defined as follows:

The closed-loop time constants are given by the eigenvalues of A_m where

$$\dot{x}_m = A_m x_m + g_m b r$$

The fastest dynamics here is given approximately by a time constant of

$$\tau = \frac{1}{\max\{|\lambda_i(A_m)|\}}$$

where λ_i represents an eigenvalue of A_m.

In addition to the dynamics of the closed-loop control system, we have to take into account the dynamics of the adaptive law as well.

We recall the Liapunov function described in equation (11)

$$V(e(t),\tilde{\alpha}(t)) = \frac{1}{2}e^T P e + \frac{1}{2}\tilde{\alpha}^T \Gamma^{-1}\tilde{\alpha}|g|$$

with the appropriate adaptive law (equation (14)), we get

$$\dot{V} = \frac{1}{2}e^T Q e \leq 0$$

The rate of adaptation may be approximately quantified by

$$\rho = \frac{|\dot{V}|}{V} = \frac{e^T Q e}{e^T P e + |g|\tilde{\alpha}^T \Gamma^{-1}\tilde{\alpha}}$$

$$\leq \frac{e^T Q e}{e^T P e}$$

$$\leq \frac{\lambda_{max}(Q)}{\lambda_{min}(P)} \stackrel{\Delta}{=} \rho_{max}$$

To increase the robustness of the adaptive controller, we have built in a deadzone option which is user selectable. Typically, the adaptive controller is allowed to adapt the control parameters until the plant output tracks the reference model fairly closely. Upon achieving this, the user may choose the deadzone option. In this option, the control gains will only be adapted if the plant output is sufficiently different from the output of the reference model (i.e. outside the deadzone). The choice of the deadzone is as follows:

Let $\theta_{threshold} = 0.0349$ (2 degrees deadzone for the position signal).

Let $\omega_{threshold} = \frac{\omega_n}{10}$ where ω_n is the natural frequency chosen for the reference model.

```
If ( ( abs ( θ - θm ) > θthreshold ) or
   ( abs ( ω - ωm ) > ωthreshold ) )
   update the control gains (alpha)
else
   do not update the control gains (alpha)
```

Another feature built into our controller is putting bounds on the control gains. This is based on the algorithm described in [3]. In cases where *a priori* bounds on the unknown plant parameters are known and where for each set of parameter values within these bounds the plant has no unstable pole-zero cancellation, we can incorporate this partial parameters knowledge in the adaptive law and prevent the parameters from drifting away. A short description of the algorithm is as follows:

Suppose we know that the set of control gains α_i, $i \in [1,n]$, where n is the number of parameters to be adapted, lies within known bounds $[\alpha_{min}, \alpha_{max}]$, and that the plant can be well controlled for all $\alpha \in [\alpha_{min}, \alpha_{max}]$.

For all $i \in [1,n]$,

if $\alpha_i > \alpha_{max_i}$

$$f_i = \alpha_i - \alpha_{max_i}$$

else if $\alpha_i < \alpha_{min_i}$

$$f_i = \alpha_i - \alpha_{min_i}$$

else

$$f_i = 0$$

The adaptive law is modified to

$$\dot{\alpha} = -\text{sign}(g)\Gamma X e^T P b - \Gamma f$$

where $f = [f_1, f_2, \ldots, f_n]^T$

4. MICROCONTROLLER-BASED ADAPTIVE CONTROLLER

An Intel 8096-based microcontroller board which was designed in-house was used to control a d.c. servo motor. A schematic of this board is shown in Figure 1.

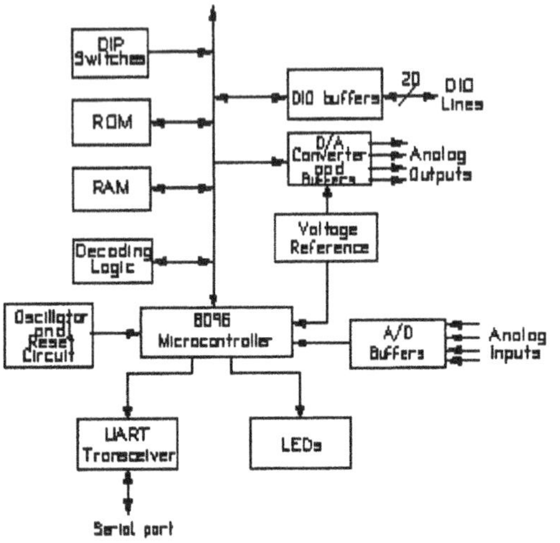

Figure 1. Functional blocks of the 8096-based microcontroller board

The advantage of the single-chip microcontroller is that it is designed for embedded real-time system applications. On-chip, it has five 8-bit I/O ports, two in-built 16-bit counters/timers and a 10-bit analog-to-digital converter with sample and hold. All these features allow the hardware to have a low chip count. In addition, the instruction set of the 8096 includes fixed-point multiplication and division. A photograph of the set-up is shown in Figure 2. The IBM PC/AT is used as a user-friendly interface.

From our initial experiments, it was found that about 15 milliseconds was needed to complete all the calculations necessary within each sampling period. This was implemented using the floating-point library provided by the Intel development system. For servo applications, a sampling time of 10 milliseconds or less is needed [2]. In order to cut down the time needed for calculations, we wrote our own floating-point library which dramatically cuts down the time needed for calculation. In our floating-point algorithm, we made use of the multiply and divide instructions provided in the 8096 instruction set. The floating-point number representation used is the IEEE single-precision format, which is the same as that used by the Intel-provided floating-point library. Four bytes are used to represent a floating point number. As the internal operations of the floating-point library are transparent to the users, one can continue to use whatever C-language programs that had been written without any modification. Table 1 shows a comparison of the times taken to do various arithmetic operations between the floating-point library provided by Intel and that we developed.

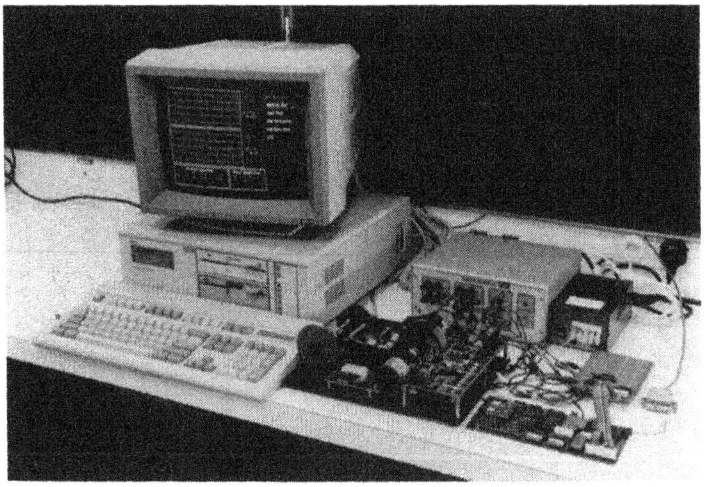

Figure 2. 8096-based microcontroller controlling a d.c. motor

Mathematical Operation	Average time taken (microseconds)	
	Using our custom library	Using Intel's library
Add	61	184
Subtract	63	195
Multiply	71	200
Divide	66	406

Table 1. Comparison of the times taken for various arithmetic operations

The microcontroller board may be used as a stand-alone controller or as a front-end controller connected to an IBM PC/AT through its serial port. The IBM PC/AT provides a friendly user-interface which allows the user to select important parameters such as the controller gains, the reference model and other run-time options. Figure 3 is a photograph of the user-interface showing the pull-down menu for setting controller parameters.

When the controller is running, the output and the control gains are plotted on an EGA monitor on-line (Figure 4). In this photograph, the top left window shows the response of the motor to a square wave input. The red signal represents the reference output θ_m. The yellow signal plots the motor position output θ. The window below this plots the control gains α_1 and α_2.

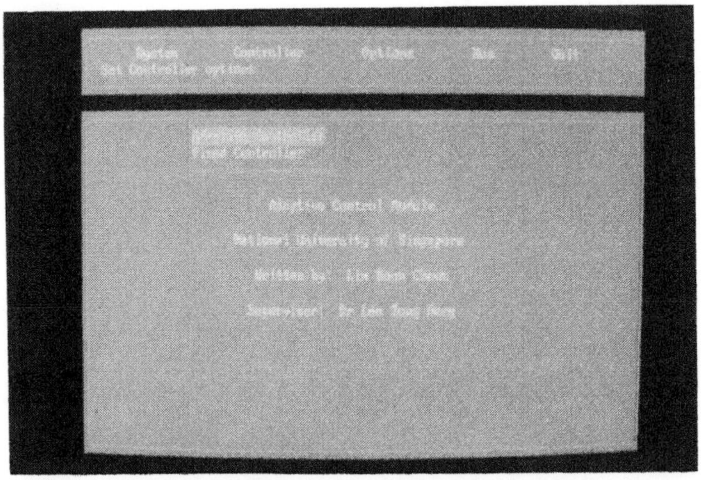

Figure 3. User-interface on the IBM PC/AT with an EGA monitor

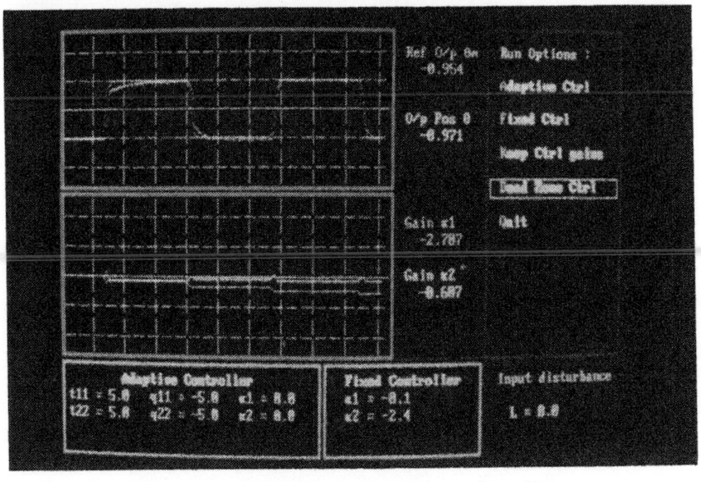

Figure 4. Controller running using the deadzone option

5. CONCLUSION

In this paper we have demonstrated the feasibility of using a relatively cheap single-chip microcontroller to implement adaptive control of a servo motor. Various measures were built in to increase the robustness of the control system. We modified the adaptive control algorithm in [1] using the partially known system approach to cut down the computations required in the control loop. In addition, guide-lines on digital implementation are also provided.

6. REFERENCES

[1]Narendra, K.S. And P. Kudva (1974), "Stable adaptive schemes for system identification and control," *IEEE Trans. on Man, Systems and Cybernetics,* SMC-4, pp. 542-560.

[2]Lim, B.C., T.H. Lee and S.K. Chan (1990), "A microcontroller-based adaptive position controller for a d.c. motor," *National University of Singapore, Department of Electrical Engineering, Control Group Technical Report.*

[3]Kreisselmeier, G., "An Approach to Stable Indirect Adaptive Control", *Automatica*, Vol. 21, No. 4, pp. 425-431, 1985.

ADAPTIVE TEMPERATURE CONTROL OF INDUSTRIAL DIFFUSION/LPCVD REACTORS[1]

H. de Waard* and W. L. de Koning**

**ASM Micro-electronics Technology Centre, P. O. Box 100, 3720 AC, Bilthoven, The Netherlands*
***Dept. of Mathematics and Informatics, Delft University of Technology, Delft, The Netherlands*

Abstract. A mathematical model describing the heat transfer characteristics of a hot-wall batch electric furnace serves as the basis for the design of a temperature controller of LQG type. Because the model is of high order, reduced-order controller design techniques are applied to satisfy constraints on on-line computations imposed by available computing power. In contrast to classical techniques for reduced-order controller design that use system-theoretic considerations, this paper adopts the optimal projection approach, thus yielding an optimal reduced-order controller. An algorithm is presented for solving the associated set of coupled (modified) Riccati and Lyapunov equations. The algorithm utilizes a homotopic continuation method. A sequential prediction error algorithm is added to estimate the parameters, making the control law adaptive. The feedback is calculated by performing a few iterations of a Riccati equation in every sample interval. Some performance results are presented.

Keywords. Temperature control; ovens; optimal projections; continuation methods; computational methods; adaptive systems.

INTRODUCTION

In the production process of making integrated circuits, diffusion/ low pressure chemical vapour deposition reactors play an important role. In these reactors, the substrates (wafers) get a heat treatment or are subjected to chemical reactions under reduced pressure. This paper deals with the design of a temperature controller for diffusion/LPCVD reactors of batch-type as depicted in Fig. 1.

In such a reactor, a batch of up to 300 150 mm wafers can be processed simultaneously. For this purpose, they are placed vertically in a boat and loaded into the furnace by means of a paddle. A process tube, usually made of quartz, provides a clean processing environment. A heating element embedded in a ceramic insulating fibre, provides the power to attain the required process temperature.

This heating element is divided into five individual zones in order to be able to control the axial temperature profile in the reactor. The temperature of each heating zone is measured by a so-called spike thermocouple and the temperature inside the process tube is measured by a 5-point thermocouple assembly, referred to as paddle thermocouple. Temperature is a critical parameter for all processes that can take place in the reactor, such as diffusion, oxidation and annealing (Schlote and Knoll, 1981). A natural requirement is that all wafers in the batch get the same heat treatment in order to achieve a good down-boat uniformity. In view of the length of the wafer load, this requires the use of a temperature controller that is capable of controlling the axial wafer temperature profile. As discussed in DeWaard and DeKoning (1990b) classical PID controllers that control *thermocouple* temperatures on a single-input single-output basis are not suitable for this purpose, because there can be a significant difference in thermocouple temperature and wafer temperature. Unfortunately, it is impossible to make direct measurements of the temperatures of the wafers in a batch in a production environment.

DeWaard and DeKoning (1990a) presented a mathematical model describing the heat transfer characteristics of the furnace. The furnace is considered to be cylindrically symmetrical and can be thought of as to consist of five constituents: insulation, heating coils, process tube, wafer load, and lids. For each of the constituents energy balances have been compiled, consisting of nonlinear partial integro-differential equations with appropriate boundary conditions. These energy balances take into account only time and axial dependency. They incorporate conductive and radiative heat transfer. This distributed parameter model is discretized in the space variable using a central-difference scheme, linearized around a steady-state solution using a multivariable Newton-Raphson algorithm, and converted into an equivalent discrete-time system.

Since the resulting model is of high order (40-100), an optimal controller based on this model will necessarily be of the same high order, thus violating constraints imposed by available computing power. Several approaches to reduced-order compensator design have been reported in the literature. In all methods the reduced-order compensator in principle is characterized by a projection on the full state-space. In the 'classical' methods the reduced-order compensator is obtained by reducing the dimension of a full-order LQG design, whereby the projection is constructed using balancing considerations. In this paper a different approach is followed (Hyland, Bernstein, and Davis, 1986), which does not require a full-order LQG design, but instead directly characterizes the quadratically optimal compensator of a given fixed order. In this approach the projection is a consequence of optimality conditions rather than system theoretic considerations. The explicit first order necessary conditions for quadratically optimal reduced-order compensation form a system of four coupled matrix equations: two modified Riccati equations and two modified Lyapunov equations, coupled by the projection matrix.

[1]This work was sponsored by the Dutch Technology Foundation (STW) under grant DWI 77.1397 (SMBT 87-1)

Motivations for the use of adaptive techniques for on-line tuning of the control law in this particular control problem are: Plant uncertainties, changing system dynamics, due for instance to the coating of the process tube during chemical vapour deposition, the frequent change of operating conditions and the effect of ageing of the thermocouples. Because the system is infinite-dimensional the adaptive control law inevitably is of restricted complexity so that the problem belongs to the class of adaptive control problems in the presence of unmodeled dynamics. Here, a sequential prediction error algorithm is used to estimate the parameters, and the control law is calculated by performing a few iterations of the associated Riccati equation in every sample interval.

After reviewing the optimal fixed-order dynamic compensation problem, a homotopy algorithm for solving the resulting optimal projection equations is presented. Next the controller is made adaptive by adding a recursive estimation scheme. Some performance results are given.

THE OPTIMAL FIXED-ORDER DYNAMIC COMPENSATION PROBLEM

Consider the system:

$$x_{i+1} = \Phi x_i + \Gamma u_i + v_i \qquad (2.1a)$$

$$y_i = Cx_i + w_i, \qquad i = 0, 1, .. \qquad (2.1b)$$

where $x_i \in R^n$ is the state, $u_i \in R^m$ the control, $y_i \in R^l$ the observation, $v_i \in R^n$ the system noise, $w_i \in R^l$ the observation noise and Φ, Γ, C are real matrices of appropriate dimensions. The processes $\{v_i\}$, $\{w_i\}$ are uncorrelated zero-mean gaussian white noise sequences with covariance $V \geq 0$ and $W > 0$ respectively. The initial condition x_0 is a gaussian stochastic variable with mean $\overline{x_0}$ and covariance P_0 and is uncorrelated with $\{v_i\}$ and $\{w_i\}$. As controller we choose the following dynamic compensator:

$$\hat{x}_{i+1} = F\hat{x}_i + Ky_i\,, \qquad (2.2a)$$

$$u_i = -L\hat{x}_i\,, \quad i = 0, 1, \ldots \qquad (2.2b)$$

characterized by the triple (F, K, L). Here $\hat{x}_i \in R^{nc}$, $n_c \leq n$. The optimal fixed-order dynamic compensation problem can now be stated as follows: Given system (2.1) find a compensator (F*,K*,L*) that minimizes the steady state performance measure

$$\sigma_\infty(F,K,L) = \lim_{N\to\infty} E\left\{\sum_{i=0}^{N-1} (x_i^T Q x_i + u_i^T R u_i)\right\}, \qquad (2.3)$$

where Q and R are real symmetric matrices of appropriate dimensions with $Q \geq 0$ and $R > 0$, and to find the minimum value $\sigma^*_\infty = \sigma_\infty(F^*,K^*,L^*)$. The closed-loop system may be described by:

$$\begin{bmatrix} x_{i+1} \\ \hat{x}_{i+1} \end{bmatrix} = \begin{bmatrix} \Phi & -\Gamma L \\ KC & F \end{bmatrix}\begin{bmatrix} x_i \\ \hat{x}_i \end{bmatrix} + \begin{bmatrix} v_i \\ Kw_i \end{bmatrix} \qquad (2.4)$$

Introducing

$$x_i' = \begin{bmatrix} x_i \\ \hat{x}_i \end{bmatrix}, \quad v_i' = \begin{bmatrix} v_i \\ Kw_i \end{bmatrix}, \quad \Phi' = \begin{bmatrix} \Phi & -\Gamma L \\ KC & F \end{bmatrix}, \quad V' = \begin{bmatrix} V & 0 \\ 0 & KWK^T \end{bmatrix}$$

gives for (2.4)

$$x'_{i+1} = \Phi' x'_i + v'_i, \qquad i = 0, 1, \ldots \qquad (2.5)$$

where $\{v'_i\}$ is a zero-mean gaussian white noise sequence with covariance V', and independent of the initial condition x_0'. System (2.5) is denoted by (Φ'). Let S^n denote the set of symmetric nxn matrices, and let $P_i \in S^{(n+nc)}$ denote the closed-loop covariance, then from (2.5)

$$P'_{i+1} = \Phi' P'_i \Phi'^T + V' \qquad (2.6)$$

If the closed-loop system is stable, the steady-state covariance P' exists as the unique positive semi-definite solution of

$$P' = \Phi' P' \Phi'^T + V' \qquad (2.7)$$

Furthermore, criterion (2.3) is finite and independent of initial conditions and can be expressed as

$$\sigma_\infty(F,K,L) = \mathrm{tr}\,(Q'P') \qquad (2.8)$$

where $Q' \in S^{(n+nc)}$ is given by

$$Q' = \begin{bmatrix} Q & 0 \\ 0 & L^T R L \end{bmatrix}$$

For this reason the admissible set of compensators C_{adm} is restricted to those compensators for which the closed-loop system is stable. Furthermore it is restricted to the set of minimal compensators because the criterion value is independent of the internal realization of (F,K,L):

$$C^m_{adm} = \left\{(F,K,L) \in C_{adm} \mid (F,K) \text{ reachable, } (F,L) \text{ observable}\right\}$$

The optimal fixed-order dynamic compensation problem can now be reformulated as to find the optimal compensator $(F^*,K^*,L^*) \in C^m_{adm}$ which minimizes (2.8) and subject to (2.7) for given value of $n_c \leq n$, and to find the minimum criterion value $\sigma^*_\infty = \sigma_\infty(F^*,K^*,L^*)$. Defining the linear transformation $A': S^{n+nc} \rightarrow S^{n+nc}$ as

$$A'X = \Phi'^T X \Phi', \qquad X \in S^{n+nc} \qquad (2.9)$$

implies (A') stable $\Leftrightarrow \rho(A') < 1$. Because the eigenvalues of A' continuously depend on (F,K,L), the set is open. Therefore, the matrix minimum principle [Athens, 1968] may be applied to find necessary conditions for the solution of the optimal compensation problem. For this purpose define the Hamiltonian H by

$$H(F,K,L,P',S') = \mathrm{tr}\left[Q'P' + (\Phi'P'\Phi'^T + V' - P')S'\right] \qquad (2.10)$$

where $S' \in S^{(n+nc)}$ is a Lagrange multiplier. Then the first-order necessary conditions are:

$$\frac{\partial H}{\partial F} = \frac{\partial}{\partial F}\mathrm{tr}\left(\Phi' P' \Phi'^T S'\right) = 0 \qquad (2.11a)$$

$$\frac{\partial H}{\partial K} = \frac{\partial}{\partial K}\mathrm{tr}\left(\Phi' P' \Phi'^T S' + V' S'\right) = 0 \qquad (2.11b)$$

$$\frac{\partial H}{\partial L} = \frac{\partial}{\partial L}\mathrm{tr}\left(\Phi' P' \Phi'^T S' + Q' P'\right) = 0 \qquad (2.11c)$$

$$\frac{\partial H}{\partial P'} = \Phi'^T S' \Phi' + Q' - S' = 0 \qquad (2.12a)$$

$$\frac{\partial H}{\partial S'} = \Phi' P' \Phi'^T + V' - P' = 0 \qquad (2.12b)$$

Partition $(n + n_c) \times (n + n_c)$ S' and P' as

$$S' = \begin{bmatrix} S_1 & S_{12} \\ S_{12}^T & S_2 \end{bmatrix}, \quad P' = \begin{bmatrix} P_1 & P_{12} \\ P_{12}^T & P_2 \end{bmatrix},$$

according to the partitioning of Φ' and define the nxn non-negative-definite matrices

$$S = S_1 - S_{12}S_2^{-1}S_{12}^T \qquad P = P_1 - P_{12}P_2^{-1}P_{12}^T$$
$$\hat{S} = S_{12}S_2^{-1}S_{12}^T \qquad \hat{P} = P_{12}P_2^{-1}P_{12}^T$$

Then the following lemma is used for stating the main result.

Lemma 2.1 (Richter and Collins, 1989). Suppose $\hat{P} \in S^n$ and $\hat{S} \in S^n$ are symmetric and nonnegative definite and rank $\hat{P}\hat{S} = n_c$. Then the following statements hold:

i) $\hat{P}\hat{S}$ is diagonalizable and has nonnegative eigenvalues

ii) The nxn matrix

$$\tau = \hat{P}\hat{S}(\hat{P}\hat{S})^{\#} \qquad (2.13)$$

is idempotent, i.e. τ is an oblique projection and

$$\text{rank } \tau = n_c \qquad (2.14)$$

iii) There exists $G, H \in R^{n_c \times n}$, and nonsingular $M \in R^{n_c \times n_c}$ such that

$$\hat{P}\hat{S} = G^T M H \qquad (2.15)$$
$$HG^T = I_{n_c} \qquad (2.16)$$

For convenience introduce the following notation:

$$\hat{W} = W + CPC^T \qquad (2.17)$$
$$\hat{R} = R + \Gamma^T S \Gamma \qquad (2.18)$$
$$\Sigma_p = \Phi PCT\, \hat{W}^{-1} (\Phi PC^T)^T \qquad (2.19)$$
$$\Sigma_s = (\Gamma^T S \Phi)^T \hat{R}^{-1} \Gamma^T S \Phi \qquad (2.20)$$
$$\Phi_p = \Phi - \Phi PC^T \hat{W}^{-1} C \qquad (2.21)$$
$$\Phi_s = \Phi - \Gamma \hat{R}^{-1} \Gamma^T S \Phi \qquad (2.22)$$
$$\tau_\perp = I_n - \tau \qquad (2.23)$$

Then the necessary conditions characterizing admissible extremals of the optimal fixed-order dynamic compensation problem are given by the following theorem.

Theorem 2.1 (Bernstein, Davis, and Hyland, 1986). Suppose $(F,K,L) \in C^m_{adm}$ solves the reduced-order dynamic compensator problem. Then there exist nxn nonnegative-definite matrices P, S, $\hat{P}$ and $\hat{S}$ such that F, K and L are given by:

$$F = H\,[\Phi - \Phi PC^T \hat{W}^{-1} C - \Gamma \hat{R}^{-1} \Gamma^T S \Phi] G^T \qquad (2.24)$$
$$K = H\,[\Phi PC^T \hat{W}^{-1}] \qquad (2.25)$$
$$L = -\,[\hat{R}^{-1} \Gamma^T S \Phi] \qquad (2.26)$$

and such that P, S, $\hat{P}$ and $\hat{S}$ satisfy

$$P = \Phi P \Phi^T - \Phi PC^T \hat{W}^{-1} (\Phi PC^T) + V + \tau_\perp \hat{P} \tau_\perp^T \qquad (2.27)$$
$$S = \Phi^T S \Phi - (\Gamma^T S \Phi)^T \hat{R}^{-1} \Gamma^T S \Phi + Q + \tau_\perp^T \hat{S} \tau_\perp \qquad (2.28)$$
$$\hat{P} = \Phi_s \tau \hat{P} \tau^T \Phi_s^T + \Sigma_p \qquad (2.29)$$
$$\hat{S} = \Phi_p^T \tau^T \hat{S} \tau \Phi_p + \Sigma_s \qquad (2.30)$$

where

$$\tau = \sum_{i=1}^{n_c} \pi_i (\Psi) \qquad (2.31)$$

For some $\Psi \in D(\hat{P},\hat{S})$ such that $(\Psi^{-1}\hat{P}\hat{S}\Psi)_{(i,i)} \neq 0$, $i = 1, \dots, n_c$ and some projective factorization G, H of τ. Here D denotes the set of contragrediently diagonalizing transformations defined for nonnegative-definite matrices T and U as $D(T,U) = \{\Psi: \Psi^{-1}T\Psi^{-T} \text{ and } \Psi^T U \Psi \text{ are diagonal}\}$

Notice that in the full-order case the optimal projection matrix τ and its factors G and H can be chosen to be the identity matrix so that (2.27) and (2.28) reduce to the standard observer and regulator Ricatti equations. Equations (2.29) and (2.30) then express the proviso that the compensator be minimal. The coupling of the equations due to the projection illustrates the nonoptimality of standard sequential controller reduction or model reduction schemes, because in the reduced-order case there is no longer separation between estimation and control operations.

CONTINUATION METHODS AND A HOMOTOPY ALGORITHM

The problem of solving (2.27-2.30) generally involves iterative schemes. In order to obtain convergence these require that a starting point be known sufficintly close to the solution. For high dimensional problems it is very difficult to find suitable starting points and numerical problems like ill-conditioning may make the actual computation impossible.

Continuation methods have the advantage that they are global in nature and do not rely upon knowledge of starting points in the domain of attraction of a particular iterative scheme. A continuation method can be used to embed the problem $F(x) = 0$ into a parameterized family of problems, $H(x, \alpha) = 0$, with $\alpha \in [0, 1]$, such that $H(x, 1) = F(x) = 0$ is the original problem and $H(x, 0)$ has a known or easily computed solution $x(0) = x_0$. Based on the above parametrization one deforms the simple problem $H(x, 0) = 0$ into the desired one $H(x, 1) = 0$, calculating the solution to the deformed problem at each stage of the deformation. Taking the derivatives of $H(x, \alpha)$ with respect to α gives the differential equation of Davidenko (Mariton and Bertrand, 1985)

$$H_x(x(\alpha), \alpha)\, \frac{dx(\alpha)}{d\alpha} + H_\alpha(x(\alpha), \alpha) = 0 \qquad (3.1)$$

where the subscripts indicate partial derivatives. Together with the initial condition $x(0) = x_0$ (3.1) defines an initial value problem which by numerical integration from 0 to 1 yields the desired solution x(1)

$$x(1) = x_0 + \int_0^1 \frac{dx(\alpha)}{d\alpha}\, dx \qquad (3.2)$$

The integration (3.2) if (3.1) yields a unique solution of $F(x) = 0$ if the following conditions are satisfied:

1) $H(x(0), 0) = 0$;
2) $x(\alpha)$ is continuously differentiable with respect to α
3) H is continuously differentiable with respect to x and α
4) H_α is continuous in α

The homotopy algorithm to be presented uses a discrete homotopy method. In contrast to the continuous homotopy methods the discrete methods do not require direct integration of Davidenko's equation, but instead partition the interval [0, 1] to obtain a finite chain of problems:

$$H(x, \alpha_k) = 0, \qquad 0 = \alpha_0 < \alpha_1 < ..... < \alpha_n = 1 \quad (3.3)$$

Starting with the known solution $x(0) = x_0$ for $H(x, 0)$, $x(\alpha_{k+1})$ is computed by an interative scheme with $x(\alpha_k)$ as the starting point.

A Homotopy Algorithm for Solution of the Discrete-Time Optimal Projection Equations

Simple algorithms for the solution of the optimal projection equations have been given by Greeley and Hyland (1988) and Hyland (1983). Richter and Collins (1989) presented a sophisticated algorithm using homotopic continuation methods. However, the algorithms given applied to the continuous time-case only. Here a simple homotopy algorithm is presented for the solution of the discrete-time optimal projection equations (2.27-2.31). Introducing the homotopy $H(x, \alpha)$, $\alpha \in [0, 1]$ as

$$\begin{aligned} & H(x, \alpha) = \\ & (X_1 - \Phi X_1 \Phi^T + \Phi X_1 C^T \hat{W}^{-1} (\Phi X_1 C^T)^T - V - \alpha^2 \tau_\perp X_3 \tau^T_\perp, \\ & X_2 - \Phi^T X_2 \Phi + (\Gamma^T X_2 \Phi)^T \hat{R}^{-1} \Gamma^T X_2 \Phi - Q - \alpha^2 \tau^T_\perp X_4 \tau_\perp, \\ & X_3 - \Phi_{X_2} [\alpha\tau + (1 - \alpha) I_n] X_3 [\alpha\tau + (1 - \alpha) I_n]^T \Phi^T_{X_2} - \Sigma_{X_1}, \\ & X_4 - \Phi^T_{X_1} [\alpha\tau + (1 - \alpha) I_n]^T X_4 [\alpha\tau + (1 - \alpha) I_n] \Phi_{X_1} - \Sigma_{X_2}) \quad (3.4) \end{aligned}$$

where I_n is the n•n identity matrix, we have the following

Theorem 3.1. $H(X, \alpha) = 0$ with $X = (P, S, \hat{P}, \hat{S})$ and $\alpha \in [0, 1]$ is a continuation process for the solution of the system of equations (2.27-2.31).

Proof: see DeWaard (1990). Notice that for $\alpha = 0$, $H(X, 0) = 0$ constitutes a set of uncoupled Riccati and Lyapunov equations corresponding to the full-order LQG problem, of which a unique solution is known to exist if the system is stabilizable and detectable.

The actual computational scheme consists of two nested loops: the outer loop increments the homotopy parameter α_k from its initial value $\alpha_0 = 0$ to its final value $\alpha_n = 1$, by fixed increments $\Delta = 1/N$. The inner loop iterates till convergence the Riccati and Lyapunov equations for fixed α_k, and using P, S, $\hat{P}$, $\hat{S}$ and τ as obtained with α_{k-1} as initial value. At the start of the algorithm τ is taken to be I_n. When convergence is obtained τ is updated and used in a next iteration. Finally, F, K and L are solved using (2.24–2.26) and the cost criterion $\sigma_\infty(F,K,L)$ is evaluated using (2.8).

CONTROLLER DESIGN FOR DIFFUSION / LPCVD REACTORS

The model of the diffusion/LPCVD reactor as derived by DeWaard and DeKoning (1990a, 1990b) consists of a set of nonlinear partial differential equations, describing the heat balances of all constituents in the furnace, i.e. the insulation, heating elements, tube, wafer load and lids (see Fig. 1). Finite differences are used to obtain a lumped parameter model, which is linearized about a steady-state solution, resulting in a continuous-time state-space description. Finally this description is converted to an equivalent discrete-time system description, yielding a model of the form (2.1). In this model x_i is a vector containing the deviations of the thermocouple temperatures from their steady-state values. The dimension n of the model is high, typically in the order of 40-100. The objective of the controller design procedure is to devise a controller that controls wafer tempratures instead of thermocouple temperatures. Furthermore, it is desired that the controller be capable of realizing a given axial temperature profile in the wafer load, and of sustaining this profile even during ramping of the set-point. In order to find the optimal control law, a cost criterion as in (2.3) is minimized that reflects the wish to weight wafer temperature set-point deviations without spending too much control effort. By appropriately choosing the off-diagonal entries in the state-weighting matrix, Q, in (2.3), an explicit penalty on wafer-to-wafer temperature deviations can be brought into the criterion, thereby effectively eliminating any undesirable temperature gradients in axial direction. A different choice of weighting matrices can be made to bring the thermocouple temperatures into the criterion instead of the wafer temperatures. This is useful for making a quick comparison between the performance of a classical PID controller and the reduced-order LQG controller, because the thermocouple temperatures are then directly available, whereas the wafer temperatures are not. For this purpose a weighting matrix Q_p is defined as

$$Q_p = \begin{pmatrix} 1 & & & & & \\ & \ddots & & & \text{paddle weighting block} & \\ & & 1 & & & \\ & & & (n-1)*un+ow & -un \cdots & -un \\ & & & -un & \ddots & \vdots \\ & & & \vdots & \ddots & -un \\ & & & -un \cdots & -un & (n-1)*un+ow \end{pmatrix} \quad (4.1)$$

The weighting matrix Q in (2.3) is now calculated as

$$Q = C^T Q_p C \quad (4.2)$$

Optimal values for the weighting parameters are found to be *un*=50, *ow*=100. The control weighting matrix R is chosen as R=diag(*inp*), where *inp*=0.0001. Figure 2 shows the simulated response of the centre paddle thermocouple to a ramped setpoint change for 5 different dimensions of the reduced-order compensator, indicated by *dpr*. The full-order model has dimension *n*=42. Figure 3 shows the corresponding flatzone degradation. This is the maximum temperature difference between thermocouples as a function of time, and is a good measure for the ability of the controller to maintain a flat profile during setpoint changes. From Fig. 3 it can be seen that there is only a slight performance decrease due to the order reduction process. Figure 4 shows the performance of a 20-dimensional controller on a real furnace, and Fig. 5 gives the corresponding flatzone degradation. Due to model inaccuracies the offset control needed to sustain a nonzero setpoint is not calculated exactly, as can be seen from Fig. 4.

The wide range and frequent change of operating conditions, and the changing system dynamics motivate the use of on-line parameter estimation schemes. Here a sequential prediction error algorithm is used to make the control system adaptive. For this purpose the reduced order compensator (2.2) is rewritten in filter form and the controller parameters are directly parameterized by a vector θ

From (2.2) and using (2.24-2.26) we can write the reduced-order compensator as

$$\hat{x}_{i+1} = \hat{\Phi}\hat{x}_i + \hat{\Gamma}u_i + K(y_i - \hat{y}_i) \tag{4.3}$$

where

$$\hat{y}_i = \hat{C}\hat{x}_i \tag{4.4}$$

and where

$$\hat{\Phi} = H\Phi G^T \; ; \; \hat{\Gamma} = H\Gamma \; ; \; \hat{C} = CG^T.$$

Defining the parameter vector θ as

$$\theta = \text{col}\,(\hat{\Phi}, \hat{\Gamma}, K, \hat{C})$$

we can estimate θ using the following recursive least squares algorithm

$$\hat{\theta}_{i+1} = \hat{\theta}_i + P_{i-1}\psi_{i-1}\left(y_i - \hat{y}_i\right) \tag{4.5}$$

$$P_{i-1} = P_{i-2} - \left(I + \psi_{i-1}^T P_{i-2}\psi_{i-1}\right)^{-1} P_{i-2}\psi_{i-1}\psi_{i-1}^T P_{i-2} \tag{4.6}$$

where

$$\psi_{i-1}^T = \frac{d}{d\theta}\left[\hat{y}_i(\theta)\right]_{\theta=\hat{\theta}_{i-1}} \tag{4.7}$$

Introducing

$$\varepsilon_i = y_i - \hat{y}_i \tag{4.8}$$

we can write the cost criterion (2.3) as

$$\sigma_\infty(F,K,L) = \lim_{N\to\infty} E\left\{\sum_{i=0}^{N-1}\left[\hat{y}_i^T Q\hat{y}_i + u_i^T R u_i + \varepsilon_i^T Q\varepsilon_i\right]\right\} \tag{4.9}$$

where Q is given by (4.2). Minimizing the last term in (4.9) is exactly the objective of the recursive least squares algorithm. Minimizing the first two terms subject to (4.3-4.4) follows by solving a Riccati equation with a dimension equal to that of the controller. Notice that this has to be done on-line, because now the weighting matrix Q depends of θ. From the solution of the Riccati equation the gain matrix L can be computed directly.

Figure 6 shows the value of the criterion function (4.9) for the simulation shown in Fig. 2 for both adaptive and non-adaptive reduced-order control. In the adaptive case the criterion function converges to zero, indicating accurate output estimation. This also provides a means to calculate the offset control needed to sustain a nonzero setpoint. This can be done by solving the offset control u_0 from

$$y_d = \hat{C}^*\left(I - \hat{\Phi}^*\right)^{-1}\hat{\Gamma}^* u_0 \tag{4.10}$$

where y_d is the desired setpoint for the thermocouples, and a * indicates the converged value. Figure 7 shows the convergence of the offset control.

CONCLUSION

A reduced-order controller has been developed for the controlling of thetemperature in industrial diffusion / LPCVD reactors. The design method adopted the novel approach of optimal projections to find an implementable low-order controller for a high dimensional system model. The performance of the resulting reduced-order compensator was shown to be close to that of the full-order LQG controller. A sequential prediction error algorithm was used to be able to control the reactor under a variety of operating conditions. This also provided a means to calculate the offset control necessary to sustain a non-zero setpoint. Current research effort is directed towards enhancing the speed of convergence of the parameters, and adding a priori information into the parameter estimation scheme. Also the robustness of the control system is still an issue. However, the optimal projection approach allows for robust controller design by adopting a stochastic parameter formulation (Bernstein and Haddad, 1987).

ACKNOWLEDGMENT

The granted facilities and valued support of ASM International is gratefully acknowledged.

REFERENCES

Athens, M. (1968). The matrix minimum principle. *Information and Control*, 11, 592-606.

Bernstein, D.S., and W.M. Haddad (1987). Optimal projection equations for discrete-time fixed-order dynamic compensation of linear systems with multiplicative white noise. *Int. J. Control,* 46, no. 1, 65-73.

Bernstein, D.S., D.C. Hyland, and L.D. Davis (1986). The optimal projection equations for reduced-order discrete-time modeling, estimation, and control. *J. Guidance*, 9, 288-293.

DeWaard, H. (1990). Reduced-order control of diffusion/LPCVD reactors using optimal projections. *Submitted for publication.*

DeWaard, H., and W.L. DeKoning (1990a). Optimal control of a diffusion / low pressure chemical vapour deposition furnace, *J. Appl. Phys.*, 67, 2264-2271.

DeWaard, H., and W.L. DeKoning (1990b). Optimal control of the wafer temperatures in diffusion / LPCVD reactors. *To appear in Automatica.*

Greeley, S.W., and D.C. Hyland (1988). Reduced-order compensation: linear-quadratic reduction versus optimal projection. *J. Guidance*, 11, 328-335.

Hyland, D.C. (1983). The optimal projection approach to fixed-order compensation: numerical methods and illustrative results. *AIAA 21st Aerospace Sciences Meeting*, Reno, Nevada, 1-10.

Mariton, M., and P. Bertrand (1985). A homotopy algorithm for solving coupled Riccati equations. *Optimal Control Applications & Methods*, 6, 351-357.Richter, S., and E.G. Collins, Jr. (1989). A homotopy algorithm for reduced order compensator design using the optimal projection equations. *Proc. 28th Conf. Dec. Contr.* , Tampa, Florida, 506-511.

Schlote, J., and D. Knoll (1981). *Exp. Techn. Phys.*, 29, 157

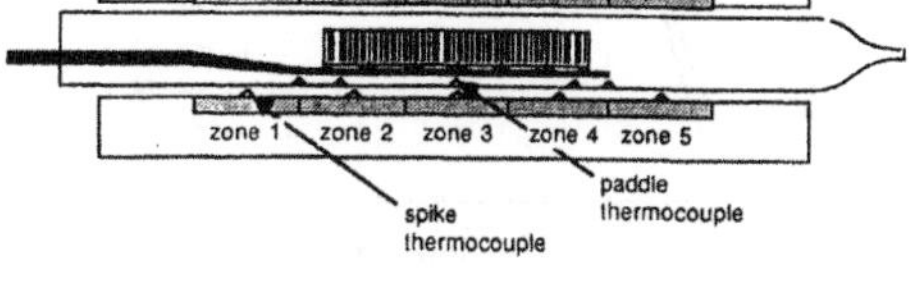

Fig. 1. Longitudinal section of a diffusion/LPCVD reactor

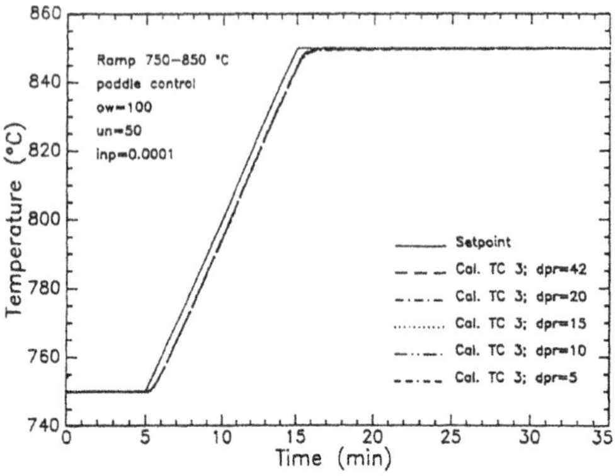

Fig. 2. Simulated response of the paddle thermocouple in zone 3 to a ramped setpoint change for a reduced-order controllers of dimension dpr.

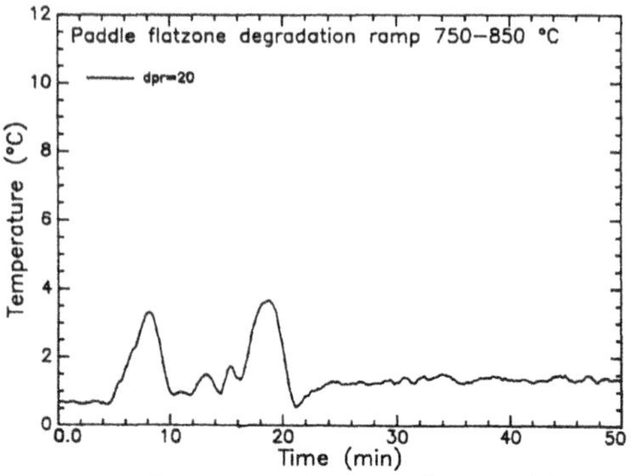

Fig. 5. Paddle flatzone degradation as a function of time, corresponding to Fig. 4.

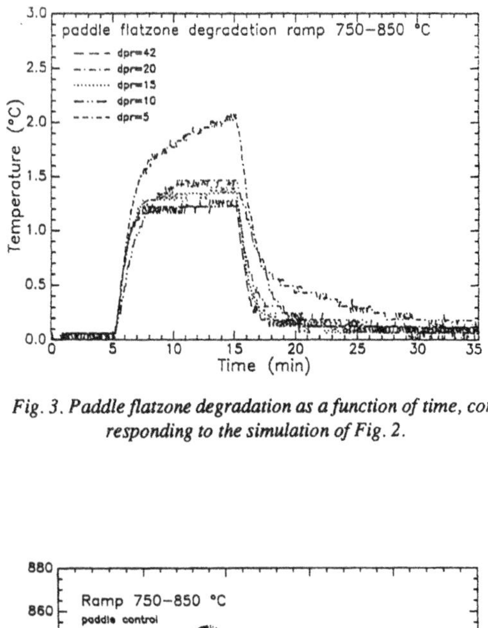

Fig. 3. Paddle flatzone degradation as a function of time, corresponding to the simulation of Fig. 2.

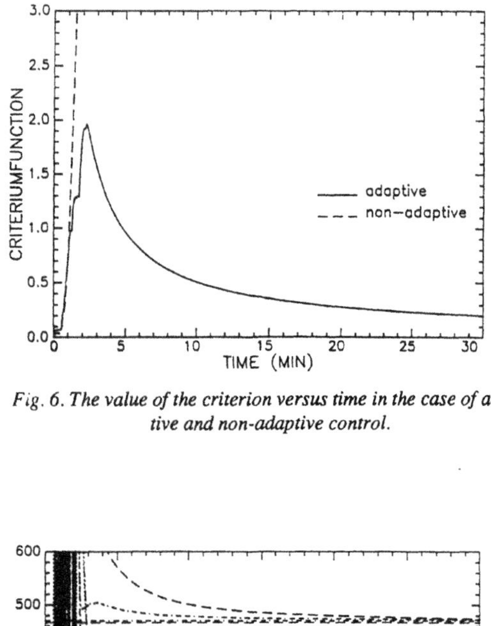

Fig. 6. The value of the criterion versus time in the case of adaptive and non-adaptive control.

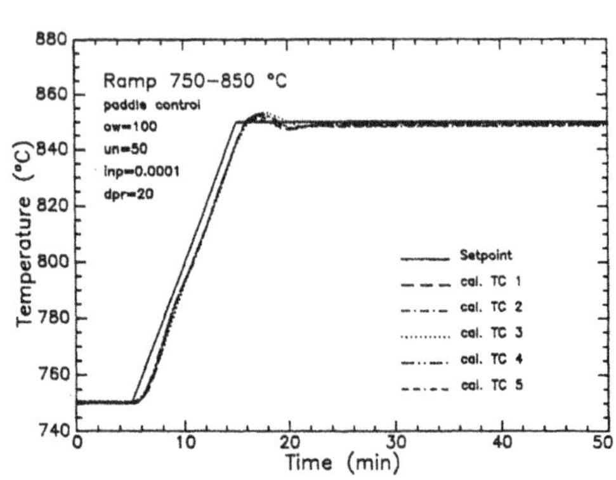

Fig. 4. Response of the paddle thermocouples to a ramped setpoint change on a real furnace, using a 20-dimensional reduced-order controller

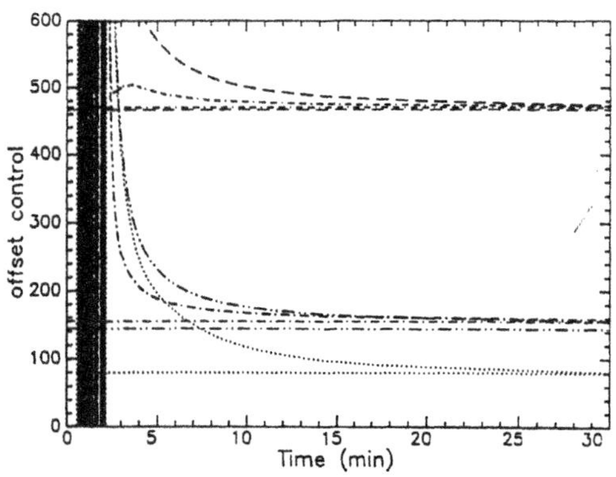

Fig. 7. The values of the offset powers, calculated according to (4.10), as a function of time.

DESIGN AND IMPLEMENTATION OF THE ADAPTIVE SYNCHRONIZING FEEDFORWARD CONTROLLER FOR TWO AXES MOTION CONTROL SYSTEMS

M. Tomizuka*, T. Kamano and T. Suzuki****

**Dept. of Mechanical Engineering, University of California, Berkeley, Berkeley, California, USA*
***Dept. of Electrical and Electronic Engineering, The University of Tokushima, Tokushima, Japan*

Abstract. In this paper, high speed motion synchronization of two d.c. motors, or motion control axes, under adaptive feedforward control is considered. The adaptive feedforward control system for one axis system consists of a proportional feedback controller, an adaptive disturbance compensator and an adaptive feedforward controller. To synchronize the motions of the two motors, a coupling controller, which responds to the synchronization error is introduced. The synchronization error is used as the adaptation error signal in the two adaptive feedforward controllers as well. When a disturbance input is applied to one axis, the motion errors appear in the undisturbed axis as well as in the disturbed axis. The motion error in the undisturbed axis is induced by the coupling controller and adaptive feedforward controller. This allows a quick removal of the synchronization error. The effectiveness of adaptive synchronizing feedforward control is demonstrated by experiment.

Keywords. Synchronizing control; Adaptive control; Feedforward control; d.c. motors; Speed control.

INTRODUCTION

One of fundamental problems in motion control systems such as a machining center is synchronization of multiple motion axes or motors. Synchronization of multiple motion axes can be achieved by either the "equal-status" approach or "master-slave" approach (Uchiyama and Nakamura, 1988). In the equal-status approach, the synchronizing controller treats multiple axes in a similar manner without favoring one axis over the other. When the dynamics are significantly different among multiple axes, the equal-status approach may not be the best because the synchronization speed of the overall system is set by the slowest axis. In a two-axes problem with significantly different dynamics between the two axes, it will make more sense to take the master-slave approach. In this case, the slow axis is under conventional servo control and acts as the master for the fast axis.

In this paper, high speed synchronization for a two-axes problem is considered from the equal-status viewpoint. To achieve high speed synchronization, the following issues must be taken into consideration: 1. dynamics of motion axes may be significantly different from one axis to another and may depend on operating conditions, and 2. motion axes must be synchronized during transient as well as at steady state. These issues are difficult to handle within a frame work of conventional feedback control theory. On the other hand, feedforward control is effective in tracking the time varying desired output signal (e.g. Tomizuka,1989). However, tracking performance under feedforward controllers depends on the accuracy of a plant model utilized in the design. When dynamic parameters such as viscous and Coulomb friction coefficients and inertia are subject to change, they must be estimated in real time and the feedforward controller must be adjusted accordingly. The objective of this paper is to explore the synchronization under adaptive feedforward control.

In the next section, we discuss an independent two-axes system, in which an adaptive feedforward controller is used in each axis independently. The feedforward controller consists of a proportional feedback controller, an adaptive disturbance compensator and an adaptive feedforward controller. Experimental results of this system for both the time varying and the constant desired output signals are shown. In the third section, we introduce a coupling controller in addition to the adaptive feedforward controller for each axis. The coupling controller responds to the synchronization error, i.e. the difference between the two motion errors. This idea is similar to the one suggested by Koren (1980) for two axes contouring systems. The synchronization error is used as the adaptation error signal in the two adaptive feedforward controllers. Experimental results are shown to demonstrate the effectiveness of the coupling controller. Conclusion is described in the fourth section.

INDEPENDENT ADAPTIVE FEEDFORWARD CONTROL

Overall Structure (Kamano et al,1990)

The overall system of a two-axes system under independent adaptive feedforward control is shown in Fig.1. The d.c. motor, of which parameters are listed in Table 1, is used as the the actuator in each axis. The load inertia, which is about ten times the moment of inertia of the motor itself, is applied to the first axis. The motor dynamics is described by

$$G_{mi}(s)=\frac{K_i}{\tau_i s+1} \qquad (i=1 \text{ or } 2) \qquad (1)$$

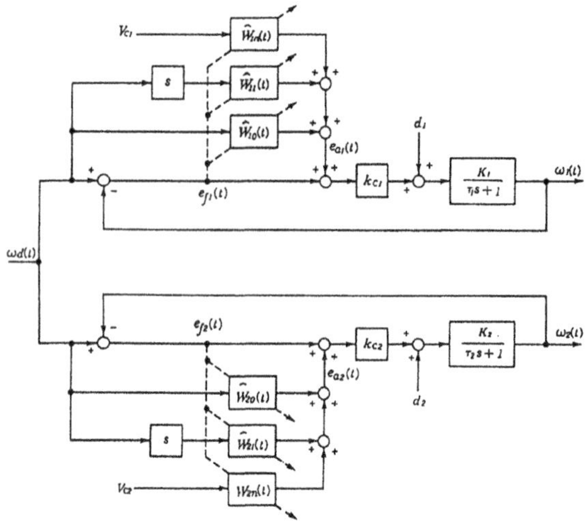

Fig.1. Independent Adaptive Feedforward Control for Two Axes System.

TABLE 1 Specifications and Parameters of Tested Motor

Specifications		
Capacity	35	W
Voltage	24	V
Current	2.5	A
Torque	1.0	kgcm
Speed	3500	rpm
Parameters		
Armature resistance	2.26	Ω
Moment of inertia	3.30×10^{-5}	kgm^2
Viscous friction torque coefficient	2.20×10^{-5}	Nm/rad/s
Back electromotive force constant	5.10×10^{-2}	V/rad/s
Torque constant	5.10×10^{-2}	Nm/A

where the input and output are the voltage and velocity, and τ_i and K_i are the time constant and gain constant, respectively. τ_i and K_i are assumed to be precisely unknown or subject to change. Notice that the load inertia on the first axis makes the two transfer functions, $G_{m1}(s)$ and $G_{m2}(s)$, significantly different.

As shown in Fig.1, the input to the motor is

$$u_i(t)=k_{ci}\{e_{fi}(t)+e_{ai}(t)\}+d_i \quad (i=1 \text{ or } 2) \qquad (2)$$

where k_{ci} is a fixed proportional control gain. $e_{ai}(t)$ is the feedforward control output defined by

$$e_{ai}(t)=\hat{W}_{i0}(t)\omega_d(t)+\hat{W}_{i1}(t)\{d\omega_d(t)/dt\} +\hat{W}_{in}(t)V_{ci} \quad (i=1 \text{ or } 2) \qquad (3)$$

where $\omega_d(t)$ is the desired output. $\omega_d(t)$ and $d\omega_d(t)/dt$ are both bounded, and V_{ci} is constant. In Eq.(3), the first two terms represent the adaptive feedforward control action for the desired output, and the third term represents that for the disturbance input. $W_i(t)$'s are adjusted by the adaptation law,

$$\frac{d\hat{W}_{i0}(t)}{dt} = \gamma\omega_d(t)e_{fi}(t) \quad (i=1 \text{ or } 2) \qquad (4)$$

$$\frac{d\hat{W}_{i1}(t)}{dt} = \gamma\{d\omega_d(t)/dt\}e_{fi}(t) \quad (i=1 \text{ or } 2) \qquad (5)$$

$$\frac{d\hat{W}_{in}(t)}{dt} = \gamma V_{ci}e_{fi}(t) \quad (i=1 \text{ or } 2) \qquad (6)$$

where γ is a positive adaptation gain. By defining,

$$\hat{\theta}_i(t)^T= [\hat{W}_{i0}(t),\ \hat{W}_{i1}(t),\ \hat{W}_{in}(t)] \quad (i=1 \text{ or } 2) \qquad (7)$$

and

$$\phi(t)^T= [\omega_d(t),\ d\omega_d(t)/dt,\ V_{ci}] \quad (i=1 \text{ or } 2) \qquad (8)$$

the adaptation law for each axis can be written as

$$\frac{d\hat{\theta}_i(t)}{dt} = \gamma\phi(t)e_{fi}(t) \quad (i=1 \text{ or } 2) \qquad (9)$$

Under the assumption that the nonadaptive feedback control loop consisting of the plant and proportional controller remains asymptotically stable, the disturbance (d_i) is constant, and the desired output and its derivative are bounded, it can be shown that the overall adaptive system is stable and the tracking error converges to zero asymptotically (see Kamano et al, 1990). Notice that when the error is zero, i.e. $e_{fi}(t)=0$, the motor are purely under feedforward control.

Experimental Results

Some experimental results are presented below for the case where the desired output during adaptation is a periodic trapezoidal function. The trapezoidal function is shaped so that the first derivative exists (see Fig.2).

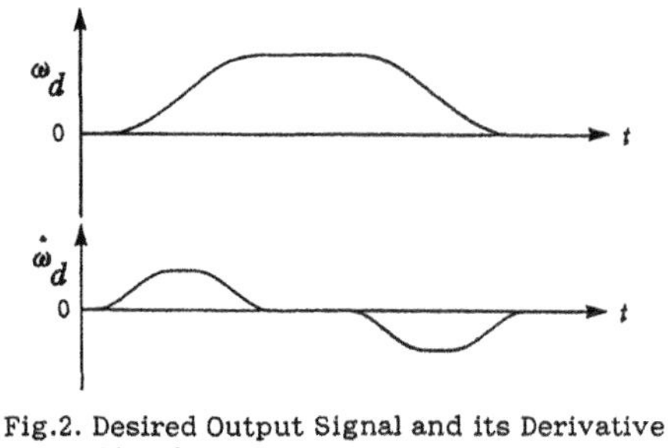

Fig.2. Desired Output Signal and its Derivative Signal.

Figures 3(a) and 3(b) show the output and error signals of the first axis and second axis in the early stage of adaptation. The outputs of the two axes approach the desired output gradually and independently. Remember the dynamics of the two axes are significantly different and there is no interaction between the two axes. The tracking error, $e_{f1}(t)$ and $e_{f2}(t)$, gradually decrease, and the output signals of the adaptive controllers for the two axes, $e_{a1}(t)$ and $e_{a2}(t)$, increase as the adaptation process progresses. The output and error signals at the twentieth period of the desired output are shown in Figs.4(a) and 4(b). The tracking error are almost zero. Therefore, the motor is mainly controlled by the output of the corresponding adaptive controller. The actual output of each axis follows the desired output well. The controller parameters are shown in Fig.5. These parameters converge to the constant values. Figure 6 shows the mean-square errors, $e_{m1}(t)$ and $e_{m2}(t)$ defined by

$$e_{mi}(t)=\frac{1}{T_r}\int_0^{T_r}\{\omega_d(t+t_k)-\omega(t+t_k)\}\,dt \qquad (10)$$

(i=1 or 2)

where T_r is the period and t_k corresponds to the beginning of the k-th period. The error $e_{m1}(t)$ and $e_{m2}(t)$ are reduced substantially by adaptive control.

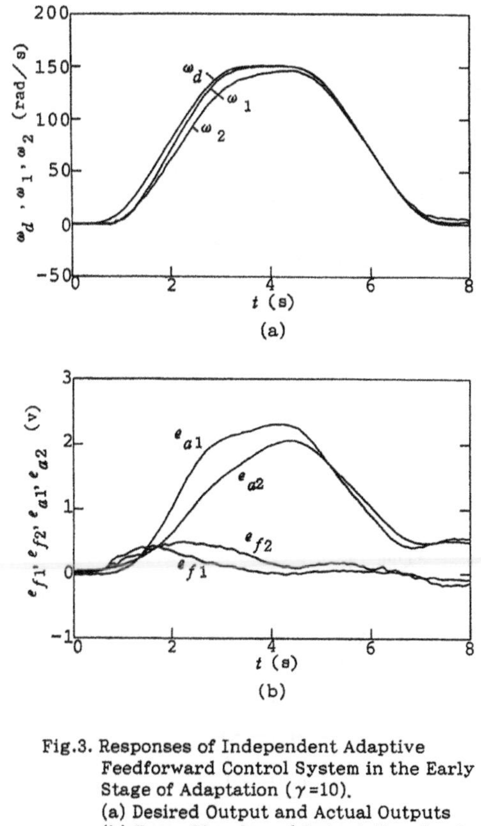

Fig.3. Responses of Independent Adaptive Feedforward Control System in the Early Stage of Adaptation (γ=10).
(a) Desired Output and Actual Outputs
(b) Error Signals and Output Signals of Feedforward Controllers

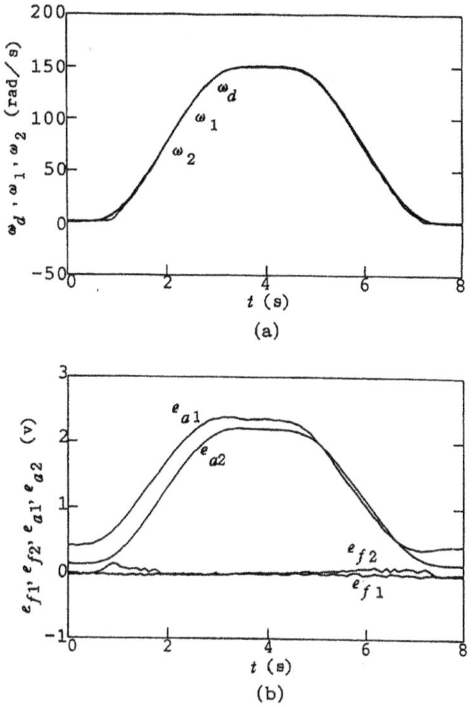

Fig.4. Responses of Independent Adaptive Feedforward Control System after Adaptation (γ=10).
(a) Desired Output and Actual Outputs
(b) Error Signals and Output Signals of Feedforward Controllers

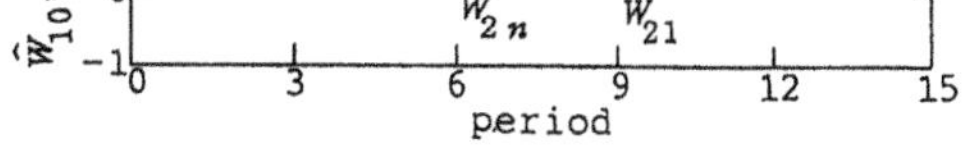

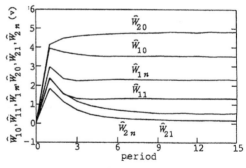

Fig.5. Responses of the Control Parameters (γ=10).

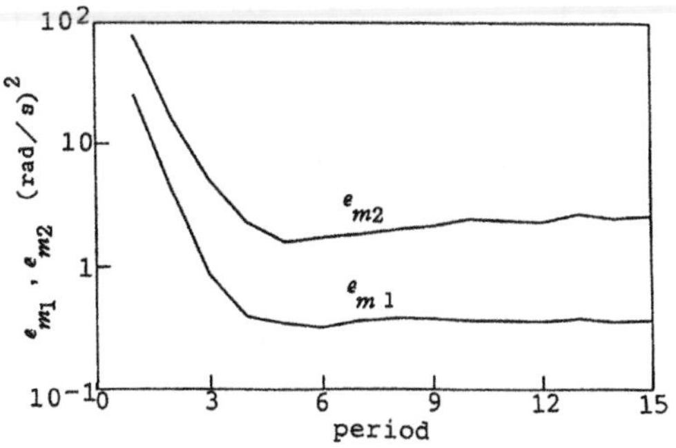

Fig.6. Mean-Square Error (γ=10).

Figures 7(a),7(b) and 7(c) show the responses for a step disturbance input, which is applied to the first axis. The desired output is constant. The actual output of the disturbed axis temporarily dips. However, since the controller parameters of the disturbed axis are modified by the adaptation algorithm, the output resumes the desired value after a short transient process. Notice that the error of the undisturbed axis remains zero, and the two axes are unsynchronized during the recovery process of the disturbed axis.

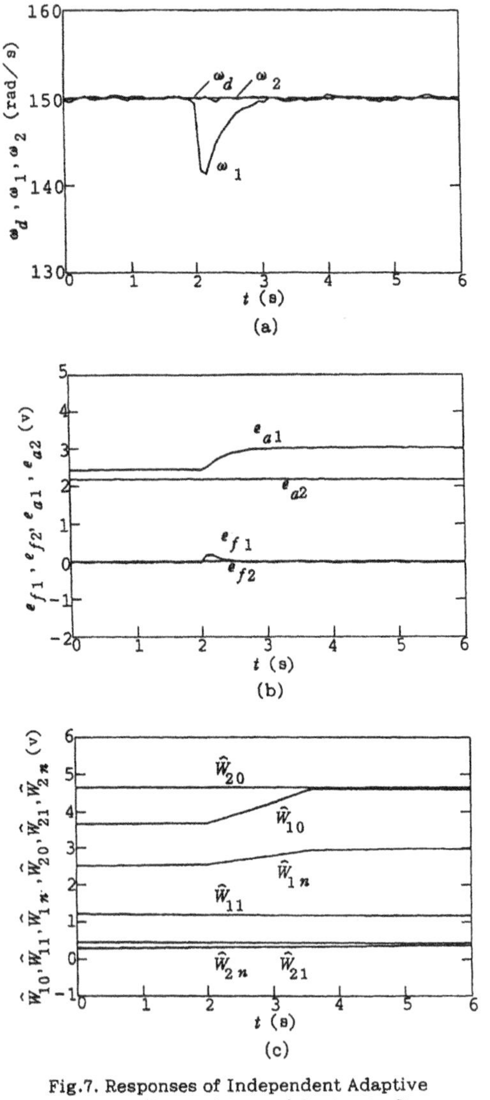

Fig.7. Responses of Independent Adaptive Feedforward Control System to Step Disturbance to the First Axis (γ=10).
(a) Desired Output and Actual Outputs
(b) Error Signals and Output Signals of Feedforward Controllers
(c) Control Parameters

ADAPTIVE SYNCHRONIZING FEEDFORWARD CONTROL

Overall Structure (Tomizuka et al,1990)

To achieve a better synchronization, we measure the synchronization error,

$$\varepsilon(t)=e_{f1}(t)-e_{f2}(t) \tag{11}$$

and modify the overall control structure in Fig.1 to the one in Fig.8. Notice that the additional coupling controller increases an effective loop gain for each axis. The parameter adaptation law is modified to

$$\frac{d\hat{\theta}_1(t)}{dt} = \gamma\phi(t)\,[e_{f1}(t)+\beta\varepsilon(t)] \tag{12}$$

$$\frac{d\hat{\theta}_2(t)}{dt} = \gamma\phi(t)\,[e_{f2}(t)-\beta\varepsilon(t)] \tag{13}$$

where the adaptation gain γ is any positive number and the positive coupling parameters β must be selected so that

$$G(s)=\begin{bmatrix} g_1 s+h_1+(1+\beta), & -\beta \\ -\beta, & g_2 s+h_2+(1+\beta) \end{bmatrix}^{-1} \tag{14}$$

is strictly positive real, where $g_1=\tau_1/k_{o1}K_1$, $h_1=1/k_{o1}K_1$, $g_2=\tau_2/k_{o2}K_2$ and $h_2=1/k_{o2}K_2$. In particular, if two axes possess identical dynamic, i.e. $\tau_1=\tau_2$, $K_1=K_2$ and $k_{o1}=k_{o2}$, β can take any positive value. Notice that the synchronization error drives the two parameter vectors, $\hat{\theta}_1(t)$ and $\hat{\theta}_2(t)$, in opposite directions, which results in faster removal of the synchronization error. Under the assumptions described in the previous section and the adequate β, it can be shown that this adaptive synchronizing feedforward system is stable and the tracking errors

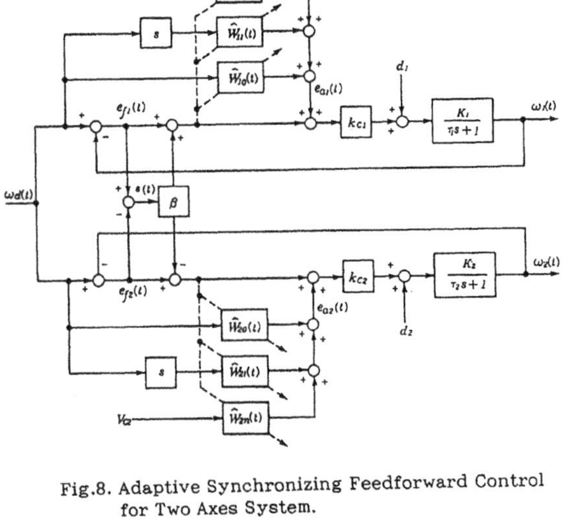

Fig.8. Adaptive Synchronizing Feedforward Control for Two Axes System.

of the two axes converge to zero asymptotically (see Tomizuka et al,1990).

Experimental Results

Figures 9(a) and 9(b) show the output and error signals of this adaptive synchronizing system for the trapezoidal desired output in the previous section. In this experiment, γ and β are set to 10 and 1.5, respectively. Although the dynamics of the two axes are significantly different, the two outputs gradually approach the desired output in a synchronized manner. The tracking errors, $e_{r1}(t)$ and $e_{r2}(t)$, gradually decrease and the outputs of the two adaptive controllers take over as the adaptation process progresses. These results imply that the coupling controller is effective for faster removal of the synchronization error. The output and error signals at the twentieth period of the desired output are shown in Figs.10(a) and 10(b). The outputs of the two axes follow the desired output well. The tracking errors are almost zero, and each axis is mainly controlled by the output of the adaptive controller. The controller parameters are shown in Fig.11. These parameters converge to the constant values. Notice that the converged values agree with those shown in Fig.5. This confirms that the primary influence of the synchronizing action is on the transient process. Figure 12 shows the mean-square error, $e_{m1}(t)$ and $e_{m2}(t)$, defined in Eq.(10). In synchronizing adaptive control, $e_{m1}(t)$ and $e_{m2}(t)$ decrease similarly because of the effect of the coupling controller.

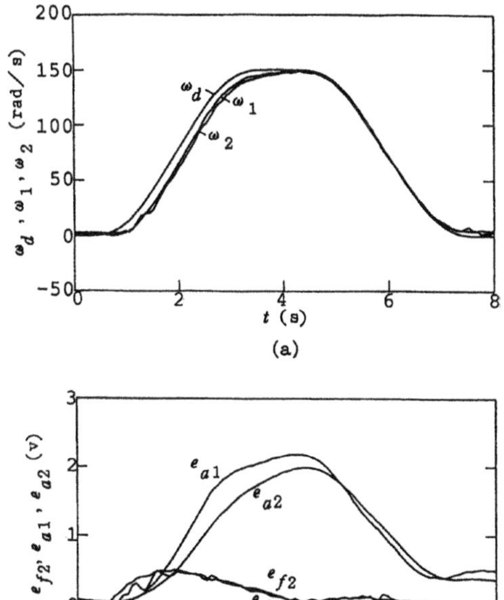

(a)

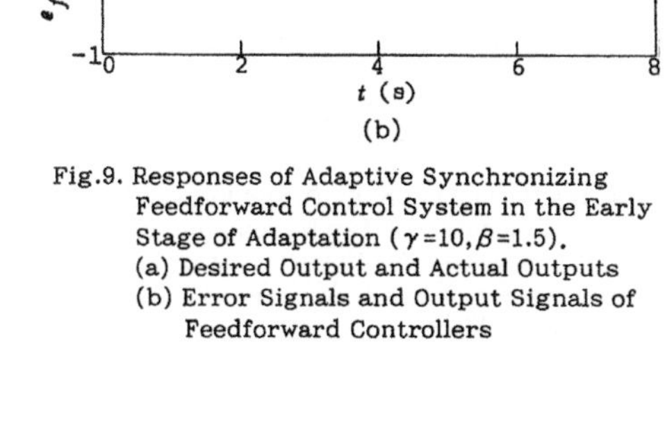

(b)

Fig.9. Responses of Adaptive Synchronizing Feedforward Control System in the Early Stage of Adaptation (γ=10,β=1.5).
(a) Desired Output and Actual Outputs
(b) Error Signals and Output Signals of Feedforward Controllers

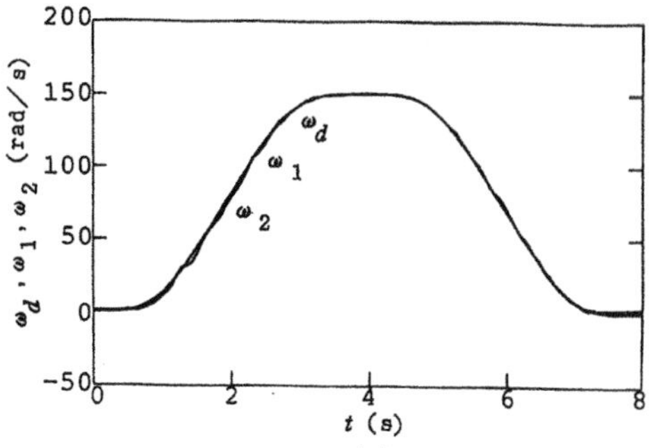

(a)

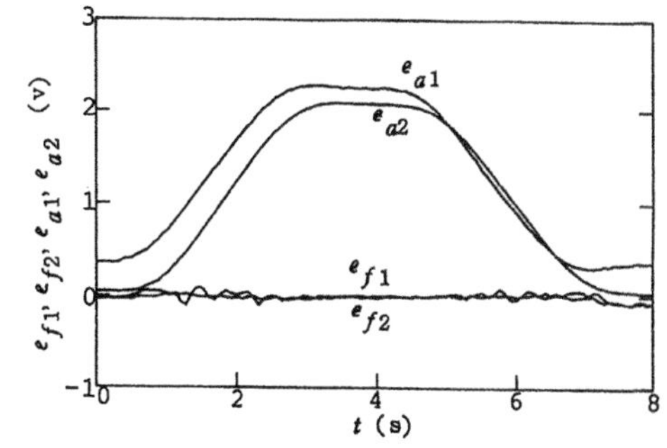

(b)

Fig.10. Responses of Adaptive Synchronizing Feedforward Control System after Adaptation (γ=10,β=1.5).
(a) Desired Output and Actual Outputs
(b) Error Signals and Output Signals of Feedforward Controllers

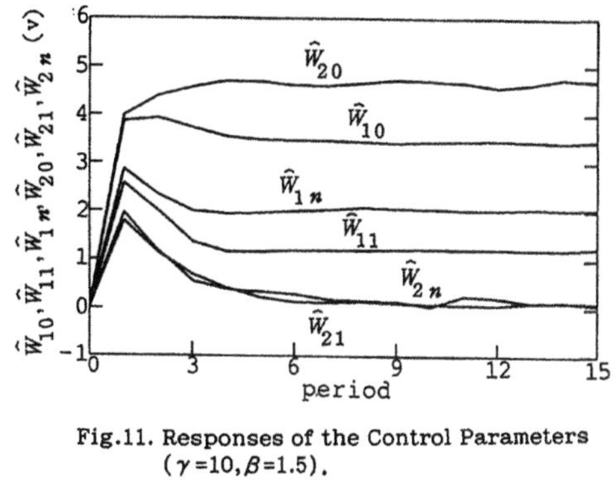

Fig.11. Responses of the Control Parameters (γ=10,β=1.5).

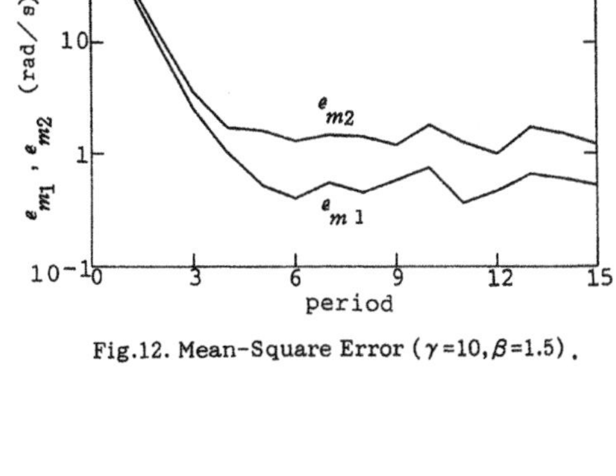

Fig.12. Mean-Square Error (γ=10,β=1.5).

Figures 13(a),13(b) and 13(c) show the responses for regulation problem when a step disturbance input is applied to the first axis. Notice that the tracking errors appear in the undisturbed axis as well as in the disturbed axis resulting in a reduced synchronizing error. Figure 14 shows the effect of the coupling parameter, β, on synchronization. As shown in Fig.14, the synchronization error is significantly reduced as the coupling parameter β is increased.

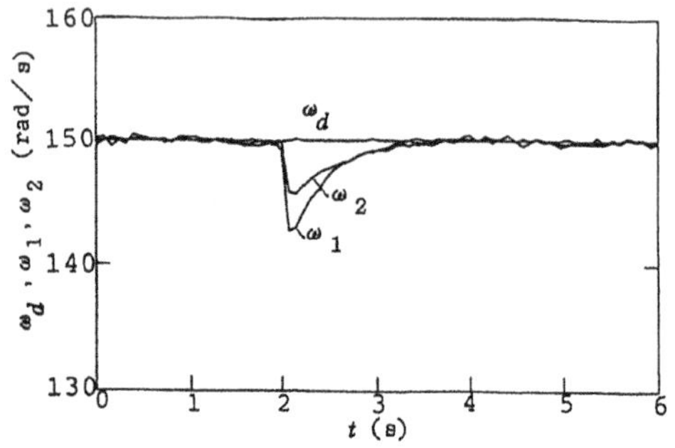

(a)

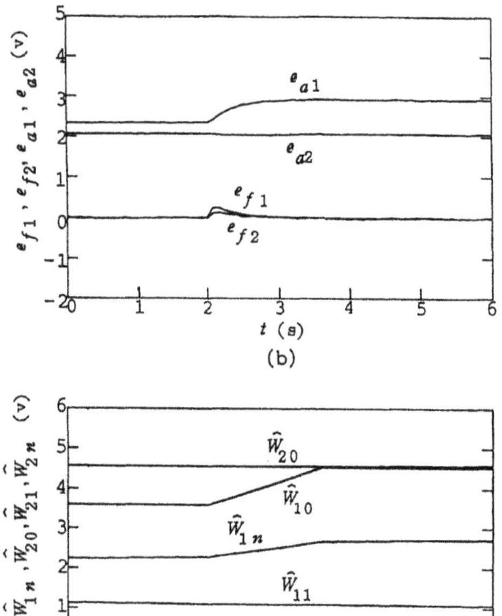

(b)

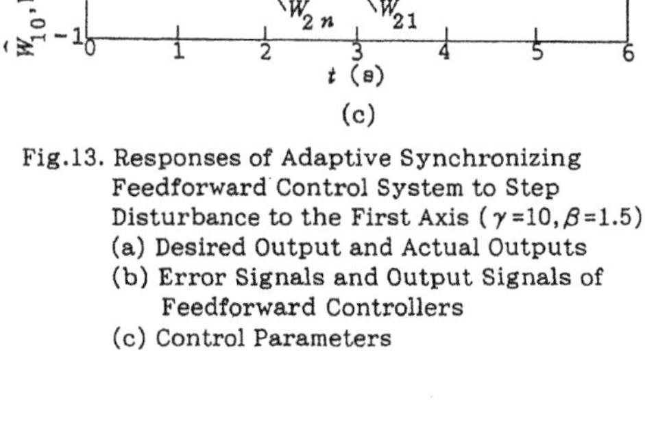

(c)

Fig.13. Responses of Adaptive Synchronizing Feedforward Control System to Step Disturbance to the First Axis (γ=10,β=1.5).
(a) Desired Output and Actual Outputs
(b) Error Signals and Output Signals of Feedforward Controllers
(c) Control Parameters

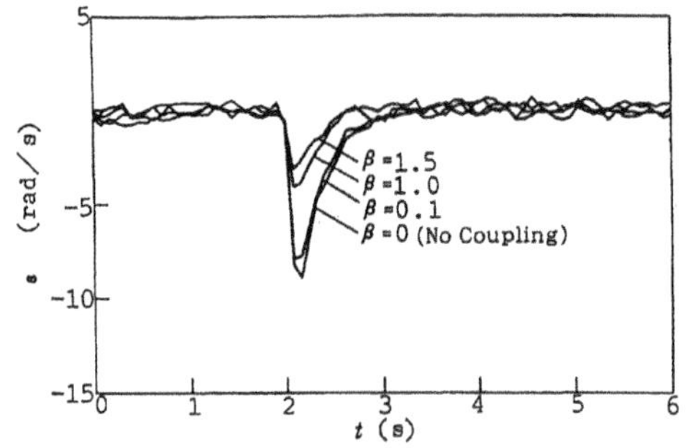

Fig.14. Effect of Coupling Parameter (β) on Synchronization Error (γ=10).

CONCLUSION

An adaptive feedforward control scheme for two-axes synchronizing control system is tested. First, independent adaptive feedforward control approach, in which the adaptive controller is used in each axis independently, is investigated. Next, the coupling controller, which results in faster removal of the synchronization error, is added to the adaptive feedforward controllers for the two axes. In this adaptive synchronizing feedforward control approach, the synchronization error drives the parameter vectors in opposite direction through the coupling controller. Experimental results demonstrate the effectiveness of the coupling controller in synchronization.

REFERENCES

Koren,Y.,(1980), Cross-Coupled Biaxial Computer Controls for Manufacturing Systems, ASME Journal of Dynamic Systems, Measurement and Control, Vol.102, pp.265-272,1980.

Tomizuka,M.,(1989), Design of Digital Tracking Controllers for Manufacturing Applications, Manufacturing Review, Vol.2,No.2,June 1989, pp.134-141 (ASME)

Uchiyama,M. and Nakamura,Y.,(1988), Symmetric Hybrid Position/Force Control of Two Cooperating Robot Manipulators, Proceedings of 1988 IEEE International Workshop on Intelligent Robots and Systems,pp.515-520, Nov.1988.

Kamano,T.,Suzuki,T.,Kuzuhara,T. and Tomizuka,M. (1990), Feedback Plus Feedforward Control for Positioning System with Progressive Wave Type Ultrasonic Motor, Proceeding of 1990 Japan-U.S.A. Symposium on Flexible Automation, Vol.II. pp.601-607, July 1990

Tomizuka,M.,Hu,J.s.,Chiu,T.C. and Kamano,T.,(1990), Synchronization of Two Motion Control Axes Under Adaptive Feedforward Control, to be presented at the 1990 ASME Winter Annual Meeting, Nov.1990

INTELLIGENT TUNING AND ADAPTIVE CONTROL FOR CEMENT RAW MEAL BLENDING PROCESS

P. Lin*, Y. S. Yun, J. P Barbier*, Ph. Babey* and P. Prevot****

* *Lafarge Coppée Recherche. BP 8, 07220 Viviers sur Rhone, France*
***Laboratoire d'Informatique Appliquee, INSA de Lyon, 20 A.V.A. Einstein, 69621 Villeurbanne, France*

ABSTRACT : The main goal of the raw meal blending control in cement industry is to maintain near the reference values and to decrease the variation of the chemical composition rejecting the disturbances. The raw materials blending process is a complex system. In this paper, a two level adaptive control policy combined with an heuristic auxiliary system is proposed for ensuring the robustness of the control system. Based upon the simulation results, two applications have been developped in Ciments Lafarge.

KEYWORDS : Adaptive control, intelligent adjustment, cement industry, raw mill.

1 - INTRODUCTION

In the manufacturing process of cement, the main raw material is crushed in the quarry and usually pre-homogenized in large stockpiles. The processed raw mill is to grind and blend the raw material with other products, and then feed to the kiln. A raw meal with a good fineness and well controlled chemical composition can improve the cement quality and the kiln operation performance. On the other hand, a good raw materials mixing can also reduce the raw meal cost, because some raw materials used for correction of the composition, such as bauxite, are very expensive. The raw materials blending process is a multivariable system (several raw materials are used) and a coupled one, as the feeder tanks do not contain chemically homogeneous raw materials. The time delay in the system are considerable in spite of X-ray fluorescence technique applied for analysing the chemical composition of the raw meal.

The main purpose of the raw meal quality control is to maintain closer to the standard set values and to decrease the variation of the chemical composition (in terms of composition moduli, as lime deficiency, silica modulus and aluminium modulus) of the raw meal by using some raw materials called "correction products". There are strong disturbances in this process. The first is the change in the main raw material composition because of geographical site changes in the quarry. The second is due to sampling and analysis techniques limits. So, in most cement plants there are only (even there are no) the static control systems for the composition control. With developing of advanced control theory and computer technology some successful dynamic control strategis have been reported (1) (2) (3). Furthermore, the problem of estimating the raw materials concentrations has been discussed by Gelb in 1974, and an on-line applicaton was carried out in 1982 (4). In this paper a new improved control policy which is verified in simulations and in experiments is presented. Based upon the simulation results, two applications have been developped in Ciments Lafarge.

2 - RAW MEAL BLENDING CONTROL PROBLEM

A simple schema of the blending control is shown in fig. 1. The raw meal which is made up of m raw materials is sampled before the silo. The oxides compositions obtained by a X-ray fluorescence analyser (RFA) are provided for the computer which calculates the new values to give to each feed flow with regard to a fixed total feed flow. There are four most important oxides: S(SiO2), A(Al2O3), F(Fe2O3) and C(CaO).

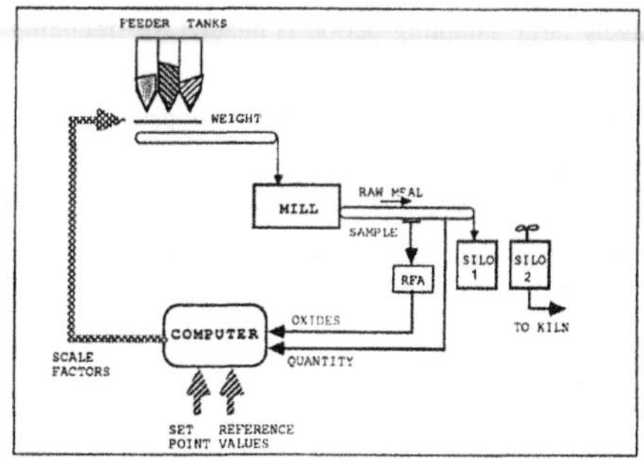

Fig.1. Simple schma of blending control

We suppose that the concentrations of the m raw materials are constant during the sampling period. The oxide compositions of the raw meal in the mill output can be expressed by:

$$\begin{pmatrix} S(t) \\ A(t) \\ F(t) \\ C(t) \end{pmatrix} = \begin{pmatrix} S_1 & S_2 & \dots & S_m \\ A_1 & A_2 & \dots & A_m \\ F_1 & F_2 & \dots & F_m \\ C_1 & C_2 & \dots & C_m \end{pmatrix} \begin{pmatrix} g_1(t) & & 0 \\ & g_2(t) & \\ & \dots & \\ 0 & & g_m(t) \end{pmatrix} \begin{pmatrix} u_1(t) \\ u_2(t) \\ \dots \\ u_m(t) \end{pmatrix} \quad (1)$$

where

S(t),A(t),F(t),C(t)	:	concentration after milling;
Si, Ai, Fi, Ci	:	concentration of the raw material i;
gi(t)	:	dynamic proprety of the raw material i passing through the mill;
ui(t)	:	weight percentage of the raw material i.

We notice that gi(t) are influenced by the physical characteristics of the raw materials and the thermal proprety of the mill. So, gi(t) are usually time-varing.

The purpose of the control is to maintain relative rates of these oxides towards their reference values. There are three specific moduli for raw meal in cement industry:

$$\text{Lime deficiency} \quad : \quad \Delta BC = \frac{280S+165A+35F-100C}{S+A+F+C}$$

$$\text{Silica modulus} \quad : \quad MS = \frac{S}{A+F} \qquad (2)$$

$$\text{Aluminium modulus} : \quad AF = \frac{A}{F}$$

The relation (2) shows the nonlinearity of the control problem.

Taking into account the disturbances to the composition of the main raw material from quarries (low frequency part), an on-line estimation is necessary. Otherwise, because of the presence of important time dalay between sampling and value knowledge (20-30mn) and the technical limits on the sampler, a relatively large sampling period is needed for reducing the measurement noise.

With the presence of a silo for homogeneizing the raw meal, the final goal of the control is to let average moduli values in the whole silo content reach desired references values, but these average moduli values in the batch silo are immeasurable.

3 - CONTROL POLICY

According to the previous description two integrated control systems are proposed:

a) static control system for ill-identified process;
b) dynamic control system for relatively stable process.

The two systems have the same structure.

Let T1 be sampling period of RFA and T2 be sampling period of raw materials feeds. We have T2 >> T1.The difference between the two control systems is that the first takes an approximate estimation for raw meal composition(every T2), the second takes a multivariable long-range prediction based upon the process model and the raw materials feeds (every T1).

The principle of the static control system has been presented in (5) (6) (7). The proposed dynamic control system contains (see Fig.2):

1). a multivariable long-range prediction;
2). a control algorithm for nonlinear optimization;
3). an estimation and adaption of the raw materials composition;
4). a self-adjustment of the reference modulus values.

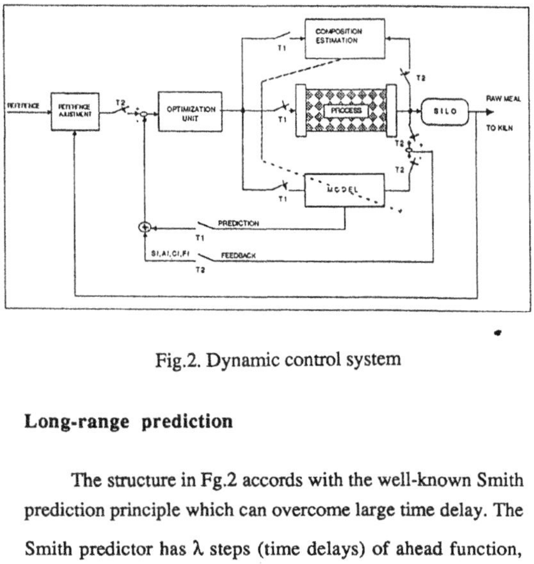

Fig.2. Dynamic control system

Long-range prediction

The structure in Fg.2 accords with the well-known Smith prediction principle which can overcome large time delay. The Smith predictor has λ steps (time delays) of ahead function, and so-called long-range predictor looks ahead for λ + P steps, which has better robustness(8). Here P is prediction horizon.

For the dynamics of the mill, the first ordre time lag approximation with delay is proved to be good in most pratical cases, i.e.

$$g_i(Z) = \frac{b'_i Z^{-\lambda_i - 1}}{1 + a_i Z^{-1}} \qquad (3)$$

so the predictive equation is derived directly from (1) and (3). Only S is taken as an example (idem for A, F, C).

At first the same time delay $\lambda = \lambda i$ ($i = 1,2,\ldots m$) is supposed. From (1), (3) and with $bi = bi'Si$, we have

$$S(k) = \sum_{i=1}^{m} y_i(k) = \sum_{i=1}^{m} [a_i y_i(k-1) + b_i u_i(k-\lambda-1)] \tag{4}$$

For the component i, λ steps of prediction are calculated by using the possessive values,

$$\tilde{y}_i(k+\lambda) = a_i \tilde{y}_i(k+\lambda-1) + b_i u_i(k-1) \tag{5}$$

As long-range prediction, more P steps predictive equations are :

$$\begin{aligned} \tilde{y}_i(k+\lambda+1) &= a_i \tilde{y}_i(k+\lambda) + b_i u_i(k) \\ \tilde{y}_i(k+\lambda+2) &= a_i^2 \tilde{y}_i(k+\lambda) + a_i b_i u_i(k) + b_i u_i(k+1) \quad (6) \\ &\ldots\ldots \\ \tilde{y}_i(k+\lambda+P) &= a_i^P \tilde{y}_i(k+\lambda) + a_i^{P-1} b_i u_i(k) + \ldots + b_i u_i(k+P-1) \end{aligned}$$

The single step control horizon is used[9], i.e.

let $ui(k) = ui(k+1) = \ldots = ui(k+P-1)$

and (6) is rewritted in vector form:

$$\tilde{y}_i = x_i u_i(k) + w_i \tilde{y}_i(k+\lambda)$$

where,

$$\begin{aligned} \tilde{y}_i &= [\tilde{y}_i(k+\lambda+1), \tilde{y}_i(k+\lambda+2), \ldots, \tilde{y}_i(k+\lambda+P)]^T \\ x_i &= [1, 1+a_i, \ldots, 1+a_i+\ldots+a_i^{P-1}]^T b_i \quad (7) \\ w_i &= [a_i, a_i^2, \ldots, a_i^P]^T \end{aligned}$$

Notice that in the right side of (7), the first term represents the contribution of future control, and the second is related to history.

The prediction of the total oxide S can be obtained from (4) and (7), and with additional feed back term:

$$\tilde{S} = X_s u + W_s \tilde{y}_s + S_f \tag{8}$$

where

$$\begin{aligned} \tilde{S} &= [\tilde{S}(k+\lambda+1), \tilde{S}(k+\lambda+2), \ldots, \tilde{S}(k+\lambda+P)]^T \\ X_s &= [x_1 \; x_2 \; \ldots \; x_m] \\ u &= [u_1(k), u_2(k), \ldots, u_m(k)]^T \\ W_s &= [w_1 \; w_2 \; \ldots \; w_m] \\ \tilde{y}_s &= [\tilde{y}_1(k+\lambda), \tilde{y}_2(k+\lambda), \ldots, \tilde{y}_m(k+\lambda)]^T \\ S_f &= [S_f(k), S_f(k), \ldots, S_f(k)]^T \end{aligned}$$

with

$$S_f(k) = S(k) - \tilde{S}(k) \tag{9}$$

$S(k)$: real analytic value (S_iO_2).

If there are different time delays, following treatment is necessary:

Let $\lambda = \min\{\lambda 1, \lambda 2, \ldots \lambda m\}$

$\Delta i = \lambda i - 1$

then $xi = [0,\ldots, 0, 1, 1+ai, \ldots, 1+ai+\ldots+ai^{P-\Delta i-1}]^T bi$

and an additional history term Xs'U' will be presented in (8)

with $Xs' = [x1' \; x2' \ldots xm']$
$U' = [u1' \; u2' \ldots um']^T$

where

$$x'_i = \begin{bmatrix} 1 & & \\ a_i & 1 & 0 \\ & a_i & \\ \ldots & & \\ & & 1 \\ a_i^{P-1} & a_i^{P-2} \ldots & a_i^{P-\Delta_i} \end{bmatrix}_{p \times \Delta_i} \cdot b_i$$

$$u'_i = [u_i(k-\Delta_i), u_i(k-\Delta_i+1), \ldots u_i(k-1)]^T_{1 \times \Delta_i}$$

Optimization problem

In order to formulate an optimization problem, we introduce a quality-cost criterion function to minimize :

$$\begin{aligned} \Phi = \min\{ & K1(\Delta BC-\Delta BC0)T(\Delta BC-\Delta BC0) + \\ & K2(MS-MSO)T(MS-MSO) + \\ & K3(AF-AF0)T(AF-AF0) + K4UThU\} \end{aligned} \tag{10}$$

where

$\Delta BC0 = [\Delta BC0, \Delta BC0, \ldots, \Delta BC0]^T$: reference modules (idem for MS0 and AF0);

$h = \text{diag}\{h1^2, h2^2, \ldots, hm^2\}$: cost vector;

K1,K2,K3,K4 : weight coefficients.

ΔBC, MS and AF are to be calculated from S, A, F and C by using (2).

The subjected constraints are:

$\sum ui(k) = 1$, for $i = 1,m$;

$ui\ min \leq ui(k) \leq ui\ max$.

Considering the nonlinearity in (10) and these conditions, a gradient searching method based on a genaralization of Davidon's algorithm is used for optimization.

Estimation of composition

The major disturbances to the process come from the variations in compositions of the raw materials. These variations can be estimated by using real ui(t) and the composition values in the raw meal obtained at interval of T2 (T2 = LT1).

Let the calculation begin at k, and the initial values on static relation between input and output of the mill substitute in (4)

$$S(k)=\sum_{i=1}^{m} y_i(k)=\sum_{i=1}^{m} u_i(k-\lambda_i-1)S_i \quad (14)$$

define $\quad Vi(k) = ui(k-\lambda i-1) \quad$ (15)

then

$$S(k+jL)=\sum_{i=1}^{m} V_i(k+jL)S_i \qquad (j=1,2,\ldots n) \quad (16)$$

where a recursive formula can be derived:

$$V_i(k+jL)= a_i^L V_i[k+(j-1)L]+ a_i^{L-1}b'_i u_i[k+(j-1)L-\lambda_i]+ a_i^{L-2}b'_i u_i[k+(j-1)L-\lambda_i+1]+\ldots+ b'_i u_i[k+jL-\lambda_i-1] \quad (17)$$

Thus the estimation equation is

$$\begin{bmatrix} S(k) \\ S(k+L) \\ \ldots \\ S(k+nL) \end{bmatrix} = \begin{bmatrix} V_1(k) & V_2(k) & \ldots & V_m(k) \\ V_1(k+L) & V_2(k+L) & \ldots & V_m(k+L) \\ & \ldots\ldots & & \\ V_1(k+nL) & V_2(k+nL) & \ldots & V_m(k+nL) \end{bmatrix} \begin{bmatrix} \hat{S}_1 \\ \hat{S}_2 \\ \ldots \\ \hat{S}_m \end{bmatrix} \quad (18)$$

The recursive least square estimation algorithm with a forgetting factor is used for this system. An explicit adaptive control loop is now formed by using the estimated values in the predictive equations.

Self-adjustment of reference values

The average composition values in the silo content which are immeasurable can be obtained by calculating:

$$\bar{S}(k')=\frac{\sum_{i=1}^{k'} M_i S(i)}{\sum_{i=1}^{k'} M_i} \qquad (k'=1,2,\ldots k_o) \quad (19)$$

where

- $S(k')$: average composition value in the silo content at k'-th sampling instant (T2);
- Mi : raw meal quality during the i-th sampling interval;
- k0 : sampling number with filling silo.

As the total flow of raw materials is quasi-constant, there are almost the same Mi = M (i = 1,2,.. k0). The equation (19) can be simplified :

$$\bar{S}(k')=\frac{\sum_{i=1}^{k'} S(i)}{k'} \quad (20)$$

A general adjustment equation is:

$$k'M\bar{S}(k')+R(k')MS_a = [k'+R(k')]\, MS_o \quad (21)$$

$$S_a = S_o + \frac{k'}{R(k')}[S_o-\bar{S}(k')] \quad (22)$$

where

- S0 : reference composition value;
- Sa : adjusted reference composition value;
- R(k') : adjustment time (time-varing).

The (21) is an oxide quantity balance equation. The first term of the left side represents the oxide quantity in silo at the k'-th instant and the second is the future one with Sa composition value into the silo during R(k') interval; the right side represents the total oxide quantity in the future k' + R(k') instant when the average oxide composition values have to be equal to the desired ones. So the adjusted reference moduli values during k'+1 sampling interval can be obtained with Sa (and Aa, Fa, Ca). It is obvious that when the silo is filled, in order to reach the reference values, the following condition is required: when k' = k0-1, R(k') = 1.

The adjustment policy in (1) is:

let R(k') = k0 - k' (23)

hence

$$S_a = S_o + \frac{k'}{k_o-k'}[S_o-\bar{S}(k')] \quad (24)$$

The disadvantage of this policy is that the initial adjusted quantity is too small and the last action is usually very violent because of a large k0 in general.

An improved policy is proposed here:

$$\text{let } R(k') = r - \frac{r-1}{k_o-2}(k'-1) \quad (25)$$

where r is a selective parameter.

Equ.(25) satisfies the required condition, but the initial adjusted point is R(1) = r. We can now easilly change r according to the silos management. Obviously, if r = k0-1, then (25) reduces to (23).

4 - HEURISTIC SUPERVISION

We have noticed that K1, K2, K3 and K4 represent the importance of ΔBC, MS, AF and raw meal cost. Usually they are consided as constant value parameters by user. If we build a three dimensional space with three axes ΔBC(t), MS(t) and AF(t), we have a cluster of points for process transition in terms of time. This cluster surrounds the reference point, and each point represents the situation at a fixed moment.

Simulation results show that with certain disturbances, it is preferable to change the relative importance of ΔBC, MS, AF and the cost, according to the real position in the space, in order to find a compromise among the moduli and to give more freedom to the operator who chooses the values.

In Lin (1988)[5], an heuristic look-up table is established from operator's experiments to adjust K1,K2, K3 and K4 in real time.

This heuristic auxiliary supervision contains other functions for ensuring the robustness of the control system:

a). test of the results of the composition estimation

Before application in the predictive equations, the results of the composition estimation should be tested and validated.

b). choice of the optimization method

In fact we have three methods in our control system[7].The first minimizes only the meal cost by subjecting to inequality constraints on chemical composition. The second is the presented gradient method. The last is an experimental method (simplified from the second one) by which the solution space is limited primarily to reduce the searching operation computational time. The choice of the method depends on the number of raw materials in use and the performance of the computer.

c). choice of the initial point for the algorithm of optimization

The choice of initial point is very important for optimization problem. A good initial point can facilitate the searching operation. This choice is taken by the heuristic system, according to the behaviour of the feeders.

5 - SIMULATION RESULTS

Before on-line industrial applications, the proposed control system is verified by a simulation example whose data come from a plant of Ciments Lafarge. There are four raw materials: quarry material (main raw material), silica, bauxite and pyrite. Their transfer functions through the mill can be written respectively as:

$$\frac{e^{-11s}}{7.32s+1}, \quad \frac{e^{-16s}}{5.67s+1}, \quad \frac{e^{-16s}}{5.67s+1}, \quad \frac{e^{-16s}}{6.85s+1}$$

The sampling and analysis time delay is about 30 mn. The sampling period T2 is 20 mn, at least. A control sampling period T1 = 2 mn is chosen. The reference values of the three moduli are:

$\Delta BC0 = -2.00$, $MS0 = 5.00$, $AF0 = 1.00$.

We choose main tuning parameters as following:

weight coefficients :

$K1 = 50; K2 = 10; K3 = 40; K4 = 1;$

prediction horizon : $P = 5;$

adjustment coefficient : $r = 6.$

When a disturbance of sine form with additional stochastic noise in introduced to S of the main raw material, the simulation results are shown in Fig.3. In comparaison with the results obtained by the static control startegy [6] shown in Fig.4. the results of the dynamic control strategy proposed in the paper are obviously better.

The results without estimation or without reference adjustment are respectively presented in Fig5 and 6. where the responses of ΔBC are given because of its sensitibity to disturbance and its importance.

Finally, the adjustment policy of (24) is used (Fig. 7). We can see that the initial small adjustment not only gives a sluggish response, but also results in the last control action larger adjusted quantity. The efficiency of the improved policy (Fig.3) is obvious.

6 - EXPERIMENTS IN CEMENT PLANTS

The biggest plant in Ciments Lafarge is Le Havre plant. With two steps raw mill process and eight raw materials in use, the control problem become very difficult. In addition, the process dynamics are time-varing. The proposed static control system has been applied to this plant. The softwares are implemented in IBM PC connected with RFA and Modcomp system as process computer. We have developed an heuristic auxiliary system to detect the process functionalities and to carry out the associated treatment procedures presented in section 4.

The Fig. 8 presents the regularity of raw meal during one week. We have observed a strong disturbance due to the change of stockpiles (sampling 34) and a change of the reference point (sampling 64).

The second application is designed to another cement plant: le Teil plant. The softwares are now implemented in a microVax computer connected with RFA and Modcomp system.

CONCLUSION

In the present paper we have designed an integrated control system for raw meal blending process in cement industry. The simulation results and some industrial applications have shown that this control policy is successful.

Because the single step control horizon is used, the derived algorithm is simple and is easily implemented. The explicit adaption schema gives flexibility to the estimation unit. In fact for our example only the composition of the main raw material was estimated, as the main raw material has a big flow (more than 90% of the total flow). About optimization, we prefer gradient method, which can not only solve the nonlinear problem, but also treat various constraints needed in the production. It is also for the last reason that an intelligent adjustment of the reference values is very important if we want to improve the control system performance.

Heuristic supervision plays an important role between automatic control theory and artificiel intelligence. Based upon the operator's experiments, an heuristic auxiliary system can increase the robustnesse of a control system.

REFERENCES

(1). Csaki F.,L.K.Keviczky and al. (1978). Simultaneous adaptive control of chemical composition, fineness and maximum quantity of ground materials at a closed circuit ball mill. 7th IFAC Congress, Helsinki.

(2). Westurlund K.T. (1983). Real-time optimization of raw materials in the cement industry. IFAC Automatica in MMM processing, Helsinki.

(3). Morant F. and P.Albertos (1988). Quality control of raw materials blending in the cement industry. 5th IFAC symposium on identification and system parameter estimation, Beijing.

(4).Mont Hubbard and Tom Dasilva (1982). Estimation of feedstream concentrations in cement raw materials blending. Automatica, Vol.18, pp595-606.

(5).Lin P., J.P.Barbier and P.Prevot (1988). Computer aided design and control for a raw mill in cement manufactory. IFAC CADCS'88, Beijing.

(6).Lin P. and al. (1989). Two-level adaptive control system for a cement raw materials mixing process. IFAC MMM'89, Buenos aires.

(7).Lin P. and al. (1988). Commande adaptative pour la correction de composition d'un broyeur de cru en cimenterie. Revue RAPA. Vol.1, N°3/1988.

(8).Cutler C.R. and B.L.Ramaker (1980). Dynamic matrix control: A computer control algorithm. J.A.C.C. preprints.

(9).Clarke D.W. and al. (1987). Generalized predictive control Part 1. The basic algorithm. Automatica, Vol.23, N°.2.

(10).Yun.Y.S.,Lin P. and al. (1990). Adaptive and predictive control of the raw meal composition in cement industry. IASTED CONTROL'90, Lugano.

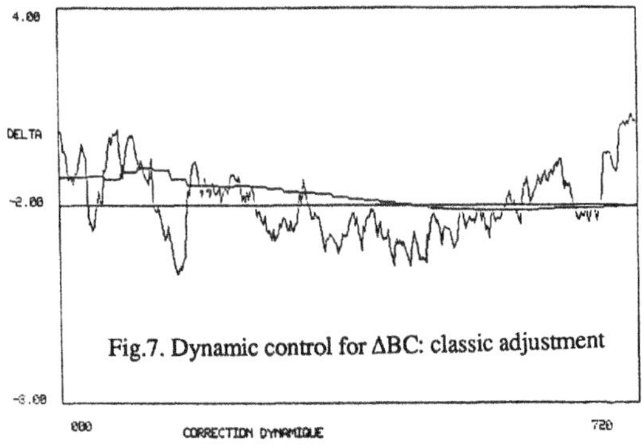

Fig.7. Dynamic control for ΔBC: classic adjustment

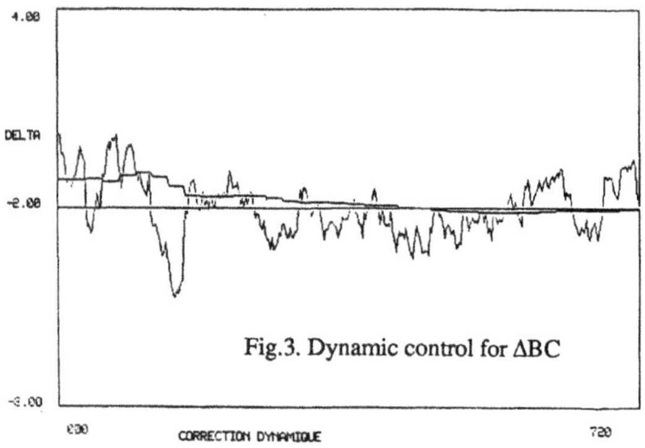

Fig.3. Dynamic control for ΔBC

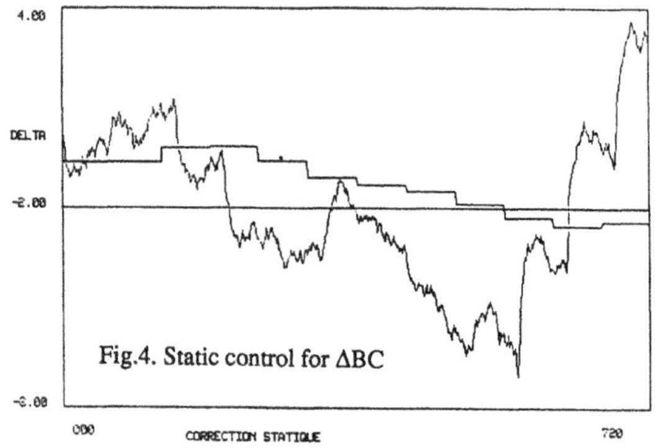

Fig.4. Static control for ΔBC

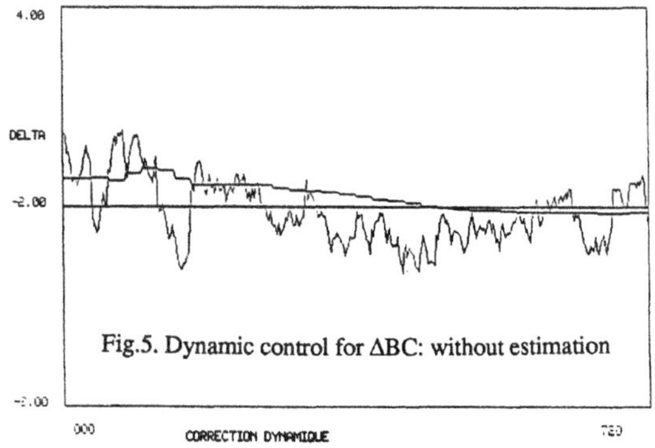

Fig.5. Dynamic control for ΔBC: without estimation

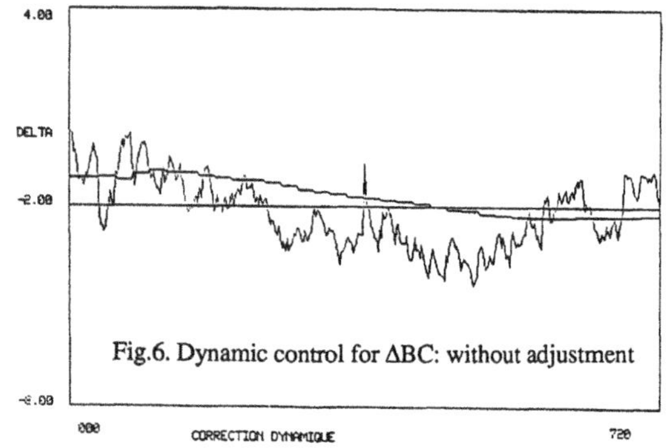

Fig.6. Dynamic control for ΔBC: without adjustment

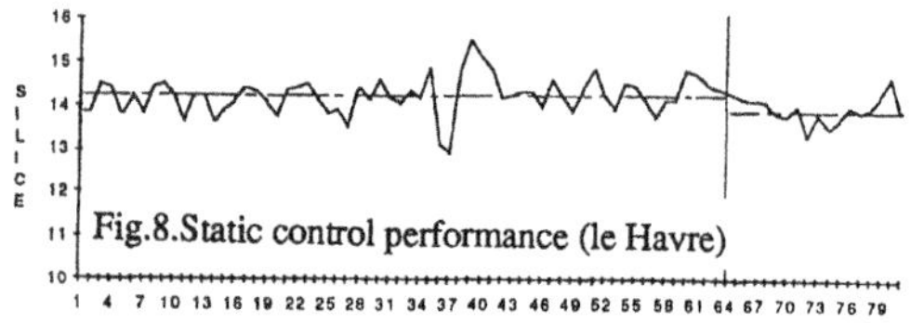

Fig.8.Static control performance (le Havre)

DYNAMIC POSITIONING SYSTEM USING SELF-TUNING CONTROL

E. A. Barros and H. M. Morishita

Dept. of Marine Engineering and Naval Architecture,
University of Sao Paulo, Caixa Postal 61548, CEP 05508, Sao Paulo, Brazil

Abstract. This paper deals with the application of a generalized minimum variance self-tuning controller to the Dynamic Positioning System (DPS) of a semi-submersible rig. In the control design the sway, surge and yaw have been considered as independent motions. There is no explicit filtering process to the high frequency motions since the rig itself may be treated as a low pass filter because it is designed to be hydrodynamically transparent. The control performance is analyzed by computer simulation, verifying influence of direction and magnitude of the wave and current. The results reveal that the self-tuning DPS keeps position and heading of rig within the accuracy requirements in spite of an existence of a small off-set.

Keywords. Adaptive control; Marine systems; minimum variance control; self-tuning regulators; dynamic positioning system.

INTRODUCTION

A dynamic positioning system (DPS) is used to maintain position and heading of an offshore industry floating vessel, mainly in deep waters where, often, it is impractical to use conventional mooring system. A DPS is designed to counteract forces and moments imposed on the vessel by wind, current and waves. It is based on the action of a set of thrusters governed by a control algorithm. It controls only three of the six modes of motion, i.e., surge, sway and yaw.
In the DPS design, the vessel motions are usually divided in two parts, the high and low frequency motions. The high ones are induced by waves, the low ones are determined by characteristics of the DPS together with the wind, wavedrift and current. The high frequency motions are oscillatory in nature.So they must be filtered away to avoid wear and tear of the thruster mechanism and to save energy.

The first generation of DPS is based on conventional PID controller with notch filter. In this case, the wave filter imposes a phase lag on the position error signal that usually restricts the allowable bandwidth of the controller. This means that the tuning of the control gains is a compromise between the effectiveness of a high frequency motion filtering and the stability margins required for satisfactory controller performance. These considerations have led to the development of a second generation of DPS based on linear quadratic theory. This generation of DPS employs Extended Kalman Filter to estimate states and parameters. But performance of EKF depends on mathematical model of the system that needs be simple to avoid excessive computational load. Balchen et alii(1980) assumed that high frequency motions were purely oscillatory and could be modeled by a second order sinusoidal oscillator with variable center frequency. Kallstrom (1983) has taken this model but has kept the frequency constant. Grimble, Patton and Wise(1980) used a fourth order wave model to high frequency motions with fixed low frequency filter gain. Fung and Grimble(1983) propose a self-tuning Kalman Filter decoupling low and high frequency motions.

A theoretical development of a semi-submersible rig model is difficult because it involves modelling of wind, wave and current action and of hydrodynamic interaction between the rig and sea water. These considerations suggest the verification of the performance of a DPS with adaptive control assuming the rig as an unknown parameter system with stochastic perturbations. This paper deals with the application of a generalized minimum variance self-tuning controller to the DPS of a semisubmersible rig. The system is described by a stochastic autoregressive moving average model with unknown coefficients. The control law is obtained by the minimization of a cost function in which system output and control effort are weighted.The application of the self-tuning controller requires the estimation of the model parameters. In this paper the implicit way is used to estimate directly the parameters of the controller by means of the recursive least-squares algorithm. A rig is designed to be hydrodynamically transparent and so it may be treated as a low pass filter.

Taking this characteristic into account the design of a DPS without explicit filtering process to the high frequency motions is attempted.

The development of the controller requires a mathematical model of the rig. The model adopted was developed by using an auto-regressive moving average for high frequency motions. For the low frequency motions maneuvering equations were used. Current and second order wave forces are described through semi empirical relationships. This paper does not emphasize wind action because the usual feedforward control to compensate the wind force has not been included. The performance of the DPS is analyzed by computer simulations with data of GVA 4000 rig, verifying influence of a direction and magnitude of the wave and current.

SELF-TUNING CONTROLLER

The generalized minimum variance controller proposed by Clark and Gawthrop(1975) is considered. As this theory is well-known only the essentials of the method are outlined here. Consider a system described by a linear difference equation:

$$A(z^{-1})y(k) = B(z^{-1})u(k-d) + C(z^{-1})v(k) + D \quad (1)$$

where y is the output; u is control signal; D is a constant; v is an uncorrelated random sequence of zero mean disturbing the system; d is the time delay; z^{-1} is the backward-shift operator, and $A(z^{-1})$, $B(z^{-1})$ and $C(z^{-1})$ are polynomials given by:

$$A(z^{-1}) = 1 + a_1 z^{-1} + ... + a_{na} z^{-na}$$

$$B(z^{-1}) = b_0 + b_1 z^{-1} + ... + b_{nb} z^{-nb}$$

$$C(z^{-1}) = 1 + c_1 z^{-1} + ... + c_{nc} z^{-nc}$$

where $b_0 \neq 0$; na, nb and nc are polynomial orders and the roots of $C(z^{-1})$ lie within the unit circle. The control law is derived by minimization of the following cost function:

$$I = E\ \{[P(z^{-1})y(k+d) - R(z^{-1})w(k)]^2 - [Q'(z^{-1})u(k)]^2\} \quad (2)$$

where w is the reference and $P(z^{-1})$, $R(z^{-1})$ and $Q(z^{-1})$ are weighing polynomials; E is the expectation operator. However, the term y(k+d) makes deduction of control law from Eq. (2) difficult. This problem may be overcome by defining the optimal predictor $y^*(k+d \mid k)$ of y(k+d) as:

$$y^*(k+j \mid k) = y(k+j) - e(k+j) \quad (3)$$

where e(k+j) is the prediction error. The optimal predictor can be obtained recursively as:

$$Cy^*(k+j \mid k) = F'_j y(k) + E_j Bu(k+j-d) + E_j(1)D \quad (4)$$

for $j \leq d$ and where $F'_j(z^{-1})$ and $E_j(z^{-1})$ are na-1 and j-1 order polynomials and their coefficients are determined by:

$$C(z^{-1}) = E_j(z^{-1})A(z^{-1}) + z^{-j}F'_j(z^{-1})$$

Now, substituing Eq. (3) into Eq. (2) and making $\partial I / \partial u(k) = 0$, it follows that the control law is given by:

$$\phi^*(k+d \mid k) = Py^*(k+d \mid k) + Qu(k) - Rw(k) = 0 \quad (5)$$

where $Q = Q' q_0 / b_0$. Considering Eq. (4) with Eq. (5) it follows:

$$C\phi^*(k+j \mid k) = Fy(k) + Gu(k) + Hw(k) + \delta \quad (6)$$

where

$$F(z^{-1}) = \Sigma\ p_j F_{d-j}(z^{-1})$$

$$G(z^{-1}) = \Sigma\ p_j z^{-j} E_{d-j}(z^{-1}) B(z^{-1}) + C(z^{-1})Q(z^{-1})$$

$$H(z^{-1}) = -C(z^{-1})R(z^{-1})$$

$$\delta = p_j E_j(1) D$$

As the control law is given by setting $^*(k+d \mid k)$ to zero at each stage it can be also obtained by:

$$Fy(k) + Gu(k) + Hw(k) + \delta = 0 \quad (7)$$

As the self-tuning control theory is developed to unknown systems it is necessary to adopt some procedure to get the control law parameters. Here the implicit approach is considered for estimating directly the coefficients of $F(z^{-1})$, $G(z^{-1})$, $H(z^{-1})$ and δ . Defining ϕ (k+d) as:

$$\phi(k+d) = Py(k+d) + Qu(k) - Rw(k)$$

and making use of Eq. (3), it follows that

$$\phi(k+d) = \phi^*(k+d \mid k) + \varepsilon(k+d) \quad (8)$$

where $\varepsilon(k+d) = \sum_{i=0}^{d-1} p_i e(k+d-i)$

Clarke and Gawthrop(1975) show that from Eq. (5) and Eq. (8) a recursive least squares algorithm can be obtained to get $\hat{F}(z^{-1})$,$\hat{G}(z^{-1})$, $\hat{H}(z^{-1})$ and $\hat{\delta}$ that are polynomials with estimated values of the coefficients of $F(z^{-1})$, $G(z^{-1})$, $H(z^{-1})$ and δ. The control law is then given by:

$$\hat{F}y(k) + \hat{G}u(k) + \hat{H}w(k) + \hat{\delta} = 0 \quad (9)$$

MODEL OF SEMI- SUBMERSIBLE RIG

The four-column semi-submersible drilling rig GVA-4000 was selected to verify the performance of adaptive DPS. In order to develop the mathematical model of the rig it is assumed that total rig motions are the sum of high and low frequency motions. Three different coordinate systems, as is shown in Fig. 1, are used to determine the dynamic equations of the rig. The first one, xyz, is fixed to the rig, and its origin is at the center of gravity of the rig. The coordinate system X0Y0 is fixed at the earth. This system is used for the integration of the low frequency equations and it defines wave, wind and current directions. The third coordinate system, Xhf-YhfZhf is fixed at the position of the centre of gravity in the low frequency motion. The axes Xhf, Yhf and Zhf are parallel to the axes XO, YO and z respectively.

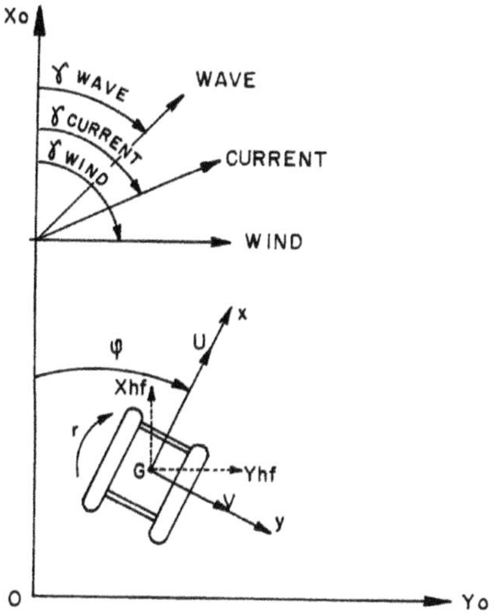

Fig. 1. The Coordinate System

High frequency motions

The high frequency motions are due to oscillatory motions of the waves and in this paper they are obtained by an auto regressive moving average model (ARMA) fed by the wave elevation for each degree of freedom. For instance, the high frequency sway motion, y_{hf} is given by:

$$y_{hf}(k) + ah_1 y_{hf}(k-1) + \ldots\ldots + ah_n y_{hf}(k-n) =$$

$$\xi(k) + bh_1\,\xi(k-1) + \ldots + bh_n\,\xi(k-n)$$

where ξ is the wave elevation. Kallstrom(1983) determined the values of coefficients and order n from scale model tests data. The parameters ah_i and bh_i were estimated with the maximum likelihood identification methods and Akaike's information criterion was applied in order to determine a suitable model order, that is four to sway, surge and yaw motions.

Wave elevation

A stochastic realization of the wave elevation may be obtained through a shaping filter fed by white noise. This filter is chosen to minimize the error between the true sea spectrum and the approximate spectrum being generated. Thus the output of the filter is the wave elevation with the desired significant wave height and peak period. Here the filter determined by Kallstrom(1983) is used that is expressed by an auto-regressive moving average model (ARMA):

$$\xi(k) + aw_1\xi(k-1) + \ldots\ldots aw_n\xi(k-n) =$$

$$v(k) + cw_1 v(k-1) \ldots\ldots + cw_n v(k-n)$$

where v is a zero mean white noise signal. The order of the filter is also four (Kallstrom,1983).

Low Frequency Motions

An ideal-fluid maneuvering model in the horizontal plane is proposed in order to represent the low frequency motions (Norrbin,1971):

$$\left(1-X_{\dot{u}}\right)\dot{u} = \left(1-Y_{\dot{v}}\right)vr + \left(Y_{\dot{v}}-X_{\dot{u}}\right)v_c r + X/m$$

$$\left(1-Y_{\dot{v}}\right)\dot{v} = \left(X_{\dot{u}}-1\right)ur + \left(Y_{\dot{v}}-X_{\dot{u}}\right)u_c r + Y/m \qquad (10)$$

$$L\left(k^2_{zz}-N_{\dot{r}}\right)\dot{r} = \left(1/L\right)\left(Y_{\dot{v}}-X_{\dot{u}}\right)\left(u-u_c\right)\left(v-v_c\right) + N/mL$$

where $X_{\dot{u}}$, $Y_{\dot{v}}$ and $N_{\dot{r}}$ are the added mass and moment in surge, sway and yaw, respectively, normalized by the use of the "bis" system (Norrbin,1971); u is the surge velocity, v is the sway velocity and r is the yaw rate; L is the length of the rig;m is the mass of the rig; k_{zz} is the radius of gyration around z axis; u_c and v_c are the components in the x and y direction of a constant homogeneous current and they are related with the earth fixed coordinate by:

$$u_c = u_{co}\sin\varphi + v_{co}\cos\varphi$$

$$v_c = v_{co}\cos\varphi - u_{co}\sin\varphi$$

where u_{co} and v_{co} are the components of the current in XO and YO directions respectively and φ is the heading of the rig. The forces X and Y and the moment N consist of three different components:

$$X = X_{visc} + X_{wdrift} + X_{thrust}$$

$$Y = Y_{visc} + Y_{wdrift} + Y_{thrust}$$

$$N = N_{visc} + N_{thrust}$$

where the visc, wdrift and thrust subscripts mean viscous, wave drift and thurst forces respectively. Low frequency motions also depend on wind efforts that are omitted here since their influence is not analyzed in this paper.

Viscous Forces. The viscous forces of GVA-4000 with a constant homogeneous current have been determined by scale model tests (Kallstrom,1983):

$$X_{visc}/m = C_{dx}V_r^2\cos\gamma r$$

$$Y_{visc}/m = C_{dy}V_r^2\sin\gamma_r$$

$$N_{visc}/mL = -0.000946V_r^2\sin(2\gamma_r) - 3.75r\,|r|$$

where V_r is the current speed relative to the rig and γ_r is the current direction relative to the rig. X_{visc} and Y_{visc} can also be expressed as (Barros, 1989):

$$X_{visc} = 0{,}0123\sqrt{\left(u_c-u\right)^2 + \left(v_c-v\right)^2}\,\left(u_c-u\right)$$

$$Y_{visc} = 0{,}0170\sqrt{(u_c-u)^2 + (v_c-v)^2}\,(v_c-v)$$

Wave Drift Forces. The wave drift forces are also obtained from scale model tests:

$$X_{wdrift}/m = 0.0021(\xi - z)^2\cos(\gamma_{wave} - \varphi)$$

$$Y_{wdrift}/m = 0.0014(\xi - z)^2\sin(\gamma_{wave} - \varphi)$$

where z is the heave and γ_{wave} is the direction of wave propagation.

Resulting Model
The position X_O, Y_O and the heading in the low frequency model are obtained by integration of the following equations:

$$\dot{X}_o = u\cos\varphi - v\sin\varphi$$

$$\dot{Y}_o = u\sin\varphi + v\cos\varphi$$

$$\dot{\varphi} = r$$

The total horizontal motions are obtained by adding the high frequency motions:

$$X_{tot} = X_O + X_{hf}$$

$$Y_{tot} = Y_O + Y_{hf}$$

$$\varphi_{tot} = \varphi + \varphi_{hf}$$

Linear Model Analysis
A state space representation for the linearized low frequency model is obtained around the values $u = v = r = 0$ and $u_c = u_{co}$ and $v_c = v_{co}$:

$$\dot{X} = AX + BU$$

$$Y = CX$$

where:
$X = [u\ v\ r\ X_O\ Y_O\ \varphi]^T$; $U = [X_{thrust}\ Y_{thrust}\ N_{thrust}]^T$

$$Y = [X_O\ Y_O\ \varphi]^T$$

$$A = \begin{bmatrix} a_{11} & a_{12} & a_{13} & 0 & 0 & 0 \\ a_{21} & a_{22} & a_{23} & 0 & 0 & 0 \\ a_{31} & a_{32} & 0 & 0 & 0 & 0 \\ 1 & 0 & 0 & 0 & 0 & 0 \\ 0 & 1 & 0 & 0 & 0 & 0 \\ 0 & 0 & 1 & 0 & 0 & 0 \end{bmatrix}$$

$$B = \begin{bmatrix} b_{11} & 0 & 0 & 0 & 0 & 0 \\ 0 & b_{22} & 0 & 0 & 0 & 0 \\ 0 & 0 & b_{33} & 0 & 0 & 0 \end{bmatrix}$$

$$C = \begin{bmatrix} 0 & 0 & 0 & 1 & 0 & 0 \\ 0 & 0 & 0 & 0 & 1 & 0 \\ 0 & 0 & 0 & 0 & 0 & 1 \end{bmatrix}$$

where a_{ij} and b_{ij} are obtained from linearization of Eq. (10) (Barros, 1989)

The low damping and absence of restoring forces in the horizontal plane are reflected by the transfer function. In the case of $u_{co} = v_{co} = 0$ it follows that:

$$H(s) = \begin{bmatrix} \frac{b_{11}}{s^2} & 0 & 0 \\ 0 & \frac{b_{22}}{s^2} & 0 \\ 0 & 0 & \frac{b_{33}}{s^2} \end{bmatrix}$$

There is a small difference in the presence of the current. For example, consider $u_{co} = 1$ m/s, $v_{co} = 0$. Neglecting the coupling terms, the transfer functions is:

$$H(s) = \begin{bmatrix} \frac{b_{11}}{s(s+0{,}021)} & 0 & 0 \\ 0 & \frac{b_{22}}{s(s+0{,}011)} & 0 \\ 0 & 0 & \frac{b_{33}}{s^2} \end{bmatrix}$$

DPS DESIGN

Three independent controllers were designed to sway, surge and yaw motions. The rig dynamics model assumed in the control desaign is described by Eq. (1) with na = 2, nb = 1 and d = 1. The control effort u is, depending on the motion, X_{thrust}, Y_{thrust} or N_{thrust}. Actually DPS has a set of propellers allocated along the rig. So it is necessary calculate control signals for each of them to get desired forces and moment. Thus, this model makes a suitable simplification that doesn't prevent further improvement. The order of polynomial $C(z^{-1})$ was also assumed as 2. The output of the transfer function $C(z^{-1})/A(z^{-1})$ fed by white noise represents the sum of high frequency motion and measurement noise. This assumption is reasonable because the amplitude of high frequency motion of the semi-submersible rig considered in this paper is of the same order as the accuracy of the position measurement system. To define the control structure the following cost function was selected:

$$I = E\{[p_o y(k+d)]^2 + [(q_o - q_1 z^{-1})u(k)]^2\}$$

The sampling period adopted is 12 s. This period was chosen considering the low frequency dynamics, whose range is 0 - 0.25 rad/s, and the maximum recommended frequency to the operation of the propellers that is 0.63 rad/s. The values of p_o, q_o and q_1 were selected by preliminary simulation tests imposing ±2° to the heading and ±5 meters to the positions.

Simulation eresults

Several simulations were made to analyze the performance of the self-tuning DPS. In this paper the influence of coefficient q_1 and the absence of high frequency motion filter are discussed in particular.

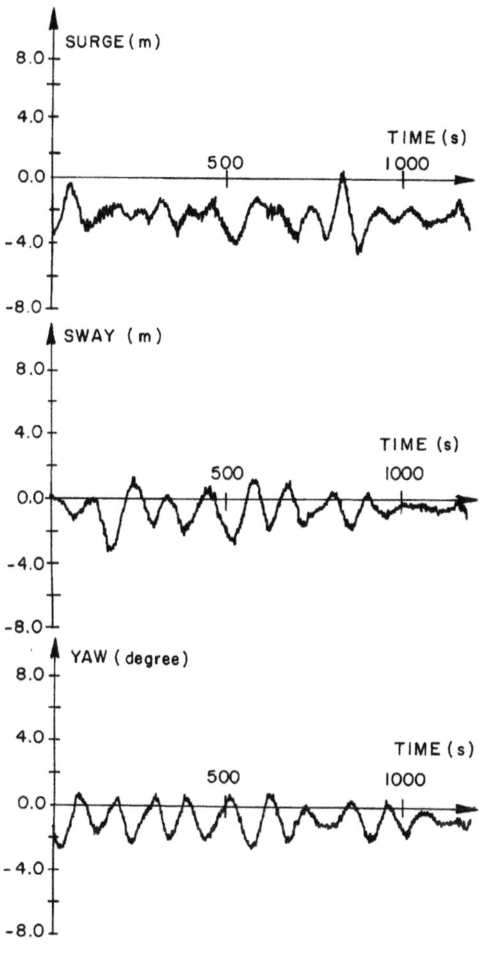

Fig. 2.1 Motions of the rig.

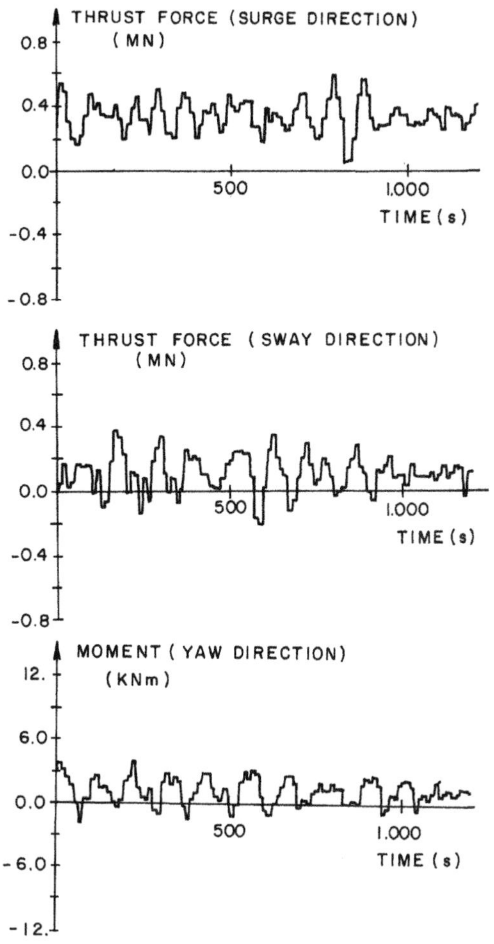

Fig. 2.2 Thruster efforts

The effect of using $q_1 \rightarrow 0$ is to increase the offset of the rig output, mainly if it is perturbed with constant environmental forces. The offset is usually eliminated by the introduction of an integral action, defining $q_1 = q_0$. However this approach failed since rig motions became oscillatory as an undamped system instead of eliminating the offset. The analysis of the closed loop transfer function by the root locus method has shown that the oscillation increases because $q_1 \rightarrow q_0$ tends to assign poles to the unit circle. An intermediate solution was obtained with $q_1 = 0.3q_0$. The test was made in accordance with the following conditions: 1) Wave with Pierson Moskowitz spectrum with significant wave height = 5.6 m and peak period = 12 s; 2) $\gamma_{wave} = 225°$; 3) $\gamma_{current} = 190°$; current velocity = 1 m/s; 4) The positions and heading are assumed to be measured with additive white noise disturbances with standard deviations of 0.5 m and 0,05° respectively.

The rig motions shown in Fig. 2.1 reveal that the self-tuning DPS keeps position and heading of rig within the accuracy requirements. As expected there is a small offset mainly in surge directions. This is due to both constant current action and $q_1 \neq q_0$. Fig. 2.2 shows that the profile of control efforts is suitable if a comparison is made with other results, for instance Kallstron (1983).

The effects of the filter abscence were also tested. A simulation without the feedback of the high frequency motion to the controller was performed, other conditions were kept the same as in previous test. The results shown in Fig. 3.1 and Fig. 3.2. are similar in comparison with Fig. 2.1 and Fig. 2.2 specially the modulation of the control signal. The sampling period selected doesn't reproduce the high frequency motion because its frequency is greater than 0.3 rad/s. But there is the aliasing effect of this signal that can cause unnecessary motion of the thrusters. This aspect needs to be analyzed further.

CONCLUSIONS

In this paper the application of self-tuning DPS was shown. The results indicate that there is some difficulty to eliminate the offset only by sear-

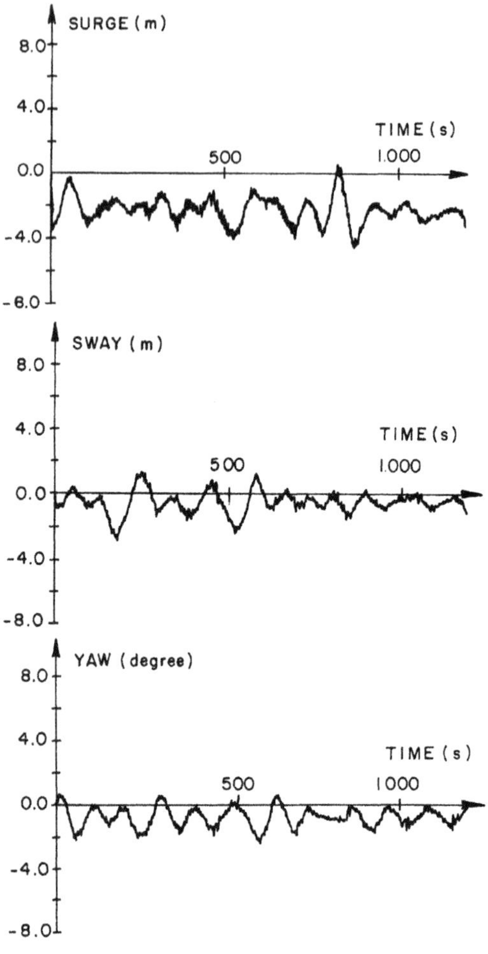

Fig. 3.1. Motions of the rig without high frequency components

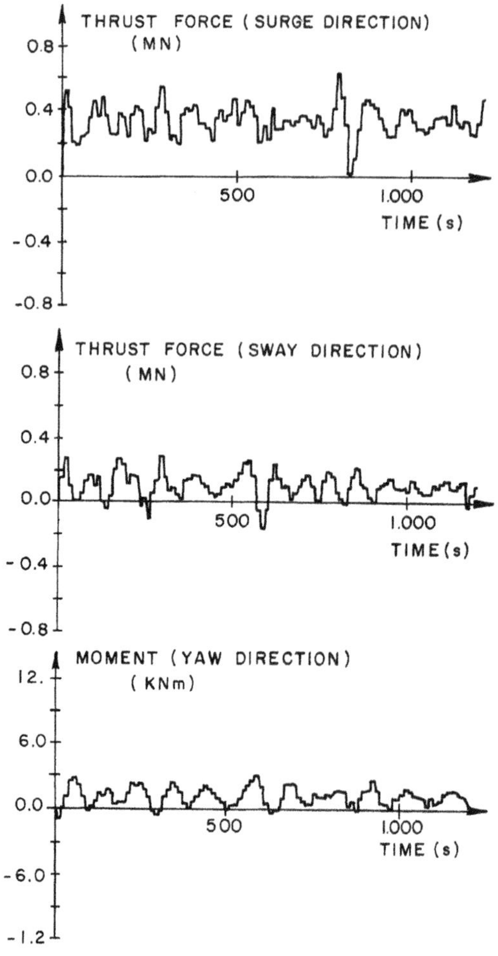

Fig. 3.2 Thruster efforts without high frequen-

ching for a suitable cost function. The low damping characteristics of the rig at low frequency motions in the horizontal plane impose limitations to the integral gain included in the cost function. The results also suggest that the control effort is not significantly affected by the filter absence for the semi-submersible rig considered. The implementation of this kind of DPS seems easier than linear quadratic control with EKF since it only requires order and time delay knowledge of the system. Thus a self-tuning regulator may be an easy way to cope with the non-linearities and the time variant characteristics of the system.

REFERENCES

Balchen, J.G. et alii (1985). Dynamic positioning system based on Kalman Filter and optimal control. Modeling, Identification and control. Oslo, 1(3), pp. 135-63.

Barros, E.A. (1989) Aplicação de um controlador auto-ajustável ao sistema de posionamento dinâmico de uma plataforma semi-submersível. Ms dissertation, University of São Paulo, São Paulo

Clarke, D.W. and P.J. Gawthrop (1975). Self-tuning controller. Proceedings of the Institution of Electrical Engineers, 22(9) London, pp. 929-934

Fung, P.T.K. and M.J. Grimble (1983). Dynamic ship positioning using a self-tuning Kalman Filter. IEEE Transactions on Automatic Control, AC-28, New York, pp 339-349.

Grimble, M.J., R.J. Patton and D.A. Wise (1980). Use of Kalman filtering techniques in dynamic ship positioning systems. IEE Proceedings. Part D:Control theory and Applications, 127(3), London pp. 93-102.

Kallstron, C.G. (1983). Mooring and dynamic positioning of a semi-submersible. A comparative simulations study. Proc. 2nd International Symposium on Ocean Engineering and Ship Handling, Gothenburg, pp.417-440.

Norbbin, N.H. (1971) Theory and observations on the use of a mathematical model for ship maneuvering in deep and confined waters. Goteborg, Statens Skeppsprovningsanstalt, (SSPA publication).

PRACTICAL ADAPTIVE LEVEL CONTROL OF A THERMAL HYDRAULIC PROCESS

J. Z. Liu*, B. W. Surgenor and J. K. Pieper****

**Dept of Thermal Power Engineering, North China Institute of Electric Power, Baoding, Hebei, PRC*
***Dept. of Mechanical Engineering, Queen's University, Kingston, Ontario, Canada, K7L 3N6*

Abstract. Presented is an extension to Clarke and Gawthrop's design for a Self-Tuning Controller (STC) to enable application to a practical process. Additions include a process monitor to oversee operation of an estimator switch, an intelligent integrator, feedforward action, nonlinear valve compensation, anti-windup reset protection, and provision for bumpless transfer. The augmented rule-based design is referred to as a Practical Self-Tuning Controller (PSTC) to differentiate it from the original design. The PSTC design is successfully applied to level control of a thermal hydraulic process. The new design demonstrates a 43 % improvement in performance when compared to a conventional well-tuned constant gain controller.

Keywords: Adaptive control; Control applications; Conventional control; Level control; Recursive least squares; Self-tuning regulators

INTRODUCTION

Adaptive control methods can provide a systematic yet flexible approach to the regulation of processes which are not well understood, are slowly time varying, or which have significant nonlinearities (Chien et al, 1985). It is expected that the adaptive system can achieve effective control and good system performance with respect to energy savings and product quality. However, even though there is much effort leading to significant advancements in theory (Åström and Wittenmark, 1989, and Seborg et al, 1986), in practical situations, adaptive controllers tend to fall short of expectation (Fernandes et al, 1989, and Åström, 1987).

A good practical adaptive control scheme will require not only a robust design method, but also a robust process parameter estimation scheme integrated together (Wittenmark and Åström, 1984, and Song et al, 1986). In this way, control can be achieved of processes with the above mentioned characteristics and also unmodelled high frequency dynamics and disturbances which commonly affect industrial processes.

A study was conducted to examine and improve the performance and robustness properties of Clarke and Gawthrop's (1975) design for a direct adaptive Self-Tuning Controller (STC) as applied to level control in a thermal hydraulic system. It was observed that optimum performance was achieved with the addition of i) an estimator switch, ii) an "intelligent" integrator iii) feedforward action, iv) nonlinear valve compensation and v) a start-up procedure with bumpless transfer from constant gain to adaptive control modes. The term Practical Self-Tuning Controller (PSTC) was adopted to highlight the new features of the STC design.

PROCESS DESCRIPTION

The thermal hydraulic process under study is illustrated in Figure 1. The apparatus is a direct contact heater similar to a power plant deaerator. The process has three control loops: level, temperature, and flow. Level is measured as the height of the column of water in a standoff tube connected to the main tank. This gives a lagging response of the measured to actual level. A pneumatically operated globe valve between a constant head source and the tank governs the inlet flow rate. The outlet delivery flow is controlled by a motor driven ball valve with the flow rate measured by a paddle wheel flowmeter. The temperature loop is not used in this study and consequently the steam supply is shut-off.

The important process characteristics are:

- There are nonlinearities in the dynamics over the range of operating conditions.

- There is significant pure dead time.
- Random input disturbances exist.
- The process is marginally stable in that the outflow is not a function of tank level.

The last point implies that the system operates with a free integrator in the forward path.

To give an indication of the dynamics of the process considered in this study, a closed loop recursive least squares parameter estimation scheme of the level control loop was implemented to yield the following nominal discrete time model (sample time, $T_s = 2.5\ s$):

$$\frac{y(q^{-1})}{u(q^{-1})} = G(q^{-1})$$

$$= \frac{0.0243\ q^{-6}\ (1 + 0.872\ q^{-1})}{1 - 1.477\ q^{-1} + 0.478\ q^{-2}} \quad (1)$$

As has become the practise, q^{-1} is used to denote the backwards-shift operator. From the model it is seen that the process has open loop poles at +1 and +0.478. That is, the process has an integrator plus first order lag structure. Based upon this identification result, it was decided to use an adaptive scheme based on a process model order of $n = 2$, number of transmission zeroes $m = 2$ and a process delay of $d = 5$ samples.

PSTC DESIGN

The process to be controlled, with input u(k), output y(k) and disturbing noise $\xi(k)$ (with a white gaussian distribution), is described in the discrete time domain by:

$$A(q^{-1})\ y(k) = q^{-d}\ B(q^{-1})\ u(k) + C(q^{-1})\ \xi(k) \quad (2)$$

$$A(q^{-1}) = 1 + a_1\ q^{-1} + \ \dots\ + a_n\ q^{-n} \quad (3)$$

$$B(q^{-1}) = b_1\ q^{-1} + \ \dots\ + b_n\ q^{-n} \quad (4)$$

$$C(q^{-1}) = 1 + c_1\ q^{-1} + \ \dots\ + c_n\ q^{-n} \quad (5)$$

where the order of the process is n. In general, the $B(q^{-1})$ and $C(q^{-1})$ polynomials will be of degree less than n. In this case, the trailing coefficients should be set to zero with no loss of generality in the process description. If it is assumed that only uncolored or uncorrelated noise enters the process ($C(q^{-1}) = 1$) then, Clarke and Gawthrop (1975) have developed the Self-Tuning Controller (STC) which, when servo action is included, gives the control law:

$$u(k) = \frac{y_r(k + d + 1) - \varphi^T(k)\ \theta(k)}{\beta_o + \rho/\beta_o} \quad (6)$$

where d is the estimated system dead time and $y_r(k)$ is the reference input or setpoint. The coefficient β_o requires an estimate of b_1, the steady-state gain for the process. In this application $\beta_o = 0.012$ and $\rho = 0.01$. The other components of the control law given by equation (6) are found from:

$$\varphi(k) = \Big(-y(k)\ \ -y(k-1)\ \dots\ -y(k-n+1)\quad u(k-1)\ u(k-2)\ \dots\ u(k-p) \Big)^T \quad (7)$$

$$\theta(k) = \Big(\alpha_0\ \alpha_1\ \dots\ \alpha_{n-1}\ \beta_1\ \beta_2\ \dots\ \beta_p \Big)^T \quad (8)$$

$$where\ \ p = n + d - 1$$

Note that $\varphi(k)$ and $\theta(k)$ are the measurement and control parameter estimate vectors, respectively. The α and β parameters of equation (8) are obtained from the solution of equation (9), which provides the transformation from the a and b parameters of equations (3) and (4).

This STC design will minimize the following cost function:

$$J(r) = \sum_{k=0}^{\infty} ([y(k) - y_r(k)]^2 + \rho u^2(k)) \quad (9)$$

given exact knowledge of the process. If $\rho = 0$ then the STC control objective of equation (9) equates to that for the Self-Tuning-Regulator (STR) design of Astrom and Wittenmark (1973).

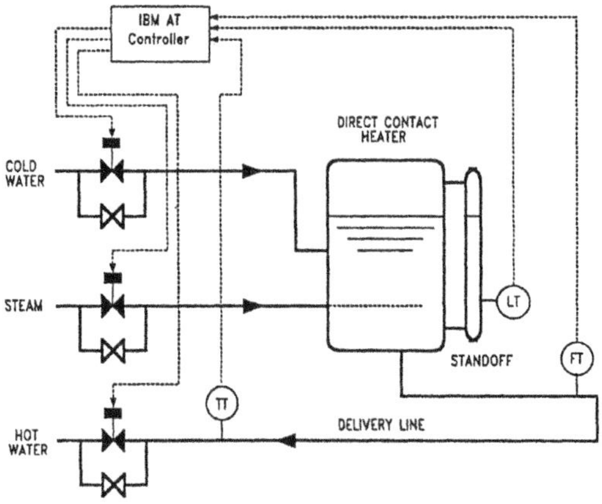

Figure 1. The thermal hydraulic process.

A recursive least squares (RLS) method is used to estimate the parameters vector $\theta(k)$ as given by the following:

$$P(0) = diag_{2n}[1000] \tag{10}$$

$$\theta(k+1) = \theta(k) + \sigma(k)\ \epsilon(k) \tag{11}$$

$$\epsilon(k) = y(k) - \varphi^T(k)\ \theta(k) \tag{12}$$

$$\sigma(k) = P(k-1)\ \varphi(k) \left[\lambda + \varphi^T(k)\ P(k-1)\ \varphi(k)\right]^{-1} \tag{13}$$

$$P(k) = \left(I_{2n} - \sigma(k)\ \varphi^T(k) \right) P(k-1)/\lambda \tag{14}$$

where σ is the adaptive gain. One can note that unbiased parameter estimates using this method result if $C(q^{-1}) = 1$ (no coloration of the input disturbance).

The Estimator Switch

To avoid unnecessary computational load on the controller, and to ensure that the covariance matrix, $P(k)$, does not "windup" or experience numerical instabilities, an estimator switch is employed. The two rules given below add hysteresis to the estimation scheme, and turn off the update mechanism in steady state conditions.

$$\begin{aligned} &RULE\ \#1 \quad IF \quad |\ y(k) - y(k-h_1)\ | > \delta_1 \\ &\qquad OR \quad |\ u(k) - u(k-h_2)\ | > \delta_2 \\ &\qquad THEN\ y(k)\ and\ u(k)\ are \\ &\qquad used\ to\ update\ \theta(k) \end{aligned} \tag{15}$$

$$\begin{aligned} &RULE\ \#2 \quad IF \quad estimator\ switch\ is\ ON \\ &\qquad THEN\ N\ samples\ are\ used \\ &\qquad for\ parameter\ updating \end{aligned} \tag{16}$$

The variables δ_1, δ_2, h_1, h_2, and N are user specified constants. Critical values for δ_1 and δ_2 are related to the inverse of the signal to noise ratio times a sensitivity factor. Experience has shown good results if this sensitivity factor is in the range 2 to 20. In this case $\delta_1 = \delta_2 = 0.15$ was used. For the comparison delays, h_1 and h_2, a value of 2 or 3 samples reduces high frequency noise effects while keeping the comparison between current data. A reasonable nominal value for the number of covariance updates is $N = 10$.

Note that when the estimator is switched off, the control law of equation (6) gives a constant gain model based proportional control. The measurement vector, $\varphi(k)$ is updated regardless of the estimator condition.

An Intelligent Integrator

Integral action is used to: 1) eliminate steady state error, 2) provide a static value for controller output, and 3) correct for nonlinearities and inaccurate process models. Disadvantages include windup and inertia problems resulting in large overshoot, oscillation and lagging effects, especially in processes with inherently long time delays or unstable open loop poles. To reduce these difficulties, "intelligent" integral action was used in this study according to:

$$e_i = y_r(k) - y(k) \tag{17}$$

$$\begin{aligned} &RULE\ \#3 \quad IF \quad y_r(k) \neq y_r(k-1) \\ &\qquad THEN\ e_i(k+j) = 0 \\ &\qquad for\ j = 0,\ 1,\ \ldots,\ N_s \end{aligned} \tag{18}$$

$$\begin{aligned} &RULE\ \#4 \quad IF\ |\ e_i(k)\ | < \delta_3 \\ &\qquad THEN \quad T_i = 2\,T_{io} \\ &\qquad ELSE \quad T_i = T_{io} \end{aligned} \tag{19}$$

With the integral action given by:

$$I(k) = I(k) + e_i(k)/T_i \tag{20}$$

$$u_c(k) = u(k) + I(k) \tag{21}$$

Again, experience has proven the best indicator for the value of the constants in the above mentioned rules. Choosing δ_3 as 1 or 2 times the inverse of the signal to noise ratio and N_s as approximately the process settling time gives good performance. T_{io} should be taken as the reset time for a well tuned constant gain PI controller. For completeness, the actual experimental values were $\delta_3 = 0.01$, $N_s = 10$ samples and $T_{io} = 60$.

Feedforward Action

Feedforward action can be used to reduce the effects of measurable disturbances. In the case of the thermal hydraulic system considered here, changes in the delivery flow rate will be propogated upstream and appear as a disturbances in the level. Knowledge of the flow rate can be used to compensate:

$$u_f(k) = u_c(k) + K_f\ Q(k) \tag{22}$$

with $Q(k)$ as the measured delivery flow and $K_f = 0.085$ as a gain.

Nonlinear Valve Compensation

There is a significant nonlinearity in the inlet water valve. A plot of the control signal versus inlet flow gives a nearly quadratic response. This behaviour can be compensated for by a type of inverting prefilter. Such a filter is accomplished by dividing the operating region into two linear parts and adjusting the control as:

$$\begin{aligned} RULE\ \#5 \quad IF \quad & u_f(k) > 2.5 \\ THEN\ & u_n(k) = 2.5 \\ & +0.75\ (u_f(k) - 2.5) \\ ELSE \quad & u_n(k) = u_f(k) \end{aligned} \tag{23}$$

This rule is applied as the final control action.

Anti-Reset Wind-Up

Another add-on feature of many practical industrial controllers is anti-reset wind-up. The integrator term, I(k), may continue to increase in magnitude, either positive or negative, due to a large process error even though the actuator is at its limit. Various anti-reset wind-up protection mechanisms are available. An inelegant but effective method is to simply stop the integrator, by setting T_i large, when the control is near the saturation limit for the actuator. Other techniques can be found in Wittenmark (1989). A simplified method was incorporated in the design considered here. First the calculated control output was sent through a software saturation limiter according to Rule #6 in parallel to the actual process.

$$\begin{aligned} RULE\ \#6 \quad IF \quad & u_n(k) > u_{max} \\ THEN \quad & u_v(k) = u_{max} \\ ELSE\ IF \quad & u_n(k) < u_{min} \\ THEN \quad & u_v(k) = u_{min} \\ ELSE \quad & u_v(k) = u_n(k) \end{aligned} \tag{24}$$

The integrator of equation (20) is updated by:

$$\begin{aligned} I(k) = I(k-1) + e_i(k)/T_i \\ + K_a\ (u_v(k-1) - u_n(k-1)) \end{aligned} \tag{25}$$

Again, for completeness, $K_a = 2$, and for the specific actuators employed, $u_{max} = 4.5\ volts$ and $u_{min} = 0.5\ volts$. It can be seen that if the control is not saturated, the final term of equation (25) is zero and the windup protection has no effect.

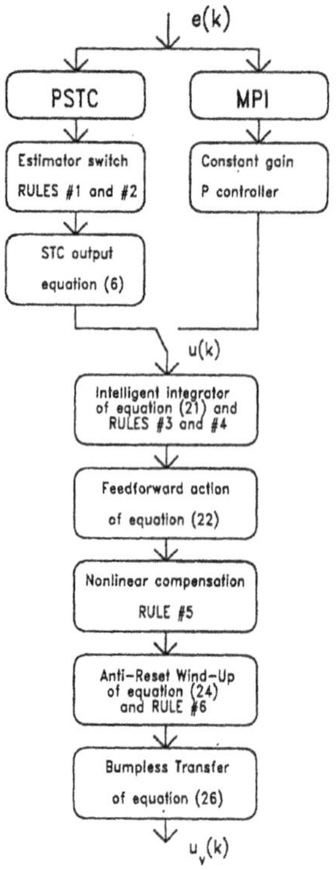

Figure 2. Control algorithm flow chart.

Bumpless Transfer and Start-up

Bumpless transfer avoids disruptions of the system when switching between control algorithms or structures. One method of achieving bumpless transfer is to set:

$$I(k_o) = K_i(u_{PI}(k_o) - u_{STC}(k_o)) \tag{26}$$

as the initial value for the STC integrator.

For the PSTC algorithm to work effectively, good estimates of the process model parameters are needed. The reasonable initial estimates allow for performance to be maintained within acceptable bounds at all times, even in the initial transients. In order for the RLS scheme to detect good process parameter estimates, the following start-up procedure was adopted:

1. Implement feedback control of the process through an appropriately simple strategy such as PI control.
2. Run the process in identification mode for 100 samples, or until proper control law parameters can be found using a PRBS excitation signal (amplitude = 0.4 v). At this point the first estimates for β_o and ρ are obtained.
3. If initial parameter estimates are available, Steps 1 and Step 2 can be run using the

constant gain minimum variance control law of equation (6) with $\theta(k) = \theta_o$, the initial estimates.

4. Implement bumpless transfer. At this point, fine tuning can be done on-line for β_o and ρ.

Finally, the full PSTC algorithm is turned on. The PSTC algorithm is summarized in Figure 2 in flow chart form.

MPI CONTROL

Multi-function PI control (MPI) is essentially classic proportional plus integral control with the addition of practical features developed for the PSTC design. Thus, the MPI design begins with a constant gain proportional controller. To this, intelligent integral action, a feedforward term and nonlinearity compensation are all incorporated for a "multi-function" design. All of these terms require proper tuning. In this study, this procedure was done by hand using a trial and error approach. The control law for MPI is:

$$u(k) = K_p(\dot{y}_r(k) - y(k)) \qquad (27)$$

followed by the application of Rules #3 to #6. The MPI control structure was tuned off line following the Ziegler-Nichols (1942). On-line tuning identified $K_p = 0.95$ and $T_{io} = 100$ as providing the best performance.

EXPERIMENTAL RESULTS

A set of disturbance rejection and setpoint change tests were conducted at three operating points: low, mid and high flow rates. To obtain quantitative measures of performance, the Integrated Time Absolute Error (ITAE) and Maximum Percentage Overshoot (MPO) were calculated. The ITAE criterion addresses the need

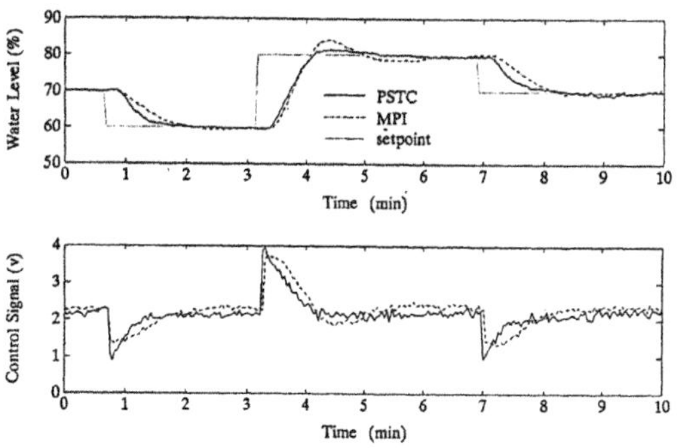

Figure 4. Response at mid flow.

to return the process to the setpoint as soon as possible, yet recognizes that early errors are unavoidable due to the process dead time.

Figure 3 compares the performance of MPI versus PSTC at low flow (15 l/min) in response to step changes in the level setpoint. Figure 4 compares the performance of MPI versus PSTC at mid range flow (20 l/min). The MPI controller was tuned at this mid range condition. The nonlinearities in the process dynamics are evident, given that the tank filling and emptying responses are seen to be quite different. The difference between the filling and emptying responses is even more pronounced at the high flow (30 l/min) condition, as illustrated in Figure 5.

In comparing the performance of MPI versus PSTC, one observes that PSTC in general shows smaller overshoots and excellent overall tracking performance. Also, the system nonlinearities are removed from the controlled variable, especially at low and mid-range flow. Comparison of the control signal and water level responses indicates a relatiively small amount of lagging effect even though the process dead time is quite long.

Table 1 summarizes the ITAE and MPO scores together with the ITAE ratio for PSTC over MPI. In all cases, PSTC matches or exceeds the performance measures for MPI. On average

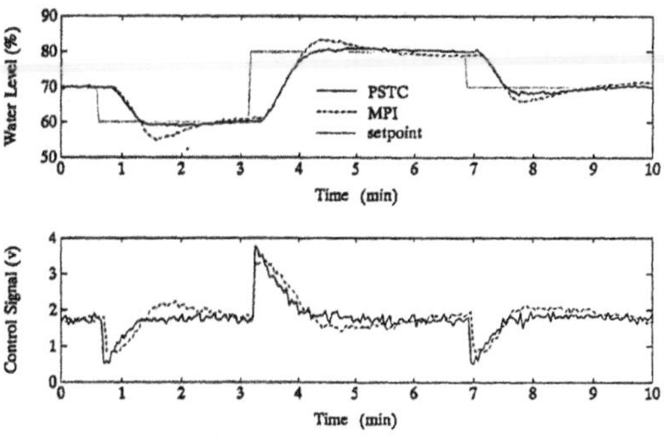

Figure 3. Response at low flow.

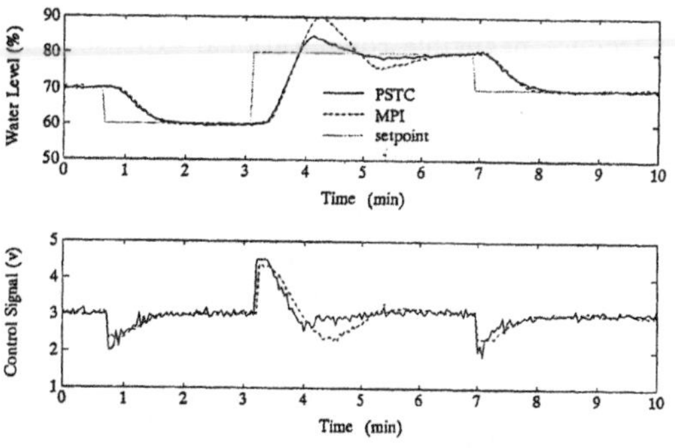

Figure 5. Response at high flow.

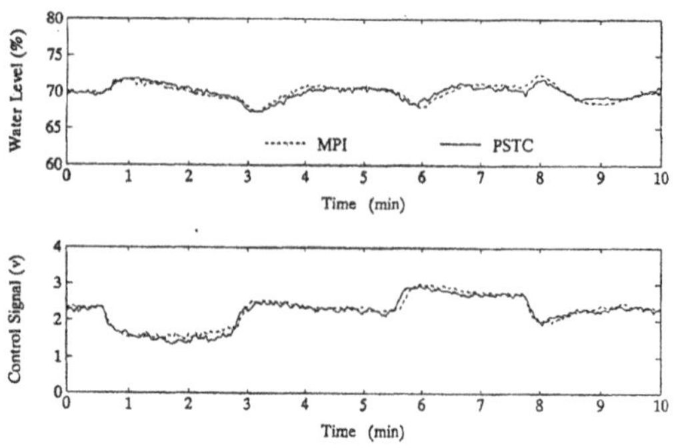

Figure 6. Response to a disturbance.

there is a reduction in ITAE of 43 %. The MPO of the PSTC is considerably smaller, indeed less than half, than that for MPI. This is primarily due to the quickness of the control action and its predictive abilities. This also indicates that the system dead time was accurately estimated in the preliminary commissioning of the control scheme.

As a test of disturbance rejection performance, the drain flow was stepped from 20 to 15 to 20 to 30 l/min at 2 min intervals. The result is given in Figure 6. In the case of both MPI and PSTC, performance is quite good even though the disturbance is quite vigorous. This is likely due to the addition of feedforward action. The IAE scores (no weighting for time) given in Table 1 show a 10 % improvement in PSTC over MPI for the disturbance.

CONCLUSIONS

This paper highlights several of the practical problems associated with direct self-tuning control schemes. The additions to the classic design combine for a new algorithm designated as PSTC. The scheme is successfully applied to level control in a thermal hydraulic process.

The rules introduced into the design and detailed in this paper are simple in nature and introduce negligible computational expense. Tuning parameters are, for the most part, intuitive and easy to select. The performance improvement is judged to be significant, with approximately a 43 % ITAE reduction over an industry accepted constant gain control scheme.

Table 1. Summary of performance results

		MPI		PSTC		ITAE
Test	Flow	ITAE	MPO	ITAE	MPO	Ratio
Setpoint	low	14.2	8.8 %	7.89	3.4 %	0.56
	mid	11.7	5.2 %	7.37	1.9 %	0.63
	high	13.8	12.6 %	8.61	6.1 %	0.51
Disturb	(IAE)	5.22		4.78		0.91

ACKNOWLEDGEMENTS

This work was supported under contract to the Manufacturing Research Corporation of Ontario (MRCO). One of the authors (JKP) wishes to acknowledge scholarship support from the Natural Sciences and Engineering Research Council of Canada (NSERC).

REFERENCES

ÅSTRÖM, K.J. (1987): "Adaptive Feedback Control," *Proceedings IEEE*, **Vol. 75**, No. 2,, pp. 185-217.

ÅSTRÖM, K.J. AND WITTENMARK, B. (1973): "On Self-Tuning Regulators," *Automatica*, **Vol. 9**, pp. 185-199.

ÅSTRÖM, K.J., AND WITTENMARK, B. (1989): *Adaptive Control*, Addison-Wesley, New York..

CHIEN, I.L., SEBORG, D.E. AND MELLICHAMP, D.A. (1985): "A Self-Tuning Controller for Systems with Unknown or Varying Time Delays," *Int. Jrnl. Control*, **Vol. 42**, No. 4, pp. 949-964.

CLARK, D.W. AND GAWTHROP, P.J. (1975): "A Self-Tuning Controller," *Proceedings IEE*, **Vol. 122**, No. 9, pp. 929-934.

FERNANDES, J.M., DE SOUZA, C.E. AND GOODWIN, G.C. (1989): "An Application of Adaptive Scheme to Fluid Level Control," *Proceedings ACC*, **Vol. 3**, Pittsburgh, June 21-23, pp. 1886-1891.

SEBORG, D.E., EDGAR, T.F. AND SHAH, S.L. (1986): "Adaptive Control Strategies for Process Control: A Survey," *AIChE Journal*, **Vol. 32**, No. 6, pp. 881-912.

SONG, H.K., SHAH, S.L. AND FISHER, D.G. (1986): "A Self-Tuning Robust Controller," *Automatica*, **Vol. 22**, No. 5, pp. 521-531.

WITTENMARK, B. AND ÅSTRÖM, K.J. (1984): "Practical Issues in the Implementation of Self-Tuning Control," *Automatica*, **Vol. 20**, No. 5, pp. 595-605.

WITTENMARK, B. (1989): "Integrators, Nonlinearities and Anti-Windup Reset for Different Control Structures," *Proceedings ACC*, **Vol. 2**, Pittsburgh, June 21-23, pp. 1679-1683.

ZIEGLER, J.G. AND NICHOLS, N.B. (1942): "Optimum Settings for Automatic Controllers," *Transactions ASME*, **Vol. 64**, No. 11, pp. 759-768.

A PROTOTYPE EXPERT SYSTEM BASED ON LAGUERRE ADAPTIVE CONTROL

Wei-Wu Zhou*, G. Dumont and B. Allison***

**Pulp and Paper Research Institute of Canada, 3800 Westbrook Mall, Vancouver, BC, V6S 2L9*
***Pulp and Paper Centre, Dept. of Electrical Engineering, University of British Columbia, Vancouver, BC, V6T 1W5*

Abstract. Part of an expert control (EC) project which uses Laguerre unstructured self-tuner (LUST) technique is presented. A commercial expert system shell G2 has been used to develop the expert knowledge for commissioning and monitoring the adaptive controller. The present prototype is developed to commissioning and monitoring the LUST adaptive controller. Simulations on process control in pulp mill give satisfactory results.

Keywords. Adaptive control, Artificial intelligence, Expert system, Expert intelligent control, Process control

INTRODUCTION

Modern adaptive control techniques provide the capability to automatically adjust control actions and thus ensure optimal performance as process operating conditions change. Although much research work on adaptive control has taken place over the past two decades and some successful applications have been developed, it is, however, still not extensively used in industry. One main obstacle toward widespread applications is a shortage of people with the specialized knowledge required for commissioning such systems. Such knowledge is required for choosing among the numerous existing control schemes, the algorithm best suited to the problem at hand. Expert system technology has shown the potential to provide the expertise necessary to help a non-expert to commission and monitor an adaptive controller.

An expert system can loosely be described as a computer program that uses stored knowledge to emulate the problem solving behaviour of human expert in some limited domain. A process control expert system is a natural extension of advanced control technology. The development of such a system has been an important research area in both expert system and process control since the early 1980s. Expert systems in process control have been initially focused on learning and self-organizing control.

In learning control systems, the controller is supposed to be able to estimate unknown information during its operation and determine an optimal control action from the estimated information. Examples of work in the area are Fu (1971) and Saridis (1977,1981). Expert systems used as off-line tools during process and control designs, in combination with a control system computer-aided design, are aimed to capture knowledge about how the process should be designed in order to fulfil vari ous objectives (control qualities, economy etc), and to capture knowledge about process configuration, control specifications and different control design techniques (*e.g.* SISO compensation, multivariable design etc). Examples of work in the area are found in Fisher (1986); Taylor (1986); Niida and Umeda (1986); MacFarlane *et al.* (1987); Lewin and Morari (1988); Birky *et al.* (1988). Other example on plant-wide control strategy planning is found in Stephanopoulos *et al.* (1987). A more well-known expert system application in process control is to use an expert system as a complement to a conventional control system for process monitoring and alarm analysis. The expert system is mainly used as an operator assistant, which packages the knowledge of experienced operators. Numerous examples of different monitoring applications can be found. A few examples are Nelson (1982) on nuclear power plants, Sakaguchi and Matsumoto (1983) on electrical power systems and Palowitch and Kramer (1985) on chemical plants. Recently, along with the availability of commercial real-time expert system shells, much attention has been paid to developing expert system to real-time process control, which is viewed as expert control (EC) or knowledge-based control (KBC). In the area the expert system is typically used on a small part of the plant (*e.g.*, single closed loop) and performed as a supervisor to commission and tune controllers, monitor process, perform fault diagnosis, and restructure control systems, etc. Examples of such work can be found in Moore *et al.* (1984); Åström *et al.* (1986); Arzen(1987); Karsai *et al.* (1987); Liu and Gertle (1987); Zhu and Dai (1988); Tzouanas *et al.* (1988); Whitlow and Debelak (1988); Basila and Cinar (1988). The work of the present paper belongs in this category. The goal in the long-run for expert control is to build an "intelligent" controller, which has incremental learning capability and can be viewed as expert intelligent control (EIC).

In this paper a prototype expert system using a Laguerre adaptive control technique, which is a part of a long-term EC project, is presented.

STRUCTURE OF EXPERT CONTROL

Expert control is a very advanced and comprehensive branch of modern control technology. It, in fact, deals with a rather wide range of disciplines and technologies in both control theory and artificial intelligence (AI). In aspect of control theory, EC may consist of all basic control algorithms, *e.g.*, PID control, auto-tuning, identification algorithms, adaptive control algorithms, etc., which constitute a toolbox of numerical algorithms for different control strategies. The numerical algorithms provide explicit optimal solutions under some certain conditions. In aspect of artificial intelligence, EC may consist of knowledge base, knowledge representation, inference engine, data base, etc., which constitute a expert "brain" for emulating "intelligent" problem solving behaviours of human expert. In this part, the knowledge is presented symbolically and the problems presented are usually fuzzy, complex and difficult numerically to be solved; instead heuristic deduction and reasoning processes are used. EC is thus a combination of modern control technologies and artificial intelligence, and is in the frontiers of modern science and technology. The structure of EC may be shown as in Fig. 1.

Control of a process basically involves two types of knowledge: control knowledge and process knowledge. Control knowledge contains knowledge of automatic control, *e.g.*, knowledge about identification and different control strategies, etc. Process knowledge contains knowledge of the process that should be controlled, *e.g.*, knowledge about normal and abnormal conditions, static and dynamic characteristics of the process, etc. This knowledge can be acquired through on-line experiments or from experienced process engineers. In Fig. 1, the blocks of numerical algorithms and controllers constitute the control knowledge, which may contain, *e.g.*, PID controllers, auto-tuning algorithms, identification and adaptive control algorithms for single-input single-output (SISO) and multi-inputs multi-outputs (MIMO) of linear systems. The blocks of knowledge base, inference engine, and database constitute the part of expert system in EC.

The knowledge base is the core of the expert system, which contains the knowledge of certain domains that has been entered directly by human experts or extracted indirectly via the process. The knowledge is formed in an ordered list of facts (related to the process), rules (reasoning conclusions), and heuristics about the domain of interest. Knowledge representation is a key issue in expert systems. Common representation forms are objects (related to facts of the domain) and rules (related to heuristic deductions and reasoning). The inference engine is the driving mechanism of expert systems. It manipulates rules from the knowledge base to form inferences and draw conclusions. The database is used to hold and represent facts (problems) about the application domain.

The implementation of an expert system for process control systems is not trivial with common symbolical language such as Lisp or Prolog. The separation between the knowledge base and inference engine has led to the development of so called expert system shells or frameworks. An expert system shell is an empty expert system without any domain knowledge. It, however, provides an inference engine and a knowledge representation structure (for building objects and rules) that can be used as a programming tool for implementation of expert systems in different applications. A real-time expert system shell G2 (Gensym Corp.) provides such a real-time framework and has been used in the work of the paper.

SYSTEM STRUCTURE

In this paper, we present preliminary results from an EC project started in 1989 in the Vancouver Laboratory, Pulp and Paper Research Institute of Canada. The initial motivation was to build an expert supervisor for commissioning and monitoring an adaptive controller. The project consists of both hardware and software developments as shown in Fig. 2 as a demonstration project.

A VAX Station 3100 running VMS and DEC Windows has been chosen as the hardware platform. The expert system shell G2 has been used for implementing the expert system part of the project. G2 is a real-time expert system shell with object-oriented knowledge representation. It provides schematics, dynamic models, and heuristics to represent the knowledge of applications. A standard interface GSI supplied along with G2 allows G2 to interface to external data sources including data acquisition equipment, databases, and other external devices. The numerical algorithms are the main part of the software development in this project, and consists of several packages of different algorithms about identification, adaptive control, PID control, and management utility. They are programmed in FORTRAN. A data communication bridge has been programmed in C to link the expert system G2 and the numerical algorithm packages. The G2 standard interface GSI provides specific functions for the communication purpose and is used to perform the development. A data acquisition equipment should be assembled to transfer data between the VAX-station and the actual controlled process. The expert system G2 and the numerical algorithm packages can run on a single computer, as in our case, or on two separate computers. The measurements obtained from the process are transferred to the main computer VAX-3100 and stored in data files. These data files constitute an external database for the expert system and provide necessary information to both the expert system and the numerical algorithm packages. Such a system structure is aimed at matching the situations of hardware equipments commonly existing in the pulp and paper industries.

LAGUERRE ADAPTIVE CONTROLLER

Several adaptive control strategies and identification methods have been selected and used in the EC project in order to fulfil different requirements in a wider applications. Considering the rather wide range of different model structures and varying dead-times we have to deal with in industrial process control systems, the Laguerre unstructured self-tuner (LUST) method (Zervos and Dumont, 1988) will first be implemented and tested. In the following a brief introduction of LUST is described.

The output of the plant $y(t)$ is described by,

$$y(t) = \sum_{i=1}^{N} c_i l_i(t) + w(t) = \underline{c}_0^T \underline{l}(t) + w(t) \qquad (1)$$

where $\underline{c}_0^T = [c_1 \quad c_2 \ldots c_N]$, $\underline{l}^T(t) = [\ l_1(t) \quad l_2(t) \quad \ldots \quad l_N(t)\ ]^T$, where the l_i's are the outputs from each Laguerre filter and $w(t)$ is the process noise.

An adaptive control scheme based on the above formulation uses the recursive least-squares (RLS) method to identify the parameter vector c. Here, we use the exponential forgetting and resetting algorithm (EFRA) (Salgado *et al.*, 1988).

Theorems proving the global convergence and stability of this scheme are presented in Zervos and Dumont (1988).

The choice of the parameter p in the Laguerre functions is not crucial. However, it influences the accuracy of the approximation of the plant dynamics as a truncated series. For a given plant, there exists an optimal p that minimizes the number of filters required to achieve a given accuracy. The chain of all-pass filters in the Laguerre network provides good representation of a time delay τ, in particular when $p = 2N/\tau$. The actual plant order has little bearing on the number of filters N. The horizon of the predictive control law is automatically adjusted on-line to assure closed-loop stability.

Based on theoretical analysis and on our industrial experience of the LUST algorithm, we know that LUST can perform an excellent control quality under the following conditions:

- proper selection of the values for
  - sampling interval T
  - Laguerre time constant p
  - number of Laguerre filters N
- sufficient excitation for convergent estimation of Laguerre spectrum gain vector c

These factors should be fulfilled when a LUST controller is used.

IMPLEMENTATION

The limitations of the LUST algoirhtm chosen imply the following constraints on the process to be controlled,

- single-input and single-output system
- open-loop stable
- well damped

The initial task is to commission the Laguerre adaptive controller. For such a commissioning, several characteristic factors of the process should initially be determined by testing the process or operator's input. In particular we require rough estimates of

T_d Delay time
T_s Response or setting time

Because we then use different rules whether the response is dominated by time delay or not, we then use the following criterion for discriminating between the two cases:

Time-delay : $T_d/\tau > 0.25$
No-delay : $T_d/\tau < 0.25$

where T_d is the estimated dead-time and τ is the dominant time constant of the process and can be calculated in a step response test that the time of the process response (without dead-time) reaches the point of 63% value of set-point.

The operator is also asked whether the process is currently under a PID control or not.

To determine above factors, a step response test is initially applied either in a close-loop (under an existing PID control) or in an open-loop (no PID control existed). Then delay time T_d, response time T_s and time constant τ are accordingly determined. Further the desired sampling interval T is calculated from the response time T_s and the time delay T_d. The number of Laguerre filters N and constant time p can then be calculated for the following two cases (Dumont, Zervos and Pageau, 1990),

- Delay representation:

$$N \geq \frac{T_d}{T} + 1$$

$$p = \frac{2}{T_d} + \frac{2}{T}$$

- No delay:

$$p = \frac{1}{\tau}$$

During the step response test (after obtaining proper values for N and p), an identification experiment is performed for initialization of the LUST controller. It is performed by sending a pseudo random binary signal (PRBS) that is added to the step signal. The Laguerre spectrum gain vector c is then estimated with open-loop signals. It means that the signals used as input signals in the estimation are the combined signals (step signal plus PRBS), and that the computed control laws $u(t)$ from the LUST algorithm are, in this case, not yet used as the input signals to the process. After passing a period to allow near convergence of the estimator, the control law $u(t)$ is shifted to be the input signal and the LUST controller is formally executed to control the process.

The procedure of commissioning is then performed in rules, *e.g.*,

- IF process-status is NO-PID-EXIST THEN in order conclude that OPEN-LOOP-STEP-RESPONSE-TEST is TRUE; and set SET-POINT to INITIAL-SET-POINT; and inform the operator that "OPEN LOOP STEP RESPONSE TEST is executing".
- IF process-status is YES-PID-EXIST THEN in order conclude that CLOSED-LOOP-STEP-RESPONSE-TEST is TRUE; and set SET-POINT to INITIAL-SET-POINT; and inform the operator that "CLOSED LOOP STEP RESPONSE TEST is executing".
- IF process-status is NO-PID-EXIST; and test-status is TEST-READY THEN in order set ADD-PRBS to 1; and set START-LAGUERRE to 1; and inform the operator that "Initialization of Laguerre adaptive controller, please wait."
- IF process-status is YES-PID-EXIST; and test-status is TEST-READY THEN in order set ADD-PRBS to 1; and set START-LAGUERRE to 1; and inform the operator that "Initialization of Laguerre adaptive controller, please wait."
- IF estimate-status is GOOD THEN in order conclude that ADAPTIVE-CONTROL-READY is TRUE; and set control-switch to 6; and inform the operator that "Laguerre adaptive control is executing !"

As mentioned previously the numerical algorithm packages (written in FORTRAN) are running parallel with the expert system G2, some control flags using digital number 0, 1, ... are thus used to pass commands between G2 and the FORTRAN packages through the GSI bridge. The number 1 and 6 appeared in above rules are these commands.

SIMULATION

An example for commissioning and monitoring the LUST controller used on a bleach-plant extraction stage in pulp mill has been simulated. A process schematic of the bleach-plant extraction stage is developed on G2 and shown in Fig. 3. The objects developed for the application have physical instances which are illustrated on the schematic, and are knowledge represented by those related attributes and rules of G2. External sensors which are related to the objects are developed on G2 in order to transfer data or messages between the external process and the expert system. For simulation purposes, the process is emulated by a theoretical model which is a first order model with time delay. The procedure of commissioning has be designed to be able to run on two cases: NO-PID-EXIST and YES-PID-EXIST, which are related to OPEN-LOOP-STEP-RESPONSE-TEST and CLOSED-LOOP-STEP-RESPONSE-TEST respectively. The control target in the process is the pH value at the exit of the first caustic extraction tower, and the control inputs are caustic flow which is manipulated by a pneumatic PI controller in inner loop. The LUST controller assembled on the computer in the outer loop is then used to control the setpoint for the inner loop. A main program written in FORTRAN which links all numerical algorithms is an external function in the C program of the GSI bridge. It is called every two seconds which is set and determined by the setting of scan interval for calling the GSI bridge from G2. The data transferred between C code and FORTRAN code are passed through the external function in C and the common data in FORTRAN. Another way to transfer data between C and FORTRAN in the VAX VMS environment is through mailbox. The knowledge of commissioning the LUST controller has been represented in both heuristics and in FORTRAN. With clear and certain objective in mind, an overall consideration for optimal programming to fulfil the target is necessary and significant. It means that if the knowledge can be easily represented in a numerical way, then it may not be necessary to use heuristics. After development of the knowledge representation for the specific process, which contains the developments of objects and rules, the expert system G2 is then ready to control the process with its filled expertise. Procedures for the commissioning and monitoring the LUST controller have been designed with guided steps. The operator is guided through the initial procedures, *i.e.* to answer the necessary questions which operator can provide (*e.g.* the process is currently under a PID control?). After answering these questions the expert system can automatically perform a correct procedure for commissioning and to monitor the performances of the control with its expert knowledge. Simulations on the application have shown a rather convincing results on expert supervisory to the commissioning procedure and monitoring capability.

CONCLUSIONS

This prototype has shown partial results of an EC project. Only the commissioning part of a LUST control using the expert system techniques has been introduced in this paper. Monitoring functions are currently being developed.

ACKNOWLEDGEMENTS

The financial support of the Science Council of British Columbia through STDF Grant #88-227 is gratefully acknowledged.

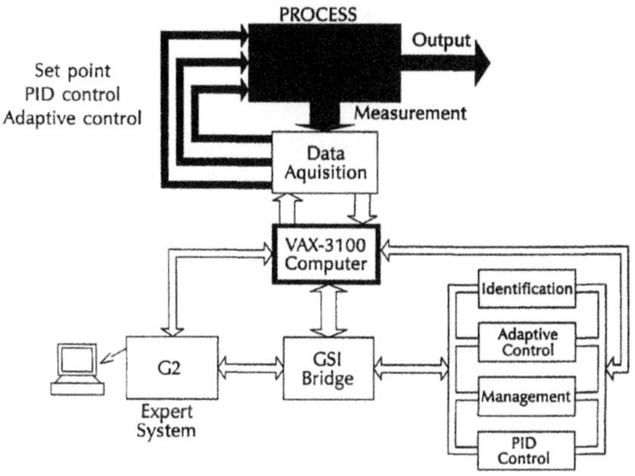

Fig. 1. Structure of Expert Control

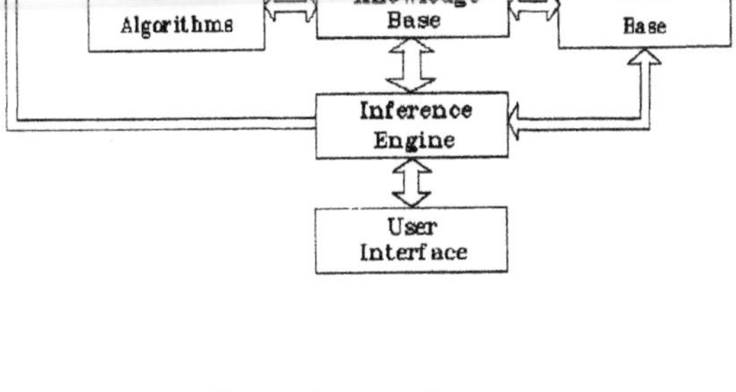

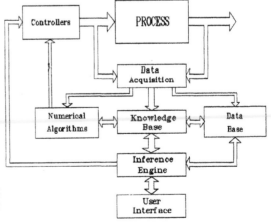

Fig. 2. System Structure

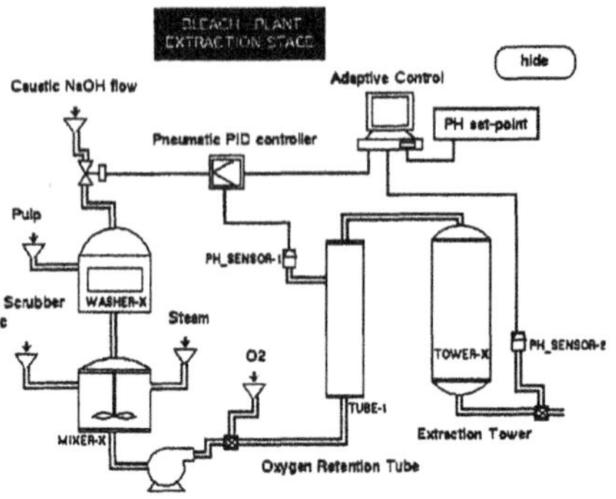

Fig. 3. Bleach-Plant Extraction Stage

REFERENCES

Åström, K.J., J.J.Anton and K.E.Arzen (1986). Expert control. Automatica. 22, 277–286.

Arzen, K.E. (1987). Realization of expert system based feedback control. Ph.D. thesis, Dept. of Automatic Control, Lund Institute of Technology, Sweden.

Basila, M.R. and A. Cinar (1988). MOBECS: Model-object based expert control system. AICHE Annual Meeting. Washington, D.C.

Birky, G.J., T.J. McAvoy and M. Modarres (1988). An Expert System for Distillation Control Design. Comput. Chem. Eng., 12(9/10).

Carmon, A. (1986). Intelligent knowledge-based system for adaptive PID controller. Journal A. 27, 133–138.

Cameron, I.T. (1986). Expert systems for hazard and operability studies of process plants. Proc. Australian Institute of Petroleum. Melbourne, pp. 1–12.

Dumont, G., C.C. Zervos and G.L. Pageau (1990). Laguerre-based adaptive control of pH in an industrial bleach plant extraction state. Automatica. 25.

Fisher, E.L. (1986). An AI-based methodology for factory design. Artificial Int., 7(4), 72–85.

Fu, K.S. (1971). Learning control system and intelligent control system — An introduction of artificial intelligent and automatic control. IEEE Trans., AC-16, 70–72.

Jackson, P. (1986). Introduction to Expert Systems. Addison-Wesley: Reading, MA.

Karsai G., E. Blokland, C. Biegl, J. Sztipanovits, K. Kawamura, N. Miyasaka and M. Inui (1987). Intelligent supervisory controller for gas distribution system. Proc. 1987 Am. Control Conf., 1353–1358.

Lewin, D.R. and M. Morari (1988). ROBEX: an expert system for robust controller synthesis. Comput. Chem. Eng., 12, 1187–1198.

Liu, K. and J. Gertler (1987). A supervisory (expert) adaptive control scheme. IFAC 10th World Congress. Munich.

MacFarlane A.G.L., G. Gruebel and J. Ackermann (1987). Future design environments for control engineering. IFAC 10th World Congress. Munich.

Moore, R.L., L.B. Hawkinson, C.G. Knickerbocker and L.M. Churchman (1984). A real-time expert system for process control. Proc. First Conf. on AI Applications. IEEE Computer Society, 529–576.

Moore R.L. and M.A. Kramer (1986). Expert system in on-line process control. In M. Morari & T.J. McAvoy, Eds., Chemical Process Control - CPC III. Elsevier, Amsterdam, pp. 839.

Nelson, W.R. (1982). REACTOR: an expert system for diagnosis and treatment of nuclear reactor accidents. Proc. of the National Conf. on Artificial Intelligence. Pittsburgh, PA, pp. 296–301.

Niida, K. and T. Umeda (1986). Process control system synthesis by an expert system. In M. Morari & T.J. McAvoy, Eds., Chemical Process Control - CPC III. Elsevier, Amsterdam.

Sakaguchi, T. and K. Matsumoto (1983). Development of a knowledge-based expert system for power system restoration. IEEE Trans. on Power Apparatus and Systems. 102, 2.

Salgado, M.E., G.C. Goodwin and R.H. Middleton (1988). Modified least squares algorithm incorporating exponential resetting and forgetting. Int. J. Control. 47(2), 477–491.

Saridis, G.N. (1977). Self-organizing control of stochastic system. Marcel, Dekker Inc. New York.

Saridis, G.N. (1981). Application of pattern recognition methods to control system. IEEE Trans., 26.

Shinskey, F.G. (1986). An expert system for the design of distillation columns. In M. Morari & T.J. McAvoy, Eds., Chemical Process Control - CPC III. Wiley, New York.

Stephanopoulos, G., J. Johnston and R. Lakshmanan (1987). An intelligent system for planning plant-wide control strategies. IFAC 10th World Congress on Automatic Control. Munich.

Taylor, J.H. (1986). Expert system for computer-aided control engineering. In M. Morari & T.J. McAvoy, Eds., Chemical Process Control - CPC III. Elsevier, Amsterdam. pp. 807.

Tzouanas, V.K., Georgakis, C., Luyben, W.L., Ungar, L.H. (1988). Expert multivariable control. Comput. Chem. Eng., 12(9/10).

Umeda, T. and K. Niida (1986). Process control system synthesis by an expert system. Control Theory Adv. Technol., 2.

Whitlow, J.E. and K.A. Debelak (1988). Expert system based process control. AICHE Annual Meeting, Washington, D.C.

Zervos, C.C. and G.A. Dumont (1988). Deterministic adaptive control based on Laguerre series representation. Int. J. Control. 48(6), 2333–2359.

Zhu, Z. and G. Dai (1988). Expert system approach of industrial process adaptive control. Proc. IFAC Workshop on Robust Adaptive Control. Newcastle.

REALIZATION OF AN EXPERT SYSTEM BASED PID CONTROLLER USING INDUSTRY STANDARD SOFTWARE AND HARDWARE ENVIRONMENTS

P. K. Yue, T. H. Lee, C. C. Hang and W. K. Ho

*Dept. of Electrical Engineering, National University of Singapore,
10 Kent Ridge Crescent (0511), Singapore*

Abstract. This paper discusses the implementation architecture of a knowledge-based PID controller capable of performing auto-tuning, fine-tuning, performance analysis and monitoring of processes in real-time. The main hardware and software requirements, and how these requirements are realized using industry standard hardware and software environments are presented. The improvisations needed to provide non-monotonic reasoning and asynchronous information handling for a knowledge-based system operating in a real-time dynamic situation are also described.

Keywords. Knowledge-based system, intelligent PID, auto-tuning, refined Ziegler-Nichols tuning, IEEE 488 bus interface, UNIX.

INTRODUCTION

The PID controller is by far the most widespread control scheme in use in industries. In spite of the availability of a host of tuning methods, it is common knowledge that most industrial PID loops are poorly tuned. One reason is that the tuning procedures are laborious and time-consuming. Another reason, as in the case of Ziegler-Nichols ultimate-cycle tuning, is that the process has to be operated near instability. Furthermore, no matter which technique is used to determine the initial PID coefficients, some further adjustment is inevitable. For processes that are time-varying and require tight control, frequent retuning and fine-tuning based on experience is necessary.

Fine-tuning is tedious and may take several weeks before the process is correctly controlled. Once the tuning is about right, the plant operator may be satisfied and reluctant to allow further testing. This is likely to result in poor control in the next upset and the operator may have to resort to manual control. On top of these, various operational issues in PID control like reset windup, bumpless transfer, performance analysis, abnormal operating conditions and alarms also have to be taken care of.(see [1])

There have been various attempts to automate controller tuning to eliminate the need of operating the plant near instability and to do away with manual fine-tuning by an expert. This paper presents an approach to intelligent PID control where the regulators are coupled to a knowledge-based system incorporating expert heuristics and experiential knowledge to automate controller tuning, performance analysis, fine-tuning and to take care of various operational issues.

The main motivation for a knowledge-based implementation is the separation of the heuristic logic from numerical control algorithms[2]. Heuristic knowledge typically appears in the form of **"if ... then ... else"** logic statements, which are of symbolic logic nature in contrast with controller algorithms that are of mathematical and numerical in nature. Conventional implementation tends to result in program codes where the numerical parts and logic are interspersed with each others. The mixture quickly causes the program to become difficult to understand and difficult to maintain modular development.

Knowledge-based approach provides a structured implementation where the heuristics and the numerics are segregated to as large a degree as possible. The logical complexity of the problem can be more effectively dealt with as the heuristics is implemented in a distinct knowledge base. An inference engine mimicking human reasoning mechanism which makes use of the knowledge base is implemented as a separate part. This facilitates rapid prototyping, incremental and modular development, logic testing and debugging. In addition, a knowledge-based system approach offers interesting potential for user interface. It provides explanation, query and tutoring facilities and permits inspection and manipulation of the contents of knowledge base during operation, e.g. modification and addition of new rules.

Real time application of knowledge-based systems in dynamic situation like process control necessitates several stipulations. This paper discusses the hardware and software requirements which must be met by knowledge-based systems; in particular, the realization of these requirements using industry standard hardware and software environment. Elaboration on how a relatively slow and non-deterministic knowledge-based system is coupled with numerical controller algorithms in an implementation architecture suitable for distributed control is presented. Communication issues in the implementation model are described. The improvisations adopted for the knowledge-based system to deal with dynamically changing incoming information and asynchronous event are also explained.

OVERVIEW OF KNOWLEDGE-BASED INTELLIGENT PID CONTROLLER

Features of Intelligent PID controller

Functionally, the intelligent PID controller can be viewed as an extension to auto-tuning and adaptive control with performance analysis and monitoring. It incorporates a noise estimator, relay tuning scheme, refined tuning scheme, performance analyzer, process monitor and on-line user interface. Figure 1 shows the overall flow chart of the intelligent PID controller. The controller has three different modes of operation: manual, tuning and PID. The operator has commands for switching between these different modes.

The procedure starts in the manual mode where the operator controls the plant manually until it reaches steady state at the desired set-point. Noise estimation procedure is then invoked. The information on plant noise is used to select a suitable relay for tuning. When tuning mode is selected, the system performs a relay experiment to obtain ultimate gain and ultimate period of the plant. At the end of the tuning mode, the PID parameters are computed and the controller will change into the PID mode.

The operator can then invoke a fine-tuning procedure, in which case, the controller will execute a routine to obtain the normalized gain of the plant and refine the PID performance. When a load disturbance comes in, the system will execute performance analysis routine and check whether retuning of the PID controller is necessary. From the PID mode, the operator could change to manual or tuning mode. It is also possible for the operator to inspect and change the content of the knowledge base and to modify and add new rules on-line.

Algorithms

The algorithms contained in the intelligent PID controller are: noise estimation, relay experiment, PID, refined PID and performance analysis. Noise estimation is performed to obtain the plant noise level. Relay feedback scheme which put the plant under on-off feedback control is used to obtain the ultimate gain and period of the plant[3] (Figure 2). The parameters of the relay are selected based on plant noise level and user-defined allowable perturbation. An alternative non-intrusive scheme using correlation technique to obtain ultimate gain and period while the process is under PID control will be implemented[4].

With the ultimate gain and ultimate period, the PID coefficients are first determined using the traditional Ziegler-Nichols tuning formula. Fine-tuning of the PID controller is realized with a refined Ziegler-Nichols tuning formula [5],[6].

$$u_c = k_c\left[(By_r - y) + \frac{1}{T_i}\int e\,dt - T_d\frac{dy_f}{dt}\right]$$

$$T_i = 0.5ut_u$$

The equation is essentially the traditional PID formula which is modified to include a constant weighting factor, ***B***, on the set-point in the proportional term to reduce overshoot. For processes with large dead-time where there are excessive undershoot and sluggish load response, the integral term is modified by a factor, ***u***, in addition to the set-point weighting. The modification factors ***B*** and ***u*** are determined using heuristics based on knowledge of the normalized gain of the plant. Controller performance analysis is done by examining the peak load error of the plant output. Based on several heuristics rules, the performance could be analyzed and the need for controller retuning could be determined.

Heuristics Rules

The heuristic rules are implemented in a knowledge base to separate them from pure numerical algorithms. The heuristics are implemented in production rules knowledge representation scheme motivated by its declarative nature. The knowledge base is written using an expert system shell, NEXPERT with a rule format

if ⟨ *conditions* ⟩ **then** ⟨ *hypothesis* ⟩
do ⟨ *actions* ⟩

which is compatible with the "**if** ... **then** ... **else**" nature of controller heuristics. A typical rule of the system is as follows:

```
(@RULE= R11
     (@LHS=
          (Yes     (normalized_gain_obtained))
          (Is      (pid.structure) ("PID"))
          (>       (normalized.gain)         (2.25))
          (<       (normalized.gain)         (15))
          (=       (acceptable.overshoot)    (20))
     )
     (@HYPO= refined_pid_control)
     (@RHS=
          (Do      (36/(27+5.0*normalized.gain))   (pid.betta))
          (Do      (ultimate.gain*0.6)        (pid.pro))
          (Do      (ultimate.period*0.5)      (pid.int))
          (Do      (ultimate.period*0.125)    (pid.der))
          (Execute   ("communicate")   (@ATOMID=pid.pro,pid.int,
                                              pid.der,pid.betta;
                                          @STRING="refinedpid";))
```

Figure 3 shows one module of the rule network of the intelligent PID controller. The heuristics rules are grouped according to their context into several modules as follows:

Noise estimation rules: Rules that oversee noise estimation and determine relay parameters.

Relay rules: Rules related to relay experiment; e.g. determining and adjusting relay parameters, determining when oscillation is steady and computing the ultimate gain and period.

Ziegler-Nichols tuning rules: Rules that compute PID coefficients using the conventional Ziegler-Nichols formula.

Fine tuning rules: Rules that refine the PID performance using refined Ziegler-Nichols formula.

PID performance analysis rules: rules that check the PID controller performance by evaluating the plant peak load error.

PID monitoring rules: Rules that handle abnormal conditions, bumpless transfer and reset windup.

Command decoding rules: Rules that handle operator commands.

HARDWARE REQUIREMENTS OF DISTRIBUTED KNOWLEDGE-BASED CONTROL SYSTEM

Overall implementation structure

Information processing system for real-time problems must, by definition, be able to process incoming information sufficiently fast to meet the time constraints imposed by the external environment. One approach is to construct hardware e.g. symbolic processor, capable of executing intelligent programs as fast as existing algorithmic programs. An alternative approach, which is more suited to existing hardware, is to use a hierarchical architecture in which a fast, precise algorithmic controller has its performance monitored and its algorithms adjusted by a flexible, heuristic, but probably slow, supervisory processor.

The implementation structure of the intelligent PID controller is shown in figure 4. The system is implemented on an HP9000/360 workstation running UNIX and an INTEL 8096-based microcontroller board built in-house.

The HP9000/360 workstation constitutes the upper layer of the hierarchy with 4 main processes running concurrently: a knowledge-based process, a graphic display process, an operator interface process and an IEEE 488 interface process. The knowledge-based process oversees and supervises several microcontroller numerical algorithms on-line. This involves starting and stopping of algorithms, calculating algorithm parameters, analyzing results from identification algorithms, reacting correctly on alarms from monitoring algorithms, etc. The graphic display and operator interface processes provide man-machine communication. The IEEE 488 communication process exchanges information with microcontroller boards.

On the lower layer of the hierachy are several microcontroller boards where various control and identification algorithms are implemented. The microcontrollers compute and send the control output to the plant.

Inter-processor Communication

Communication interface between different processors in a hierarchical architecture must supports many system throughput requirements. IEEE 488 standard is adopted for several reasons. The bus interface functions are comprehensive for the intended application, convenient and relatively easy to implement with the given hardware. The UNIX operating system on the HP9000/360 provides system calls for the IEEE 488 interface. Dedicated IEEE 488 chips that implement the IEEE 488 functions are available for easy connection to microcontrollers. Parallel interface also provides higher data transfer rates in addition to richer interface functions in comparison with a serial interface like RS232C.

IEEE 488 bus interface is a byte-serial bit-parallel asynchronous data transfer standard using 3-wire handshake interlock. The standard allows up to a maximum of 15 devices on one contiguous bus. Each bus connector has complementing female and male parts that allows one to be piggy-backed onto the other. This facilitates daisy-chaining of devices on the bus into a star or linear network.

Devices on the bus can assume the functions of controller, talker and listener. A controller exercises control over all interface functions on the bus. It is responsible for issuing commands to the bus. A talker is a device authorized to place data on the bus and a listener is a device authorized to accept all device-dependent messages from the bus. In this implementation, the upper layer HP9000/360 is assigned as the bus controller. The distributed microcontrollers connected to the bus are assigned to be a talker or a listener by the controller.

Figure 5 shows the information flow diagram of the intelligent PID controller. A dedicated process on the HP9000/360 oversees communication with the distributed microcontrollers through the IEEE 488 interface and acts as the controller of the bus. Information are obtained from a particular microcontroller by addressing it as a talker and HP9000/360 as listener. Upon being addressed as a talker, the microcontroller sends the plant input, output and set-point error values to the bus interface. These values are read and logged into the shared memory on the HP9000/360. In return, operator and knowledge-based system commands are sent to the microcontrollers. In this case, the microcontrollers are addressed as listeners and HP9000/360 as a talker.

Apart from providing the basic functions of controller, talker and listener, the IEEE 488 standard also specifies the following five secondary interface functions imperative in a distributed control system.

Remote/Local: This function provides the capability for a microcontroller on the bus to select between two sources of input information. Local corresponds to front panel controls and remote to the input information from the bus.

Service Request: This function permits a microcontroller to asynchronously request service from the bus interface controller.

Device Clear: This function allows a microcontroller to be initialized to a pre-defined initial state.

Device Trigger: This function permits a microcontroller to have its designer-specified operation initiated by any talker on the bus.

Serial/Parallel Poll: This function allows the bus interface controller to poll microcontrolllers attached to the bus either one by one or as a group for status report.

Development effort is sharply reduced with these functions provided by the bus standard. In a serial interface like the RS232, these functions would have to be explicitly implemented in software.

SOFTWARE REQUIREMENTS FOR KNOWLEDGE-BASED CONTROL

Concurrent processes implementation

In knowledge-based control, the coupling of expert system and numerical algorithms must meet certain real-time demands. Rule execution in a knowledge-based system is essentially a large search problem which is inherently slow and non-deterministic. On the other hand, numerical control algorithms must be able to compute an output within a fixed sampling interval. It cannot be halted or delayed by a knowledge-based system searching among different rules. As a result, the usual practice of coupling an expert system and numerical algorithms, that is, by making an expert system call external numerical routines or by embedding an expert system in conventional procedural languages, is not suitable in this application. The solution is to implement the two differing parts as communicating concurrent processes with different priorities. The numerical algorithms must be given highest priority.

The knowledge-based system, graphic display, IEEE 488 communication and operator interface are implemented as concurrent processes in HP9000/360 workstation under the UNIX operating system. The UNIX multiprocessing structure provides an excellent tool for organizing this application. The UNIX process environment has a hierarchical tree structure. A process may start other child processes to accomplish subtasks within an application or concurrently execute another program. The following UNIX system calls for process creation and manipulation are used in this application.

fork, vfork: the basic process creation primitive used to create new process by duplicating the calling process.

exec: a family of system calls that transform a process by overlaying its memory space with a new program.

exit: used to terminate a process.

Figure 6 shows the process tree structure of the implementation. The knowledge-based system is implemented as the parent process. Immediately upon execution, it spawns 3 child processes: graphic display, user interface and IEEE 488 communication. This is accomplished by first duplicating the knowledge-based system process by calling ***vfork*** and then overlaying the new duplicated child process with a new program accordingly using *exec*. Several numerical algorithmic processes will be spawned and terminated as and when it is necessary by the knowledge-based system to perform various calculation.

Priority-based pre-emptive scheduling

In this multi-process implementation, it is vital that the process scheduling policy be priority-based and pre-emptive. Priority-based means that a more important process can be assigned a priority higher than other processes, so that the important process will be executed before other processes. Pre-emptive means that the high priority process can interrupt or pre-empt the execution of a lower priority process. Unfortunately, the scheduling policy of the UNIX operating system dynamically adjusts process priorities, favouring interactive processes with light CPU usage at the expense of those using CPU heavily. This is unacceptable in this application, as computation intensive numerical algorithms processes must be given higher priority than I/O intensive user interface processes.

Traditional UNIX provides users some control of priorities with the *nice(2)* system call, but the *nice* value is only one factor in the scheduling formula. As a result, it is difficult to guarantee that one process has a priority greater than another process. The UNIX operating system on HP9000/360, known as HP-UX overcome this problem by adding a new range of priorities from 0 to 127, called real-time priorities, with 0 being the highest priority. These real-time priorities do not fluctuate over time and can be set with the *rtprio(2)* system call. The real-time stringent processes like IEEE 488 communication and numerical algorithms processes are given a real time priority of 1. The knowledge-based process is given priority 2 and the user interface processes are given priority 3.

Memory locking of processes

A second important feature in a real-time system is the ability to lock a process in memory so that it can execute without waiting to be paged in from the disc. In UNIX operating system, processes are not normally locked in memory; they are paged in from disc as needed. The time required to page in a process, or page in one of more pages of a process can range from several milliseconds to several seconds, which violates the response time requirements of many real-time applications.

To avoid unexpected swapping or paging, the executable code and data of a process can be locked into the system memory using the UNIX system call *plock(2)*. In this implementation, all processes are locked into memory for immediate execution.

Inter-process communication

The inter-process information exchange within the HP9000/360 is accomplished by two UNIX inter-process communication facilities: signals and shared memory. A signal is essentially a software interrupt. A process can send a signal to inform another process that an event has occurred, and then the other process can immediately enter its interrupt handler to respond to the event. However, the disadvantage of using signals is that they pass little or no data. To overcome this, shared memory is used as an additional communication tool. Two or more processes can attach the same segment of memory to their data space and then write to and read from it. As shared memory constitutes a critical region where uncontrolled access can cause data inconsistency, semaphore synchronization is used to mediate access. Table 1 below lists the various system calls provided by UNIX that permit easy operations to create, gain access, get status information or set control values of inter-process communication facilities.

Inter-process communication facility	Associated system calls
Using shared memory for high-bandwidth communication	shmget(2) shmctl(2) shmop(2)
Synchronizing access to resources with semaphores	semget(2) semctl(2) semop(2)
Using signals as an asynchronous notification mechanism	sigvector(2) signal(2) kill(2)

Table 1. UNIX inter-process communication system calls

In this implementation, plant input, output and set-point values are each stored in a four-kilobytes shared memory buffer with a pointer indicating the most recent value. This buffer keeps a reasonably long historical record of plant data. Older data could be stored in a file on the system hard disk. These data are used for trend plotting, inferencing and various calculations. In addition, several shared memory buffers are set up for commands and parameters passing between different processes. For instance, the knowledge-based process has a dedicated communication shared memory buffer where asynchronous information from external environment (e.g. the operator or microcontrollers) is channeled through. An asynchronous software interrupt is used to notify the knowledge-based process on the arrival of information.

Nonmonotonic reasoning

Process control environment is dynamic and changing. As new information comes in, the basis for some drawn conclusions may no longer be valid. To maintain true facts in such non-monotonic environment, the system must be able to retract these previous conclusions and resolve contention of new facts contradicting old facts. Many truth maintenance systems have been proposed: extended logic, meta-system handling of non-monotonicity, data validity time tag, etc. However, the theoretical approaches to non-monotonic reasoning have yet to mature. The usual approach taken in applications is to use various ad hoc methods to circumvent the problems. The method adopted in this implementation is to explicitly ensure that the database is consistent for each case. Consider the following two rules when the conditions or antecedents are implemented as boolean.

```
(@RULE= R1
        (@LHS=
                (Yes    (manual_control))
        )
        (@HYPO= process_under_manual)
        (@RHS=
                (Execute        ("manual_control_action"))
        )
)

(@RULE= R2
        (@LHS=
                (Yes    (pid_control))
        )
        (@HYPO= process_under_pid)
        (@RHS=
                (Execute        ("pid_control_action"))
        )
```

When ***manual_control*** is volunteered as true, the first rule fires. The hypothesis ***process_under_manual*** becomes true and the corresponding action to control the plant under manual is executed. At a later time, when ***pid_control*** is volunteered as true, the second rule fires with the hypothesis ***process_under_pid*** becoming true. The plant is then put under PID control. Note that unless the old data ***manual_control*** is explicitly declared to be false in the second rule, the condition of ***manual_control*** and other hypotheses related to it remain true. This will cause logical inconsistency in the inference system. To overcome this problem, the same set of rules are implemented in a different way where the conditions are represented as a string value.

```
(@RULE= R1
        (@LHS=
                (Is     (control_mode)  ("manual"))
        )
        (@HYPO= process_under_manual)
        (@RHS=
                (Execute        ("manual_control_action"))
        )
)

(@RULE= R2
        (@LHS=
                (Is     (control_mode)  ("pid"))
        )
        (@HYPO= process_under_pid)
        (@RHS=
                (Execute        ("pid_control_action"))
        )
```

Note that when the ***control_mode*** changes from ***"manual"*** to ***"pid"***, both rules will be affected with one hypothesis becoming true and the other becoming false. The above is an example where non-monotonicity can be overcome by a different rule representation. In certain instances, additional rules have to created to achieve the same purpose.

Asynchronous event handling in knowledge-based system

Real-time control are influenced by asynchronous events in the external environment and information have to accepted as and when they happen to arrive. A knowledge-based system must be able to process these incoming events. However, the information input by any asynchronous mechanism should not create new data or set new data values because this will create an inconsistent inference state and corrupt the working memory. Instead, these information must be set up as an internal queue to be processed by the inference engine when it is in a stable inference state. The expert system, NEXPERT, used to implement the knowledge-based system allows the users to install a queue polling handler which will be called by the inference engine when it is in a stable inference state. The procedure must be installed in the initialization part of the knowledge-based program by calling the following system library:

```
NXP_SetHandler( NXP_PROC_POLLING,
                        Polling_procedure, (Char*)0)
```

CONCLUSIONS

This investigation shows that implementation of an intelligent PID controller for real-time process control is feasible using a hierarchical architecture with separate processors. The numerical controller algorithms is implemented with a dedicated microcontroller for each control loop and the supervisory knowledge-based system in a workstation environment. The communication issues between separate processors can be successfully solved using the IEEE 488 bus which provides comprehensive interface functions. The real-time knowledge-based system can be implemented on the UNIX platform. With the exception of real-time priority system call ***rtprio(2)*** which is unique to HP-UX, all other process management and inter-process communication system calls are defined in the UNIX System V protocol. The hierarchical implementation is also amenable for a distributed control system. As UNIX is a general operating system rapidly becoming available on a wide range of computer hardware, the main hardware and software elements used in this implementation can be easily replicated in other computer systems with X-Windows and a full-function IEEE 488 card. With this portability, moving the implementation to a new system requires less man-months of efforts.

The knowledge-based implementation approach offers several distinct advantages. Separation of heuristics and numerics not only facilitates prototype development, debugging and maintenance. It is also relatively easy to extend the system with new functionality by introducing new algorithms and logic in an incremental and modular manner. In addition, expert system technique provide interesting possibilities in user interface. Embedded knowledge can be inspected and modified on-line. In comparison, outdated knowledge in conventional programs is difficult to handle. Often the users are unaware of its existence and significant programming effort is required for updating.

Experience from testing the intelligent PID controller shows that the PID and PI tuning is accurate over a much larger range of process dynamics. Figure 7 shows a session of relay-experiment, Ziegler-Nichols tuning and fine-tuning. Through set-point weighting and factor modification of the integral term in a refined Ziegler-Nichols formula, the PID controller is tuned to give both good set-point response as well as good load disturbance response. Consequently, the need for manual fine-tuning and human expertise is largely eliminated. In future, the system can be extended with additional functionality like dead-time compensation, feed-forward, pole placement and incorporation of other control schemes like GMV, GPC, etc. The knowledge-based system can also be extended with cooperative reasoning capability in a blackboard architecture.

REFERENCES

[1] Astrom, K. J., J. J. Anton, and K. E. Arzen (1986). Expert Control. Automatica, Vol. 22, No. 3, pp 277-286.

[2] Arzen, K. E. (1987). Realization of Expert System Based Feedback Control. Dept. of Automatic Control, Lund Inst. of Technology. Doctoral Dissertation. LUTFD2/TFRT-1029.

[3] Astrom, K. J. (1982). Ziegler-Nichols auto-tuners. Dept. of Automatic Control, Lund Inst. of Technology, Rep. LUTFD2/TFRT-3167.

[4] Hang, C. C., and K. K. Sin. (1988). On-line Auto-tuning of PID Controllers Based on Cross Correlation. Proceedings of 14th Annual Conference of IEEE Industrial Electronics Society, pp 441-446.

[5] Hang. C. C., and K. J. Astrom, (1987) Refinements of the Ziegler-Nichols Tuning Formula for PID Auto-tuners. Dept. of Automatic Control, Lund Inst. of Technology. Rep. LUTFD2/TFRT-7371.

[6] Hang. C. C., K. J. Astrom, and W. K. Ho.(1989). Refinements of the Ziegler-Nichols Tuning Formula. Dept. of Electrical Engineering, National University of Singapore. Rep. CI-89-9.

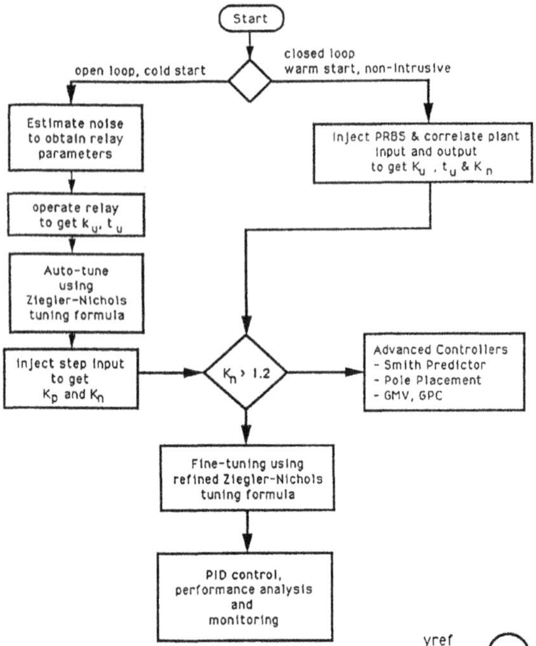

Notations:
K_u, t_u: ultimate gain and period
K_p : plant static gain
$K_n = K_u K_p$: plant normalized gain

Figure 1. Overall Flowchart of Intelligent PID Controller

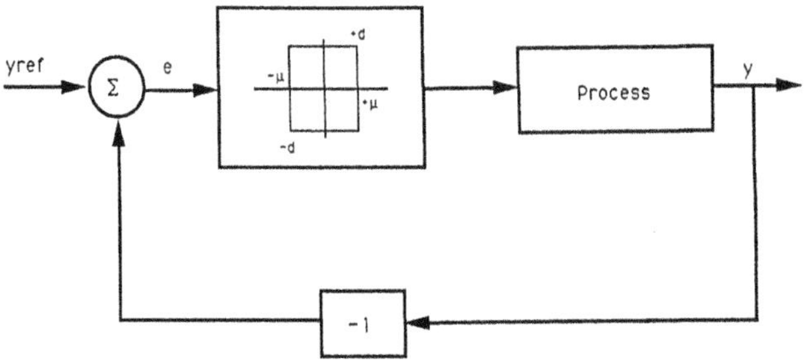

Figure 2. Relay Feedback

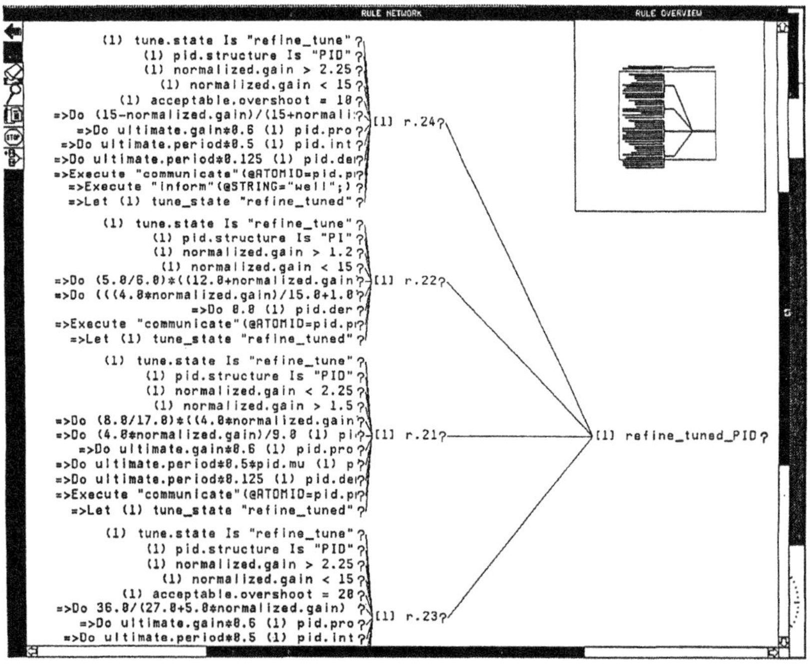

Figure 3. Rule Network of Fine-tuning Heuristics

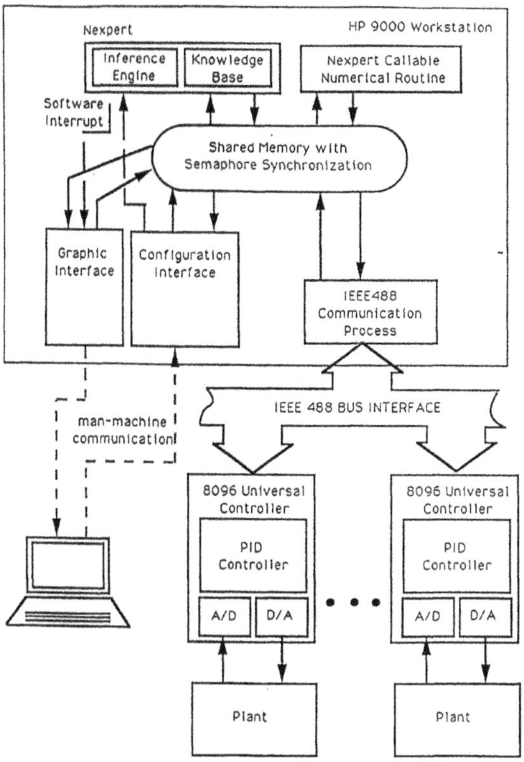

Figure 4. Implementation Architecture of Intelligent PID Controller

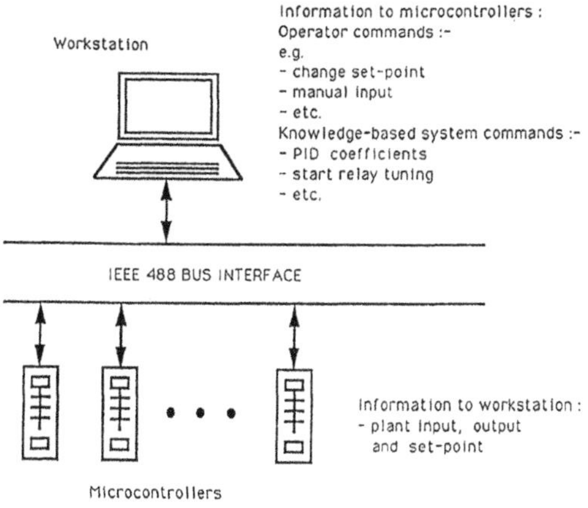

Figure 5. Information of between supervisory workstation and distributed microcontrollers

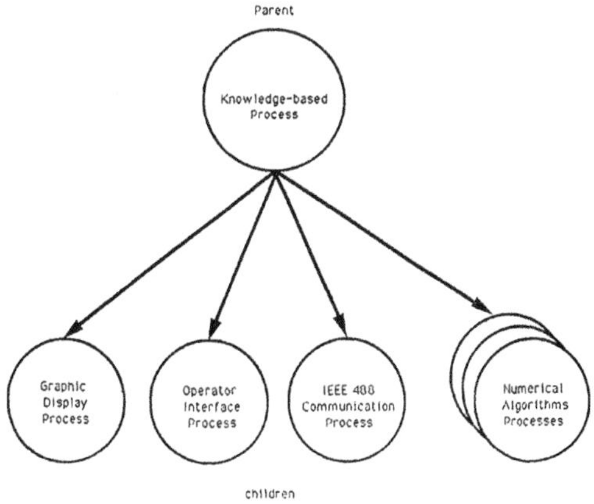

Figure 6. Process Tree of Intelligent PID Controller

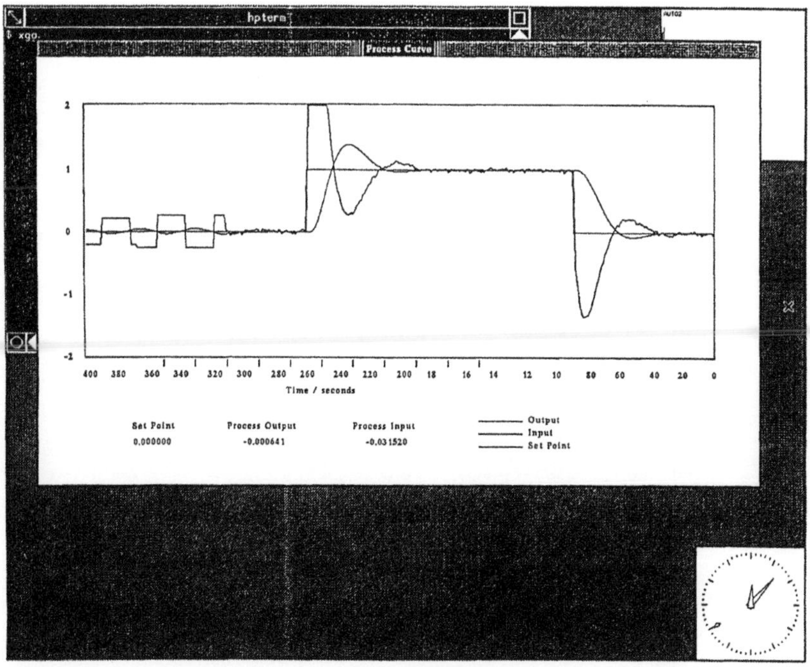

Figure 7. A Session with Intelligent PID controller

ROBUST ADAPTIVE CONTROL SYSTEM BASED ON EXTENDED LEAKY INTEGRATION METHOD

H. Ohmori and A. Sano

Dept. of Electrical Engineering, Keio University, 3-14-1 Hiyoshi, Kohoku-ku, Yokohama 223, Japan

Abstract. The main objective of this paper is to propose a discrete-time model reference robust adaptive control system of a plant in the presence of bounded disturbances. This adaptive control system has a new robust adaptive law whose features are: one is that it belongs to the category of leaky integration method with multiple regularization parameters. It is shown how to theoretically determine these regularization parameters which are regarded as σ in the extended σ-modification to multiple parameters. Furthermore it is one of techniques to assure persistent excitation regardless of the size of disturbances. The other is that, upper bound of disturbances is not necessary to be known *a priori* to design the adaptive control system.

Keywords. Model reference adaptive control; robustness; discrete time systems; least-squares estimation; eigenvalues

INTRODUCTION

The model reference adaptive control systems(MRACS) are designed under the assumption that the plant dynamics are exactly those of one member of a specified class of models. In practice, the true plant is not perfectly described by any model in the given class, then robust stability and performance is very important for the practical applicability of adaptive control laws. Unfortunately, it has been shown(Rohrs and others, 1982) that bounded disturbances or unmodeled stable dynamics can make the adaptive control system unstable.

Several modified adaptive control laws have been proposed (Ortega and Tang, 1989) to achieve the robust stability and performance against the presence of unmodeled dynamics and disturbances. These methods are divided into two major approaches: (i) the leaky integration approach which includes the fixed σ-modification(Ioannou and Kokotovic, 1983, 1984; Johnson, 1988), the error-weighted modification (Narendra and Annaswamy, 1986; Rey and Johnson, 1987) and the switching σ-modification(Ioannou and Tsakalis, 1986; Hsu and Costa, 1987). (ii) the error dead zone modification approach(Peterson and Narendra, 1982; Kreisselmeier and Narendra, 1982; Goodwin and Sin, 1984; Ortega and Lozano-Leal, 1987) which includes relative dead zone method(Kreisselmeier and Anderson, 1986), etc.

Both approaches are based on the same basic idea that since the regressor and the output signal is violated by the effects of the unmodeled dynamics and/or disturbances, the adaptive control law abandons the action of the exactly parameter adjustment. In the leaky integration approach, introducing the leaky integration(Johnson, 1988) in the adaptive control law suggests that the adaptive control law with this integration should have the permissive of the parameter adjustment. This approach requires the knowledge of a bound on the norm of the desired controller parameters. In the dead zone approach, a dead zone is introduced in the adaptive control law such that adaptation is stopped when the augmented error becomes smaller than a certain threshold. It means that the dead zone approach has also the permissive of parameter adjustment. In order to choose the size of the dead zone appropriate, an upper bound on norm of the desired constant controller parameter vector and/or the disturbances must be known.

It is also well known that the condition for exponential convergence of the estimated parameter is that the measured vector signal is persistently exciting(Anderson, 1982, 1985), which can be used to establish robustness and stability in the presence of bounded disturbances and/or unmodeled dynamics. The persistently excitation condition implies the invertibility of a certain time-varying matrix which is made from regressors. In the presence of unmodeled dynamics and/or disturbances, since input signal is strongly correlated, then persistently excitation condition is not satisfied, that is, the problem of loss of identifiablity due to feedback control loop, of what we call ill-posed problem has occurred. A well-known method for overcoming the ill-posed problem is the Tikhonov regularization method(Tikhonov and Arsenin, 1977) in which a regularization parameter is added into the diagonal elements of the ill-conditioned matrix(Hoerl and Kennard, 1970; Draper and Nostrand, 1979).

It is shown that the regularization parameter can be regarded as σ in σ-modification(Ohmori and Sano, 1989). Consider the following system which described by the regression form in the presence of disturbances.

$$y(k) = \phi^T(k)\theta_* + d(k), \qquad k = 1, 2, \cdots \quad (1)$$

Here $\phi(k)$ is regressor, θ_* is unknown constant parameter vector and $y(k)$ is available signal.

We will consider the loss function denoted by

$$J(\hat{\theta}, k) = \frac{1}{2}\frac{\left(y(k) - \hat{\theta}^T(k)\phi(k)\right)^2}{1 + \phi^T(k)\phi(k)}$$

$$+\frac{1}{2}\rho(k)\left(\hat{\theta}(k)-\theta_0(k)\right)^2 \quad (2)$$

where $\hat{\theta}(k)$ is the parameter estimate vector, $\rho(k)$ is referred to a regularization parameter. The gradient descent of $J(\hat{\theta},k)$ results in the updating the parameter estimate vector:

$$\begin{aligned}\hat{\theta}(k+1) &= \hat{\theta}(k)+\frac{\mu\phi(k)\varepsilon(k)}{1+\phi^T(k)\phi(k)} \\ &\quad -\mu\rho(k)[\hat{\theta}(k)-\theta_0(k)] \end{aligned} \quad (3)$$

where

$$\varepsilon(k) \equiv y(k)-\phi^T(k)\hat{\theta}(k), \quad 0<\mu<2 \quad (4)$$

If $\rho(k)=0$, Eq.(3) is known as normalized LMS.

We now look at two special forms of the algorithm by the difference of the choice of $\theta_0(k)$. First interesting observation is that if we let $\theta_0(k)=\mathbf{0}$, then the algorithm above reduces to the leaky integration approach:

$$\begin{aligned}\hat{\theta}(k+1) &= (1-\mu\rho(k))\hat{\theta}(k) \\ &\quad +\frac{\mu\phi(k)\varepsilon(k)}{1+\phi^T(k)\phi(k)} \end{aligned} \quad (5)$$

If the regularization parameter $\rho(k)$ is chosen as

$$\rho(k) = \sigma/\mu, \quad 0<\sigma<1$$

then Eq.(5) is the fixed σ-modification. Let $\rho(k)$ be chosen as

$$\rho(k) = \frac{\gamma/\mu|\varepsilon(k)|}{\kappa+\gamma|\varepsilon(k)|}, \quad \gamma,\kappa>0$$

then Eq.(5) is the error-weighted modification.

Second choice for $\theta_0(k)=\hat{\theta}(k+1)$, then the Eq.(3) becomes the dead-zone approach:

$$\begin{aligned}\hat{\theta}(k+1) &= \hat{\theta}(k) \\ &\quad +\frac{\mu}{1-\mu\rho(k)}\frac{\phi(k)\varepsilon(k)}{1+\phi^T(k)\phi(k)} \end{aligned} \quad (6)$$

where

$$\rho(k)=\begin{cases}\infty, & \text{if} \quad |\varepsilon(k)|<2\bar{d} \\ 0, & \text{if} \quad |\varepsilon(k)|\geq 2\bar{d},\end{cases}$$

$$|d(k)|<\bar{d}, \quad {}^{\forall}k$$

The main objective of this paper is to propose a discrete-time model reference robust adaptive control system containingrobust adaptive law with multiple regularization parameters. The proposed robust adaptive law is derived from minimizing a generalized criterion which includes multiple regularization parameters(Hemmerle, 1975) and an *a priori* estimate. In this paper, the existence of theoretically optimal multiple regularization parameters is clarified, in the sense of minimizing the error between true controller parameters and estimated parameters. These optimal parameters, however, involve the unknown plant parameters and disturbances. Hence, introducing the suboptimal regularization parameters in the sense of minimizing the upper bound on parameter error, we show the robust adaptive law including only accessible data signals.

This adaptive law has the following features:

(i) It belongs to the category of leaky integration method with multiple σ parameters;

(ii) The upper bound of disturbances is not necessary to be known *a priori* to design the adaptive control system.

SYSTEM DESCRIPTION AND PROBLEM STATEMENT

Consider the following single-input, single-output(SISO), linear-time-invariant(LTI) system,

$$y(k) = G(z)u(k)+v(k) \quad (7)$$

where

$$\begin{aligned} G(z) &\equiv \frac{z^{-L}B(z^{-1})}{A(z^{-1})}, \quad L\equiv n-m\geq 1 \\ A(z^{-1}) &\equiv 1+\sum_{i=1}^{n} a_i z^{-i}, \quad a_i\in\mathcal{R}^1, \\ B(z^{-1}) &\equiv \sum_{i=0}^{m} b_i z^{-i}, \quad b_i\in\mathcal{R}^1, \quad (b_0\neq 0). \end{aligned}$$

where $\{y(k)\}$ and $\{u(k)\}$ are the measured output and input, respectively, $\{v(k)\}$ represents the effect of disturbances. z^{-1} is the unit delay operator. We will make the following assumptions.

(A1) Structure of the parametric part
The order n and delay time L are known. Furthermore $A(z^{-1})$ and $B(z^{-1})$ are relatively prime.

(A2) Bounded disturbance
The disturbance $\{v(k)\}$ has an upper bound $\bar{v}$, ($|v(k)|<\bar{v},{}^{\forall}k$), which is not necessary to be known *a prior.*

(A3) Nonminimum phase system
The zeros of the polynomial $B(z^{-1})$ lie strictly inside the unit circle.

The desired output $\{y_M(k)\}$ satisfies the following reference model:

$$y_M(k) = G_M(z)r(k) \quad (8)$$

where

$$\begin{aligned} G_M(z) &\equiv \frac{z^{-L_M}B_M(z^{-1})}{A_M(z^{-1})}, \quad L_M\equiv n_M-m_M \\ A_M(z^{-1}) &\equiv 1+\sum_{i=1}^{n_M} a_{Mi} z^{-i}, \quad a_{Mi}\in\mathcal{R}^1, \\ B(z^{-1}) &\equiv \sum_{i=0}^{m_M} b_{Mi} z^{-i}, \quad b_{Mi}\in\mathcal{R}^1. \end{aligned}$$

We also impose the following additional constraints.

(A4) $A_M(z^{-1})$ is asymptotically stable (having no zeros for $|z| \geq 1$).

(A5) The delay in the reference model, L_M, should be greater than or equal to the delay, L, in the plant, i.e. $L_M \geq L$.

(A6) $\{r(k)\}$ is uniformly bounded.

Now it is problem considered in this paper to design the adaptive control system with extended leaky integration algorithm so that the adjusted controller parameter $\hat{\theta}(k)$ tracks the model-matching parameter θ_* as closely as possible against disturbances.

CONTROLLERS STRUCTURE

This section presents the controllers structure for minimum phase plants in the presence of disturbances. We introduce any asymptotically stable polynomial $D(z^{-1})$ which is described by

$$D(z^{-1}) \equiv 1+\sum_{i=1}^{n} d_i z^{-i}, \quad d_i \in \mathcal{R}^1. \quad (9)$$

Since $A(z^{-1})$ and z^{-L} are relatively prime, there exist uniquely polynomials $R(z^{-1})$ and $S(z^{-1})$ which can satisfy following Diophantine equation in polynomials:

$$D(z^{-1}) = A(z^{-1})R(z^{-1}) + z^{-L}S(z^{-1}) \quad (10)$$

where

$$R(z^{-1}) \equiv 1+\sum_{i=1}^{L-1} r_i z^{-i}, \quad r_i \in \mathcal{R}^1,$$

$$S(z^{-1}) \equiv \sum_{i=0}^{n-1} s_i z^{-i}, \quad s_i \in \mathcal{R}^1.$$

Furthermore by defining

$$\begin{aligned} B_R(z^{-1}) &\equiv R(z^{-1})B(z^{-1}) - b_0 \\ &\equiv \sum_{i=1}^{n-1} b_{Ri} z^{-i}, \quad b_{Ri} \in \mathcal{R}^1, \end{aligned}$$

the plant Eq.(7) can be rewritten as

$$D(z^{-1})y(k+L) = \theta_*^T \phi(k) + d(k+L) \quad (11)$$

where

$$\begin{aligned} \theta_* &\equiv \left[b_0, \ \bar{\theta}_*^T\right]^T \\ &\equiv \left[b_0, b_{R1}, \cdots, b_{R(n-1)}, s_0, s_1, \cdots, s_{n-1}\right]^T, \\ \phi(k) &\equiv \left[u(k), \ \bar{\phi}^T(k)\right] \\ &\equiv [u(k), u(k-1), \cdots, u(k-n+1), \\ &\quad\ y(k), y(k-1), \cdots, y(k-n+1)]^T, \\ d(k) &\equiv R(z^{-1})A(z^{-1})v(k). \end{aligned}$$

The adaptive controllers structure is given by

$$\begin{aligned} u(k) &= \frac{1}{\hat{b}_0(k)} \Big[z^L D(z^{-1})G_M(z^{-1})r(k) \\ &\quad -\hat{\bar{\theta}}^T(k)\bar{\phi}(k)\Big] \end{aligned} \quad (12)$$

or

$$D(z^{-1})y_M(k+L) = \hat{\theta}^T(k)\phi(k) \quad (13)$$

where

$$\hat{\theta}(k) \equiv \left[\hat{b}_0(k), \hat{\bar{\theta}}^T(k)\right]^T \in \mathcal{R}^{2n}.$$

Defining the output error between the reference model and the plant as

$$e(k) \equiv y(k) - y_M(k), \quad (14)$$

and using adaptive control input Eq.(12), the output error is obtained as follows:

$$\begin{aligned} D(z^{-1})e(k+L) &= \psi^T(k)\phi(k) \\ &\quad + d(k+L) \end{aligned} \quad (15)$$

where $\psi(k) \equiv \theta_* - \hat{\theta}(k)$ is the parameter error vector.

In next section, we will introduce the extended leaky integration adaptive algorithm even if in the presence of disturbances.

EXTENDED LEAKY INTEGRATION METHOD

Combining Eq.(11) from the time instant $k - N + 1$ to k into the vector-matrix form gives

$$D(z^{-1})\boldsymbol{y}(k) = \boldsymbol{\Phi}(k-L)\theta_* + \boldsymbol{d}(k) \quad (16)$$

where

$$\begin{aligned} \boldsymbol{y}(k) &\equiv [\ y(k-N+1), \ \cdots, \ y(k)\ ]^T \in \mathcal{R}^N \\ \boldsymbol{\Phi}(k) &\equiv [\ \phi(k-N+1), \ \cdots, \ \phi(k)\ ]^T \in \mathcal{R}^{N\times 2n} \\ \boldsymbol{d}(k) &\equiv [\ d(k-N+1), \ \cdots, \ d(k)\ ]^T \in \mathcal{R}^N. \end{aligned}$$

We require $2n \leq k \leq N$. Multiplying both sides of Eq.(16) by $\boldsymbol{\Phi}^T(k-L)/N$ from the left, we can obtain

$$\boldsymbol{h}(k) = \boldsymbol{R}(k-L)\theta_* + \boldsymbol{w}(k) \quad (17)$$

where

$$\begin{aligned} \boldsymbol{h}(k) &\equiv \boldsymbol{\Phi}^T(k-L)\left[D(z^{-1})\boldsymbol{y}(k)\right]/N \in \mathcal{R}^{2n}, \\ \boldsymbol{w}(k) &\equiv \boldsymbol{\Phi}^T(k-L)\boldsymbol{d}(k)/N \in \mathcal{R}^{2n}, \\ \boldsymbol{R}(k) &\equiv \boldsymbol{\Phi}^T(k)\boldsymbol{\Phi}(k)/N \in \mathcal{R}^{2n\times 2n}. \end{aligned}$$

In order to determine the parameter estimate $\hat{\theta}(k)$ which is estimated from the data on the observation interval $[k-N+1, k]$, we will consider the general form of the weighted criterion denoted by

$$J(\hat{\theta},k) \equiv \frac{1}{N}\|D(z^{-1})y(k) - \Phi(k-L)\hat{\theta}(k)\|^2 + \left(\hat{\theta}(k)-\theta_0\right)^T Q(k)\left(\hat{\theta}(k)-\theta_0\right) \quad (18)$$

where θ_0 is an *a prior* estimate of the true controller parameter θ_*, and $Q(k)$ is referred to a regularization matrix. Note that Eq.(18) is extension to the least squares type criterion of Eq.(2). The ordinary least squares criterion has $Q(k) = O$ and k=N in Eq.(18). We will investigate the effects of inducing the both θ_0 and $Q(k)$ in Eq.(18).

The parameter estimate $\hat{\theta}(k)$ that minimizes the criterion function Eq.(18) is given by

$$\hat{\theta}(k) = F(k)^{-1}g(k) \quad (19)$$

where

$$F(k) \equiv R(k) + Q(k), \quad (20)$$
$$g(k) \equiv h(k) + Q(k)\theta_0. \quad (21)$$

We take a specified form of $Q(k)$ as

$$Q(k) \equiv V(k)P(k)V(k)^T, \quad (22)$$
$$P(k) \equiv \mathrm{diag}\left[\ \rho_1(k), \ \cdots\ , \rho_{2n}(k)\ \right] \quad (23)$$

where $\rho_i(k)$ are the regularization parameters and $V(k) \equiv \left[\ v_1(k), \ \cdots\ , v_{2n}(k)\ \right]$ where $\{v_i(k)\}$ are the normal eigenvectors of $R(k)$, i.e.

$$R(k) \equiv V(k)\Lambda(k)V(k)^T, \quad (24)$$
$$\Lambda(k) \equiv \mathrm{diag}\left[\ \lambda_1(k), \ \cdots\ , \lambda_{2n}(k)\ \right] \quad (25)$$

where $\lambda_i(k)$ is the eigenvalue of $R(k)$ corresponding to the eigenvector $v_i(k)$. Hence the solution Eq.(19) reduces to

$$\hat{\theta}(k) = \left\{\sum_{j=1}^{2n}\frac{v_j(k)v_j(k)^T}{\rho_j(k)+\lambda_j(k)}\right\}\{h(k)+Q(k)\theta_0\}$$
$$= \left\{\sum_{j=1}^{2n}\frac{v_j(k)v_j(k)^T}{\rho_j(k)+\lambda_j(k)}\right\}\{h(k)+\rho_j(k)\theta_0\}. \quad (26)$$

Using Eqs.(19), (20) and (21) gives

$$\begin{aligned}\psi(k) &= \theta_* - F^{-1}(k)(h(k)+Q(k)\theta_0)\\ &= F^{-1}(k)\left[F(k)\theta_* - h(k) - Q(k)\theta_0\right]\\ &= F^{-1}(k)\left[Q(k)\psi_0 + R(k)\theta_* - h(k)\right]\\ &= F^{-1}(k)\left[Q(k)\psi_0 - w(k)\right] \quad (27)\end{aligned}$$

where

$$\psi_0 \equiv \theta_* - \theta_0.$$

Remark 1: If $Q(k) = O$, i.e. in the case of the ordinary least squares estimate, from Eq.(27) we get

$$\psi(k) = R^{-1}(k)w(k). \quad (28)$$

From this expression, we note that when $R(k)$ has small eigenvalues, the parameter error increases(Adachi and Sano, 1986). The follwing theorem gives the multiple regularization parameters to be determined optimally.

Theorem 1: (Optimal Regularization Parameters) *The following parameter error criterion*

$$I(k) \equiv \|\psi(k)\|^2 \quad (29)$$

is minimal for the regularization parameters $\rho_i^0(k)$ such that

$$\rho_i^0(k) = \frac{v_i^T(k)w(k)}{v_i^T(k)\psi_0}, \quad \text{for } i = 1,2,\cdots,2n. \quad (30)$$

Proof:

$$\frac{\partial}{\partial Q}\|Q(k)\psi_0 - w(k)\|^2 = (Q(k)\psi_0 - w(k))\,\psi_0^T = O.$$

Hence, $Q(k)\psi_0\psi_0^T = w(k)\psi_0^T$. Using Eqs.(22) and (23), we can obtain Eq.(30).

Since the theoretically optimal regularization parameters involve the unknown true parameters θ_* and unknown $w(k)$ which is affected by disturbances, it can not be used actually for the regularization parameters. Therefore we will show the suboptimal method which can overcome the above difficulties.

In order to derive the suboptimal regularization parameters based on accessible data, we will introduce the upper bound of parameter error criterion.

Lemma 1: *From Eqs. (27) and (29), we can get the following relation.*

$$I(k) \equiv \|\psi(k)\|^2 \le I_1(k;\theta_*) + \bar{I}_2(k;\eta) \equiv \bar{I}(k;\theta_*,\eta) \quad (31)$$

where

$$I_1(k;\theta_*) \equiv 2\sum_{j=1}^{2n}\left\{\frac{\rho_j(k)}{\lambda_j(k)+\rho_j(k)}\right\}^2 \cdot\left\{v_j^T(k)(\theta_*-\theta_0)\right\}^2, \quad (32)$$

$$\bar{I}_2(k;\eta) \equiv 2\sum_{j=1}^{2n}\frac{\lambda_j(k)}{\{\lambda_j(k)+\rho_j(k)\}^2}\eta^2(k) \quad (33)$$

$$\eta^2(k) \equiv \frac{\|d(k)\|^2}{N}. \quad (34)$$

Proof: See Ohmori and Sano(1989).

Remark 2: Note that $I_1(k;\theta_*)$ increases monotonically with respect to the regularization parameter, and $\bar{I}_2(k;\eta)$ decreases monotonically with respect to the regularization parameter. So we can see that there are the regularization parameters which can attain to minimize $\bar{I}(k;\theta_*,\eta)$.

We will find the suboptimal regularization parameters $\rho_i^{sub}(k)$ which minimize the new criterion, the upper bound $\bar{I}(k;\boldsymbol{\theta}_*)$ instead of $I(k)$ with respect to $\rho_j(k)$ using only accessible data.

Since $\bar{I}(k;\boldsymbol{\theta}_*,\eta)$ in Eq.(31) involves the unknown true controller parameter $\boldsymbol{\theta}_*$ and $\eta(k)$ from the unknown disturbances, then it is impossible to evaluate $\bar{I}(k;\boldsymbol{\theta}_*,\eta)$ actually. Therefore using the estimated value, we shall replace the true parameter $\boldsymbol{\theta}_*$ in Eq.(32) and $\eta^2(k)$ in Eq.(33) as follows,

$$\bar{\boldsymbol{\theta}}(k,\{\mu_m\}) = \left\{\sum_{j=1}^{2n} \frac{\boldsymbol{v}_j(k)\boldsymbol{v}_j(k)^T}{\mu_j(k)+\lambda_j(k)}\right\} \bar{g}(k,\{\mu_m\}) \quad (35)$$

$$\begin{aligned}\bar{\eta}^2(k,\{\mu_m\}) &= \frac{1}{N}\|D(z^{-1})y(k) - \boldsymbol{\Phi}(k-L)\bar{\boldsymbol{\theta}}(k,\{\mu_m\})\|^2 \\ &+(\bar{\boldsymbol{\theta}}(k,\{\mu_m\}) - \boldsymbol{\theta}_0)^T \\ &\boldsymbol{V}(k)\boldsymbol{M}(k)\boldsymbol{V}(k)^T(\bar{\boldsymbol{\theta}}(k,\{\mu_m\}) - \boldsymbol{\theta}_0) \quad (36)\end{aligned}$$

where

$$\begin{aligned}\bar{g}(k,\{\mu_m\}) &\equiv \boldsymbol{h}(k) + \boldsymbol{V}(k)\boldsymbol{M}(k)\boldsymbol{V}(k)^T\boldsymbol{\theta}_0 \\ \boldsymbol{M}(k) &\equiv \mathrm{diag}[\mu_1(k),\cdots,\mu_{2n}(k)].\end{aligned}$$

Since the optimum of $\{\rho_j(k)\}$ are not yet obtained, we should employ other regularization parameters $\{\mu_m\}$ for calculating the estimates Eq.(35). Then, using Eqs.(31), (32) and (33), we consider the following upper bound of the estimated parameter error $\bar{\boldsymbol{\theta}}(k,\{\mu_m\}) - \hat{\boldsymbol{\theta}}(k)$ described by

$$\begin{aligned}\hat{\bar{I}}(k;\bar{\boldsymbol{\theta}},\bar{\eta}) &\equiv I_1(k;\bar{\boldsymbol{\theta}}) + \bar{I}_2(k;\bar{\eta}) \\ &= 2\sum_{j=1}^{2n} \frac{\rho_j^2(k)\{\boldsymbol{v}_j^T(k)(\bar{\boldsymbol{\theta}}(k,\{\mu_m(k)\}) - \boldsymbol{\theta}_0)\}^2}{\{\rho_j(k)+\lambda_j(k)\}^2} \\ &+2\sum_{j=1}^{2n} \frac{\lambda_j(k)}{\{\rho_j(k)+\lambda_j(k)\}^2}\bar{\eta}^2(k,\{\mu_m\}). \quad (37)\end{aligned}$$

Now the procedure of determining the regularization parameters $\{\rho_j(k)\}$ and $\{\mu_m(k)\}$, such that $\rho_j(k) = \mu_j(k)$ is satisfied for $j = 1,\cdots,2n$, is summarized as follows: First, for specified $\{\mu_m(k)\}$, calculate the suboptimal $\rho_j^{sub}(k,\{\mu_m(k)\})$ so that Eq.(37) may be minimized, as

$$\rho_j^{sub}(k,\{\mu_m\}) = \frac{\bar{\eta}^2(k,\{\mu_m\})\{\mu_j(k)+\lambda_j(k)\}^2}{\{\boldsymbol{v}_j(k)^T\boldsymbol{h}(k) - \lambda_j(k)\boldsymbol{v}_j(k)^T\boldsymbol{\theta}_0\}^2} \quad (38)$$

Secondly solve the next equation

$$\rho_j^{sub}(k,\{\mu_m\}) = \mu_j(k) \qquad j = 1,\cdots,2n \quad (39)$$

and finally, by using this solution, determine the optimum as $\rho_j^{sub}(k) = \mu_j(k)$.

We will show each procedures to determine the regularization parameter $\{\rho_j^{sub}(k)\}$ for two cases, (a) $\eta^2(k) \le \bar{d}^2$ is known; (b) $\eta^2(k)$ is unknown.

(a) The case that $\bar{d}$ is known.
In this case, the equation Eq.(39) can be solved analytically, we can obtain the following theorem.

Theorem 2: (Suboptimal Regularization Parameters Based on Accessible Data)

Let $\alpha_j(k)$ be denoted by

$$\alpha_j(k) = \frac{\{\boldsymbol{v}_j(k)^T\boldsymbol{h}(k) - \lambda_j(k)\boldsymbol{v}_j(k)^T\boldsymbol{\theta}_0\}^2}{\bar{d}^2} \quad (40)$$

Then the regularization parameters $\rho_j^{sub}(k)$ are given as:

i) *If $\alpha_j(k) \ge 4\lambda_j(k)$, then*

$$\rho_j^{sub}(k) = \frac{\alpha_j(k) - 2\lambda_j(k) - \sqrt{\beta_j(k)}}{2}$$

$$\beta_j(k) \equiv \alpha_j(k)\big(\alpha_j(k) - 4\lambda_j(k)\big)$$

ii) *If $\alpha_j(k) < 4\lambda_j(k)$, then $\rho_j^{sub}(k) = \infty$*

The case of $\rho_j^{sub}(k) = \infty$ implies that the j-th eigenvalue should be discarded. As described in Theorem 1, the theoretically optimum $\rho_j^o(k)$ obtained by using the true $\boldsymbol{\theta}$ never becomes infinity except when $\lambda_j(k) = 0$. On the other hand, the result in Theorem 2 is very interesting in that the optimal $\rho_j^{sub}(k)$ based on the only available input-output data possibly becomes infinity when $\alpha_j(k) < 4\lambda_j(k)$ is satisfied.

(b) The case that $\eta^2(k)$ is unknown.
In this case, we should take the iterative way, as a result, the algorithm can be summarized as follows:

Main Adaptation algorithm:

(Step1) For number of iteration $\ell = 0$, set the initial value $\mu_m^{(0)} = 10^{-8}$, for $m = 1,2,\cdots,2n$.

(Step2) Calculate $\bar{\eta}^2(k,\{\mu_m^{(\ell)}\})$ in Eq.(36) replacing μ_m with $\mu_m^{(\ell)}$, for $m = 1,2,\cdots,2n$.

(Step3) Calculate $\rho_j^{sub}(k,\{\mu_m^{(\ell)}\})$ by Eq.(38), and correct $\mu_j^{(\ell+1)}$ to $\rho_j^{sub}(k,\{\mu_m^{(\ell)}\})$, for $j = 1,2,\cdots,2n$.

(Step4) If it holds that $|\mu_j^{(\ell+1)} - \mu_j^{(\ell)}| \le \delta$ (δ: threshold number), or $\mu_j^{(\ell+1)} > 10^{32}$, then go to Step 5. Otherwise, let $\ell = \ell + 1$ and return Step 2.

(Step5)
Substituting $\rho_j^{sub}(k,\{\mu_m^{(\ell_*)}\})$ into Eq.(26), calculate the estimated controller parameters as follows,

$$\begin{aligned}\hat{\boldsymbol{\theta}}(k) &= \left\{\sum_{j=1}^{2n} \frac{\boldsymbol{v}_j(k)\boldsymbol{v}_j(k)^T}{\rho_j^{sub}(k)+\lambda_j(k)}\right\}\boldsymbol{h}(k) \\ &+\left\{\sum_{j=1}^{2n} \frac{\rho_j^{sub}(k)\boldsymbol{v}_j(k)\boldsymbol{v}_j(k)^T}{\rho_j^{sub}(k)+\lambda_j(k)}\right\}\boldsymbol{\theta}_0\end{aligned}$$

This adaptation algorithm has the following two important properties, one is that if $y(k)$ is bounded, $\hat{\theta}(k)$ is also bounded. Because when the absence of persistent excitation i.e. $^\exists j$, such that $\lambda_j(k) = 0$, $\rho_j^{sub}(k)$ becomes ∞, so $\hat{\theta}(k)$ also becomes *a priori* estimates θ_0, which prevent from bursting of adaptive parameters. The other is that this algorithm can be realized without requiring additional information regarding the plant or the disturbance.

CONCLUSIONS

We have shown how to theoretically determine the regularization parameters regarded as σ in σ-modification. Main idea is that introducing the suboptimal regularization parameters in the sense of minimizing the upper bound on parameter error. Then we have shown the robust adaptive law including only accessible data signals. The major advantage of using the multiple variable regularization parameters is that it is not necessary to know *a prior* knowledge of the upper bound of disturbances. Furthermore, it is one of techniques to assure persistent excitation regardless of the size of disturbance.

REFERENCES

Adachi, S., and A. Sano (1986). Least squares estimation of impulse response in consideration on illcondition of input autocovariance matrix. *Trans. SICE, Vol.22*, pp.1156-1161(in Japanese).

Anderson, B.D.O. (1982). Exponential Convergence and Persistent Excitation. *Proc. of the 21st IEEE Conference on Decision and Control*, Orlando, FL, pp.12-17.

Anderson, B.D.O. (1985). Adaptive Systems, Lack of Persistency of Excitation and Bursting Phenomena. *Automatica, Vol.21*, No.3, pp.247-258.

Draper, N.R., and R.C. van Nostrand (1979). Ridge regression and James-Stein estimation. *Technometrics, Vol.21*, pp.451-466.

Goodwin, C.G., and K.S. Sin (1984). *Adaptive Filtering Prediction and Control*, Englewood Cliffs, NJ:Printice-Hall, pp.88-91.

Hemmerle, W.J. (1975). An explicit solution for generalized ridge regression. *Technometrics, Vol.17*, pp.309-314.

Hoerl, A.E., and R.W. Kennard (1970). Ridge regression: biased estimation for nonorthogonal problems. *Technometrics, Vol.12*, pp.55-67.

Hsu, L., and R. R. Costa (1987). Adaptive control with discontinuous σ -factor and saturation for improved robustness. *Int. J. Control, Vol.45*, No.3, pp.843-859.

Ioannou, P.A., and P.V.Kokotovic (1983). *Adaptive systems with reduced models* , Springer-Verlag.

Ioannou, P.A., and P.V.Kokotovic (1984). Robust redesign of adaptive control. *IEEE Trans. Automat. Contr., Vol. AC-29*, No.3, pp.202-211.

Ioannou, P.A., and K. Tsakalis (1986). A Robust Discrete-time adaptive controller. *Proc. of 25th Conf. on Decision and Control*, Athens, Greece, December, pp.838-843.

Johnson Jr., C.R. (1988). *Lectures on Adaptive Parameter Estimation*, Englewood Cliffs, NJ:Printice-Hall, pp.160-163.

Kreisselmeier, G., and K.S.Narendra (1982). Stable Model Reference Adaptive Control in the Presence of Bounded Disturbances. *IEEE Trans. Automat. Contr., Vol. AC-27*, No.6, pp.1169-1175.

Kreisselmeier, G., and B.D.O.Anderson (1986). Robust Model Reference Adaptive Control. *IEEE Trans. Automat. Contr., Vol. AC-31*, No.2, pp.127-133.

Narendra, K.S., and A.M. Annaswamy (1986). Robust adaptive control in the presence of bounded disturbances. *IEEE Trans. Automat. Contr., Vol. AC-31*, No.4, pp.306-315.

Ohmori, H., and A. Sano (1989). A New Adaptive Law Using Regularization Parameters for Robust Adaptation. *Proc. of the 28th Conference on Decision and Control*, Tampa, Florida, December, pp.1563-1565.

Ortega, R., and R. Lozano-Leal (1987). A Note on Direct Adaptive Control of Systems with Bounded Disturbances. *Automatica, Vol.23*, No.2, pp.253-254.

Ortega, R., and Yu Tang (1989). Robustness of Adaptive Controllers - a Survey. *Automatica, Vol.25*, No.5, pp.651-677.

Peterson, B.B., and K.S.Narendra (1982). Bounded Error Adaptive Control. *IEEE Trans. Automat. Contr., Vol. AC-27*, No.6, pp.1161-1168.

Rey, G.J., D.R. Johnson, Jr. and S. Dasguputa (1987). Tuning leakage for robust adaptive control. *Proc. of 26th Conf. on Decision and Control*, Los Angeles, December, pp.1660-1665.

Rohrs, C.E., L. Valavani, M. Athand, and G, Stein (1982). Robustness of adaptive control algorithms in the presence of unmodeled dynamics. *Proc. 21th IEEE Conf.Decision Contr.*, Orlando, FL, pp.3-11.

Sano, A., H. Ohmori, and M. Furuya (1989). Least square type of adaptive filter with optimally. stabilized convergency based on eigenstructred approach. *Proc. IFAC Symp. Adaptive Systems and Signal Processing*, Glasgow, UK, April, pp.471-476.

Tikhonov, A.N., and V.Y. Arsenin (1977). *Solutions of Ill-Posed Problems*, Wiley.

Copyright © IFAC Intelligent Tuning and
Adaptive Control, Singapore 1991

AN INTELLIGENT NONLINEAR ADAPTIVE MINIMUM-VARIANCE CONTROLLER

H.-U. Flunkert and H. Unbehauen

Dept. of Electrical Engineering, Ruhr-University Bochum, D-4630 Bochum 1, Germany

Summary

The aim of this paper is to control stochastically disturbed nonlinear plants. For the solution an adaptive control strategy will be described, the identification of which is based on a discrete-time nonlinear model structure. The optimal nonlinear model structure of the plant will be identified by an algorithm which selects the statistically significant terms of the model structure automatically. Based on this model an indirect parameter-adaptive controller will be designed.

1. Problem formulation

The dynamic behaviour of most industrial processes is nonlinear. Nevertheless, if there are only minor changes around an operating point, a linear model would be sufficient for describing its dynamic behaviour. However, if the operating conditions change in a broad range and if these changes are rapid ones, the corresponding control action should be based on an adaptive controller scheme using either an explicit or implicit parameter estimation.
However, during the transient phase between different operating points an optimal control action can not be expected in this case because of the adaptation of the controller. One possibility to improve this situation is to base the controller design on a nonlinear plant model. Such a controller has to fulfil two demands: (i) The static and dynamic nonlinear plant behaviour must be identified and be used for the controller design; (ii) The controller must be able to adapt itself to parameter changes of the plant, i.e. it has to be an adaptive one. The following example will clarify this problem.
Fig. 1 shows the block diagram of a nonlinear plant, consisting of a linear dynamic and nonlinear static partial system. The output signal is superimposed by a stochastic disturbance. A corresponding controller has now the task

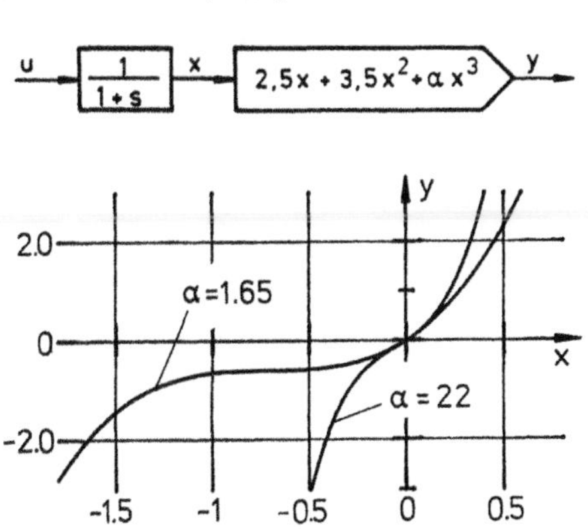

Fig. 1. Block diagram and output signal of a special nonlinear plant for α=1,65 and α=22

of optimally tracking the controlled signal with respect to the desired reference signal and at the same time to provide minimum variance of the controlled signal in the presence of stochatic disturbances. Furthermore the controllers must be adaptive to changes of the plant, due to variation of the parameter α.
It is easy to understand that under these conditions the nonlinear process dynamics can only be described adequately by a nonlinear model, on which also the corresponding adaptive minimum variance (MV)-controller design must be based. In order to guarantee an unbiased parameter estimation the included recursive least squares (RLS)-algorithm is activated only if a sufficient excitation of the algorithm is available. For this reason, the estimation algorithm is fitted with an additional intelligent switching criterion, which provides that the excited signals can lead to minimal variance of the parameter vector, and hence also of the controlled signal of the nonlinear process.
Up to now there have been only a few contributions to the design of adaptive controllers on the basis of nonlinear plant models. For special applications nonlinear models have been used to design new self-tuning controllers, as e.g. in [1 to 3]. A more general approach, containing already a broad class of nonlinear single-input, single-output systems had been proposed in [4]. Another possibility to deal with nonlinear systems is a multi-model approach, where the nonlinear system is approximated by a number of linear ones in different operating points and the adaptive controller is designed according to such a model [5]. Up to now there exists no approach which covers a broader range of nonlinearities than those dealt with in [4]. This paper, therefore, tries to include a very general nonlinear model structure within the design of adaptive controllers using an intelligent algorithm for the structure detection. The paper is organized as follows: In section 2 the pre-identification procedure is discussed. Section 3 describes the derivation of the control law. In section 4 the final parameter estimation and the switching conditions will be presented. Section 5 demonstrates the efficiency of the new approach by simulation studies.

2. Pre-identification and structure determination

A general problem of identifying nonlinear systems consists in their inherent complexity and the resulting difficulty to derive an appropriate mathematical model for control purposes. For example, in order to describe already a very simple nonlinear processes adequately, a high number of parameters is necessary. Therefore, nonlinear system identification does not only consist of parameter estimation but also structure determination has to be performed. This is especially true when a very general nonlinear model structure using the Kolmogorov-Gabor polynomial approach [6] is applied. In this case the following discrete-time relation between input sequence u(k) and output sequence y(k) results.

$$y(k)=\bar{y}+\sum_{i=0}^{n} b_i\, u(k-i)+\sum_{j=0}^{n}\sum_{j=i}^{n} b_{ij}\, u(k-i)k(k-j)$$

$$+\ldots+\sum_{i=0}^{n}\ldots\sum_{v=p}^{n}\sum_{q=v}^{n} b_{i\ldots q}\, u(k-i)\ldots u(k-q)$$

$$+\sum_{i=1}^{m} a_i y(k-i) + \sum_{i=1}^{m}\sum_{j=1}^{m} a_{ij}\, y(k-i)\, y(k-j)$$
$$+\ldots+ \sum_{i=0}^{m}\sum_{j=1}^{m} c_{ij}\, u(k-i)\, y(k-j)$$
$$+\ldots+ \sum_{i=0}^{n}\sum_{j=1}^{m}\ldots\sum_{v=p}^{m}\sum_{q=v}^{m} c_{ij\ldots q}\, u(k-i)\, y(k-j)\ldots y(k-q) \qquad (1)$$

In this equation q represents the degree of the polynomial and $\bar{y}$ is a bias term. If the sytem includes a discrete dead-time $d=\tau_t/T$ (where d is an integer multiple of the sampling time T, and τ_t is the dead time), then the input sequence $u(k-i)$ has to be replaced by $u(k-i-d)$ for $i=1,\ldots,n$. The integers m and n represent the maximal backward shifts of the input, respectively the output sequence. The statistically significant terms of eq. (1) can be determined by using an approach [7] based on a statistical test procedure. Thus a reduced optimal model structure can be obtained. As this test procedure had been described extensively several times elsewhere [7 to 9] it will not be discussed here in detail. However, in order to demonstrate the operation of this structure selection procedure the above example, already, shown in Fig. 1, will briefly be discussed again.

This process with $\alpha=1.65$ had been excited by a test signal over the whole operating range. Then a mathematical model with n=m=q=3 was determined using the measured input and output signals as well as eq. (1). Without structure selection this model would have 84 terms. However, the above mentioned statistical test procedure led to the following optimal model structure:

$$y(k)=\bar{y}+b_1 u(k-1)+b_2 u(k-2)+a_1 y(k-1)$$
$$+a_2 y(k-2)+a_{12} u^2(k-1)u(k-2)+c_{231} u(k-2)u(k-3)y(k-2)$$
$$+c_{333}u(k-3)y(k-3)y(k-3)+c_{133} u(k-1)y(k-3)y(k-3). \qquad (2)$$

As can easily be seen from Fig. 1 the step response of the system does not contain an initial step and, therefore, the term with b_0 is not included in eq. (2).

In general eq. (1) can also be written in the form

$$y(k)=\alpha_0+\alpha_1 u(k)+\ldots+\alpha_q u^q(k), \qquad (3)$$

where the parameters α_i $(i=1,2,\ldots,q)$ contain the previous values of the input and output sequences as well as their ascending powers and α_0 is a bias. According to this structure eq. (2) can also be written in the form

$$y(k)=\nu_0+\nu_1 u(k-1) + \nu_2 u^2(k-1) \;. \qquad (4)$$

3. Control law

As mentioned above, the aim of this paper is to derive a control scheme for stochastically disturbed nonlinear plants. According to that we use an optimal nonlinear structure of the plant model and a performance index minimising the variance of the output signal. To have influence on the control action a weighting of the manipulated variable is introduced. Following these conditions a control law for the nonlinear Minimum-Variance (MV)-controller will be presented.

Using the performance index of Clarke and Hastings-James [10]

$$I^*=[(y^*(k+1\,|\,k)-w(k))^2 + r u^2(k)] \stackrel{!}{=} \text{Min} \qquad (5)$$

and the corresponding partial derivatives

$$\frac{\partial^2 I^*}{\partial u(k)} = 0 = a_0+a_1 u(k)+\ldots+a_{2q-1}u^{2q-1}(k) \qquad (6)$$

and

$$\frac{\partial^2 I^*}{\partial u^2(k)} > 0 \qquad (7)$$

then the optimal control sequence

$$u_{opt}(k) \text{ for } k=0,1,2,\ldots$$

can be determined. Hereby $y^*(k+1\,|\,k)$ is the optimal prediction of the controlled sequence, w(k) is the reference and $ru^2(k)$ represents the weighted control action. According to eq. (1) the parameters a_j $(j=0,\ldots,2q-1)$ contain again the previous values of the input and output sequences and the system coefficients. The second partial derivation, eq. (7), is a necessary condition because of the nonlinear model structure and because of the possible existence of several real solutions. However, there exists only one optimal solution $u_{opt}(k)$ for $k=0,1,\ldots$. It is well known that the weighting factor $r\neq 0$ in eq. (5) results in a remaining control error $e(k)=w(k)-y(k)\neq 0$. One possibility to avoid this, would be to introduce an integral action into the loop, where the integrator according to Fig. 2 can be considered as an extension of the process model. The discrete transfer function of the integrator

$$\frac{U_I(z)}{U(z)} = \frac{1}{T_i}\,\frac{1}{1-z^{-1}}$$

with the integration time constant $T_i=1$ provides in the time-domain

$$u_I(k) = u(k) - u_I(k-1). \qquad (9)$$

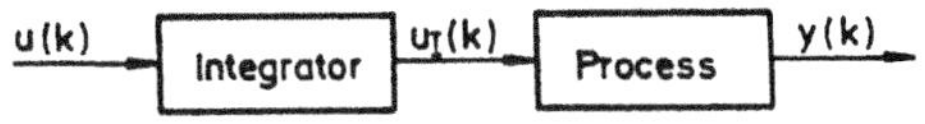

Fig. 2. Extended process model

Thus we obtain

$$y(k) = f\,[u_I(k)] \qquad (10)$$

with

$$u_I(k) = g\,[u(k)] \qquad (11)$$

and therefore

$$y(k) = f\,[g\,[u(k)]]. \qquad (12)$$

The optimal control sequence

$$u_{Iopt}(k) \quad \text{for } k=0,1,2,\ldots$$

can directly be obtained in analogy to eqs. (5) to (7) without explicit calculation of u(k).

4. Parameter estimation and switching conditions

The nonlinear process model according to eq. (1) can be transformed to a vector form, linear in the parameters, with z+1 variables for the purpose of parameter estimation [7]. Taking together the terms of equal polynomial order, the signal vector

$$\mathbf{m}^T(k) = [v_0(k),\ v_1(k),\ v_2(k),\ldots,v_z(k)] \qquad (13)$$

and the parameter vector

$$\mathbf{p}^T = [\Theta_0,\ \Theta_1,\ \Theta_2,\ \ldots\ ,\Theta_z] \qquad (14)$$

can be introduced. The sequences $v_i(k)$ are given by

$$\left.\begin{aligned}
v_0(k) &= 1\\
v_1(k) &= u(k)\\
&\vdots\\
v_{n+1}(k) &= u(k-n)\\
&\vdots\\
v_{m+n+1}(k) &= y(k-m)\\
v_{m+n+2}(k) &= u(k)u(k)\\
&\vdots\\
v_z(k) &= y^q(k-m)
\end{aligned}\right\} . \qquad (15)$$

θ_0 corresponds to the bias term $\bar{y}$ of the model. Thus eq. (1) can be written as [7]

$$y(k) = \mathbf{m}^T(k)\ \mathbf{p}. \qquad (16)$$

Introducing the estimated parameter vector $\hat{\mathbf{p}}$ and the equation error

$$e(k) = y(k) - \mathbf{m}^T(k)\hat{\mathbf{p}} \qquad (17)$$

and defining it as an uncorrelated stochastic variable with expectation value equal to zero and variance σ^2, i.e. white noise, then the minimization of the summation of the squared equation error over N measuring points gives a consistent 'Least Squares'-estimation (LS) $\hat{\mathbf{p}}$ of the parameter vector [11]. As new measurements in adaptive systems are always available $\hat{\mathbf{p}}$ can be obtained by the well-known recursive LS-version (RLS).
However, to obtain an unbiased parameter estimation $\hat{\mathbf{p}}$, where the elements of this vector are represented by the estimated parameters $\hat{\alpha}_i$, this algorithm is applied only if sufficient excitation of the input and output signals of the controlled process is available. This can be provided by an intelligent switching condition which tests if the present signal excitation can really lead to a minimal variance of the parameter error vector $\tilde{\mathbf{p}} = \hat{\mathbf{p}} - \mathbf{p}$, using the matrix

$$E[\tilde{\mathbf{p}}\ \tilde{\mathbf{p}}^T] = \sigma^2\mathbf{P}. \qquad (18)$$

To fulfil the condition of minimal variance of the RLS-algorithm, the covariance matrix **P** must be invertable and positive definite. Based on this preliminary conditions two criteria for identifiability of the controlled process can be formulated as follows: By introducing the scalars

$$\gamma_1(k) = [\det \mathbf{P}(k)]^{-1} \qquad (19a)$$

and

$$\gamma_2(k) = [\mathrm{trace}\ \mathbf{P}(k)]^{-1} \qquad (19b)$$

the switching variables

$$K_1(k) = \frac{\gamma_1(k) - \gamma_1(k-1)}{\gamma_1(k) + \gamma_1(k-1)} \qquad (20a)$$

and

$$K_2(k) = \frac{\gamma_2(k) - \gamma_2(k-1)}{\gamma_2(k) + \gamma_2(k-1)} \qquad (20b)$$

can be used for the activation of the RLS-estimation. Only in the case that either $K_1(k)>0$ or $K_2(k)>0$ the estimation is switched on, if the variance of the parameter vector according to eq. (4) decreases. The basic schematic block diagram of this new adaptive control scheme is given in Fig. 3.

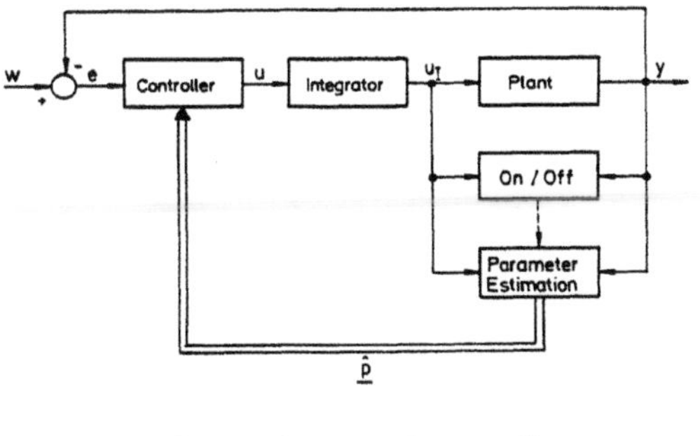

Fig. 3. Block diagram of the adaptive controller

5. Simulation studies

The process used for the simulation studies is the same as introduced already in Fig. 1. Using the adaptive control structure of Fig. 3 and a sampling interval of 0.1 sec., Fig. 4 shows the simulation result during a period of 1000 samples. At the sampling time k=500 the value of α is changed from 1.65 to 22. A white noise sequence of the normalised maximal amplitude of 0.1 was superimposed to the output sequence. The process model is given by eq. (2). The reference sequence w(k) was changed over the whole operating condition between the normalized extremal values of -1.2 and 1.2.

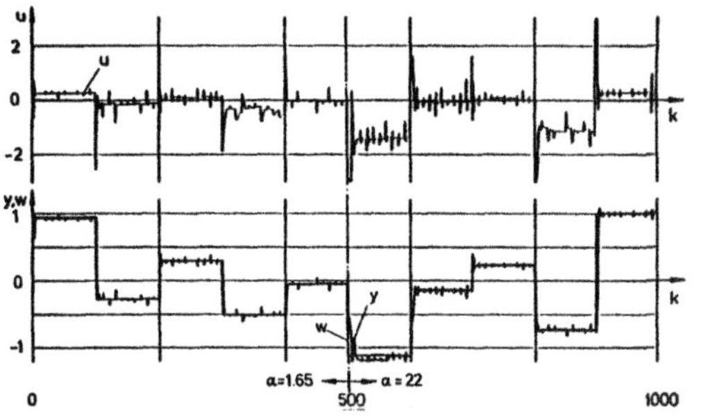

Fig. 4. Dynamic behaviour of the control action u(k), the process output y(k) and the reference sequence w(k)

From Fig. 4 we can see the good tracking behaviour of y(k) after a short initial adaptation phase. Further changes of the operating conditions do not require adaptation of the controller parameters in the case of convergence of the parameters of the nonlinear plant model. This is in contrary to linear adaptive control laws, where also in case of changing operating conditions adaptation is necessary. After changes of the plant parameter α a short adaptation phase again is necessary.

Conclusions

This paper presents a new adaptive control scheme based on reduced nonlinear process models. In this scheme the adaptation is stopped for low excitation of the process and by increasing excitation the adaptation is again activated. This new adaptive control scheme is based on an identification procedure with included model selection and an indirect parameter-adaptive controller. The already available positive simulation results and practical experiences advise to perform the structure determination on-line, i.e. parallel in time to the control action.

References

[1] Ambumani, K., L. Patnaik and I. Sarma (1981). Self-tuning minimum-variance control of nonlinear systems of the Hammerstein model. IEEE Trans. Aut. Contr., AC-26, pp. 959-961.

[2] Lachmann, K. (1982). Parameter-adaptive control of a class of nonlinear processes. Proc. 6th IFAC-Symposium on Identification and System Parameter Estimation, Arlington (VA), pp. 372-378.

[3] Dochain, D. and G. Bastin (1984). Adaptive identification and control algorithms for nonlinear bacterial grooth system. Automatica, 20, pp. 621-644.

[4] Agarwal, M. and D. Seborg (1985). Self-tuning controllers for nonlinear systems. Proceed. IFAC-Symposium on Adaptive Control of Chemical Processes, Frankfurt a. M., pp. 151-157.

[5] Jedner, U. and H. Unbehauen (1988). Intelligent adaptive control for a class of time-varying systems. Proceed. 12th IMACS World Congress on "Scientific Computation", Paris, pp. 123-126.

[6] Eykhoff, P. (1974). System Identification. John Wiley & Sons, New York.

[7] Kortmann, M. and H. Unbehauen (1988). A model

structure selection algorithm in the identification of multivariable nonlinear systems with application to a turbo-generator set. Pepr. 12th IMACS World Congress on "Scientific Computation", Paris, pp. 536-539.

[8] Kortmann, M. and H. Unbehauen (1988). Two algorithms for model structure determination of nonlinear dynamic systems with applications to industrial processes. IFAC Symposium on "Identification'88" Peking.

[9] Haber, R and H. Unbehauen (1988). Structure identification of nonlinear dynamic systems - A survey on input-output approaches. IFAC Symposium on "Identification'88" Peking.

[10] Clarke, D. and R. Hastings-James (1971). Design of digital controllers for randomly disturbed systems. Proceed, IEE 118, pp. 1503-1506.

[11] Unbehauen H. (1985). Regelungstechnik III. Vieweg-Verlag Braunscheig.

SELF-TUNING ADAPTIVE CONTROL BASED ON A NEW PARAMETER ESTIMATION METHOD

P. H. Thoa*, N. T. Loan and H. H. Son****

**Institute of Computer Science, Lieu Giai, Ba dinh, Hanoi, Vietnam*
***Institute of Theoretical Physics, P.O. Box 429 Bo Ho, Hanoi 10000, Vietnam*

Abstract. Self-tuning adaptive control on the basis of a new and efficient parameter estimation method is presented in this paper. The approach to parameter estimation is based on the Marchuc variational method and optimal filtering theory. General formulas and an equation system for parameter estimation are developed and a simple algorithm for system parameter estimation is then given in this paper. Using this estimator, an adaptive regulator is derived on the basis of linear quadratic Gaussian theory. Simulation results are presented to demonstrate the effectiveness of the self-tuning adaptive control based on the proposed parameter estimation method.

Keywords. Adaptive control; parameter estimation; self-tuning regulators; optimal filtering; linear optimal control; discrete time systems; variational method.

INTRODUCTION

In many practical applications of industrial process controllers, it is difficult to determine the parameters of the controller, since the dynamics of the process and its disturbances are unknown. The process parameters thus have to be estimated, and it is desirable to have a regulator which tunes its parameters on-line. A popular approach for such controller design is the self - tuning adaptive control (Åström, 1983; Åström, 1987 ; Åström and coworkers, 1977). In this approach the process is represented by linear discrete time models so that the controller is inherently digital. The self-tuning regulator has recently obtained much attention, because it is flexible and easy to implement with microprocessors.

It is assumed that the process to be controlled can be described by

$$y(k) = \sum_{i=1}^{n} a_i y(k-i) + \sum_{i=1}^{n} b_{i-1} u(k-d-i+1) + w(k) \qquad (1)$$

where u(k) and y(k) denote the system input and output, w(k) is a disturbance affecting the system and d is a positive integer specifying the time delay.

The purpose of self-tuning regulator is to control the system (1) with unknown but constant parameters a_i and b_{i-1}, $i = 1,\dots,n$. The regulator can also be applied to systems with slowly varying parameters.

The class of regulators can be thought of as composed of three parts: a parameter estimator, a linear controller and a block which determines the controller parameters from the estimated parameters. There are many possible self-tuning regulators for the system (1) depending on parameter estimation techniques and the controller design that are used. For the estimator there are different estimation schemes which have been used, for example, least-squares, extended and generalized least-squares , stochastic approximation, instrumental variables, maximum likelihood method... (Åström and Eykhoff, 1971; Ljung and Söderström , 1983). They are all based on different recursive schemes of estimating the parameters of a prediction model. In this paper a new parameter estimation method used for self-tuning adaptive control is developed on the basis of the variational method (Marchuc, 1977; Loan and Son, 1987, 1990; Thoa and Son , 1990) and the optimal filtering theory.

The paper is organized as follows. Firstly, we begin with some notations and definitions for the variational method and use this method and the optimal filtering theory to develope a new parameter estimation method. A general algorithm for system parameter estimation, which reduces in the special case to least-squares algorithm, is then proposed in the paper. The estimated parameters are used in the next section for controller design on the basis of linear quadratic Gaussian theory. Finally, some simulation results are illustrated to verify the effectiveness of the self-tuning adaptive control based on the proposed parameter estimation method.

VARIATIONAL METHOD

The variational method is originally suggested by Marchuc (1977) to solve the deterministic inverse problem of mathematical physics (IPMP). Let $\Phi(x)$ be a process satisfying the following equation written in the operator form:

$$L_\alpha \Phi(x) = q_\beta(x), \quad \Phi(0) = \Phi_0, \quad x \in D \subset R^n \qquad (2)$$

where L is some linear operator, x is a set of all variables (time, space,...), $\Theta = (\alpha^T, \beta^T)^T$ is a vector of unknown parameters and q(x) is a source distribution.

Introduce the conjugate (adjoint) equation corresponding to Eq. (2) :

$$L^*\Phi_p^*(x) = p(x) \qquad (3)$$

where $p(x) \in \mathcal{L}_n^2(D)$ and L^* is a conjugate operator for L satisfying the following equation:

$$\langle L\lambda , \mu \rangle_H = \langle \lambda , L^*\mu \rangle_H \qquad (4)$$

for all $\lambda, \mu \in H$.
Here $\langle \cdot , \cdot \rangle_H$ denotes the inner product in the Hilbert space $H = \mathcal{L}_n^2(D)$.

Let the measurements be given by the measurement functional

$$J_p[\Phi] \doteq \langle \Phi , p \rangle_H = \int_D \Phi(x)p(x)dx \qquad (5)$$

and define

$$J_p[\Phi] = J_q[\Phi_p^*] \quad \text{or} \quad \langle \Phi , p\rangle_H = \langle \Phi_p^*, q\rangle_H \qquad (6)$$

The IPMP is to determine L (or Θ) using the measurement functional $J_p[\Phi]$ (Marchuc, 1977).

<u>Theorem 1</u>. Let Φ' be the solution of Eq. (2) with $L \doteq L'$ and $q \doteq q'$. Then the following relation holds for the variations δJ_p, δL and δq:

$$\delta J_p = - \langle \Phi_p^* , \delta L\Phi' \rangle + \langle \Phi_p^* , \delta q \rangle \qquad (7.a)$$

with $\delta J_p = J_p' - J_p$, $\delta L = L' - L$ and $\delta q = q' - q$.

<u>Proof.</u> See appendix 1.

Assume that the variation δL is sufficiently small , instead of Eq. (7.a) we can use the following

$$\delta J_p = - \langle \Phi_p^* , \delta L\Phi \rangle + \langle \Phi_p^* , \delta q \rangle \qquad (7.b)$$

In the special case $q' = q$, i.e. $\delta q = 0$, we obtain

$$\delta J_p = - \langle \Phi_p^* , \delta L\Phi \rangle \qquad (8)$$

The formula (8) is called the formula of small perturbances theory obtained by Marchuc (1977).

We now consider the general case, when $\Phi(x)$ in Eq. (2) is a random process and Φ_0 may be a random value. For simplicity we assume that

$$q_\beta(x) = q^\circ_\beta(x) + w(x) \qquad (9)$$

where $w(x)$ is a disturbance affecting the system.

Propose that the output of system (2) can be observed and given by

$$z(x) = \Phi(x) + v(x) \qquad (10)$$

here $v(x)$ is random observation error.

In this case Eq. (7) can not be used to estimate the variations δL, δq (or $\delta\alpha$, $\delta\beta$) since Φ is a random field and it is therefore impossible to calculate the both hands of Eq. (7). To overcome this difficulty we shall estimate Φ by solving the filtering problem (2) and (10) with $\Theta = \Theta^\nu$

<u>Theorem 2</u>. Let Φ_Θ be the solution of the filtering problem (2) and (10), and $\hat{\Phi}_\Theta$ be solution of filtration problem with Θ^ν. If δL and $\delta L\varepsilon_\Theta$ are sufficiently small , $\varepsilon_\Theta \doteq \Phi_\Theta - \hat{\Phi}_\Theta$, $\delta L = L' - L$, $\delta q = q' - q$, then the following relation holds for the variations δL, δq and $\delta\hat{J}_p$:

$$\delta\hat{J}_p = - \langle \Phi_p^* , \delta L\hat{\Phi}_\Theta \rangle + \langle \Phi_p^* , \delta q \rangle + \xi \qquad (11)$$

with $\xi \doteq J_p[\varepsilon_\Theta] = \langle \varepsilon_\Theta, p \rangle$; $\delta\hat{J}_p \doteq J'_p - \hat{J}_p$; $\hat{J}_p \doteq J_p[\hat{\Phi}_\Theta]$

and the relation between δL, δq and $\delta\hat{Z}_p$, $\delta\hat{Z}_p \doteq Z'_p - \hat{J}_p$ can be represented by

$$\delta\hat{Z}_p = - \langle \Phi_p^* , \delta L\hat{\Phi}_\Theta \rangle + \langle \Phi_p^* , \delta q \rangle + \mathfrak{Z} \qquad (12)$$

with $\mathfrak{Z} = \Upsilon + \xi$; $\Upsilon \doteq \langle p, v \rangle$; $\delta\hat{Z}_p \doteq Z_p - \hat{J}_p$; $Z_p = \langle p, z \rangle$

<u>Proof.</u> See appendix 2.

The formulas (11) and (12) are called the general formulas for system parameter estimation.

PARAMETER ESTIMATION

<u>Parameter Estimation Equation System</u>

In order to estimate the parameters a_i, b_{i-1}, $i = 1, \ldots ,n$, of Eq. (1) by using the formulas (11) or (12) we rewrite Eq. (1) in the form of Eq. (2):

$$[1 - \sum_{i=1}^{n} a_i B^i]y(k) = \sum_{i=1}^{n} b_{i-1}B^{i-1}u(k-d) + w(k) \qquad (13)$$

where B denotes the backward shift operator, i.e. $B^i y(k) = y(k-i)$.
The output observation of system (13) is given in general case by

$$z(k) = y(k) + v(k) \qquad (14)$$

here $v(k)$ is random observation error.

Corresponding to the process model (13) the following prediction model will be used:

$$[1 - \sum_{i=1}^{n} \hat{a}_i(k-1)B^i]\tilde{y}(k) = \sum_{i=1}^{n} \hat{b}_{i-1}(k-1)B^{i-1}u(k-d) + w(k) \qquad (15)$$

By defining that

$$L' = 1 - \sum_{i=1}^{n} a_i B^i, \quad q' = \sum_{i=1}^{n} b_{i-1}B^{i-1}u(k-d) + w(k) \qquad (16.a)$$

$$\Phi' = y(k) , \quad \Phi = \tilde{y}(k)$$

$$L = 1-\sum_{i=1}^{n}\hat{a}_i(k-1)B^i, \quad q = \sum_{i=1}^{n}\hat{b}_{i-1}(k-1)B^{i-1}u(k-d) + w(k) \quad (16.b)$$

and using the variational method we obtain

$$\delta L = -\sum_{i=1}^{n}\delta a_i(k)B^i, \quad \delta q = \sum_{i=1}^{n}\delta b_{i-1}(k)B^{i-1}u(k-d) \quad (17)$$

where $\delta a_i(k) = a_i - \hat{a}_i(k-1)$,
$\delta b_{i-1}(k) = b_{i-1} - \hat{b}_{i-1}(k-1)$.

For the problem (1), let $D = \{0, \pm1, \pm2, \ldots\}$ and without loss of generality by defining that

$$\langle \lambda, \mu \rangle_H \doteq \sum_{-\infty}^{\infty}\lambda(l)\mu(l) = \sum_{l=1}^{k}\lambda(l)\mu(l) \quad (18)$$

with $\lambda(l) = 0$, $\mu(l) = 0$ for all $l \le 0$ and $l > k$ we obtain from the formula (12):

$$\delta\hat{Z}_p = \sum_{l=1}^{k}y^*(l)\sum_{i=1}^{n}\delta a_i(k)B^i\hat{y}(l) + \sum_{l=1}^{k}y^*(l)\sum_{i=1}^{n}\delta b_{i-1}(k)B^{i-1}u(l-d) + \mathfrak{Z} \quad (19.a)$$

with

$$\mathfrak{Z} = \xi + \Upsilon, \quad \xi = \sum_{l=1}^{k}p(l)\varepsilon(l), \quad \Upsilon = \sum_{l=1}^{k}p(l)v(l), \text{ and}$$

$$\varepsilon(l) = y(l) - \hat{y}(l)$$

where $\hat{y}(k)$ is the solution of the optimal filtering problem (15) with the system observations (14).

For the considered system (1) we assume that the exact observation $z(k) = y(k)$ is obtained, i.e. $v(k) = 0$ in Eq. (14). In this case we can use the formula (7.a) for parameter estimation directly. It follows then

$$\delta\hat{J}_p = \sum_{l=1}^{k} y^*(l)\sum_{i=1}^{n} \delta a_i(k)B^iy(l) + \sum_{l=1}^{k}y^*(l)\sum_{i=1}^{n}\delta b_{i-1}(k)B^{i-1}u(l-d) + \xi \quad (19.b)$$

where $y^*(k)$ is the solution of the conjugate equation corresponding to Eq. (15):

$$L^*y^*(k) = p(k) \quad (20)$$

L^* is the conjugate operator for L

__Lemma__. The conjugate operator L^* for L defined in Eq. (16.b) is

$$L^* = 1 - \sum_{i=1}^{n} \hat{a}_i(k-1) F^i \quad (21)$$

here F^i is the forward shift operator, i.e.

$$F^iy(k) = y(k+i).$$

__Proof.__ See appendix 3.

Let $p_j(l)$, a function of $l \in D$, belong to H, i.e. $p_j(l) = 0$ for all $l \le 0$ and $l > k$. The conjugate equation is then given from Eq. (20) and Eq. (21):

$$y^*_{pj}(l) = \hat{a}_1(k-1)y^*_{pj}(l+1) + \ldots + \hat{a}_n(k-1)y^*_{pj}(l+n) + p_j(l) \quad (22)$$

with $l = k, \ldots, 1$ and $y^*_{pj}(k+1) = y^*_{pj}(k+2) = \ldots = 0$ and the measurement functionals J_{pj} and $\hat{J}_{pj}$ can be constructed as

$$J_{pj} = \langle p_j, y\rangle_H = \sum_{l=1}^{k} p_j(l)y(l)\ ;\ j = 1,\ldots,m \quad (23.a)$$

$$\hat{J}_{pj} = \langle p_j, \hat{y}\rangle_H = \sum_{l=1}^{k} p_j(l)\hat{y}(l)\ ;\ j = 1,\ldots,m \quad (23.b)$$

By that way we can obtain m functionals J_{pj}, $\hat{J}_{pj}$ and the parameter estimation equation system can be given from Eq. (19.b).

Denoting that

$$\delta\hat{J} = (\delta\hat{J}_{p1}, \delta\hat{J}_{p2}, \ldots, \delta\hat{J}_{pm})^T$$

and

$$\delta\Theta = (\delta a_1(k),\ldots, \delta a_n(k), \delta b_0(k),\ldots, \delta b_{n-1}(k))^T$$

the parameter estimation equation system can be written in the form:

$$\delta\hat{J} = R\,.\,\delta\Theta + \xi \quad (24)$$

where R is a (m x 2n)-matrix whose elements are

$$r_{j,i} = \sum_{l=1}^{k}y^*_{pj}(l)y(l-i); \quad r_{j,n+i} = \sum_{l=1}^{k}y^*_{pj}(l)u(l-d-i+1) \quad (25)$$

with $i = 1,\ldots,n$; $j = 1,\ldots,m$ and ξ is a random variable vector.

Solving Eq. (24) by some method (optimal filtering, stochastic approximation, least-squares, ...) we obtain the parameter variation vector $\delta\hat{\Theta}(k)$ and the estimated parameters can computed as

$$\hat{\Theta}(k) = \hat{\Theta}(k-1) + \delta\hat{\Theta}(k) \quad (26)$$

The parameter estimation equation system derived here may be of minimal order which is equal the number of unknown parameters to be estimated. This leads to a simple estimation algorithm which can be used for the estimation of unknown process parameters.

__General Parameter Estimation Algorithm__

For instance, let us know a priori

$$\Theta = \Theta(0) + \delta\Theta\ ,\ E(\delta\Theta) = 0\ ,\ E(\delta\Theta.\delta\Theta^T) = M_\Theta \qquad (27)$$

Choose an approximate value for $\Theta(0)$, for example, $\Theta(0) = E(\Theta)$.

Step 1. Solve the optimal filtering problem (15) and (14). Its solution will be the couple $(\hat{y}(k), P(k))$ where $\hat{y}(k)$ is the optimal estimated value for $\tilde{y}(k)$ and $P(k)$ is the error covariance matrix for $\hat{y}(k)$.

Step 2. Form the estimation equation system (24) by the aid of m constructed functionals (23). To do that, it is necessary to solve m conjugate equations determined by Eq. (22) with p_j, $j = 1, \ldots, m$ and to compute the statistical characteristics of sequence (ξ_j).

Step 3. Solve the estimation equation system (24) by one of filtering methods or other estimation methods. Its solution is the estimated parameter variation vector $\delta\hat{\Theta}(k)$.

Step 4. Compute the new value of parameters by the formula (26).
Repeat the above procedure at each time instant k.

A Simple Parameter Estimation Algorithm

In the special case, $v(k) = 0$, it is not necessary to solve the filtering problem (15) and (14) in step 1. The algorithm then is very simple.

Let the system (1) be represented by the following form ($v(k) = 0$)

$$y(k) = \Phi(k-1)\Theta + w(k) \qquad (28)$$

where

$$\Phi(k-1) = [y(k-1), \ldots, y(k-n), u(k-d), \ldots, u(k-d-n+1)]$$

$$\Theta = [a_1, a_2, \ldots, a_1, b_0, b_1, \ldots, b_{n-1}]^T$$

The Eq. (28) can be written in the operator form:

$$L\ y(k) = q_\beta(k) \qquad (29)$$

with $L = 1$, $q_\beta(k) = \Phi(k-1)\Theta$ and $\beta = \Theta$.

Thus, we have to estimate $\beta = \Theta$ using the formula (7.a).
Choosing $p(l) = 1$ for $l = k$ and $p(l) = 0$ for $l \neq k$, we obtain

$$J(y) = \sum_{l=1}^{k} p(l)y(l) = y(k) \qquad (30)$$

and the conjugate equations are then given by

$$y^*(l) = p(l) = \begin{cases} 1 & \text{for } l = k \\ 0 & \text{for } l \neq k \end{cases} \qquad (31)$$

The left-hand side of Eq. (7.a) becomes

$$\delta J = J - \hat{J} = y(k) - \hat{y}(k) \qquad (32)$$

with $\hat{y}(k) = \Phi(k-1)\hat{\Theta}(k-1)$.

From the right-hand side of Eq. (7.a) we obtain

$$\delta J = \langle y^*, \delta q \rangle + \xi = \langle y^*, \Phi\,\delta\Theta \rangle + \varepsilon$$

$$= \sum_{l=1}^{k} y^*(l)\Phi(k-1)\,\delta\Theta(k) + \xi$$

or

$$\delta J = \Phi(k-1)\,\delta\Theta(k) + \varepsilon \qquad (33)$$

with $\delta\Theta(k) = \Theta - \hat{\Theta}(k-1)$, $\xi = \sum_{l=1}^{k} p(l)\ w(l) = w(k)$

The parameter estimation problem reduces then simply to solving of Eq. (33) for $\delta\Theta(k)$ and to computing the new value of parameters by Eq. (26).
Let the criterion be chosen as to minimize the loss function:

$$V = \sum_{l=1}^{k} (\delta J - \Phi(l-1)\,\delta\hat{\Theta}(l))^2 \qquad (34.a)$$

It follows then

$$V = \sum_{l=1}^{k} (y(l) - \Phi(l-1)\hat{\Theta}(l))^2 \qquad (34.b)$$

where $\hat{\Theta}(l) = \hat{\Theta}(l-1) + \delta\hat{\Theta}(l)$

We see that Eq. (34.b) is the well-known least-squares criterion.
Solve Eq. (33) with the criterion (34) we obtain the least-squares solution for Θ of the system (1). This implies that the proposed parameter estimation method reduces in this case to the well- known least-squares method (Åström and Eykhoff, 1971; Åström and co-workers, 1977).

ADAPTIVE REGULATOR

To control the system (1) the performance criterion is chosen in the general case as

$$J_k(u) = E(\|y(k+d) - y_d(k+d)\|_Q^2 + \|u(k)\|_R^2 \ /\Phi(k-1)) \qquad (35)$$

where $\|\cdot\|_R$ indicates the norm with weight R, i.e., $\|u\|_R^2 = R\,u^2$ and $y_d(.)$ describes the desired output value. The expectation operation E is conditioned on the available measurements up to and including time k-1:

$$\Phi(k-1) = [y(k-1), \ldots, y(k-n), u(k-d), \ldots, u(k-d-n+1)]^T.$$

The control which minimizes Eq. (35) is determined by the following equation (Borisson, 1979; Koivo and Guo, 1983):

$$Ru(k) + b_0 Q[\hat{y}(k+d/k-1) - y_d(k+d)] = 0 \qquad (36)$$

where $\hat{y}(k+d/k-1)$ denotes the optimal predicted

value (in the sense of least- squares error) of $y(k+d)$ given the past measurements and controls up to and including the time $(k-1)$. The r-step ahead predicted value may be calculated recursively for $r = 0, 1, \ldots, d$ by:

$$\hat{y}(k+r/k-1) = \Phi(k-1+r/k-1)\,\hat{\Theta}(k-1) \tag{37}$$

where

$$\Phi(k-1+r/k-1) = [\,\hat{y}(k-1+r/k-1), \ldots, \hat{y}(k/k-1), y(k-1), \ldots, y(k+r-n), u(k-d+r), \ldots, u(k-d-n+1+r)\,] \tag{38}$$

SIMULATION

__Example 1__. To demonstrate the behaviour of the derived parameter estimator a first order system will be used:

$$y(k) = ay(k-1) + bu(k-1)$$

where the numerical values are $\Theta^* = (a^*, b^*)^T = (0.5,\ 1)^T$.
The initial values of the parameters are assumed to be $\Theta^\circ = (\ 0.1,\ 0.3\ )^T$.
Choose $p_j(l) = 0.5$ for $l = k-2j$ or $l = k-(2j+1)$; $j = 0, 1$; $l = 1, \ldots, k$; else put $p_j(l) = 0$.
The simulation results for the proposed algorithm are illustrated and compared with those of the well-know recursive least-squares method It is shown that for a sinusoidal signal input $u(k) = \sin(k)$ the LS estimates converge slowly to the true process parameters (Fig. 1). But using the proposed estimation method the estimated parameters quickly converge to the true values of the process parameters ($k=4$ in Fig.1)

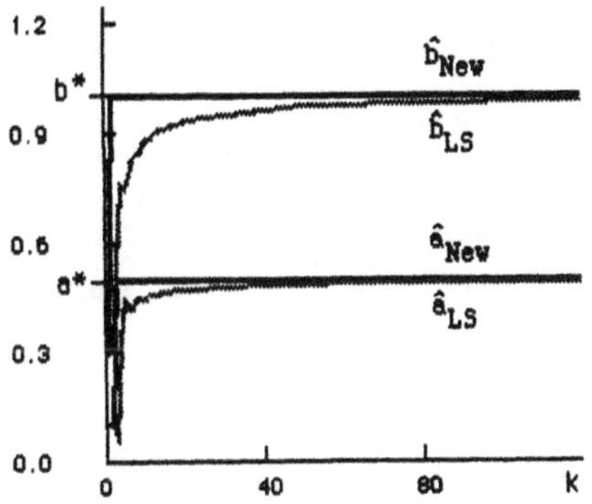

Fig. 1. Estimated parameters in example 1.

In this case we obtain the following simulation results at the sampling instant $k = 500$:
Least-Squares method:

$$\hat{\Theta}_{LS} = (\ 0.498975,\ 0.997252\ )^T \quad \text{for } k = 500$$

Proposed Method:

$$\hat{\Theta}_{New} = (\ 0.500013,\ 0.999999\ )^T \quad \text{for } k = 4$$

__Example 2__. Suppose that it is necessary to control the speed of a DC servomotor with unknown loads. The autoregressive model is chosen for $n = 2, d = 1, a_1 = 0$:

$$y(k) = a_2y(k-2) + b_0u(k-1) + b_1u(k-2) + w(k) \tag{39}$$

where the true parameters are $\Theta^* = (0.919, 3.45, 3.308)^T$ and $w(k)$ is an independent zero - mean additive white noise with variance $\delta_w^2 = 1$.
The initial values of the parameters are assumed to be $\Theta^\circ = (\ 0.5,\ 5.3,\ 2.1\ )^T$.
In this case we use the proposed algorithm to estimate the unknown motor parameter $\Theta = (\ a_2, b_0, b_1\ )^T$. The estimated parameters $\hat{\Theta}(k)$ are then used for the controller design on the basis of Eq.(36). The following control law is then chosen:

$$u(k) = [y_d(k+1) - \hat{a}_2(k)y(k-1) - \hat{b}_1(k)u(k-1)]/\hat{b}_0(k) \tag{40}$$

where $y_{d(k)}$ is chosen as the output of a reference model described by

$$y_d(k+1) = 1.9y_d(k) - 0.905y_d(k-1) + 0.01r(k-1)$$

Here $r(k)$ denotes the reference input. (41)

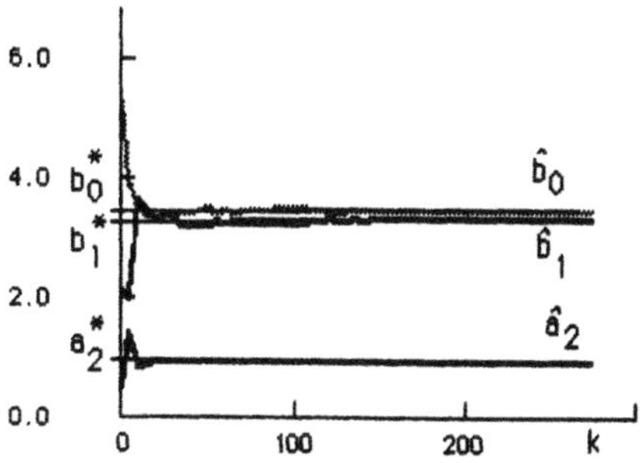

Fig. 2. Estimated parameters in example 2.

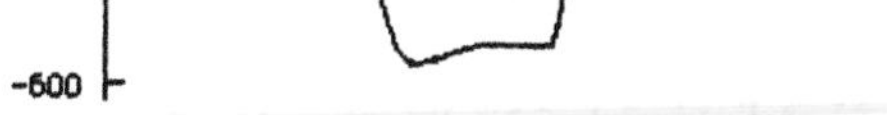

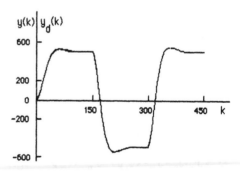

Fig. 3. Response for adaptive algorithm in example 2.

The simulation results are presented in Fig.2 and Fig.3. It is shown in Fig.2 that the estimated parameters quickly converge to the true values of motor parameters. The output of the self-tuning adaptive system $y(k)$ follows exactly the desired trajectory $y_d(k)$ (Fig.3).

CONCLUSION

This paper has presented the self-tuning adaptive control on the basis of a new and efficient parameter estimation method. Using the

Marchuc variational method and optimal filtering theory, general formulas and an equation system for parameter estimation were developed and a system parameter estimation algorithm was then proposed in the paper. Using this estimator, an adaptive regulator was derived on the base of linear quadratic Gaussian theory. Simulation results have shown that the estimated parameters quickly converged to the true process parameters and they have verified the effectiveness of the self - tuning adaptive control based on the proposed parameter estimation method.

Another interesting key problems, such as stability, convergence analysis and optimal choosing of function p(k), will be done by the authors in a future work.

APPENDIX 1 - PROOF OF THEOREM 1

Let L and q in Eq.(2) become L' and q' due to variations of α and β respectively, i. e. $L' = L + \delta L$ and $q' = q + \delta q$, then

$$L'\Phi' = q' \tag{42}$$

Multiplying scalarly Eq.(42) with Φ_p^*, Eq.(3) with Φ' and then substracting one from another we obtain from the left-hand side

$$\langle\Phi_p^*, L'\Phi'\rangle - \langle\Phi', L^*\Phi_p^*\rangle = \langle\Phi_p^*, L'\Phi'\rangle - \langle L\Phi', \Phi_p^*\rangle =$$
$$\langle\Phi_p^*, L'\Phi' - L\Phi'\rangle = \langle\Phi_p^*, \delta L\Phi'\rangle \tag{43}$$

and from the right-hand side

$$\langle\Phi_p^*, q'\rangle - \langle\Phi', p\rangle = \langle\Phi_p^*, q + \delta q\rangle - J_p' =$$
$$\langle\dot{\Phi}_p^*, q\rangle + \langle\Phi_p^*, \delta q\rangle - J' = J_p + \langle\Phi_p^*, \delta q\rangle - J_p'$$
$$= -\delta J_p + \langle\Phi_p^*, \delta q\rangle \tag{44}$$

The following relation is then obtained from Eq.(43) and Eq.(44):

$$\delta J_p = -\langle\Phi_p^*, \delta L\Phi'\rangle + \langle\Phi_p^*, \delta q\rangle \tag{45}$$

This implies that theorem 1 is proved.

APPENDIX 2 - PROOF OF THEOREM 2

The left- hand side of Eq.(7.b) can be written as

$$\delta J_p = J'_p - J_p = J'_p - \langle\Phi, p\rangle = J'_p - \langle\hat{\Phi}_\Theta, p\rangle - \langle\varepsilon_\Theta, p\rangle =$$
$$J'_p - \hat{J}_p - J_p[\varepsilon_\Theta] = \delta\hat{J}_p - J_p[\varepsilon_\Theta] \tag{46}$$

It follows from the right-hand side of Eq.(7.b) that

$$-\langle\Phi_p^*, \delta L\Phi\rangle + \langle\Phi_p^*, \delta q\rangle =$$
$$-\langle\Phi_p^*, \delta L(\hat{\Phi}_\Theta + \varepsilon_\Theta)\rangle + \langle\Phi_p^*, \delta q\rangle \tag{47}$$

If $\delta L\varepsilon_\Theta$ is sufficiently small, then we obtain from Eq.(46) and Eq.(47):

$$\delta J_p = \delta\hat{J}_p - J_p[\varepsilon_\Theta] = -\langle\Phi_p^*, \delta L\hat{\Phi}_\Theta\rangle + \langle\Phi_p^*, \delta q\rangle$$
$$\text{or } \delta\hat{J}_p = -\langle\Phi_p^*, \delta L\hat{\Phi}_\Theta\rangle + \langle\Phi_p^*, \delta q\rangle + J_p[\varepsilon_\Theta] \tag{48}$$

The last formula implies that Eq.(11) is proved. By the same way we can establish the relation (12).

APPENDIX 3 - PROOF OF LEMMA

It is necessary to show that with $\forall\ \lambda, \mu \in H$

$$\langle L\lambda, \mu\rangle_H = \langle\lambda, L^*\mu\rangle_H \tag{49}$$

According to Eq.(16.b) we have from the left-hand side of Eq. (49):

$$\langle(1-\sum_{i=1}^{n}\hat{a}_i(k-1)B^i)\lambda, \mu\rangle_H = \sum_{l=1}^{k}\lambda(l)\mu(l) -$$
$$\hat{a}_1(k-1)\sum_{l=2}^{k}\lambda(l-1)\mu(l) - \ldots - \hat{a}_n(k-1)\sum_{l=n+1}^{k}\lambda(l-n)\mu(l) \tag{50}$$

On the other hand, from the right- hand side of Eq. (49) we have

$$\langle\lambda, (1-\sum_{i=1}^{n}\hat{a}_i(k-1)F^i)\mu\rangle_H = \sum_{l=1}^{k}\lambda(l)\mu(l) -$$
$$\hat{a}_1(k-1)\sum_{l=1}^{k-1}\lambda(l)\mu(l+1) - \ldots - \hat{a}_n(k-1)\sum_{l=1}^{k-n}\lambda(l)\mu(l+n) \tag{51}$$

We see that Eq.(50) is equal Eq.(51) in pairs, and it implies that Eq.(49) is proved.

REFERENCES

Åström, K.J. (1983). Theory and applications of adaptive control - a survey. Automatica, 19, 471-486.

Åström, K.J. (1987). Adaptive feedback control. Proc. of the IEEE, 75, 185-215.

Åström, K.J., and P.Eykhoff (1971). System identification - a survey. Automatica, 7, 123.

Åström, K.J., U. Borisson, L. Ljung, and B. Wittenmark (1977). Theory and application of self - tuning regulators. Automatica, 13, 457-476.

Borisson, U. (1979). Self-tuning regulators for a class of multivariable systems. Automatica, 15, 209-215.

Koivo, A.J., and T.H. Guo(1983). Adaptive linear controller for robotic manipulators. IEEE Trans. Aut. Control, AC-28, 162-171.

Loan, N. T., and H. H. Son (1987). Variational method for inverse problems of mathematical physics & its application with optimal filtering for solving the inverse problems. Thesises of III Vietnam Conference on Physic, Hanoi.

Loan, N.T., and H.H. Son (1990). Adaptive parameter identification method in controlled contamination industries systems. Proc. 5th World Filtration Congress, Vol.3, 221-229, Acropolis, Nice, France.

Ljung, L., and T. Söderström (1983). Theory and Practice of Recursive Identification, MIT Press, Cambridge.

Marchuc, G.I.(1977). Methods of computational mathematics (in Russian). Nauka, Moskow.

Thoa, P.H., and H.H. Son (1990). A new system parameter identification method using sampled input - output measurement data. Proc. 2nd National Conference on Metrology, Hanoi.

ON THE DESIGN OF THE UNIFIED PREDICTIVE CONTROLLER

A. R. M. Soeterboek, H. B. Verbruggen and P. P. J. van den Bosch

Delft University of Technology, Dept. of Electrical Engineering, Control Laboratory, P.O. Box 5031, 2600 GA Delft, The Netherlands

Abstract

In this paper the design parameters of the Unified Predictive Controller are discussed. It is shown that separate tuning of the servo and regulator behavior can be obtained by a special choice of the design parameters. Rule of thumb methods are presented that can be used in order to obtain initial settings for the design parameters.

Introduction

Over the last ten years many predictive controllers have been proposed [3,4,8]. Although all predictive controllers are based on the same concept, there are significant differences among predictive controllers in the literature (see, for example, [4]). This is due to the fact that, within the concept of predictive control, many different approaches can be used to design a predictive controller and, therefore, it can be quite hard to select which controller to use in a particular situation. In order to get an insight into the different approaches used in the literature to obtain a predictive controller, a unified approach has been proposed: the most commonly used approaches to predictive controller design are united into the Unified Predictive Controller (UPC) [6,8,9]. Relations with other controller design methods such as minimum-variance, pole-placement and dead-beat control are discussed in [8]. Due to this unified approach many well-known predictive controllers and classical controllers, each having its own features, can be obtained simply by selecting the parameters of the UPC controller. Some theoretical results concerning the design parameters of the UPC controller are presented in [8]. A detailed description of the design of the UPC controller can be found in [9].
The UPC controller has many design parameters among which polynomials. The purpose of this paper is to gain an insight into the influence of the UPC design parameters on the performance and robustness of the closed-loop system and to formulate rule of thumb methods on how to select the design parameters of the controller assuming that a model of the process to be controlled is available.

Unified Predictive Control

All predictive controllers are based on the predictive control strategy illustrated in Figure 1. Suppose the current time is $t = k$ and $u(k)$, $y(k)$ and $w(k)$ denote the controller output, the process output and the desired process output at $t = k$, respectively. Further, define

$$\boldsymbol{u} = [u(k), \cdots, u(k+H_p-1)]^T$$

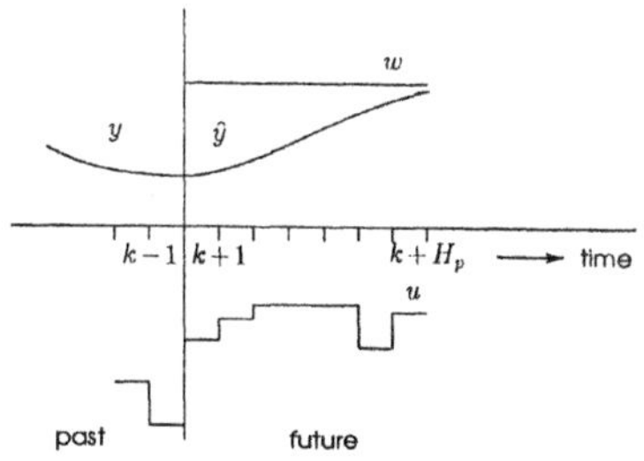

Figure 1: The predictive control strategy.

$$\hat{\boldsymbol{y}} = [\hat{y}(k+1), \cdots, \hat{y}(k+H_p)]^T$$
$$\boldsymbol{w} = [w(k+1), \cdots, w(k+H_p)]^T$$

where H_p is called the prediction horizon and the symbol $\hat{}$ denotes estimation. Then, a predictive controller calculates such a future controller output sequence $\boldsymbol{u}$ that the predicted output of the process $\hat{\boldsymbol{y}}$ is 'close' to the desired process output $\boldsymbol{w}$. This desired process output is often called the reference trajectory. The first element of the controller output sequence ($= u(k)$) is used to control the process. At the next sample, the whole procedure is repeated using the latest measured information. This is called the *receding horizon* principle.

The process output is predicted by using a model of the process. Any model that describes the relation between the input and the output of the process can be used. Hence, not only ARMAX (Auto-Regressive Moving-Average eXogenous) models can be used, but also step and impulse response models, state-space models and nonlinear models. Further, if the process is subject to disturbances, a disturbance or noise model can be added to the process model thus allowing the effect of disturbances on the predicted process output to be taken into account. In the UPC controller the following model is used to predict the process output over the prediction horizon.

$$y(k) = \frac{q^{-d}B(q^{-1})}{A(q^{-1})}u(k-1) + \frac{C(q^{-1})}{D(q^{-1})A(q^{-1})}e(k) \quad (1)$$

in which q^{-1} is the backward shift operator, d is the time delay in samples ($d \geq 0$) and $e(k)$ is a discrete white noise sequence with zero mean. A, B, C and D are polynomials in q^{-1} with degree n_A, n_B, n_C and n_D, respectively. Further, A, C and D are monic. The model (1) unifies familiar process models such as ARX (Auto-Regressive eXogenous), ARMAX, ARIMAX (Auto-Regressive Integrated Moving-Average), FIR (Finite Impulse Response) and FSR (Finite Step Response) models.
The parameters A, B and d can be estimated by using a

least-squares method yielding $\hat{A}$, $\hat{B}$ and $\hat{d}$. Because the polynomials C and D are in practice difficult to estimate they are often used as design parameters where $\hat{C}$ is usually denoted as T [2]. The model that is used to predict the process output then becomes:

$$y(k) = \frac{q^{-\hat{d}}\hat{B}(q^{-1})}{\hat{A}(q^{-1})}u(k-1) + \frac{T(q^{-1})}{\hat{A}(q^{-1})\hat{D}(q^{-1})}e(k) \quad (2)$$

This model is used to predict $y(k+i)$ for all $i \geq d+1$ (note that prediction of $y(k+1), \ldots, y(k+d)$ makes no sense because they cannot be influenced by u). In [9] a detailed discussion can be found on how to predict $y(k+i)$.

In order to define how well the predicted process output tracks the reference trajectory, a criterion function J is used. In the UPC controller the following criterion function is minimized subject to the constraint (4):

$$J = \sum_{i=H_s}^{H_p} [P\hat{y}(k+i) - Rw(k+i)]^2 + $$

$$+\rho \sum_{i=1}^{H_p-d} \left[\frac{Q_n}{Q_d}u(k+i-1)\right]^2 \quad (3)$$

$$N\Delta^{\beta}u(k+i-1) = 0 \qquad 1 \leq H_c < i \leq H_p - d \quad (4)$$

where $\Delta = 1 - q^{-1}$. The parameters H_p, H_s and H_c are called the prediction horizon, the minimum cost horizon and the control horizon, respectively. Further, P, R, Q_n, Q_d and N are polynomials in q^{-1}. Finally, β is an integer variable ≥ 1 and ρ is a weighting factor ≥ 0. Now the controller output sequence u is obtained by minimization of J with respect to u. It is shown in [9] that if the criterion is quadratic and the model is linear, an analytical solution to the minimization problem exists. In this case, the control law is given by:

$$\mathcal{R}u(k) = -\mathcal{S}y(k) + \mathcal{T}w(k+H_p) \quad (5)$$

where the polynomials $\mathcal{R}$, $\mathcal{S}$ and $\mathcal{T}$ are a complex function of the parameters of the model (2) and the UPC design parameters. Figure 2 shows the control law (5) in a block diagram. From Figure 2 the following closed-loop transfer

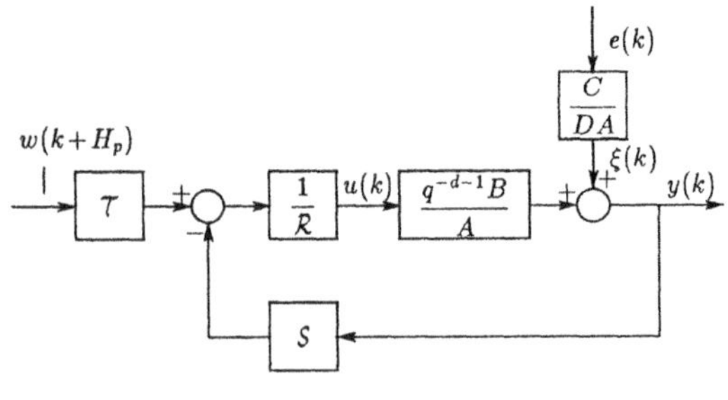

Figure 2: The closed-loop system

function can easily be calculated:

$$y(k) = \frac{q^{-d-1}B\mathcal{T}w(k+H_p) + \frac{C\mathcal{R}}{D}e(k)}{A\mathcal{R} + q^{-d-1}B\mathcal{S}} \quad (6)$$

The UPC controller has many design parameters to be selected by the designer. Next it is shown that although UPC has more design parameters than any other predictive controller, tuning of UPC is relatively simple.

Tuning of the UPC Design Parameters

Q_n, β and $\hat{D}$

Q_n, β and $\hat{D}$ must satisfy certain conditions in order to prevent steady-state errors [8]. Suppose the reference trajectory is of type r ($r = 1$ denotes a constant trajectory, $r = 2$ denotes a triangular trajectory and so on) and the disturbance $\xi(k)$ is of type p. Further, introduce δ and γ, where δ is the number of factors Δ in $\hat{D}$ and γ is the number of factors Δ in Q_n. It has been shown in [8] that steady-state errors do not occur at the sampling instants if the following conditions are satisfied:

$$\delta + n \geq \max(p, r) \quad (7)$$

$$\gamma + n \geq \max(p, r) \quad (8)$$

$$\beta + n \geq \max(p, r) \quad (9)$$

in which n is the number of integrators of the process. Therefore, the following defaults can be used for Q_n, β and $\hat{D}$:

$$\beta = \max(\max(p, r) - n, 1)) \quad (10)$$

$$Q_n = \Delta^{\beta} \quad (11)$$

$$\hat{D} = \Delta^{\beta}$$

ρ and Q_d

The parameters ρ and Q_d realize, together with Q_n, controller output weighting. In order to attenuate undesired frequencies in the controller output, ρ can be used together with Q_n and Q_d. The polynomials Q_n and Q_d must be chosen such that the frequencies that must be attenuated are passed by the filter Q_n/Q_d. All other frequencies must not be passed by this filter. However, one must take care that the number of factors Δ in Q_n satisfies (8) in order to prevent steady-state errors. The value for ρ must be determined by using Bode plots and/or simulations.
Usually ρ is used as a tool to decrease the controller output variance: increasing ρ makes the controller output variance smaller. In many predictive controllers $\beta = 1$ and $Q_n = \Delta$ in order to avoid steady-state errors due to constant trajectories and disturbances for processes without integrators. The polynomial Q_d is usually equal to 1. Consequently, controller increments are weighted. It has been shown in [7] that by doing so, the closed-loop system can become badly damped or even unstable for a certain range of ρ. By weighting the controller output directly (hence, $Q_n = Q_d = 1$) these undesirable effects did not occur. However, in this case, steady-state errors can occur for constant trajectories and disturbances. It was shown in [7] that, as a compromise, $Q_n = \Delta$ and $Q_d = 1 - 0.95q^{-1}$ can be chosen. Then, (8) is satisfied and the higher frequencies in $u(k)$ are equally weighted by ρ. In the case of non-constant disturbances or reference trajectories Q_n must be chosen according to (11) while Q_d can be chosen as:

$$Q_d = (1 - 0.95q^{-1})^{\beta} \quad (12)$$

However, still it is quite difficult to choose a value for ρ. Therefore, it is argued in, for example, [2] to use $\rho = 0$ by default and to obtain the desired behavior of the control system by using the other design parameters (see below). However, a small value for ρ can (and sometimes must) be used in order to improve the numerical robustness of the controller. For example, if $H_c > H_p - H_s + 1$, ρ must be chosen different from zero in order to obtain a unique solution to the minimization problem.

H_p, H_c and H_s

The prediction, control and minimum cost horizon play an important role in determining the desired behavior of the closed-loop system. It has been shown in [8] that a pole-placement controller is obtained if $H_p \geq n_A + n_B + d + 1$, $H_c = n_A + 1$, $H_s = n_B + d + 1$, $\rho = 0$, $N = P$ and $R = P(1)$. The closed-loop poles are determined by P. The N polynomial is not considered as a tuning parameter. For this reason, subsequently, N is taken equal to P. If the input/output behavior of the process is estimated correctly (hence, $\hat{A} = A$, $\hat{B} = B$ and $\hat{d} = d$) and the reference trajectory is equal to the set point Sp, the closed-loop transfer function (6) can be written as [8]:

$$y(k) = \frac{q^{-d-1}B(q^{-1})P(1)}{P(q^{-1})B(1)}Sp + \frac{A(q^{-1})R(q^{-1})}{P(q^{-1})T(q^{-1})}\xi(k) \quad (13)$$

where $\xi(k)$ is a disturbance acting on the output of the process (see Figure 2). Equation (13) clearly shows that the zeros of the process are not cancelled. Consequently, processes with an unstable inverse (such as non-minimum phase processes) can be controlled. Equation (13) also shows that by choosing appropriate values for P, minimum-variance, dead-beat and mean-level controllers can be obtained [8]. Obviously, the desired servo-behavior can be defined by P. However, (13) shows that the regulator behavior is affected by P too.

Examination of the criterion (3) shows that in the case of set point changes H_p need not be chosen larger than the settling time of $Py(k)$ which, as is shown later in this paper, is equal to the settling time of the closed-loop system when $P = 1$. In general, the open-loop system will respond slower to set point changes than the closed-loop system obtained with $P = 1$. Therefore, if the process is stable and well damped, a default value for H_p is: $H_p = \text{int}(t_s(5\%)/T_s)$, where $t_s(5\%)$ is the 5% settling time of the process' step response, T_s is the sampling period and int(.) is a function that converts a real value into an integer. If the process is badly damped or unstable, then H_p must be related to the expected settling time of the closed-loop system. If, as a rule of thumb, the sampling period is selected $10 - 20$ times smaller than the settling time of the closed-loop system [1], this results in $H_p = 10 - 20$. It has been shown in [6] that by choosing H_p as described above, the sensitivity of the closed-loop system to changes in H_p is small.

Usually, predictive controllers are not operated as pole-placement controllers resulting in other values for H_c and H_s. The influence of H_s on the closed-loop system has been discussed in, among others, [4]. There, it is argued that increasing H_s usually makes the system respond slower to set point changes and disturbances. However, by choosing H_s greater than $d + 1$, the tracking error in the near future is not included in the criterion function. For minimum phase processes this usually does not cause problems. However, in the case the process is non-minimum phase, the influence of H_s on the system is quite unpredictable as is shown by the following example.

Example 1 .

Settings: $H_p = 25$, $\beta = 1$, $N = P = 1$, $R = 1$, $T = 1$, $\hat{D} = 1$, $\rho = 10^{-4}$, $Q_n = \Delta$, $Q_d = 1$

Process:

$$H(s) = \frac{-4(s-1)}{s^2 + 1.6s + 4} \overset{T_s=0.2s}{\Longrightarrow}$$

$$H(z^{-1}) = \frac{-0.5954z^{-1}(1 - 1.2269z^{-1})}{1 - 1.5910z^{-1} + 0.7261z^{-2}} \quad (14)$$

Parameters: $H_s = 1, \ldots, 25$ and $H_c = 1, 2$

The process used in this example is an underdamped non-minimum phase process. Note that ρ has been chosen different from zero in order to obtain a unique solution to the minimization problem.
In the case $H_c = 1$, simulations have shown that the closed-loop system is hardly affected by H_s. For all values of H_s, the controller is approximately equal to a mean-level controller. The rise and settling time as a function of H_s if $H_c = 2$ are shown in Figure 3. The figure shows quite

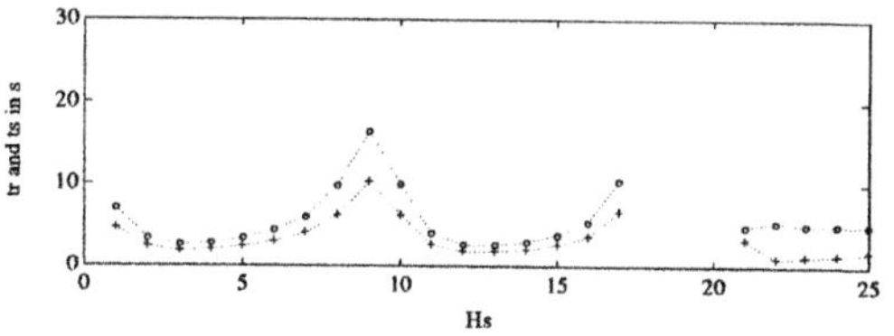

Figure 3: Rise ('+') and settling time ('o') as a function of H_s with $H_c = 2$.

a remarkable behavior of t_r and t_s. Increasing H_s from 1 till 3, makes the system faster. Further increasing H_s makes the system respond slower to set point changes. If $H_s = 18, \ldots, 20$, t_r and t_s are not defined because for these values of H_s, the closed-loop system is unstable. This is clearly illustrated by Figure 4 which shows a root-locus plot as a function of H_s. The root locus also shows that for

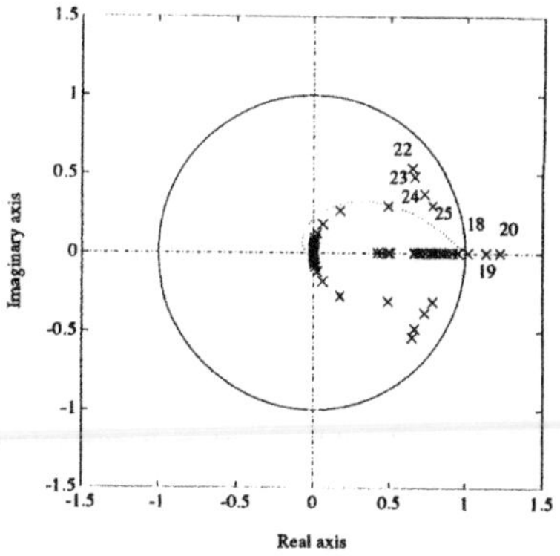

Figure 4: Root locus as a function of H_s with $H_s = 1, \ldots, 25$ and $H_c = 2$.

$H_s = 22, \ldots, 25$, the closed-loop poles are badly damped resulting in a large overshoot (> 30%). Only in the case $H_s = 21$ a good response is obtained. The behavior of the closed-loop system is very sensitive to the choice of H_s: choosing $H_s = 20$ yields an unstable closed-loop system. Choosing $H_s = 22$ yields a closed-loop system which step response shows an overshoot of 101%.
To conclude: by increasing H_s, the tracking error in the

near future is not taken into account in the criterion function. Especially in the case of a non-minimum phase process this can result in an unpredictable behavior of the closed-loop system when H_s is changed. A small change in H_s can even make the closed-loop system move from nicely damped to unstable or vice versa. Therefore, in our opinion, H_s should not be used as a tuning parameter. H_s can best be chosen equal to $d+1$. Only in the case UPC is operated as a pole-placement controller, H_s can be chosen different from $d+1$ [8].

The control horizon H_c can be chosen equal to n_A+1 as is required to obtain pole-placement control. However, many simulations using a wide variety of processes [6] have shown that choosing $H_c = n_A$ yields a compromise between robustness and performance of the closed-loop system.

The T polynomial

It has been shown in [2,6,9] that the T polynomial plays an important role in determining the robustness and the regulator behavior of the closed-loop system. In this paper the stability robustness of the closed-loop system is considered only. The gain margin (gm) and the phase margin are used as measures of the system's robustness. The phase margin is translated into a time delay margin (dm). The larger the gain and time delay margin, the more robust the control system to variations in the DC gain and the time delay of the process. Both gm and dm can be calculated from the Nyquist plot of the loop transfer function H_L: $H_L = q^{-\hat{d}-1}\hat{B}S/\hat{A}R$.
In [9] it has been shown that if $\hat{A} = A$, $\hat{B} = B$ and $\hat{d} = d$, T does not have effect on the servo behavior of the system. In order to obtain a rule of thumb on how to select T consider the following theorem which has been proven in [9]:

Theorem 1 *If $PT = \hat{A}\hat{D}$, then $S = 0$.*

The theorem implies that choosing $PT = \hat{A}\hat{D}$ involves that feedback disappears. Hence, the closed-loop system is stable if the process is stable. As a result, the stability robustness of the control system with respect to changes in DC gain and the time delay is infinite. Now, for stable processes, a logical choice for T and $\hat{D}$, is:

$$\begin{aligned} T &= \hat{A}(1-\mu q^{-1})^{\beta} \\ \hat{D} &= P\Delta^{\beta} \end{aligned}$$

with $0 \le \mu \le 1$. Note that the number of factors Δ in $\hat{D}$ still satisfies (7) and that P has been made a factor of $\hat{D}$. It is shown later in the paper that there is another reason for doing this. For small values of μ, the gain and delay margins have certain values. For $\mu \to 1$, the (stability) robustness of the closed-loop system is guaranteed to be infinite by Theorem 1. Therefore, it can be expected that increasing μ results in an increasing robustness of the system. This is illustrated by Figure 5a which shows the gain and delay margin as a function of μ ($\mu = 0,\ldots,1$) for the non-minimum phase process (14) with settings $H_p = 25$, $H_c = 2$, $H_s = 1$, $\rho = 0$, $\beta = 1$, $P = 1$ and $\hat{D} = \Delta$. The dotted line in Figure 5 indicates the value 1. The figure clearly shows that the robustness of the system increases monotonically with μ.

Another choice for T proposed by Clarke [2] is: $T = (1-\mu q^{-1})^{n_A}$ with $0 \le \mu \le 1$. As a default $\mu = 0.8$ is suggested. Simulations have shown that by using this choice for T the robustness usually does not increase monotonically with μ. Moreover, the range of the gain margin especially is

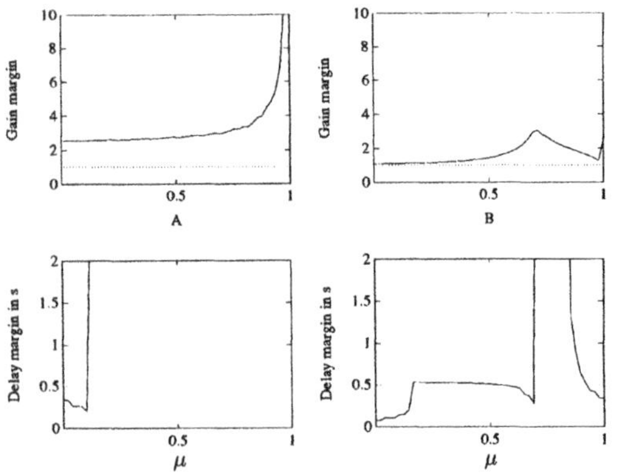

Figure 5: Robustness criteria as a function of μ with $T = \hat{A}(1-\mu q^{-1})$ (Figure A) and with $T = (1-\mu q^{-1})^2$ (Figure B).

small compared to that using $T = \hat{A}(1-\mu q^{-1})^{\beta}$. For example, Figure 5b shows the stability margins as a function of μ with $T = (1-\mu q^{-1})^{n_A}$ for the non-minimum phase process (14). The settings of the UPC controller were equal to those used in the previous example. Clearly, the robustness range is quite small. Further, for $\mu = 0$ the stability margins are very small: $gm = 1.07$ and $dm = 0.089$s. Increasing μ does not result in a monotonically increasing robustness. Therefore, the choice $T = \hat{A}(1-\mu q^{-1})^{\beta}$ is to be preferred. However, it is shown in [9] that if the process is correctly estimated, the roots of T appear as closed-loop poles. Therefore, $T = \hat{A}(1-\mu q^{-1})^{\beta}$ cannot be used if the process is unstable or badly damped. Then, $T = (1-\mu q^{-1})^{n_A}$ can be used instead.

So far, the influence of T on the robustness of the closed-loop system has been discussed. It is well known that T also influences the system's regulator behavior [2]. Simulations have shown [6] that using either $T = \hat{A}(1-\mu q^{-1})^{\beta}$ or $T = (1-\mu q^{-1})^{n_A}$ makes the closed-loop system respond slower to disturbances if μ is increased. If UPC is operated as a pole-placement controller this is clearly shown by (13). As a result, a large controller output variance caused by disturbances can be made smaller by increasing μ. However, by doing so, in most cases the process output variance increases. If $\mu = 1$, feedback no longer exists and disturbances are no longer rejected. Only if the process is disturbed by measurement noise this is a good policy. Simulations have shown that using $T = \hat{A}$ as a default usually results in a compromise between the regulator behavior and robustness.

The P polynomial

In this section the influence of P on the regulator behavior and the robustness of the closed-loop system is discussed. In the case a pole-placement controller is selected, this influence is easy to analyze. Now the closed-loop transfer function is, if the process is correctly modeled, given by (13) and P can be used to define the desired servo behavior of the closed-loop system. However, (13) also shows that P affects the transfer function from the disturbance $\xi(k)$ to $y(k)$ and hence the regulator behavior of the system.

From the design point of view it may be desirable to make the regulator behavior and the robustness independent of P. Then P can be used to define the servo behavior of the closed-loop system while T defines its regulator behavior and its robustness. Two ways to realize this are:

1. Use $P = 1$ in the criterion function (3) and generate the reference trajectory by filtering the set point by R/P.

2. Use P as a factor of $\hat{D}$. It is shown in [6] that by doing so, P does not have influence on the regulator behavior and the robustness of the control system. In the case P is not a factor of $\hat{D}$ the effect of P on the robustness and the regulator behavior of the closed-loop system is identical to that of T [6].

Considering the servo behavior of the closed-loop system the following theorem, which has been proven in [6], can be stated:

Theorem 2 *If $A = \hat{A}$, $B = \hat{B}$, $d = \hat{d}$, $N = P$ and P is a factor of Q_n, then the closed-loop poles are given by the roots of P and the poles that would have been obtained if $P = 1$ had been used.*

Theorem 2 shows that, as the servo behavior of the closed-loop is concerned, the roots of P are added to the set of closed-loop poles obtained when using $P = 1$. If, for example, a dead-beat controller is selected the closed-loop poles are all in the origin. By choosing P different from 1, the roots of P are added to the set of closed-loop poles which in this case results in the closed-loop poles being equal to the roots of P. Then, by tuning P, the desired response time of the process to set point changes can be obtained. In general, selecting a desired response by means of P is possible only if the closed-loop poles determined by P dominate the closed-loop poles that are obtained when $P = 1$.
Remark: choosing P a factor of $\hat{D}$, the robustness and regulator behavior become independent of P. Moreover, together with $T = \hat{A}(1 - \mu q^{-1})^{\beta}$, it also ensures that the stability margins are infinite if $\mu \to 1$ (see the previous section).

Design of P and R using a single parameter

In the case the desired closed-loop system is known in terms of poles and zeros the choices for R and P are straightforward. However, it may also be desirable to define the closed-loop servo behavior using a single parameter. For example, the rise time of the response to a step wise change in the set point.
If a first-order trajectory is desired, R and P can easily be calculated: $R = 1 - \alpha$ and $P = 1 - \alpha q^{-1}$. The parameter α can be related to the desired rise time t_r of the trajectory by using: $\alpha = e^{-2.3T_s/t_r}$. The disadvantage of using a first-order trajectory is that it is quite fast in the beginning and reaches slowly its steady-state value (in theory in an infinite time). This typically results in a large controller output after a set point change and a large settling time.
In order to overcome this problem, a second-order trajectory can be used. Now two parameters must be specified: the damping ratio ς and the natural frequency ω_n. The damping ratio can be fixed at $\varsigma = 0.7$. In this case the overshoot is approximately 4.6% and the natural frequency coincides with the desired bandwidth of the closed-loop system. The natural frequency ω_n can be related to the peak time t_p of the step response: $\omega_n = \pi/(t_p\sqrt{1-\varsigma^2})$. Hence, given the desired peak time of the closed-loop system and assuming $\varsigma = 0.7$, ω_n can easily be calculated. Now, the reference model is known in the s-domain. Transforming this continuous model to the discrete domain using a zero-order hold yields for P:

$$P = 1 + p_1 q^{-1} + p_2 q^{-2} \tag{15}$$

with $p_1 = -2e^{-\varsigma\omega_n T_s}\cos(\omega_n T_s\sqrt{1-\varsigma^2})$ and $p_2 = e^{-2\varsigma\omega_n T_s}$. It is common practice not to include the zero and the unit delay of the discretized reference model into the model that is used to define the characteristics of the closed-loop system. As a result, R is given by:

$$R = P(1) = 1 + p_1 + p_2 \tag{16}$$

By using one of the methods mentioned above, the desired speed of the closed-loop system can be adjusted by using a single parameter: the rise or peak time of the closed-loop step response. The following example illustrates the use of a first and second-order reference trajectory.

Example 2 .	
Settings:	$H_p = 4$, $H_s = 2$, $H_c = 3$, $\beta = 1$, $R = P(1)$, $T = 1$, $\hat{D} = 1$, $\rho = 0$, $N = P$, $Q_n = 1$, $Q_d = 1$
Process:	$H(s) = \dfrac{5}{(4s+1)(5s+1)} \overset{T_s=0.5s}{\Longrightarrow}$ $H(z^{-1}) = \dfrac{0.029z^{-1}(1+0.928z^{-1})}{(1-0.882z^{-1})(1-0.905z^{-1})}$
Parameters:	P

The P (and N) polynomials were selected such that both reference trajectories have a rise time of 3.5s. Figure 6a shows the response in the case a first-order reference trajectory is selected ($P = 1 - 0.72q^{-1}$). Figure 6b shows the use of a second-order trajectory ($P = 1 - 1.398q^{-1} + 0.540q^{-2}$). Both responses have the same rise time. However, the second-order trajectory makes the system settle faster to the set point. Further, the controller output is smoother compared to the one using a first-order trajectory. For this reason a second-order reference trajectory is suggested using (15) and (16) to define P and R.

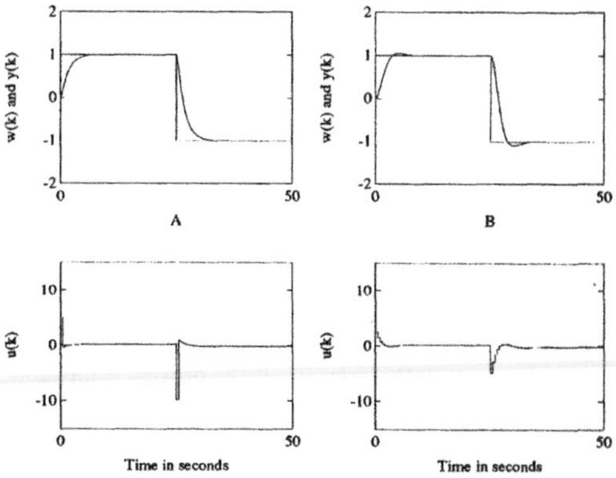

Figure 6: The use of a first-order (Figure A) and a second-order reference trajectory (Figure B).

Rule of Thumb methods

Based on the results shown above, rules of thumb can be formulated that provide (initial) settings for the UPC design parameters. Because there is a slight difference between the rules of thumb for well-damped and badly-damped processes (among which unstable processes), they are discussed separately.

Rule of thumb methods for well-damped processes
In this section the rules of thumb are formulated for processes with well-damped poles. Processes containing integrators are also considered as such.

$H_p = \text{int}(t_s(5\%)/T_s)$ where $t_s(5\%)$ is the 5% settling time of the step response of the (continuous) process. Simulations have shown that this value for H_p yields a low sensitivity of the closed-loop system to changes in H_p.

$H_s = d + 1$. It was shown that other values for H_s sometimes cause unpredictable results. Only in the case UPC is operated as a pole-placement controller, other values for H_s can be considered.

$H_c = n_A$. This value for H_c has shown to yield a compromise between robustness and performance.

$\beta = \max(\max(p, r) - n, 1)$ where r and p denote the type of the reference trajectory and the disturbances and n denotes the number of integrators in the process.

$\rho = 0$. For most applications, the weighting factor ρ can be taken equal to zero. Then, Q_n and Q_d can be taken equal to 1.

$\hat{D} = P\Delta^{\beta}$. Together, with the choice of β (Q_n does not play a role because $\rho = 0$), steady-state errors do not occur. Further, by making P a factor of $\hat{D}$, P does not affect the servo-behavior and the robustness of the closed-loop system.

$T = \hat{A}$. This value for T can be used only if the poles of the model are well damped. Usually a compromise between robustness and regulator behavior is obtained.

$P = 1 + p_1 q^{-1} + p_2 q^{-2}$ where p_1 and p_2 are calculated by specifying the desired peak time of the closed-loop system. The reference trajectory is in this case equal to the set point (hence, $w = [Sp, \cdots, Sp]^T$).

Further, $R = P(1)$ and $N = P$.

Rule of thumb methods for badly-damped processes
Most of the rules mentioned above for well-damped processes can also be used for badly-damped processes. Only those that are different are discussed here.

H_p. If the process is stable but badly damped, utilizing the rule of thumb mentioned in the previous section can yield an extremely large H_p. If the process is unstable, this rule can no longer be used. In such a case H_p must be selected according to the (expected) settling time of the closed-loop system.

$T = (1 - \mu q^{-1})^{n_A}$ with $\mu = 0.8$. Because the roots of T appear as poles in the closed-loop system using $T = \hat{A}$ is not possible. The choice $T = (1 - \mu q^{-1})^{n_A}$, however, can also be used for well-damped processes.

Usually, UPC tuned by the above-mentioned rule of thumb methods yields a response that is quite acceptable. However, in some situations fine tuning may be desired. The following parameters are considered as initialization parameters: Q_n, Q_d, β, H_s, $\hat{D}$ and H_p. These parameters have either less influence on the closed-loop system (e.g. H_p) or their influence is hard to predict (e.g. H_s). The parameters Q_n, Q_d, β and $\hat{D}$ are obtained by applying theorems concerning the steady-state behavior and the regulator behavior. The parameters T, P, ρ and H_c can be used as possible fine tuning parameters (remember that $R = P(1)$ and $N = P$). The whole design procedure can be realized in an expert system [5].

Conclusions

In this paper the design parameters of the Unified Predictive Controller have been discussed. It has been shown that separate tuning of the servo and regulator behavior can be obtained by choosing particular settings for P and T. Then, P can be used to define the servo behavior and T can be used to define the regulator behavior and the robustness of the closed-loop system. Further, a simple procedure for selecting P has been presented. It has been shown that a second-order polynomial for P is to be preferred. Rule of thumb methods have been proposed that can be used to obtain initial settings for the UPC design parameters. Usually, the resulting response is quite acceptable. However, in some situations fine tuning may be desired.

References

[1] Åström, K.J. and B. Wittenmark (1984). *Computer Controlled Systems.* Prentice-Hall, Inc.

[2] Clarke, D.W. and C. Mohtadi (1989). "Properties of Generalized Predictive Control". *Automatica.* Vol.25, No.6, pp. 859-875.

[3] Clarke, D.W., C. Mohtadi and P.S. Tuffs (1987). "Generalized Predictive Control-Parts I & II". *Automatica,* Vol.23, No.2, pp. 137-160.

[4] Keyser, R.M.C. de, Ph. G.A. van de Velde and F.A.G. Dumortier (1988). "A Comparative Study of Self-adaptive Longe-range Predictive Control Methods". *Automatica.* Vol.24, No.2, pp. 149-163.

[5] Krijgsman, A.J., H.B. Verbruggen and P.M. Bruijn (1991). "Knowledge-Based Tuning and Control". To be presented at the IFAC symposium on Intelligent Tuning and Adaptive Control. Singapore.

[6] Soeterboek, A.R.M. (1990). *Predictive Control - A Unified Approach.* Ph.D. Thesis, Delft University of Technology, the Netherlands.

[7] Soeterboek, A.R.M., H.B. Verbruggen and P.P.J. van den Bosch (1989). "Unified Predictive Control - Analysis of Design Parameters". *Proceedings of the IASTED International Symposium on Modelling, Identification and Control.* Grindelwald, Switzerland.

[8] Soeterboek, A.R.M., H.B. Verbruggen, P.P.J. van den Bosch and H. Butler (1990). "Adaptive Predictive Control - A Unified Approach". *Proceedings of the Sixth Yale Workshop on Adaptive and Learning Systems.* Yale University, New Haven, U.S.A.

[9] Soeterboek, A.R.M., H.B. Verbruggen, P.P.J. van den Bosch and H. Butler (1990). "On the Unification of Predictive Control Algorithms". *Proceedings of the 29th IEEE Conference on Decision and Control.* Honolulu (HI), U.S.A.

A COMPARATIVE STUDY OF SOME MULTIVARIABLE PI CONTROLLER TUNING METHODS

J. T. Tanttu and J. Lieslehto

Tampere University of Technology, Control Engineering Laboratory, P.O. Box 527, SF-33101 Tampere, Finland

Abstract.

In this paper four different multivariable PI controller tuning methods are compared. Multivariable PI controller is designed using all four methods for a two input two ouput state state space model. The resulted designs are then compared using frequency domain performance and robustness criteria and time domain simulations

Key words. PID control; Multivariable control systems

INTRODUCTION

The tuning of scalar (single input single output) PID controllers has received new attention in the 80's. Several survey papers and monographs have been published about the subject (Clarke, 1986). The combination of adaptive methods with PID controller tuning has also led to several schemes for automatic tuning of scalar PID controllers (Åström and Hägglund, 88). Although several different methods for the tuning of multivariable PI(D) controllers have also been suggested: Solheim (1974), Davison (1976), Seraji and Tarokh (1977), Penttinen and Koivo (1980), Jussila and Koivo (1987), systematic comparison of different design methods for multivariable PI(D)-controllers has received less interest (Yu, 1989).

In this paper four different multivariable PI controller tuning methods are compared. The methods included are: the tuning method proposed by Davison (1976), an extension of that proposed by Penttinen and Koivo (1980), another modification of Davison tuning presented in Maciejowski (1989) and finally a newly developed method (Lieslehto, 1990) which is based on the IMC tuning ideas (Rivera, Morari, Skogestad, 1986)

Multivariable PI controller is designed using all the listed methods for a state space model proposed by Kouvaritakis (1978). The resulted designs are then compared using the new frequency domain performance and robustness criteria (Maciejowski, 1989) and also with time domain simulations.

1. THE COMPARED TUNING METHODS

The multivariable PI controller is described with the transfer function matrix

$$K(s) = K_p + K_i \frac{1}{s} \tag{1.1}$$

where K_p and K_i are the gain matrices of the proportional and integral parts of the controller respectively. The task of any tuning method is to determine these matrices based on some kind of process model or experimental measurements. In this paper the starting point is a stable transfer function matrix model

$$y(s) = G(s)\, u(s) \tag{1.2}$$

We will assume that the number of inputs of the system equals the number of outputs so that G(s) is an *mxm* rational matrix.

Davison tuning

Davison (1976) proposed a PI controller where the tuning matrices are selected as follows

$$K_p = \delta\, G^{-1}(0) \tag{1.3a}$$

$$K_i = \varepsilon\, G^{-1}(0) \tag{1.3b}$$

if the inverse exists. Here

$$K = G^{-1}(0)$$

is called the ***rough tuning matrix*** and δ and ε are the ***fine tuning parameters*** of the method. In this method the rough tuning matrix is the inverse of the steady state gain of the process which may be determined experimentally for stable processes. Usually the fine tuning parameters are in the interval (0 , 1).

Penttinen - Koivo tuning

Assuming that the transfer matrix G(s) in (1.2) is strictly proper there exists a (minimal) state space realization for system (1.2) of the form

$$\dot{x} = Ax + Bu \tag{1.4a}$$

$$y = Cx \tag{1.4b}$$

The transfer function matrix has the series expansion

$$G(s) = \frac{CB}{s} + \frac{CAB}{s^2} + \frac{CA^2B}{s^3} + \dots \tag{1.5}$$

Now Penttinen and Koivo (1980) proposed that the proportional part of the controller to be selected as

$$K_p = \delta\,(CB)^{-1} \qquad (1.6)$$

if the required inverse exists. This selection aims at decoupling the system at high frequencies (c.f. Maciejowski, p.142). The rough tuning matrix in (1.6) can also be determined graphically from the open-loop step responses of the system (Penttinen and Koivo, 1980). The tuning matrix for the I-part of the controller is selected as in Davison tuning (1.3) which using the state space model (1.4) becomes

$$K_i = \varepsilon\,(-CA^{-1}B)^{-1} \qquad (1.7)$$

The fine tuning parameters ε and δ are selected as in the Davison method.

Maciejowski tuning

When the compensator (1.6) is used then

$$G(s)\,K_p \rightarrow \frac{\delta I}{s} \text{ as } |s| \rightarrow \infty \qquad (1.8)$$

The problem is however that usually (1.8) holds for frequencies that are much above the desired bandwidth of the system. So Maciejowski (1989) suggests that instead of (1.6)

$$K_p \approx \delta\, G^{-1}(j\omega_b) \qquad (1.9)$$

should be used where $\omega_b \approx$ the desired bandwidth of the system.

Lieslehto tuning

Rivera, Morari & Skogestad (1986) proposed a PID controller tuning method based on the Internal Model Control (IMC) principle. What they actually showed was that for several low order transfer functions IMC-method yields a PI(D)-type of controller. Lieslehto (1990) has developed a multivariable tuning method based on this idea. In this method a SISO PI(D) controller is first designed for each element $g_{ij}(s)$ of the transfer function matrix G(s) separately yielding two parameters the proportional gain k_{pij} and the integral gain k_{iij}. Then the tuning matrix for the MIMO P-controller is computed as follows

$$K_p = \begin{bmatrix} 1/k_{p11} & 1/k_{p12} & \cdots & 1/k_{p1m} \\ 1/k_{p21} & 1/k_{p22} & \cdots & 1/k_{p2m} \\ \vdots & \vdots & & \vdots \\ 1/k_{pm1} & 1/k_{pm2} & \cdots & 1/k_{pmm} \end{bmatrix}^{-1} \qquad (1.10)$$

and for the I-part in the similar way. In this method there is only one tuning parameter supplied by the user. This tuning parameter is approximately the inverse of the desired bandwidth of the closed loop system (in scalar case) and it is used at the first stage of the design. Since the IMC method yields PI(D)-type of controllers only for simple transfer functions (no more than two poles and one zero, no delays) higher order transfer functions are first approximated by low order ones in this method. For a more detailed description of the algorithm see (Lieslehto, 1990).

2. FREQUENCY DOMAIN COMPARISON CRITERIA

Let us study the closed loop configuration of Figure 2.1. Then the output of the system becomes

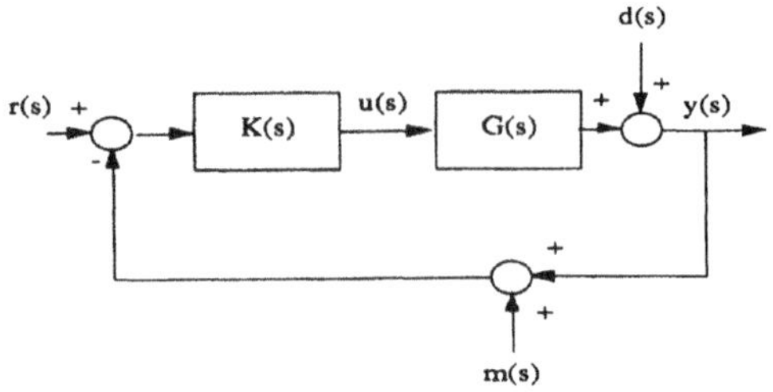

Figure 2.1 Standard closed loop configuration for one degree of freedom control.

$$y(s) = S(s)\,d(s) + T(s)\,r(s) - T(s)\,m(s) \qquad (2.1a)$$

$$S(s) = [I + G(s)K(s)]^{-1} \qquad (2.1b)$$

$$T(s) = [I + G(s)K(s)]^{-1}G(s)K(s) \qquad (2.1c)$$

The transfer function matrix S(s) from load disturbance d(s) to output y(s) is usually called the *sensitivity function* and the transfer function T(s) from reference signal r(s) (or measurement noise m(s)) to output y(s) is the *closed loop transfer function matrix* of the system. Based on this configuration several performance and robustness criteria may be stated (c.f. Ford & al, 1988; Maciejowski, 1989).

1) From the *load disturbance attenuation* point of view the largest singular value of the sensitivity function should be as small as possible (especially at frequencies below the desired bandwidth)

$$\bar{\sigma}\,[S(j\omega)] \le \rho_1 \quad \omega \le \omega_1 \qquad (2.2a)$$

$$\bar{\sigma}\,[S(0)] \le \rho_2 \qquad (2.2b)$$

2) For the *noise attenuation* the largest singular value of the closed loop transfer function matrix should be as small as possible for frequencies above the desired bandwidth of the closed loop system.

$$\bar{\sigma}\,[T(j\omega)] \le \rho_3 \quad \omega \ge \omega_3 \qquad (2.3)$$

3) To minimize the *tracking error*

$$\bar{\sigma}\,[T(j\omega) - I] \le \rho_4 \quad \omega \le \omega_4 \qquad (2.4)$$

4) To minimize the *control activity*

$$\bar{\sigma}\,[(I + KG)^{-1}K] \qquad (2.5)$$

should be made as small as possible.

5) Let the true transfer function of the plant

$$G_t(s) = G(s) + \Delta_a(s)$$

where $\Delta_a(s)$ is an unstructured additive perturbation. Then the largest singular value of $\Delta_a(s)$ for which the closed loop system

(2.1) remains stable is obtained as

$$\bar{\sigma}(\Delta_a) = \frac{\underline{\sigma}\,[(I + GK)]}{\bar{\sigma}(K)} \qquad (2.6)$$

6) Similarly for input multiplicative perturbations

$$G_i(s) = G(s)[I + \Delta_i(s)]$$

$$\bar{\sigma}(\Delta_i) = \underline{\sigma}\,[(I + (KG)^{-1}] \qquad (2.7)$$

7) and for output multiplicative perturbations

$$G_o(s) = [I + \Delta_o(s)]G(s)$$

$$\bar{\sigma}(\Delta_o) = \underline{\sigma}\,[(I + (GK)^{-1}] \qquad (2.8)$$

3. THE SIMULATION EXAMPLE

The following example proposed by Kouvaritakis (1978)

$$A = \begin{bmatrix} -3 & 0 & -2 \\ -2 & -1 & -2 \\ -1 & 1 & -2 \end{bmatrix} \qquad B = \begin{bmatrix} 6 & -5 \\ 4 & -2 \\ -4 & 5 \end{bmatrix} \qquad (3.1a)$$

$$C = \begin{bmatrix} 16 & -13 & 11 \\ 6 & -5 & 4 \end{bmatrix} \qquad (3.1b)$$

was used in the simulations. For this example

$$K_0 = G^{-1}(0) = [-CA^{-1}B]^{-1} = \begin{bmatrix} 5.0 & -12 \\ 3.0 & -7.5 \end{bmatrix} \qquad (3.2)$$

However the matrix

$$CB = \begin{bmatrix} 0 & 1 \\ 0 & 0 \end{bmatrix} \qquad (3.3)$$

needed in (1.6) is not invertible. Peltomaa and Koivo (1983) suggest that in this case the tuning rule

$$K_p = \delta\,(CB + CAB)^{-1} \qquad (3.4)$$

should be used. For our example (3.1)

$$(CB + CAB)^{-1} \approx \begin{bmatrix} -1.00 & 2.75 \\ -0.50 & 1.25 \end{bmatrix} \qquad (3.5)$$

Selecting the fine tuning parameters $\varepsilon = \delta = 0.1$ the following-tuning matrices are obtained for the Davison tuning

$$K_p = K_i \approx \begin{bmatrix} 0.50 & -1.2 \\ 0.30 & -0.75 \end{bmatrix} \qquad (3.6)$$

and for Penttinen - Koivo - Peltomaa tuning

$$K_p \approx \begin{bmatrix} -0.100 & 0.275 \\ -0.050 & 0.125 \end{bmatrix} \quad K_i \approx \begin{bmatrix} 0.50 & -1.2 \\ 0.30 & -0.75 \end{bmatrix} \qquad (3.7)$$

The inverse of the tright hand side of the Maciejowski tuning equation (1.9) is a complex matrix. For the compensator design we need a *real approximation* of it. The result

$$K_1 \approx G^{-1}(0.1j) \approx \begin{bmatrix} 4.88 & -11.7 \\ 2.92 & -7.32 \end{bmatrix} \qquad (3.8)$$

was obtained using the ALIGN function in the MATLAB MFD toolbox (Ford & al, 1988). This result is very near the inverse of the steady state gain of the system thus yielding a PI controller that is almost identical to Davison tuning (3.6). So instead of (3.8) the matrix

$$K_2 \approx G^{-1}(1.0j) \approx \begin{bmatrix} 2.37 & -9.32 \\ 1.27 & -5.69 \end{bmatrix} \qquad (3.9)$$

was used.

Thus for the Maciejowski tuning the gain matrices become

$$K_p = 0.1\ K_2 \approx \begin{bmatrix} 0.237 & -0.932 \\ 0.127 & -0.569 \end{bmatrix} \qquad (3.10a)$$

$$K_i = 0.1\ K_0 \approx \begin{bmatrix} 0.50 & -1.2 \\ 0.30 & -0.75 \end{bmatrix} \qquad (3.10b)$$

The Lieslehto tuning with the fine tuning parameter corresponding to closed loop bandwidth equal to 0.1 rad/s yields the tuning parameters

$$K_p \approx \begin{bmatrix} -0.100 & 0.275 \\ -0.050 & 0.125 \end{bmatrix} \quad K_i \approx \begin{bmatrix} 0.50 & -1.2 \\ 0.30 & -0.75 \end{bmatrix} \qquad (3.11)$$

4. FREQUENCY DOMAIN COMPARISONS

In the Figure 4.1 the largest singular value of the sensitivity function is displayed for all the proposed tunings of section 3. As was explained in section 2 it is an index of the ability of the compensated system to attenuate load type disturbances.

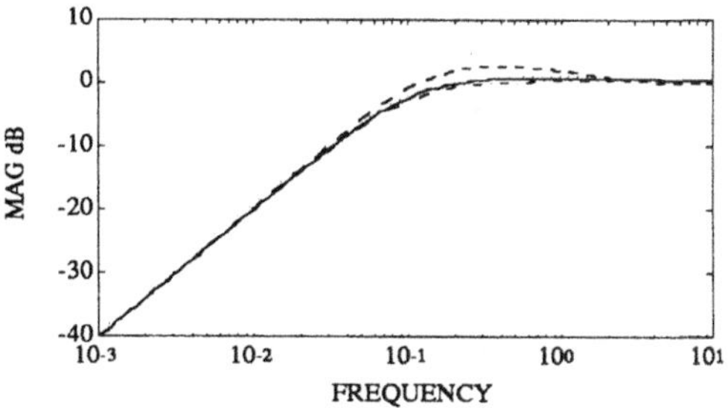

Figure 4.1 Load disturbance attenuation index for Davison (—), Penttinen (- -), Lieslehto (⁻) and Maciejowski (- · -) tuning.

It is also an index of the servo properties of the control system since for the control configuration of Figure 2.1

$$S(s) + T(s) = I \qquad (4.1)$$

and thus

$$\bar{\sigma}\,[S(j\omega)] = \bar{\sigma}\,[T(j\omega) - I] \qquad (4.2)$$

It seems that the disturbance attenuation properties of the four tunings are almost identical. However, in the interval $0.1 \leq \omega \leq 1.0$ Penttinen tuning seems to be poorer than the three others.

In the Figure 4.2 the noise attenuation indices for the four designs are displayed.

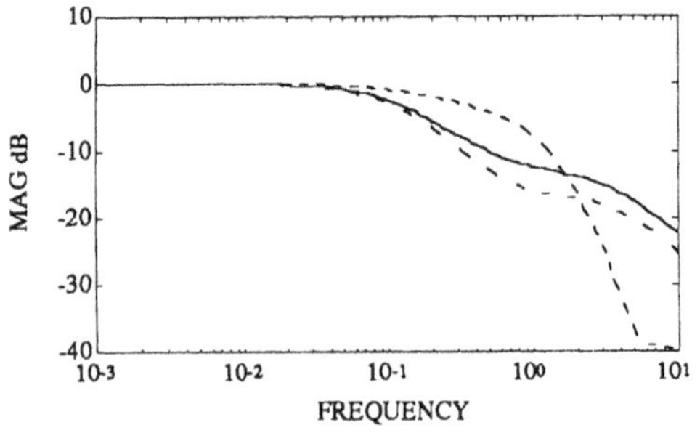

Figure 4.2 Noise attenuation index for Davison (—), Penttinen (- -), Lieslehto (-) and Maciejowski (- · -) tuning.

At intermediate frequencies 0.1 ≤ ω ≤ 1.0 Penttinen design seems again to be the poorest. At high frequencies, however, Penttinen tuning performs best then comes Lieslehto tuning. The high frequency noise attenuation properties of the Davison and Maciejowski tuning are almost identical, Maciejowski tuning being all the time slightly better.

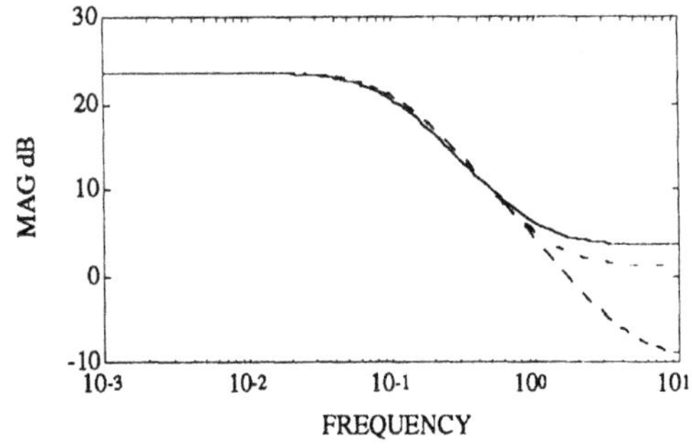

Figure 4.3 Control energy distribution index for Davison (—), Penttinen (- -), Lieslehto (-) and Maciejowski (- · -) tuning.

Figure 4.3 displays the control activity at the plant inputs u caused by an impulse at the reference signal r. In this respect the four designs seem to be almost identical at low and intermediate frequencies. The control activity at high frequencies is significantly smaller for Lieslehto and Penttinen designs than for Maciejowski and Davison designs.

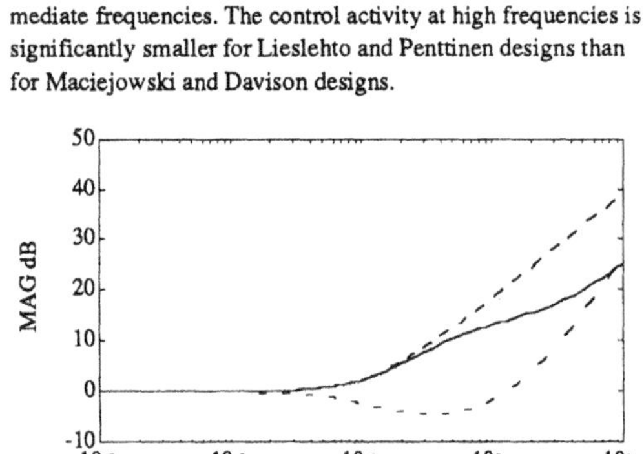

Figure 4.4 Maximal input multiplicative perturbation (2.7) for Davison (—), Penttinen (- -), Lieslehto (-) and Maciejowski (- · -) tuning.

The maximal input multiplicative perturbation index (Figure 4.4) of the Maciejowski design is significantly poorer than other designs especially at intermediate frequencies. For the output multiplicative perturbation index the situation was almost vice versa Maciejowski design being the best at low and intermediate frequencies. At high frequencies the order was Penttinen, Lieslehto, Maciejowski and Davison from the best to the worst.

In the Figures 4.5a - d the singular values of the closed loop transfer function matrix T(s) are depicted for each tuning. As Maciejowski (1989, pp. 83 - 84) points out, for good design the largest and smallest singular value of the closed loop transfer function matrix should be close to each other and he suggests as a design criterion to minimize

$$\Delta = (\omega_{by} - \omega_{ty}) \tag{4.3a}$$

where ω_{by} and ω_{ty} are defined in the following way

$$\bar{\sigma}\, T(j\omega_{by}) = \frac{\bar{\sigma}\, T(0)}{\sqrt{2}} \tag{4.3b}$$

$$\underline{\sigma}\, T(j\omega_{ty}) = \frac{\underline{\sigma}\, T(0)}{\sqrt{2}} \tag{4.3c}$$

i.e. to minimize the difference between the best case and worst case "bandwidth" of the system.

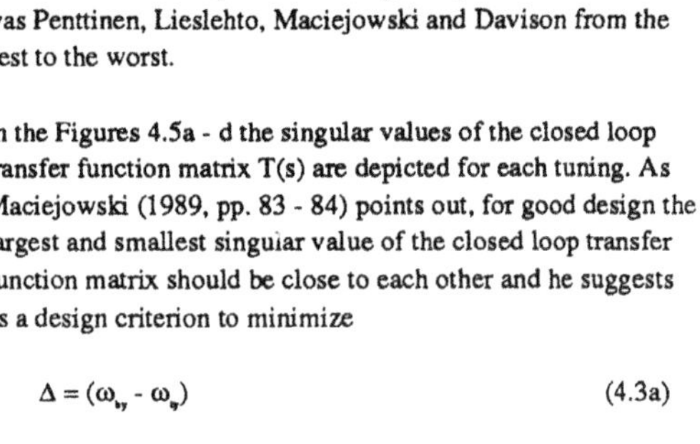

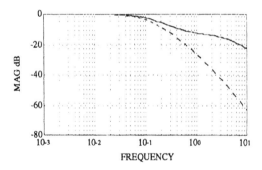

Figure 4.5a Singular values of the closed loop transfer function matrix - Davison tuning.

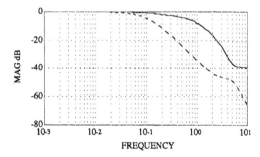

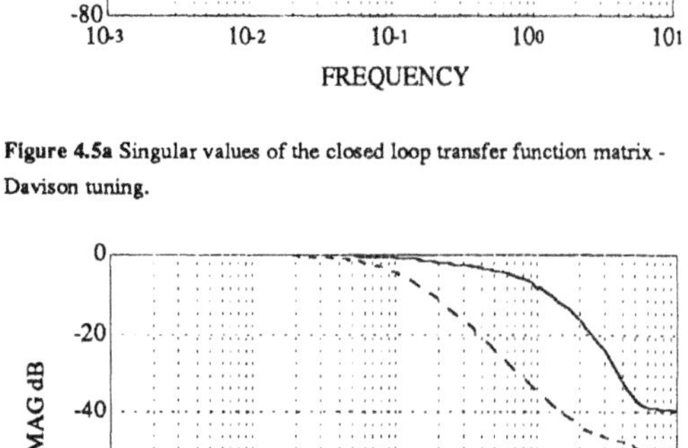

Figure 4.5b Singular values of the closed loop transfer function matrix - Penttinen tuning.

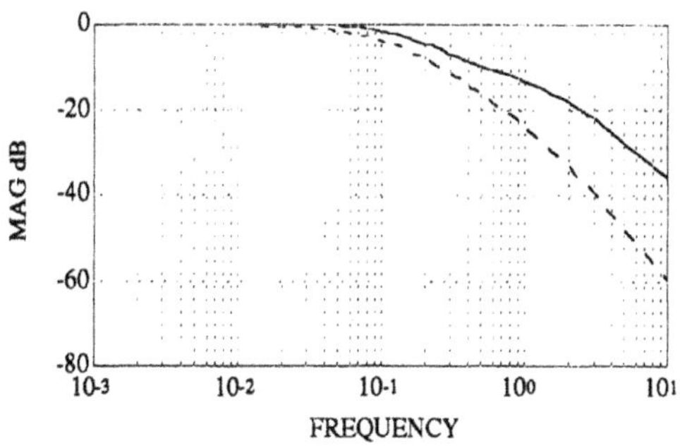

Figure 4.5c Singular values of the closed loop transfer function matrix - Lieslehto tuning.

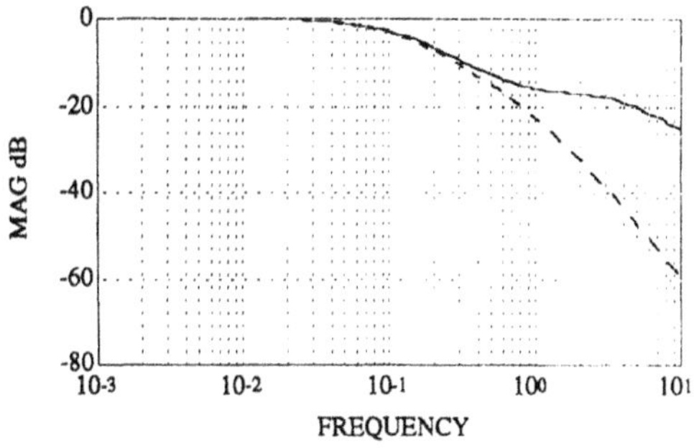

Figure 4.5d Singular values of the closed loop transfer function matrix Maciejowski tuning.

If the criterion (4.3) is used Maciejowski design is superior to the other although also the singular values of Lieslehto tuning remain quite near each other at all frequencies.

5. TIME DOMAIN COMPARISONS

In the first time domain experiment a unit step was used as the reference signal for the output y_2. The results of the experiment are depicted in the Figures 5.1a - 5.1d. It should be noted that digitally implemented PI(D) controllers with sampling period $h = 0.25$ were used in all the simulations of this section.

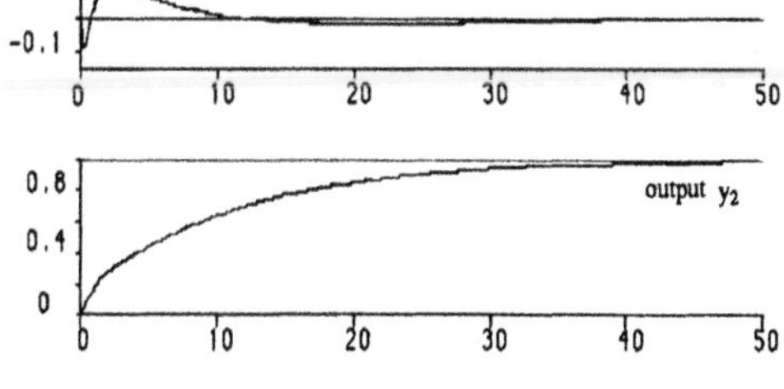

Figure 5.1a Response of the system (3.1) when a unit step reference signal was applied for the second loop (r_2) - Davison tuning.

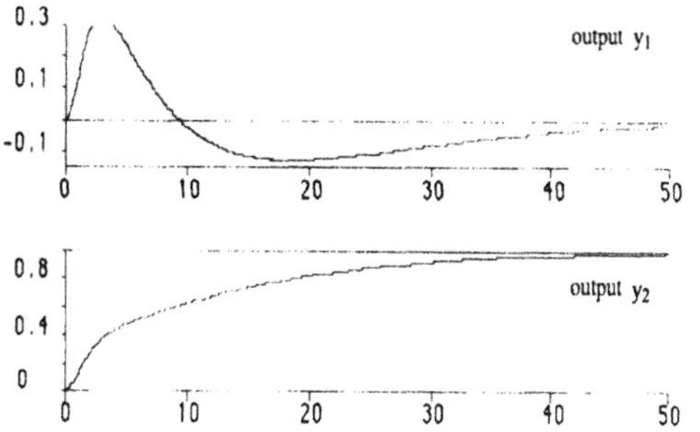

Figure 5.1b Response of the system (3.1) when a unit step reference signal was applied for the second loop (r_2) - Penttinen tuning.

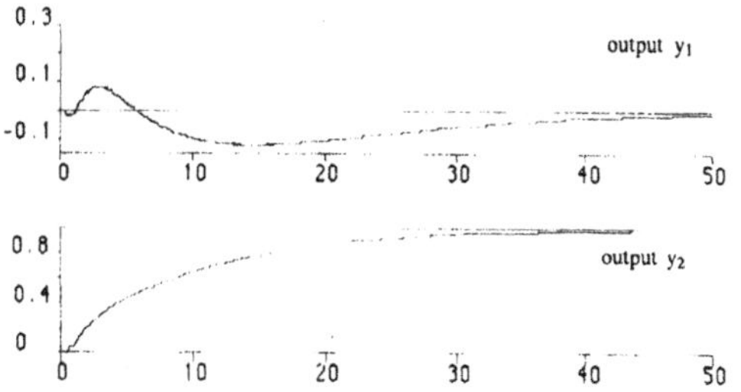

Figure 5.1c Response of the system (3.1) when a unit step reference signal was applied for the second loop (r_2) - Lieslehto tuning.

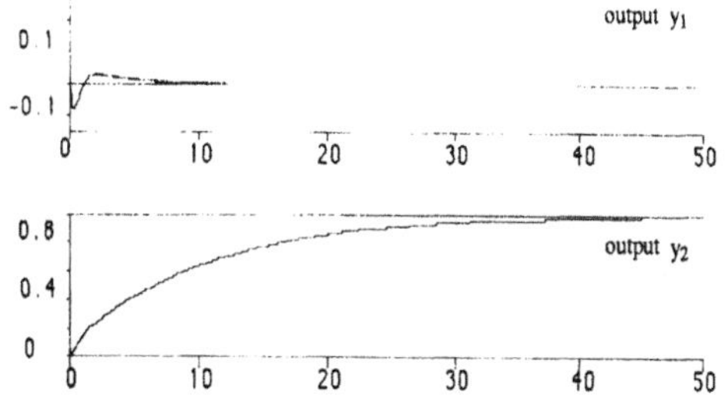

Figure 5.1d Response of the system (3.1) when a unit step reference signal was applied for the second loop (r_2) - Maciejowski tuning.

As was to be expected the rise time of the second output (y_2) is roughly the same for all the designs (since fine tuning of the controllers was made to achieve roughly the same bandwidth ≈ 0.1 rad/s) being about 20 s for each tuning. However, from interaction reduction point of view the controllers behave quite differently. Maciejowski tuning has least interactions then come Davison and Lieslehto designs, Penttinen tuning being the worst in this respect.

In the second time domain experiment a unit disturbance was added to the first output (y_1) at the time $t = 0$. Also in this experiment all the controllers attenuate the effect of the disturbance on the first output component (y_1) to 10% in about 20 seconds. But again the disturbance decoupling properties (the effect of the disturbance on y_2) of the designs differ significantly Maciejowski and Lieslehto tunings being far better than Davison and Penttinen tunings.

In the third time domain experiment an actuator gain change was simulated by changing the real transfer matrix G(s) so that

$$G'(s) = G(s)\, \mathrm{diag}[0.99,\ 1.0] \qquad (5.1)$$

which means that the gain of the actuator for the first input (u_1) decreases 1%. Otherwise the experiment is similar to the first one. The responses for Davison tuning are displayed in the Figure 5.2.

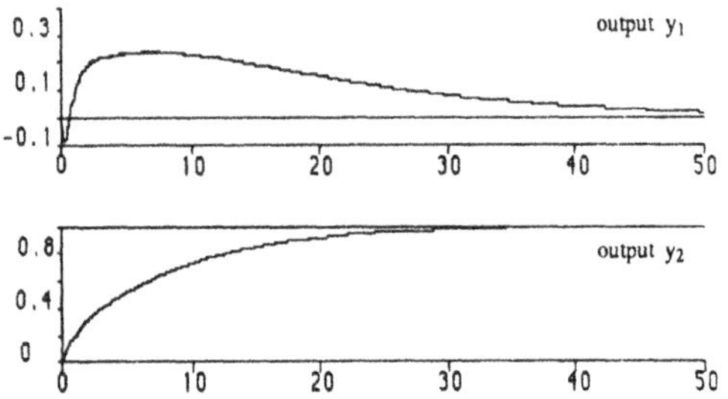

Figure 5.2 Step response when gain of the actuator of first loop is reduced 1% - Davison tuning.

It is obvious from the results of the third experiment that the performance robustness of none of the designs is very good. Lieslehto tuning seems to be the least sensitive in this respect but for all the designs their decoupling properties deteriorate considerably due to the small change in the actuator gain.

6. CONCLUSIONS

In this paper four different multivariable PI controller tuning methods have been compared using frequency domain performance indices and time domain simulations. Although only one rather simple state space model was used several interesting issues arose. The frequency domain indices proved to be quite useful in assessing the disturbance and noise attenuation properties and especially the robustness of the closed loop systems but they did not give information about interaction reduction or disturbance decoupling properties of the designs. The latter two important aspects of the controller designs were more easily assessed in time domain simulation experiments.

Further work is carried on to include also optimization based PI controller designs into the comparisons. Also another test example with input delays will be used and work is also in progress to apply these tuning methods to a laboratory scale pilot process.

Acknowledgement

The authors thank Mr Terho Jussila for useful discussions concerning this paper.

7. REFERENCES

Åström, K.J. and **T. Hägglund** (1988). *Automatic tuning of PID controllers*, ISA, Research Triangle Park, N.C.

Clarke, D. (1986). Automatic tuning of PID regulators, *Expert Systems and Optimisation in Process Control*, Technical Press, Aldershot, England.

Davison, E. J.(1976). Multivariable tuning regulators: The feedforward and robust control of general servo-mechanism problem. *IEEE Trans. Aut. Control* AC-21, 35-47.

Ford, M. B. , **J.M. Maciejowski** and **J. M. Boyle** (1988). *Multivariable frequency domain toolbox - User's Guide*. GEC, Leicestershire, England.

Jussila, T.T. and **H.N. Koivo**.(1987). Tuning of multivariable PI-controllers for unknown delay-differential systems. *IEEE Trans. Aut. Control* AC-32, 364-368.

Kouvaritakis, B. (1978). Gain margins and root locus: asymptotic behaviour in multivariate design. *Int. J. Control* Vol. 25, 33-62.

Lieslehto, J. Tanttu, J. T. and **H.N. Koivo** (1990). An expert system for the multivariable controller design. *In this conference.*

Maciejowski, J.M. (1989). *Multivariable feedback design*, Addison-Wesley, Wokingham, England.

Morari, M. and **E. Zafiriou** (1989). *Robust process control*, Prentice-Hall, Englewood Cliffs, N.J.

Peltomaa , A. and **H.N. Koivo** (1983). Tuning of multivariable discrete-time PI controller for unknown systems. *Int. J. Control* Vol 38, 735-745.

Penttinen, J. and **H.N. Koivo** (1980). Multivariable tuning regulators for unknown systems. *Automatica* Vol 16, 393-398.

Pohjolainen, S. and **T. Mäkelä** (1988). How to improve a given linear controller using H - methods, 12th IMACS World Congress, Paris.

Rivera, D.E., M. Morari and **S. Skogestad** (1986). Internal model control 4. PID controller design, *Ind.Eng.Chem.Process Des. Dev.*, 25, 252-256.

Seraji, H. and **M. Tarokh** (1977). Design of PID controllers for multivariable systems. Int J. Control, Vol 26, No 1, 75-83.

Solheim, O.A. (1974). Ein optimaler mehrgrössen-PI-regler, Regelungstechnik, Heft 9.

Yu, S. W. (1989). An optimal design technique for multivariable SISO controllers, AIChE J. 35(11), 1903-1906.

PREDICTIVE CONTROL OF NONLINEAR PROCESSES

A. R. M. Soeterboek*, H. B. Verbruggen*, J. M. Wissing* and A. J. Koster**

**Delft University of Technology, Dept. of Electrical Engineering, Control Laboratory, P.O. Box 5031, 2600 GA Delft, The Netherlands*
***Royal Gist-brocades N.V., Research and Development, Dept. PRA, P.O. Box 1, 2600 MA Delft, The Netherlands*

Abstract

In this paper a predictive controller for a class of nonlinear processes is proposed. An application to a highly nonlinear process, a pH process, is discussed by simulations and real-time experiments. It is shown that the nonlinear predictive controller yields a robust and well performing control system.

Introduction

The concept of predictive control is not new. It has been introduced in the late 70s by Richalet [4] and Cutler et al. [3]. Nowadays many predictive control schemes are known. For example, GPC [2] and UPC [5,6,7]. Most of the predictive controllers that have been developed use a linear model of the process to calculate the predictions of the process output. Further, in order to obtain the predictive control law, in all predictive controllers a quadratic criterion, weighting the predicted tracking error and controller output, is minimized. Because the criterion is quadratic and the model is linear, this minimization can be performed analytically yielding a linear control law.
For some processes a linear model cannot describe its properties in an acceptable way. An example of such a process is a pH process. The concept of predictive control is, however, not limited to linear processes. However, by using a nonlinear model the criterion function cannot be minimized analytically and hence a nonlinear optimization method is required. In this paper a nonlinear predictive controller will be presented that can be used to control a class of nonlinear processes. Further, an application to a highly nonlinear pH process is discussed. A nonlinear model of this process will be used to calculate the predictions. Simulation and experimental results are presented showing the performance and robustness of the closed-loop system.
The paper is organized as follows. First the nonlinear predictive controller is discussed followed by a description of the pH process. Next some simulation and experimental results are presented followed by some conclusions.

Predictive Control

The way predictive controllers operate for a single-input single-output system is illustrated by Figure 1. Suppose,

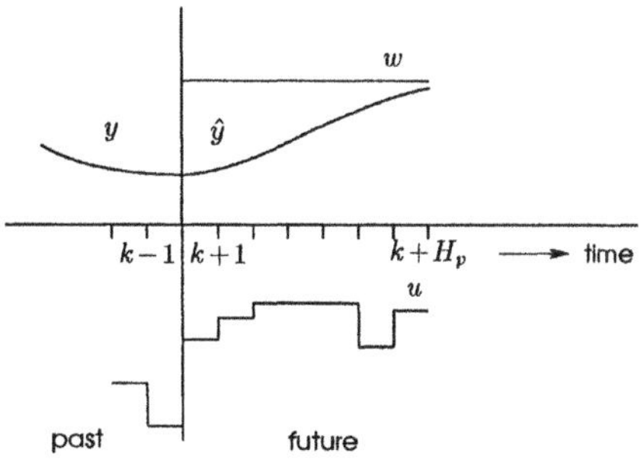

Figure 1: The predictive control concept.

the current time is $t = k$ and $u(k)$, $y(k)$ and $w(k)$ denote the controller output, the process output and the desired process output at $t = k$, respectively. Further, define

$$\begin{aligned} \boldsymbol{u} &= [u(k), \cdots, u(k+H_p-1)]^T \\ \hat{\boldsymbol{y}} &= [\hat{y}(k+1), \cdots, \hat{y}(k+H_p)]^T \\ \boldsymbol{w} &= [w(k+1), \cdots, w(k+H_p)]^T \end{aligned}$$

where H_p is called the prediction horizon and the symbol $\hat{}$ denotes estimation. Then, a predictive controller calculates such a future controller output sequence $\boldsymbol{u}$ that the predicted output of the process $\hat{\boldsymbol{y}}$ is 'close' to the desired process output $\boldsymbol{w}$. This desired process output is often called the reference trajectory which can be an arbitrary sequence of points. The first element of the controller output sequence that is determined in this way ($= u(k)$) is used to control the process. At the next sample, the whole procedure is repeated based on the latest measured information. This is called the *receding horizon* principle. The process output is predicted by utilizing a model of the process. Any model that describes the relation between the input and the output of the process can be used. Hence, linear as well as nonlinear models can be used.
In order to define how well the predicted process output tracks the reference trajectory, a criterion function J is used. Typically, such a criterion function is a function of $\hat{\boldsymbol{y}}$, $\boldsymbol{w}$ and $\boldsymbol{u}$. Now the controller output sequence $\boldsymbol{u}$ is obtained by minimization of J with respect to $\boldsymbol{u}$. Then, $\boldsymbol{u}$ is optimal with respect to the criterion that is minimized.
It is shown in [5] that if the criterion is quadratic and the model is linear, there is an analytical solution to the minimization problem. In this paper a nonlinear process is controlled using a nonlinear model. Then, an analytical solution to the optimization problem no longer exists.

Prediction of the Process Output

As shown above the output of the process must be predicted over the prediction horizon. For this purpose a model of the process is used. In this paper the process is assumed to be described by:

$$y(k) = f(x(k)) \tag{1}$$

$$x(k) = \frac{q^{-d}B}{A}u(k-1) + \frac{C}{\Delta}e(k) \tag{2}$$

where where d is the time delay of the process ($d \geq 0$), f is a function which inverse f^{-1} exists, $e(k)$ is a discrete white noise sequence with zero mean and Δ is the differencing operator: $\Delta = 1 - q^{-1}$. Further, A, B and C are polynomials in the back-ward shift operator q^{-1} of degree n_A, n_B and n_C, respectively. The parameters of the process and the function f are assumed to be time invariant.

As shown above, predictions of the process output at $t = k + i$ are required for $i \geq d + 1$ (prediction of $y(k+1), \ldots, y(k+d)$ makes no sense because the controller output sequence cannot influence these values). For this purpose an i-step-ahead predictor is used. It has been shown (see, for example, [5]) that the minimum-variance i-step-ahead predictor predicting $x(k+i)$ is for $i \geq d+1$ given by:

$$\hat{x}(k+i) = \frac{E_i B \Delta}{C} u(k+i-d-1) + \frac{F_i}{C} x(k) \tag{3}$$

where the polynomials E_i (of degree $i-1$) and F_i can be obtained by solving the following Diophantine equation:

$$\frac{C}{\Delta A} = E_i + q^{-i}\frac{F_i}{\Delta A} \tag{4}$$

Further, the time delay d and the polynomials A, B and C are replaced by those of the model thus applying the certainty equivalence principle. In many cases it is difficult to estimate C. Instead of estimating C, $\hat{C}$ is often selected by the designer. The $\hat{C}$ polynomial is usually called T and is often used as a tuning parameter to influence the regulator behavior and the robustness of the closed-loop system [2,6]. This yields for (3):

$$\hat{x}(k+i) = \frac{E_i \hat{B} \Delta}{T} u(k+i-\hat{d}-1) + \frac{F_i}{T} x(k) \qquad i \geq \hat{d}+1 \tag{5}$$

where E_i and F_i are calculated using (4) with A and C replaced by their estimates. Prediction of the process output is now realized by using the following relation for $i \geq \hat{d}+1$:

$$\hat{y}(k+i) = \hat{f}(\hat{x}(k+i))$$

where $\hat{f}$ is an estimate of the function f. Further, because $x(k)$ cannot be measured, the inverse of $\hat{f}$ is used to obtain an estimate of $x(k)$ from the measured $y(k)$:

$$x(k) = \hat{f}^{-1}(y(k))$$

Note: by making Δ a factor of the denominator of the noise term in (2), integral action is introduced in the control loop thus preventing steady-state errors due to constant disturbances and/or constant reference trajectories [7].

For convenience, the prediction model (5) is for $i = d + 1, \ldots, H_p$ written in a matrix notation:

$$\hat{\boldsymbol{x}} = \boldsymbol{\Gamma u} + \boldsymbol{\Psi s} + \boldsymbol{k} \tag{6}$$

where $\boldsymbol{\Gamma}$ and $\boldsymbol{\Psi}$ are matrices of appropriate dimensions build up of the elements of E_i, F_i, $\hat{A}$ and $\hat{B}$. The vectors $\hat{\boldsymbol{x}}$ and $\boldsymbol{s}$ are given by:

$$\hat{\boldsymbol{x}} = [\hat{x}(k+\hat{d}+1), \ldots, \hat{x}(k+H_p)]^T$$

$$\boldsymbol{s} = [x(k), x(k-1), \cdots, u(k-1), u(k-2), \cdots]^T$$

and $\boldsymbol{k}$ is given by $\boldsymbol{k} = \boldsymbol{K}_c \boldsymbol{c}$, where the matrix $\boldsymbol{K}_c$ depends on $\hat{A}$, T and Δ and $\boldsymbol{c}$ is a vector of appropriate dimension given by $\boldsymbol{c} = [c(k), c(k-1), \cdots]^T$, in which $c(k) = \dfrac{\hat{A}y(k) - \hat{B}u(k-\hat{d}-1)}{T}$. For a detailed discussion on predictors for linear processes, see [5].

The Criterion Function

The criterion function that is minimized in most predictive controllers is a quadratic function weighting tracking error and controller output. In this paper a simple criterion function weighting tracking error only is used:

$$J = \sum_{i=d+1}^{H_p} (\hat{y}(k+i) - w(k+i))^2 \tag{7}$$

where $w(k+i)$ is equal to the set point for all i.

It is well known (see, for example, [6]) that minimization of (7) yields a minimum-variance controller independent of H_p. Since minimum-variance control can hardly ever be used to control a real process, controller output weighting is required. One way to realize this is by including the controller outputs in the criterion function that is minimized and using a weighting factor to determine which of the objectives (minimization of tracking error or minimization of the controller output variance) is the most important. However, it has been shown in [6] that it is quite hard to select an appropriate value for this weighting factor a priori. For this reason, another way of weighting the controller output is introduced by Cutler and Ramaker [3]. They proposed the principle of the control horizon: the controller outputs are calculated over the *control horizon*, all other controller outputs over the prediction horizon are kept equal to the last calculated controller output and thus remain constant. The control horizon is denoted by H_c. The optimization problem can now be formulated as: minimize (7) with respect to $u(k), \ldots, u(k+H_c)$ subject to the constraint:

$$\Delta u(k+i-1) = 0 \qquad 1 \leq H_c < i \leq H_p - d \tag{8}$$

It is shown in [7] that by choosing H_c different from $H_p - d$, control laws other than minimum-variance control can be obtained. Making H_c smaller results in a less active controller output. For $H_c = 1$, a mean-level controller is obtained which yields a robust but usually slow control system (see, [2]).

In [5] it is shown that minimization of the criterion function (7) subject to the constraint (8) can be performed analytically if the model of the process is linear. However, in this paper the model is assumed to be nonlinear thus making an analytical solution to the minimization problem impossible. Hence, a nonlinear (iterative) optimization method must be used to minimize the criterion function.

Minimization of the Criterion Function

In searching for an optimum, most iterative optimization methods make use of the following algorithm.

Algorithm 1 Basic algorithm of an iterative optimization method.

Step 1: Initialize $\bar{u} \Longrightarrow \bar{u}_0$ and $n = 0$.

Step 2: If stop criterion satisfied Then
Optimal solution found.
End If

Step 3: $\bar{u}_{n+1} = \bar{u}_n + \lambda_n d_n$ (9)
$n = n + 1$
Goto Step 2

where: $\bar{u}_n = [u(k), \cdots, u(k+H_c)]^T$. Further, n denotes the iteration number ($n \geq 0$), d_n denotes the search direction at iteration n and $\lambda_{n,opt}$ is a scalar denoting the optimal step size in the search direction. Because the search direction is determined by d_n, $\lambda_{n,opt}$ is positive. The way d_n and $\lambda_{n,opt}$ are calculated distinguishes optimization algorithms. In this paper the steepest-descent method [1] is utilized and hence the search direction is given by:

$$d_n = -g_n = -\frac{\partial J}{\partial \bar{u}_n}$$

where g_n is the gradient of J with respect to $\bar{u}_n$. The optimal step size in the search direction is obtained by using a line search method. Many such methods are available. However, in this paper the golden-section method is used [1].

In any optimization problem conditions are required that determine whether or not an optimum has been found and the iterative optimization algorithm can be stopped. If the search direction d_n is equal to zero an optimum has been found. In practice the following condition must be satisfied for a solution to the optimization problem:

$$\max_j |d_j| < \epsilon \quad \text{for } j = 1, \ldots, H_c \tag{10}$$

where d_j is the jth element of d_n and ϵ is a small positive value determining the accuracy of the solution.

The starting point of the optimization plays an important role in the number of iterations that is required to find an optimal solution with a certain accuracy. In general, the closer the starting point to the optimum, the faster an optimum with accuracy ϵ is found. One way to initialize $\bar{u}$ is to use the solution calculated in the previous sample. Let $\bar{u}_{k-1}$ be the vector $\bar{u}$ at $t = k - 1$ and let $u_{k-1,j}$ be the jth element of $\bar{u}_{k-1}$. Then, the following algorithm can be used to initialize $\bar{u}$ at $t = k$.

$$\begin{aligned} u_{0,j} &= u_{k-1,j+1} \qquad j = 1, \ldots, H_c - 1 \\ u_{0,j} &= u_{k-1,j} \qquad j = H_c \end{aligned} \tag{11}$$

where $u_{0,j}$ denotes the jth element of $\bar{u}_0$ at $t = k$.

In the case $k = 0$, the solution to the optimization problem in the previous sample is not available. Then, for example, $\bar{u}_0 = 0$ can be used instead.

Calculation of the Gradient

In this section the calculation of the gradient is discussed considering the nonlinear process described by (1) and (2). The gradient becomes:

$$g_n = \frac{\partial J}{\partial \bar{u}_n} = 2\frac{\partial \hat{y}_n}{\partial \bar{u}_n}(\hat{y}_n - w) \tag{12}$$

Further, by using (8), the relation between u_n and $\bar{u}_n$ is given by $u_n = M\bar{u}_n$, where:

$$M = \begin{bmatrix} 1 & 0 & \cdots & 0 \\ 0 & 1 & \ddots & \vdots \\ \vdots & \ddots & \ddots & 0 \\ 0 & \cdots & 0 & 1 \\ 0 & \cdots & 0 & 1 \\ \vdots & & \vdots & \vdots \\ 0 & \cdots & 0 & 1 \end{bmatrix} \begin{matrix} \left.\vphantom{\begin{matrix}1\\1\\1\\1\end{matrix}}\right\} H_c \\ \left.\vphantom{\begin{matrix}1\\1\\1\end{matrix}}\right\} H_p - H_c \end{matrix}$$

The prediction model (6) can therefore be written as: $\hat{x}_n = \Gamma M\bar{u}_n + \Psi s + k$. Further, because

$$\frac{\partial \hat{y}(k+i)}{\partial u(k+j-1)} = \frac{\partial \hat{y}(k+i)}{\partial \hat{x}(k+i)}\frac{\partial \hat{x}(k+i)}{\partial u(k+j-1)}$$

and $\frac{\partial \hat{x}_n}{\partial \bar{u}_n} = (\Gamma M)^T$, the gradient (12) becomes:

$$g_n = 2M^T\Gamma^T F_n(\hat{y}_n - w) \tag{13}$$

where:

$$F_n = \text{diag}\left(\frac{\partial \hat{y}_n(k+d+1)}{\partial \hat{x}_n(k+d+1)}, \ldots, \frac{\partial \hat{y}_n(k+H_p)}{\partial \hat{x}_n(k+H_p)}\right)$$

Note that the matrix F_n changes during the iterative search, the matrix ΓM remains constant.

The nonlinear predictive controller now becomes:

Algorithm 2 The nonlinear predictive controller.

Step 1: Initialize $\bar{u} \Longrightarrow \bar{u}_0$ and $n = 0$. Measure $y(k)$ and calculate $x(k) = f^{-1}y(k)$.

Step 2: While no optimal solution found Do
Determine search direction d_n
If $\max_j |d_j| < \epsilon$ Then
Optimal solution found.
Else
Calculate $\lambda_{opt,n}$
$\bar{u}_{n+1} = \bar{u}_n + \lambda_{opt,n} d_n$
$n = n + 1$
End If
End While
Use $u(k)$ to control the process.
Wait until next sample
Goto Step 1.

Next an application of the above mentioned algorithm is discussed on a pH process.

Description of the pH Process

In the chemical industry the control of the acidity of a solution, indicated by the pH value, is common practice. Chemical reactions and physical operations flourish best at certain pH values, while the pH of waste water from a plant is often subject to environmental regulations.

In most cases the pH-control takes place in a vessel which receives an influent of variable acidity and flow rate. The pH can be controlled by adding a stream of concentrated reagent. The level in the vessel is kept constant so that the effluent (the stream out of the vessel) equals the sum of the influent and the reagent stream. The contents of the vessel is stirred. This configuration is characterized as a Continuous Stirred Tank Reactor (CSTR) and shown in Figure 2. The pH of the effluent is measured and compared with the

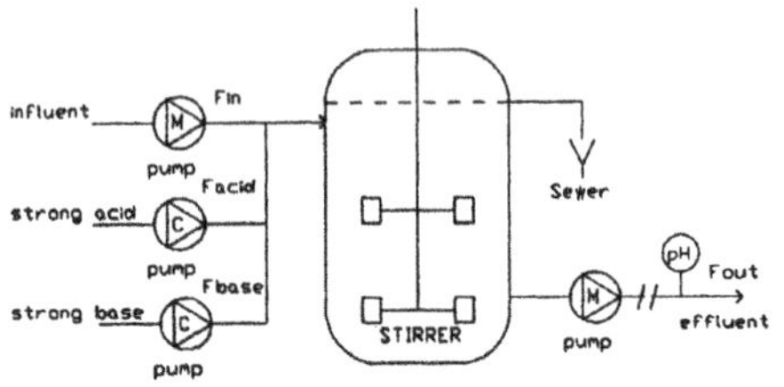

Figure 2: Configuration of the pH process.

required set point. As a function of the difference, the flow rate of the reagent stream is controlled.
Problems arise from the severe nonlinearity of the pH process which is clearly illustrated by the shape of the titration curve of a specific solution. An example of a titration curve is shown in Figure 3. The titration curve shows the pH as

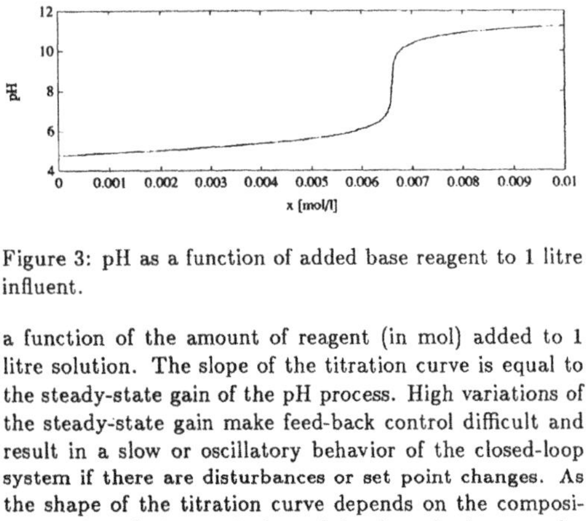

Figure 3: pH as a function of added base reagent to 1 litre influent.

a function of the amount of reagent (in mol) added to 1 litre solution. The slope of the titration curve is equal to the steady-state gain of the pH process. High variations of the steady-state gain make feed-back control difficult and result in a slow or oscillatory behavior of the closed-loop system if there are disturbances or set point changes. As the shape of the titration curve depends on the composition of the solution, variations of the shape in time are also possible. In some simple cases it is possible to calculate the titration curve [8], but in general the titration curve has to be determined experimentally.
Another problem is the pH measurement. In the application described here, there is a time delay between the outlet of the vessel and the pH measurement as a result of the position of the pH-sensor.

The Process Model

In order to construct a model to be used by the predictive controller, the so called 'effective base/acid concentration' x is defined. It denotes the amount of reagent (in mol) that is added to 1 litre solution. For the influent, x_{in} follows from:

$$x_{in} = (F_{base}M_{base} - F_{acid}M_{acid})/(F_{acid} + F_{base} + F_{in})$$

where M_{acid} and M_{base} are the molarity of the acid and base reagent in mol/l, respectively. Further, F_{acid} and F_{base} are the flow of the acid and base reagent in l/s, respectively. Finally, F_{in} is the influent in l/s.
The dynamics of the vessel can be described by a first-order system:

$$\frac{dVx}{dt} = (F_{in} + F_{acid} + F_{base})x_{in} - F_{out}x \tag{14}$$

where V is the volume of the vessel (in l) and F_{out} is the effluent (in l/s). In this case V is constant and therefore $F_{in} + F_{acid} + F_{base} = F_{out} = F$. Now, (14) can be simplified into $dx/dt = F(x_{in} - x)/V$. This yields for the transfer function in the s-domain:

$$X(s) = \frac{1}{s\tau + 1}X_{in}(s) \tag{15}$$

where $\tau = V/F$. The pH of the solution is defined by the titration curve f: pH $= f(x)$. Usually, f must be determined experimentally. The titration curve used during the simulations and experiments is shown in Figure 3.
Further, because the pH is measured in the effluent, the measurements are delayed. Since the titration curve is assumed to be static, the time delay T_d (in seconds) can be included in the first-order process described by (15). Now the pH process can be described by:

$$\begin{aligned} X(s) &= \frac{e^{-T_d s}}{s\tau + 1}X_{in}(s) \\ \text{pH}(s) &= f(X(s)) \end{aligned}$$

where x_{in} is calculated by the computer. The flow of the reagent stream is then obtained by multiplication of x_{in} by the the influent flow. A positive value of x_{in} can be obtained by adding base reagent only, while a negative value for x_{in} can be obtained by adding the acid reagent to the influent.
Remark: by assuming a static titration curve, changes in the composition of the influent are not taken in account.

Alternative approach to controlling pH processes

Another approach to control a pH process is to use the inverse titration curve to linearize the process. In combination with a linear controller, the closed-loop system becomes as shown in Figure 4. However, in spite of the fact

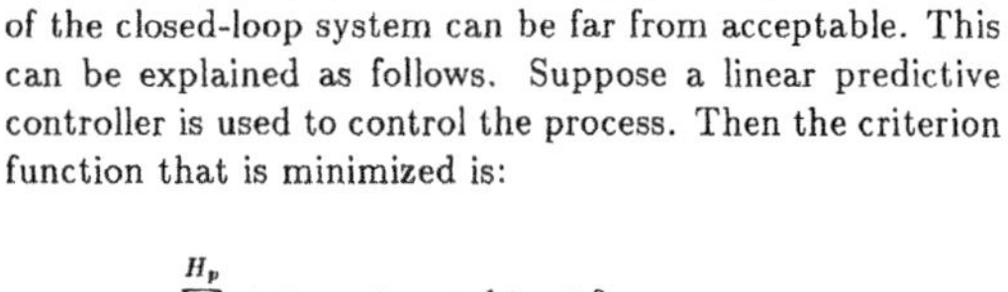

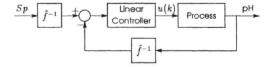

Figure 4: Using a linear controller for pH control.

that the resulting system is linear (if $f = \hat{f}$), the behavior of the closed-loop system can be far from acceptable. This can be explained as follows. Suppose a linear predictive controller is used to control the process. Then the criterion function that is minimized is:

$$J = \sum_{i=d+1}^{H_p} (\hat{x}(k+i) - f^{-1}(Sp))^2$$

Minimization of J taking into account the control horizon then results in a good control behavior for $x(k)$ with possibly a small overshoot during set point changes. However, the resulting pH response can be far from acceptable because small changes in x can result in large changes in the pH (see Figure 3). For example, when a small overshoot occurs during a set point change, the overshoot of the pH can be extremely large depending on the set point. By using the approach proposed in this paper, the difference between the actual pH en the pH set point is included directly in the criterion function:

$$J = \sum_{i=d+1}^{H_p} (\widehat{\text{pH}}(k+i) - Sp))^2$$

Obviously, much better results can be expected.

Simulated and Experimental Results

Experimental set up

The experimental set up is shown in Figure 2. The experiments were carried out in a stirred vessel with a contents of 10 litres ($V = 10$l). A buffer solution (0.0066 mol sodium-acetate and 0.0066 mol acetic acid per litre) was fed to the vessel at 100 l/h. This results in a time constant of 360s. In the outlet of the vessel the pH was measured and fed to a computer (a PC/AT) on which the nonlinear predictive controller was implemented. Step responses showed that the time delay of the process is about 7s. This time delay could be varied by changing the position of the pH sensor. Based on the measured pH and the pH set point the required x_{in} was calculated by the nonlinear predictive controller described above. In order to obtain positive values as well as negative values for x_{in}, two hose pumps were used to add either sulphuric acid (0.044 N) or sodium hydroxide (0.044 N) to the vessel. If the calculated value for x_{in} is positive the base pump was driven while the acid pump was driven for negative values of x_{in}. The capacity of the hose pumps was such that a maximum of 0.02 mol of reagent could be added to 1 litre of influent if the influent flow is 100 l/h.

The titration curve of the buffer solution is shown in Figure 3. This curve has been determined experimentally. It is implemented by using a table of 128 points in combination with linear interpolation in between the points.

In order to be able to judge the regulator performance of the control system, pulse and step disturbances were added to the process. Pulse disturbances were added to the process by adding during 10s acid or base reagent to the influent at a rate of 1 mol/h. Further, step disturbances were added to the process by changing the influent flow from 100 l/h to 70 l/h and vice versa. Note, however, that this also makes the time constant of the process change.

The servo behavior of the control system was judged by making a number of set point changes over a large pH range.

Tuning and Simulations

The nonlinear predictive controller as discussed above has several tuning parameters: H_p, H_c, T and the sampling period T_s. As a rule of thumb, the sampling period can be chosen 2-4 times smaller than the rise time of the closed-loop system. Because the response time of the pH as a function of x_{in} can be very fast due to the nonlinear relation between x_{in} and the pH, the sampling period must be chosen relatively small. During the experiments $T_s = 1$s has been used.

For linear predictive controllers it is well-known that in order to obtain a robust control system, H_c must be chosen small and H_p large (see, for example, [6]). Further, by choosing $T = (1 - \mu q^{-1})^2$ (with $0 \leq \mu \leq 1$), the effect of disturbances on the controller output can be decreased by increasing μ. Also, the system's robustness increases as μ increases (see, [2] and [6]). During the experiments the following settings were used: $H_c = 1$, $H_p = 16$ and $T = (1 - 0.91q^{-1})^2$. For linear processes this yields a robust but usually slow closed-loop system with smooth control actions. Further, the reaction of the controller output to disturbances is rather smooth. These settings also result in an acceptable behavior of the nonlinear system as is shown by the Figures 5a and 5b. Figure 5a shows a simulated response with the nonlinear process and the nonlinear predictive controller for different set points. Clearly, the response is quite acceptable: there are no oscillations and overshoot does not occur. That the control system is highly nonlinear is illustrated by fact that the controller actions required to bring the pH from pH = 9 to pH = 7.5 are different from those to bring the pH from pH = 7.5 to pH = 6. Figure 5b shows the effect of pulse and step disturbances on the control system at pH = 6.7. Pulse disturbances occurred at $t = 500$s and $t = 950$s while step disturbances occurred at $t = 1250$s and $t = 1650$s. Clearly, these disturbances are rejected nicely. The large peaks occurring especially during pulse disturbances cannot be made smaller. This is mainly caused by the time delay of the process. When a disturbance occurs, the controller is simply too late to prevent a large deviation of the pH from the set point.

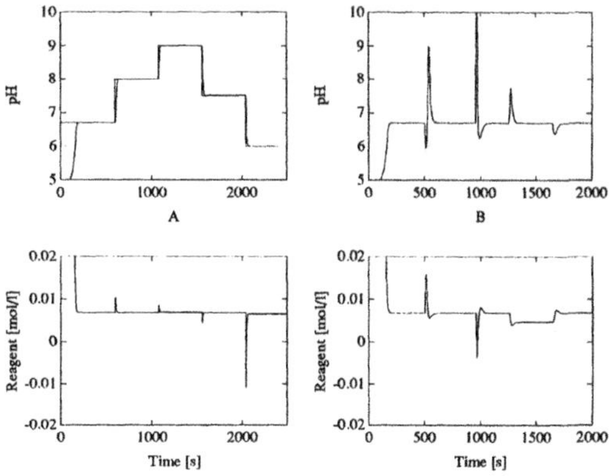

Figure 5: Figure A shows (in simulation) the response to different set-point changes. Figure B shows the regulator behavior at pH = 6.7.

Experimental Results

In this section a number of experimental results are reported that were obtained by using the set up as shown in Figure 2. Figure 6 shows the regulator behavior of the control system for two different pH set points, namely pH = 7.5 and pH = 6.7, respectively. At $t = 500$s and $t = 950$s, the control system is disturbed by pulse disturbances. Clearly, they are rejected nicely in a way similar to those during the simulations (see Figure 5b). At $t = 1250$s the influent flow was changed from 100 l/h to 70 l/h resulting in a step wise disturbance and a change in the time constant of the first-order process. At $t = 1650$s the influent flow was changed to its original value (= 100 l/h). Due to the integral action of the controller, step-wise disturbances do

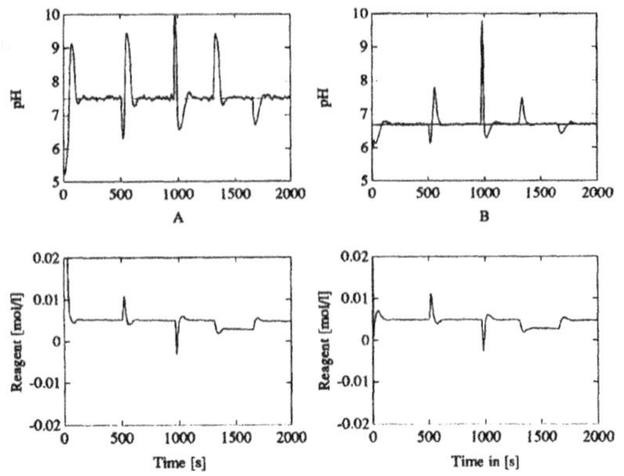

Figure 6: The regulator behavior at pH = 7.5 (Figure A) and at pH = 6.7 (Figure B).

not yield steady-state errors. The Figures 6a and 6b show that for both pH set points, step and pulse disturbances are rejected nicely although the steady-state gains for pH = 7.5 and pH = 6.7 are very different (see Figure 3).

Figure 7a shows the servo behavior of the closed-loop system as it was measured during the experiments. Comparing Figure 7a with Figure 5a shows that, although the responses are similar, overshoot occurred during the experiment while this was not the case during the simulations. This is probably caused by the presence of a mismatch between the model and the process. Nevertheless, it is our believe that the response shown in Figure 7a is quite good considering the severe nonlinearity of the process.

Finally, the robustness of the control system with respect to variations in the time delay has been examined by changing the position of the pH sensor. The time delay of the model (= 7s) has not been adapted thus resulting in a mismatch between the real time delay and that of the model. Figure 7b shows the response at pH = 6.7 in the case the time delay of the process is changed to 13s (hence, the time delay is increased by a factor 1.9). Pulse disturbances were added at t = 500s and at t = 950s. Although, the response is not as good as that shown in Figure 6b, it is still quite acceptable despite of the large time delay mismatch.

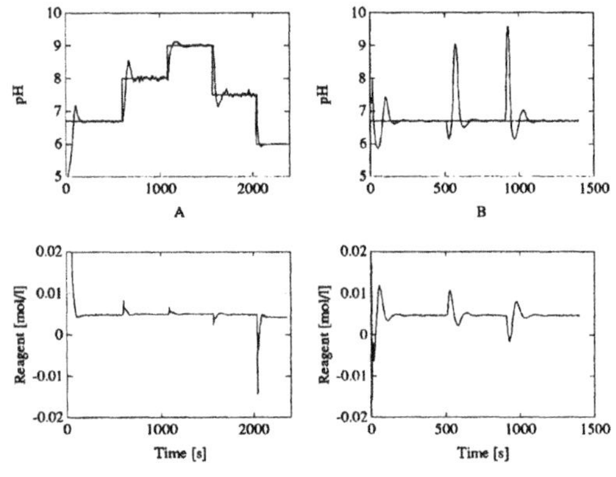

Figure 7: The servo behavior of the experimental setup (Figure A) and the regulator behavior in the presence of model mismatch (Figure B).

Conclusions

In this paper a nonlinear predictive controller has been proposed that is able to control a class of nonlinear processes. Attention has been paid especially to solving the resulting nonlinear optimization problem. By means of simulations and experiments on a test set up, it has been shown that a pH process, a highly nonlinear process, can well be controlled by using the proposed nonlinear predictive controller. Therefore, it is our believe that predictive control is a good candidate for controlling a class of nonlinear processes. However, a model of the process including its nonlinearities must be available.

References

[1] Bazaraa, M.S. and C.M. Shetty (1979). *Nonlinear Programming - Theory and Algorithms.* John Wiley & Sons, Inc.

[2] Clarke, D.W., C. Mohtadi and P.S. Tuffs (1987). "Generalized Predictive Control-Parts I & II". *Automatica*, Vol.23, No.2, pp. 137-160.

[3] Cutler, C.R. and B.L. Ramaker (1980). "Dynamic Matrix Control - A Computer Control Algorithm". *Proceedings JACC*, San Francisco, U.S.A.

[4] Richalet, J., A. Rault, J.L. Testud and J. Papon (1978). "Model predictive heuristic control: applications to industrial processes". *Automatica*, Vol.14, No.5, pp. 413-428.

[5] Soeterboek, A.R.M., H.B. Verbruggen, P.P.J. van den Bosch and H. Butler (1990). "On the Unification of Predictive Control Algorithms". To be published in the Proceedings of the 29th IEEE Conference on Decision and Control. Honolulu (HI), U.S.A.

[6] Soeterboek, A.R.M., H.B. Verbruggen and P.P.J. van den Bosch (1991). "On the Design of the Unified Predictive Controller". To be published in the Proceedings of the IFAC Symposium on Intelligent Tuning and Adaptive Control (ITAC 91). Singapore.

[7] Soeterboek, A.R.M., H.B. Verbruggen, P.P.J. van den Bosch and H. Butler (1990). "Adaptive Predictive Control - A Unified Approach". To be published in the Proceedings of the Sixth Yale Workshop on Adaptive and Learning Systems. Yale University, New Haven, U.S.A.

[8] Westerlund, T., B. Skrifvard and S. Karrila (1985). "On the uniqueness in pH calculations". *Chemical Engineering Science*, Vol.40, pp. 973-976.

ENHANCED ROBUSTNESS FOR PITCH POINTING FLIGHT USING SLIDING MODE CONTROL

M. E. Penati*, G. Bertoni* and C. Chen**

**Electronics, Computer Science and System Theory Dept., University of Bologna, Viale Risorgimento 2, 40136, Bologna, Italy*
***Dept. of Automatic Control, Nanjing Aeronautical Institute, Nanjing, PRC*

Abstract: This paper proposes a methodology of combining eigenstructure assignment, feedforward gain with sliding mode control technique to realize an enhanced robust pitch pointing flight control system. This control system can track step input commands without error and it turns out to be consistently robust against parameter variations and external disturbances. The design methodology is illustrated by application to an AFTI/F–16 aircraft.

Keywords: C.C.V. (A.C.T.), V.S.S., Sliding Mode, Decoupling, Robustness

1 INTRODUCTION

Advanced aircraft such as Control Configured Vehicles (C.C.V.) provide the capability for implementing unconventional flight modes for bombing, strafing and air–to–air combat. In longitudinal dynamics these modes are pitch pointing, vertical translation and direct lift control; correspondingly, in lateral dynamics they are yaw pointing, lateral translation and direct sideforce. The control objectives for these advanced modes include the ability to command a chosen variable without significantly affecting another specified variable; for example, in pitch pointing mode, pitch attitude command must change the pitch attitude without changing the flight path angle. Hence, these modes are characterized by the decoupling of controlled variables; in general, this kind of decoupling is static.

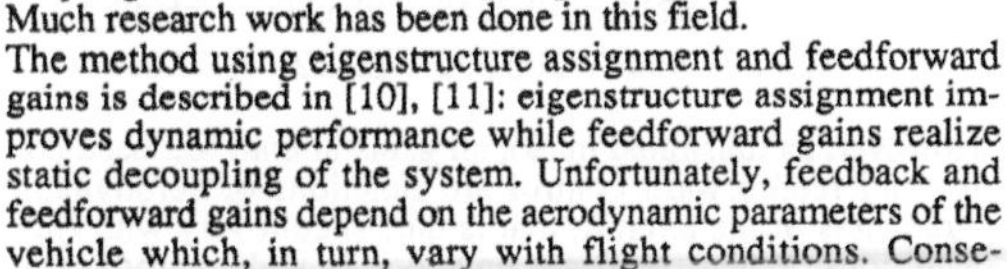

Fig. 1. Pitch pointing flight

In order to realize the decoupling of the controlled variables for the longitudinal dynamics of a C.C.V., the flaperons and the elevator form a set of redundant control surfaces capable of decoupling normal control forces and pitching moments.
Much research work has been done in this field.
The method using eigenstructure assignment and feedforward gains is described in [10], [11]: eigenstructure assignment improves dynamic performance while feedforward gains realize static decoupling of the system. Unfortunately, feedback and feedforward gains depend on the aerodynamic parameters of the vehicle which, in turn, vary with flight conditions. Consequently, when flight conditions change, the performance of the flight control system degrades.
This paper proposes an enhanced robust pitch pointing flight control system which utilizes sliding mode control of Variable Structure Systems (V.S.S.). The design procedure is illustrated by the application to an AFTI/F–16 aircraft. This system is simple and practical, it has good decoupling capability and robustness.

2 PITCH POINTING FLIGHT CONTROL SYSTEM

The dynamics of an aircraft with multiple control surfaces is described by means of a system of linear time–invariant differential equations of the type:

$$\dot{\mathbf{x}} = \mathbf{A}\mathbf{x} + \mathbf{B}\mathbf{u} \quad (1)$$

where:

$\mathbf{x} \in R^n \qquad \mathbf{u} \in R^m \qquad \text{rank}[\mathbf{B}] = m$

The eigenstructure assignment procedure, which, as already mentioned in the Introduction, allows the design of a suitable performance control system, can be stated as follows: a state feedback law is determined so that to the closed–loop system an arbitrary self–conjugate set of eigenvalues together with any permissible associated set of eigenvectors and/or generalized eigenvectors is assigned [5]. As known [2], when in a controllable and observable system of the type (1) output feedback is used, the number of closed–loop eigenvalues which can be assigned are given by max (m, r) and the eigenvectors which can be partially assigned are still given by max (m, r) while the entries which can arbitrarily be chosen in each vector are min (m, r).
If all the eigenvalues can arbitrarily be placed by means of a suitable feedback law $\mathbf{u} = \mathbf{K}\mathbf{x}$, hence also the desired damping and rise time can be obtained.
If the measurements of all the states are available we can write:

$$\dot{\mathbf{x}} = \mathbf{A}_c\,\mathbf{x} + \mathbf{B}\mathbf{u} \quad (2)$$

being:

$$\mathbf{A}_c = \mathbf{A} + \mathbf{B}\mathbf{K}$$

Matrix $\mathbf{K}$ is to be chosen so to assign eigenvalues such to imply a suitable damping and a suitable raising time and eigenvectors so as to decouple as far as possible the pitch angle and the flight path angle.
Let us show, now, how to obtain the steady state decoupling of as many outputs as there are controls or, in other words, how to force the steady state outputs:

$$\mathbf{y}_0 = \mathbf{H}\mathbf{x}_0 \quad (3)$$

to follow the input command $\mathbf{u}_c$, that is how to get:

$$\mathbf{y}_0 = \mathbf{u}_c \quad (4)$$

where $\mathbf{H}$ is a $(m \times n)$ matrix.
In the case of a flight control system, the aircraft controlled variables (such as pitch attitude angle ϑ or flight path angle γ) must track the pilot's command $\mathbf{u}_c$. When the pilot's command is a step input, this is a problem of a tracking system with nonzero set points [6].
From (3) and (4) it follows:

$$\mathbf{u}_c = \mathbf{H}\mathbf{x}_0 \quad (5)$$

Hence, the problem to be solved is to find, as a function of $\mathbf{u}_c$, a constant input $\mathbf{u}_0$ such that, at the steady state, relation (5) holds. From (2) it follows that $\mathbf{x}_0$ and $\mathbf{u}_0$ must be related by:

$$\mathbf{0} = \mathbf{A_c\,x_0 + B u_0} \tag{6}$$

hence:

$$\mathbf{x_0} = -\mathbf{A_c}^{-1}\mathbf{B u_0} \tag{7}$$

Substituting (7) in (5) we get:

$$\mathbf{u_c} = -\mathbf{H A_c}^{-1}\mathbf{B u_0}$$

Hence, if the $\mathbf{HA_c}^{-1}\mathbf{B}$ matrix is invertible, we get:

$$\mathbf{u_0} = -\,[\mathbf{HA_c}^{-1}\mathbf{B}]^{-1}\,\mathbf{u_c} \tag{8}$$

Consequently, the overall input which imposes both a static decoupling and an assigned dynamics is given by:

$$\mathbf{u} = \mathbf{Kx} - [\mathbf{HA_c}^{-1}\mathbf{B}]^{-1}\,\mathbf{u_c} = \mathbf{Kx} + \mathbf{Du_c} \tag{9}$$

where:

$$\mathbf{D} = -\,[\mathbf{HA_c}^{-1}\mathbf{B}]^{-1} \tag{10}$$

is the feedforward gain.
In other words, the control law is divided into two parts: one is the feedback control, the other is the feedforward one that leads to the desired static decoupling of the system.
Such a system, adopting both feedback control and feedforward control, possesses a static decoupling capability and a satisfactory dynamical response.
When we use only the pitch attitude command ϑ_c and keep flight path angle command $\gamma_c = 0$, we can get the movement of pitch attitude of the aircraft without changing the flight path angle (pitch pointing mode); alternatively, when $\vartheta_c = 0$, we can command the flight path angle without changing the pitch attitude (vertical translation). Obviously, if this were required, we could also command both ϑ_c and γ_c.
As an example, let's take the AFTI/F–16 flight control system. This example has been considered in many papers and will also be used in this one in order to compare the performance of a static decoupling system with the performance of the same system having the type of control we will propose in the next section.
The model of AFTI/F–16 will be described by the short period approximation equations augmented by control actuator dynamics (elevator and flaperon). The variables of the system (1) are the following:

$$\mathbf{x} = \begin{vmatrix} \gamma \\ q \\ \alpha \\ \delta_e \\ \delta_f \end{vmatrix} \quad \begin{matrix} \text{flight path angle} \\ \text{pitch rate} \\ \text{angle of attack} \\ \text{elevator deflection} \\ \text{flaperon deflection} \end{matrix} \tag{11}$$

$$\mathbf{u} = \begin{vmatrix} \delta_{ec} \\ \delta_{fc} \end{vmatrix} \quad \begin{matrix} \text{elevator deflection command} \\ \text{flaperon deflection command} \end{matrix} \tag{12}$$

When the flight condition corresponds to a Mach number $M = 0.6$ and to an altitude $h = 3{,}000$ ft, the parameters of the system can be assumed as:

$$\mathbf{A} = \begin{vmatrix} 0 & 0 & 1.3411 & 0.169 & 0.2518 \\ 0 & -0.8694 & 43.223 & -17.251 & -1.5766 \\ 0 & 0.9934 & -1.3411 & -0.169 & -0.2518 \\ 0 & 0 & 0 & -20 & 0 \\ 0 & 0 & 0 & 0 & -20 \end{vmatrix}$$

$$\mathbf{B} = \begin{vmatrix} 0 & 0 \\ 0 & 0 \\ 0 & 0 \\ 20 & 0 \\ 0 & 20 \end{vmatrix} \tag{13}$$

Assuming that the variables to be decoupled are the pitch attitude angle ϑ (given by $\vartheta = \gamma + \alpha$) and the flight path angle γ, the output matrix $\mathbf{H}$ becomes:

$$\mathbf{H} = \begin{vmatrix} 1 & 0 & 1 & 0 & 0 \\ 1 & 0 & 0 & 0 & 0 \end{vmatrix} \tag{14}$$

while the vector of the steady state values (which the variables ϑ and γ are supposed to assume) can be labeled as:

$$\mathbf{u_c} = \begin{vmatrix} \vartheta_c \\ \gamma_c \end{vmatrix} \tag{15}$$

Moreover, let us suppose that, for the closed–loop system, the desired eigenvalues are the following:

$$\begin{vmatrix} \lambda_1 \\ \lambda_2 \\ \lambda_3 \\ \lambda_4 \\ \lambda_5 \end{vmatrix} = \begin{vmatrix} -5.6 + j\,4.2 \\ -5.6 - j\,4.2 \\ -1 \\ -19 \\ -19.5 \end{vmatrix} \tag{16}$$

while the desired eigenvectors are the following:

$$\mathbf{v}_1 = \begin{vmatrix} 0 \\ 1 \\ -0.9286 \\ -5.13 \\ 8.36 \end{vmatrix} \quad \mathbf{v}_2 = \begin{vmatrix} 0 \\ -9.5 \\ 1 \\ 0.1286 \\ -5.16 \end{vmatrix} \quad \mathbf{v}_3 = \begin{vmatrix} 1 \\ 0 \\ -1 \\ -2.80 \\ 3.23 \end{vmatrix}$$

$$\mathbf{v}_4 = \begin{vmatrix} -0.0057 \\ 1.07 \\ -0.0508 \\ 1 \\ 0 \end{vmatrix} \quad \mathbf{v}_5 = \begin{vmatrix} -0.0137 \\ 0.0601 \\ 0.0106 \\ 0 \\ 1 \end{vmatrix} \tag{17}$$

Consequently, the matrices $\mathbf{K}$ and $\mathbf{D}$ (which give a steady state decoupled system having the assigned eigenvalues and the assigned eigenvectors) are given by:

$$\mathbf{K} = \begin{vmatrix} 3.25 & 0.8911 & 7.1162 & -0.5264 & -0.084 \\ -6.1 & -0.8977 & -10.0296 & 0.4206 & 0.1055 \end{vmatrix} \tag{18}$$

$$\mathbf{D} = \begin{vmatrix} -2.88 & -0.367 \\ 2.02 & 4.08 \end{vmatrix} \tag{19}$$

Consequently, the closed–loop state matrix $\mathbf{A_c}$ is given by:

$$\mathbf{A_c} = \begin{vmatrix} 0 & 0 & 1.3411 & 0.169 & 0.2518 \\ 0 & -0.8694 & 43.223 & -17.251 & -1.5766 \\ 0 & 0.9934 & -1.3411 & -0.169 & -0.2518 \\ 65 & 17.8214 & 142.3248 & -30.5288 & -1.679 \\ -122 & -17.9544 & -200.592 & 8.412 & -17.89 \end{vmatrix} \tag{20}$$

However, as illustrated in (9), all gains in both feedback control and feedforward control are functions of aerodynamic parameters of the aircraft, that is they should vary with flight conditions. When the flight envelope is wide, aerodynamic parameters change intensively and hence, if we use fixed–gain, we can not obtain a *clean* pitch attitude.
As an example, let us consider the equations of the aircraft motion under a different flight condition, i.e., $M = 0.9$, $h = 20{,}000$ ft. In this case [12] the equations are given by:

$$\begin{aligned} \dot\gamma &= 1.6187\,\alpha + 0.16625\,\delta_e + 0.19\,\delta_f \\ \dot q &= -\,0.7704\,q + 32.9618\,\alpha - 22.5382\,\delta_e - 4.2615\,\delta_f \\ \dot\alpha &= 0.997\,q - 1.6187\,\alpha - 0.16625\,\delta_e - 0.1904\,\delta_f \end{aligned} \tag{21}$$

In figures 2 and 3 a comparison between the behavior of the system in the two different conditions of flight is made. As shown, in the second case there is a difference between the pitch angle command ($\vartheta_c = 2$) and the flight path angle com-

mand ($\gamma_c = 1$) on the one hand and the actual angles on the other. We could utilize the gain–schedule method, but the control system would be quite complicated.
This is why it could be convenient to solve the problem by means of a method which guarantees a suitable robustness. This will be done in the next section using the sliding mode model–following method.

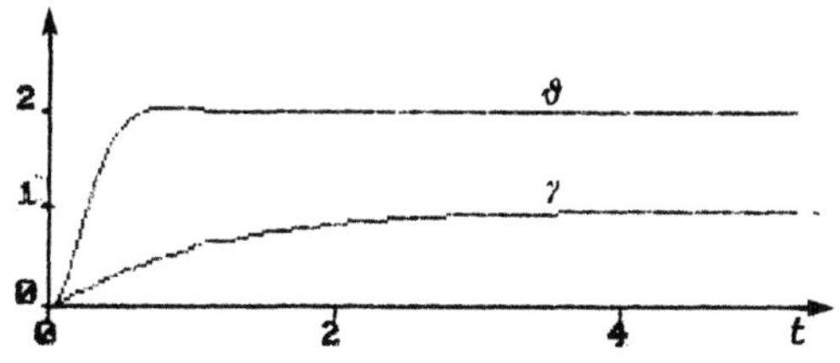

Fig. 2

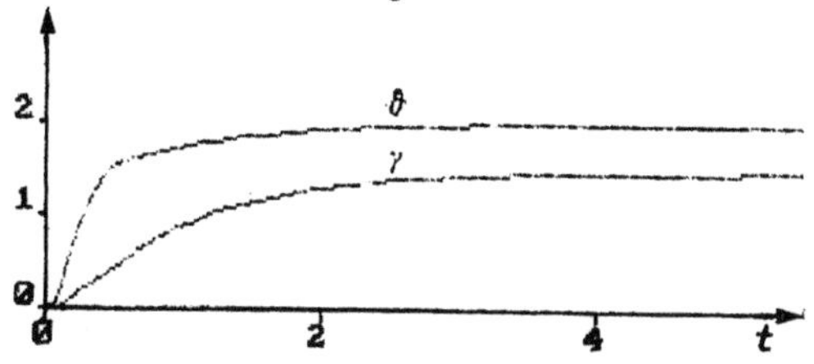

Fig. 3

3 VARIABLE STRUCTURE DESIGN

Recently, V.S.S. have received considerable attention. A V.S.S. is characterized by a state feedback control structure which switches as the system state crosses certain switching (discontinuity) surfaces in the state space. One of the central features of V.S.S. are the sliding modes. This kind of behavior occurs when the system state repeatedly crosses and immediately re–crosses a certain manifold of the switching surfaces that is, in a certain sense, it is forced to slide on this manifold. While in the sliding mode the system is constrained to lie within a certain subspace of the full state space so that it turns out to be equivalent to a system of lower order, called *equivalent system*. This equivalent system has the very interesting characteristic of being insensible to some parameter variations and external disturbances as shown in the Appendix.
In the case we are studying the problem to solve is to determine a sliding mode control which will force system (1) to satisfy the following constraints:

$$e_1 = (\gamma + \alpha) - \vartheta_c = \vartheta - \vartheta_c = 0 \qquad (22)$$
$$e_2 = \gamma - \gamma_c = 0$$

Unfortunately, in this case, as it is easy to verify, such a control does not exist, that is, there is no sliding mode control which can force the system to slide on the intersection of the two surfaces.
However, a way to solve the problem is to apply the model–following method using the sliding control (see Appendix); to this aim, let us suppose the plant is represented by a system like (A.21), $\mathbf{A}_p$ and $\mathbf{B}_p$ being two matrices coinciding with matrices $\mathbf{A}$ and $\mathbf{B}$ given by (13). Besides, the model is represented by a system like (A.22), where $\mathbf{A}_m$ is a matrix coinciding with matrix $\mathbf{A}_c$ given by (20) and $\mathbf{B}_m$ a matrix coinciding with matrix $\mathbf{B}$ given by the second of the two matrices in (13).
Since the problem is to determine the control $\mathbf{u} = \mathbf{u}_0$ which will force system (A.33) to have the steady state input (5), the overall system can be written as follows:

$$\begin{aligned}
\dot{\mathbf{x}}_{m1} &= \mathbf{A}_{m11}\,\mathbf{x}_{m1} + \mathbf{A}_{m12}\,\mathbf{x}_{m2} + \mathbf{v}\\
\dot{\mathbf{x}}_{m2} &= \mathbf{A}_{m21}\,\mathbf{x}_{m1} + \mathbf{A}_{m22}\,\mathbf{x}_{m2} + \mathbf{B}_{m2}\,\mathbf{u}_0\\
\dot{\mathbf{e}}_1 &= \mathbf{A}_{v11}\,\mathbf{x}_{m1} + \mathbf{A}_{v12}\,\mathbf{x}_{m2} + \mathbf{A}_F\,\mathbf{e}_1 + \mathbf{v}\\
\dot{\mathbf{v}} &= \mathbf{E}\,\mathbf{e}_1
\end{aligned} \qquad (23)$$

$$\mathbf{u}_c = \mathbf{H}_1\,\mathbf{x}_{m01} \qquad (24)$$

where:

$$\mathbf{A}_F = [(\mathbf{A}_{m11} - \mathbf{A}_{v11}) - (\mathbf{A}_{m12} - \mathbf{A}_{v12})\,\mathbf{F}]$$

$\mathbf{x}_{m01}$ is the steady state value of $\mathbf{x}_{m1}$ and $\mathbf{H}_1$ is made up of the first three columns of matrix $\mathbf{H}$ (whose the last two columns are made of zeros). In order to determine $\mathbf{u}_0$ let us consider the first two equations of (23) in the steady state case. We can write:

$$\mathbf{0} = \mathbf{A}_c\,\mathbf{x}_{m0} + \mathbf{B}_b\,\mathbf{w} \qquad (25)$$

where we define:

$$\mathbf{w} = \begin{vmatrix} \mathbf{v} \\ \mathbf{u}_0 \end{vmatrix} \qquad \mathbf{B}_b = \begin{vmatrix} \mathbf{I} & \mathbf{0} \\ \mathbf{0} & \mathbf{B}_{m2} \end{vmatrix} \qquad (26)$$

It follows:

$$\mathbf{x}_{m0} = -\mathbf{A}_b\,\mathbf{w} \qquad (27)$$

where:

$$\mathbf{A}_b = \mathbf{A}_c^{-1}\mathbf{B}_b \qquad (28)$$

By partitioning, in the usual form, the matrix $\mathbf{A}_b$, vector (27) can be written as:

$$\begin{aligned}
\mathbf{x}_{m01} &= -\mathbf{A}_{b11}\,\mathbf{v} - \mathbf{A}_{b12}\,\mathbf{u}_0\\
\mathbf{x}_{m02} &= -\mathbf{A}_{b21}\,\mathbf{v} - \mathbf{A}_{b22}\,\mathbf{u}_0
\end{aligned} \qquad (29)$$

By premultiplying the first equation in (29) by $\mathbf{H}_1$ and taking into account (24) we get:

$$\mathbf{u}_0 = -[\mathbf{H}_1\,\mathbf{A}_{b12}]^{-1}\,\mathbf{u}_c - [\mathbf{H}_1\,\mathbf{A}_{b12}]^{-1}\,\mathbf{H}_1\,\mathbf{A}_{b11}\,\mathbf{v}$$

which is:

$$\mathbf{u}_0 = -\mathbf{B}_u\,\mathbf{u}_c - \mathbf{B}_v\,\mathbf{v} \qquad (30)$$

where:

$$\mathbf{B}_u = [\mathbf{H}_1\,\mathbf{A}_{b12}]^{-1} \qquad \mathbf{B}_v = \mathbf{B}_u\,\mathbf{H}_1\,\mathbf{A}_{b11} \qquad (31)$$

Therefore, the overall system becomes:

$$\begin{aligned}
\dot{\mathbf{x}}_{m1} &= \mathbf{A}_{m11}\,\mathbf{x}_{m1} + \mathbf{A}_{m12}\,\mathbf{x}_{m2} + \mathbf{v}\\
\dot{\mathbf{x}}_{m2} &= \mathbf{A}_{m21}\,\mathbf{x}_{m1} + \mathbf{A}_{m22}\,\mathbf{x}_{m2} - \mathbf{B}_{m2}\,\mathbf{B}_v\,\mathbf{v} = \mathbf{B}_{m2}\,\mathbf{B}_u\,\mathbf{u}_c\\
\dot{\mathbf{e}}_1 &= \mathbf{A}_{v11}\,\mathbf{x}_{m1} + \mathbf{A}_{v12}\,\mathbf{x}_{m2} + \mathbf{A}_F\,\mathbf{e}_1 + \mathbf{v}\\
\dot{\mathbf{v}} &= \mathbf{E}\mathbf{e}_1
\end{aligned} \qquad (32)$$

and, hence, its state matrix turns out to be:

$$\mathbf{A}_T = \begin{vmatrix} \mathbf{A}_{m11} & \mathbf{A}_{m12} & \mathbf{0} & \mathbf{I} \\ \mathbf{A}_{m21} & \mathbf{A}_{m22} & \mathbf{0} & -\mathbf{B}_{m2}\,\mathbf{B}_v \\ \mathbf{A}_{v11} & \mathbf{A}_{v12} & \mathbf{A}_F & \mathbf{I} \\ \mathbf{0} & \mathbf{0} & \mathbf{E} & \mathbf{0} \end{vmatrix}$$

Consequently, if matrix $\mathbf{A}_T$ has eigenvalues with negative real parts, the steady state errors revert to zero, i.e., the model–following turns out to be perfect even though the difference matrices $\mathbf{A}_{v11}$ e $\mathbf{A}_{v12}$ are not null. As shown in figures 4 and 5 for the two different flight conditions shown above, these matrices affect the behavior of the system only during the transients, but not during the steady state.

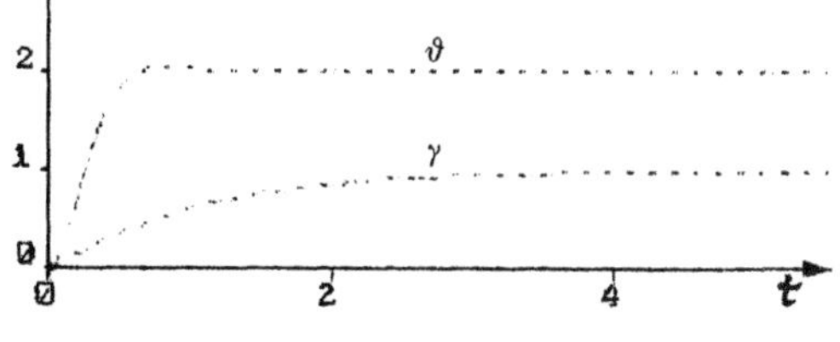

Fig. 4

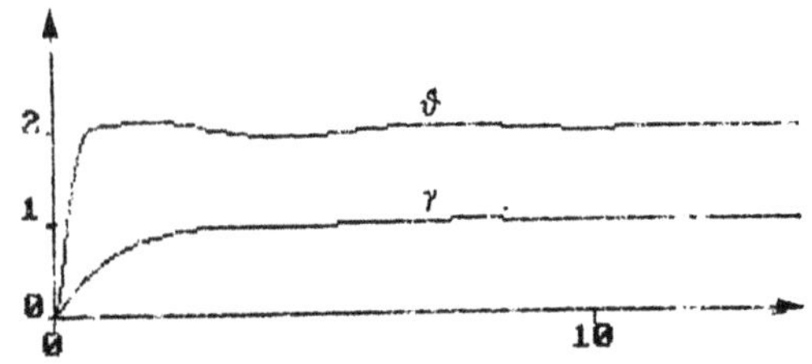

Fig. 5

4 CONCLUSION

This paper has attempted to describe an application of V.S.S. to the design of the longitudinal decoupled mode controller for the AFTI/F–16. The design methodology combines eigenstructure assignment, feedforward gains and variable structure model–following technique. The resulting system possesses the ability of tracking step input commands without error. The zero steady state tracking error means that the system is insensitive to parameter variations.

APPENDIX

Invariance conditions for sliding mode equations

a) Output disturbances

Let us consider a system:

$$\dot{\mathbf{x}} = \mathbf{A}\mathbf{x} + \mathbf{B}\mathbf{u} + \mathbf{D}\mathbf{f} \tag{A.1}$$

where **x**, **A**, **B** and **u** are vectors and matrices with the usual dimensions, while **f** is the column vector ($r \times 1$) of the output disturbances, **D** being the corresponding input matrix ($n \times r$).
As known [4], supposing the sliding surface is given by:

$$\mathbf{C}\mathbf{x} = \mathbf{0} \tag{A.2}$$

in order for the disturbances not to affect the sliding mode equations, it must be:

$$[\mathbf{I} - \mathbf{B}\,(\mathbf{C}\mathbf{B})^{-1}\,\mathbf{C}]\,\mathbf{D} = \mathbf{0} \tag{A.3}$$

that is:

$$\rho[\mathbf{B}\;\;\mathbf{D}] = \rho[\mathbf{B}] \tag{A.4}$$

(A.4) is the well known Drazenoviç invariance condition which, however, does not allow an easy physical interpretation. Fortunately enough, this condition can be expressed in a different form which turns out to be more practical and easy to understand.
To this aim and without loss of generality, we can suppose that the equations of the system given by (A.1) can be put in the partitioned form:

$$\begin{aligned} \dot{\mathbf{x}}_1 &= \mathbf{A}_{11}\,\mathbf{x}_1 + \mathbf{A}_{12}\,\mathbf{x}_2 + \mathbf{D}_1\,\mathbf{f} \\ \dot{\mathbf{x}}_2 &= \mathbf{A}_{21}\,\mathbf{x}_1 + \mathbf{A}_{22}\,\mathbf{x}_2 + \mathbf{D}_2\,\mathbf{f} + \mathbf{B}_2\,\mathbf{u} \end{aligned} \tag{A.5}$$

where:

$$\mathbf{x} = \begin{vmatrix} \mathbf{x}_1 \\ \mathbf{x}_2 \end{vmatrix} \qquad \mathbf{A} = \begin{vmatrix} \mathbf{A}_{11} & \mathbf{A}_{12} \\ \mathbf{A}_{21} & \mathbf{A}_{22} \end{vmatrix}$$
$$\mathbf{D} = \begin{vmatrix} \mathbf{D}_1 \\ \mathbf{D}_2 \end{vmatrix} \qquad \mathbf{B} = \begin{vmatrix} \mathbf{0} \\ \mathbf{B}_2 \end{vmatrix} \tag{A.6}$$

In this case, relation (A.2) becomes:

$$\mathbf{C}_1\,\mathbf{x}_1 + \mathbf{C}_2\,\mathbf{x}_2 = \mathbf{0} \tag{A.7}$$

where:

$$\mathbf{C} = |\;\mathbf{C}_1 \;\; \mathbf{C}_2\;| \tag{A.8}$$

On the other hand, the existence of a sliding mode means that relation (A.7) is satisfied; hence, from this relation itself we can obtain:

$$\mathbf{x}_2 = -\mathbf{F}\mathbf{x}_1 = 0 \tag{A.9}$$

where we defined:

$$\mathbf{F} = \mathbf{C}_2^{-1}\,\mathbf{C}_1 \tag{A.10}$$

Substituting (A.9) in the first equation of (A.5) we get:

$$\dot{\mathbf{x}}_1 = (\mathbf{A}_{11} - \mathbf{A}_{12}\,\mathbf{F})\,\mathbf{x}_1 + \mathbf{D}_1\,\mathbf{f} \tag{A.11}$$

Consequently, in order for the sliding mode equations not to be affected by the disturbances, it must hold that:

$$\mathbf{D}_1 = \mathbf{0} \tag{A.12}$$

As can immediately be verified, this condition is completely equivalent to condition (A.3). From (A.12) it can be easily seen that the disturbances do not affect the sliding mode equations only if the terms that take into account the disturbances act on the same equations as the controls do.
As far as the sliding mode control is concerned, it is well known that, in the case of a linear system, the control components depend on the position of the state vector with respect to the switching surfaces and must coincide with one of the following two functions:

$$\begin{aligned} u_i^+ &= \beta_i^+\,\mathbf{x}_1 \qquad i = 1, \ldots, m \\ u_i^- &= \beta_i^-\,\mathbf{x}_1 \qquad i = 1, \ldots, m \end{aligned} \tag{A.13}$$

where the row vectors β_i^+ and β_i^- depend on the system matrices **A**, **B**, **C**. However, it must be emphasized that the conditions which must be satisfied by β_i^+ and β_i^- so as to have a sliding mode control are inequalities. This means that, at each instant of time, components u_i^+ and u_i^- must be greater (smaller) than some suitable values $\overline{u}_i^+$ and $\overline{u}_i^-$. Consequently, if the range spanned by the maximum and minimum values of the disturbances entering the system is known, it is possible, in general, to choose vectors β_i^+ and β_i^- so that the terms of the second member of (A.13) (even if added to the disturbances) are always greater or smaller than $\overline{u}_i^+$ or $\overline{u}_i^-$. In other words, if the disturbances remain within a predetermined range, it is still possible to find a control which forces the system to slide on the sliding surface. Obviously, the larger the bounds of the disturbances the greater the power required to implement the needed control. The situation will turn out to be particularly critical in the neighborhood of the origin, since in this case the terms in the second member of (A.13) revert to zero whatever values have been assigned to β_i^+ and β_i^-; consequently, small disturbances can also stop the sliding mode.

b) Disturbances due to state matrix parameter variations

Let us consider a system of the type:

$$\dot{\mathbf{x}} = (\mathbf{A} + \mathbf{A}_v)\,\mathbf{x} + \mathbf{B}\mathbf{u} \tag{A.14}$$

where $\mathbf{A}_v$ is the ($n \times n$) matrix which takes into account the parameter variations; in order for the influence of these kinds of disturbances on sliding mode equations to be null, it must hold that [4]:

$$[\mathbf{I} - \mathbf{B}\,(\mathbf{C}\mathbf{B})^{-1}\,\mathbf{C}]\,\mathbf{A}_v = \mathbf{0} \tag{A.15}$$

that is:

$$\rho[\mathbf{B}\;\;\mathbf{A}_v] = \rho[\mathbf{B}] \tag{A.16}$$

(A.16) is another Drazenoviç invariance condition and, like (A.4), it can be expressed in a different and more practical form.
To this aim and without loss of generality, we can suppose that the equations of the system given by (A.14) can be put in the partitioned form:

$$\begin{aligned} \dot{\mathbf{x}}_1 &= \mathbf{A}_{11}\,\mathbf{x}_1 + \mathbf{A}_{12}\,\mathbf{x}_2 + \mathbf{A}_{v11}\,\mathbf{x}_1 + \mathbf{A}_{v12}\,\mathbf{x}_2 \\ \dot{\mathbf{x}}_2 &= \mathbf{A}_{21}\,\mathbf{x}_1 + \mathbf{A}_{22}\,\mathbf{x}_2 + \mathbf{A}_{v21}\,\mathbf{x}_1 + \mathbf{A}_{v22}\,\mathbf{x}_2 + \mathbf{B}_2\,\mathbf{u} \end{aligned} \tag{A.17}$$

where we defined:

$$\mathbf{A}_v = \begin{vmatrix} \mathbf{A}_{v11} & \mathbf{A}_{v12} \\ \mathbf{A}_{v21} & \mathbf{A}_{v22} \end{vmatrix} \tag{A.18}$$

so that, taking (A.9) into account, the sliding mode equations can be written in the following form:

$$\dot{x}_1 = [(A_{11} + A_{v11}) - (A_{12} + A_{v12})\, F]\, x_1 \qquad (A.19)$$

Consequently, in order for the sliding mode equations not to be affected by the disturbances due to parameter variations, it must hold that:

$$A_{v11} = 0 \qquad A_{v12} = 0 \qquad (A.20)$$

Condition (A.20) coincides with condition (A.15); from (A.20) it is easy to understand that sliding mode equations are not affected by parameter variations only if the terms which take into account these variations act on the same equations as the controls do.
On the other hand, since the eigenvalues assigned by matrix F to matrix $[A_{11} - A_{12} F]$ have negative real parts, matrix $[(A_{11} + A_{v11}) - (A_{12} + A_{v12})\, F]$, provided the terms of the variation matrices are not too large, can also maintain eigenvalues with negative real parts. This means that the sliding mode equations may maintain dumped solutions even if disturbances due to parameter variations occur; obviously, variables x_1 and x_2 revert to zero with a different dynamics.
As far as the sliding mode control is concerned, things are quite similar to the previous case since, in general, the values of vectors β_i^+ and β_i^- in (A.13) can be chosen so as to satisfy the proper inequalities no matter what the values of the terms of matrix A_v are (provided they are bound within certain limits). Obviously, no critical case in the neighborhood of the origin can occur.

c) Disturbances on model-following systems

Let us consider a system whose plant is described by:

$$\dot{x}_p = A_p\, x_p + B_p\, u_p \qquad (A.21)$$

where x_p and u_p are, respectively, n and m dimension vectors, while A_p and B_p are matrices of proper dimensions; as far as the model to be followed is concerned, let us suppose it is represented by:

$$\dot{x}_m = A_m\, x_m + B_m\, u_m \qquad (A.22)$$

where x_m and u_m are, respectively, n and m dimension vectors, while A_m and B_m are matrices of proper dimensions.
Defining the error variable as:

$$e = x_m - x_p \qquad (A.23)$$

and defining:

$$A_v = A_m - A_p \qquad (A.24)$$

the overall error equations become:

$$\dot{e} = A_m\, e - B_p\, u_p + A_v\, x_p + B_m\, u_m \qquad (A.25)$$

As known [9] [13], in order for a sliding mode control u_p to exist, so as to the steady state error e may be null (and hence the perfect model-following may exist) the following conditions are sufficient:

$$\rho[B \;\; A_v] = \rho[B] \qquad \rho[B \;\; B_m] = \rho[B] \qquad (A.26)$$

In this case, again, making assumptions similar to the ones given in (A.6) and (A.18) and defining:

$$\begin{vmatrix} e_1 \\ e_2 \end{vmatrix} = \begin{vmatrix} x_{m1} - x_{p1} \\ x_{m2} - x_{p2} \end{vmatrix} \qquad (A.27)$$

the model equations (A.22) and the sliding mode error equations can be written in the following form:

$$\dot{x}_{m1} = A_{m11}\, x_{m1} + A_{m12}\, x_{m2}$$

$$\dot{x}_{m2} = A_{m21}\, x_{m1} + A_{m22}\, x_{m2} + B_{m2}\, u \qquad (A.28)$$

$$\dot{e}_1 = A_{m11}\, e_1 + A_{m12}\, e_2 + A_{v11}\, x_{p1} + A_{v12}\, x_{p2} \qquad (A.29)$$

On the other hand, assuming that in this case relation (A.10) also holds, we can write:

$$e_2 = -Fe_1 = 0 \qquad (A.30)$$

so that, taking (A.27) into account, (A.29) becomes:

$$\dot{e}_1 = A_F\, e_1 + A_{v11}\, x_{m1} + A_{v12}\, x_{m2} \qquad (A.31)$$

where:

$$A_F = [(A_{m11} - A_{v11}) - (A_{m12} - A_{v12})\, F]$$

Consequently, we can conclude that, in order for equation (A.31) not to be affected by the disturbances due to the differences between the system state matrix and the model state matrix it must be that:

$$A_{v11} = 0 \qquad A_{v12} = 0 \qquad (A.32)$$

As it is easy to verify, this condition coincides with the first one in (A.26). Moreover, from (A.32) we can see that, in order for the system to follow the model perfectly, the variations of the matrices must only affect the components which appear in the equations where the controls are taken into account. If this condition is not satisfied, the matrix variations could represent a serious shortcoming for the exploitation of the sliding mode model-following method as a means of implementing a robust control, i.e., a parameter variation insensitive control.
A way to limit this shortcoming is to implement a further input v in the model equations (and, consequently, in the error equations); hence the overall system turns out to be:

$$\dot{x}_{m1} = A_{m11}\, x_{m1} + A_{m12}\, x_{m2} + v$$

$$\dot{x}_{m2} = A_{m21}\, x_{m1} + A_{m22}\, x_{m2} + B_{m2}\, u$$

$$\dot{e}_1 = A_{v11}\, x_{m1} + A_{v12}\, x_{m2} + A_F\, e_1 + v \qquad (A.33)$$

$$\dot{v} = Ee_1$$

As can easily be understood, v is a $((n - m) \times 1)$ vector giving the integral of the error while E is a $((n - m) \times (n - m))$ diagonal matrix; in order not to cosiderably modify the error dynamics assigned by matrix F when all the terms of A_{v11} and A_{v12} are zero, the eigenvalues of E must have negative real parts smaller than the ones of the eigenvalues of matrix $[A_{m11} - A_{m12} F]$. Obviously, in order to obtain steady state zero errors even when condition (A.32) is not satisfied (that is notwithstanding the presence of forcing terms in the error equations), the state matrix of system (A.33) must have eigenvalues with negative real parts.

REFERENCES

[1] Ambrosino, G., G. Celentano and F. Garofalo (1984). Variable Structure Model Reference Adaptive Control Ststems. *Int. J. Control, 39 N° 6.*

[2] Andry, A.N., E.Y. Shapiro and J.C. Chung (1983). Eigenstructure Assignment for Linear System. *IEEE Trans. on Aerospace Electronic Systems, AES-19.*

[3] Dorling, C.M. and A.S.I. Zinober (1986). Two Approaches to Hyperplane Design in Multivariable Variable Structure Control Systems. *Int. J. Control, 44 N° 1.*

[4] Draženović, D. (1969). The Invariance Conditions in Variable Structure Systems. *Automatica, 5,* 287÷295.

[5] Fahmy, M.M. and J. O'reilly (1982). On Eigenstructure Assignment in Linear Multivariable Systems. *IEEE Trans. on AC, 27 N° 3.*

[6] Kwakernaak, H. and R. Sivan (1972). In John Wiley (Ed.), *Linear Optimal Control System,* New York.

[7] Mudge, S.K. and R.J. Patton (1988). Analysis of The Technique of Robust Eigenstructure Assignment with Application to Aircraft Control. *IEE Proceedings, 4.*

[8] Mudge, S.K. and R.J. Patton (1988). Enhanced Assessment of Robustness for an Alircraft's Sliding Mode Controller. *Journal of Guidance, Control and Dynamics, 11 N° 6.*

[9] Penati, M.E. and G. Bertoni (1989). Decoupling and Model-Following Control in Variable Structure System with Linear Plane. *IFAC/IFORS/IMACS Symposium on*

Large Scale Systems.

[10] Sobel, K.M. and E.Y. Shapiro (1985). A Design Methodology for Pitch Pointing Flight Control Systems. *Journal of Guidance, Control and Dynamics, 8 N°2.*

[11] Sobel, K.M. and E.Y. Shapiro (1985). Eigenstructure Assignment for Design of Multimode Flight Control Systems. *IEEE Control System Magazine.*

[12] Speyer, J.L., J.E. White, R. Douglas and D.G. Hull (1984). Multi–Input/Multi–Output Controller Design for Longitudinal Decoupled Aircraft Motion. *Journal of Guidance, Control and Dynamics, 7 N° 6.*

[13] Young, K.K.D. (1978). Design of Variable Structure Model–Following Control Systems. *IEEE Trans. on AC, 23.*

A DESIGN OF NONLINEAR MODEL REFERENCE ADAPTIVE CONTROL AND THE APPLICATION TO THE LINK MECHANISM

S. Okubo

Dept. of Mechanical System Engineering, Faculty of Engineering, Yamagata University, Yonezawa, Yamagata, Japan

Abstract. This paper shows a design of nonlinear MRACS and the proof of gloval stability. Nonlinear systems which are dealed in this paper havethe property of norm constraints. We design nonlinear MRACS by combineing nonlinear model following control systems and parameter adjustable laws of Kleisselmeier type. In this method parameter errors converge to zero exponentially under the condition that identification signals are persistently exciting. As a useful application we show a design of MRACS to link mechanism of robot manipulators.

Keywords. Model reference adaptive control systems; nonlinear control systems; model following control systems; global stability; link mechanism.

1. INTRODUCTION

Model reference adaptive control system (MRACS) has been developed since Monopoli (1974) presented. In case of single input and single output many applications are performed. However the researches of multi input and multi output MRACS or nonlinear MRACS are rare. Morse (1980) succeeded the proof of global stability as sigle input and single output linear case. Though the applications of Morse method to multi input and multi output or nonlinear case are not adequate, Morse method is based on a direct method and scalar gain. In case of multi input and multi output case , numerator gain becomes a matrix, an indirect method must be adopted. In this paper the design of nonlinear and multi input multi output MRACS is shown. The nonlinear element f(v(t)) is such that the norm is a power function of $\| v(t) \|$,that is $\| f(v(t)) \| \leqq \alpha + \beta \| v(t) \|^{\gamma}$. The internal stability of nonlinear MRACS can be proved (S.Okubo 1989,1990). Also we apply this control method to the nonlinear link mechanism and show usefulness of this paper's method.

2. DESIGN OF NONLINEAR MODEL FOLLOWING CONTROL SYSTEM

A controlled object is described in (1), (2),(3)and a model is given in (4),(5).

$$\dot{x}(t)=Ax(t)+Bu(t)+B_f f(v(t))+d(t) \quad (1)$$
$$v(t)=C_f x(t)+d_v(t) \quad (2)$$
$$y(t)=Cx(t)+d_o(t) \quad (3)$$

$$\dot{x}_m(t)=A_m x_m(t)+B_m r_m(t) \quad (4)$$
$$y_m(t)=C_m x_m(t) \quad (5)$$

Where $x(t) \in R^{n}$, $u(t) \in R^{\ell}$, $v(t) \in R^{\ell_f}$, $f(v(t)) \in R^{\ell_f}$, $d(t) \in R^{n}$, $d_o(t) \in R^{\ell}$, $d_v(t) \in R^{\ell_f}$, $x_m(t) \in R^{n_m}$, $r_m(t) \in R^{\ell_m}$ and $y_m(t) \in R^{\ell}$. Available states are y(t),v(t). On the other hand internal state x(t) is not available. Output y(t) is controlled value and v(t) is measurement output. The function form of $f(\cdot)$ is known. So f(v(t)) is available. The linear part of controlled object (C,A,B) is controllable and observable Zero points of $C(pI-A)^{-1}B$ are stable. Nonlinear function f(v(t)) satifies the next condition.

$$\| f(v(t)) \| \leqq \alpha + \beta \| v(t) \|^{\gamma} \quad (6)$$

Where $\alpha \geqq 0$、 $\beta \geqq 0$、 $\gamma \geqq 0$、 $\| \cdot \|$ is Euclidean norm. Disturbance d(t),do(t) are bounded and satisfy (7).

$$D_d(p)d(t)=0、 D_d(p)d_o(t)=0 \quad (7)$$

$D_d(p)$ is a scalar characteristic polynomial of disturbances. Where $\partial D_d(p)=n_d, \Gamma(D_d(p))=1$. $d_v(t)$ is general bounded distubance which adds to measurement output. Output error is given in (8).

$$e(t)=y(t)-y_m(t) \quad (8)$$

The representations of state equations are described as followings to obtain input-output relations of controlled object and model.

$$D(p)y(t)=N(p)u(t)+N_f(p)f(v(t))+w(t) \quad (9)$$
$$D_m(p)y_m(t)=N_m(p)r_m(t) \quad (10)$$
$$w(t)=C\,adj(pI-A)d(t)+D(p)d_o(t) \quad (11)$$

The aim of control system design is to obtain a control law which makes output error zero and keeps internal states bounded. Control input u(t) is calculated like as (16) using polynomial algorithm (12) of stable polynomial $T(p)D_m(p)$.

$$T(p)D_m(p)=D_d(p)D(p)R(p)+S(p) \quad (12)$$
$$T(p)D_m(p)e(t)=0 \quad (13)$$

$$T(p)D_m(p)e(t)=D_d(p)D(p)R(p)y(t)+S(p)y(t) -T(p)N_m(p)r_m(t) \quad (14)$$

From these results,

$$\begin{aligned} e(t)=Q(p)N_r/\{T(p)D_m(p)\}[&u(t) \\ &+N_r^{-1}Q(p)^{-1}\{D_d(p)R(p)N(p)-Q(p)N_r\}u(t) \\ &+N_r^{-1}Q(p)^{-1}S(p)y(t) \\ &+N_r^{-1}Q(p)^{-1}D_d(p)R(p)N_f(p)f(v(t)) \\ &-N_r^{-1}Q(p)^{-1}T(p)N_m(p)r_m(t)] \end{aligned} \quad (15)$$

The next control law is obtained from $e(t)=0$.

$$\begin{aligned} u(t)=&-N_r^{-1}Q(p)^{-1}\{R(p)D_d(p)N(p)-Q(p)N_r\}u(t) \\ &-N_r^{-1}Q(p)^{-1}S(p)y(t) \\ &-N_r^{-1}Q(p)^{-1}D_d(p)R(p)f(v(t))+u_m(t) \end{aligned} \quad (16)$$

There are following relations in each polynomial matrices using above calculations. $C(pI-A)^{-1}B=N(p)/D(p)$, $C(pI-A)^{-1}B_f=N_f(p)/D(p)$, $D(p)=\det(pI-A)$, $\partial_{ri}N(p)=\sigma_i$, $\Gamma_r(N(p))=N_r$, $|N_r| \neq 0$, $\partial_{ri}N_f(p)=\sigma_{fi}$, $\partial_{ri}N_m(p)=\sigma_{mi}$, $T(p)$ is $\partial T(p)=\rho$ $(\rho \geq n_d+2n-n_m-1-\sigma_i)$, $\Gamma(T(p))=1$ and a stable scalar polynomial. $Q(p)$ is $\partial_{ri}(Q(p))=\rho+n_m-n+\sigma_i$, $\Gamma_r(Q(p))=I$ and a stable polynomial matrix. The external signal $u_m(t)$ is given in (17).

$$u_m(t)=N_r^{-1}Q(p)^{-1}T(p)N_m(p)r_m(t) \quad (17)$$

For using no derivatives of signals in control input $u(t)$, next constraints of polynomial degree must be satisfied.

$$n_m-\sigma_{mi} \geqq n-\sigma_i \ (i=1\sim \ell),\ \sigma_i \geqq \sigma_{fi}(i=1\sim \ell),$$
$$\rho \geqq n_d+2n-n_m-1-\sigma_i (i=1\sim \ell) \quad (18)$$

Therefore $e(t)$ satisfies (19) and converges to zero.

$$T(p)D_m(p)e(t)=0 \quad (19)$$
$$e(t)\to 0 \quad (t\to\infty) \quad (20)$$

3. PROOF OF BOUNDEDNESS OF INTERNAL STATES

Control input $u(t)$ can be described as following using internal states.

$$u(t)=-H_1\zeta_1(t)-E_2y(t)-H_2\zeta_2(t)-E_3f(v(t)) -H_3\zeta_3(t)+u_m(t) \quad (21)$$

$$\dot{\zeta}_1(t)=F_1\zeta_1(t)+G_1u(t) \quad (22)$$

$$\dot{\zeta}_2(t)=F_2\zeta_2(t)+G_2y(t) \quad (23)$$

$$\dot{\zeta}_3(t)=F_3\zeta_3(t)+G_3f(v(t)) \quad (24)$$

Where $|pI-F_i|=|Q(p)|$ $(i=1,2,3)$, there are next relations between system matrices and transfer function matrices.

$$N_r^{-1}Q(p)^{-1}\{D_d(p)R(p)N(p)-Q(p)N_r\} =H_1(pI-F_1)^{-1}G_1 \quad (25)$$
$$N_r^{-1}Q(p)^{-1}S(p)=E_2+H_2(pI-F_2)^{-1}G_2 \quad (26)$$
$$N_r^{-1}Q(p)^{-1}D_d(p)R(p)N_f(p)=E_3+H_3(pI-F_3)^{-1}G_3 \quad (27)$$

From above considerations, the state space representation of an overall system can be obtained.

$$\frac{d}{dt}\begin{bmatrix} x(t) \\ \zeta_1(t) \\ \zeta_2(t) \\ \zeta_3(t) \end{bmatrix} = \begin{bmatrix} (A-BE_2C) & -BH_1 & -BH_2 & -BH_3 \\ -G_1E_2C & (F_1-G_1H_1) & -G_1H_2 & -G_1H_3 \\ G_2C & 0 & F_2 & 0 \\ 0 & 0 & 0 & F_3 \end{bmatrix} \cdot \begin{bmatrix} x(t) \\ \zeta_1(t) \\ \zeta_2(t) \\ \zeta_3(t) \end{bmatrix} + \begin{bmatrix} B_f-BE_3 \\ -G_1E_3 \\ 0 \\ G_3 \end{bmatrix} f(v(t)) + \begin{bmatrix} Bu_m(t)+d(t)-BE_2d_o(t) \\ G_1u_m(t)-G_1E_2d_o(t) \\ G_2d_o(t) \\ 0 \end{bmatrix} \quad (28)$$

$$v(t)=C_fx(t)+d_v(t) \quad (29)$$
$$y(t)=Cx(t)+d_o(t) \quad (30)$$

These state equation system is arranged to the next way to be set

$$z(t)^T=(x(t)^T, \xi_1(t)^T, \xi_2(t)^T, \xi_3(t)^T).$$

$$\dot{z}(t)=A_sz(t)+B_sf(v(t))+d_s(t) \quad (31)$$
$$v(t)=C_sz(t)+d_v(t) \quad (32)$$

The contents of $A_s, B_s, C_s, d_s(t)$ are clear from (28),(29). The boundedness of internal states is that $z(t)$ is bounded.

First of all characteristic polynomial is calculated in the next equation.

$$|pI-A_s|=|Q(p)|^2\ V_s(p)\ T(p)^{\ell}\ D_m(p)^{\ell} \quad (33)$$

Where $V_s(p)$ is zero point polynomial of $C(pI-A)^{-1}B = U(p)^{-1}V(p)$ (left coprime decomposition), that is $V_s(p)=|V(p)|/|N_r|$. From (33), A_s is a stable system matrix.

For the next the nonlinear element $f(v(t))$ is considered, $f(v(t))$ is a function having a proper of (6). From (6), $f(v(t))$ is bounded as $v(t)$ is bounded. The proofs of boundedness are different depending on γ. The proof case of $0\leqq\gamma<1$ is easy, but the case of $\gamma\geqq 1$ is complicated with adding other conditions. For the first time, the case of $0\leqq\gamma<1$ is treated.

To set $V(t)$ a quadratic form of $z(t)$, $\dot{V}(t)$ is given in (35).

$$V(t)=1/2z(t)^TPz(t) \quad (34)$$

$$\dot{V}(t)=-1/2z(t)^TQz(t)+z(t)^TPB_sf(v(t)) +z(t)^TPd_s(t) \quad (35)$$

Where P, Q are symmetric positive definite matrices, $A_s{}^TP+PA_s=-Q$ is satisfied.

$\dot{V}(t)$ satisfies the next inequality as $f(v(t))$ satisfies (6) and $d_s(t)$ is bounded.

$$\dot{V}(t)=-\mu\|z(t)\|^2+M_1\|z(t)\|^{\gamma+1}+M_2\|z(t)\| \leqq -q_1\ V(t)+q_2 \quad (36)$$

Where $\mu=(1/2)\lambda_{min}(Q)>0$, M_1, M_2, q_1, q_2 are positive. As the result of (37) is obtained, $z(t)$ is bounded.

$$V(t)=q_2/q_1(1-\exp(-q_1t))+V(0)\exp(-q_1t) \leqq q_2/q_1+V(0)<\infty \quad (37)$$

For the next time, the case of $\gamma \geqq 1$ is considered. In this case the next condition is satisfied.

(condition 1)

A transfer function from $f(v(t))$ to $v(t)$,that is $H(p)$ is positive real.

$H(p)=C_s(pI-A_s)^{-1}B_s$ is positive real.

(condition 2)

$$v(t)^T f(v(t)) \leqq \alpha_1 - \beta_1 \| v(t) \|^{\gamma_1}$$

is satisfied and $\alpha_1 \geqq 0, \beta_1 > 0, \gamma_1 > \gamma \geqq 1$ can be set.

When the above two conditions are satisfied, the boundedness can be proved. The relation of $v(t)$ and $f(v(t))$ is given as a next equation.

$$v(t)=C_s(pI-A_s)^{-1}B_s f(v(t))+C_s(pI-A_s)^{-1}d_s(t) +d_v(t)=\underline{C_s}(pI-\underline{A_s})^{-1}\underline{B_s} f(v(t))+\underline{d_v}(t) \quad (38)$$

Where $(\underline{C_s},\underline{A_s},\underline{B_s})$ is a minimum realization of (C_s,A_s,B_s). Therefore $(\underline{C_s},\underline{A_s},\underline{B_s})$ is controllable and observable. $\underline{d_v}(t)=C_s(pI-A_s)^{-1}d_s(t)+d_v(t)$ is a bounded signal becouse A_s is a stable system matrix. In case that the external signal appears positively, an analysis of boundedness becomes complicated. So the elimination of external signals from a state space representation is necessary.

$$\underline{v}(t)=v(t)-\underline{d_v}(t) \quad (39)$$

$$\underline{v}(t)=H(p)\underline{f}(\underline{v}(t)), H(p)=\underline{C_s}(pI-\underline{A_s})^{-1}\underline{B_s} \quad (40)$$

$$\underline{f}(\underline{v}(t))=f(v(t))=f(\underline{v}(t)+\underline{d_v}(t)) \quad (41)$$

To change (40) to a state space representation, (42) can be obtained.

$$\dot{\underline{z}}(t)=\underline{A_s}\,\underline{z}(t)+\underline{B_s}\,\underline{f}(\underline{v}(t)),\ \underline{v}(t)=\underline{C_s}\,\underline{z}(t) \quad (42)$$

For the system of (42), a quadratic form of (43) is set, and the time derivative of $V(t)$ is calculated.

$$V(t)= 1/2\ \underline{z}(t)^T P\ \underline{z}(t) \quad (43)$$

$$\dot{V}(t)= -1/2\ \underline{z}(t)^T Q\ \underline{z}(t) + \underline{z}(t)^T P\ \underline{B_s}\ \underline{f}(\underline{v}(t)) \quad (44)$$

Where $H(p)$ is positive real, the lemma of positive real is used.

$$P\,\underline{A_s} + \underline{A_s}^T P = -Q,\ \underline{B_s}^T P = \underline{C_s},$$
$$P = P^T > 0, \quad Q = Q^T \geqq 0 \quad (45)$$

Using a result of (45), $\dot{V}(t)$ becomes the following.

$$\begin{aligned}\dot{V}(t) &\leqq \underline{v}(t)^T\underline{f}(\underline{v}(t))\\ &=(\underline{v}(t)+\underline{d_v}(t)-\underline{d_v}(t))^T f(\underline{v}(t)+\underline{d_v}(t))\\ &\leqq \alpha_1-\beta_1 \| \underline{v}(t)+\underline{d_v}(t) \|^{\gamma_1}\\ &\quad + \alpha_2 + \beta_2 \| \underline{v}(t)+\underline{d_v}(t) \|^{\gamma}\\ &\leqq \alpha_3-\beta_3 \| \underline{v}(t) \|^{\gamma_1} \end{aligned} \quad (46)$$

Where $\alpha_3 \geqq 0, \beta_3 > 0, \gamma_1 > \gamma \geqq 1$ can be set. For the first, it will be shown that $\underline{z}(t)$ has no finite escape time. From (46),

$$\begin{aligned}V(t) &\leqq V(0)+\int_0^t (\alpha_3-\beta_3 \| \underline{v}(t) \|^{\gamma_1})\,dt\\ &\leqq V(0)+\alpha_3 t\end{aligned} \quad (47)$$

From the above and (43),the next relation is satisfied.

$$\| \underline{z}(t) \| \leqq (\alpha_4 t+\beta_4)^{1/2},\ (\alpha_4 \geqq 0, \beta_4 \geqq 0) \quad (48)$$

This means that $\underline{z}(t)$ has no finite escape time. Therefore $\underline{v}(t)$ has no finite escape time from $\underline{v}(t)=\underline{C_s}\,\underline{z}(t)$. In the next, the boundedness of $\underline{z}(t)$ will be shown using reductio ad absurdum. To suppose that $\underline{v}(t)$ is divergent as $t \to \infty$, $\underline{z}(t)$ is also divergent becouse of $\underline{v}(t)=\underline{C_s}\,\underline{z}(t)$. Namely

$$\| \underline{v}(t) \| \to \infty,\quad \| \underline{z}(t) \| \to \infty \quad (t \to \infty) \quad (49)$$

Then the next inequality is satisfied and $M_3 \geqq 0$ exits.

$$\dot{V}(t) \leqq \alpha_3-\beta_3 \| \underline{v}(t) \|^{\gamma_1} \leqq \alpha_3-\beta_3 M_3^{\gamma_1} \leqq 0 \quad (50)$$

As $v(t)$ has no finite escape time and is divergent as $t \to \infty$, there is a finite time T like as $\| \underline{v}(t) \| \geqq M_3$ $(t \geqq T)$ for $t \geqq T$. So considering $V(t)$, $\dot{V}(t)$ for $t \geqq T$, (51) is satisfied and $V(t)$ is non-increasing as $t \geqq T$. Therefore

$$V(t) \leqq V(T)=1/2\ \underline{z}(T)^T\ P\ \underline{z}(T) < \infty \quad (t \geqq T) \quad (51)$$

This is contradiction to (49). So $\underline{v}(t)$ is bounded. From the facts of $\| \underline{v}(t) \| < \infty$ and $v(t)=\underline{v}(t)+\underline{d_v}(t)$, $v(t)$ is bounded.

$$\| v(t) \| < \infty \quad (52)$$

Continuously $f(v(t))$ is bounded from boundedness of $v(t)$. In system equation (31), $f(v(t)), d_s(t)$ are bounded and A_s is a stable system matrix , it concludes that internal state $z(t)$ is bounded.

$$\| z(t) \| < \infty \quad (53)$$

The above results are summarized in a next theorem.

<<Theorem 1>>

In the nonlinear system, $z(t) \in R^n, v(t) \in R^{\ell_f}$, $f(v(t)) \in R^{\ell_f}$, $d(t) \in R^n$, $d_v(t) \in R^{\ell_f}$, A_s is a stable system matrix, and disturbances of $d(t), d_v(t)$ are bounded.

$$\dot{z}(t)=A_s z(t)+B_s f(v(t))+d(t) \quad (54)$$

$$v(t)=C_s z(t)+d_v(t) \quad (55)$$

A nonlinear element $f(v(t))$ is satisfied in (56).

$$\| f(v(t)) \| \leqq \alpha + \beta \| v(t) \|^{\gamma} \quad (56)$$

Where α, β, γ are $\alpha \geqq 0, \beta \geqq 0, \gamma \geqq 0$. Then the two below conditions are satisfied, $z(t)$ and $v(t)$ are bounded.

(condition 1) $H(p)=C_s(pI-A_s)^{-1}B_s$ is positive real.

(condition 2) $v(t)^T f(v(t)) \leqq \alpha_1 - \beta_1 \| v(t) \|^{\gamma_1}$ is satisfied and $\alpha_1 \geqq 0, \beta_1 > 0, \gamma_1 > \gamma \geqq 1$ are set. 』

$H(p)$ of (condition 1) can be changed to (59) by elimination $u(t)$ from (57) and (58) by setting $y(t) \to y_m(t)$.

$$y(t)=C(pI-A)^{-1}Bu(t)+C(pI-A)^{-1}B_f f(v(t)) +C(pI-A)^{-1}d(t)+d_o(t) \quad (57)$$

$$v(t)=C_f(pI-A)^{-1}Bu(t)+C_f(pI-A)^{-1}B_f f(v(t)) +C(pI-A)^{-1}d(t)+d_v(t) \quad (58)$$

$$H(p)=C_s(pI-A_s)^{-1}B_s =C_f(pI-A)^{-1}B_f -C_f(pI-A)^{-1}B\{C(pI-A)^{-1}B\}^{-1}C(pI-A)^{-1}B_f \quad (59)$$

Furthermore using a block transformation of an inverse matrix, H(p) is given in (60).

$$H(p)=[C_f \ 0]\begin{bmatrix} pI-A & -B \\ -C & 0 \end{bmatrix}^{-1}\begin{bmatrix} B_f \\ 0 \end{bmatrix} \quad (60)$$

4. NONLINEAR MODEL REFERENCE ADAPTIVE CONTROL SYSTEM AND THE GLOBAL STABILITY

In this section a design of nonlinear model reference adaptive control for a controlled object described from (1) to (5) using a parameter adjustable law by Kreisselmeier and a proof of the global stability will be shown in case of $0 \leqq \gamma < 1$.

$$\| f(v(t)) \| \leqq \alpha + \beta \| v(t) \|^{\gamma} \quad (61)$$

Where α, β, γ are $\alpha \geqq 0$, $\beta \geqq 0$, $0 \leqq \gamma < 1$. The other conditions are following .

(i) n, σ_i, n_d are known.
(ii) Zero points of $C(pI-A)^{-1}B$ are stable, and $|N_r| \neq 0$ is satisfied.
(iii) $\sigma_i \geqq \sigma_{if}, n_m-\sigma_{mi} \geqq n-\sigma_i$
(iv) $r_m(t)$ is P.E.
(v) y(t),v(t) are available and f(v(t)) is known.

Control input u(t) is given in (62).

$$u(t)= -N_r^{-1}Q(p)^{-1}\{R(p)D_d(p)N(p)-Q(p)N_r\}u(t) -N_r^{-1}Q(p)^{-1}S(p)y(t) -N_r^{-1}Q(p)^{-1}D_d(p)R(p)f(v(t)) +N_r^{-1}Q(p)^{-1}T(p)N_m(p)r_m(t)$$

$$= C_r r_h(t)-H_1\zeta_1(t)-\{E_2y(t)+H_2\zeta_2(t)\} -\{E_3f(v(t))+H_3\zeta_3(t)\}$$

$$= [C_r,-H_1,-E_2,-H_2,-E_3,-H_3]\zeta(t) = \Theta\zeta(t) \quad (62)$$

Where $\zeta(t)$ is given in (63).

$$\zeta(t)^T=[r_h(t)^T,\zeta_1(t)^T,y(t)^T,\zeta_2(t)^T, f(v(t))^T,\zeta_3(t)^T] \quad (63)$$

$\zeta_1(t),\zeta_2(t),\zeta_3(t)$ are state variable filters.

$$\dot{\zeta}_1(t)=F_1\zeta_1(t)+G_1u(t) \quad (64)$$

$$\dot{\zeta}_2(t)=F_2\zeta_2(t)+G_2y(t) \quad (65)$$

$$\dot{\zeta}_3(t)=F_3\zeta_3(t)+G_3f(v(t)) \quad (66)$$

$\xi(t)$ is defined as following.

$$u_p(t)=u(t)/(T(p)Dm(p))=\Theta\xi(t) \quad (67)$$

$$\xi(t)^T=[\{Q(p)^{-1}y(t)\}^T,\xi_1(t)^T, \{y(t)/(T(p)D_m(p))\}^T,\xi_2(t)^T, \{f(v(t))/(T(p)D_m(p))\}^T,\xi_3(t)^T] \quad (68)$$

$$\dot{\xi}_1(t)=F_1\xi_1(t)+G_1\{u(t)/(T(p)D_m(p))\} \quad (69)$$

$$\dot{\xi}_2(t)=F_2\xi_2(t)+G_2\{y(t)/(T(p)D_m(p))\} \quad (70)$$

$$\dot{\xi}_3(t)=F_3\xi_3(t)+G_3\{y(t)/(T(p)D_m(p))\} \quad (71)$$

In parameter adjustable law , the exponential decay law of Kreisselmeier(1977) is used. To set $\underline{\Theta}(t)$ as estimation of Θ, the following constraction is performed.

$$J(t)=\int_0^t \{\underline{\Theta}(t)\xi(\tau)-u_p(\tau)\}^T \cdot\{\underline{\Theta}(t)\xi(\tau)-u_p(\tau)\} e^{-q(t-\tau)} d\tau \rightarrow \min \quad (72)$$

$$\dot{\underline{\Theta}}(t)= -\mu/2\,\partial J(t)/\partial\underline{\Theta}(t) \quad (73)$$

$$\dot{\Omega}(t)= -q\Omega(t)+\xi(t)\xi(t)^T \quad (74)$$

$$\dot{\Lambda}(t)= -q\Lambda(t)+u_p(t)\xi(t)^T \quad (75)$$

$$\dot{\underline{\Theta}}(t)= -\mu\{\underline{\Theta}(t)\Omega(t)-\Lambda(t)\} \quad (76)$$

Where $q>0, \mu>0, \Omega(t)\in R^{N\ell \times N\ell}$, $\Lambda(t)\in R^{\ell \times N\ell}$, to set $\Delta\underline{\Theta}(t)=\underline{\Theta}(t)-\Theta$ and to remark

$$\Lambda(t)=\Theta\Omega(t),$$

$$\Delta\dot{\underline{\Theta}}(t)= -\mu\Delta\underline{\Theta}(t)\Omega(t) \quad (77)$$

$$W(t)=Tr\{\Delta\underline{\Theta}(t)\Delta\underline{\Theta}(t)^T\} \quad (78)$$

$$\dot{W}(t)=-2Tr\{\Delta\underline{\Theta}(t)\Omega(t)\Delta\underline{\Theta}(t)^T\} \quad (79)$$

Signal $\xi(t)$ is satisfied in (80).

$$\int_t^{t+T} \xi(\tau)\xi(\tau)^T d\tau \geqq kI>0 ,t\geqq 0,T>0 \quad (80)$$

It means that internal signal is persistently exciting. Then

$$\Omega(t+T)\geqq e^{-qT}\int_t^{t+T} \xi(\tau)\xi(\tau)Td\tau \geqq ke^{-qT} I \quad (81)$$

is satisfied. From (79) and (81),(82) is organized.

$$\dot{W}(t)\leqq -2\mu k e^{-qT} W(t) \quad (82)$$

From this , (119) is consisted

$$\|\Delta\underline{\Theta}(t)\| \leqq \|\Delta\underline{\Theta}(0)\| e^{-\mu k e^{-qT} t} \quad (83)$$

Then the next result is obtained.

$$\|\Delta\underline{\Theta}(t)\| \rightarrow 0 \quad (t\rightarrow\infty) \quad (84)$$

(83) equation means that estimated value $\underline{\Theta}(t)$ converges to true value Θ whether $\xi(t)$ is bounded or not. The control input is used in (85).

$$u(t)=\underline{\Theta}(t)\zeta(t)=\Theta\zeta(t)+\Delta\underline{\Theta}(t)\zeta(t) \quad (85)$$

Using the above preparations, a proof of the global stability is performed. The overall behaviour is described as the following.

$$\dot{x}(t)=Ax(t)+Bu(t)+B_f f(v(t))+d(t)$$
$$y(t)=Cx(t)+d_o(t)$$
$$v(t)=C_f x(t)+d_v(t)$$
$$u(t)= -(H_1+\Delta\underline{H}_1(t))\zeta_1(t) -(E_2+\Delta\underline{E}_2(t))y(t) -(H_2+\Delta\underline{H}_2(t))\zeta_2(t)$$

$$
\begin{aligned}
&-(E_3+\Delta \underline{E}_3(t))f(v(t)) \\
&-(H_3+\Delta \underline{H}_3(t))\zeta_3(t) \\
&+(C_r+\Delta \underline{C}_r(t))r_h(t)
\end{aligned}
$$

$$\dot{\zeta}_1(t)=F_1\zeta_1(t)+G_1u(t)$$

$$\dot{\zeta}_2(t)=F_2\zeta_2(t)+G_2y(t)$$

$$\dot{\zeta}_3(t)=F_3\zeta_3(t)+G_3f(v(t)) \tag{86}$$

(86)is summarized as (87).

$$
\begin{aligned}
\dot{z}(t)= &A_sz(t)+B_sf(v(t)) \\
&+\Delta \underline{A_s}(t)z(t)+\Delta \underline{B_s}f(v(t))+d_s(t)
\end{aligned} \tag{87}
$$

$$v(t)= C_sz(t)+d_v(t) \tag{88}$$

Where the contents of each variables become as next way .

$$
A_s = \begin{bmatrix} (A-BE_2C) & -BH_1 & -BH_2 & -BH_3 \\ -G_1E_2C & (F_1-G_1H_1) & -G_1H_2 & -G_1H_3 \\ G_2C & 0 & F_2 & 0 \\ 0 & 0 & 0 & F_3 \end{bmatrix},
$$

$$
\Delta \underline{A_s}(t) = -\begin{bmatrix} B \\ G_1 \\ 0 \\ 0 \end{bmatrix}
\times \begin{bmatrix} \Delta \underline{E_2}(t)C, & \Delta \underline{H_1}(t), & \Delta \underline{H_2}(t), \Delta \underline{H_3}(t) \end{bmatrix},
$$

$$
B_s = \begin{bmatrix} B_f-BE_3 \\ -G_1E_3 \\ 0 \\ G_3 \end{bmatrix}, \quad \Delta \underline{B_s}(t) = -\begin{bmatrix} B \\ G_1 \\ 0 \\ 0 \end{bmatrix} \Delta \underline{E_3}(t),
$$

$$
d_s(t)=
\begin{bmatrix} BC_rr_h(t)-BE_2d_o(t)+d(t) \\ G_1C_rr_h(t)-G_1E_2d_o(t) \\ G_2d_o(t) \\ 0 \end{bmatrix}
+ \begin{bmatrix} B\Delta \underline{C_r}(t)r_h(t)-B\Delta \underline{E_2}(t)d_o(t) \\ G_1\Delta \underline{C_r}(t)r_h(t)-G_1\Delta \underline{E_2}(t)d_o(t) \\ 0 \\ 0 \end{bmatrix},
$$

$$C_s= \begin{bmatrix} C_f,0,0,0 \end{bmatrix} \tag{89}$$

$d_s(t)$is finite signal from the above contents. The aim is to proof of boundedness of $z(t)$ in (87) and (88). Firstly a characteristic equation of A_s is obtained in (93).

$$|pI-A_s| = |Q(p)|^2 V_s(p)T(p)^{\ell} D_m(p)^{\ell} \tag{90}$$

$V_s(p)$ is an invariant zero polynomial of $C(pI-A)^{-1}B$, and this is a stable polynomial, so the characteristic equation of A_s is stable. Error parameters $\Delta \underline{A_s}(t), \Delta \underline{B_s}(t)$ decay exponentially becouse $\| \Delta \underline{\Theta}(t) \|$ decays exponentially.

$$\| \Delta \underline{A_s}(t) \| \leqq C_3 \| \Delta \underline{\Theta}(t) \| \leqq C_4 e^{-\nu t} \tag{91}$$

$$\| \Delta \underline{B_s}(t) \| \leqq C_5 \| \Delta \underline{\Theta}(t) \| \leqq C_6 e^{-\nu t} \tag{92}$$

Where

$$\nu = \mu k e^{-qT} \tag{93}$$

Such a fact that parameter error decays exponentially is important for the proof of global stability. The time derivative of a qurdratic form of (94) will be calculated.

$$V(t)=1/2z(t)^TP_sz(t) \tag{94}$$

$$
\begin{aligned}
\dot{V}(t)=&-1/2z(t)^TQ_sz(t)+z(t)^TP_sB_sf(v(t)) \\
&+z(t)^TP_s\Delta \underline{A_s}(t)z(t) \\
&+z(t)^TP_s\Delta \underline{B_s}(t)f(v(t))+z(t)^TP_sd_s(t)
\end{aligned} \tag{95}
$$

Where $P_sA_s+A_s{}^TP_s= -Q_s$ and $P_s>0$, $Q_s>0$ are performed. To remark $\| \Delta \underline{A_s}(t) \|$ and $\| \Delta \underline{B_s}(t) \|$, $\dot{V}(t)$ becomes as following.

$$
\begin{aligned}
\dot{V}(t) \leqq &-q_1V(t)+z(t)^TC_s{}^Tf(v(t)) \\
&+q_2\|z(t)\|^2 e^{-\nu t} \\
&+q_3\|z(t)\| \|f(v(t))\| e^{-\nu t} + q_4\|z(t)\| \\
\leqq &-q_1V(t)+v(t)^Tf(v(t))+\|d_v(t)\| \|f(v(t))\| \\
&+q_2\|z(t)\|^2 e^{-\nu t} + q_3\|z(t)\|(\alpha+\beta\|v(t)\|)^{\gamma} \\
&\cdot e^{-\nu t} + q_4\|z(t)\| \\
\leqq &-q_1V(t)+\alpha_3+\beta_3\|v(t)\|^{\gamma_1}+q_5(\alpha_4+\beta_4\|v(t)\|)^{\gamma} \\
&+q_6V(t)e^{-\nu t} + q_3\|z(t)\|(\alpha+\beta\|v(t)\|)^{\gamma} e^{-\nu t} \\
&+q_4\|z(t)\| \\
\leqq &-q_7V(t)+q_8V(t)e^{-\nu t} + q_9
\end{aligned} \tag{96}
$$

On the above equation, q_i is positive. By solving the inequality of (96), the next result can be obtained.

$$
\begin{aligned}
V(t) &\leqq e^{\frac{q_8}{\nu}(1-e^{-\nu t})}[V(0)e^{-q_7t} + q_9/q_7(1-e^{-q_7t})] \\
&\leqq e^{\frac{q_8}{\nu}}(V(0) + q_9/q_7)<\infty
\end{aligned} \tag{97}
$$

So $z(t)$ is bounded.

$$\|z(t)\|<\infty \tag{98}$$

5.THE APPLICATION TO THE LINK MECHANISM

Dynamic equations of multi-link mechanism of robot manipulators are given in (99).

$$J(\theta)\ddot{\theta}+X(\theta)\dot{\theta}^2+M^TDM\dot{\theta}+Z(\theta)=M^Tu(t)+d(t) \tag{99}$$

Where

$\theta= (\theta_1, \theta_2, \cdots\cdots, \theta_n)^T$,
$\theta^2=(\theta_1{}^2, \theta_2{}^2, \cdots\cdots, \theta_n{}^2)^T$,
$J(\theta)=\{L_{ij}\cos(\theta_i-\theta_j)\}+\mathrm{diag}(\tilde{I}_i)$,
$X(\theta)=\{L_{ij}\sin(\theta_i-\theta_j)\}$,
$D=\mathrm{diag}(D_1,D_2, \cdots\cdots,D_n)$,

$Z(\theta)=-[G_1\sin\theta_1, G_2\sin\theta_2, \cdots, G_n\sin\theta_n]$ (100)

L_{ij}, G_i are given in mass of link(m_i), length of link(ℓ_i), gravity constant g. $\tilde{I}_i$ is an inertia moment of actuator. D_i is a damping factor. θ_i is an angular of link, u(t) is input torque, and d(t) is disturbance.

$$M=\begin{bmatrix} 1 & & 0 \\ -1 & 1 & \\ \cdot & \cdot & \cdot \\ 0 & -1 & 1 \end{bmatrix} \quad (n\times n) \qquad (101)$$

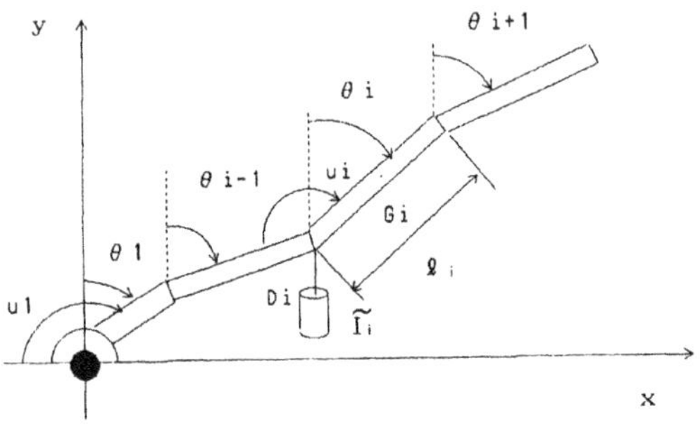

Fig.1 Schmatic diagram of multi-link mechanism

In the special case of single nonlinear link a dynamic equation is given as the following.

$$J\ddot{\theta}(t)+D\dot{\theta}(t)-mg\ell/2\cdot\sin(\theta(t))=u(t)+d(t) \qquad (102)$$

We set $x_1(t)=\theta(t), x_2=\dot{\theta}(t), y(t)=\theta(t)$, $f(v(t))=\sin(v(t))$, then eq.(102) becomes the form of eq.(1),(2).

$$\dot{x}(t)=\begin{bmatrix} 0 & 1 \\ 0 & -D/J \end{bmatrix} x(t)+\begin{bmatrix} 0 \\ 1/J \end{bmatrix} u(t) + \begin{bmatrix} 0 \\ mg\ell/2J \end{bmatrix} \sin(v(t))+\begin{bmatrix} 0 \\ 1/Jd(t) \end{bmatrix} \qquad (103\ a)$$

$$v(t)=\begin{bmatrix} 1 & 0 \end{bmatrix} x(t),\quad y(t)=\begin{bmatrix} 1 & 0 \end{bmatrix} \qquad (103\ b)$$

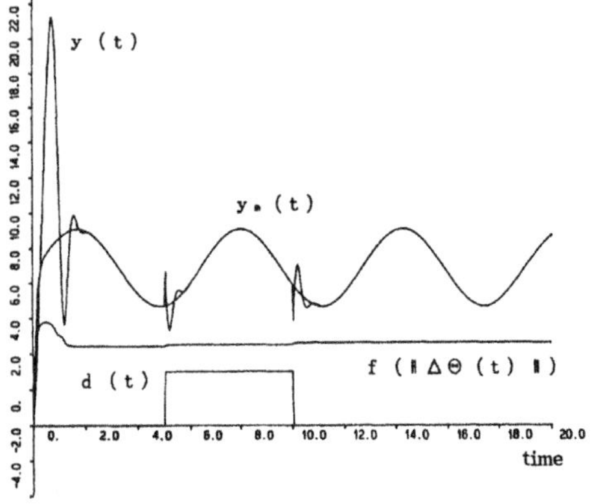

Fig.2 System responses of nonlinear adaptive control system

Reference model is given in (104).

$$y_m(t)=1/(p+\alpha)^2\ r_m(t),\quad r_m(t)=500(\sin(t)+4) \qquad (104)$$

$T(p)=(p+\alpha)^2, Q(p)=(p+\alpha)^2, q=0.01, \lambda=10, \alpha=15$, $d(t)=3.0$

We show a result of simulation in Fig.2. $F(\|\Delta\underline{\Theta}(t)\|)$ shows a behavior of convergence of parameters.

$$F(\|\Delta\underline{\Theta}(t)\|)=\{\|\Delta\underline{\Theta}(t)\|/\|\Theta\|-(1-\delta)\}\times 10+1 \qquad (105)$$

$$\underline{\Theta}(0)=\Theta\times\delta,\quad \delta=0.8 \qquad (106)$$

We can confirm that output signal follows the reference signal in Fig.2.

6. CONCLUSION

In this paper the design of the nonlinear model reference adaptive control system of multi input and multi output is showed and the proof of the global stability is performed. The structure of MRACS is to add parameter adjustable law to MFCS. The non-linear element is treated in case of power function of state norm. When nonlinear index γ is less than 1, the global stability can be proved. Furthermore we showed the application of the nonlinear model reference control system to the link mechanism and the usefulness of this control method.

REFERENCES

R.V.Monopoli(1974). Model Reference Adaptive Control with an Augmented Error Signal. IEEE Trans. Aut. Control, AC-19-5, 474-484

A.S.Morse(1980).Global Stability of Parameter Adaptive Control Systems, IEEE Trans. Aut. Control, AC-25-3, 433-439

I.D.Landau and H.M.Silveira(1979). A Stability Theorem with Applications to Adaptive Control, IEEE Trans. Aut. Control, AC-24-2,305-312

G.Kreisselmeier(1977). Adaptive Observers with Exponential Rate of Convergence, IEEE Trans. Aut. Control, AC-22-1,2-8

B.O.D.Anderson and J.B.Moore(1968). Algebraic Structure of Generalized Positive Real Matrices,SIAM J.Control, 6-4, 615-624

S.Okubo(1990). Robust Control of Multi Link Robot Manipulators Based on Model Following Control Method,Proceedings of 1st Int. Symp. Measurement and Control in Robotics,Houston,USA, G2.1.1-G2.1.7

S.Okubo(1989).A Design of Nonlinear MRACS and the Proof of Stability, Preprints of an IFAC Symp. on Adaptive Systems in Control and Signal Processing,Glasgow,UK ,525-530

S.Okubo(1990).Globally Stable MRACS for Multi-Input and Multi-Output Nonlinear Systems,Trans.Society of Instrumentation and Control Engineers(in Japan),26-1, 46-53(Japanese)

Richard P.Paul(1986). Robot Manipulators,The MIT Press

THE DESIGN AND APPLICATION OF EXPERT INTELLIGENCE CONTROL SYSTEM OF AUTOMATIC COMPENSATION EXPOSURE ENERGY

Chu Xuedao

Institute of Automation, QuFu Normal University, QuFu, Shan-dong Province, PRC

Abstract: In this paper ,based on the discrete state space optimization model developed by the author , an algorithm for solving exposure energy set points optimization problem by using heuristic search technique is presented. According to the proposed algorithm a computer control system for automatic compensation exposure energy is designed and implemented. It is used in the SBK-II fast-starting Iodine Gallium Lamp (IGL) printer. The running result shows that the proposed system is both reliable and stable , and the plate exposing qaulity is improved.

Keywords. expert intelligence control system (EICS) ; heuristic search technique; automatic compensation exposure energy(ACEE)

INTRODUCTION

Offset printing is now using advanced printing techniques at home and abroad , and the plate exposing is key process in offset plate making. The qaulity of a plate exposing actually relies on more or less of light energy absorbed by a plate. The requirement of more or less of light energy absorbed by a plate is different for different types of printing plate and different temperature of developer , and the result of plate exposing can be tested only after the plate is developed. Therefore , the control of qaulity of plate exposing is a black box control, and plate exposing time only depends on experience at present , that is, plate exposing time is from manual adjustment. In process of plate exposing,the exposure of a plate is a complex phsical and chemical reaction for which it is difficult to establish a mathematical model. In oder to improve the qaulity of plate exposing , the discrete optimization mathematical model of trailling optimally set points of exposure energy was established by author. The control for plate exposing qaulity is changed into one for plate exposing time by the convertor of light-electricity-frequency developed by us. We have designed an EICS , through heuristic optimization algorithm, the optimal trail of exposure energy for the set points have been realized in the process of plate exposing.

THE DYNAMICAL MATHEMATICAL MODEL OF ACEE

In the process of plate exposing , the plate exposing time is different with the varying of voltage of light souree and light flux from IGL. The main function of the dynamical optimization model is that it provide the decision information for the printer to realize the optimal trail of set points of exposure energy according to the observable values(Light flux , electricity and voltage , type of printing plate and plate exposing time) . Based on the characteristic of computer control , the discrete optimization mathematical model of the ACEE was established by author. It is as follows:

$$\min_{U_{sp}\in\Omega_t} J = \left\| W(i) - \int_0^T p(x(t),u(t),t)\,dt \right\|^2 , \quad (1)$$

subject to:

$$x(k+1) = A(x(k),u(k),k)\,x(k) + B(x(k),u(k),k)\,u(k) + C(x(k),u(k),k)\,e(k) , \quad (2)$$

$$x(0) = x_0 , \quad u(k) = F(T,k)$$

$$|p(x(t),u(t),t) - p^*| < \varepsilon$$

$$\Omega_t = [v_{min} , v_{max}] ,$$

where w(i) is set point of exposure energy , $\int_0^T p(x(t),u(t),t)dt$ is light energy absorbed by a plate in time interval [0,T]; u(k), x(k) are voltage and electricity on the light source respectively, {e(k)} is white noise random process; F(T,k) is a sectioned linear function; Ω_t is allowed limits of set points of voltage on the light source.

OPTIMAL CONTROL POLICY OF ACEE

The ACEE is actually, according to the type of printing plate and the temperature of developer, that the set points of exposure energy determined by experts experience and experiments are trailled optimally by light energy absorbed from a plate under the case of voltage fluctuating movement and the varying of light flux. Because parameters tested in the process of plate exposing are related with control scheme, and the control policy should synthesize the control for the plate exposing time according to the habit of operation, it is difficlt for solving above problem by classical optimization approach. We use the convertor of light-electricity-frequency to change the energy of ultraviolet light into the sum of frequency, and thus the optimization problem for set points of exposure energy is changed into the optimization for plate exposing time. Based on the dynamical model above, we have designed an EICS and taken a heuristic optimizing algorithm to realize the optimal trail for set points of exposure energy.

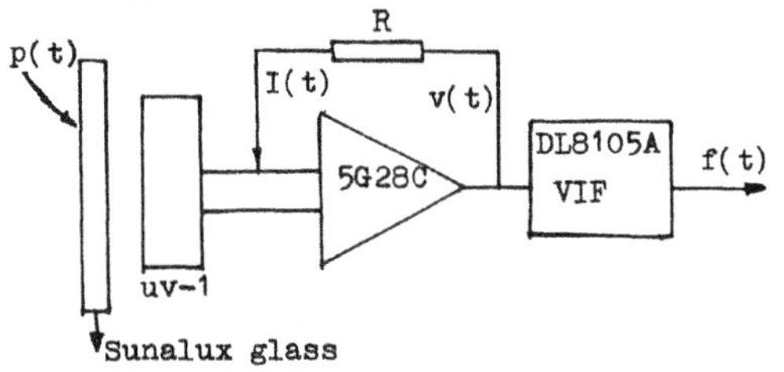

Fig.1 Converting principle figure of the light-electricity-frequency

In the process of plate exposing, light sent from the IGL, at first, passes the sunalux glass, and then it is converted into electricity signal through violet photosensitive diode uv-1 and amplifing of D.C amplifier 5G28C. Soppose that radiating light flux of ultraviolet light on unit area is p(t) , (p(t)=p(x(t),u(t),t)) , converting rate of light-electricity of the uv-1 is α, and resistance of amplifier is R, then relation among the p(t), electricity I(t) and voltage v(t) are as follows:

$$I(t)=\alpha p(t) \qquad (3)$$

$$v(t)=R\,I(t) \qquad (4)$$

In Fig.1, the DL8105A VIF is voltage—frequency convertor with the function converting voltage signal v(t) into pulse signal f(t), sopposing converting rate is β, then

$$f(t)=\beta v(t) \qquad (5)$$

from (3),(4),(5) yield that

$$f(t)=\alpha\beta R\,p(t) \qquad (6)$$

Form (6) shows the linear relation between the pulse frequency f(t) and p(t). Suppose the plate exposing time is T seconds(T is natural number),then exposure energy of ultraviolet light is

$$W=\int_0^T \alpha\beta R\,p(t)dt=\int_0^T f(t)dt \qquad (7)$$

Not losing generality, let the interval of taking sample is $\Delta t_i=t_i-t_{i-1}=1$ second , $(i=1,2,\ldots n,\ \ t_0=0,\ \ t_n=T)$, then

$$F=\sum_{i=1}^{n} f(t_i)\,\Delta t_i=\sum_{i=1}^{n} f(t_i)\doteq\int_0^T f(t)dt$$

that is, $F=\sum_{i=1}^{n} f(t_i) \doteq W$ (8)

The form (8) shows that the exposure energy was changed into the sum of frequency , and thus the optimization algorithm for the set points of exposure energy was changed into one for the set time.

The heuristic optimization algorithm are as following recurisive process:

1) datum steady working case of the plate exposing process belongsing to is decided according to the peer of the selected steady working case set with the types of printing plate, temperature of developer, light flux and the tested case of electricity and voltage of light source.
2) according to the datum steady working case, select the set pointt w(i) of the first exposing by relative heuristic knowledge, and let initial mark i=0 .
3) according to the mathematical model, compute the mean of light flux p(t), in every sample interval by the tested parameters.
4) compute the performance index by the

form $J=\| w(i)-\sum_{i=1}^{n} f(i) \|^2$.

5) if $J=J_{min}$, then best plate exposing time corresponding to datum steady working case is derived, and the optimization algorithm is over.
6) if $J \neq J_{min}$, according to the information of R (R is defined by (9)), transfer proper heuristic regulation, and let $i \Leftarrow i+1$, turn 3), and make continually optimization search.

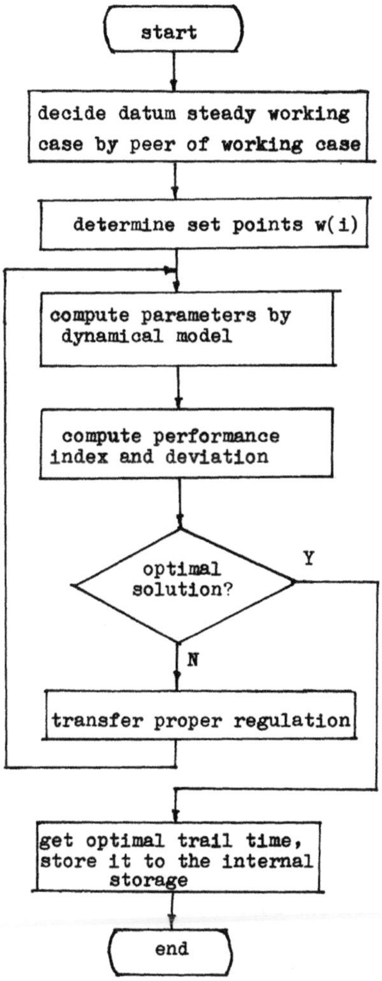

Fig.2 The principle diagram of heuristic optimization algorithm

The restricted deviation R shows the satisfied case of restricting conditions in the optimization problem. For the optimization algorithm (1),(2), R is defined as follows:

$$R \triangleq \begin{bmatrix} r_1 \\ r_2 \\ r_3 \end{bmatrix} = \begin{bmatrix} G(p(k)-p^*) \\ G(u(k)-u^*) \\ G(t(k)-T^*) \end{bmatrix} \qquad (9)$$

$$G(x) = \begin{cases} 0 & |x| \le \varepsilon \\ x-\varepsilon\,\mathrm{sign}(x) & |x| > \varepsilon \end{cases}$$

$$\mathrm{sign}(x) = \begin{cases} 1 & x > 0 \\ -1 & x < 0 \end{cases}$$

For the optimization algorithm above, the heuristic regulation sets play a key role in the whole algorithm, that is, it guides the seraching direction of set points w(i). The EICS which we designed takes mixed knowledge structure, that is, the heuristic regulation set is formmed by the method which link outer knowledge based on experts experience with inner knowledge based on dynamical model. Some parts of heuristic regulation are as following table1 and the table 2 .

Table1: some part of heuristic regulation based on experts experience

order of regulation	condition	action
1	use A-type ps-plate	energy set point of first exposing is w(A)
2	if voltage of I&L is too high or low	shut power source
3	second exposing	time is a bit shorter than first exposing
⋮	⋮	⋮

Table2: some part of heuristic regulation based on dynamical model

order of regulation	condition			action	
	r_1	r_2	r_3	$\Delta u(k)$	$\Delta p(k)$
1	0	0	0	0	0
2	0	0	1	0	0
3	0	1	0	0	1
4	0	1	1	0	1
5	1	0	0	1	0
⋮	⋮			⋮	⋮

Those elements in the regulation set are discribed by producing regulation with the following structure,

IF(condition) THEN(action)

The initial set points of the first exposing in optimization process is decided mainly by outer knowledge, that is, it gives a better search starting point, the searching direction of trailling set point in optimization process is determined by the inner knowledge, and fine searching results is produced through the combination of the outer and inner knowledge, and thus it strengthens the ability of solving optimization problem.

The control principle that exposure energy trailled optimally set points is as following figure3:

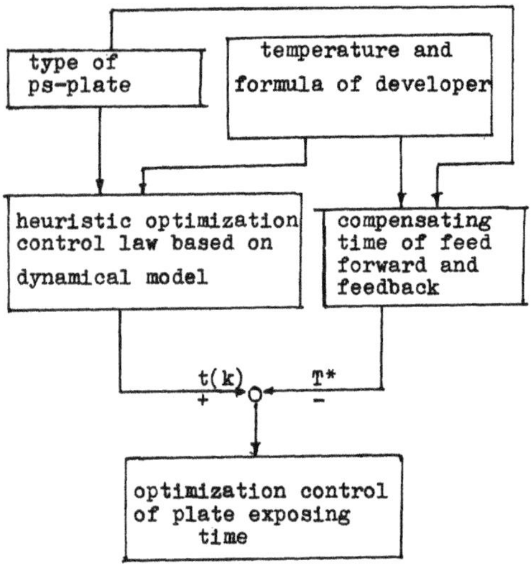

Fig.3

THE DESIGN OF COMPUTER CONTROL SYSTEM

The EICS of compensating automatically exposure energy have been used in the SBK-II fast-starting IGL printer. The diagram of computer control system of automatic compensation exposure energy is as following figure4:

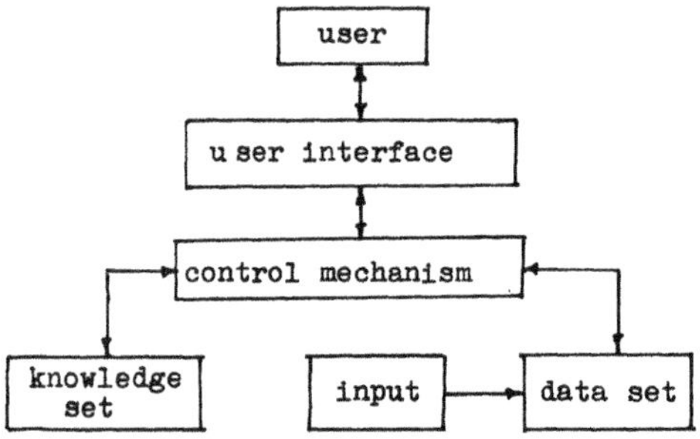

Fig.4

The main computer of control mechanism is 8098 single chip processor produced by Intel company, and A/D conversion is realized by light—electricity — freqency convertor. A simple user interface is used in system design, that is, the computer asks for user to input and answers yes or no, and user inputs parameters of types of ps-plate, formula and temperature of developer etc. according to the answer. The reliability and robustness are considered in the sowftware design, and it possess the function of automatic fault diagnosis, processing alarm, promising disturbance switchover of control form(from the first exposing to second exposing), and the manual interference under the unusual condition.

CONCLUSION

The EICS of compensating automatically exposure energy was used SBK-II fast-starting IGL printer designed by ourself in 1988 . Through over two years online running, it shows that this system is both reliable and accurate enough for exposure energy control, and it is simple for operating. Because of curbbing blindness in control of manual set points, it improved the qaulity of plate exposing.

In october, 1988, the SBK-II printer was appraised that EICS of automatic compensation exposure energy arrived international advanced level. In September, 1989, SBK-II IGL printer with fine qaulity won championship in international tenders which came from 103 factories of producing machine from 13 countries, and thus products penetrates into international markets.

REFERENCES

Astrom, K.J., Anton, T.J., and Arzer, K.E. (1986). Expert control. Automatica, 22, 277-286.

Herrod, R.A., and Papas, B.(1985). Artificial intelligence moves into industrial and process control. I&CS, vol.58, No.3, 45-52 .

Chu Xuedao, etc. A closed-loop self-adaptive fluid infusion monitoring and controlling system for the resuscitation of burn patients. 8th IFAC/IFOR Symposium on Identification and System parameter estimation preprints, vol.3, 1756-1757 .

Zhou Qijian, Li Zuxou, and Chen Menyou . (1987). Intelligence control and its prospect. Information and Control. vol.16,No.2, 38-45 .

Zhou Qijian, and Chen Menyou.(1987). Intelligence control of PH process. Information and Control, vol.16, No.2, 1-5 .

He Min, and Lu Yongzai. (1988). Mixed-type Intelligence regulator. Information and Control,vol.17, No.4, 2-4 .

A DIRECT ADAPTIVE POLE ASSIGNMENT CONTROL SCHEME FOR NON NECESSARILY STABLY INVERTIBLE DISCRETE-TIME PLANTS

G. Bartolini and A. Ferrara

Dept. of Communication, Computer and System Sciences, University of Genova, Via Opera Pia, 11 A - 16145 Genova, Italy

Abstract. This paper deals with discrete-time adaptive control of possibly non stably invertible linear SISO plants. The control objective is that of solving an explicit adaptive pole assignment problem without either introducing the so-called augmented error signal, regardless of the relative degree of the plant, or avoiding any identification procedure. The control scheme here proposed presents a number of positive aspects, namely, a true parallelism between the model and the plant, simpler polynomial equations in the underlying linear structure and the possibility of ignoring the plant time delay. However, its main feature concerns its general applicability, since, unlike classical schemes, it does not require the cancellation of the polynomial containing the zeros of the plant, but that of a manipulable polynomial, so that it is possible to make use of recent results obtained in the research area of polynomial robustness, under suitable and not unrealistic assumptions on the system to be controlled, in order to guarantee the minimum phase requirement for the zeros involved in the cancellation, even in presence of a partially unknown non stably invertible plant.

Keywords. Model reference adaptive control; discrete-time systems; nonminimum-phased systems; robustness.

1. INTRODUCTION

Ever since the seventies, the problem of designing differentiator-free adaptive controller for unknown linear time-invariant single input-single output plants has intensively been dealt with (see, Monopoli, 1974, and Narendra and Valavani, 1978). In particular, the basic issue of proving the stability of the overall adaptive control scheme has been faced and solved for both continuous-time, (Narendra, Lin and Valavani, 1980), and discrete-time, (Narendra and Lin, 1980), systems. Yet, the resulting control structures are somewhat complex to be implemented and, unfortunately, require the system to be stably invertible. In this paper, the more general case in which the system is possibly non stably invertible is addressed. More precisely, we consider a single input-single output discrete deterministic linear time-invariant plant. The control objective is that of solving an explicit adaptive pole assignment problem without making use of the so-called augmented error signal, independently of the relative degree of the plant. We also aim at avoiding any explicit identification procedure. In practice, in this paper we present a first attempt to extend to non stably invertible discrete-time plants the work and results presented in (Bartolini and Ferrara, 1990).

Having chosen a simpler control objective (pole assignment instead of complete transfer function assignment), we have the opportunity to devise a control scheme which turns out to be simpler and more flexible than the classical augmented error signal method. More precisely, the controller structure proposed is characterized by the fact that a true parallelism between the model and the plant is retained, even in the case of high relative degree. Further, the underlying linear structure is described by very simple polynomial equations and the degree of complexity of the adaptation mechanism does not differ from that present in conventional proposals. However, there is an interesting feature which can be exploited so as to guarantee a more general applicability than classical model reference schemes. In fact, unlike them, the controller structure proposed does not require the cancellation of $B(z)$, namely the polynomial which contains the zeros of the plant, but that of a manipulable polynomial. For this reason, introducing a suitable set of assumptions on the plant, it is possible to relate the control problem stated above with a robust stabilization problem in terms of interval polynomials, and apply recent results obtained in the research area of polynomial robustness in order to guarantee the minimum phase requirement for the roots of the polynomial involved in the cancellation. In fact, what we want to highlight is the possibility, ensured by the scheme proposed, of supporting the adaptive control mechanism by means of an off-line pre-stabilization procedure, so that any drawback related to nonminimum phase behaviours is circumvented.

It is worth noting that the avoidance of the minimum phase requirement for the plant numerator turns out to be highly desirable in practice, since it has been observed that the pulse transfer function obtained discretizing a continuous minimum phase system with a sample-and-hold device can present zeros outside the unit disk. That is, a continuous-time plant whose zeros are all contained in the left half plane does not necessarily produce a system with zeros inside the unit disk when sampled, as made clear in (Astrom, Hagander and Sternby, 1984). The occurence of such a situation is far from being unusual, even in the case of plants with low relative degree. For instance, continuous-time plants with relative degree greater than 2, have always a tendency to give rise to sampled systems with nonminimum phase zeros, this phenomenon usually taking place when the sampling period is small. A possible way to counteract this drawback is that of using the so-called δ-operator, as indicated in (Middleton and Goodwin, 1986). Yet, in this paper we prefer to remain within the classical z-transform environment, to make our work consistent and comparable with existing literature.

This paper is organized as follows. In the next section, we present the basic design philosophy and solve the underlying control problem in the case in which all the plant parameters are assumed to be perfectly known. The discrete adaptive control scheme is presented in Section 3 and its global asymptotic stability is briefly discussed. Section 4 is devoted to the investigation of the particular characteristics of the adaptive control scheme proposed which lead to the statement of the robust stabilization problem. Finally, in Section 5, some comments conclude the paper.

2. PROBLEM STATEMENT

In this work we consider a single input-single output discrete linear time-invariant plant described by the following finite difference equation

$$A(z)y_p(j) = B(z)z^{-d}\, u(j) \qquad (1)$$

where

$$A(z)= z^n+a_1z^{n-1}+ \ldots +a_n$$
$$B(z)= b_0z^n+b_1z^{n-1}+ \ldots +b_n$$

that is, the discrete-time version of a continuous-time plant with a possible time delay, as often encountered in industrial applications. The plant considered satisfies the following assumptions: i) $A(z)$ has zeros inside the unit disk, i.e., $|z| < 1$; ii) the plant order "d+n" and the relative degree "d" are known; iii) the sign of coefficient b_0 is known. Note that the zeros of $B(z)$ are not assumed to lie within the unit disk. Moreover, to make this plant tractable in terms of Schur stability of interval polynomials some further assumptions, which are obviously unnecessary to cope with the adaptive control problem, are required. Let us express the uncertain knowledge about the plant in terms of perturbations acting on each of its coefficients. Then, if we denote with $q \in \Re^{2n+1}$ a vector of perturbations such that $q_i^- \leq q_i \leq q_i^+$, for $i=1,\ldots, 2n+1$, with q_i^-, q_i^+ known limits, we can make explicit the influence of such a vector on the plant polynomials, i.e., we can rewrite (1) as

$$A(z, q)y_p(j) = B(z, q)z^{-d}\, u(j) \qquad (2)$$

where

$$A(z, q)= z^n+a_1(q)z^{n-1}+ \ldots +a_n(q)$$
$$B(z, q)= b_0(q)z^n+b_1(q)z^{n-1}+ \ldots +b_n(q)$$

In the general case treated in literature, the plant coefficients $a_i(q)$, $b_i(q)$ are assumed to be given affine linear functions of any component of the vector q. In this paper, we limit ourselves to considering the case in which each perturbation element q_i acts on a single coefficient, that is $a_i(q)=a_{i0}+q_i$, $i=1,\ldots, n$, $b_i(q) = b_{i0}+q_{i+n+1}$, $i=0,\ldots, n$.

Our aim is that of controlling this plant so as to fulfil a pole assignment requirement. In particular, in order to simplify the following formulas and with no loss of generality, all the desired poles are assumed to be placed at the origin of the z-plane. Hence, the behaviour required to the plant can be represented by the reference model

$$z^{n+d}y_p(j) = B(z)r(j) \qquad (3)$$

where $r(j)$ is a preassigned reference sequence which is assumed to be uniformly bounded, and we have chosen $A_m(z)=z^{n+d}$.

The linear structure underlying the control scheme proposed when we deal with the case of perfectly known plant (i.e., we refer to (1) with $A(z)$, $B(z)$ known polynomials), can be described by the following parametrization. We denote with

$$\theta_p^{*T}=[F^*_{11},\ldots, F^*_{1(n+d)}, F^*_{21},\ldots, F^*_{2(n+d)}, F^*_{30}, F^*_{31},\ldots, F^*_{3(n+d)}] \qquad F^*_{30}=1 \qquad (4)$$

the vector of the coefficients F_{ki}^*, for $k=1,...,3$, of the numerators of three discrete state variable filters with input $u(j)$, $v_1(j)$, $r(j)$, respectively, and characteristic polynomial $D(z)=z^{n+d}$, and with

$$\Phi_p^T(j)=[-u_p(j-1),\ldots,-u_p(j-n-d),\ -v_1(j-1),\ldots,-v_1(j-n-d),\ r(j),\ldots,r(j-n-d)] \qquad (5)$$

the vector containing the relevant signals, with

$$v_1(j)=y_p(j)+C(z)u_p(j) \qquad (6)$$

where $C(z)=N_c(z)/D_c(z)$ is in general a feedforward compensator whose utility will become apparent in the sequel, dealing with robustness issues. However, for the sake of simplicity, we now set $C(z)=1$. When the plant is perfectly known, the parameter vector θ_p^* can be determined so as to achieve the specified control objective. To this end, let us set

$$u_m^*(j)=y_m^*(j)=F_{30}^*r(j)+\sum_{t=1}^{n+d}F_{3t}^*r(j-t) \qquad (7)$$

$$y_1^*(j)=\sum_{t=1}^{n+d}F_{1t}^*u_p(j-t) \qquad (8)$$

$$y_2^*(j)=\sum_{t=1}^{n+d}F_{2t}^*v_1(j-t) \qquad (9)$$

As a tracking error we consider the sequence

$$v(j)=y_m^*(j)-y_p^*(j)-u_p^*(j)=y_m^*(j)-v_1^*(j)=0, \qquad (10)$$

the signal fed back into the plant being

$$u_p^*(j)=u_m^*(j)-y_1^*(j)-y_2^*(j)=\theta_p^{*T}\Phi_p(j). \qquad (11)$$

Therefore, solving the control problem in the known parameters case has just the meaning of searching for a fixed vector θ_p^* such that the tracking error is identically zero. More precisely, we can introduce the following problem.

<u>Problem 1.</u> Find polynomials $F_k(z)$, $k:1,..., 3$, such that

i) $y_p(j)=\frac{B(z)}{A_m(z)}r(j)$ (i.e., the pole assignment requirement);

ii) set $y_m(j)=T_1(z)r(j)$ and $v_1(j)=T_2(z)r(j)$, then $T_1(z)=T_2(z)$;

iii) the transfer functions between $u_m^*(j)$ and $y_m^*(j)$, and between $u_p^*(j)$ and $v_1(j)$ are strictly positive real.

Δ

Clearly, points ii)-iii) essentially refer to the adaptive case.

A possible solution to Problem 1 can be characterized by the following choice of polynomials $F_k(z)$, for $k=1,...,3$.

$$F_1(z)=A'(z)+B(z)-D(z)$$
$$F_2(z)=A_m(z)-A'(z) \qquad (12)$$
$$F_3(z)=A'(z)+B(z)$$

where $A'(z)=z^dA(z)$, and $A_m(z)$ and $D(z)$ are chosen so that the ratio $D(z)/A_m(z)$ is strictly positive real ($D(z)/A_m(z)=1$, in our case). In fact, using Mason's rule, the transfer function between $y_p(j)$ and $r(j)$ results

$$T(z)=\frac{F_3(z)\,B(z)}{A'(z)(D(z)+F_1(z))+F_2(z)(A'(z)+B(z))} \qquad (13)$$

while $T_1(z)$ and $T_2(z)$ are respectively given by

$$T_1(z)=\frac{F_3(z)}{D(z)}\cdot\frac{D(z)}{A_m(z)}=\frac{F_3(z)}{A_m(z)} \qquad (14)$$

$$T_2(z)=\frac{A'(z)+B(z)}{A'(z)(D(z)+F_1(z))+F_2(z)(A'(z)+B(z))} \qquad (15)$$

By substitution, it is easy to verify that the proposed choice for the coefficients of polynomials $F_k(z)$'s actually constitutes a possible solution to Problem 1, provided that the ratio $D(z)/A_m(z)$ is strictly positive real, as assured by assumption. Note that, in this case, the relevant Bezoutian equation is $A'(z)(D(z)+F_1(z))+F_2(z)(A'(z)+B(z))=(A'(z)+B(z))A_m(z)$. This equation can be solved assuming $\deg(F_1(z))=\deg(F_2(z))=n+d-1$. Further, the filter $F_3(z)/D(z)$ can be written as $(A'(z)+B(z))/D(z)=1+(A'(z)+B(z)-D(z))/D(z)=1+F_1(z)/D(z)$. Therefore, it is possible to characterize the proposed scheme by means of a single parameter vector which contains the coefficients of $F_1(z)$ and $F_2(z)$ only. Clearly, this fact constitutes an advantage in designing the adaptation mechanism, since it makes the complexity of the proposed scheme, in terms of parameter sets to be tuned, comparable with that of conventional schemes.

Hence, the solution to the above problem can be denoted by a fixed parameter vector θ_p^*, chosen as

$$F_{30}^*=1$$

$$F_{1j}^*=F_{3j}^*=\{\ a_j,\ \text{if } j<d;\ \ a_j+b_{j-d},\ \text{if } d\le j,\ \text{with } a_j=0,\ \text{if } j>n,\ j=1,\ldots,n+d\ \} \qquad (16)$$

$$F_{2j}^*=\{\ -a_j,\ \text{for } j=1,\ldots,n;\ 0,\ \text{for } j=n+1,\ldots,n+d\ \}$$

When considering the adaptive case, the controller structure previously described is retained, as will be outlined in the next section.

3. THE ADAPTIVE CONTROLLER

In this section we refer to the problem of adaptively controlling the system introduced in (2), in order to meet the requirements listed in the statement of Problem 1, assuming the plant to be partially known, i.e., to be uncertain in the structured way indicated by the assumptions. For the time being, let us assume that no problem related to nonminimum phase behaviours arises (i.e., possible failures are supposed to be avoided by means of the pre-stabilization procedure we are going to sketch in the reminder of the paper), and depict the adaptive mechanism which enables us to force the tracking error to zero as required. To this end, we proceed by referring to the parametrization introduced in the previous section. When the plant parameters are unknown, the same structure for the regulator as in (11) is maintained, of course assuming the parameters to be time-varying, i.e.,

$$u_p(j)=\theta_p^T(j)\Phi_p(j) \tag{17}$$

Since we can write $u_p(j) = [\ \theta_p^* + \theta_p(j) - \theta_p^*]^T\ \Phi_p(j)$, the control sequence $u_p(j)$ becomes

$$u_p(j)=u_p^*(j)+\Delta\theta_p^T(j)\Phi_p(j) \tag{18}$$

where $\Delta\theta_p^T(j)$ denotes the parameter error. By construction, the output $v(j)$ does not depend on $r(j)$. So, only the effect of the disturbance $\tilde{u}_p(j)=\Delta\theta_p^T(j)\Phi_p(j)$ must be taken into account on deriving the error equation associated with the control scheme. Therefore, the traking error is given by

$$v(j)=-\hat{\Delta}\theta_p^T(j)\hat{\Phi}_p(j) \tag{19}$$

where

$$\hat{\Delta}\theta_p^T(j)=[\Delta F_{11}(j),\dots,\ \Delta F_{1(n+d)}(j),\ \Delta F_{21}(j),\dots,\ \Delta F_{2(n+d)}(j)]$$

with $\Delta F_{ki}(j)=F_{ki}(j)-F_{ki}^*$, for $k=1,2;\ i=1,\ \dots,\ n+d$, and

$$\hat{\Phi}_p^T(j)=[-u_p(j-1),\dots,\ -u_p(j-n-d),\ -v_1(j-1),\dots,\ -v_1(j-n-d)].$$

The error equation (19) defines a linear system with output $v(j)$ and input $\hat{\Delta}\theta_p^T(j)\hat{\Phi}_p(j)$. Then, its transfer function is strictly positive real. Therefore, adaptation can take place according to one of the various adaptation mechanisms proposed in literature, since the overall stability is directly guaranteed. For instance, we can implement an adaptation mechanism of proportional(P)-plus-integral(I) type, that is

$$\theta_p(j)=\theta_p^P(j)+\theta_p^I(j) \tag{20}$$

where $\theta_p^I(j)=\theta_p^I(j-1)+\Gamma\hat{\Phi}_p(j)v(j)$, with Γ being a $2(n+d)\cdot 2(n+d)$ arbitrary positive definite matrix, and $\theta_p^P(j)=\alpha\Gamma\hat{\Phi}_p(j)v(j)$, with α suitably chosen to preserve stability. Note that, in order to avoid the algebraic loop intrinsic in the proportional component of the adaptation mechanism, the following physically realizable algorithm, which is equivalent to (20), can be used

$$\theta_p(j)=\frac{1}{1+\alpha\hat{\Phi}_p^T(j)\Gamma\hat{\Phi}_p(j)}\theta_p^I(j) \tag{21}$$

A positive aspect is surely the fact that the proof of stability for the present control scheme, even when the relative degree is greater than one, keeps being identical, since the scheme, unlike the augmented error one, does not require any modification.

4. ROBUST STABILIZATION OF THE INVERSE CONTROL LAW W.R. TO v_1

In this section we will make clear the possibility of introducing an off-line pre-stabilization step for the polynomial involved in the cancellation. First, let us recall that, like many techniques for the design of control systems, most classical adaptive control schemes require the cancellation of the process zeros. On the contrary, in our case, because of the presence of a feedforward path in parallel with the plant, it turns out that the polynomial involved in the cancellation is not simply $B(z)$, but the manipulable polynomial $[A'(z)+B(z)]$. Thus, let us distinguish between some different situations which are liable to occur in practice, on using the scheme introduced. By assumption $A'(z)$ is Schur. Of course, it can happen to have $[A'(z)+B(z)]$ Schur as well. If this is the case, as long as a suitable parametrization for the system is selected, the approach presented in (Bartolini and Ferrara, 1990) could be used to solve the adaptive pole assignment problem without taking any pre-stabilization step. On the other hand, even if the "nominal" polynomial $[A'(z)+B(z)]^*$, i.e., that obtainable in the case of absence of uncertainty, is Schur, we cannot guarantee a trouble-free cancellation for any polynomials in the family considered.

Thus, we are going to consider a rational feedforward compensator of general type, as the one indicated in (6), and substitute the parallel connection of the plant and this compensator, characterized by the transfer function $[B(z)/z^dA(z)+C(z)]$, to the previous one. It is easy to show that this fact implies a number of modifications in the control structure: i) $D(z)=z^{n+d}N_c(z)$; ii) $D(z)/A_m(z)=N_c(z)/D_c(z)$; iii) the adaptation mechanism must be suitably modified. Yet, the main result of this rearrangement is the fact that the polynomial which is cancelled is not $[A'(z)+B(z)]$ any more, but

$$\Delta(z,\ q) \triangleq B(z,\ q)D_c(z)+z^dA(z,\ q)N_c(z). \tag{22}$$

This is a polytope of polynomials, whose stability must be checked in order to validate our adaptive control scheme.

However, as it is well known, for discrete-time

interval polynomials the attainment of Kharitonov-like results, i.e., concerning the stability of a rectangle of polynomials, is by no means straightforward. A possibility is that of assuming that uncertainty only affects a subset of the coefficients of the discrete-time polynomial in order to have robust stability of the entire family implied by that of the extreme polynomials, as outlined in (Hollot and Bartlett, 1986) and (Barmish, 1989). Another approach is that presented in the work by Mansour, Kraus and Anderson (1988) in which an analog of the strong Kharitonov theorem is derived for discrete systems, in that making it possible to check stability of a relatively small number of extreme polynomials even for that sort of systems (such a number increasing with the interval polynomial order). In this paper we prefer to make reference to this second line, since the assumption of a-priori knowledge on the coefficients of the manipulable polynomial which needs stabilizing would be in contrast with the salient features of the framework within which we operate, i.e., that of adaptive control. Nevertheless, the number of assumptions we are bound to introduce in order to set out our problem in robustness terms are somehow unusual compared to the ones generally required in standard adaptive control. They are however quite realistic and often met in many practical situations, in which it is not unnatural to have information about the nominal behaviour of the plant and the range of variation of the operating point.

Moreover, up to now the theory of polynomial robustness does not suggest a simple way for robustly stabilizing a system in presence of parametric perturbations of general type. Making improvement in this area is however out of the scope of this paper. What we want to highlight is the fact that an unsolvable problem in the field of adaptive control can be solved, under a suitable form of a-priori knowledge about plant parameters, using results coming from another research area. More precisely, we can think of devising an off-line iterative procedure (e.g., starting with a proportional compensator, testing stability, possibly increasing the complexity of such a compensator, and so on) to robustly stabilize the polynomial $\Delta(z, q)$. In other words, in order to disregard nonminimum phase behaviours, the following problem requires to be faced.

Problem 2. Find a compensator of $C(z)$ such that the polynomial (22) is Schur for all $q_i \in [q_i^-, q_i^+]$, $i=1,\dots,2n+1$.

Δ

Some peculiarities of the present problem with respect to the conventional stabilization problem deserve to be explicitly remarked:

1) The location of the assigned poles must be chosen in order to acquire some specific behaviour in time (for instance, damping or convergence rate requirements), while the placement of the zeros simply needs to obey a simple stabilization criterion, since only boundedness of the control law is concerned.
2) Some particular situations are frequently encountered which can be solved by using standard techniques. For example, if the zeros which are outside the unit disk lie within a disk with radius λ_z, and the poles are enclosed in a disk with radius λ_p, then a constant compensator

$$C(z) = k, \qquad k \leq \frac{(1+\lambda_z)^n}{(1-\lambda_p)^n\, b_0^+} \tag{23}$$

suffices for robust stability of the relevant polynomial.

5. CONCLUSIONS

In this paper a discrete time adaptive controller for linear SISO plants has been presented which solves a pole assignment control problem, guaranteeing the convergence to zero of the tracking error. The control philosophy also includes an off-line robust stabilization step, so that non stably invertible plants can be dealt with as well. The possibility, under suitable assumptions, of assuring Schur stability of the polynomial which is cancelled by the controller is allowed by the peculiar structure proposed and surely makes it largely applicable in practice.

REFERENCES

Astrom, K. J., O. Hagander, and J. Sternby (1984). Zeros of sampled systems. *Automatica*, 20, 31-38.

Barmish, B. R. (1989). An extreme point result for robust stability of discrete-time interval polynomials. *Proc. of the 28th CDC*, Tampa, Florida.

Bartolini, G., and A. Ferrara (1990). A simplified direct approach to the problem of adaptive pole assignment. *Proc. of the Joint Conf. New Trends in Systems Theory*, Genova, Italy.

Hollot, C. V., and A. C. Bartlett (1986). Some discrete-time counterparts to Kharitonov's stability criterion for uncertain systems. *IEEE Trans. Automat. Contr.*, 31, 355-356.

Mansour, M., F. Kraus, and B. D. O. Anderson (1988). Strong Kharitonov theorem for discrete systems. *Proc. of the 27th CDC*, Austin, Texas.

Middleton, R. H., and G. C. Goodwin (1986). Improved finite word length characteristics in digital control using delta operators. *IEEE Trans. Automat. Contr.*, 31, 1015-1021.

Monopoli, R. V. (1974). Model reference adaptive control with an augmented error signal. *IEEE Trans. Automat. Contr.*, 19, 474-484.

Narendra, K. S., and L. S. Valavani (1978). Stable adaptive controller design- direct control. *IEEE*

Trans. Automat. Contr., 23, 570-583.
Narendra, K. S., Y. H. Lin, and L. S. Valavani (1980). Stable adaptive controller design, part II: proof of stability. *IEEE Trans. Automat. Contr.*, 25, 440-448.
Narendra, K. S., and Y. H. Lin (1980). Stable discrete adaptive control. *IEEE Trans. Automat. Contr.*, 25, 456-461.

A SELF-TUNING FEEDFORWARD COMPENSATOR

R. Devanathan

School of Electrical and Electronic Engineering, Nanyang Technological Institute, Nanyang Avenue, Singapore 2263

Abstract In this paper, the development of a self-tuning feedforward compensator is discussed. The technique is to witness the uncompensated load response and identify the system transfer functions from it and set the compensator for ideal compensation. The compensator is then tuned by a trial and error method so as to account for the modelling error. The transfer functions corresponding to the manipulated and the load variables are assumed to be modelled by first order dead time systems. The self-tuning technique is verified using simulation on a distributed control system.

Keywords:- Feedforward control, compensator, self-tuning, pattern recognition, lead-lag compensator, dead time compensation.

INTRODUCTION

Feedforward control systems (Fig.1) have found wide use in the chemical process industry (Bristol and Hansen, 1987, Luyben and Buckley, 1977, MacDonald, McAvoy and Tits, 1987, Nisenfeld and Miyasaki, 1973, Shinskey, 1979, 1988). Typically, these control systems can provide quick correction to load upsets in the short term while the feedback control systems (with the integral action) provide the long term trim to the feedforward action. The steady state corrections provided by the feedforward control systems are sufficient for many processes. However, those processes which experience frequent load upsets require the dynamic feedforward correction as well. Also, those processes which tend to change their dynamic characteristics with load necessitate on-line tuning of the dynamic compensator of the feedforward control system so as to obtain the full benefits of the feedforward action.

Morari and Zafiriou (1989) have proposed an internal model control (IMC) based feedforward controller. The practical implementation of such advanced algorithms require a certain level of control-theoretic knowledge of the field personnel. Also, often, a relatively large number of parameters need adjustment in the case of the advanced algorithms thus precluding the full realisation of their potential in less knowledgeable environments. In contrast, a simple lead-lag compensator (Shinskey, 1988), when correctly tuned, is known to reduce the integrated absolute error (IAE) of the process load response by a factor of 10 or more from that experienced under feedback control alone. Also, the number of parameters needing adjustment in a lead-lag compensator is only two. Even for those processes, which require a dead time compensation, a two step procedure can be followed. The first step implements the dead time compensation followed by the second step which corresponds to the lead-lag compensation. The lead-lag compensator not only compensates for the differential lags in the manipulation and the load paths but also (by the fact that a negative lead is a good approximation to dead time) can also account for the residual dead time compensation error. This paper discusses the development of a self-tuning feedforward dynamic compensator incorporating the pattern recognition technique.

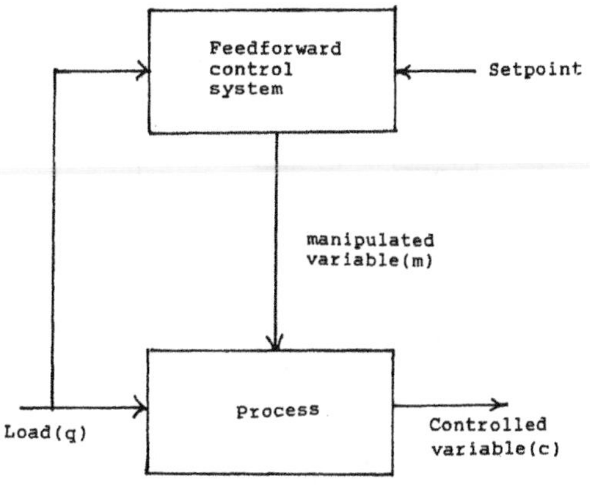

Fig.1 Schematic of Feedforward Control System

To summarise the rest of the paper, the following section classifies and discusses the uncompensated load response patterns. This is followed by the theory behind the self-tuning compensator both for dead time balanced systems and for dead time compensation systems. Implementation of the self-tuning compensator is discussed next followed by simulation studies to test the performance of the self-tuning algorithm. Finally a section on conclusions highlights the important results of the paper.

UNCOMPENSATED LOAD RESPONSE

Consider the feedforward compensation action given in Shinskey (1979,1988) as shown in Fig.2. The symbols g_m and g_q correspond to the transfer functions of the dynamic elements in the path of the manipulated variable (m) and the load variable (q) respectively. The symbol g_d stands for the transfer function of the dynamic compensator placed in the path of the manipulated variable as shown in Fig. 2. For the industrial processes, each of the transfer functions g_m and g_q can be modelled in combinations of apparent dead time (T_{dm}, T_{dq}) and dominant lag (T_m, T_q). Nine possible uncompensated responses have been catalogued by Shinskey (1988) (see Fig.3) based on the relative values of the dead time and the dominant lag on the side of both the load and the manipulated variables. Class I processes are dead time balanced. Class II processes correspond to the case when $T_{dm} < T_{dq}$. Class III processes correspond to the reverse case, viz., $T_{dm} > T_{dq}$. The class types a, b and c are defined in terms of the dominant lags. The type 'b' processes are dominant lag balanced. The type 'a' corresponds to the case $T_m < T_q$ and the type 'c' corresponds to the case $T_m > T_q$. The class Ib corresponds to the case when the dominant lags and the dead times are both balanced and hence does not need any compensation. Classes Ia and Ic can be completely compensated for by a lead-lag compensator. Class II processes strictly need a dead time compensation in addition to the lead-lag compensation. But due to the fact that a negative lead is a good approximation to a dead time (Shinskey, 1988), these classes can also be approximately compensated for by a lead-lag compensator alone. Class III processes cannot be compensated for completely since the dead time in the manipulated variable is too long.

The uncompensated load responses given in Fig. 3 fall into two broad categories, viz., (i) the patterns where the load response settles out asymptotically without crossing the setpoint line and (ii) those where the response curve crosses the setpoint line before settling out. The patterns IIIa and IIc in Fig. 3 correspond to the classification (ii) while the rest of the patterns in Fig. 3 correspond to the classification (i) above. Notice that the patterns IIIa and IIc under classification (ii) merely differ from each other in the polarity of the signal. Similarly , the patterns Ia and Ic also differ only in the polarity of the signal. The analysis under 'Dead time balanced systems' below apply to the patterns Ia and Ic. The results of same section can also be extended to the patterns IIa and IIb as an approximation to the dead time compensation. Arguments for the patterns Ia, IIa and IIb also apply to the patterns Ic, IIIb and IIIc except that the lead and lag time constants of the lead-lag compensator will be interchanged subject to the limit of the lead action. The patterns IIc and IIIa need dead time compensation and they are considered under the section 'Dead Time Compensation' below.

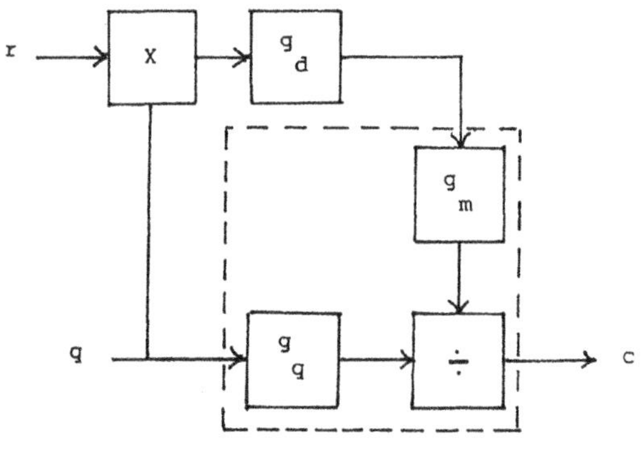

Fig.2 Feedforward Control Block Diagram

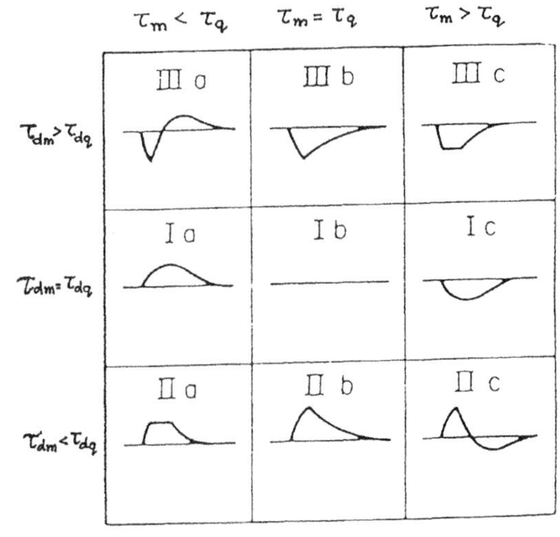

Fig.3 Uncompensated Load Response Patterns

FEEDFORWARD COMPENSATOR

Dead Time Balanced System

In a dead time balanced system, the transfer functions of the elements in the manipulated variable and the load variable paths can be assumed as follows.

$$g_m = 1 / (1 + s T_m) \quad (1)$$

$$g_q = 1 / (1 + s T_q) \quad (2)$$

The transfer function of the compensator can be assumed to be

$$g_d = (1 + s G T) / (1 + s T) \quad (3)$$

where T is the lag time constant of the compensator and G corresponds to the gain of the compensator at higher frequencies.

For small disturbances around a fixed operating point, assuming perfect steady state feedforward cancellation, the normalized uncompensated dynamic load response (see Fig. 4) is given by

$$dc_u(t) = \exp (-t/ T_q) - \exp (-t/ T_m) \quad (4)$$

The process time constants T_m and T_q are generally unknown and have to be estimated from the uncompensated load response. Then perfect dynamic compensation can be achieved, within the modelling error, as per the result of the following theorem.

Theorem 1

Consider the system of Fig.2 where g_m, g_q and g_d are given by Eq.(1),(2) and (3) respectively. The perfect feedforward dynamic compensation is achieved when

$$T = T_q = T_p (x-1)/(x \ln x) \quad (5)$$

and

$$G = x \quad (6)$$

where x is related to the uncompensated peak response dc_{up} as in

$$dc_{up} = x^{\{x/(1-x)\}} - x^{1/(1-x)} \quad (7)$$

and T_p corresponds to the time to peak of the uncompensated load response.

Proof:- See Appendix-I.

Dead Time Compensation

In Fig. 3, the patterns IIa and IIb belong to the same classification as the pattern Ia. This means that, for the patterns IIa and IIb, the evaluation of the lead-lag compensator parameters T and G can follow the technique mentioned above under 'Dead Time Balanced Systems' However, starting from these values of T and G, further trial and error tuning of the lead-lag compensator may be necessary to account for the imbalance in the dead times in these cases. The trial and error tuning procedures are discussed in the following section. The patterns IIIb and IIIc are similar to patterns IIa and IIb except that the polarity is reversed and similar arguments as above apply.

For the pattern IIc, consider the uncompensated load response given in Fig.5. The dead times T_{dm} and T_{dq} can be determined as shown in the diagram. The dead time compensation is then given by

$$T_d = T_{dq} - T_{dm} \quad (8)$$

The dead time compensation having been carried out approximately, the next load response, in this case, will then resemble the pattern Ic which can be handled using the technique outlined under the previous section on 'Dead Time Balanced Systems'.

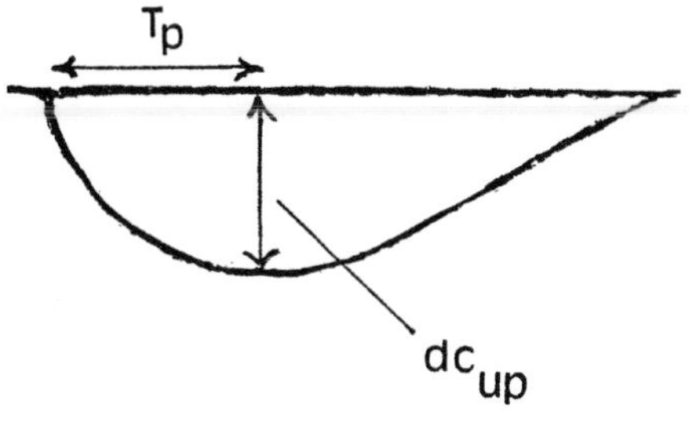

Fig.4 Uncompensated Response- Dead Time Balanced System

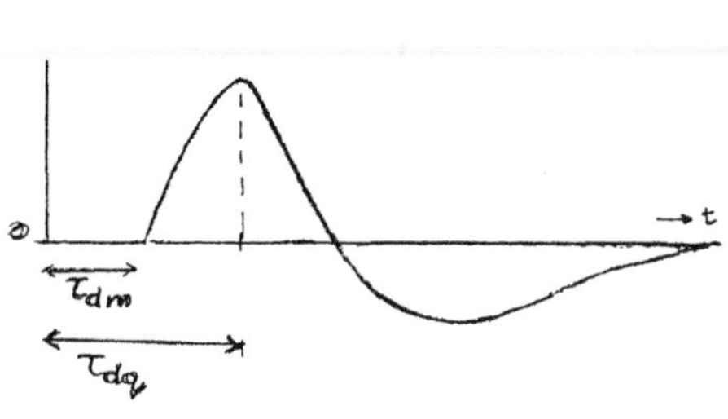

Fig.5 Uncompensated Load Response-Dead Time Compensation System

For the case IIIa , dead time compensation is impossible.Only lead-lag compensation is carried out.The result in this case is contained in the following theorem.

Theorem 2

Consider the system of Fig.2 where g_m, g_q and g_d are represented by first order dead time systems such that $T_{dm} > T_{dq}$ and $T_m < T_q$.For ideal lead-lag compensation of the differential time lags, the parameters T and G of Eq.(3) are then given by

$$G = x = T_o / (T_o + T_d) \qquad (9)$$

and

$$T = T_p (x -1)/ x \ln x \qquad (10)$$

where T_o, T_d and T_p are as shown in the uncompensated load response of Fig.6.

Proof:- See Appendix-II.

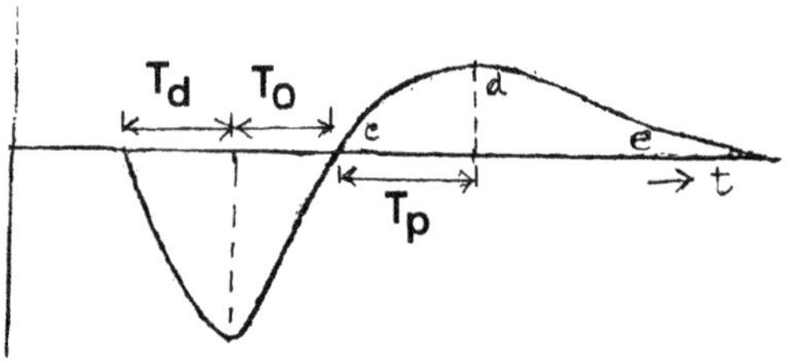

Fig.6 Uncompensated Load Response - Case IIIa

Trial and Error Tuning

After the initial settings have been made on the feedforward compensator as outlined in the last section , the load response may still exhibit dynamic error due to reasons of unmodelled dynamics or error in the evaluation of the initial values of the compensator parameters.To reduce this error, a two step approach is used.The first step is to drive the integrated error (IE) to zero followed by the second step which attempts to drive the integrated absolute error (IAE) to zero or minimum.

From Eq.(1), (2) and (3),the IE for a compensated system can be shown (Shinskey, 1988) to be

$$IE = (T_m - T_q) - T(G-1) , \; G > 1$$

$$= T(1-G) - (T_q - T_m) , \; G < 1 \qquad (11)$$

Referring to Fig. 7 (i) which corresponds to a compensated load response, the IE in this case is positive.To reduce IE to zero, will mean, for G>1, that T has to be increased and for G < 1, that T has to be decreased.A zero IE will result in a compensated load response such as that given in Fig.7(ii).In order to reduce the integral absolute error (IAE) to zero, for the case of Fig.7(ii), one has to increase G while at the same time keeping the IE at or near zero by assuring that the product T(G-1) or ((1-G)T) remains constant.This process is repeated until a tolerance limit on the peak error is reached or the minimum absolute error is reached.

SYSTEM IMPLEMENTATION

Two approaches hve been taken for the implementation of the feedforward compensator, viz., the expert system approach and the distributed control system approach.These are described below.

Expert System Approach

In this approach, the system used is as shown in Fig.8.The feedforward control system of Fig.2 is simulated and run on a FOX-3 computer (The Foxboro Company,1982) using the Foxboro Control Package (FCP).FCP is a high level language used for the configuration of control system strategies.The blocks of FCP correspond to the emulation of different analog elements, such as, PID (Proportional, Integral and Derivative),LL (lead-lag) , DT (Dead time) etc.The blocks are configured to form a control scheme.A programme called Task which is written in the Foxboro Process Basic (FPB) helps to activate the scheme for real time control and to output system data to the IBM/PC /XT via a RS-232 link.

The expert system residing in the IBM/ PC/ XT monitors the system response as accessed by the Task programme running on FOX-3.Based on the data in the database

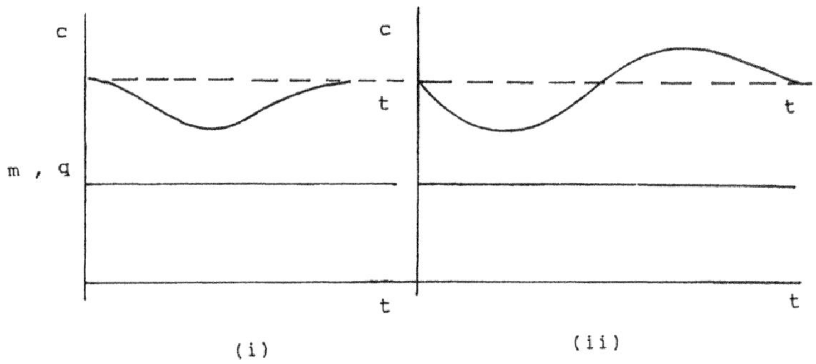

Fig.7.Trial and Error Tuning

which is derived from the system response and using the rules and facts specified in the knowledge base ,the expert system makes a decision to tune the compensator by a certain amount.This decision is communicated to the system simulation (running on FOX-3) via the RS-232 link.The language employed for the expert system is Turbo-Prolog developed by Borland International for the IBM/PC and compatible computers (Borland International Inc., Townsend, 1987).

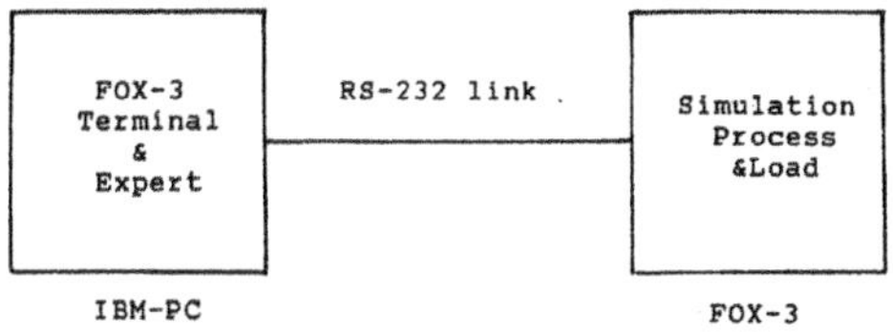

Fig.8 Expert System Approach to Self-tuning Compensator

DCS Approach

In this approach, the feedforward control system and the dynamic compensator as well as the self-tuning feature are implemented in the Foxboro's Intelligent Automation Series (I/A series) workstation (The Foxboro Company,1989).The simulation is as per the plan shown in Fig.9.The I/A series uses the compound concept wherein the different analog functional blocks are emulated.In addition,there is provision for freely mixing continuous blocks , the sequence blocks and the calculator blocks.The sequence blocks allow a high level batch language programme to be written into it.The calculator blocks are useful in computation and in recognising the different patterns shown in Fig.3.Typically, different peaks of the waveform shown and their times of occurences are noted.The independent and the dependent blocks take care of the logistics aspect of the problem.The other continuous blocks simply simulate the feedforward control system of Fig.2.

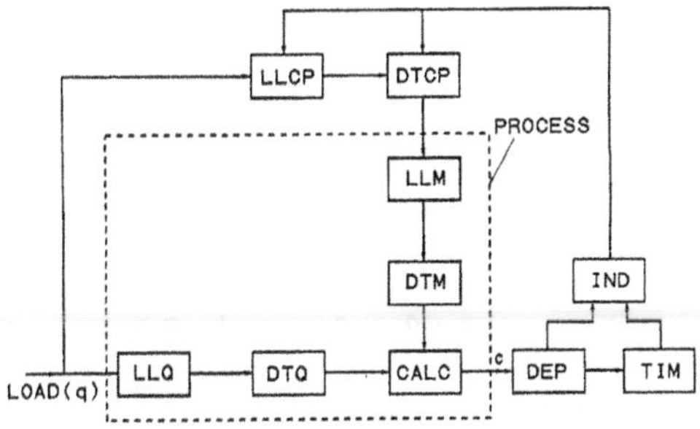

Fig.9 Distributed Control System Approach to Self-tuning Compensator

SIMULATION RESULTS

Dead Time Balanced Systems

Figure 10 shows the results of simulations numbering 1 to 10 conducted using the expert system approach.Table 1 gives the lag values used for the simulations.For all cases considered, as shown in Fig. 10 , the algorithm is such that the compensator parameters TG and T tend to converge to the true values of the lags T_m and T_q respectively.The different lines shown in Fig.10 tend to converge to the point (1,1).The peak error encountered in each case is less than or equal to 1 % for a 10% change in load.

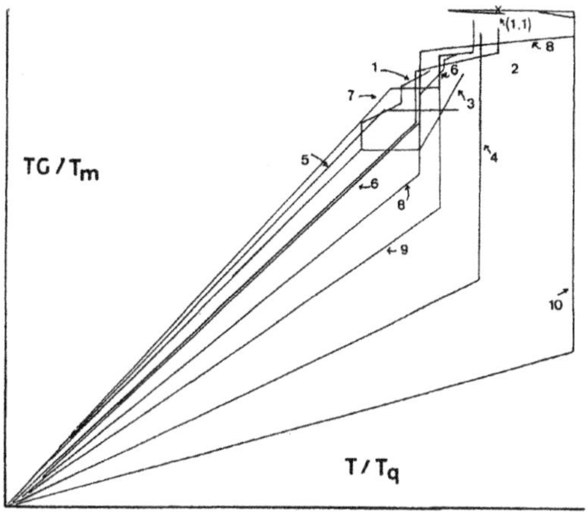

Fig.10.Convergence of Self-tuning Compensator - Dead Time Balanced System

Table 1 System Parameters for Simulation in Fig.10

No.	T_q	T_m	No.	T_q	T_m
1	0.1	0.2	6	0.2	0.5
2	0.1	0.4	7	0.5	0.2
3	0.2	0.1	8	0.1	1.0
4	0.1	4.0	9	0.1	2.0
5	2.0	0.1	10	0.1	10.0

Dead Time Compensation Systems

Figure 11 shows the convergence of the algorithm in the case when dead time compensation is added to the lead-lag compensation.Peak positive (E_1) error and

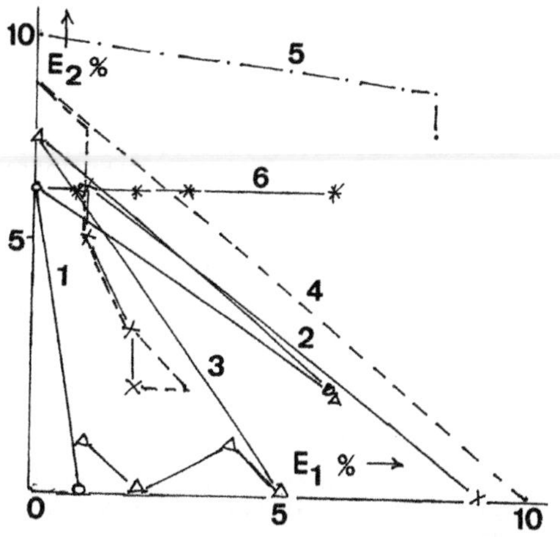

Fig.11 Convergence of Self-tuning Compensator-Dead Time Compensation System

peak negative (E_2) error values are plotted in Fig.11 for the different simulations numbering 1 to 6.The system parameters for the simulations numbering 1 to 6 together with the corresponding case, as in Fig.3, are given in Table 2.Fig.11 shows that in the case of simulations 1 and 3 (case IIc) the peak error converges to the tolerance limit of 1% . For simulations 2 and 4 (case IIa) the limit is reached when the IAE reaches a minimum. In the case of simulations 5 and 6 (cases IIIc and IIIa respectively), the minimum IAE error obtainable is large.The latter is understandible since the dead time in the path of the manipulated variable is too long.

Table 2.System Parameters for Simulations in Fig.11

No.	T_{dm_m}	T_{m_m}	T_{dq_m}	T_{q_m}	T_{d_m}	T_m	G	Case (Fig.3)
1	0	0.5	0.5	0.05	0.48	0.1	6.5	IIc
2	0	0.1	0.5	0.5	-	1.25	-0.1	IIa
3	0.5	1.0	1.5	0.1	0.99	0.12	10.0	IIc
4	0.5	0.1	1.5	1.0	-	3.05	0.01	IIa
5	0.5	1.0	0.2	0.1	-	0.08	43	IIIc
6	1.0	0.1	0	1.0	-	0.03	6.4	IIIa

CONCLUSIONS

This paper has described the development of a self-tuning dynamic feedforward compensator.The algorithm has been verified using computer simulations.It appears that the algorithm works well in all cases except when the dead time in the manipulated variable is too long.The latter case is more a limitation of the control system rather than of the compensator itself.

Acknowledgement

The author would like to thank Nanyang Technological Institute for the facilities provided towards the research reported in this paper.

References

E.H.Bristol, and P.D. Hansen. (1987). Moment Projection Feedforward Control Adaptation. Proc.1987 American Control Conference,pp. 1755-1762.

Borland International Inc. Turbo-Prolog Owner's Handbook,Calif.,U.S.A.

W.L.Luyben and P.S.Buckley,(1977) .A Proportional Lag-Level Controller. InTech.

K.A.MacDonald,T.J.McAvoy and A.Tits(1986). Optimal Averaging Level Control. AIChE J.

M.Morari and E.Zafiriou (1989). Robust Process Control,Prentice Hall,pp.131-135.

A.E.Nisenfeld and R.K.Miyasaki (1979). Applications of Feedforward Control to Distillation Control, Automatica, 9, p.319.

F.G.Shinskey, (1979). Process Control Systems,McGraw-Hill, New York,N.Y.,pp. 178-186.

F.G. Shinskey, (1988). Process Control Systems,McGraw-Hill,New York,N.Y., pp. 269-281.

The Foxboro Co.(1982) FOX-3 System User's Guide.

The Foxboro Company (1989). Integrated Configuration,Intelligent Automation Series (I/A series) Documentation.

C.Townsend (1987). Introduction to Turbo-Prolog.Tech.Publ.Singapore.

Appendix-I

Proof of Theorem 1

The time to peak (T_p) for the response of Eq.(4) is given by

$$T_p = [\ln (T_q / T_m)] / [(1/T_m) - (1/T_q)] \quad (1.1)$$

If we now define T'_m and T'_q as

$$T'_m = T_m / T_p \quad (1.2)$$

$$T'_q = T_q / T_p \quad (1.3)$$

then from Eq.(1.1), it follows that

$$T'_q = (x - 1)/(x \ln x) \quad (1.4)$$

where

$$x = T'_m / T'_q = T_m / T_q \quad (1.5)$$

dc_{up} is then given by

$$dc_{up} = x^{x/(1-x)} - x^{1/(1-x)} \quad (1.6)$$

For perfect dynamic compensation,

$$T = T_q \quad (1.7)$$

and

$$G = x \quad (1.8)$$

From Eq.(1.3),(1.4) and (1.7), Eq.(5) follows immediately.Hence the result.

Appendix-II

Proof of Theorem 2

The uncompensated load response of Fig.6 (case IIIa in Fig.3) is given by

$$dc(t) = e^{-\{(t-T_{dq})u(t-T_{dq})\}/T_q} - e^{-\{(t-T_{dm})u(t-T_{dm})\}/T_m}, t>=0 \quad (2.1)$$

where u(.) is an unit step function. Put

$$t_1 = t - T_{dm} \quad (2.2)$$

Eq.(2.1) becomes for $t_1 >= T_{dm}$ as

$$dc(t_1) = e^{-(t_1+T_d)/T_q} - e^{-t_1/T_m} \quad (2.3)$$

Putting $dc(T_o) = 0$ in Eq.(2.3), we get,

$$x = T_m / T_q = T_o/(T_o + T_d) \quad (2.4)$$

The 'cde' part of the curve of Fig.6 corresponds to the case of differential lags only (the dead times having elapsed) with a common load disturbance .Hence it corresponds to the curve of Fig.4 and Eq.(5) applies where T_p corresponds to the time to peak as shown in Fig.6.Thus Eq.(10) results and hence the result.

MODEL REFERENCE ADAPTIVE CONTROL FOR NON-MINIMUM PHASE SYSTEM BY PERIODIC FEEDBACK

Y. Miyasato

The Institute of Statistical Mathematics, 4-6-7 Minami-Azabu, Minato-ku, Tokyo, Japan 106

Abstract : In most of the studies of model reference adaptive control (MRAC), the controlled systems are confined to minimum phase systems, since the MRAC techniques utilize the control laws involving cancellations of zeros of the systems. In this paper, we propose a design method of model reference adaptive control systems for non-minimum phase systems. The poles and zeros of the controlled systems are relocated by the periodic feedback control laws with multirate sampling, and no cancellation of zeros occurs in our method. It is shown that even if the unstable zeros exist, the output error converges to zero asymptotically, while the control input remains uniformly bounded. Some simulation results for the non-minimum phase system, also show the effectiveness of the proposed method.

Keywords : adaptive control; model reference adaptive control; non-minimum phase system; periodic feedback; multirate sampling.

INTRODUCTION

In most of the studies of model reference adaptive control (MRAC), the controlled systems are confined to minimum phase systems, since the MRAC techniques utilize the control laws involving cancellations of zeros of the systems (Goodwin, Ramadge and Caines,1980; Goodwin and Sin,1984). It makes the scope of application of MRAC too restrictive, for non-minimum phase discrete-time systems can often appear. For example, when continuous-time systems with relative degree greater than two, are sampled at a fast rate, those give rise to non-minimum phase discrete-time systems (Åström, Hagander and Sternby,1984). Hence, the study of MRAC for non-minimum phase systems is of great importance.

In the present paper, we propose a design method of MRAC for non-minimum phase systems. The poles and zeros of the controlled system are relocated by the periodic feedback control with multirate sampling (Chammas and Leondes, 1978a 1978b; Kaczorek, 1985; Hagiwara and Araki, 1988; Ortega and Kreisselmeier, 1990), and no cancellation of zeros occurs in our method. It is shown that even if unstable zeros exist, the output error converges to zero asymptotically, while the control input remains uniformly bounded. Finally, some simulation results for the non-minimum phase system also show the effectiveness of the proposed method.

PROBLEM STATEMENT

We consider a continuous-time, single-input, single-output linear system, called the controlled system, described as follows:

$$y(t) = G(s)u(t), \tag{1}$$

where the degree of $G(s)$ is assumed to be known (denoted as n). The control problem to be solved in this paper is stated as follows: Given an unknown system described by (1), and a known reference signal described by

$$y_M(iT) \quad (i = 0,\ 1,\ 2,\ \cdots), \tag{2}$$

which is defined at the time instant $t = 0,\ T,\ 2T,\ \cdots$, determine a suitable control input $u(t)$ such that the discrete-time asymptotic model-following is achieved, i.e.,

$$y(iT) \rightarrow y_M(iT) \quad (i \rightarrow \infty). \tag{3}$$

Usually, in order to solve that problem, we construct the discrete-time model of (1) based on the sampling method where the input and output of the system are sampled at the same instant (see Figure 1); the discretized model is represented as follows:

$$y(iT) = G_T(z)u(iT), \tag{4}$$

$$G_T(z) = (1 - z^{-1})Z_T L^{-1}\{\frac{G(s)}{s}\}, \tag{5}$$

where L is a Laplace-transform and Z_T is a $z-$transform with the sampling period T.

However, that discrete-time model might have unstable zeros when the relative degree of $G(s)$ is greater than 2 and the sampling period T is small. In that case, the MRAC method which involves cancellations of unstable zeros, cannot be utilized.

In order to avoid those cancellations of zeros, we choose the multirate sampling method (see Figure 2), where the input is discretized with the sampling period $t = T/n$, while the output is discretized with the sampling period $t = T$, and determine $u(t)$ $(u(jT/n) : j = 0, 1, 2, \cdots)$ which achieves discrete-time asymptotic model following (3).

DESCRIPTION OF MULTIRATE SAMPLING SYSTEM

Let (c, A, b) $(A \in \Re^{n\times n},\ b,\ c \in \Re^{n})$ be a discrete-time model based on the usual sampling method with the sampling period $\frac{T}{n}$, that is,

$$c^T(zI - A)^{-1}b = G_{T/n}(z) \tag{6}$$

Then the multirate sampling system (Figure 2) is represented in the followings:

$$\begin{aligned} x_{j+1} &= Ax_j + bu_j, & (7)\\ y_{in} &= c^T x_{in}, & (8)\\ &(j = 0,\ 1,\ 2,\ \cdots;\ i = 0,\ 1,\ 2,\ \cdots), \end{aligned}$$

where

$$\begin{aligned} x_j &\equiv x(\frac{jT}{n}), & (9)\\ u_j &\equiv u(\frac{jT}{n}), & (10)\\ y_{in} &\equiv y(iT). & (11) \end{aligned}$$

With the notations

$$\begin{aligned} X(i) &\equiv x_{in}, & (12)\\ y(i) &\equiv y(iT) = y_{in}, & (13)\\ y_M(i) &\equiv y_M(iT), & (14) \end{aligned}$$

we can obtain the following state-space representations by rearranging (7) and (8).

$$\begin{aligned} X(i+1) &= A^n X(i) + BU(i), & (15)\\ y(i) &= c^T X(i), & (16) \end{aligned}$$

where

$$\begin{aligned} B &\equiv [A^{n-1}b,\ A^{n-2}b,\ \cdots,\ Ab,\ b], & (17)\\ U(i) &\equiv [u_{in},\ u_{in+1},\ \cdots,\ u_{in+n-1}]^T. & (18) \end{aligned}$$

For notational convenience, let

$$\begin{aligned} u_k(i) &\equiv u_{in+k-1} \quad (k = 1,\ 2,\ \cdots,\ n), & (19)\\ U(i) &= [u_1(i),\ u_2(i),\ \cdots,\ u_k(i)]^T, & (20) \end{aligned}$$

then the controlled system (15), (16), (20), is considered a discrete-time system with the $n-$inputs $(u_1(i), \cdots, u_n(i))$, and the single output $(y(i))$.

PERIODIC FEEDBACK

The controlled system (15), (16), (20), can be seen as the $n-$inputs, single output system described by

$$y(i) = A(z^{-1})y(i) + \sum_{k=1}^{n} B_k(z^{-1})u_k(i), \tag{21}$$

where z is a shift operator, such as $zy(i) = y(i+1)$, and

$$\begin{aligned} A(z^{-1}) &\equiv a_1 z^{-1} + \cdots + a_n z^{-n}\\ &= 1 - z^{-n}\det(zI - A^n), & (22)\\ B_k(z^{-1}) &\equiv b_{k1}z^{-1} + \cdots + b_{kn}z^{-n}\\ &= z^{-n}c^T \mathrm{adj}(zI - A^n)A^{n-k}b. & (23) \end{aligned}$$

We consider the periodic feedback control (Chammas and Leondes, 1978a 1978b; Kaczorek, 1985) described by

$$u_k(i) = g_k y(i) + h_k y_M(i+1). \tag{24}$$

From (21) and (24), it follows that the close-loop system becomes

$$\begin{aligned} y(i) &= A(z^{-1})y(i) + \sum_{k=1}^{n} g_k B_k(z^{-1})y(i)\\ &\quad + \sum_{k=1}^{n} h_k B_k(z^{-1})y_M(i+1). & (25) \end{aligned}$$

If there exist the feedback parameters g_k, h_k such that the following equations hold,

$$\begin{aligned} \sum_{k=1}^{n} g_k B_k(z^{-1}) &= -A(z^{-1}), & (26)\\ \sum_{k=1}^{n} h_k B_k(z^{-1}) &= z^{-1}, & (27) \end{aligned}$$

then the discrete-time model following (3) is attained.

The next lemma is concerned with the condition of the existance of $\{g_k\}$ and $\{h_k\}$ satisfying (26) and (27), respectively.

Lemma 1 *Assume that (A, b) is controllable, and that (c, A^n) is observable. Then for any $p(z^{-1})$*

$$p(z^{-1}) = p_1 z^{-1} + \cdots + p_n z^{-n}, \quad (p_i \in \Re) \tag{28}$$

there exists a unique set $\{l_k \in \Re;\ k = 1,\ 2,\ \cdots, n\}$ such that

$$\sum_{k=1}^{n} l_k B_k(z^{-1}) = p(z^{-1}). \tag{29}$$

That is, the following holds under those controllability and observability conditions.

$$\det B_0 \neq 0, \tag{30}$$

where

$$B_0 \equiv \begin{bmatrix} b_{11} & \cdots & b_{n1}\\ \vdots & \vdots & \vdots\\ b_{1n} & \cdots & b_{nn} \end{bmatrix} \quad (\in \Re^{n\times n}). \tag{31}$$

(Proof)
From the assumption that (c, A^n) is observable, it follows that there exist a non-singular matrix T such that

$$\begin{aligned} \overline{A^n} &\equiv T^{-1}A^nT\\ &= \begin{bmatrix} * & 1 & 0 & \cdots & 0\\ * & 0 & 1 & \cdots & 0\\ \vdots & \vdots & \cdots & \vdots & \vdots\\ * & 0 & \cdots & 0 & 1\\ * & 0 & \cdots & 0 & 0 \end{bmatrix}, & (32)\\ \overline{c}^T &\equiv c^T T = [1\,0 \cdots 0]. & (33) \end{aligned}$$

(Observable canonical form)
Using those representations (32) and (33), $B_k(z^{-1})$ is rewritten in the form

$$B_k(z^{-1}) = z^{-n}c^T \mathrm{adj}(zI - A^n)A^{n-k}b$$
$$= [z^{-1}, z^{-1}, \cdots, z^{-n}]T^{-1}A^{n-k}b. \quad (34)$$

Therefore, in order that (29) holds, it is sufficient for l_k $(k = 1, 2, \cdots, n)$ to be chosen for given $p(z^{-1})$ in the following :

$$l = B^{-1}Tp, \quad (35)$$

where

$$l = [l_1, l_2, \cdots, l_n]^T, \quad (36)$$
$$p = [p_1, p_2, \cdots, p_n]^T. \quad (37)$$

Since (A, b) is controllable, it follows that B^{-1} exists and that l is determined uniquely.

(Q.E.D.)

(Remark)
The assumptions in Lemma 1 assert that the controlled system (1) must be observable when it is discretized with the sampling period T, and controllable with the sampling period $\frac{T}{n}$.

The following lemma is given concerned with the observability condition of (c, A^n).

Lemma 2 *A must be cyclic because of the controllability condition of (A, b). For cyclic A, it is shown that*
1) when A has distinct eigenvalues,
(c, A^n) is observable $\Leftrightarrow$ (c, A) is obeservable.
2) otherwise,
if multiple eigenvalues of $A \neq 0$, then
(c, A^n) is observable $\Leftrightarrow$ (c, A) is observable.

The proof is omitted.

ADAPTIVE CONTROL BY PERIODIC FEEDBACK

In the present section, we construct an adaptive control system, where the unknown system parameters $\{a_i\}, \{b_i\}$ are estimated recursively, and the control parameters $\{g_k\}, \{h_k\}$ are adjusted automatically. For convenience' sake, we describe the controlled system in the following regression form.

(Controlled system)

$$y(i) = \theta^T \phi(i-1), \quad (38)$$

where

$$\theta^T \equiv [a_1, \cdots, a_n, b_{11}, \cdots, b_{1n}, \cdots, b_{n1}, \cdots, b_{nn}]^T \quad (39)$$
$$\phi(i-1) \equiv [y(i-1), \cdots, y(i-n), u_1(i-1), \cdots, u_1(i-n), \cdots, u_n(i-1), \cdots, u_n(i-n)]^T. \quad (40)$$

We consider the adaptive identifier, the adaptive law (Goodwin, Ramadge and Caines, 1980; Goodwin and Sin, 1984) and the control law in the followings:

(Adaptive identifier)

$$\hat{y}(i) = \hat{\theta}(i-1)^T \phi(i-1), \quad (41)$$

where $\hat{\theta}(i-1)$ is a current estimate of θ at the time instant $(i-1)$.

(Adaptive law) (method I)

$$\hat{\theta}(i) = \hat{\theta}(i-1) + \alpha(i)\frac{\phi(i-1)}{1 + \| \phi(i-1) \|^2}\epsilon(i) \quad (42)$$
$$\epsilon(i) = y(i) - \hat{y}(i) \quad (43)$$
$$0 < \alpha(i) < 2 \quad (44)$$

(Control law)

$$u_k(i) = \hat{g}_k(i)y(i) + \hat{h}_k y_M(i+1), \quad (45)$$

where

$$\hat{g}(i) = -\hat{B}_0(i)^{-1}\hat{a}(i), \quad (46)$$
$$\hat{h}(i) = \hat{B}_0(i)^{-1}e_1, \quad (47)$$
$$\hat{g}(i) \equiv [\hat{g}_1(i), \cdots, \hat{g}_n(i)]^T, \quad (48)$$
$$\hat{h}(i) \equiv [\hat{h}_1(i), \cdots, \hat{h}_n(i)]^T, \quad (49)$$
$$\hat{a}(i) \equiv [\hat{a}_1(i), \cdots, \hat{a}_N(i)]^T, \quad (50)$$
$$e_1 \equiv [1, 0, \cdots, 0]^T, \quad (51)$$
$$\hat{B}_0(i) \equiv \begin{bmatrix} \hat{b}_{11}(i) & \cdots & \hat{b}_{n1}(i) \\ \vdots & \vdots & \vdots \\ \hat{b}_{1n}(i) & \cdots & \hat{b}_{nn}(i) \end{bmatrix}. \quad (52)$$

The control law is given by replacing in (24), g_k and h_k by their current estimates.

The next theorem is our main result.

Theorem 1 *Consider the contolled system (15), (16),(20), with the adaptive law (42), (43), (44), and the control law (45), (46), (47), (48), (49), (50), (51), (52). Suppose that (c, A^n) is observable and (A, b) is controllable. The adaptive gain $\alpha(i)$ is chosen such that the following holds:*

$$| \det \hat{B}_0(i) | \neq 0. \quad (53)$$

Then, it follows that for any bounded $y_M(i)$, all signals in the resulting control system are uniformly bounded, and that the tracking error converges to zero asymptotically.

$$\lim_{i \to \infty} \{y_M(i) - y(i)\} = 0 \quad (54)$$

(Remark)
This theorem holds even if some of the zeros of the system are unstable (that is, the non-minimum phase system).

(Proof)

The following are derived as the well-known results (Goodwin, Ramadge and Caines, 1980; Goodwin and Sin, 1984):

1) $\lim_{i\to\infty} \frac{\epsilon(i)^2}{1+\|\phi(i-1)\|^2} = 0$

2) $\hat{\theta}(i)$is bounded.

3) $\lim_{i\to\infty} \| \hat{\theta}(i) - \hat{\theta}(i-N) \| = 0 \quad (0 \le N < \infty)$

We can also show the following results using 2), 3) and Eq. (53).

2') $\hat{g}(i)$, $\hat{h}(i)$ are bounded.

3') $\lim_{i\to\infty} \| \hat{g}(i) - \hat{g}(i-N) \| = 0$

$\lim_{i\to\infty} \| \hat{h}(i) - \hat{h}(i-N) \| = 0 \quad (0 \le N < \infty)$

Next, we consider the relationship between the output error $e(t)$ and the identification error $\epsilon(i)$. It is shown that the relationship is described as follows:

$$\begin{aligned}\epsilon(i) &= e(i) \\ &-\sum_{k=1}^{n}\sum_{m=2}^{n} \hat{b}_{kl}(i-1)\{\hat{g}_k(i-m) \\ &-\hat{g}_k(i-1)\}y(i-m) \\ &-\sum_{k=1}^{n}\sum_{m=2}^{n} \hat{b}_{kl}(i-1)\{\hat{h}_k(i-m) \\ &-\hat{h}_k(i-1)\}y_M(i+1-m). \end{aligned} \tag{55}$$

Then, using 2') and 3'), it follows that

$$\begin{aligned}\lim_{i\to\infty} &\frac{\hat{b}_{k1}(i-1)\{\hat{g}_k(i-m) - \hat{g}_k(i-1)\}}{\{1+ \| \phi(i-1) \|^2\}^{1/2}} \cdot \\ &\cdot y(i-m) \\ &= 0, \end{aligned} \tag{56}$$

$$\begin{aligned}\lim_{i\to\infty} &\frac{\hat{b}_{k1}(i-1)\{\hat{h}_k(i-m) - \hat{h}_k(i-1)\}}{\{1+ \| \phi(i-1) \|^2\}^{1/2}} \cdot \\ &\cdot y_M(i+1-m) \\ &= 0, \end{aligned} \tag{57}$$

$$(1 \le k \le n,\ 2 \le m \le n)$$

and we get

$$\lim_{i\to\infty} \frac{e(i)}{\{1+ \| \phi(i-1) \|^2\}^{1/2}} = 0. \tag{58}$$

From the control law (45), and 2'), it follows that

$$| u_k(i) | \le M_1 + M_2 | y(i) |, \quad (0 < M_1, M_2 < \infty) \tag{59}$$

and this implies

$$\| \phi(i-1) \|^2 \le M_3 + M_4 \sup_{m\ge 0}\{e(i-m)^2\}. \tag{60}$$

$$(0 < M_3, M_4 < \infty)$$

Therefore, using Lemma 3-1 (Goodwin, Ramadge and Caines, 1980), we show the uniform boundedness of $\phi(i)$, and the convergence of the output error $e(t)$.

(Q.E.D.)

The adaptive gain $\alpha(i)$ satisfying (53), can be chosen in the following way.

Lemma 3 *We describe the update of $\hat{B}_0(i)$ as follows:*

$$\hat{B}_0(i) = \hat{B}_0(i-1) + \alpha(i)V(i), \tag{61}$$

where $V(i)$ is constructed with the corresponding elements of the second term in (42). Then, in order that $\det \hat{B}_0(i) \ne 0$ $(i \ge 0)$, it is sufficient for $\alpha(i)$ to be chosen in the followings:

i) $\det \hat{B}_0(0) \ne 0$

ii) $0 < \alpha(i) < 2$ *and*

$\frac{1}{\alpha(i)} \ne$ eigenvalues of

$\{-\hat{B}_0(i-1)^{-1}V(i)\}. \quad (i \ge 1)$

(Proof)

The proof is similar to that of Lemma 9-1 (Goodwin, Ramadge, and Caines, 1980). Rewriting (61), we get

$$\hat{B}_0(i) = \hat{B}_0(i-1)\{I + \alpha(i)\hat{B}_0(i-1)^{-1}V(i)\}, \tag{62}$$

and it follows that

$$\begin{aligned}&\det \hat{B}_0(i) \\ &= \det \hat{B}_0(i-1) \det\{I + \alpha(i)\hat{B}_0(i-1)^{-1}V(i)\} \\ &= \det \hat{B}_0(i-1)\{\alpha(i)\}^n \\ &\quad \cdot \det[\{I/\alpha(i)\} + \hat{B}_0(i-1)^{-1}V(i)]. \end{aligned} \tag{63}$$

Then i) and ii) follow immediately.

(Q.E.D.)

We can also choose the least-squares-type adaptive law (Landau, 1981).

Theorem 2 *The same control law is adopted. But the adaptive law is different from (42), (43), and (44).*

(Adaptive law) (method II)

$$\hat{\theta}(i) = \hat{\theta}(i-1) - \frac{P(i-1)\phi(i-1)}{1 + \phi(i-1)^T P(i-1)\phi(i-1)}\epsilon(i) \tag{64}$$

$$\epsilon(i) = y(i) - \hat{y}(i) \tag{65}$$

$$\begin{aligned}P(i) &= \frac{1}{\lambda_1(i)}[P(i-1) \\ &- \frac{\lambda_2(i)P(i-2)\phi(i-1)\phi(i-1)^T P(i-1)}{\lambda_1(i) + \lambda_2(i)\phi(i-1)^T P(i-1)\phi(i-1)}] \end{aligned} \tag{66}$$

$$P(0) = P(0)^T > 0 \tag{67}$$

$$0 < \lambda_1(i) \le 2, \quad 0 \le \lambda_2(i) < 2 \tag{68}$$

The adaptive gains $\lambda_1(i)$, $\lambda_2(i)$ and $\hat{B}_0(0)$ are chosen in the following ways:

i) $\det \hat{B}_0(0) \ne 0$

ii) $\lambda_1(i)$, $\lambda_2(i)$ satisfying (68), are chosen such that

$$\text{eigenvalue of } \{\hat{B}_0(i)^{-1}V(i, \lambda_1(i), \lambda_2(i))\} \ne -1, \tag{69}$$

where $V(\cdot,\cdot,\cdot)$, which is defined by

$$V(i, \lambda_1(i), \lambda_2(i)) = \hat{B}_0(i) - \hat{B}_0(i-1), \quad (70)$$

is constructed with the corresponding elements of the second term in (64). Then, under the same controllability and observability conditions, the resulting control system is uniformly bounded, and the output error converges to zero asymptotically.

The proof is similar to that of Theorem 1, and then omitted here.

SIMULATION RESULTS

A numerical simulation study is performed to show the effectiveness of the proposed method. Let us consider a continuous-time system described as follows:

$$G(s) = \frac{1}{(s+1)^3}. \quad (71)$$

The control problem is to make the sampled output of the system (71), i.e., $y(iT)$, follow the prescribed reference output $y_M(iT)$.

$$y(iT) \rightarrow y_M(iT) \quad (72)$$

The discretized model based on the usual sampling method where the input and output are sampled with the same period T, is given as follows:

$$G_T(z) = \frac{B_{3T}(z)}{\{z - \exp(-T)\}^3}, \quad (73)$$

where

$$B_{3T}(z) \rightarrow \frac{T^3}{6} B_3(z) \ (T \rightarrow 0), \quad (74)$$

$$B_3(z) = z^2 + 4z + 1. \quad (75)$$

The roots of $B_3(z) = 0$ are called limiting zeros (Åström, Hagander and Sternby,1983). In this case, one unstable zero occurs for sufficiently small T. The region of T where one unstable zero exists, is written as follows:

$$0 \leq T < 1.8399 \quad (76)$$

The sampling period T is chosen as follows:

$$T = 1. \quad (77)$$

Then, the poles and zeros of the resulting discrete-time system become

poles : 0.367879 (stable)

zeros : −0.123776 (stable), −1.79896 (unstable).

In our method, the sampling period of the input is $T/3 (= 1/3)$, while the output is sampled with the period $T = 1$. The reference signal, and design parameters are chosen as follows:

$$y_M(i) = \sin(2\pi i/50), \quad (78)$$

$$\hat{B}_0(0) = I \ (\in \Re^{3\times 3}), \quad (79)$$

$$\lambda_1(i) = \lambda_2(i) = 1 \ \text{(method II)}, \quad (80)$$

$$P(0) = 10^6 \ \text{(method II)}. \quad (81)$$

Figure 3 shows the result where the method II (Theorem 2) is used. On the other hand, the usual MRAC method which involves cancellations of zeros, could not achieve discrete-time model-following (72) in this example.

CONCLUSION

In the present paper, we propose a design method for constructing discrete-time model reference adaptive control systems for non-minimum phase systems, where the poles and zeros are relocated by periodic feedback control with multirate sampling, and no cancellation of zeros ocuurs. Some simulation study for non-minimum phase systems, also show the effectiveness of the proposed method.

References

[1] Goodwin, G.C., Ramadge, P.J. and Caines, P.E. (1980). Discrete-time multivariable adaptive control. *I.E.E.E. Trans. Autom. Control,* **25**, 449-456.

[2] Goodwin, G.C. and Sin, K.S.(1984). *Adaptive Filtering, Prediction, and Control.* Prentice-Hall, New Jersey.

[3] Åström, K.J., Hagander, P. and Sternby, J.(1983). Zeros of sampled systems, *Automatica,* **20**, 31-38.

[4] Chammas, A.B. and Leondes, C.T.(1978a). On the design of linear time invariant systems by periodic output feedback Part I. Discrete-time pole assignment. *Int. J. Control,* **27**, 885-894.

[5] Chammas, A.B. and Leondes, C.T.(1978b). On the design of linear time invariant systems by periodic output feedback Part II. Output feedback controllability. *Int. J. Control,* **27**, 895-903.

[6] Kaczorek, T.(1985). *Pole placement for linear discrete-time systems by periodic output feedbacks.* Systems & Control Letters, **6**, 267-269.

[7] Hagiwara, T. and Araki, M.(1988). Design of stable state feedback controller based on the multirate sampling of the plant output. *I.E.E.E. Trans. Autom. Control,* **33**, 812-819.

[8] Ortega, R. and Kreisselmeier, G.(1990). Discrete-time model reference adaptive control for continuous-time systems using generalized sampled-data hold functions. *I.E.E.E. Trans. Autom. Control,* **35**, 334-338.

[9] Landau, Y.D.(1979) Adaptive Control, The Model Reference Approach. Marcel Dekker, New York.

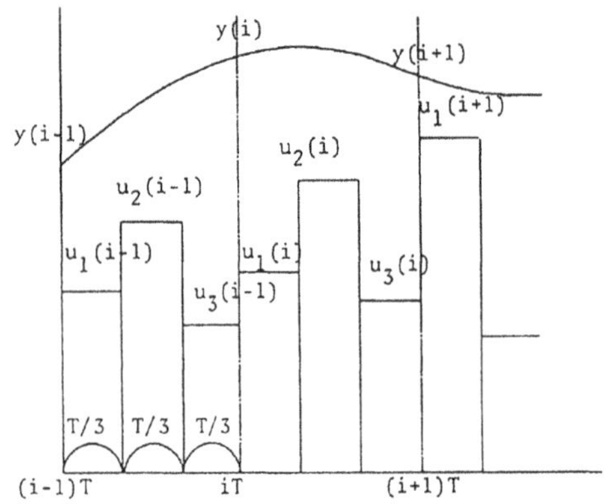

Fig. 2 Multirate sampling (n=3)

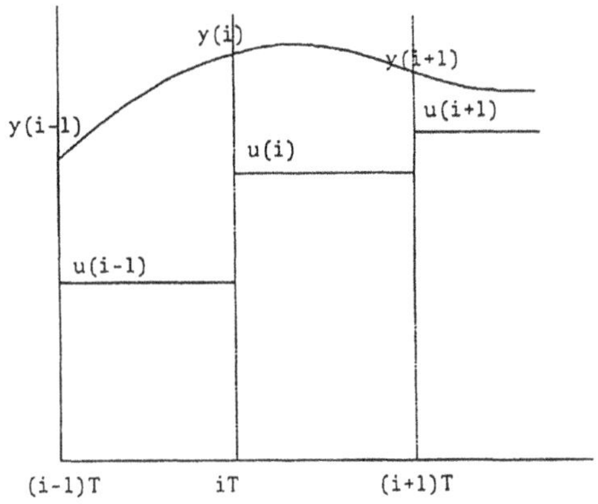

Fig. 1 Usual sampling

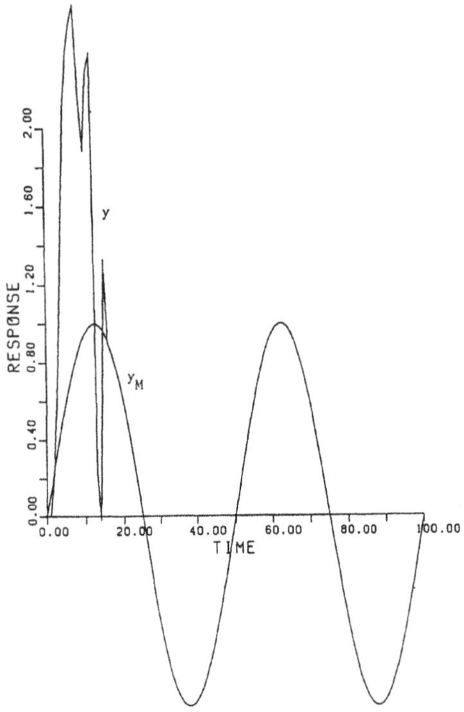

Fig. 3-1 Simulation results (method II) Outputs

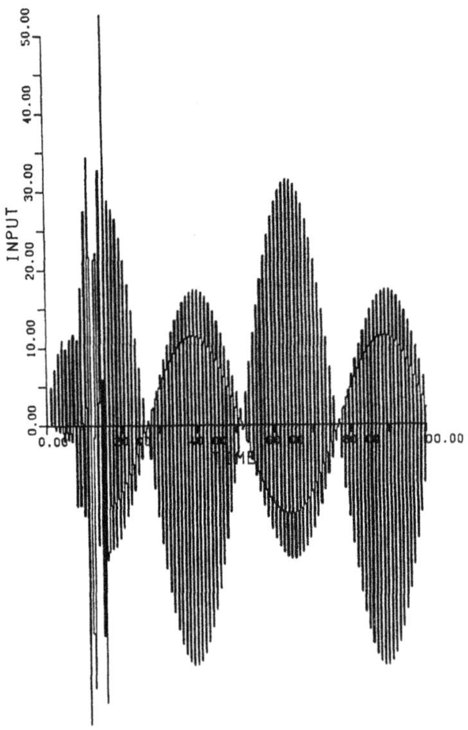

Fig. 3-2 Simulation results (method II) Inputs

SMART SYNTHESIS OF A PID CONTROLLER

S. E. Sallé* and K. J. Åström**

**Laboratoire d'Automatique de Grenoble, B.P.n° 46, F-38 402 Saint-Martin d'Hères, France*
***Dept. of Automatic Control, Lund Institute of Technology, P.O. Box 118, S-221 00 Lund, Sweden*

Abstract. A new auto-tuner for PID controllers is described. It combines two commonly taken approaches : analysis of process transient responses and estimation of the process critical point through a relay feedback experiment. The system selects the most appropriate controller among a PI, a PID or a PI regulator coupled with a Smith predictor. The controller parameters are based on estimation of the normalised process dead-time and modified Ziegler-Nichols rules. The system is implemented on a Sun workstation and uses a real-time expert system developed with Muse. The paper describes the methods used and gives simulation results.

Key Words. PID control; normalised dead-time; relay feedback; intelligent control, real-time expert system.

1. INTRODUCTION

Despite many advances in control theory, most industrial control loops are still based on PID controllers. Significant efforts have lately been devoted to the automatic tuning of such regulators ; a discussion on this subject can be found e.g. in [*Åström & Hägglund 1988*]. Two ideas are commonly employed for automatic tuning. One approach is based on an open-loop or closed-loop transient response analysis, as in the Foxboro "Exact" controller [*Bristol 1977*]. A second approach is based on analysis of the process response under relay feedback as in the Satt Control controller ECA 40. Both systems use traditional tuning methods of PID controllers in the Ziegler-Nichols spirit [*Ziegler & Nichols 1942*].

The two approaches are combined in this paper. A relay feedback experiment gives an estimation of the process critical point and allows the design of a crude PI controller. Then a pattern recognition is performed for a closed-loop and an open-loop step response. The results of the experiments provide a rough model of the process: dead-time, apparent time-constant, static gain, order (first order or not). Then the controller parameters are selected based on the refined Ziegler-Nichols tuning formula presented in [*Hang & Åström 1988*].

The paper is organised as follows: The statement of the problem is first given. Experiences of many different methods tested, modified and used are described in section 3 and simulations are presented in section 4. Comparison of the proposed controller with two other controllers is presented in section 4 before the conclusions in section 5.

2. PROBLEM DESCRIPTION

A description of a new generation of PID auto-tuners is given in [*Åström et al 1989*]. These systems have reasoning capabilities which help them to "smartly" select the parameters of a controller. Based on knowledge of the normalised process dead-time θ (which is defined as the ratio of the process apparent time-delay L over the process apparent time constant T as shown in figure 2.1, i.e. $\theta = L/T$), the difficulty of controlling the treated process is estimated and some heuristics rules allow to select among a PI, a PID or a PI controller coupled with a Smith predictor. In [*Hang & Åström 1988*], this normalised dead-time is also used in order to refine the Ziegler-Nichols tuning formula. To build such an auto-tuner, the main problem is to estimate the normalised dead-time. Several ways to do this are discussed.

Within this study, the process is assumed to have one input and one output, to be linear and stable (no integrator), to have a global monotone step response except possibly at the beginning and to give stable oscillations under relay-feedback. The class of systems considered is roughly similar to the one considered in the classical works on Ziegler-Nichols tuning. Such systems can be characterised by three parameters : static gain Kp, apparent dead-time L and apparent time constant T (cf figure 2.1).

The approach used in this paper consists of determining the critical point and the parameters Kp, L and T in an autonomous manner. requiring very little a-priori knowledge of the process. These data are then used with a refined Ziegler-Nichols method.

The operator is asked the range of the process rise-time, the magnitude order of the static process gain, and the minimum and maximum values of the process input. The approximate value of the process rise-time is necessary to check the steady state of the process (which is assumed to be stable) before performing the relay experiment. In commercial controllers, the static process gain is often of the order of one. It is one task of the process engineer to set the static gain of the entire process (actuator + process + sensor) in the range of 0.2 and 5. Some a-priori settings of the system assume the same hypothesis. If this hypothesis is not true, a rough estimation is required in order to reschedule the processed data (process input, output and reference). An anti-windup device for the integral part of the developed controller is included. It requires knowledge of the maximal values of the process input.

After this preliminary stage, the operator brings the process manually to the desired operating point and three experiments are performed. After the steady state of the process is checked, a relay experiment is performed. During the first two oscillations, the relay parameters can be changed in order to get a significant variation of the process output of reasonable size, typically three times the noise level of the process output. After this phase, the period and magnitude of the oscillations are measured to get an idea of the process critical point. At the same time, estimations are made of the maximum process delay and of first order and second order models of the process, based on analysis of the wave form. A PI controller, based on the Ziegler-Nichols tuning formula, is designed with the obtained information. This regulator will be used later as a safety regulator.

Then a closed loop step response is performed and a precise estimation of the static process gain is obtained. This experiment is followed by an open loop step response which gives a first order model with time delay of the process. In case of too large process errors, the safety PI is switched on. The experiment is stopped when the process has reached 63% of the final value. The PI controller is then used to bring the process rapidly into steady state at the operating point initially fixed by the operator. Different treatments are also performed during the step responses. They are described in the third section.

System Architecture

The system consist of procedural algorithms written in C supervised by a knowledge based system. Such an architecture is described in [*Åström et al 1986*] or in [*Årzén 1987*]. The algorithms are monitored by an expert system developed with Muse (a development tool for real time expert systems which is briefly described in [*Sallé 1989*]). The use of an expert system gives a much cleaner auto-tuner implementation than a procedural language would give, mainly because of the clear separation of algorithms and logic. Another benefit due to the modularity of the expert system is the ease of changing or incrementing the associated logic. The real-time features of Muse are also helpful. The two different computer-processes are linked through a Unix socket. Muse supports this communication.

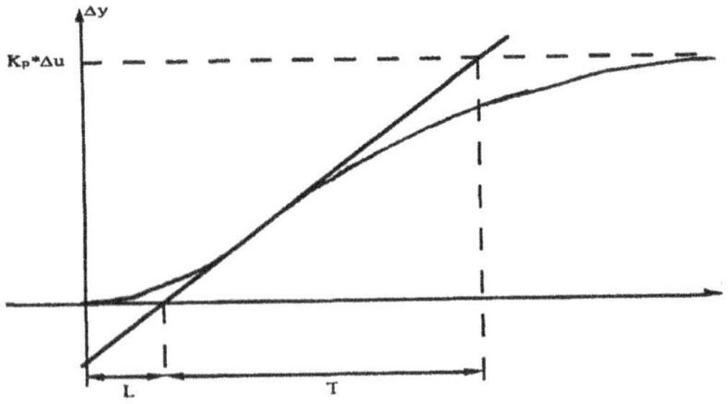

Figure 2.1 Typical step response of considered processes.

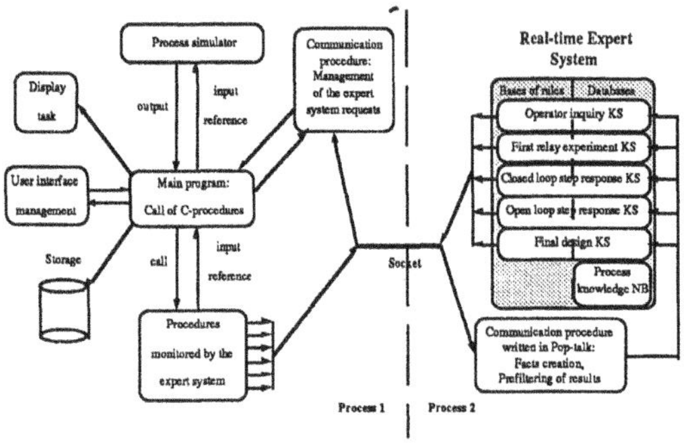

Figure 2.2 System architecture.

The Expert System

Muse from Cambridge Consultants in U.K. is a toolkit for embedded real-time artificial intelligence. It consists of an integrated environment for knowledge representation with an object structure and databases. The Pop-talk language is the central component of this package. Pop-talk is implemented in C and is derived from the Pop series of languages. It has been extended to support object-oriented programming. It also provides a stack-based language that combines list-processing elements with a block-structured syntax as in C or Pascal. On the top of the basic object language, a frame (or schema) system is built which includes multiple inheritance, methods, relations and demons.

Muse applications range from a simple expert system with a single database and rule-set to a complete blackboard system with many knowledge sources (KS) and databases (DB) which co-operate to solve the problem. This possibility is used in the developed system since the set of developed rules is split into five knowledge sources. Each knowledge source is associated with a precise task : Operator interaction, performing the relay experiment, performing the closed-loop and the open-loop step response, computing and testing the final controller. All the knowledge acquired is stored in a notice board.

The present version of the expert system comprises 70 rules. Only one knowledge source is active at a time. Its activation is done in a defined order. This is implemented using objects which have a slot containing the present level of reasoning. The task of each KS consists of bringing the system from one state to another state. To do this, each KS fires the desired procedures, performed tests on the process and elaborates conclusions. Intermediate facts are created in the DB of the active KS, but the final conclusions and results obtained by this KS are stored in the separate notice board. We think that it is a good way to structure the system. For example, to restart a task, the affected KS can be reset without destroying the knowledge already acquired on the process.

Procedure Libraries

The procedures monitored by the expert system are written in C. They include a PID regulator, digital filters, an implementation of a relay with hysteresis with automatic adjustment of its amplitude and hysteresis, relay oscillation analysis, statistic computation routines, pattern recognition routines, etc. These procedures also incorporate a process simulator, a user interface and display tasks. The expert system gives initial parameters to the procedures, it receives and stores the obtained results. The main procedures will be further described in section 3.

Each request made by the expert system is associated with a procedure. A request, sent by the expert system on the socket, is characterised by the first part of the message. The socket is polled and its content is read and analysed through an "if ... else if ..." structure in the communication procedure. The parameters of the request are read and the associated procedure is initialised. The main programme calls the procedure until it is completed or if Muse asks it to be cancelled.

3. METHODS

The different methods used in the system will now be described. Their concepts are shortly explained. It is shown how they are implemented. It is also attempted to point out their advantages and disadvantages based on our experience with the system.

Testing For Steady State

Before performing an experiment, the system checks if the process is in steady state. This is done by investigating if the process output is close to a constant and if the variations around this value are almost constant. The precondition for the test is knowledge about the time scale of the process. This is based either on the process rise time Tr, as given by the operator, or the relay oscillation period Tu. Three time intervals of length Tr/3 or Tu are used in the test. The mean value and the standard deviation of the measured output and the control error are estimated on the first and on the last interval. Depending on the desired confidence, the variation of the mean (and standard deviation) estimates must be less than 5% or 10% (10% or 20%). The mean value of the error must also be less than the deviation in the measured variable and the process error must be approximately zero. The thresholds are set to allow very small measurement noise (typically less than 0.01%). The permissible variations of the standard deviation estimates are larger than the ones allowed for the mean values due to the estimators' characteristics. The measurement noise is determined as the difference of the maximum and minimum values of the process output. These simple tests have been found, in the authors' experiments, very effective to determine if the process output is in steady state.

Relay Tuning

The idea of relay tuning is to introduce a relay in the feedback loop. For a large class of processes, there will be a limit cycle oscillation. The amplitude and the period of the oscillations give information about the process dynamics that can be used to compute the appropriate controller parameters. The idea of using a relay for tuning purposes is described in details in [*Aström & Hägglund 1988*]. The process information obtained from a relay experiment is essentially knowledge about one point on the open-loop Nyquist curve of the process. In many cases (if the system has a phase lag of at least π at high frequencies), this point is at or close to the intersection of the Nyquist curve with the negative real axis. This point is traditionally described in terms of ultimate gain Ku, and ultimate period Tu. Ziegler and Nichols gave a method to determine PID parameters based on knowledge of this critical point.

When a limit cycle oscillation is established, the output is a periodic signal with period Tue which is close to Tu. If d represents the relay amplitude and ε the relay hysteresis, it follows from a Fourier series expansion that the first harmonic of the relay output has the amplitude $4d/\pi$. If the amplitude (maximum value minus minimum value) of the relay input (which is the control error) is 2a, the ultimate gain is thus approximately given by:

$$K_u \approx \frac{4\,d}{\pi\,a} \quad \text{since } a \approx \frac{4\,d}{\pi}\left|G(i\omega_u)\right| \text{ and } K_u = \frac{1}{\left|G(i\omega_u)\right|}.$$

The advantages of this method are:

* it requires little a-priori knowledge of the process,
* the estimation method is easy to implement since it is based on counting and comparisons only,
* it is easy to control the oscillation amplitude by an appropriate feedback choice of the relay characteristics d and ε,
* it is robust to measurement noise and the use of a well chosen hysteresis improves its robustness.

Three parameters associated with the relay experiment must be set : the relay amplitude, the relay hysteresis and the desired amplitude of the output oscillations. The value of the relay hysteresis is determined from the measurement noise : more noise gives a larger hysteresis in order to avoid erroneous relay switching. A minimal value of the hysteresis is prescribed to avoid problems with fast relay oscillations as e.g. with first order processes without time delay. The value of the desired oscillation amplitude is also determined from the noise level. As a protection against processes with very high static gain and short time delay, during the first half period of the oscillation, the relay amplitude is increased linearly from zero until either it reaches a default amplitude (which is presently set to 0.2), or the error signal exceeds the desired amplitude.

The amplitude d and the hysteresis ε may be adjusted separately to obtain the desired oscillation amplitude. Notice that an analysis based on harmonic balance gives:

$$\arg\left[G(i\omega_{uc})\right] \approx -\pi + \arcsin\left(\frac{\varepsilon}{a}\right)$$

The ratio ε/a determines how far from the negative real axis the estimated point is. A possible adjustment law consists in fixing the desired phase lag and to adjust ε. It is not presently done.

It typically takes from 4 to 7 half periods to reach a steady state limit cycle oscillation. The actual number depends on whether the relay parameters are adjusted or not. The oscillation is said to be stable when the relative variation between the last two peak-to-peak amplitudes is less than 10%. The oscillation analysis takes two periods. During the first period, the peak amplitudes and the oscillation half periods are measured. During the second period, regularly spaced points are selected on the oscillation curve in order to analyse the wave form (this is explained in the next section).

Wave Form Analysis

The shape of the oscillation under relay feedback may be used to estimate a process model [*Åström & Hägglund 1988*]. The idea of this method is that the process input and output are periodic signals. Regularly spaced samples $y_0, y_1, \ldots, y_{2n-1}$ are chosen as shown in figure 3.2. Based on Z-transform properties, it can be shown that the coefficient of the standard input/output model $A(q)\ y(k) = B(q)\ u(k)$ can be determined from the equation:

$$A(z)\ D(z) + z^r\ B(z)\ E(z) = z^r\ (z^n + 1)\ Q(z)$$

where $r = \deg A(q) - \deg B(q)$, d is the relay amplitude and

$$E(z) = z^n + z^{n-1} + \ldots + z,$$

$$D(z) = (y_r\ z^n + y_{r+1}\ z^{n-1} + \ldots + y_{r+n-1}\ z)\ .\ d$$

The polynomial Q corresponds to initial conditions which give a steady state periodic output. For example, with $n = 3$, a model $G(s) = K_p\ e^{-s\ L} / (1 + s\ T)$ can be determined. The calculations and formula can be found in [*Åström & Hägglund 1988*].

Remarks: This method works very well for a first order process if there is **little process noise**. In discrete time implementations, it is also important that the times associated with the 2n regularly spaced samples correspond exactly to a measured sample, unless the sampling period is very small compared to the process oscillation period. If this is not the case, the errors of the estimated continuous time parameters may be very large (from 10% up to 50%). The method works well for thermal processes where the noise level is small and because such processes are well approximated by first order dynamics with time delays. The method does not work well at high noise levels. To get good estimates in such a case, it is necessary to increase the relay amplitude. The determination of high order models has proven to be difficult. With n equal to five, it was found to be very difficult to determine second order models.

Minimum phase systems with a relative degree of one is an interesting class of systems that can be successfully approximated by a first order system. A controller can be designed using special techniques such as a constant high gain proportional feedback. In theory, such processes can be detected by examining the value of the process output derivative at the relay oscillations extreme point. Such an analysis is performed in the system. The slopes before and after the extreme point are estimated based on 3, 5 or 7 points of the curve (depending on the noise level). The mean values of the slopes computed on each extreme are sent to the expert system. Based on the amplitude of the variation of these values and on the comparisons with the other results, the expert system deduces if the system belongs to this class. If it is the case, special rules for synthesising the controller are used.

The mean value of the times required to reach the extreme is also sent to the expert system. It is used to estimate the maximum process time delay.

Method of Moments

This method is described in [*Åström & Hägglund 1988*]. It consists of estimating a process model from the values of the transfer function and its derivatives at $\omega = 0$. If the process model is:

$$G(s) = \frac{K_p e^{-sL} Q(s)}{P(s)} = \frac{K_p e^{-sL} \prod_{i=1}^{q} (1 + q_i s)^{m_i}}{\prod_{i=1}^{p} (1 + p_i s)^{n_i}},$$

an expression of $G^{(n)}(0)$ can be established [*Sallé 1990*] and:

$$G(0) = K_p,\ \frac{dG}{ds}(0) = -\ G(0) * \left[L + \sum_{i=1}^{p} n_i p_i - \sum_{i=1}^{q} m_i q_i\right], \text{ etc} \ldots$$

The value of the transfer function and its derivate at $s = 0$ can be computed from the following equations:

$$Y(0) = G(0)*U(0),\ \frac{dY}{ds}(0) = \frac{dG}{ds}(0)*U(0) + G(0)*\frac{dU}{ds}(0), \text{ etc} \ldots$$

and from the following integrals:

$$U(0) = \int_0^\infty u(t)\,dt,\ U^{(n)}(0) = \int_0^\infty t^n u(t)\,dt \text{ and } Y(0) = \int_0^\infty y(t)\,dt,\ Y^{(n)}(0) = \int_0^\infty t^n y(t)\,dt$$

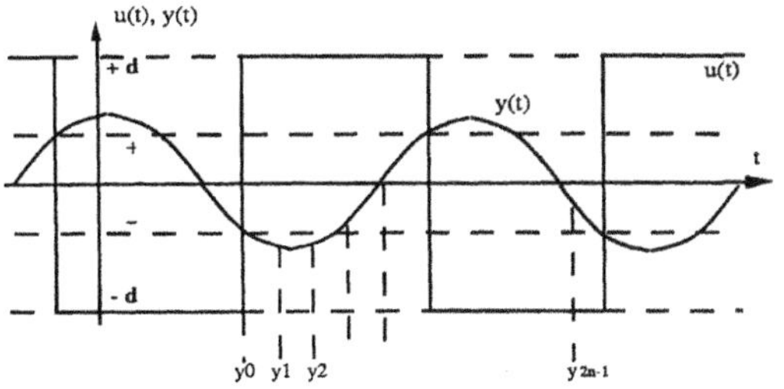

Figure 3.1 Input and output given by a process under relay feedback.

This method can be used with arbitrary input signals. However, for the computations of the integrals to be truncated in time, the input and the output level must be the same at the beginning and at the end of the experiment.

This method works perfectly in theory. It should be robust to noise since it is based on computations of integrals. However, the estimation of a simple first order model with time delay requires the computation of G"(0) and Y"(0). Due to the t^2 term, these computations led to numerical problems for long transients. Due to errors in estimation of G"(0), the results are poor even at low noise levels. We have had good experience in estimating process gain Kp and the sum T + L for first order systems. Combined with the estimates of Ku and Tu from the relay experiment, this gives a good method for determining the parameters Kp, T and L of a first order system. This method is used during the relay experiment and the two step responses. The results are sent to the expert system and their accuracies are estimated based on the estimated measurement noise and on the value of the input integral.

Pattern Recognition

Information about the process can also be deduced from pattern recognition of transient responses. In this way, it is possible to determine the static process gain and the dead-time from a closed loop step response. It is also possible to detect a non-minimum phase system. This is done by checking that the closed loop system is in steady state. A change in the set point is then performed. When the process is in steady state, the static gain is evaluated as the ratio of the variation of the process output mean values and the variation of the process input mean values. The difference between the first significant variation of the input and the first significant variation of the output gives a rough estimate of the process dead-time.

When the estimation of the static process gain is performed, the PI is turned off and an open loop step response is performed. The amplitude of the process input step is computed in order to get the same final value of the process output as before the closed loop step response. A maximum error limit is set based on estimation of measurement noise. As soon as the error is greater than this value, the PI controller is turned on and the experiment is stopped. It is restarted when the process is considered again to be in steady state. A rough estimate of the process dead-time is computed and a non-minimum phase characteristic is detected. The inflection point of the step response is determined as the point with maximum slope. This slope is estimated based on 3, 5 or 7 points of the process output curve, depending on the noise level. This experiment is finished when the output has reached 63% of its final value and the maximum slope has been reached several points before. The PI controller is then switched on to bring the system to steady state at the desired level. The apparent dead-time (cf figure 2.1) is estimated and two estimates of the apparent time constant are computed. One is based on the slope computation (cf figure 2.1) and the other on the time it takes to reach 63% of the final value. Notice that this operation is quite safe since the process output is confined to a predetermined band when the loop is opened.

Special calculations are done to estimate a precise discrete time first order model of the process. The step response of such a process has no inflection point since the slope is maximal at the beginning of the response. Since we are working in discrete time and using more than two points for the estimation of the slope, a correction must therefore be done [*Sallé 1990*].

An Asymmetrical Relay

The relay experiment essentially gives information about one point on the Nyquist curve. This is equivalent to two parameters Ku and Tu. It is possible to determine a reasonable PID controller based on this implementation. To obtain a fine tuned PID controller, it is however useful to know a third parameter e.g. the static process gain. This can be determined from open or closed loop step tests as it has been discussed. It would, however, be highly desirable to determine all parameters from a single experiment. This can be

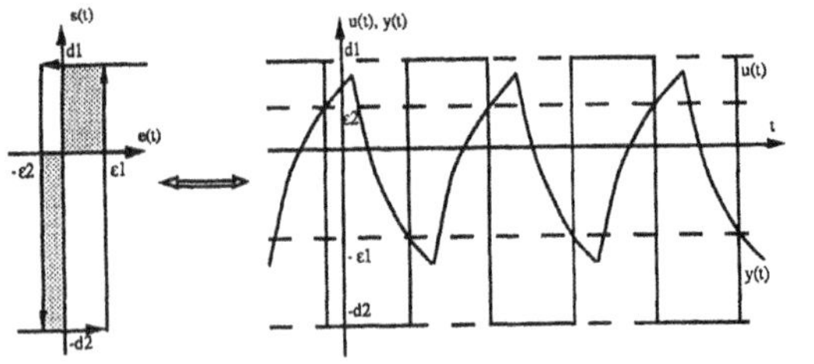

Figure 3.2 An asymmetrical relay characteristic function and an associated time response.

obtained by using an asymmetrical relay as the one shown in figure 3.2. Depending on the placement of the origin in the rectangle of the relay characteristic function, three main types of asymmetrical relays can be defined. The first kind of asymmetrical relay consists of keeping the vertical axis in the middle of the rectangle and shifting the horizontal axis towards the negative or positive values. This kind of relay is equivalent to taking a symmetric relay and to adding a bias to the relay input. Another type of asymmetrical relay consist of shifting both vertical and horizontal axes in the same direction. The last type of asymmetrical relay (which is displayed in figure 3.2) consists of shifting both vertical and horizontal axes in the two opposite directions. This last type has been chosen since, when tested with several processes, it gave the most symmetric (in time) oscillations and the biggest absolute value of the input integral (this is a necessary condition to rely on the results given by the method of moments). The first type gives a small value of the input integral and the second type gives the most asymmetrical (in time) oscillations. The usefulness of using two different hysteresis is shown in [*Sallé 1990*] in the case of a first order process.

The static gain may be obtained either as the ratio of the mean values of input and output or with the method of moments. An analysis based on harmonic balance gives an estimation of the ultimate gain. The equations used in the developed system are fully described in [*Sallé 1990*]. Experiments with many different processes showed that the estimates of the critical point obtained with an asymmetrical relay were quite similar to those obtained with a symmetric relay. Moreover, when the measurement noise is not too large, the estimate of the static gain is accurate. A drawback of this method is that it requires symmetric oscillations if a normal relay is used. Should it not be the case due to, for instance, a nonlinear process, the previous reasoning is erroneous. In [*Hang & Åström 1987*], it is shown how the standard experiment may be modified to obtain symmetrical oscillations.

The method presented here may be employed when a process identification is performed by using a pseudo random binary sequence. Usually, this sequence is centred around zero. If its mean value differs from zero and if the final output is close to the initial output, a precise estimation of the static process gain may be obtained by the method of moments.

Detection of Outliers

Outliers are particularly detrimental for controllers with derivative action. A simple scheme to detect outliers is therefore introduced. It just consist of detecting measurements that are incompatible with the next and previous measurement values. If at time t a process output is larger than expected (e.g. the output at time t-1 plus two or three times the measurement noise level), the output at time t+1 is compared to the output at time t-1. If these data are almost equal (e.g. their difference is smaller than the measurement noise level), a false measure is considered to be detected and the process input at time t+1 is set to the opposite of the input at time t. This simple device is very effective as it will be shown in section 4. It is especially true if the process has higher order dynamics and a rather long dead-time.

Combination of the Different Approaches

The expert system gathers all the results obtained from the C procedures.It analyzes them to find a consistent process model.

The two models obtained by the method of moments are evaluated. The static process gain is first estimated by comparing the available estimations. The estimate given by the closed loop step response normally gives the best accuracy and emphasis is put on it. Depending on the experimental conditions, an estimation of $G'(0)$ and $G''(0)$ may be also obtained. These data give an idea about the process time delay and the sum of the process time constants.

The process time delay is then estimated as the time when the output starts to change after a variation in the process input or as the apparent time delay (cf figure 2.1). The latter may be larger particularly if the process is non-minimum phase. Such a feature may be detected at this stage.

Based on the slope analysis of the relay oscillations, on the comparison of the different models, and on their estimated time delays, the process is classified as being well approximated by a first order process (type 1) or not well approximated (type 2). The process time constant is then estimated, depending on the type of the process. The estimate of the ultimate gain Kue is recomputed in the case of type 1 processes in order to reduce the committed error on this data (which may be bigger than 50%).

An estimate of the normalised process dead-time θ and the normalised process gain κ (which is defined as Kp*Ku) is then obtained. These two parameters are dimensionless and experiments have shown that processes with a small θ or a large κ are easy to control and processes with a large θ or a small κ are difficult to control [*Åström et al 1989*].

The final controller follows the recommendations given in [*Åström et al 1989*]. Set point weighting is introduced to reduce the overshoot. A derivative part is added to the controller when the noise level is said to be low and when θ is smaller than one.The parameters of the controller are based on the refined Ziegler-Nichols tuning formula stated in [*Hang & Åström 1988*].

Computation of the Regulator Parameters

Depending on the value of the normalised process dead-time θ, four cases are considered for the synthesis of the controller. The used formula are given without any explanations. The interested readers must refer to the relevant references in order to get further details. T_{ue} represents the relay oscillation period, K_{ue} the estimate of the ultimate process gain. The factor N introduced in order to filter the derivative action is arbitrarily set to 8. The set point weighting consists of a ß factor introduced before the reference signal in the proportional term of the controller.

$\theta \le 0.15$:

In this case, the process is declared to be easy to control and rather well approximated by a first order process. Since Ziegler-Nichols tuning may not give the best results in this case, a new way of designing a PI controller is tested. This design tries to take profit of the available power in the following way. In the case of a first order process (with the transfer function Kp/(1 + T*p)) coupled with a PI controller (with the transfer function Kr + 1/(Ti*p)), the poles of the closed loop transfer function can be fixed such that they correspond to a relative damping of 0.707 and a natural frequency ω_n equal to n/T rd/s. The factor n measures how fast the closed loop is. Some calculations leads to:

$$Kr = \frac{n\sqrt{2} - 1}{Kp} \text{ and } Ti = T\frac{n\sqrt{2} - 1}{n^2}.$$

The settings we chose, are that a set point change of an amplitude equal to half the peak-to-peak amplitude of the relay oscillations, creates an immediate, and often maximum change, in the process input equal to 70% of the available power. This last quantity is estimated as the difference between the present process input and the closest input extreme value. These requirements give a value of Kr since it may be assumed that, in case of a set point change, the initial variation of the control signal is mainly due to the proportional part of the controller. From knowledge of Kp and Kr, the factor n is deduced. Its value is then restricted to the interval [0.5 ; 3]. Eventually, Ti is computed since both T and n are known. The factor ß is set to one.

A derivative part is added if the estimated noise level is said to be low (smaller than 0.005). The time derivative follows the recommendation given in [*Åström et al 1989*]: Td = Ti/8.

$0.15 < \theta \le 0.6$:

This is the prime application area for PID controllers with Ziegler-Nichols tuning.

If the noise level is smaller than 0.01, a PID controller is used and its parameters are given by the well known relations:

Kr = 0.6*Kue, Ti = 0.5*Tue and Td = Ti/4.

If the noise level is greater than 0.01, a PI controller with the following parameters is used:

If $\theta \le 0.4$ then $\mu = -1.2*\theta*\theta + 1.3*\theta - 0.11$

else $\mu = -0.15*\theta*\theta + 0.33*\theta + 0.11$.

If $\theta \le 0.4$ then $\Omega = 0.16*\theta^{-0.87}$ else $\Omega = 0.25*\theta^{-0.37}$.

Kr = μ*Kue, Ti = Ω*Tue and Td = 0.

ß is set to $0.1+5*\theta/3$ if $\theta \le 0.3$ and to $0.3+\theta$ otherwise.

$0.6 < \theta \le 1.0$:

Ziegler-Nichols tuning becomes less useful. It is recommended to introduce some dead-time and possibly feedforward compensation devices.

If the noise level is smaller than 0.01, a PID controller is used:

Kr = 0.6*Kue, Ti = 0.5*(1.5 - 0.8*θ)*Tue and Td = Tue/8.

If the noise level is greater than 0.01, a PI controller with the following parameters is used:

$\mu = -0.15*\theta*\theta + 0.33*\theta + 0.11$, $\Omega = 0.25*\theta^{-0.37}$.

Kr = μ*Kue

Ti = Ω*Tue and Td = 0

The ß factor is set to $1.6 - \theta$ if $\theta \le 0.8$ and to 0.8 otherwise.

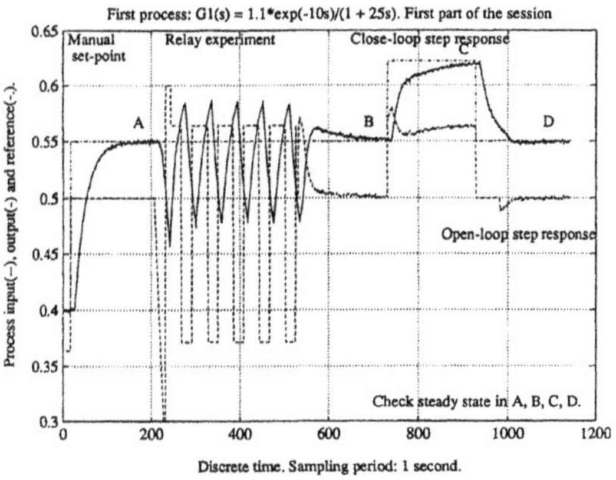

Figure 4.1 Process input, output and reference obtained during an entire session with the developed system.

$1.0 \le \theta$:

The process is said to have a long dead-time. It is then essential to introduce some dead-time and possibly feedforward compensation devices. This can be done for instance with a pole placement or a Smith predictor. The latter structure has been chosen. The parameters of the PI controller used with this structure are given by the relations obtained with a very small θ. However, due to the process characteristic, the factor n, related to the closed loop's speed, is restricted to the interval [0.5 ; 1.5]. Notice that no extensive test of this way of setting the PI parameters has been performed.

4. SIMULATION RESULTS

Figure 4.1 shows a typical experiment performed with the developed system. The four different phases previously described can easily be identified. A first order process with $Kp = 1.1$, $L = 10$ s and $T = 25$ s is used. The estimated value of θ is 0.395 which must be compared with the theoretical value 0.4. The estimated value of κ is 4.63 which must be compared with the theoretical value 4.59. During all the showed simulations, a Gaussian centred white noise with a standard deviation of 0.001 is added to the process output.

The next two tables display the results given by the developed system obtained with two different processes. Their transfer functions are equal to:

$$G_1(s) = \frac{K_{p1} e^{-Ls}}{1 + T_1 s} = \frac{1.1\, e^{-Ls}}{1 + 25\, s} \text{ and } G_2(s) = \frac{K_{p2} e^{-Ls}}{(1 + T_2 s)^2} = \frac{0.85\, e^{-Ls}}{(1 + 20\, s)^2}$$

The tables report the results obtained for different values of L. The last column states if the tested process is considered as a first order process or not. The errors committed on the estimation of the process normalised delay are bigger with the second process than with the first process. This may be explained by the fact that this data has no theoretical value in the latter case. Indeed, the dominant time constant may be defined in several ways: based on the largest slope, on the time when the output reaches 63% or 95% of its final value, etc... The theoretical values of θ reported in the table and in the next curves use the time when the output reaches 63% of its final value.

Comparison With Two Other Controllers

Two controllers are designed at the end of the tuning procedure. The first regulator is based on a pure application of Ziegler-Nichols tuning formula. The theoretical values of the process ultimate gain and ultimate period are given by the operator. These values can be computed or estimated since the true model of the process is known. A PI structure is selected if the measurement noise is greater than 0.01 or if the estimate normalised process dead-time is greater than one. If this is not the case, a PID controller is selected.

The second regulator is based on the rules used in the Satt control auto-tuner ECA 40 or ECA 400 as it is exposed in [*Hägglund & Åström 1989*]. If, based on the relay experiment, the process is classified as a first order system without dead-time or if the measurement noise is larger than 0.05 and the normalised process dead-time less than one, a PI controller based on the relations $Kr = 0.5*Kue$ and $Ti = 4*Tue/(2\pi)$ is used. If the normalised process dead-time is bigger than one, a PI controller satisfying the relations $Kr = 0.25*Kue$ and $Ti = 1.6*Tue/(2\pi)$ is used. Otherwise, the Satt controller has a PID structure. Its

TABLE 1: Estimates obtained with the first process.

L	θ	$\hat{\theta}$	K_u	$\hat{K}_u$	ω_u	$\hat{\omega}_u$	$\hat{K}_p$	$\hat{T}$	$\hat{L}$	d°
1	0.04	0.05	36.3	35.3	1.6	1.365	1.12	25.85	1.17	1
3	0.12	0.21	12.5	3.92	0.55	0.185	1.075	23.05	4.77	>1
5	0.2	0.19	7.72	8.29	0.338	0.343	1.08	26.0	4.86	1
7	0.28	0.25	5.70	6.66	0.25	0.249	1.053	27.95	6.89	1
10	0.4	0.395	4.17	4.24	0.179	0.176	1.093	25.75	10.15	1
15	0.6	0.62	2.99	2.915	0.125	0.1225	1.093	24.7	15.35	1
20	0.8	0.78	2.40	2.44	0.098	0.0985	1.10	25.25	19.75	1
25	1	0.94	2.06	2.15	0.0812	0.081	1.10	26.4	24.85	1
30	1.2	1.12	1.83	1.90	0.0697	0.071	1.10	25.91	29.1	1
40	1.6	1.15	1.545	1.88	0.055	0.0595	1.10	30.3	34.8	1
50	2	1.46	1.38	1.63	0.0458	0.0495	1.10	29.87	43.70	1

TABLE 2: Estimates obtained with the second process.

L	θ	$\hat{\theta}$	K_u	$\hat{K}_u$	ω_u	$\hat{\omega}_u$	$\hat{K}_p$	$\hat{T}$	$\hat{L}$	d°
1	0.2	0.35	47.84	4.94	0.314	0.089	0.844	36.1	12.7	>1
3	0.25	0.41	16.5	4.09	0.18	0.082	0.844	32.7	13.5	>1
5	0.3	0.35	10.2	3.7	0.139	0.076	0.84	35.2	12.4	>1
7	0.36	0.55	7.52	3.32	0.116	0.071	0.84	30.9	17.1	>1
10	0.44	0.38	5.51	5.72	0.096	0.122	0.838	38.4	14.6	1
15	0.58	0.7	3.96	2.33	0.077	0.054	0.848	33.7	23.6	>1
20	0.72	1.1	3.185	2.09	0.065	0.047	0.844	30.3	33.25	>1
25	0.86	0.99	2.725	1.89	0.057	0.042	0.848	34.1	33.7	>1
30	1.0	1.2	2.42	1.76	0.051	0.039	0.843	32.2	29.5	>1
40	1.28	1.03	2.05	2.6	0.043	0.05	0.844	39.4	40.9	1
50	1.56	1.42	1.83	1.55	0.037	0.03	0.84	38.9	55.4	>1

parameters are such that, by introducing this regulator in the control loop, the point $G(i\omega_{ue})$ estimated on the process Nyquist curve is moved to the point

$$G(i\omega_{ue})*G_{PID}(i\omega_{ue}) = 0.5*e^{-i\frac{3\pi}{4}}$$

This design method can be viewed as a combination of phase and amplitude margin specification. The value of the ratio of Ti over Td is fixed to 6.25. The factor ß is set to zero.

The behaviour of the closed loop system is compared during four experiments : a set point change, a set of false measures, a linear variation of the process reference and a load disturbance. Figure 4.2 displays a typical test-phase performed with the three different controllers. The four experiments have the same length. The set of false measures is equal for the three tested controllers. A false measure has a probability equal to 5% to occur ; the disturbance added to the output is uniformly distributed in the interval [-0.2 ; 0.2]. During these experiments, the sums of the absolute values of the process error are computed.

The three controllers performances tested with three different processes are displayed in the curves in figure 4.3 to 4.5. The theoretical normalised dead time is reported on the x-axis and a weighted sum of the absolute value of the control error on the

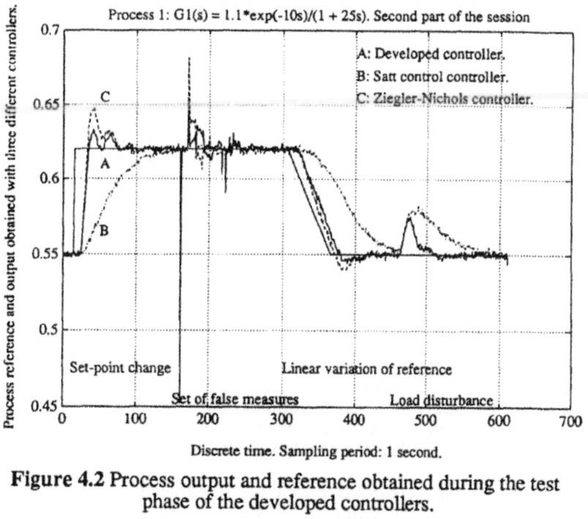

Figure 4.2 Process output and reference obtained during the test phase of the developed controllers.

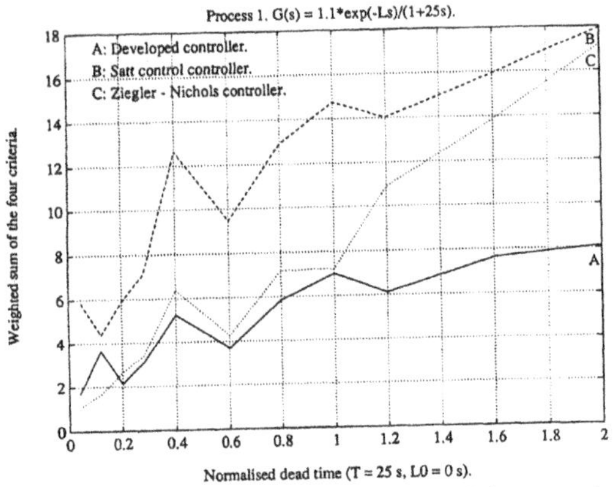

Figure 4.3: Test of the three controllers with the process 1.

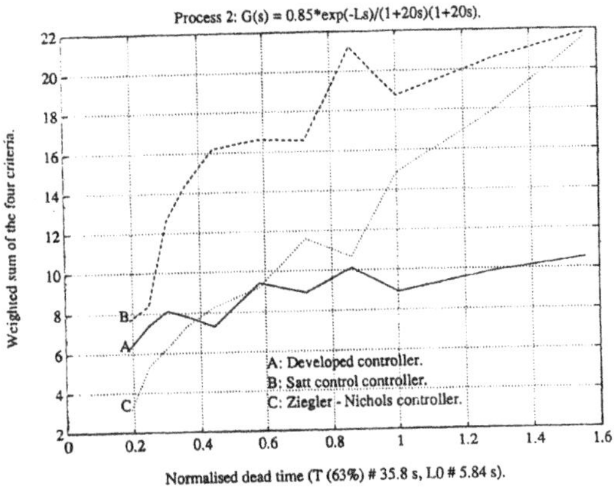

Figure 4.4: Test of the three controllers with the process 2.

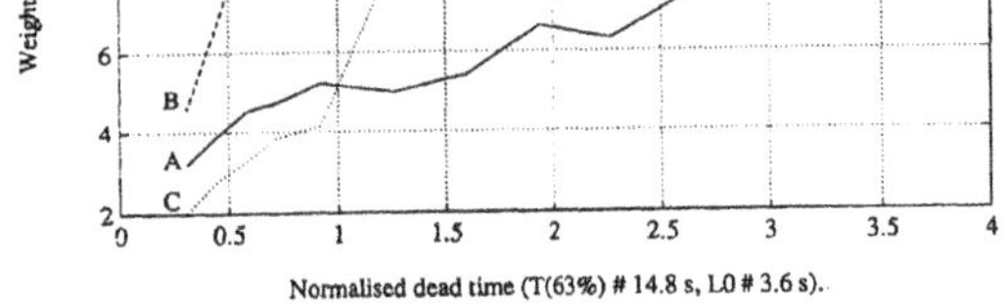

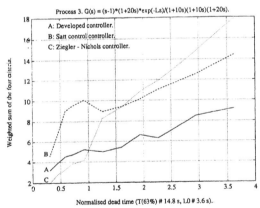

Figure 4.5: Test of the three controllers with the process 3.

y-axis: $y(\theta) = (1+\mu)\cdot \sum_{exp.1} |e_k| + \sum_{exp.2} |e_k| + \sum_{exp.3} |e_k| + \sum_{exp.4} |e_k|$

where μ is the overshoot obtained during the set point change experiment. The contribution of each experiment to the final value of $y(\theta)$ is roughly 40%, 10%, 30% and 20%.
For small values of θ, it can be seen that the "ideal" Ziegler-Nichols controller is better than the developed controller. However, it can be argued that:
1. This is an ideal controller since it uses the true ultimate point.
2. In this case, the value of μ is much bigger (typically 0.6 compared to 0.2 or 0) than for the two other controllers.
3. The process input variations given to obtain a constant process output are much larger than those given by the two other controllers.

5. CONCLUSIONS

This paper has described a possible architecture of a system for expert control and its use to obtain an improved PID tuner. The expert system structure is similar to the one described in [*Årzén 1987*]. Our system is, however, built up of commercial components as the expert system shell Muse, procedures in C and Unix sockets.

Muse has some very nice features such as the blackboard facilities, the object-oriented programming, the downloading facilities. However, it is closer to a general programming environment than an expert system shell with all the advantages and disadvantages this implies. Two of these drawbacks are that it takes some time to learn the system and the built-in real-time facilities are less developed than in other similar tools.

The expert system has the following ingredients: relay oscillations analysis, open and closed loop step responses. The advantages of characterising process dynamics by three parameters suggested in [*Åström et al 1989*] have clearly been demonstrated. A novel feature introduced is the asymmetrical relay. By using such a device, it is possible to obtain three parameters from one single experiment. This is worth further works.

REFERENCES

Årzén, K.E. (1987): "Realization of Expert System Based Feedback Control", PhD dissertation, Coden: LUFTD2/TFRT-1029, Dpt of Automatic Control, Lund Inst. of Technology.
Åström, K.J., Anton, J.J., Årzén, K.E. (1986): "Expert Control", Automatica, 22, nº 3, pp. 277-286.
Åström, K.J., & Hägglund, T. (1988): "A new autotuning method", IFAC Workshop on Adpative Control of Chemical processes, 17-19 August, Copenhagen, Denmark
Åström, K.J., & Hägglund, T. (1988): "Automatic Tuning of PID Controllers", ISA, Research Triangle Park, NC, USA.
Åström, K.J., Hang, C.C., Persson, P. (1989): "Towards Intelligent PID Control", IFAC Workshop on Artificial Intelligence in Real-Time Control, 19-21 Sept., Shenyang, PRC.
Bristol (1977): "Pattern Recognition: An alternative to Parameter Identification in Adaptive Control", Automatica 13., pp. 197 - 202.
Hägglund, T. & Åström, K.J. (1989): "An Industrial Adaptive PID Controller", Int. Symposium on Adaptive Systems in Control and Signal Processing, 19-21 April, Glasgow, UK.
Hang, C.C. and Åström, K.J. (1988): "Pratical aspects of PID auto-tuners based on relay feedback", IFAC Workshop on Adpative Control of Chemical processes, 17-19 Aug., Copenhagen.
Hang, C.C. and Åström, K.J. (1988): "Refinements of the Ziegler-Nichols Tuning Formula for PID Auto-tuners", Proc. ISA Annual Conference, Houston, USA, pp. 1021-1030.
Sallé, S.E. and Årzén, K.E. (1989): "A Comparison Between Three Development Tools for Real-Time Expert-Systems: Chronos, G2 and Muse", Proc. of CACSD, Tampa, Florida-USA.
Sallé, S.E. (1990): "Sur l'utilisation des systèmes experts dans l'automatique", PhD dissertation. Laboratoire d'Automatique de Grenoble, France.
Ziegler, J.G. & Nichols, N.B. (1942): "Optimum Settings for Automatic Controllers", Trans. of the ASME, 64, pp. 759-768.

KNOWLEDGE-BASED TUNING AND CONTROL

A. J. Krijgsman, H. B. Verbruggen and P. M. Bruijn

Delft University of Technology, Dept. of Electrical Engineering, Control Laboratory, P.O. Box 5031, 2600 GA Delft, The Netherlands

Abstract

This article presents an in-line knowledge-based supervisory and tuning system for predictive control. The set up for the system is discussed and a summary is given of the control algorithm and the parameters to be tuned and initialized. After an initialisation phase in which certain parameters are set in accordance to the data required by the system, fine-tuning is performed in-line and the control system is brought to a state in which the control requirements are fulfilled. The results of the system, tuning a predictive controller, are shown in a real-time experiment.

1 Introduction

Nowadays there is a strong tendency towards the use of plant-wide control systems, which operate on tactical, managerial, scheduling, operational and control level for production and process control. Such a multilayered system is based on several knowledge sources. Each knowledge source operates on its own level of decision making. So it is clear that for a complete control system an increasing number of people from different disciplines are involved. All these people have their own background and expertise and their different demands and requirements concerning the overall system behavior. The system should provide the various users with advisory and consultancy tools, based on the different kinds of specialized knowledge.

Between the levels of automation in the system a lot of quantitative information has to be processed, but also a lot of qualitative information has to be handled. Expert systems should operate on the level of qualitative information processing with emphasis on explanation facilities and knowledge translation. Each level has its own kind of knowledge, thus knowledge passing between the various levels of automation becomes more and more important. Therefore the system should provide facilities to pass global information to the various levels of automation. At each level a more specialized knowledge source is used which operates on more detailed information.

When we classifiy the application areas of expert sytems in automation systems we can distinguish systems which are based mainly on static or on dynamic information. Another division could be based on classification requirements or supervision and control problems. It is clear that because of the time critical situation the highest demands are put on the use of expert and other intelligent systems on the supervisory and control level in which dynamic information is involved. In these systems the expertise of process operators and control systems designers is mixed with the time-varying information obtained directly from the process by the measurements.

This paper is concerned with the lower levels of automation at which the knowledge-based and intelligent system should be implemented in a real-time environment. The most promising applications are found in the areas of:

- alarm monitoring, diagnosis and handling
- supervisory and adaptive control
- modeling of the operator
- intelligent and direct expert control

The application of expert systems for direct expert control (replacement of a conventional or advanced controller by an expert system directly controlling the process) and the application of fuzzy control algorithms for modelling the operator (to mimic the most succesful control strategies), will not be discussed in this paper. This article focuses on the use of expert systems to tune an advanced control algorithm on-line. The expert system should form the natural link between the demands and wishes of the operator and the actual implementation and parameters of the control algorithm. That means that the following tasks are involved in such a system:

- off-line knowledge acquisition facilities to initiate the control algorithm, and in-line explanation and advisory facilities to inform the control engineer/operator about the overall behaviour of the controlled process and to indicate whether the control requirements are fulfilled or could be strengthened or should be weakened.
- monitoring of the measured variables to be used in the control algorithm. Monitoring and evaluation of the performance of the controller. Decisions should be taken whether parameters of the controller should be adapted.
- tuning of some of the parameters of the algorithm depending on the requirements which should be met (fine-tuning) or switching on or off some parameters in case of emergency and deterioration of the process behaviour.

In this way artificial intelligence methods can provide additional assistance to the control engineer/operator and enhance the performance of the on-line controlled process.

The outline of this article is as follows: section 2 describes the motivation to use a knowledge-based tuner. Section 3 gives a short description of the controller algorithm which has to be tuned. Section 4 gives some insight in the set up of the knowledge base and section 5 gives some implementation details of this system. In section 6 an experiment is described where the tuner is used. The article ends with some conclusions about the use of a knowledge-based tuner.

2 Motivation

In the actual practical implementation of control systems the algorithms are only a minor part of the code in a control system. Apart from the man-machine interface the major part of the code in a control system is actually the logic that surrounds the control algorithm. This logic takes care of switches between manual and automatic control, bumpless parameter changes and anti-windup in simple controllers. In more complex controllers it also handles the supervision of automatic tuning and adaptation. It is also a common experience that the effort required to implement and debug this code is significant. Since the supervision code is easily expressed in logic it is a natural candidate for use in an expert system.

There are several devices on the market that attempt to help a user to tune a controller. Examples are the Supertuner and the Protuner. These devices typically carry out some type of system identification from plant experiments and then give the recommended controller tuning. Related techniques are used in some of the single loop controllers with automatic tuning. (See Bristol and Kraus [2], Kraus and Myron [4] and Åström and Hägglund [1]). Although these devices are useful it is clear that the tuning of a controller is not always uniquely determined by the process dynamics. It also depends on the purpose of control. A typical example is level control where the purpose can be tight level control as well as surge tank operation when it is desired that the level swings over the full range. From this viewpoint it appears reasonable to have a more sophisticated system for tuning and advice that considers also the purpose of control. It would also be highly desirable to handle design data in such a system because sometimes good tuning parameters can be computed from design data. It is also clear that applications in control system design and in automatic tuning are closely related.

In advanced controllers such as adaptive and predictive control algorithms a number of parameters have to be preset by a control engineer and to be tuned or supervised by an intelligent system. A few parameters are, however, strongly related to the system requirements and should be tuned by the operator. These parameters represent the key parameters in a control loop design, like bandwidth, overshoot and noise reduction. They can be interpreted as the controller knobs of the controller and are translated to parameters inside the controller algorithm which can be completely irrelevant for the user. A knowledge-based system forms the natural link between the idea of the control system designer as to how the system should behave, and the mathematical description of the process. Using the correct calculations and heuristics it sets the right parameters of a controller or chooses the right controller configuration.

In the last decade a number of advanced adaptive and predictive control algorithms have been developed. They can be easily implemented in a real-time environment on the level just above the direct digital control level. The number of practical applications of these algorithms is however rather restricted because:

- few attention has been paid to simple tuning rules for the relatively large number of parameters to be set either a priori or during process operation;
- the robustness of the controllers is sometimes rather poor because of the lack of sufficient and appropriate jacketing of the controller algorithm.

It is very important to pay attention to these points, and provisions should be provided by a supervisory and tuning system to support the control engineer and operator.
In this article a tuning device for such a predictive control algorithm, the Unified Predictive Controller (UPC) [7,5], is described.

The UPC is a very powerful control algorithm uniting a large number of predictive control algorithms presented in literature in the past decade. By choosing the right set of parameters a specific algorithm can be chosen exhibiting certain advantages and disadvantages for a given process. This feature can be interpreted as a Control-Law Selector. The selection can be performed off-line during the initialization phase. Reconfiguration can be performed on-line depending on the results of the Control-System Assessor, which evaluates the performance of the system and compares the results with the requirements set by the operator. Fine-tuning of a number of parameters is performed in order to adjust the parameters such that the control requirements are fulfilled. An additional option involves the redefinition of the requirements via a dialogue with the operator when the requirements can not be met or could be made tighter.

3 The Unified Predictive Controller

The Unified Predictive Controller (UPC) [6,7] belongs to the class of predictive control algorithms. All predictive controllers are based on the prediction of the future behavior of the process output as is illustrated by Figure 1. Suppose the current time is $t = k$ and $u(k)$, $y(k)$ and $w(k)$

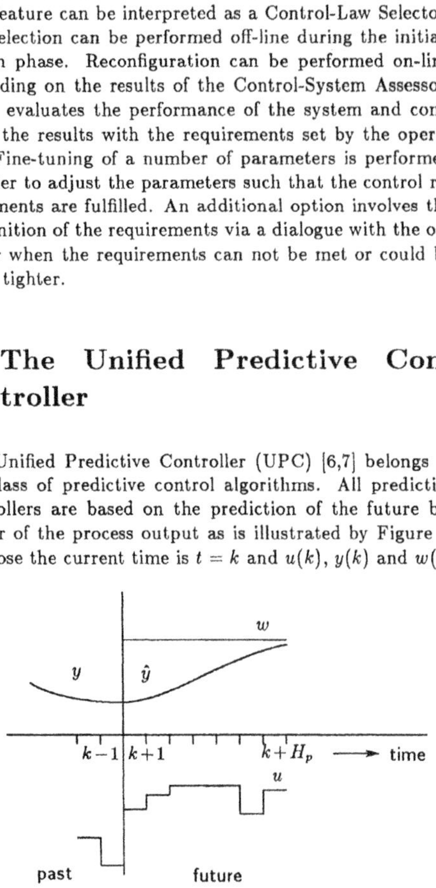

Figure 1: The predictive control strategy.

denote the controller output, the process output and the desired process output at $t = k$, respectively. Further, define

$$\begin{aligned} \boldsymbol{u} &= [u(k), \cdots, u(k+H_p-1)]^T \\ \hat{\boldsymbol{y}} &= [\hat{y}(k+1), \cdots, \hat{y}(k+H_p)]^T \\ \boldsymbol{w} &= [w(k+1), \cdots, w(k+H_p)]^T \end{aligned}$$

where H_p is called the prediction horizon and the symbol $\hat{}$ denotes estimation. Then, a predictive controller calculates such a future controller output sequence $\boldsymbol{u}$ that the predicted output of the process $\hat{\boldsymbol{y}}$ is 'close' to the desired process output w. This desired process output is often called the reference trajectory. The first element of the controller output sequence ($= u(k)$) is used to control the process. At the next sample, the whole procedure is repeated using the latest measured information. This is called the *receding horizon* principle.

The process output is predicted by using a model of the process. In UPC the following model is used for this purpose.

$$y(k) = \frac{q^{-d}B(q^{-1})}{A(q^{-1})}u(k-1) + \frac{C(q^{-1})}{D(q^{-1})A(q^{-1})}e(k)$$

in which q^{-1} is the backward shift operator, d is the time delay in samples ($d \geq 0$) and $e(k)$ is a discrete white noise sequence with zero mean. A, B, C and D are polynomials in q^{-1} with degree n_A, n_B, n_C and n_D, respectively. Further, A, C and D are monic. Note that the model unifies familiar process models such as ARIMAX (Auto-Regressive Integrated Moving-Average) and FIR (Finite Impulse Response) models. The parameters A, B and d can be estimated by using a least-squares method yielding $\hat{A}$, $\hat{B}$ and $\hat{d}$. The polynomials C and D are usually not estimated but used as design parameters [7]. Often $\hat{C}$ is denoted as T while $\hat{D}$ denotes the estimated D polynomial being a design polynomial.

A criterion function is used in order to define how well the predicted process output tracks the reference trajectory. In the UPC controller the following criterion function is minimized subject to the constraint (2):

$$J = \sum_{i=H_s}^{H_p} [P\hat{y}(k+i) - Rw(k+i)]^2 +$$

$$+\rho \sum_{i=1}^{H_p-d} \left[\frac{Q_n}{Q_d}u(k+i-1)\right]^2 \qquad (1)$$

$$N\Delta^{\beta}u(k+i-1) = 0 \qquad 1 \leq H_c < i \leq H_p - d \qquad (2)$$

where $\Delta = 1 - q^{-1}$. The parameters H_p, H_s and H_c are called the prediction horizon, the minimum cost horizon and the control horizon, respectively. Further, P, R, Q_n, Q_d and N are polynomials in q^{-1}. Finally, β is an integer variable ≥ 1 and ρ is a weighting factor ≥ 0. Now the controller output sequence $\mathbf{u}$ is obtained by minimization of J with respect to $\mathbf{u}$. It is shown in [7] that if the criterion is quadratic and the model is linear, there is an analytical solution to the minimization problem.

As a result of the unified approach, UPC has many design parameters among which polynomials. In [5] rule of thumb methods are discussed that can be used to select initial settings of all parameters. These methods are based on a model of the process and the requirements on the closed-loop system such as the rise time of the step response and the steady-state behavior. These rule of thumb methods can be summarized as follows:

$H_p = \mathrm{int}(t_s(5\%)/T_s)$ where $t_s(5\%)$ is the 5% settling time of the step response of the (continuous) process, T_s is the sampling period and int(.) is a function that converts a real value into an integer.

$H_s = d + 1$.

$H_c = n_A$.

$\beta = \max(\max(p,r) - n, 1)$ where r and p denote the type of the reference trajectory and the disturbances ($r = 1$ denotes a constant trajectory, $r = 2$ denotes a triangular trajectory and so on) and n denotes the number of integrators in the process.

$\rho = 0$. For this value for ρ, Q_n and Q_d can be taken equal to 1.

$\hat{D} = P\Delta^{\beta}$.

$T = \hat{A}$.

$P = 1 - \alpha q^{-1}$ where $\alpha = e^{-2.3T_s/t_r}$ and t_r is the desired rise time of the closed-loop system. The reference trajectory is in this case equal to the set point (hence, $w = [Sp, \cdots, Sp]^T$). Note that in [5] a second-order trajectory is suggested. However, for simplicity in this paper a first-order trajectory is used.

$R = P(1)$.

$N = P$.

It has been shown in [5] that by using the above-mentioned rule of thumbs methods, the servo behavior and the regulator behavior can be tuned independently by P and T, respectively.
Selecting the UPC design parameters by using the rules of thumb usually yields a response that is quite acceptable. However, in some situations fine tuning may be desired especially if the process is different from the model. It is argued in [5] that the following parameters can be considered as fine tuning parameters: P, T, ρ and H_c. Note, however, that by changing P the parameters R, N and $\hat{D}$ also change.

4 Knowledge-based tuning

In the previous section it was described how the Unified Predictive Controller (UPC) is designed. A large number of parameters has been defined which are difficult to select by a novice user of the algorithm. However, based on many experiments and backed by the knowledge of relationships between process and controller parameters, much information has been built up about the tuning of the controller parameters for different situations. This knowledge is partly used to intialize the controller and partly used to supervise and tune the controller on-line. This supervision logic is based on shallow reasoning (heuristics) and deep reasoning (design) as well. To develop an intelligent tuning device we have to develop a knowledge-base which contains deep as well as shallow knowledge.

One of the main problems in using expert system technology in control is the time dependance of the information. It is essential that during the tuning phase of the system all decisions and conclusions are stored, such that the system can learn from its behaviour. Therefore the system has been extended with state information of the process. State information about the previous inference cycles and the decisions made are stored in a state. This state information is necessary to avoid problems when using the knowledge base, cyclical actions would be initiated without improving the performance. A hierarchical knowledge model (see figure 2) is developped to implement this reasoning method.

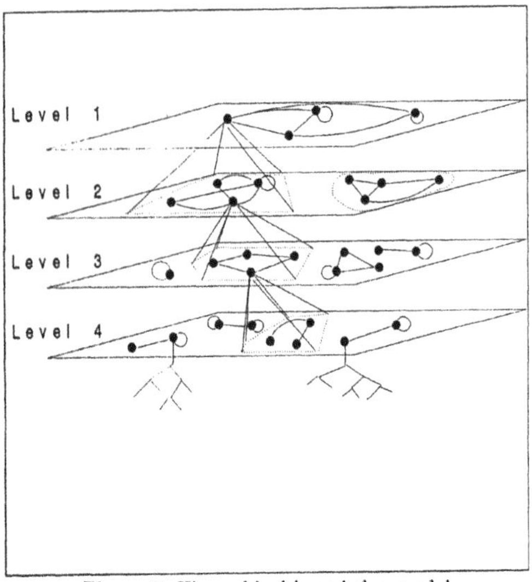

Figure 2: Hierarchical knowledge model

At each level the system focuses on another subject. Level 1 concentrates on the decisions whether the system meets basic stability requirements. If so the system proceeds with the actual tuning and if not so this problem is solved immediately. Level 2 decides about the strategy for the tuning procedure of which requirement has to be focussed on (overshoot, speed, accuracy etc.). The other levels are used to decide about the actual tuning algorithm. At the lowest level of reasoning a decision is made about the change of a specific UPC parameter, coupled to an external routine. In every inference cycle the system can use the state information to decide whether the previous action of the tuner has been succesfull or not.

Knowledge sources

The knowledge base consists of several knowledge sources which can focus on a specific area or parameters. Such a knowledge source contains the knowledge to solve a very specific problem. One can imagine that knowledge which is very appropriate to solve stability problems is not suitable for optimization of the controller parameters. The main knowledge sources are:

- **defaults:** this knowledge source provides initial values to the controller algorithm according to some knowledge about the values of process parameters. When no information about the process is available a default set of parameters for a worst case situation is used;
- **stability:** this knowledge source focuses on stability. Whenever stability problems arise, detected by the Control-System Assessor, they have to be solved immediately. The first decision which has to be made is whether a previous change of a parameter caused the problem or not. If this is true this change will be cancelled. When the problems are not solved quickly a major change in some of the essential controller parameters is made to increase the stability and robustness of the control loop;
- **requirements:** this knowledge source contains the knowledge about the relation between the controller parameters and the required performance given by the user. Most of the requirements are given in the time domain (overshoot, settling time, rise time etc). This knowledge source is split into modules, related to the mode of operation (setpoint tracking or distubance rejection);
- **optimization and redefinition:** this knowledge source can be used when all requirements are fulfilled and an even better performance is possible.

In general we can say that the knowledge sources have been developed to mimic the way the actual designer of UPC would focus on some specific problem area and tune the UPC. This tuning procedure can be seen as an optimization procedure which is orchestrated by an expert system.

Knowledge acquisition

The knowledge for the UPC tuner has been obtained by using a very generally applied technique: interviewing the expert which developed the UPC-algorithm. An extensive interview gave us some insight in how the expert would tune the controller under certain circumstances. Moreover, a number of experiments on different processes has supported this expertise. In several refinement steps this knowledge was updated until there was very little difference in the actions obtained by the expert and the expert system. The total knowledge base consists of approximately 160 rules.

Control-System Assessor

The knowledge base uses very specific process information like overshoot, rise time, settling time, variances etc. Therefore several external routines were written to collect this information from the (simulated) process. As soon as the information is available and the requirements given by the user are not met, the expert system is triggered to solve the problem using the actual process information.

Control system requirements

The requirements of the user are given to the system via a dialogue or a knowledge source. The requirements are given for servo as well as regulator behaviour:

- maximum overshoot
- maximum rise time
- maximum settling time
- maximum output variance

Note that most of these values are given in the time domain which is closely related to the way an operator 'thinks' about a process.

Tuning parameters

In order to meet the requirements given by the user some of the UPC parameters can be tuned. It is possible to tune all parameters of the algorithm, but the knowledge of the expert showed us that only some crucial parameters of the algorithm are used to change the overall performance of the system. The parameters which can be tuned are: H_c, ρ and polynomials $T(q^{-1})$ and $P(q^{-1})$. The parameters of the model are kept constant during the tuning procedure.

5 Implementation

The actual expert system tuner of the UPC has been implemented in a real-time environment: MUSIC (MUltipurpose SImulation and Control) [3]. This is a block oriented program for simulation and control. MUSIC is used to simulate the process and to implement the external routines for the expert system to gather information about the process. After simulation the simulated process can be replaced by the real process.

The knowledge base is implemented using the commercially available expert system shell NEXPERT. This is an object oriented program with a very enhanced user interface to edit and update the knowledge base. The expert system can be accessed by using the *callable interface*, which can access every function of NEXPERT. The communication

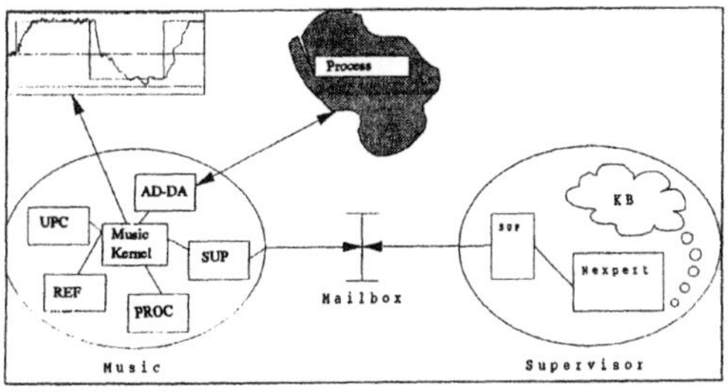

Figure 3: Communication via mailboxes

is set up using a mailbox mechanism, see figure 3. Both tasks, expert system tuner and the control loop, can send information to this mailbox and read information from it. The separation of expert system and control loop guarantees the real-time behaviour, which is handled by MUSIC in this set up. The simulation task is used to trigger the expert system in an event driven way. As soon as information is available a logical flag is set in the simulation. The information is sent to the mailbox and triggers a demon at the NEXPERT side. As soon as this demon is triggered a knowledge session is started to reason about the information. The user interface consists of two parts. The first part is the operator and display window of MUSIC. Via this window parameters of the algorithm and the process can directly be set. The second part of the user interface is the expert system window. This window prompts for the essential information from the user about the process and his demands and requirements for the performance of the system.

6 Experiments

The complete tuning configuration which has been described in the previous chapter has been tested on several processes, with different characteristics. In this section an example is given of such a tuning experiment.
The process to be controlled is a fairly simple linear process, described by the transfer function:

$$H(s) = \frac{Ke^{-T_d s}}{(s\tau_1+1)(s\tau_2+1)} = \frac{e^{-4s}}{(10s+1)(8s+1)} \tag{3}$$

The tuning experiment now starts by a session in which the system asks for a priori information of the process. When this information is not available default settings are used. We provided the system with an initial set of information:

1. the open loop process is stable
2. the open loop process has no integration
3. the sampling time is $1s$
4. the reference signal is a block signal
5. the open loop process is well damped
6. there is measurement noise
7. the process exhibits delay time

The requirements of the user to the controller are given in the time domain. The servo performance requirements are criteria related to a step response:

$$\begin{pmatrix} \text{maximum overshoot} \\ \text{maximum rise time} \end{pmatrix} \le \begin{pmatrix} 12\% \\ 8s \end{pmatrix} \tag{4}$$

The expert system uses the initial set of process information and the appropriate knowledge source to produce an initial parameter set for the UPC controller. Some of these parameters were determined using a second-order model close to the process to be controlled. This model is also used as prediction model by the UPC. The model chosen is:

$$H(z^{-1}) = z^{-1}\frac{0.0290z^{-1}+0.0269z^{-2}}{1-1.79z^{-1}+0.799z^{-2}} \tag{5}$$

Note that the only mismatch between the model and process is a difference in time delay. The complete set of initial parameters for the UPC is chosen as follows:

$$\begin{pmatrix} H_p \\ H_c \\ H_s \\ \rho \\ \beta \\ T(q^{-1}) \\ R(q^{-1}) \\ P(q^{-1}) \\ \hat{D}(q^{-1}) \\ Q_n(q^{-1}) \\ Q_d(q^{-1}) \end{pmatrix} = \begin{pmatrix} 14 \\ 1 \\ 2 \\ 0 \\ 1 \\ (1-1.79q^{-1}+0.799q^{-2}) \\ 0.394 \\ (1-0.606q^{-1}) \\ (1-0.606q^{-1})(1-q^{-1}) \\ 1 \\ 1 \end{pmatrix} \tag{6}$$

Determining $t_s(5\%)$ and using rules of thumb (see section 3) H_p was determined to be 41. However, experiments show that a value of 14 is sufficient, because the closed-loop system is much faster. To increase the robustness of the system the value of H_c was chosen equal to 1. Further, in order to show the tuning procedure, t_r was selected equal to 4.6s. In figure 4 the step responses are given using this initial set of parameters. The obtained results of the control loop with respect to the user requirements are:

$$\begin{pmatrix} \text{overshoot} \\ \text{rise time} \end{pmatrix} = \begin{pmatrix} 20\% \\ 10s \end{pmatrix} \tag{7}$$

These results do not meet the requirements thus the fine-tuning procedure is started. It is decided that to meet the requirements it is useful to vary the control horizon H_c and the polynomial $P(q^{-1})$. Actually the tuning reduces to the variation of two parameters: H_c and t_r. In figure 5 the results of this tuning experiment are shown. At the end of the tuning procedure the controller has the following set of parameters:

$$\begin{pmatrix} H_c \\ R(q^{-1}) \\ P(q^{-1}) \end{pmatrix} = \begin{pmatrix} 2 \\ 0.154 \\ (1-0.846q^{-1}) \end{pmatrix} \tag{8}$$

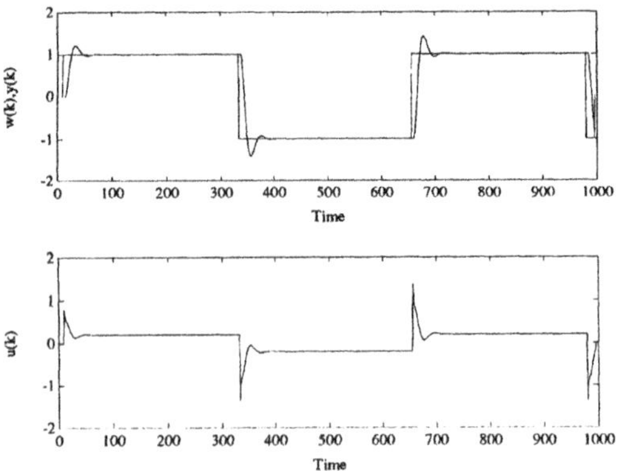

Figure 4: UPC control with default settings

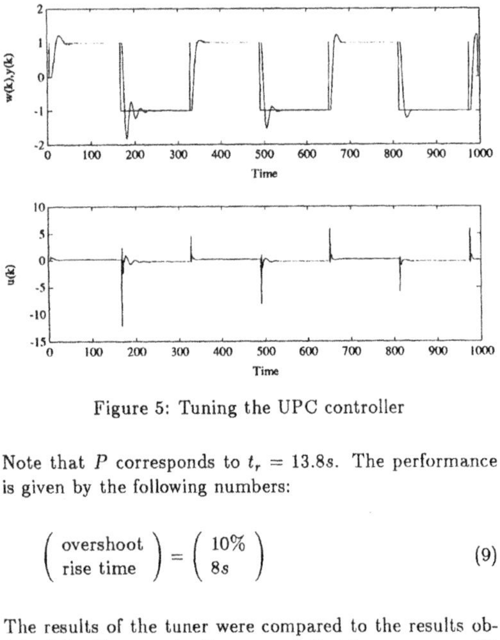

Figure 5: Tuning the UPC controller

Note that P corresponds to $t_r = 13.8s$. The performance is given by the following numbers:

$$\begin{pmatrix} \text{overshoot} \\ \text{rise time} \end{pmatrix} = \begin{pmatrix} 10\% \\ 8s \end{pmatrix} \tag{9}$$

The results of the tuner were compared to the results obtained by the expert. It appeared that he came up with a set of parameters which were comparable and lead to almost the same results. Note that for the given requirements there is no unique solution.

7 Conclusions

The use of complex control schemes in practice is limited due to the difficulty of tuning the algorithms. The gap between the expertise of the designer of the control scheme and the user of these complex schemes, due to the lack of experience and design knowledge, can be bridged by knowledge-based systems.

A knowledge-based system has been developed containing several knowledge sources, each with their own area of attention. In order to cope with time dependent information all state information is stored in the knowledge base.

A tuner for a predictive control algorithm is presented, leading to a very satisfying behaviour for various processes. The results of the knowledge-based tuner are very satisfactory and are comparable to the results gained by an experienced expert working for many years in this field.

Acknowledgement

The authors want to thank A.R.M. Soeterboek for his collaboration on this paper with respect to the Unified Predictive Controller.

References

[1] Åström, K.J. and T. Hägglund (1988), *Automatic Tuning of PID Regulators,* Instrument Society of America, Research, Triangle Park, N.C.

[2] Bristol, E.H. and T.W. Kraus (1984), *Life with Pattern Adaptation,* Proc. 1984 American Control Conference, San Diego, CA, pp.888-892.

[3] Cser J., P.M. Bruijn and H.B. Verbruggen (1986). *MUSIC: a tool for simulation and real-time control.* 4th IFAC/IFIP Symposium on Software for Computer Control SOCOCO'86, Graz, Austria, 20-23 May.

[4] Kraus T.W. and T.J. Myron (1984), *Self-tuning PID controller uses pattern recognition approach,* Control Engineering Magazine, June.

[5] Soeterboek, A.R.M., H.B. Verbruggen and P.P.J. van den Bosch (1991). "On the Design of the Unified Predictive Controller". To be published in the Proceedings of the IFAC Symposium on Intelligent Tuning and Adaptive Control (ITAC 91). Singapore.

[6] Soeterboek, A.R.M., H.B. Verbruggen, P.P.J. van den Bosch and H. Butler (1990). "Adaptive Predictive Control - A Unified Approach". *Proceedings of the Sixth Yale Workshop on Adaptive and Learning Systems.* Yale University, New Haven, U.S.A.

[7] Soeterboek, A.R.M., H.B. Verbruggen, P.P.J. van den Bosch and H. Butler (1990). "On the Unification of Predictive Control Algorithms". *Proceedings of the 29th IEEE Conference on Decision and Control.* Honolulu (HI), U.S.A.

AUTHOR INDEX

KEYWORD INDEX

Printed and bound by CPI Group (UK) Ltd, Croydon, CR0 4YY
03/10/2024
01040323-0016